Graduate Texts in Mathematics 107

Springer
New York
Berlin
Heidelberg
Barcelona
Hong Kong
London
Milan
Paris
Singapore
Tokyo

Graduate Texts in Mathematics

1 TAKEUTI/ZARING. Introduction to Axiomatic Set Theory. 2nd ed.
2 OXTOBY. Measure and Category. 2nd ed.
3 SCHAEFER. Topological Vector Spaces. 2nd ed.
4 HILTON/STAMMBACH. A Course in Homological Algebra. 2nd ed.
5 MAC LANE. Categories for the Working Mathematician. 2nd ed.
6 HUGHES/PIPER. Projective Planes.
7 SERRE. A Course in Arithmetic.
8 TAKEUTI/ZARING. Axiomatic Set Theory.
9 HUMPHREYS. Introduction to Lie Algebras and Representation Theory.
10 COHEN. A Course in Simple Homotopy Theory.
11 CONWAY. Functions of One Complex Variable I. 2nd ed.
12 BEALS. Advanced Mathematical Analysis.
13 ANDERSON/FULLER. Rings and Categories of Modules. 2nd ed.
14 GOLUBITSKY/GUILLEMIN. Stable Mappings and Their Singularities.
15 BERBERIAN. Lectures in Functional Analysis and Operator Theory.
16 WINTER. The Structure of Fields.
17 ROSENBLATT. Random Processes. 2nd ed.
18 HALMOS. Measure Theory.
19 HALMOS. A Hilbert Space Problem Book. 2nd ed.
20 HUSEMOLLER. Fibre Bundles. 3rd ed.
21 HUMPHREYS. Linear Algebraic Groups.
22 BARNES/MACK. An Algebraic Introduction to Mathematical Logic.
23 GREUB. Linear Algebra. 4th ed.
24 HOLMES. Geometric Functional Analysis and Its Applications.
25 HEWITT/STROMBERG. Real and Abstract Analysis.
26 MANES. Algebraic Theories.
27 KELLEY. General Topology.
28 ZARISKI/SAMUEL. Commutative Algebra. Vol.I.
29 ZARISKI/SAMUEL. Commutative Algebra. Vol.II.
30 JACOBSON. Lectures in Abstract Algebra I. Basic Concepts.
31 JACOBSON. Lectures in Abstract Algebra II. Linear Algebra.
32 JACOBSON. Lectures in Abstract Algebra III. Theory of Fields and Galois Theory.
33 HIRSCH. Differential Topology.
34 SPITZER. Principles of Random Walk. 2nd ed.
35 ALEXANDER/WERMER. Several Complex Variables and Banach Algebras. 3rd ed.
36 KELLEY/NAMIOKA et al. Linear Topological Spaces.
37 MONK. Mathematical Logic.
38 GRAUERT/FRITZSCHE. Several Complex Variables.
39 ARVESON. An Invitation to C^*-Algebras.
40 KEMENY/SNELL/KNAPP. Denumerable Markov Chains. 2nd ed.
41 APOSTOL. Modular Functions and Dirichlet Series in Number Theory. 2nd ed.
42 SERRE. Linear Representations of Finite Groups.
43 GILLMAN/JERISON. Rings of Continuous Functions.
44 KENDIG. Elementary Algebraic Geometry.
45 LOÈVE. Probability Theory I. 4th ed.
46 LOÈVE. Probability Theory II. 4th ed.
47 MOISE. Geometric Topology in Dimensions 2 and 3.
48 SACHS/WU. General Relativity for Mathematicians.
49 GRUENBERG/WEIR. Linear Geometry. 2nd ed.
50 EDWARDS. Fermat's Last Theorem.
51 KLINGENBERG. A Course in Differential Geometry.
52 HARTSHORNE. Algebraic Geometry.
53 MANIN. A Course in Mathematical Logic.
54 GRAVER/WATKINS. Combinatorics with Emphasis on the Theory of Graphs.
55 BROWN/PEARCY. Introduction to Operator Theory I: Elements of Functional Analysis.
56 MASSEY. Algebraic Topology: An Introduction.
57 CROWELL/FOX. Introduction to Knot Theory.
58 KOBLITZ. p-adic Numbers, p-adic Analysis, and Zeta-Functions. 2nd ed.
59 LANG. Cyclotomic Fields.
60 ARNOLD. Mathematical Methods in Classical Mechanics. 2nd ed.
61 WHITEHEAD. Elements of Homotopy Theory.

(continued after index)

Peter J. Olver

Applications of Lie Groups to Differential Equations

Second Edition

Springer

Peter J. Olver
School of Mathematics
University of Minnesota
Minneapolis, MN 55455
USA

With 10 illustrations.

Mathematics Subject Classifications (1991): 22E70, 34-01, 70H05

Library of Congress Cataloging-in-Publication Data
Olver, Peter J.
Applications of Lie groups to differential equations / Peter J. Olver.—2nd ed.
p. cm.—(Graduate texts in mathematics; 107)
Includes bibliographical references and indexes.
ISBN 0-387-95000-1
1. Differential equations. 2. Lie groups. I. Title. II. Series.
QA372.055 1993
515'.35—dc20 92-44573

Printed on acid-free paper.

First softcover printing, 2000.

Production managed by Jim Harbison, manufacturing supervised by Vincent Scelta.
Typeset by Asco Trade Typesetting Ltd., Hong Kong.
Printed and bound by Hamilton Printing Co., Castleton on Hudson, New York.
Printed in the United States of America.

9 8 7 6 5 4 3 2 1

ISBN 0-387-95000-1 Springer-Verlag New York Berlin Heidelberg SPIN 10755712

Preface to the First Edition

This book is devoted to explaining a wide range of applications of continuous symmetry groups to physically important systems of differential equations. Emphasis is placed on significant applications of group-theoretic methods, organized so that the applied reader can readily learn the basic computational techniques required for genuine physical problems. The first chapter collects together (but does not prove) those aspects of Lie group theory which are of importance to differential equations. Applications covered in the body of the book include calculation of symmetry groups of differential equations, integration of ordinary differential equations, including special techniques for Euler–Lagrange equations or Hamiltonian systems, differential invariants and construction of equations with prescribed symmetry groups, group-invariant solutions of partial differential equations, dimensional analysis, and the connections between conservation laws and symmetry groups. Generalizations of the basic symmetry group concept, and applications to conservation laws, integrability conditions, completely integrable systems and soliton equations, and bi-Hamiltonian systems are covered in detail. The exposition is reasonably self-contained, and supplemented by numerous examples of direct physical importance, chosen from classical mechanics, fluid mechanics, elasticity and other applied areas. Besides the basic theory of manifolds, Lie groups and algebras, transformation groups and differential forms, the book delves into the more theoretical subjects of prolongation theory and differential equations, the Cauchy–Kovalevskaya theorem, characteristics and integrability of differential equations, extended jet spaces over manifolds, quotient manifolds, adjoint and co-adjoint representations of Lie groups, the calculus of variations and the inverse problem of characterizing those systems which are Euler–Lagrange equations of some variational problem, differential operators, higher Euler operators and the variational complex, and the general theory of Poisson structures, both for finite-dimensional Hamiltonian systems as well as systems of evolution equations, all of which have direct bearing on the symmetry analysis of differential equations. It is hoped that after reading this book, the reader will, with a minimum of difficulty, be able to readily apply these important group-theoretic methods to the systems of differential equations he or she is interested in, and make new and interesting deductions concerning them. If so, the book can be said to have served its purpose.

A preliminary version of this book first appeared as a set of lecture notes, distributed by the Mathematical Institute of Oxford University, for a graduate seminar held in Trinity term, 1979. It is my pleasure to thank the staff of Springer-Verlag for their encouragement for me to turn these notes into book form, and for their patience during the process of revision that turned out to be far more extensive than I originally anticipated.

Preface to the Second Edition

For the second edition, I have corrected a number of misprints and inadvertent mathematical errors that found their way into the original version. More substantial changes are the inclusion of a simpler proof of Theorem 4.26 due to Alonso, [1], and the omission of the false (at least in the form stated in the first edition) Theorem 5.22 on the commutativity of generalized symmetries. Also, I have corrected some of the exercises and added several new ones. Hopefully this now eliminates all of the major (and almost all of the minor) mistakes. The one substantial addition to the second edition is a short presentation of the calculus of pseudo-differential operators and their use in Shabat's theory of formal symmetries, which provides a powerful, algorithmic method for determining the integrability of evolution equations.

The years since the appearance of the original edition of the book have witnessed a remarkable explosion of research, both pure and applied, into symmetry group methods in differential equations, proceeding at a pace well beyond my expectations. Innumerable papers, as well as several substantial textbooks devoted to the subject of symmetry and differential equations, have appeared in the literature. The former are too numerous to try to list here, although I have added a few of the more notable contributions to the list of references and have correspondingly updated the historical notes at the end of each chapter. Of the latter, I recommend the books of Bluman and Kumei, [2], and Stephani [3], on symmetry methods, and Zharinov, [1], on the geometrical theory of differential equations. There has also been a lot of activity in the development of computer algebra (symbolic manipulation) computer programs to (partially) automate the determination of symmetry groups of differential equations. A good survey of the available codes, as of 1991, including a discussion of their strengths and weaknesses, can be found in the paper of Champagne, Hereman, and Winternitz, [1].

I would like to acknowledge, with gratitude, Ian Anderson, Ken Driessel, Darryl Holm, Niky Kamran, John Maddocks, Jerry Marsden, Sascha Mikhailov, and Alexei Shabat, who offered valuable comments and suggestions for improving the first edition. Finally, I should reiterate my thankfulness and love to my wife, Cheri, and children, Pari, Sheehan, and Noreen, for their continued, all-important love and support!

Acknowledgments

Let me first express my gratitude to the National Science Foundation for supporting my summer research during the period in which this book was being written.

It is unfortunately impossible to mention all those colleagues who have, in some way, influenced my mathematical career. However, the following people deserve an especial thanks for their direct roles in aiding and abetting the preparation of this book (needless to say, I accept full responsibility for what appears in it!).

Garrett Birkhoff—my thesis advisor, who first introduced me to the marvellous world of Lie groups and expertly guided my first faltering steps in the path of mathematical research.

T. Brooke Benjamin, and the staff of the Mathematical Institute at Oxford University—who encouraged me to present the first version of this material as a seminar during Trinity term, 1979 and then typed up as lecture notes.

Willard Miller, Jr.—who encouraged me to come to Minnesota and provided much needed encouragement during the preparation of the book, including reading, criticizing and suggesting many improvements on the manuscript.

David Sattinger—who first included what has become Sections 2.2–2.4 in his lecture notes on bifurcation theory, and provided further encouragement after I came to Minnesota.

Ian Anderson—who played an indispensible role in the present development of the variational complex and higher order Euler operators of Section 5.4, who helped with the historical material, and who read, criticized and helped improve the manuscript of the book as it appeared.

Yvette Kosmann-Schwarzbach—for having the time and patience to read the entire manuscript, and for many valuable comments, criticisms and corrections.

Darryl Holm, Henry Hermes and Juha Pohjanpelto—whose many comments and suggestions on the manuscript led to much-needed improvements.

Phillip Rosenau—whose problems and suggestions never failed to get me thinking.

Debbie Bradley, Catherine Rader and Kaye Smith—who selflessly and wholeheartedly took on the arduous task of typing the manuscript, and did such a wonderful job.

My mother, Grace Olver, and father, Frank Olver—for their unfailing support and influence in my life.

And, of course, my wife, Chehrzad Shakiban, and children, Parizad, Sheehan and Noreen—whose love, patience and understanding were essential to it all!

Table of Contents

Introduction

When beginning students first encounter ordinary differential equations they are, more often than not, presented with a bewildering variety of special techniques designed to solve certain particular, seemingly unrelated types of equations, such as separable, homogeneous or exact equations. Indeed, this was the state of the art around the middle of the nineteenth century, when Sophus Lie made the profound and far-reaching discovery that these special methods were, in fact, all special cases of a general integration procedure based on the invariance of the differential equation under a continuous group of symmetries. This observation at once unified and significantly extended the available integration techniques, and inspired Lie to devote the remainder of his mathematical career to the development and application of his monumental theory of continuous groups. These groups, now universally known as Lie groups, have had a profound impact on all areas of mathematics, both pure and applied, as well as physics, engineering and other mathematically-based sciences. The applications of Lie's continuous symmetry groups include such diverse fields as algebraic topology, differential geometry, invariant theory, bifurcation theory, special functions, numerical analysis, control theory, classical mechanics, quantum mechanics, relativity, continuum mechanics and so on. It is impossible to overestimate the importance of Lie's contribution to modern science and mathematics.

Nevertheless, anyone who is already familiar with one of these modern manifestations of Lie group theory is perhaps surprised to learn that its original inspirational source was the field of differential equations. One possible cause for the general lack of familiarity with this important aspect of Lie group theory is the fact that, as with many applied fields, the Lie groups that do arise as symmetry groups of genuine physical systems of differential equations are often not particularly elegant groups from a purely mathematical viewpoint, being neither semi-simple, nor solvable, nor any of the other special classes of Lie groups so popular in mathematics. Moreover, these groups often act nonlinearly on the underlying space (taking us outside the domain of representation theory) and can even be only locally defined, with the transformations making sense only for group elements sufficiently near the identity. The relevant group actions, then, are much closer in spirit to Lie's original formulation of the subject in terms of local Lie groups acting on open subsets of Euclidean space, and runs directly counter to the modern tendencies towards abstraction and globalization which have enveloped much of present-day Lie group theory. Historically, the applications of Lie groups to differential equations pioneered by Lie and Noether faded into obscurity just as the global, abstract reformulation of differential geometry and Lie group theory championed by E. Cartan gained its ascendency in the mathematical community. The entire subject lay dormant for nearly half a century until G. Birkhoff called attention to the unexploited applications of Lie groups to the differential equations of fluid mechanics. Subsequently, Ovsiannikov and his school began a systematic program of successfully applying these methods to a wide range of physically important problems. The last two decades have witnessed a veritable explosion of research activity in this field, both in the applications to concrete physical systems, as well as extensions of the scope and depth of the theory itself. Nevertheless, many questions remain unresolved, and the full range of applicability of Lie group methods to differential equations is yet to be determined.

Roughly speaking, a symmetry group of a system of differential equations is a group which transforms solutions of the system to other solutions. In the classical framework of Lie, these groups consist of geometric transformations on the space of independent and dependent variables for the system, and act on solutions by transforming their graphs. Typical examples are groups of translations and rotations, as well as groups of scaling symmetries, but these certainly do not exhaust the range of possibilities. The great advantage of looking at continuous symmetry groups, as opposed to discrete symmetries such as reflections, is that they can all be found using explicit computational methods. This is not to say that discrete groups are not important in the study of differential equations (see, for example, Hejhal, [1], and the references therein), but rather that one must employ quite different methods to find or utilize them. Lie's fundamental discovery was that the complicated nonlinear conditions of invariance of the system under the group transformations could, in the case of a continuous group, be replaced by equivalent, but far simpler, *linear* conditions reflecting a form of "infinitesimal" invariance of the system under the generators of the group. In almost every physically important system of differential equations, these infinitesimal symmetry conditions—the so-called defining equations of the symmetry group of the system—can be explicitly solved in closed form and thus the most general continuous symmetry group of the system can be explicitly determined. The entire procedure consists of rather mechanical computations, and, indeed, several symbolic manipulation computer programs have been developed for this task.

Once one has determined the symmetry group of a system of differential equations, a number of applications become available. To start with, one can directly use the defining property of such a group and construct new solutions to the system from known ones. The symmetry group thus provides a means of classifying different symmetry classes of solutions, where two solutions are deemed to be equivalent if one can be transformed into the other by some group element. Alternatively, one can use the symmetry groups to effect a classification of families of differential equations depending on arbitrary parameters or functions; often there are good physical or mathematical reasons for preferring those equations with as high a degree of symmetry as possible. Another approach is to determine which types of differential equations admit a prescribed group of symmetries; this problem is also answered by infinitesimal methods using the theory of differential invariants.

In the case of ordinary differential equations, invariance under a one-parameter symmetry group implies that we can reduce the order of the equation by one, recovering the solutions to the original equation from those of the reduced equation by a single quadrature. For a single first order equation, this method provides an explicit formula for the general solution. Multi-parameter symmetry groups engender further reductions in order, but, unless the group itself satisfies an additional "solvability" requirement, we may not be able to recover the solutions to the original equation from those of the reduced equation by quadratures alone. If the system of ordinary differential equations is derived from a variational principle, either as the Euler–Lagrange equations of some functional, or as a Hamiltonian system, then the power of the symmetry group reduction method is effectively doubled. A one-parameter group of "variational" symmetries allows one to reduce the order of the system by two; the case of multi-parameter symmetry groups is more delicate.

Unfortunately, for systems of partial differential equations, the symmetry group is usually of no help in determining the general solution (although in special cases it may indicate when the system can be transformed into a more easily soluble system such as a linear system). However, one can use general symmetry groups to explicitly determine special types of solutions which are themselves invariant under some subgroup of the full symmetry group of the system. These "group-invariant" solutions are found by solving a reduced system of differential equations involving fewer independent variables than the original system (which presumably makes it easier to solve). For example, the solutions to a partial differential equation in two independent variables which are invariant under a given one-parameter symmetry group are all found by solving a system of ordinary differential equations. Included among these general group-invariant solutions are the classical similarity solutions coming from groups of scaling symmetries, and travelling wave solutions reflecting some form of translational invariance in the system, as well as many other explicit solutions of direct physical or mathematical importance. For many nonlinear systems, these are the *only* explicit, exact solutions which are available, and, as such, play an important role in both the mathematical analysis and physical applications of the system.

In 1918, E. Noether proved two remarkable theorems relating symmetry groups of a variational integral to properties of its associated Euler–Lagrange equations. In the first of these theorems, Noether shows how each one-parameter variational symmetry group gives rise to a conservation law of the Euler–Lagrange equations. Thus, for example, conservation of energy comes from the invariance of the problem under a group of time translations, while conservation of linear and angular momenta reflect translational and rotational invariance of the system. Chapter 4 is devoted to the so-called classical form of Noether's theorem, in which only the geometrical types of symmetry groups are used. Noether herself proved a far more general result and gave a one-to-one correspondence between symmetry groups and conservation laws. The general result necessitates the introduction of "generalized symmetries" which are groups whose infinitesimal generators depend not only on the independent and dependent variables of the system, but also the derivatives of the dependent variables. The corresponding group transformations will no longer act geometrically on the space of independent and dependent variables, transforming a function's graph point-wise, but are nonlocal transformations found by integrating an evolutionary system of partial differential equations. Each one-parameter group of symmetries of a variational problem, either geometrical or generalized, will give rise to a conservation law, and, conversely, every conservation law arises in this manner. The simplest example of a conserved quantity coming from a true generalized symmetry is the Runge–Lenz vector for the Kepler problem, but additional recent applications, including soliton equations and elasticity, has sparked a renewed interest in the general version of Noether's theorem. In Section 5.3 we prove a strengthened form of Noether's theorem, stating that for "normal" systems there is in fact a one-to-one correspondence between *nontrivial* variational symmetry groups and *nontrivial* conservation laws. The condition of normality is satisfied by most physically important systems of differential equations; abnormal systems are essentially those with nontrivial integrability conditions. An important class of abnormal systems, which do arise in general relativity, are those whose variational integral admits an infinite-dimensional symmetry group depending on an arbitrary function. Noether's second theorem shows that there is then a nontrivial relation among the ensuing Euler–Lagrange equations, and, consequently, nontrivial symmetries giving rise to only trivial conservation laws. Once found, conservation laws have many important applications, both physical and mathematical, including existence results, shock waves, scattering theory, stability, relativity, fluid mechanics, elasticity and so on. See the notes on Chapter 4 for a more extensive list, including references.

Neglected for many years following Noether's prescient work, generalized symmetries have recently been found to be of importance in the study of nonlinear partial differential equations which, like the Korteweg–de Vries equation, can be viewed as "completely integrable systems". The existence of infinitely many generalized symmetries, usually found via the recursion operator methods of Section 5.2, appears to be intimately connected with the possibility of linearizing the system, either directly through some change of variables, or, more subtly, through some form of inverse scattering method. Thus, the generalized symmetry approach, which is amenable to direct calculation as with ordinary symmetries, provides a systematic means of recognizing these remarkable equations and thereby constructing an infinite collection of conservation laws for them. (The construction of the related scattering problem requires different techniques such as the prolongation methods of Wahlquist and Estabrook, [1].) A systematic method for determining evolution equations having recursion operators, and hence classifying "integrable" systems, is provided by the theory of formal symmetries.

A number of the applications of symmetry group methods to partial differential equations are most naturally done using some form of Hamiltonian structure. The finite-dimensional formulation of Hamiltonian systems of ordinary differential equations is well known, but in preparation for the more recent theory of Hamiltonian systems of evolution equations, we are required to take a slightly novel approach to the whole subject of Hamiltonian mechanics. Here we will de-emphasize the use of canonical coordinates (the p's and q's of classical mechanics) and concentrate instead on the Poisson bracket as the cornerstone of the subject. The result is the more general concept of a Poisson structure, which *is* easily extended to include the infinite-dimensional theory of Hamiltonian systems of evolution equations. An important special case of a Poisson structure is the Lie–Poisson structure on the dual to a Lie algebra, originally discovered by Lie, and more recently used to great effect in geometric quantization, representation theory, and fluid and plasma mechanics. In this general approach to Hamiltonian mechanics, conservation laws can arise not only from symmetry properties of the system, but also from degeneracies of the Poisson bracket itself. In the finite-dimensional set-up, each one-parameter Hamiltonian symmetry group allows us to reduce the order of a system by two. In its modern formulation, the degree of reduction available for multi-parameter symmetry groups is given by the general theory of Marsden and Weinstein, which is based on the concept of a momentum map to the dual of the symmetry Lie algebra. In more recent work, there has been a fair amount of interest in systems of differential equations which possess not just one, but *two* distinct (but compatible) Hamiltonian structures. For such a "bi-Hamiltonian system", there is a direct recursive means of constructing an infinite hierarchy of mutually commuting flows (symmetries) and consequent conservation laws, indicating the system's complete integrability. Most of the soliton equations, as well as some of the finite-dimensional completely integrable Hamiltonian systems, are in fact bi-Hamiltonian systems.

Underlying much of the theory of generalized symmetries, conservation laws, and Hamiltonian structures for evolution equations is a subject known as the "formal calculus of variations", which constitutes a calculus specifically devised for answering a wide range of questions dealing with complicated algebraic identities among objects such as the Euler operator from the calculus of variations, generalized symmetries, total derivatives and more general differential operators, and several generalizations of the concept of a differential form. The principal result in the formal variational calculus is the local exactness of a certain complex—called the "variational complex"—which is in a sense the proper generalization to the variational or jet space context of the de Rham complex from algebraic topology. In recent years, this variational complex has been seen to play an increasingly important role in the development of the algebraic and geometric theory of the calculus of variations. Included as special cases of the variational complex are:

(1) a solution to the "inverse problem of the calculus of variations", which is to characterize those systems of differential equations which are the Euler–Lagrange equations for some variational problem;
(2) the characterization of "null Lagrangians", meaning those variational integrals whose Euler–Lagrange expressions vanish identically, as total divergences; and
(3) the characterization of trivial conservation laws, also known as "null divergences", as "total curls".

Each of these results is vital to the development of our applications of Lie groups to the study of conservation laws and Hamiltonian structures for evolution equations. Since it is not much more difficult to provide the proof of exactness of the full variational complex, Section 5.4 is devoted to a complete development of this proof and application to the three special cases of interest.

Although the book covers a wide range of different applications of Lie groups to differential equations, a number of important topics have necessarily been omitted. Most notable among these omissions is the connection between Lie groups and separation of variables. There are two reasons for this: first, there is an excellent, comprehensive text—Miller, [3]—already available; second, except for special classes of partial differential equations, such as Hamilton–Jacobi and Helmholtz equations, the precise connections between symmetries and separation of variables is not well understood at present. This is especially true in the case of systems of linear equations, or for fully nonlinear separation of variables; in neither case is there even a good definition of what separation of variables really entails, let alone how one uses symmetry properties of the system to detect coordinate systems in which separation of variables is possible. I have also not attempted to cover any of the vast area of representation theory, and the consequent applications to special function theory; see Miller, [1] or Vilenkin, [1]. Bifurcation theory is another fertile ground for group-theoretic applications; I refer the reader to the lecture notes of Sattinger, [1], and the references therein. Applications of symmetry groups to numerical analysis are given extensive treatment in Shokin, [1], and Dorodnitsyn, [1]. Applications to control theory can be found in van der Schaft, [1], and Ramakrishnan and Schaettler, [1]. See Maeda, [1], and Levi and Winternitz, [2], for applications to difference and differential-difference equations. Extensions of the present methods to boundary value problems for partial differential equations can be found in the books of Bluman and Cole, [1], and Seshadri and Na, [1], and to free boundary problems in Benjamin and Olver, [1]. Although I have given an extensive treatment to generalized symmetries in Chapter 5, the related concept of contact transformations introduced by Lie has not been covered, as it seems much less relevant to the equations arising in applications, and, for the most part, is subsumed by the more general theory presented here; see Anderson and Ibragimov, [1], Bluman and Kumei, [2], and the references therein for these types of transformation groups. Finally, we should mention the use of Lie group methods for differential equations arising in geometry, including, for example, motions in Riemannian manifolds, cf. Ibragimov, [1], or symmetric spaces and invariant differential operators associated with them, cf. Helgason, [1], [2].

Notes to the Reader

The guiding principle in the organization of this book has been so as to enable the reader whose principal goal is to apply Lie group methods to concrete problems to learn the basic computational tools and techniques as quickly as possible and with a minimum of theoretical diversions. At the same time, the computational applications have been placed on a solid theoretical foundation, so that the more mathematically inclined reader can readily delve further into the subject. Each chapter following the first has been arranged so that the applications and examples appear as quickly as feasible, with the more theoretical proofs and explanations coming towards the end. Even should the reader have more theoretical goals in mind, though, I would still strongly recommend that they learn the computational techniques and examples first before proceeding to the general theory. It has been said that it is far easier to abstract a general mathematical theory from a single well-chosen example than it is to apply an existing abstract theory to a specific example, and this, I believe, is certainly the case here.

For the reader whose main interest is in applications, I would recommend the following strategy for reading the book. The principal question is how much of the introductory theory of manifolds, vector fields, Lie groups and Lie algebras (which has, for convenience, been collected together in Chapter 1 and Section 2.1), really needs to be covered before one can proceed to the applications to differential equations starting in Section 2.2. The answer is, in fact, surprisingly little. Manifolds can for the most part be thought of locally, as open subsets of a Euclidean space $\mathbb{R}^m$ in which one has the freedom to change coordinates as one desires. Geometrical symmetry groups will just be collections of transformations on such a subset which satisfy certain elementary group axioms allowing one to compose successive symmetries, take inverses, etc. The key concept in the subject is the infinitesimal generator of a symmetry group. This is a vector field (of the type already familiar in vector calculus or fluid mechanics) on the underlying manifold or subset of $\mathbb{R}^m$ whose associated flow coincides with the one-parameter group it generates. One can regard the entire group of symmetries as being generated in this manner by composition of the basic flows of its infinitesimal generators. Thus a familiarity with the basic notation for and correspondence between a vector field and its flow is the primary concept required from Chapter 1. The other key result is the infinitesimal criterion for a system of algebraic equations to be invariant under such a group of transformations, which is embodied in Theorem 2.8. With these two tools, one can plunge ahead into the material on differential equations starting in Section 2.2, referring back to further results on Lie groups or manifolds as the need arises.

The generalization of the infinitesimal invariance criterion to systems of differential equations rests on the important technique of "prolonging" the group transformations to include not only the independent variables and dependent variables appearing in the system, but also the derivatives of the dependent variables. This is most easily accomplished in a geometrical manner through the introduction of spaces whose coordinates represent these derivatives: the "jet spaces" of Section 2.3. The key formula for computing symmetry groups of differential equations is the prolongation formula for the infinitesimal generators in Theorem 2.36. Armed with this formula (or, at least the special cases appearing in the following example) and the corresponding infinitesimal invariance criterion, one is ready to compute the symmetry groups of well-nigh any system of ordinary or partial differential equations which may arise. Several illustrative examples of the basic computational techniques required are presented in Section 2.4; readers are also advised to try their hands at some additional examples, either those in the exercises at the end of Chapter 2, or some system of differential equations of their own devising.

At this juncture, a number of options as to what to pursue next present themselves. For the devotee of ordinary differential equations, Section 2.5 provides a detailed summary of the basic method of Lie for integrating these equations using symmetry groups. See also Sections 4.2 and 6.3 for the case of ordinary differential equations with some form of variational structure, either in Lagrangian or Hamiltonian form. Those interested in determining explicit group-invariant solutions to partial differential equations can move directly on to Chapter 3. There the basic method for computing these solutions through reduction is outlined in Section 3.1 and illustrated by a number of examples in Section 3.2. The third section of this chapter addresses the problem of classification of these solutions, and does require some of the more sophisticated results on Lie algebras from Section 1.4. The final two sections of Chapter 3 are devoted to the underlying theory for the reduction method, and are not required for applications, although a discussion of the important Pi theorem from dimensional analysis does appear in Section 3.4.

The reader whose principal interest is in the derivation of conservation laws using Noether's theorem can move directly from Section 2.4 to Chapter 4, which is devoted to the "classical" form of this result. A brief review of the most basic concepts required from the calculus of variations is presented in Section 4.1. The introduction of symmetry groups and the basic infinitesimal invariance criterion for a variational integral is the subject of Section 4.2, along with the reduction procedures available for ordinary differential equations which are the Euler–Lagrange equations for some variational problem. The third section introduces the general notion of a conservation law. Here the treatment is more novel; the guiding concept is the correspondence between conservation laws and their "characteristics", although the technically complicated proof of Theorem 4.26 can be safely omitted on a first reading. Once one learns to deal with conservation laws in characteristic form, the statement and implementation of Noether's theorem is relatively straightforward.

Beginning with Chapter 5, a slightly higher degree of mathematical sophistication is required, although one can still approach much of the material on generalized symmetries and conservation laws from a purely computational point of view with only a minimum of the Lie group machinery. The most difficult algebraic manipulations have been relegated to Section 5.4, where the variational complex is developed in its full glory for the true aficionado. Incidentally, Section 5.4, along with Chapter 7 on Hamiltonian structures for evolution equations are the only parts of the book where the material on differential forms in Section 1.5 is used to any great extent. Despite their seeming complexity, the proofs in Section 5.4 are a substantial improvement over the current versions available in the literature.

Chapter 6 on finite-dimensional Hamiltonian systems is by-and-large independent of much of the earlier material in the book. Up through the reduction method for one-parameter symmetry groups, not very much of the material on Lie groups is required. However, the Marsden–Weinstein reduction theory for multi-parameter symmetry groups does require some of the more sophisticated results on Lie algebras from Sections 1.4 and 3.3. Chapter 7 depends very much on an understanding of the Poisson bracket approach to Hamiltonian mechanics adopted in Chapter 6, and, to a certain extent, the formal variational calculus methods of Section 5.4. Nevertheless, acquiring a basic agility in the relevant computational applications is not that difficult.

The exercises which appear at the end of each chapter vary considerably in their range of difficulty. A few are fairly routine calculations based on the material in the text, but a substantial number provide significant extensions of the basic material. The more difficult exercises are indicated by an asterisk; one or two, signaled by a double asterisk, might be better classed as miniature research projects. A number of the results presented in the exercises are new; otherwise, I have tried to give the most pertinent references where appropriate. Here references have been selected more on the basis of direct relevance for the problem as stated rather than on the basis of historical precedence.

At the end of each chapter is a brief set of notes, mostly discussing the historical details and references for the results discussed there. While I cannot hope to claim full historical accuracy, these notes do represent a fair amount of research into the historical roots of the subject. I have tried to determine the origins and subsequent history of many of the important developments in the area. The resulting notes undoubtedly reflect a large number of personal biases, but, I hope, may provide the groundwork for a more serious look into the fascinating and, at times, bizarre history of this subject, a topic which is well worth the attention of a true historian of mathematics. Although I have, for the most part, listed what I determined to be significant papers in the historical development of the subject, owing to the great duplication of efforts over the decades, I have obviously been unable to provide an exhaustive listing of all the relevant references. I sincerely apologize to those authors whose work does play a significant role in the development, but was inadvertently missed from these notes.

CHAPTER 1

Introduction to Lie Groups

Roughly speaking, a Lie group is a "group" which is also a "manifold". Of course, to make sense of this definition, we must explain these two basic concepts and how they can be related. Groups arise as an algebraic abstraction of the notion of symmetry; an important example is the group of rotations in the plane or three-dimensional space. Manifolds, which form the fundamental objects in the field of differential geometry, generalize the familiar concepts of curves and surfaces in three-dimensional space. In general, a manifold is a space that locally looks like Euclidean space, but whose global character might be quite different. The conjunction of these two seemingly disparate mathematical ideas combines, and significantly extends, both the algebraic methods of group theory and the multi-variable calculus used in analytic geometry. This resulting theory, particularly the powerful infinitesimal techniques, can then be applied to a wide range of physical and mathematical problems.

The goal of this chapter is to provide the reader with a relatively quick and painless introduction to the theory of manifolds and Lie groups in a form that will be conducive to applications to differential equations. No prior knowledge of either group theory or differential geometry is required, but a good background in basic analysis (i.e. "advanced calculus"), including the inverse and implicit function theorems, will be assumed. Of necessity, proofs of most of the "hard" theorems in Lie group theory will be omitted; references can be found in the notes at the end of the chapter.

Throughout this chapter, I have tried to strike a balance between the local coordinate picture, in which a manifold just looks like an open subset of some Euclidean space, and the more modern global formulation of the theory. Each has its own particular uses and advantages, and it would be a mistake to emphasize one or the other unduly. The applications-oriented reader might question the inclusion of the global framework here, since admittedly most of the applications of the theory presented in this book take place on open subsets of Euclidean space. Suffice it to say that the geometrical insight and understanding offered by the general notion of a manifold amply repays the relatively slight effort required to gain familiarity with the definition. However, if the reader still remains unconvinced, they can replace the word "manifold" wherever it occurs by "open subset of Euclidean space" without losing too much of the flavour or range of applicability of the theory. With this approach, they should concentrate on the sections on local Lie groups (which were, indeed, the way Lie himself thought of Lie groups) and use these as the principal objects of study.

The first section gives a basic outline of the general concept of a manifold, the second doing the same for Lie groups, both local and global. In practice Lie groups arise as groups of symmetries of some object, or, more precisely, as local groups of transformations acting on some manifold; the second section gives a brief look at these. The most important concept in the entire theory is that of a vector field, which acts as the "infinitesimal generator" of some one-parameter Lie group of transformations. This concept is fundamental for both the development of the theory of Lie groups and the applications to differential equations. It has the crucial effect of replacing complicated nonlinear conditions for the symmetry of some object under a group of transformations by easily verifiable linear conditions reflecting its infinitesimal symmetry under the corresponding vector fields. This technique will be explored in depth for systems of algebraic and differential equations in Chapter 2. The notion of vector field then leads to the concept of a Lie algebra, which can be thought of as the infinitesimal generator of the Lie group itself, the theory of which is developed in Section 1.4. The final section of this chapter gives a brief introduction to differential forms and integration on manifolds.

1.1. Manifolds

Throughout most of this book, we will be primarily interested in objects, such as differential equations, symmetry groups and so on, which are defined on open subsets of Euclidean space $\mathbb{R}^m$. The underlying geometrical features of these objects will be independent of any particular coordinate system on the open subset which might be used to define them, and it becomes of great importance to free ourselves from the dependence on particular local coordinates, so that our objects will be essentially "coordinate-free". More specifically, if $U \subset \mathbb{R}^m$ is open and $\psi: U \to V$, where $V \subset \mathbb{R}^m$ is open, is any diffeomorphism, meaning that ψ is an infinitely differentiable map with infinitely differentiable inverse, then objects defined on U will have equivalent counterparts on V. Although the precise formulae for the object on U and its

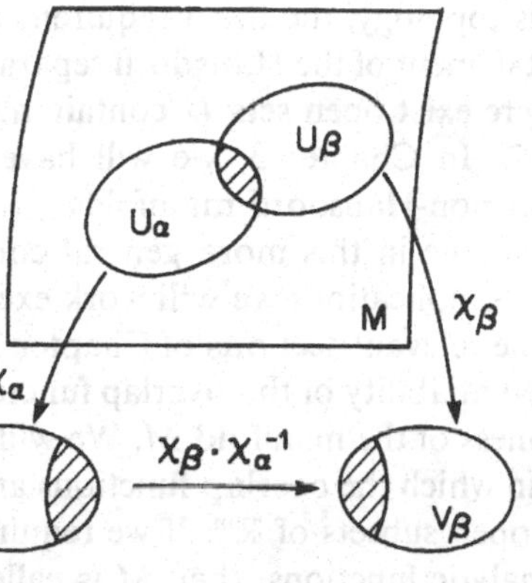

Figure 1. Coordinate charts on a manifold.

counterpart on V will, in general, change, the essential underlying properties will remain the same. Once we have freed ourselves from this dependence on coordinates, it is a small step to the general definition of a smooth manifold. From this point of view, manifolds provide the natural setting for studying objects that do not depend on coordinates.

Definition 1.1. An *m-dimensional manifold* is a set M, together with a countable collection of subsets $U_\alpha \subset M$, called *coordinate charts*, and one-to-one functions $\chi_\alpha \colon U_\alpha \to V_\alpha$ onto connected open subsets $V_\alpha \subset \mathbb{R}^m$, called *local coordinate maps*, which satisfy the following properties:

(a) The coordinate charts cover M:

$$\bigcup_\alpha U_\alpha = M.$$

(b) On the overlap of any pair of coordinate charts $U_\alpha \cap U_\beta$ the composite map

$$\chi_\beta \circ \chi_\alpha^{-1} \colon \chi_\alpha(U_\alpha \cap U_\beta) \to \chi_\beta(U_\alpha \cap U_\beta)$$

is a smooth (infinitely differentiable) function.

(c) If $x \in U_\alpha$, $\tilde{x} \in U_\beta$ are distinct points of M, then there exist open subsets $W \subset V_\alpha$, $\tilde{W} \subset V_\beta$, with $\chi_\alpha(x) \in W$, $\chi_\beta(\tilde{x}) \in \tilde{W}$, satisfying

$$\chi_\alpha^{-1}(W) \cap \chi_\beta^{-1}(\tilde{W}) = \varnothing.$$

The coordinate charts $\chi_\alpha \colon U_\alpha \to V_\alpha$ endow the manifold M with the structure of a topological space. Namely, we require that for each open subset $W \subset V_\alpha \subset \mathbb{R}^m$, $\chi_\alpha^{-1}(W)$ be an open subset of M. These sets form a *basis* for the topology on M, so that $U \subset M$ is open if and only if for each $x \in U$ there is a neighbourhood of x of the above form contained in U; so $x \in \chi_\alpha^{-1}(W) \subset U$ where $\chi_\alpha \colon U_\alpha \to V_\alpha$ is a coordinate chart containing x, and W is an open subset of V_α. In terms of this topology, the third requirement in the definition of a manifold is just a restatement of the Hausdorff separation axiom: If $x \neq \tilde{x}$ are points in M, then there exist open sets U containing x and $\tilde{U}$ containing $\tilde{x}$ such that $U \cap \tilde{U} = \varnothing$. In Chapter 3, we will have occasion to drop this property and consider non-Hausdorff manifolds. Many of the results of the other chapters remain true in this more general context, but as this introduces some technical complications we will work exclusively with Hausdorff manifolds except in the relevant sections of Chapter 3.

The degree of differentiability of the overlap functions $\chi_\beta \circ \chi_\alpha^{-1}$ determines the degree of smoothness of the manifold M. We will be primarily interested in *smooth manifolds*, in which the overlap functions are smooth, meaning C^∞, diffeomorphisms on open subsets of $\mathbb{R}^m$. If we require the overlap functions $\chi_\beta \circ \chi_\alpha^{-1}$ to be real analytic functions, then M is called an *analytic manifold*. Most classical examples of manifolds are in fact analytic. Alternatively, we can weaken the differentiability requirements and consider C^k-*manifolds*, in which the overlap functions are only required to have continuous derivatives up to order k. Many of our results hold under these weaker differentiability requirements, but to avoid keeping track of precisely how many continuous derivatives are needed at each stage, we simply stick to the case of smooth or, occasionally, analytic manifolds. The weakening of our differentiability hypotheses is left to the interested reader. We begin by illustrating the general definition of a manifold with a few elementary examples.

Example 1.2. The simplest m-dimensional manifold is just Euclidean space $\mathbb{R}^m$ itself. There is a single coordinate chart $U = \mathbb{R}^m$, with local coordinate map given by the identity: $\chi = \mathbb{1}: \mathbb{R}^m \to \mathbb{R}^m$. More generally, any open subset $U \subset \mathbb{R}^m$ is an m-dimensional manifold with a single coordinate chart given by U itself, with local coordinate map the identity again. Conversely, if M is any manifold with a single global coordinate chart $\chi: M \to V \subset \mathbb{R}^m$, we can identify M with its image V, an open subset of $\mathbb{R}^m$.

Example 1.3. The unit sphere

$$S^2 = \{(x, y, z): x^2 + y^2 + z^2 = 1\}$$

is a good example of a nontrivial two-dimensional manifold realized as a surface in $\mathbb{R}^3$. Let

$$U_1 = S^2 \setminus \{(0, 0, 1)\}, \qquad U_2 = S^2 \setminus \{(0, 0, -1)\}$$

be the subsets obtained by deleting the north and south poles respectively. Let

$$\chi_\alpha: U_\alpha \to \mathbb{R}^2 \simeq \{(x, y, 0)\}, \qquad \alpha = 1, 2,$$

be stereographic projections from the respective poles, so

$$\chi_1(x, y, z) = \left(\frac{x}{1-z}, \frac{y}{1-z}\right), \qquad \chi_2(x, y, z) = \left(\frac{x}{1+z}, \frac{y}{1+z}\right).$$

It can be easily checked that on the overlap $U_1 \cap U_2$,

$$\chi_1 \circ \chi_2^{-1}: \mathbb{R}^2 \backslash \{0\} \to \mathbb{R}^2 \backslash \{0\}$$

is a smooth diffeomorphism, given by the inversion

$$\chi_1 \circ \chi_2^{-1}(x, y) = \left(\frac{x}{x^2 + y^2}, \frac{y}{x^2 + y^2} \right).$$

The Hausdorff separation property follows easily from that of $\mathbb{R}^3$, so S^2 is a smooth, indeed analytic, two-dimensional manifold. The unit sphere is a particular case of the general concept of a surface in $\mathbb{R}^3$, which historically provided the principal motivating example for the development of the general theory of manifolds.

Example 1.4. An easier example is the unit circle

$$S^1 = \{(x, y): x^2 + y^2 = 1\},$$

which is similarly seen to be a one-dimensional manifold with two coordinate charts. Alternatively, we can identify a point on S^1 with its angular coordinate θ, where $(x, y) = (\cos\theta, \sin\theta)$, with two angles being identified if they differ by an integral multiple of 2π.

The Cartesian product

$$T^2 = S^1 \times S^1$$

of S^1 with itself is a two-dimensional manifold called a *torus*, and can be though of as the surface of an inner tube. (See Example 1.6.) The points on T^2 are given by pairs (θ, ρ) of angular coordinates, with two pairs being identified if they differ by integral multiples of 2π. In other words, (θ, ρ) and $(\tilde{\theta}, \tilde{\rho})$ describe the same point on T^2 if and only if

$$\theta - \tilde{\theta} = 2k\pi \quad \text{and} \quad \rho - \tilde{\rho} = 2l\pi,$$

for integers k, l. Thus T^2 can be covered by three coordinates charts

$$U_1 = \{(\theta, \rho): 0 < \theta < 2\pi, 0 < \rho < 2\pi\},$$
$$U_2 = \{(\theta, \rho): \pi < \theta < 3\pi, \pi < \rho < 3\pi\},$$
$$U_3 = \{(\theta, \rho\}: \pi/2 < \theta < 5\,\pi/2, \pi/2 < \rho < 5\,\pi/2\}.$$

The first overlap function is

$$\chi_1 \circ \chi_2^{-1}(\theta, \rho) = \begin{cases} (\theta, \rho), & \pi < \theta < 2\pi, \quad \pi < \rho < 2\pi, \\ (\theta - 2\pi, \rho), & 2\pi < \theta < 3\pi, \quad \pi < \rho < 2\pi, \\ (\theta, \rho - 2\pi), & \pi < \theta < 2\pi, \quad 2\pi < \rho < 3\pi, \\ (\theta - 2\pi, \rho - 2\pi), & 2\pi < \theta < 3\pi, \quad 2\pi < \rho < 3\pi \end{cases}$$

on the intersection $U_1 \cap U_2$, which is the set of all (θ, ρ) with neither θ nor ρ being an integral multiple of π. More generally, an m-dimensional torus is given by the m-fold Cartesian product $T^m = S^1 \times \cdots \times S^1$ of S^1 with itself.

In general, if M and N are smooth manifolds of dimension m and n respectively, then their Cartesian product $M \times N$ is easily seen to be a smooth $(m+n)$-dimensional manifold. If $\chi_\alpha: U_\alpha \to V_\alpha \subset \mathbb{R}^m$, and $\tilde{\chi}_\beta: \tilde{U}_\beta \to \tilde{V}_\beta \subset \mathbb{R}^n$ are coordinate charts on M and N respectively then their Cartesian products

$$\chi_\alpha \times \tilde{\chi}_\beta: U_\alpha \times \tilde{U}_\beta \to V_\alpha \times \tilde{V}_\beta \subset \mathbb{R}^m \times \mathbb{R}^n \simeq \mathbb{R}^{m+n}$$

provide coordinate charts on $M \times N$. The verification of the requirements of Definition 1.1 for $M \times N$ are left to the reader.

Change of Coordinates

Besides the basic coordinate charts $\chi_\alpha: U_\alpha \to V_\alpha$ given in the definition of M, one can always adjoin many additional coordinate charts $\chi: U \to V \subset \mathbb{R}^m$, subject to the requirement that they be *compatible* with the given charts. This means that for each α, $\chi \circ \chi_\alpha^{-1}$ is smooth on the intersection $\chi_\alpha(U \cap U_\alpha)$. Thus, restriction of a given set of local coordinates χ_α to a smaller chart $\tilde{U}_\alpha \subset U_\alpha$ will also be a valid coordinate chart. An additional possibility is to compose a given local coordinate map $\chi_\alpha: U_\alpha \to V_\alpha$ with any diffeomorphism $\psi: V_\alpha \to \tilde{V}_\alpha$ of $\mathbb{R}^m$. Such a diffeomorphism is referred to as a *change of coordinates*. Since both χ_α and $\psi \circ \chi_\alpha$ are equally valid local coordinates on the chart U_α, any property of M, or object defined on M, must be independent of any particular choice of local coordinates. (Of course, the explicit formulae for the given object may change when going from one coordinate chart to another, but the intrinsic characterization of the object remains coordinate-free.) If we choose to define an object on a manifold using its formula in a given coordinate chart, we must then check that the definition is actually independent of the particular coordinates used. This will require an investigation into how the object behaves under changes of coordinates. Often, as computations are most easily done in local coordinates, the choice of a special coordinate chart in which the object of interest takes a particularly simple form will enable us to considerably simplify many of these computations. The use of this basic technique will become clearer as we continue.

Often one expands the collection of coordinate charts to include all those compatible with the defining charts. The resulting collection, called a *maximal collection* of charts or *atlas* on M, still satisfies the basic properties (a), (b), (c) of Definition 1.1 (but, of course, is no longer countable!). The easy details of proving that two charts, compatible with the defining charts, are mutually compatible, are left to the reader.

Usually, in talking about local coordinates on a manifold, we will dispense with explicit reference to the map χ_α defining the local coordinate chart, and speak as if the local coordinate expressions were identical with the corresponding points on the manifold itself. Thus, we will say "let $x = (x^1, \ldots, x^m)$ be local coordinates on M", which, more precisely, means that there is a local coordinate chart $\chi_\alpha: U_\alpha \to V_\alpha$, with $U_\alpha \subset M$ open, $V_\alpha \subset \mathbb{R}^m$ open, and such that each p in U_α has local coordinates $x = \chi_\alpha(p)$. Since χ_α is one-to-one, we can clearly identify p with its local coordinate expression x. By the compatibility condition, we know that $y = (y^1, \ldots, y^m)$ are also local coordinates if and only if on the overlap of the two coordinate charts there is a diffeomorphism $y = \psi(x)$ defined on an open subset of $\mathbb{R}^m$ relating the two coordinates. For example, in the case of the circle S^1, the angle $-\pi < \theta = \arctan(y/x) < \pi$ is a local coordinate on $S^1 \setminus \{(-1, 0)\}$. The ratio $\rho = y/x$ is a local coordinate on $S^1 \cap \{x > 0\}$. On the overlap, the change of coordinates is given by $\rho = \tan\theta$, which is a diffeomorphism from the interval $(-\pi/2, \pi/2)$ to $\mathbb{R}$.

Maps Between Manifolds

If M and N are smooth manifolds, a map $F: M \to N$ is said to be *smooth* if its local coordinate expression is a smooth map in every coordinate chart. In other words, for every coordinate chart $\chi_\alpha: U_\alpha \to V_\alpha \subset \mathbb{R}^m$ on M and every chart $\tilde{\chi}_\beta: \tilde{U}_\beta \to \tilde{V}_\beta \subset \mathbb{R}^n$ on N, the composite map

$$\tilde{\chi}_\beta \circ F \circ \chi_\alpha^{-1}: \mathbb{R}^m \to \mathbb{R}^n$$

is a smooth map wherever it is defined (i.e. on the subset $\chi_\alpha[U_\alpha \cap F^{-1}(\tilde{U}_\beta)]$). In other words, a smooth map is of the form $y = F(x)$, where F is a smooth function on the open subsets giving local coordinates x on M and y on N.

Example 1.5. An easy example is provided by the map $f: \mathbb{R} \to S^1$, $f(t) = (\cos t, \sin t)$. In terms of the angular coordinate θ on S^1, f is a linear function: $\theta = t \bmod 2\pi$, and so is clearly smooth.

Example 1.6. For a less trivial example we show how the torus T^2 can be mapped smoothly into $\mathbb{R}^3$. Define $F: T^2 \to \mathbb{R}^3$ by

$$F(\theta, \rho) = ((\sqrt{2} + \cos\rho)\cos\theta, (\sqrt{2} + \cos\rho)\sin\theta, \sin\rho).$$

Then F is clearly smooth in θ and ρ, and one-to-one. The image of F is the toroidal surface in $\mathbb{R}^3$ given by the single equation

$$x^2 + y^2 + z^2 + 1 = 2\sqrt{2(x^2 + y^2)}.$$

Thus T^2 can be realized as a surface in $\mathbb{R}^3$. The local coordinates (θ, ρ) on T^2 serve as a parameterization of the image in $\mathbb{R}^3$.

The Maximal Rank Condition

Definition 1.7. Let $F: M \to N$ be a smooth mapping from an m-dimensional manifold M to an n-dimensional manifold N. The *rank* of F at a point $x \in M$ is the rank of the $n \times m$ Jacobian matrix $(\partial F^i/\partial x^j)$ at x, where $y = F(x)$ is expressed in any convenient local coordinates near x. The mapping F is of *maximal rank* on a subset $S \subset M$ if for each $x \in S$ the rank of F is as large as possible (i.e. the minimum of m and n).

The reader can easily check that the definition of the rank of F at x does not depend on the particular local coordinates chosen on M or on N. For example, the rank of $F(x, y) = xy$ on $\mathbb{R}^2$ is 1 at all points except the origin (0, 0) since its Jacobian matrix $(F_x, F_y) = (y, x)$ is nonzero except at $x = y = 0$. (Here and elsewhere, subscripts denote derivatives, so $F_x = \partial F/\partial x$, etc.)

Theorem 1.8. *Let $F: M \to N$ be of maximal rank at $x_0 \in M$. Then there are local coordinates $x = (x^1, \ldots, x^m)$ near x_0, and $y = (y^1, \ldots, y^n)$ near $y_0 = F(x_0)$ such that in these coordinates F has the simple form*

$$y = (x^1, \ldots, x^m, 0, \ldots, 0), \quad \textit{if } n > m,$$

or

$$y = (x^1, \ldots, x^n), \qquad \textit{if } n \leqslant m.$$

This theorem is an easy consequence of the implicit function theorem—see Boothby, [1; Theorem II.7.1] for the proof. It is the first illustration of our contention that one can significantly simplify objects (in this case functions) on manifolds through a judicious choice of local coordinates.

Submanifolds

The previous examples of surfaces in $\mathbb{R}^3$—the sphere and the torus—are special cases of the general notion of a submanifold. Naïvely, given a smooth manifold M, a submanifold $N \subset M$ should be a subset which is also a smooth manifold in its own right. However, this preliminary definition can be interpreted in several fundamentally different ways, so we need to be more careful. There are also several methods of describing submanifolds, either implicitly by the vanishing of some smooth functions, as was the case with the sphere, or parametrically by some local parametrization, as we did initially with the torus. Both methods are very useful; we begin though with the latter, which leads to a more general notion of submanifolds.

Definition 1.9. Let M be a smooth manifold. A *submanifold* of M is a subset $N \subset M$, together with a smooth, one-to-one map $\phi: \tilde{N} \to N \subset M$ satisfying the maximal rank condition everywhere, where the *parameter space* $\tilde{N}$ is some other manifold and $N = \phi(\tilde{N})$ is the image of ϕ. In particular, the dimension of N is the same as that of $\tilde{N}$, and does not exceed the dimension of M.

The map ϕ is often called an *immersion*, and serves to define a parametrization of the submanifold N. Often such a submanifold is referred to as an *immersed submanifold*, to emphasize the difference between this definition and other notions of submanifold. In this book, the term "submanifold" without qualifications always refers to "immersed submanifold" as in the above definition. The maximal rank condition is needed to ensure that N does not have singularities. For instance, the function $\phi(t) = (t^2, t^3)$ is a smooth map from $\mathbb{R}$ to $\mathbb{R}^2$, but the image of ϕ is the curve $y^2 = x^3$, which has a cusp at (0, 0). The Jacobian matrix is $\dot{\phi}(t) = (2t, 3t^2)$, which is not of maximal rank at $t = 0$, indicating the appearance of a singularity in the image of ϕ.

The following series of examples will indicate some of the possibilities for submanifolds which are allowed by this definition. As the reader will see, although the maximal rank condition does have the effect of eliminating singularities like cusps, general submanifolds can still exhibit rather bizarre properties.

Example 1.10. In all of these examples of submanifolds, the parameter space $\tilde{N} = \mathbb{R}$ is the real line, with $\phi\colon \mathbb{R} \to M$ parameterizing a one-dimensional submanifold $N = \phi(\mathbb{R})$ of some manifold M.

(a) Let $M = \mathbb{R}^3$. Then

$$\phi(t) = (\cos t, \sin t, t)$$

defines a circular helix spiralling up the z-axis. Here ϕ is clearly one-to-one, and $\dot{\phi} = (-\sin t, \cos t, 1)$ never vanishes, so the maximal rank condition holds.

(b) Let $M = \mathbb{R}^2$, and

$$\phi(t) = ((1 + e^{-t}) \cos t, (1 + e^{-t}) \sin t).$$

Then as $t \to \infty$, N spirals in to the unit circle $x^2 + y^2 = 1$. Similarly, $\tilde{\phi}(t) = (e^{-t} \cos t, e^{-t} \sin t)$ defines a logarithmic spiral at the origin.

(c) Let $M = \mathbb{R}^2$ again, and consider the map

$$\hat{\phi}(t) = (\sin t, 2 \sin(2t)).$$

Then $\hat{\phi}$ parametrizes a figure eight, which is a curve with self-intersections; namely $\hat{\phi}(t) = (0, 0)$ whenever t is an integral multiple of π. By slightly modifying this example, for instance

$$\phi(t) = (\sin(2 \arctan t), 2 \sin(4 \arctan t)),$$

we can arrange that the parametrization is one-to-one, with the curve passing through the origin just once. The maximal rank condition holds everywhere. The image of ϕ is again the figure eight, so we have a parametrization of a submanifold with "apparent" self-intersections, even though the immersion ϕ is one-to-one. Note that the same figure eight can be parametrized in a

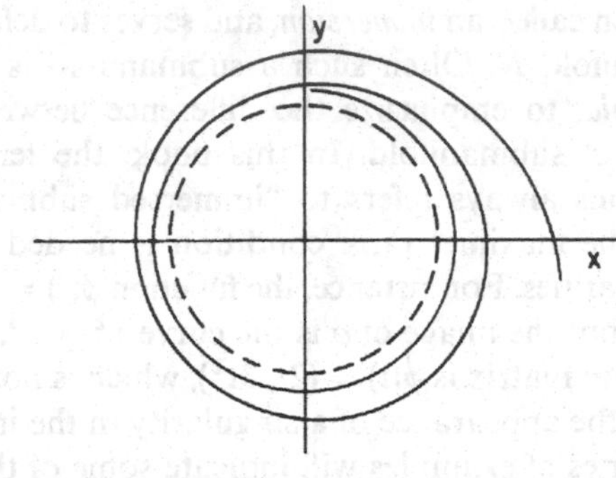

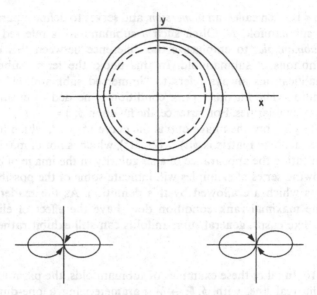

Figure 2. Examples of submanifolds.

different, inequivalent way:

$$\tilde{\phi}(t) = (-\sin(2 \arctan t), 2 \sin(4 \arctan t)).$$

The image of $\tilde{\phi}$ is the same, but the composition $\phi \circ \tilde{\phi}^{-1}: \mathbb{R} \to \mathbb{R}$ is *not* a continuous map!

This example shows that in general we must specify not only the subset $N \subset M$, but also the immersion $\phi: \tilde{N} \to M$ in order to define a submanifold unambiguously.

(d) Let $M = T^2$ be the two-dimensional torus with angular coordinates (θ, ρ). Let $\phi: \mathbb{R} \to T^2$ be the curve $\phi(t) = (t, \omega t)$, where ω is some fixed real number and the coordinates are taken modulo integral multiples of 2π as before. Note that $\dot{\phi} = (1, \omega)$, so the maximal rank condition is satisfied. If $\omega = p/q$ is a rational number, then ϕ is not one-to-one; indeed $\phi(t + 2\pi q) = \phi(t)$, so the image of ϕ is a closed curve on T^2. It can be realized as a one-dimensional manifold by using $\tilde{N} = S^1$ as the parametrizing manifold, with $\tilde{\phi}(\theta) = (q\theta, \omega q\theta)$, $\theta \in S^1$ (provided p and q have no common factors). If ω is irrational, ϕ itself is one-to-one and the image curve $N = \phi(\mathbb{R})$ can, without too much difficulty, be shown to a dense submanifold of T^2 whose closure is the entire torus. (See Boothby, [1; p. 86] for the details.) An analogous example can be constructed in $\mathbb{R}^3$ by following ϕ with the map $F: T^2 \to \mathbb{R}^3$ given in Example 1.6. Thus for ω irrational,

$$\phi(t) = ((\sqrt{2} + \cos \omega t) \cos t, (\sqrt{2} + \cos \omega t) \sin t, \sin \omega t)$$

parametrizes a one-dimensional submanifold of $\mathbb{R}^3$ whose closure is the entire two-dimensional toroidal surface $x^2 + y^2 + z^2 + 1 = 2\sqrt{2(x^2 + y^2)}$.

Regular Submanifolds

The latter two examples in 1.10 are perhaps more pathological than what one might wish to consider as submanifolds. Although, as we will see, there are good reasons for retaining Definition 1.9 as the general definition of a submanifold, it is also helpful to distinguish a class of examples, the regular or embedded submanifolds, which corresponds perhaps more accurately to one's intuitive notion of submanifold.

Definition 1.11. A *regular submanifold* N of a manifold M is a submanifold parametrized by $\phi: \tilde{N} \to M$ with the property that for each x in N there exist arbitrarily small open neighbourhoods U of x in M such that $\phi^{-1}[U \cap N]$ is a connected open subset of $\tilde{N}$.

In Example 1.10, (a) and (b) are both regular submanifolds, whereas (c) and (d) (for irrational ω) are not. In case (c), any neighbourhood U of $(0, 0)$ will contain both the piece of the curve passing through $(0, 0)$ together with the two "ends" of the curve coming back to the origin. In other words, $\phi^{-1}[U]$ consists of at least three disjoint open intervals $(-\infty, a)$, (b, c), $(d, +\infty)$ with $a < b < 0 < c < d$. Similarly in case (d), if ω is irrational and U is any open subset of T^2, then $\phi^{-1}[U]$ consists of an infinite collection of disjoint open intervals.

As a consequence of the Implicit Function Theorem 1.8 we obtain a local-coordinate characterization of regularity.

Lemma 1.12. *An n-dimensional submanifold $N \subset M$ is regular if and only if for each $x_0 \in N$ there exist local coordinates $x = (x^1, \ldots, x^m)$ defined on a neighbourhood U of x_0 such that*

$$N \cap U = \{x: x^{n+1} = \cdots = x^m = 0\}.$$

Such a coordinate chart is called a *flat coordinate chart* on M. Note that, in view of this lemma, for regular submanifolds $N \subset M$ we can dispense with the parametrizing manifold $\tilde{N}$ and just treat N as a manifold in its own right. Namely the flat local coordinates $x = (x^1, \ldots, x^m)$ on $U \subset M$ induce local coordinates, namely $\tilde{x} = (x^1, \ldots, x^n)$, on $U \cap N$. The parametrization thereby is replaced by the natural inclusion $N \subset M$.

Implicit Submanifolds

Instead of defining a surface S in $\mathbb{R}^3$ parametrically, an alternative method is to define it *implicitly* by the vanishing of a smooth function:

$$S = \{F(x, y, z) = 0\}.$$

If we assume that the gradient $\nabla F = (F_x, F_y, F_z)$ never vanishes on S, then by the implicit function theorem, at each point (x_0, y_0, z_0) in S we can solve for one of the variables x, y or z in terms of the other two. Thus if $F_z(x_0, y_0, z_0) \neq 0$, there is a neighbourhood U_α of (x_0, y_0, z_0) such that in U_α, S is given as the graph $z = f(x, y)$ of some smooth function f defined on an open subset $\tilde{V}_\alpha \subset \mathbb{R}^2$. This permits us to define a local coordinate chart on S by projecting along the z-axis; in other words, set $\tilde{U}_\alpha = S \cap U_\alpha$, with $\chi_\alpha\colon \tilde{U}_\alpha \to \tilde{V}_\alpha$, $\chi_\alpha(x, y, z) = (x, y)$. Similar constructions apply if F_y or F_x is non-zero. On the overlap $\tilde{U}_\alpha \cap \tilde{U}_\beta$, if $\tilde{U}_\alpha$ is given by $z = f(x, y)$, so $\chi_\alpha(x, y, z) = (x, y)$, and $\tilde{U}_\beta$ by $y = h(x, z)$, say, so $\chi_\beta(x, y, z) = (x, z)$, then

$$\chi_\beta \circ \chi_\alpha^{-1}(x, y) = \chi_\beta(x, y, f(x, y)) = (x, f(x, y)),$$

which is clearly smooth with smooth inverse $\chi_\alpha \circ \chi_\beta^{-1}(x, z) = (x, h(x, z))$. Thus S is a two-dimensional submanifold of $\mathbb{R}^3$. This motivates the general concept of an *implicitly defined submanifold.*

Theorem 1.13. *Let M be a smooth m-dimensional manifold, and $F\colon M \to \mathbb{R}^n$, $n \leqslant m$, be a smooth map. If F is of maximal rank on the subset $N = \{x\colon F(x) = 0\}$, then N is a regular, $(m - n)$-dimensional submanifold of M.*

The proof of this theorem follows easily from the implicit function theorem using arguments similar to the above case of surfaces in $\mathbb{R}^3$. Indeed, Theorem 1.8 says that we can choose local coordinates $x = (x^1, \ldots, x^m)$ on M near each $x_0 \in N$ such that $F(x) = (x^1, \ldots, x^n)$. Thus, in terms of these coordinates, $N = \{x^1 = \cdots = x^n = 0\}$, and so the x's provide the flat local coordinates for N near x_0. Moreover, the latter $m - n$ components $(x^{n+1}, \ldots, x^m)$ then provide local coordinates on N itself. In particular, this proves that N is a regular submanifold. Note especially that we do not require that the rank of F be maximal everywhere on M—this condition is only needed on the subset N where F vanishes. If, however, F is of maximal rank everywhere, then *every* level set of F, $\{x\colon F(x) = c\}$, is a regular $(m - n)$-dimensional submanifold of M.

For example, $F(x, y, z) = x^2 + y^2 + z^2 - 2\sqrt{2(x^2 + y^2)}$ is of maximal rank everywhere on $\mathbb{R}^3$ except on the z-axis (where it is not even smooth) and the circle $\{x^2 + y^2 = 2, z = 0\}$. The level sets $\{(x, y, z)\colon F(x, y, z) = c\}$ are tori for $-2 < c < \sqrt{2} - 2$, and like spheres with indented dimples on the z-axis for $c \geqslant \sqrt{2} - 2$. For $c = -2$, the level set is the circle $\{x^2 + y^2 = 2, z = 0\}$, on which the gradient of F vanishes. This example shows the importance of both the differentiability and the maximal rank conditions for the validity of the theorem.

Curves and Connectedness

A *curve* C on a smooth manifold M is parametrized by a smooth map $\phi\colon I \to M$ where I is a subinterval of $\mathbb{R}$. In local coordinates, C is defined by m functions $x = \phi(t) = (\phi^1(t), \ldots, \phi^m(t))$. Note that we are *not* requiring that

ϕ be one-to-one—so a curve can have self-intersections, or be of maximal rank—so a curve can have singularities like cusps. In consequence, curves are more general than one-dimensional submanifolds. A particularly degenerate curve occurs when $\phi(t) \equiv x_0$ for all t, for some fixed x_0, so C consists of just one point. A *closed curve* is one whose endpoints coincide: $\phi(a) = \phi(b)$, with $I = [a, b]$, a closed interval.

A topological space is *connected* if it cannot be written as the disjoint union of two open sets. Since any manifold looks locally like Euclidean space, it is not difficult to prove that any connected manifold is *pathwise connected*, meaning that there is a smooth curve joining any pair of points. For our purposes, it will be very useful to impose, from the outset, the requirement that all manifolds under consideration are connected.

Blanket Hypothesis. *Unless explicitly stated otherwise, all manifolds (submanifolds, etc.) are assumed to be connected.*

This will avoid constantly restating the connectedness condition in the statement of our results.

A manifold M is *simply-connected* if every closed curve $C \subset M$ can be continuously deformed to a point. This is equivalent to the existence of a continuous map

$$H: [0, 1] \times [0, 1] \to M$$

such that $H(t, 0) = x_0$ for all $0 \leqslant t \leqslant 1$, while $H(t, 1)$, $0 \leqslant t \leqslant 1$ parametrizes C. For example, $\mathbb{R}^m$ is simply-connected, while $\mathbb{R}^2 \setminus \{0\}$ is not, as there is no way to continuously contract the unit circle to a point without passing through the origin. (On the other hand, $\mathbb{R}^m \setminus \{0\}$ *is* simply connected for $m \geqslant 3$.) If M is any manifold, there exists a simply-connected *covering manifold* $\pi: \tilde{M} \to M$, where the covering map π is onto and a local diffeomorphism. For example, the simply-connected cover of the unit circle S^1 is the real line $\mathbb{R}$ with covering map $\pi(t) = (\cos t, \sin t)$, $t \in \mathbb{R}$.

1.2. Lie Groups

At first sight, a Lie group appears to be a somewhat unnatural marriage between on the one hand the algebraic concept of a group, and on the other hand the differential-geometric notion of a manifold. However, as we shall soon see, this combination of algebra and calculus leads to powerful techniques for the study of symmetry which are not available for, say, finite groups.† We begin by recalling the definition of an abstract group.

† Witness, for instance, the recent complete classification of finite simple groups (Gorenstein, [1]); the corresponding problem for Lie groups was solved before the turn of the century.

Definition 1.14. A *group* is a set G together with a group operation, usually called multiplication, such that for any two elements g and h of G, the product $g \cdot h$ is again an element of G. The group operation is required to satisfy the following axioms:

(1) *Associativity*. If g, h and k are elements of G, then

$$g \cdot (h \cdot k) = (g \cdot h) \cdot k.$$

(2) *Identity Element*. There is a distinguished element e of G, called the identity element, which has the property that

$$e \cdot g = g = g \cdot e$$

for all g in G.

(3) *Inverses*. For each g in G there is an inverse, denoted g^{-1}, with the property

$$g \cdot g^{-1} = e = g^{-1} \cdot g.$$

Before proceeding to Lie groups, we discuss a couple of elementary examples of groups which give some idea of the features which distinguish Lie groups from more general types of groups.

Example 1.15. (a) Let $G = \mathbb{Z}$, the set of integers, with addition being the group operation. Clearly associativity is satisfied, the identity element is 0 and the "inverse" of an integer x is $-x$.

(b) Similarly $G = \mathbb{R}$, the set of real numbers, is also a group with addition serving as the group operation. Again 0 is the identity, and $-x$ the inverse of the real number x. In both of these cases the group operation is commutative: $g \cdot h = h \cdot g$ for $g, h \in G$. Such groups are called *abelian*; they form only a small subclass of the full range of possibilities for groups.

(c) Let $G = \mathrm{GL}(n, \mathbb{Q})$, the set of invertible $n \times n$ matrices with rational numbers for entries. The group operation is given by matrix multiplication. The identity element is, of course, the identity matrix I, the inverse of a matrix A is the ordinary matrix inverse, which again has rational entries.

(d) Similarly, $\mathrm{GL}(n, \mathbb{R})$, the set of all invertible $n \times n$ matrices with real entries is a group under matrix multiplication, the identity and inverse being the same as in the previous example. For brevity, we will usually denote the *general linear group* $\mathrm{GL}(n, \mathbb{R})$ by just $\mathrm{GL}(n)$.

The distinguishing feature of a Lie group is that it also carries the structure of a smooth manifold, so the group elements can be continuously varied. Thus, in each of the above pairs of examples of groups, the second case is a Lie group since it is also a smooth manifold. For $\mathbb{R}$, the manifold structure is clear. As for the general linear group, it can be identified with the open subset

$$\mathrm{GL}(n) = \{X : \det X \neq 0\}$$

of the space $\mathbf{M}_{n\times n}$ of all $n \times n$ matrices. But $\mathbf{M}_{n\times n}$ is isomorphic to $\mathbb{R}^{n^2}$, the coordinates being the matrix entries x_{ij} of X. Thus $\mathrm{GL}(n)$ is also an n^2-dimensional manifold. In both cases the group operation is smooth (indeed analytic). This leads to the general definition of a Lie group.

Definition 1.16. An *r-parameter Lie group* is a group G which also carries the structure of an r-dimensional smooth manifold in such a way that both the group operation

$$m: G \times G \to G, \qquad m(g, h) = g \cdot h, \qquad g, h \in G,$$

and the inversion

$$i: G \to G, \qquad i(g) = g^{-1}, \qquad g \in G,$$

are smooth maps between manifolds.

Example 1.17. Here we discuss a couple of examples of Lie groups besides the two already presented.

(a) Let $G = \mathbb{R}^r$, with the obvious manifold structure, and let the group operation be vector addition $(x, y) \mapsto x + y$. The "inverse" of a vector x is the vector $-x$. Both operations are clearly smooth, so $\mathbb{R}^r$ is an example of an r-parameter abelian Lie group.

(b) Let $G = \mathrm{SO}(2)$, the group of rotations in the plane. In other words

$$G = \left\{ \begin{pmatrix} \cos\theta & -\sin\theta \\ \sin\theta & \cos\theta \end{pmatrix} : 0 \leq \theta < 2\pi \right\},$$

where θ denotes the angle of rotation. Note that we can identify G with the unit circle

$$S^1 = \{(\cos\theta, \sin\theta): 0 \leq \theta < 2\pi\}$$

is $\mathbb{R}^2$, which serves to define the manifold structure on $\mathrm{SO}(2)$. If we include reflections we obtain the orthogonal group

$$\mathrm{O}(2) = \{X \in \mathrm{GL}(2): X^T X = \mathrm{I}\}.$$

It has the manifold structure of two disconnected copies of S^1.

(c) More generally, we can consider the group of *orthogonal* $n \times n$ *matrices*

$$\mathrm{O}(n) = \{X \in \mathrm{GL}(n): X^T X = \mathrm{I}\}.$$

Thus $\mathrm{O}(n)$ is the subset of $\mathbb{R}^{n^2}$ defined by the n^2 equations

$$X^T X - \mathrm{I} = 0,$$

involving the matrix entries x_{ij} of X. It can be shown that precisely $\frac{1}{2}n(n+1)$ of these equations, corresponding to the matrix entries on or above the diagonal, are independent and satisfy the maximal rank condition everywhere on $\mathrm{O}(n)$. Thus, by Theorem 1.13, $\mathrm{O}(n)$ is a regular submanifold of

GL(n) of dimension $\frac{1}{2}n(n-1)$. Moreover, matrix multiplication and inversion remain smooth maps when restricted to O(n), hence O(n) is a Lie group in its own right. The *special orthogonal group*

$$\mathrm{SO}(n) = \{X \in \mathrm{O}(n) \colon \det X = +1\},$$

being the connected component of the identity of the orthogonal group, is also an $\frac{1}{2}n(n-1)$-parameter Lie group for the same reasons. (A simpler proof of these facts will appear shortly.)

(d) The group T(n) of upper triangular matrices with all 1's on the main diagonal is an $\frac{1}{2}n(n-1)$-parameter Lie group. As a manifold T(n) can be identified with the Euclidean space $\mathbb{R}^{n(n-1)/2}$ since each matrix is uniquely determined by its entries above the diagonal. For instance, in the case of T(3), we identify the matrix

$$\begin{pmatrix} 1 & x & z \\ 0 & 1 & y \\ 0 & 0 & 1 \end{pmatrix} \in \mathrm{T}(3)$$

with the vector (x, y, z) in $\mathbb{R}^3$. However, except in the special case of T(2), T(n) is *not* isomorphic to the abelian Lie group $\mathbb{R}^{n(n-1)/2}$. In the case of T(3), the group operation is given by

$$(x, y, z) \cdot (x', y', z') = (x + x', y + y', z + z' + xy'),$$

using the above identification. This is not the same as vector addition—in particular, it is not commutative. Thus a fixed manifold may be given the structure of a Lie group in more than one way.

A Lie group *homomorphism* is a smooth map $\phi \colon G \to H$ between two Lie groups which respects the group operations:

$$\phi(g \cdot \tilde{g}) = \phi(g) \cdot \phi(\tilde{g}), \qquad g, \tilde{g} \in G.$$

If ϕ has a smooth inverse, it determines an *isomorphism* between G and H. In practice, we will not distinguish between isomorphic Lie groups. For example, the Lie group $\mathbb{R}^+$ consisting of all positive real numbers, with ordinary multiplication being the group operation, is isomorphic to the additive Lie group $\mathbb{R}$. The exponential function $\phi \colon \mathbb{R} \to \mathbb{R}^+$, $\phi(t) = e^t$, provides the isomorphism. For all practical purposes, $\mathbb{R}$ and $\mathbb{R}^+$ are the same Lie group. (In fact, up to isomorphism there are only two connected one-parameter Lie groups: $\mathbb{R}$ and SO(2).)

If G and H are r- and s-parameter Lie groups, then their Cartesian product $G \times H$ is an $(r+s)$-parameter Lie group with group operation

$$(g, h) \cdot (\tilde{g}, \tilde{h}) = (g \cdot \tilde{g}, h \cdot \tilde{h}), \qquad g, \tilde{g} \in G, \qquad h, \tilde{h} \in H,$$

which is easily seen to be a smooth map in the product manifold structure. Thus, for example, the tori T^r are all Lie groups, being r-fold Cartesian products of the Lie group $S^1 \simeq \mathrm{SO}(2)$. The group law on T^2, for instance, is given in terms of the angular coordinates (θ, ρ) by addition modulo integer multiples of 2π:

$$(\theta, \rho)\cdot(\theta', \rho') = (\theta + \theta', \rho + \rho') \bmod 2\pi.$$

Note that each torus T^r is a connected, compact, abelian r-parameter Lie group, and, in fact, is the only such Lie group up to isomorphism.

Our blanket assumption on manifolds that they be connected also carries over to the Lie groups we consider in this book. Thus *unless explicitly stated otherwise, all Lie groups are assumed to be connected.* For instance, the orthogonal groups $\mathrm{O}(n)$ are not connected, while the special orthogonal groups $\mathrm{SO}(n)$ are connected Lie groups. By restricting our attention to connected Lie groups, we are consciously excluding discrete symmetries, such as reflections, from consideration and concentrating on symmetries, like rotations, which can be continuously connected to the identity element in the group. There are, of course, important applications of discrete groups to differential equations, but these lie outside the scope of this book. (Technically speaking, without the assumption of connectivity, both Examples 1.15(a) and 1.15(c) are Lie groups, being totally disconnected zero-dimensional manifolds. However, none of the infinitesimal techniques vital to Lie group theory have any relevance there, and so we are justified in excluding them.) The general linear group $\mathrm{GL}(n)$ can be shown to consist of two connected components: $\mathrm{GL}^+(n) = \{X : \det X > 0\}$, which is itself a Lie group, and $\mathrm{GL}^-(n) = \{X : \det X < 0\}$. More generally, if G is any (not necessarily connected) Lie group, the connected component of the identity G^+ will always be a Lie group of the same dimension, and we will always concentrate on this part of G, the other components of G being obtained from G^+ via a discrete subgroup of elements.

Lie Subgroups

Often Lie groups arise as subgroups of certain larger Lie groups; for example, the orthogonal groups are subgroups of the general linear groups of all invertible matrices. In general we will be interested only in subgroups of Lie groups which can be considered as Lie groups in their own right. The proper definition of a Lie subgroup is modelled on that of an (immersed) submanifold.

Definition 1.18. A *Lie subgroup H* of a Lie group G is given by a submanifold $\phi: \tilde{H} \to G$, where $\tilde{H}$ itself is a Lie group, $H = \phi(\tilde{H})$ is the image of ϕ, and ϕ is a Lie group homomorphism.

For example, if ω is any real number, the submanifold

$$H_\omega = \{(t, \omega t) \bmod 2\pi : t \in \mathbb{R}\} \subset T^2$$

is easily seen to be a one-parameter Lie subgroup of the toroidal group T^2. If ω is rational, then H_ω is isomorphic to the circle group SO(2), and forms a closed, regular subgroup of T^2, while if ω is irrational, then H_ω is isomorphic to the Lie group $\mathbb{R}$, and is dense in T^2. Thus Lie subgroups of Lie groups do not have to be regular submanifolds. However, for many applications there is one very simple method of testing whether a subgroup is a regular Lie subgroup.

Theorem 1.19. *Suppose G is a Lie group. If H is a closed subgroup of G, then H is a regular submanifold of G and hence a Lie group in its own right. Conversely, any regular Lie subgroup of G is a closed subgroup.*

Note that we need only check that H is a subgroup of G and is closed as a subset of G in order to conclude that H is a regular Lie subgroup. This circumvents the problem of actually proving that H is a submanifold. In particular, if H is a subgroup defined by the vanishing of a number of (continuous) real-valued functions

$$H = \{g: F_i(g) = 0, i = 1, \ldots, n\},$$

then H is automatically a Lie subgroup of G; we do not need to check the maximal rank conditions on the F_i! (Of course, to find the dimension of H we need to determine how many of the F_i are independent.) Thus, for example, the orthogonal group O(n) is a Lie group, being given by the n^2 equations

$$A^T A = \mathrm{I}, \qquad A \in \mathrm{GL}(n).$$

Another important example is the *special linear group*

$$\mathrm{SL}(n) \equiv \mathrm{SL}(n, \mathbb{R}) \equiv \{A \in \mathrm{GL}(n): \det A = 1\},$$

which is an $(n^2 - 1)$-dimensional subgroup, given by the vanishing of a single function $\det A - 1$.

Local Lie Groups

Often we are not interested in the full Lie group, but only in group elements close to the identity element. In this case we can dispense with the abstract manifold theory and define a local Lie group solely in terms of local coordinate expressions for the group operations.

Definition 1.20. An *r-parameter local Lie group* consists of connected open subsets $V_0 \subset V \subset \mathbb{R}^r$ containing the origin 0, and smooth maps

$$m: V \times V \to \mathbb{R}^r,$$

defining the group operation, and

$$i: V_0 \to V,$$

defining the group inversion, with the following properties.

(a) *Associativity.* If $x, y, z \in V$, and also $m(x, y)$ and $m(y, z)$ are in V, then

$$m(x, m(y, z)) = m(m(x, y), z).$$

(b) *Identity Element.* For all x in V, $m(0, x) = x = m(x, 0)$.

(c) *Inverses.* For each x in V_0, $m(x, i(x)) = 0 = m(i(x), x)$.

If we write $x \cdot y$ for $m(x, y)$, and x^{-1} for $i(x)$, then the above axioms translate into the usual group axioms, except that they are not necessarily defined everywhere. Thus $x \cdot y$ makes sense only for x and y sufficiently near 0. Associativity says that $x \cdot (y \cdot z) = (x \cdot y) \cdot z$ whenever both sides of this equation are defined. The identity element of the group is the origin 0. Finally, x^{-1} again is defined only for x sufficiently near 0, and $x \cdot x^{-1} = 0 = x^{-1} \cdot x$ for such x's.

Example 1.21. Here we present a nontrivial example of a local (but not global) one-parameter Lie group. Let $V = \{x: |x| < 1\} \subset \mathbb{R}$ with group multiplication

$$m(x, y) = \frac{2xy - x - y}{xy - 1}, \qquad x, y \in V.$$

A straightforward computation verifies the associativity and identity laws for m. The inverse map is $i(x) = x/(2x - 1)$, defined for $x \in V_0 = \{x: |x| < \frac{1}{2}\}$. Thus m defines a local one-parameter Lie group.

One easy method of constructing local Lie groups is to take a global Lie group G and use a coordinate chart containing the identity element. Less trivial is the fact that (locally) every local Lie group arises in this fashion. In other words, every local Lie group is locally isomorphic to a neighbourhood of the identity of some global Lie group G.

Theorem 1.22. *Let $V_0 \subset V \subset \mathbb{R}^r$ be a local Lie group with multiplication $m(x, y)$ and inversion $i(x)$. Then there exists a global Lie group G and a coordinate chart $\chi: U^* \to V^*$, where U^* contains the identity element, such that $V^* \subset V_0$, $\chi(e) = 0$, and*

$$\chi(g \cdot h) = m(\chi(g), \chi(h))$$

whenever $g, h \in U^*$, and

$$\chi(g^{-1}) = i(\chi(g))$$

whenever $g \in U^*$. Moreover, there is a unique connected, simply-connected Lie group G^* having the above properties. If G is any other such Lie group, there exists a covering map $\pi: G^* \to G$ which is a group homomorphism, whereby G^* and G are locally isomorphic Lie groups. (G^* is called the simply-connected covering group *of G.)*

Example 1.23. The only connected, simply-connected one-parameter Lie group is $\mathbb{R}$, so the local Lie group of Example 1.21 must coincide with some coordinate chart containing 0 in $\mathbb{R}$. Indeed, if we let $\chi: U^* \to V^* \subset \mathbb{R}$, where

$$\chi(t) = t/(t-1), \qquad t \in U^* = \{t < 1\},$$

then we easily see that

$$\chi(t+s) = m(\chi(t), \chi(s)) = \frac{2\chi(t)\chi(s) - \chi(t) - \chi(s)}{\chi(t)\chi(s) - 1},$$

$$\chi(-t) = i(\chi(t)) = \frac{\chi(t)}{2\chi(t) - 1},$$

where defined, so χ satisfies the requirements of the theorem.

Once we know that such a global Lie group exists, we can essentially reconstruct it from knowledge of just the neighbourhood of the identity determining the local Lie group.

Proposition 1.24. *Let G be a connected Lie group and $U \subset G$ a neighbourhood of the identity. Also, let $U^k \equiv \{g_1 \cdot g_2 \cdot \ldots \cdot g_k : g_i \in U\}$ be the set of k-fold products of elements of U. Then*

$$G = \bigcup_{k=1}^{\infty} U^k.$$

In other words, every group element $g \in G$ can be written as a finite product of elements of U.

As shown in Exercise 1.26 this follows directly from the connectedness of G. A similar result holds for connected local Lie groups as well.

Local Transformation Groups

In practice, Lie groups arise most naturally not as abstract, self-contained entities, but rather concretely as groups of transformations on some manifold M. For instance, the group SO(2) arises as the group of rotations in the plane $M = \mathbb{R}^2$, while GL(n) appears as the group of invertible linear transformations on $\mathbb{R}^n$. In general a Lie group G will be realized as a group of transformations of some manifold M if to each group element $g \in G$ there is associated a map from M to itself. It is important not to restrict our attention solely to linear transformations. Moreover, the group may act only locally, meaning that the group transformations may not be defined for all elements of the group nor for all points on the manifold.

Definition 1.25. Let M be a smooth manifold. A *local group of transformations* acting on M is given by a (local) Lie group G, an open subset $\mathcal{U}$, with

$$\{e\} \times M \subset \mathcal{U} \subset G \times M,$$

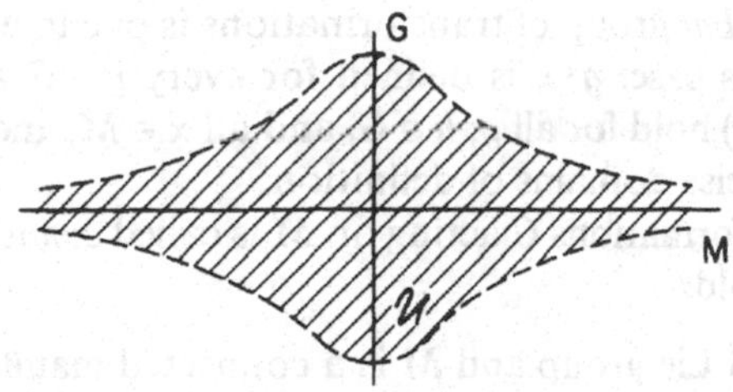

Figure 3. Domain for a local transformation group.

which is the domain of definition of the group action, and a smooth map $\Psi: \mathscr{U} \to M$ with the following properties:

(a) If $(h, x) \in \mathscr{U}$, $(g, \Psi(h, x)) \in \mathscr{U}$, and also $(g \cdot h, x) \in \mathscr{U}$, then

$$\Psi(g, \Psi(h, x)) = \Psi(g \cdot h, x).$$

(b) For all $x \in M$,

$$\Psi(e, x) = x.$$

(c) If $(g, x) \in \mathscr{U}$, then $(g^{-1}, \Psi(g, x)) \in \mathscr{U}$ and

$$\Psi(g^{-1}, \Psi(g, x)) = x.$$

(Note that except for the assumption of the form of the domain $\mathscr{U}$, part (c) follows directly from parts (a) and (b).)

For brevity, we will denote $\Psi(g, x)$ by $g \cdot x$, and the conditions of this definition take the simpler form:

$$g \cdot (h \cdot x) = (g \cdot h) \cdot x, \qquad g, h \in G, \quad x \in M, \tag{1.1}$$

whenever both sides of this equation make sense,

$$e \cdot x = x \quad \text{for all} \quad x \in M, \tag{1.2}$$

and

$$g^{-1} \cdot (g \cdot x) = x, \qquad g \in G, \quad x \in M, \tag{1.3}$$

provided $g \cdot x$ is defined. As a consequence of (1.3), we see that each group transformation is a diffeomorphism where it is defined.

Note that for each x in M, the group elements g such that $g \cdot x$ is defined form a local Lie group

$$G_x \equiv \{g \in G: (g, x) \in \mathscr{U}\}.$$

Conversely, for any $g \in G$, there is an open submanifold

$$M_g \equiv \{x \in M: (g, x) \in \mathscr{U}\}$$

of M where the transformation given by g is defined. In certain cases, the only group element which acts on all of M might be the identity element. At the other extreme, a *global* group of transformations is one in which we can take $\mathcal{U} = G \times M$. In this case, $g \cdot x$ is defined for every $g \in G$ and every $x \in M$. Thus (1.1), (1.2), (1.3) hold for all $g, h \in G$, and all $x \in M$, and there is no need to worry about precise domains of definition.

A group of transformations G acting on M is called *connected* if the following requirements hold:

(a) G is a connected Lie group and M is a connected manifold;
(b) $\mathcal{U} \subset G \times M$ is a connected open set; and
(c) for each $x \in M$, the local group G_x is connected.

As with manifolds and Lie groups, we make the blanket restriction that *unless explicitly stated otherwise, all local groups of transformations are assumed to be connected, in the above restricted sense.* These connectivity requirements help us avoid several technical complications when we come to discuss infinitesimal methods and invariants. They can always be realized by suitably shrinking the domain of definition $\mathcal{U}$ of the group action.

Orbits

An *orbit* of a local transformation group is a minimal nonempty group-invariant subset of the manifold M. In other words, $\mathcal{O} \subset M$ is an orbit provided it satisfies the conditions

(a) If $x \in \mathcal{O}$, $g \in G$ and $g \cdot x$ is defined, then $g \cdot x \in \mathcal{O}$.
(b) If $\tilde{\mathcal{O}} \subset \mathcal{O}$, and $\tilde{\mathcal{O}}$ satisfies part (a) then either $\tilde{\mathcal{O}} = \mathcal{O}$, or $\tilde{\mathcal{O}}$ is empty.

In the case of a global transformation group, for each $x \in M$ the orbit through x has the explicit definition $\mathcal{O}_x = \{g \cdot x \colon g \in G\}$. For local transformation groups, we must look at products of group elements acting on x:

$$\mathcal{O}_x = \{g_1 \cdot g_2 \cdot \ldots \cdot g_k \cdot x \colon k \geqslant 1, g_i \in G \text{ and } g_1 \cdot g_2 \cdot \ldots \cdot g_k \cdot x \text{ is defined}\}.$$

As we will see, the orbits of a Lie group of transformations are in fact submanifolds of M, but they may be of varying dimensions, or may not be regular. We distinguish two important subclasses of group actions.

Definition 1.26. Let G be a local group of transformations acting on M.

(a) The group G acts *semi-regularly* if all the orbits $\mathcal{O}$ are of the same dimension as submanifolds of M.
(b) The group G acts *regularly* if the action is semi-regular, and, in addition, for each point $x \in M$ there exist arbitrarily small neighbourhoods U of x with the property that each orbit of G intersects U in a pathwise connected subset.

Note that in particular, if G acts regularly on M then each orbit of G is a regular submanifold of M. However, the regularity condition on the group action is much stronger than this last statement, as Exercise 1.8 will bear out.

A group action is called *transitive* if there is only one orbit, namely the manifold M itself. Clearly any transitive group of transformations acts regularly. In most of our applications, the most interesting group actions will *not* be transitive.

Example 1.27. *Examples of Transformation Groups.*

(a) The group of *translations* in $\mathbb{R}^m$: Let $a \neq 0$ be a fixed vector in $\mathbb{R}^m$, and let $G = \mathbb{R}$. Define

$$\Psi(\varepsilon, x) = x + \varepsilon a, \qquad x \in \mathbb{R}^m, \quad \varepsilon \in \mathbb{R}.$$

This is readily seen to give a global group action. The orbits are straight lines parallel to a, so the action is regular with one-dimensional orbits.

(b) Groups of *scale transformations*: Let $G = \mathbb{R}^+$ be the multiplication group. Fix real numbers $\alpha_1, \alpha_2, \ldots, \alpha_m$, not all zero. Then $\mathbb{R}^+$ acts on $\mathbb{R}^m$ by the scaling transformations

$$\Psi(\lambda, x) = (\lambda^{\alpha_1} x^1, \ldots, \lambda^{\alpha_m} x^m), \qquad \lambda \in \mathbb{R}^+, \quad x = (x^1, \ldots, x^m) \in \mathbb{R}^m.$$

The orbits of this action are all one-dimensional regular submanifolds of $\mathbb{R}^m$, except for the singular orbit consisting of just the origin $\{0\}$. For instance, in the special case of $\mathbb{R}^2$ with $\Psi(\lambda, (x, y)) = (\lambda x, \lambda^2 y)$, the orbits are halves of the parabolas $y = kx^2$ (corresponding to either $x > 0$ or $x < 0$), the positive and negative y-axes, and the origin. In general, this scaling group action is regular on the open subset $\mathbb{R}^m \backslash \{0\}$. These group actions arise in the dimensional analysis of partial differential equations and historically provided the main impetus behind the development of the general theory of group-invariant solutions of differential equations.

(c) An action similar to the following comes up in the study of the heat equation. Let $M = \mathbb{R}^2$, $G = \mathbb{R}$ and consider the map

$$\Psi(\varepsilon, (x, y)) = \left(\frac{x}{1 - \varepsilon x}, \frac{y}{1 - \varepsilon x}\right),$$

which is defined on

$$\mathcal{U} = \left\{(\varepsilon, (x, y)) \colon \varepsilon < \frac{1}{x} \text{ for } x > 0, \text{ or } \varepsilon > \frac{1}{x}, \text{ for } x < 0\right\} \subset \mathbb{R} \times \mathbb{R}^2.$$

To show that this is indeed a group action, we must check condition (a) of Definition 1.25:

$$\begin{aligned}
\Psi(\delta, \Psi(\varepsilon, (x, y))) &= \Psi\left(\delta, \left(\frac{x}{1 - \varepsilon x}, \frac{y}{1 - \varepsilon x}\right)\right) \\
&= \left(\frac{x/(1 - \varepsilon x)}{1 - \delta x/(1 - \varepsilon x)}, \frac{y/(1 - \varepsilon x)}{1 - \delta x/(1 - \varepsilon x)}\right) \\
&= \left(\frac{x}{1 - (\delta + \varepsilon) x}, \frac{y}{1 - (\delta + \varepsilon) x}\right) \\
&= \Psi(\delta + \varepsilon, (x, y))
\end{aligned}$$

wherever defined. Note that this local group action has no global counterpart on $\mathbb{R}^2$; indeed $|\Psi(\varepsilon, (x, y))| \to \infty$ as $\varepsilon \to 1/x$ for $x \neq 0$. The orbits of the action consist of the straight rays emanating from the origin, and the origin itself. The action is regular on the punctured plane $\mathbb{R}^2\backslash\{0\}$.

(d) The "*irrational flow*" on the torus: Let $G = \mathbb{R}$ and M be the two-dimensional torus T^2. Let ω be a fixed real number. Using the angular coordinates (θ, ρ) on T^2 we define a global group action

$$\Psi(\varepsilon, (\theta, \rho)) = (\theta + \varepsilon, \rho + \omega\varepsilon) \quad \bmod 2\pi.$$

The orbits of G are easily seen to all be one-dimensional submanifolds of T^2, so G acts semi-regularly in all cases. If ω is a rational number, the orbits are closed curves, and the action is regular. On the other hand, if ω is irrational, each orbit is a dense submanifold of T^2. This is the simplest example of a semi-regular group action which is not regular.

1.3. Vector Fields

The main tool in the theory of Lie groups and transformation groups is the "infinitesimal transformation". In order to present this, we need first to develop the concept of a vector field on a manifold. We begin with a discussion of tangent vectors. Suppose C is a smooth curve on a manifold M, parametrized by $\phi: I \to M$, where I is a subinterval of $\mathbb{R}$. In local coordinates $x = (x^1, \ldots, x^m)$, C is given by m smooth functions $\phi(\varepsilon) = (\phi^1(\varepsilon), \ldots, \phi^m(\varepsilon))$ of the real variable ε. At each point $x = \phi(\varepsilon)$ of C the curve has a *tangent vector*, namely the derivative $\dot\phi(\varepsilon) = d\phi/d\varepsilon = (\dot\phi^1(\varepsilon), \ldots, \dot\phi^m(\varepsilon))$. In order to distinguish between tangent vectors and local coordinate expressions for points on the manifold, we adopt the notation

$$\mathbf{v}|_x = \dot\phi(\varepsilon) = \dot\phi^1(\varepsilon)\frac{\partial}{\partial x^1} + \dot\phi^2(\varepsilon)\frac{\partial}{\partial x^2} + \cdots + \dot\phi^m(\varepsilon)\frac{\partial}{\partial x^m} \tag{1.4}$$

for the tangent vector to C at $x = \phi(\varepsilon)$. On first encounter, this notation may look rather strange, but its usefulness and naturalness will be amply demonstrated throughout this book. For the moment, the reader can view the symbols $\partial/\partial x^i$ just as "place holders" for the components $\dot\phi^i(\varepsilon)$ of the tangent vector $\mathbf{v}|_x$, or, equivalently, as a special "basis" of tangent vectors corresponding to the coordinate curves whose local coordinate expressions are $x + \varepsilon e_i$, e_i being the i-th basis vector of $\mathbb{R}^m$. Later we will see how each $\partial/\partial x^i$ does indeed correspond to a partial differential operator.

For example, the helix

$$\phi(\varepsilon) = (\cos\varepsilon, \sin\varepsilon, \varepsilon)$$

in $\mathbb{R}^3$, with coordinates (x, y, z), has tangent vector

$$\dot\phi(\varepsilon) = -\sin\varepsilon\frac{\partial}{\partial x} + \cos\varepsilon\frac{\partial}{\partial y} + \frac{\partial}{\partial z} = -y\frac{\partial}{\partial x} + x\frac{\partial}{\partial y} + \frac{\partial}{\partial z}$$

at the point $(x, y, z) = \phi(\varepsilon) = (\cos\varepsilon, \sin\varepsilon, \varepsilon)$.

Two curves $C = \{\phi(\varepsilon)\}$ and $\tilde{C} = \{\tilde{\phi}(\theta)\}$ passing through the same point

$$x = \phi(\varepsilon^*) = \tilde{\phi}(\theta^*)$$

for some ε^*, θ^*, have the same tangent vector if and only if their derivatives agree at the point:

$$\frac{d\phi}{d\varepsilon}(\varepsilon^*) = \frac{d\tilde{\phi}}{d\theta}(\theta^*). \tag{1.5}$$

It is not difficult to see that this concept is independent of the local coordinate system used near x. Indeed, if $x = \phi(\varepsilon) = (\phi^1(\varepsilon), \ldots, \phi^m(\varepsilon))$ is the local coordinate expression in terms of $x = (x^1, \ldots, x^m)$ and $y = \psi(x)$ is any diffeomorphism, then $y = \psi(\phi(\varepsilon))$ is the local coordinate formula for the curve in terms of the y-coordinates. The tangent vector $\mathbf{v}|_x = \dot{\phi}(\varepsilon)$, which has the formula (1.4) in the x-coordinates, takes the form

$$\mathbf{v}|_{y=\psi(x)} = \sum_{j=1}^{m} \frac{d}{d\varepsilon}\psi^j(\phi(\varepsilon))\frac{\partial}{\partial y^j} = \sum_{j=1}^{m}\sum_{k=1}^{m} \frac{\partial \psi^j}{\partial x^k}(\phi(\varepsilon))\frac{d\phi^k}{d\varepsilon}\frac{\partial}{\partial y^j} \tag{1.6}$$

in the y-coordinates. Since the Jacobian matrix $\partial\psi^j/\partial x^k$ is invertible at each point, (1.5) holds if and only if

$$\frac{d}{d\varepsilon}\psi(\phi(\varepsilon^*)) = \frac{d}{d\theta}\psi(\tilde{\phi}(\theta^*)),$$

which proves the claim. Note that (1.6) tells how a tangent vector (1.4) behaves under the given change of coordinates $y = \psi(x)$.

The collection of all tangent vectors to all possible curves passing through a given point x in M is called the *tangent space* to M at x, and is denoted by $TM|_x$. If M is an m-dimensional manifold, then $TM|_x$ is an m-dimensional vector space, with $\{\partial/\partial x^1, \ldots, \partial/\partial x^m\}$ providing a basis for $TM|_x$ in the given local coordinates. The collection of all tangent spaces corresponding to all points x in M is called the *tangent bundle* of M, denoted by

$$TM = \bigcup_{x \in M} TM|_x.$$

These tangent spaces are "glued" together in an obvious smooth fashion, so that if $\phi(\varepsilon)$ is any smooth curve then the tangent vectors $\dot{\phi}(\varepsilon) \in TM|_{\phi(\varepsilon)}$ will vary smoothly from point to point. This makes the tangent bundle TM into a smooth manifold of dimension $2m$.

For example, if $M = \mathbb{R}^m$, then we can identify the tangent space $T\mathbb{R}^m|_x$ at any $x \in \mathbb{R}^m$ with $\mathbb{R}^m$ itself. This stems from the fact that the tangent vector $\dot{\phi}(\varepsilon)$ to a smooth curve $\phi(\varepsilon)$ can be realized as an actual vector in $\mathbb{R}^m$, namely $(\dot{\phi}^1(\varepsilon), \ldots, \dot{\phi}^m(\varepsilon))$. Another way of looking at this identification is that we are identifying the basis vector $\partial/\partial x^i$ of $T\mathbb{R}^m|_x$ with the standard basis vector e_i of $\mathbb{R}^m$. The tangent bundle of $\mathbb{R}^m$ is thus a Cartesian product $T\mathbb{R}^m \simeq \mathbb{R}^m \times \mathbb{R}^m$. If S is a surface in $\mathbb{R}^3$, then the tangent space $TS|_x$ can be identified with the usual geometric tangent plane to S at each point $x \in S$. This again uses the identification $T\mathbb{R}^3|_x \simeq \mathbb{R}^3$, so $TS|_x \subset T\mathbb{R}^3|_x$ is a plane in $\mathbb{R}^3$.

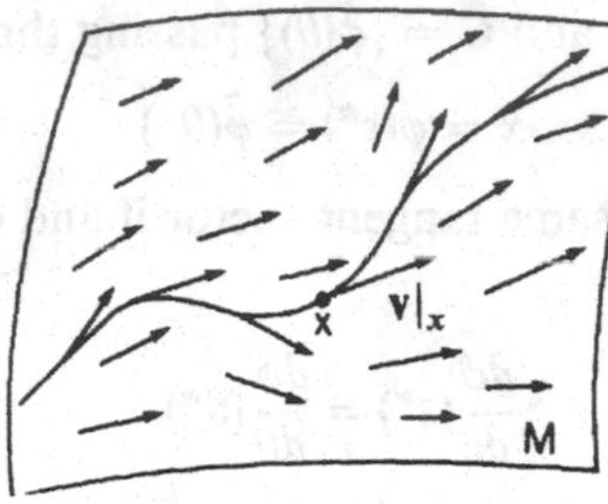

Figure 4. Vector field and integral curve on a manifold.

A *vector field* $\mathbf{v}$ on M assigns a tangent vector $\mathbf{v}|_x \in TM|_x$ to each point $x \in M$, with $\mathbf{v}|_x$ varying smoothly from point to point. In local coordinates $(x^1, \ldots, x^m)$, a vector field has the form

$$\mathbf{v}|_x = \xi^1(x)\frac{\partial}{\partial x^1} + \xi^2(x)\frac{\partial}{\partial x^2} + \cdots + \xi^m(x)\frac{\partial}{\partial x^m},$$

where each $\xi^i(x)$ is a smooth function of x. (Technically, we should put the symbol $|_x$ on each $\partial/\partial x^i$ to indicate in which tangent space $TM|_x$ it lies, but this should be clear from the context.) A good physical example of a vector field is the velocity field of a steady fluid flow in some open subset $M \subset \mathbb{R}^3$. At each point $(x, y, z) \in M$, the vector $\mathbf{v}|_{(x,y,z)}$ would be the velocity of the fluid particles passing through the point (x, y, z).

An *integral curve* of a vector field $\mathbf{v}$ is a smooth parametrized curve $x = \phi(\varepsilon)$ whose tangent vector at any point coincides with the value of $\mathbf{v}$ at the same point:

$$\dot{\phi}(\varepsilon) = \mathbf{v}|_{\phi(\varepsilon)}$$

for all ε. In local coordinates, $x = \phi(\varepsilon) = (\phi^1(\varepsilon), \ldots, \phi^m(\varepsilon))$ must be a solution to the autonomous system of ordinary differential equations

$$\frac{dx^i}{d\varepsilon} = \xi^i(x), \qquad i = 1, \ldots, m, \tag{1.7}$$

where the $\xi^i(x)$ are the coefficients of $\mathbf{v}$ at x. For $\xi^i(x)$ smooth, the standard existence and uniqueness theorems for systems of ordinary differential equations guarantee that there is a unique solution to (1.7) for each set of initial data

$$\phi(0) = x_0. \tag{1.8}$$

This in turn implies the existence of a unique *maximal* integral curve $\phi: I \to M$ passing through a given point $x_0 = \phi(0) \in M$, where "maximal" means that it is not contained in any longer integral curve; i.e. if $\tilde{\phi}: \tilde{I} \to M$ is any other integral curve with the same initial value $\tilde{\phi}(0) = x_0$, then $\tilde{I} \subset I$ and

$\tilde{\phi}(\varepsilon) = \phi(\varepsilon)$ for $\varepsilon \in \tilde{I}$. Note that if $\mathbf{v}|_{x_0} = 0$, then the integral curve through x_0 is just the point $\phi(\varepsilon) \equiv x_0$ itself, defined for all ε.

We note that if $\mathbf{v}$ is any smooth vector field on a manifold M, and $f(x)$ is any smooth, real-valued function defined for $x \in M$, then $f \cdot \mathbf{v}$ is again a smooth vector field, with $(f \cdot \mathbf{v})|_x = f(x)\mathbf{v}|_x$. In local coordinates, if $\mathbf{v} = \sum \xi^i(x)\partial/\partial x^i$, then $f \cdot \mathbf{v} = \sum f(x)\xi^i(x)\partial/\partial x^i$. If f never vanishes, the integral curves of $f \cdot \mathbf{v}$ coincide with the integral curves of $\mathbf{v}$, but the parametrizations will differ. For instance, the integral curves for $2\mathbf{v}$ will be traversed twice as fast as those of $\mathbf{v}$, but otherwise will be the same subsets of M.

Flows

If $\mathbf{v}$ is a vector field, we denote the parametrized maximal integral curve passing through x in M by $\Psi(\varepsilon, x)$ and call Ψ the *flow* generated by $\mathbf{v}$. Thus for each $x \in M$, and ε in some interval I_x containing 0, $\Psi(\varepsilon, x)$ will be a point on the integral curve passing through x in M. The flow of a vector field has the basic properties:

$$\Psi(\delta, \Psi(\varepsilon, x)) = \Psi(\delta + \varepsilon, x), \qquad x \in M, \tag{1.9}$$

for all $\delta, \varepsilon \in \mathbb{R}$ such that both sides of equation are defined,

$$\Psi(0, x) = x, \tag{1.10}$$

and

$$\frac{d}{d\varepsilon}\Psi(\varepsilon, x) = \mathbf{v}|_{\Psi(\varepsilon, x)} \tag{1.11}$$

for all ε where defined. Here (1.11) simply states that $\mathbf{v}$ is tangent to the curve $\Psi(\varepsilon, x)$ for fixed x, and (1.10) gives the initial conditions for this integral curve. The proof of (1.9) follows easily from the uniqueness of solutions to systems of ordinary differential equations; namely as functions of δ both sides of (1.9) satisfy (1.7) and have the same initial conditions at $\delta = 0$. If $\mathbf{v}$ is the velocity vector field of some steady state fluid flow, the integral curves of $\mathbf{v}$ are the stream lines followed by the fluid particles, and the flow $\Psi(\varepsilon, x)$ tells the position of a particle at time ε which started out at position x at time $\varepsilon = 0$.

Comparing the first two properties (1.9), (1.10) with (1.1), (1.2), we see that the flow generated by a vector field is the same as a local group action of the Lie group $\mathbb{R}$ on the manifold M, often called a *one-parameter group of transformations*. The vector field $\mathbf{v}$ is called the *infinitesimal generator* of the action since by Taylor's theorem, in local coordinates

$$\Psi(\varepsilon, x) = x + \varepsilon\xi(x) + O(\varepsilon^2),$$

where $\xi = (\xi^1, \ldots, \xi^m)$ are the coefficients of $\mathbf{v}$. The orbits of the one-parameter group action are the maximal integral curves of the vector field $\mathbf{v}$. Conversely, if $\Psi(\varepsilon, x)$ is any one-parameter group of transformations acting on

M, then its infinitesimal generator is obtained by specializing (1.11) at $\varepsilon = 0$:

$$\mathbf{v}|_x = \frac{d}{d\varepsilon}\bigg|_{\varepsilon=0} \Psi(\varepsilon, x). \tag{1.12}$$

Uniqueness of solutions to (1.7), (1.8) guarantees that the flow generated by $\mathbf{v}$ coincides with the given local action of $\mathbb{R}$ on M on the common domain of definition. Thus there is a one-to-one correspondence between local one-parameter groups of transformations and their infinitesimal generators.

The computation of the flow or one-parameter group generated by a given vector field $\mathbf{v}$ (in other words, solving the system of ordinary differential equations) is often referred to as *exponentiation* of the vector field. The suggestive notation

$$\exp(\varepsilon\mathbf{v})x \equiv \Psi(\varepsilon, x)$$

for this flow will be adopted in this book. In terms of this exponential notation, the above three properties can be restated as

$$\exp[(\delta + \varepsilon)\mathbf{v}]x = \exp(\delta\mathbf{v})\exp(\varepsilon\mathbf{v})x \tag{1.13}$$

whenever defined,

$$\exp(0\mathbf{v})x = x, \tag{1.14}$$

and

$$\frac{d}{d\varepsilon}[\exp(\varepsilon\mathbf{v})x] = \mathbf{v}|_{\exp(\varepsilon\mathbf{v})x}. \tag{1.15}$$

for all $x \in M$. (In particular, $\mathbf{v}|_x$ is obtained by evaluating (1.15) at $\varepsilon = 0$.) These properties mirror the properties of the usual exponential function, justifying the notation.

Example 1.28. *Examples of Vector Fields and Flows.*

(a) Let $M = \mathbb{R}$ with coordinate x, and consider the vector field $\mathbf{v} = \partial/\partial x \equiv \partial_x$. (In the sequel, we will often use the notation ∂_x for $\partial/\partial x$ to save space.) Then

$$\exp(\varepsilon\mathbf{v})x = \exp(\varepsilon\partial_x)x = x + \varepsilon,$$

which is globally defined. For the vector field $x\partial_x$ we recover the usual exponential

$$\exp(\varepsilon x\partial_x)x = e^{\varepsilon}x,$$

since it must be the solution to the ordinary differential equation $\dot{x} = x$ with initial value x at $\varepsilon = 0$.

(b) In the case of $\mathbb{R}^m$, a constant vector field $\mathbf{v}_a = \sum a^i \partial/\partial x^i$, $a = (a^1, \ldots, a^m)$ exponentiates to the group of translations

$$\exp(\varepsilon\mathbf{v}_a)x = x + \varepsilon a, \qquad x \in \mathbb{R}^m,$$

in the direction a. Similarly, a linear vector field

$$\mathbf{v}_A = \sum_{i=1}^{m} \left(\sum_{j=1}^{m} a_{ij} x^j \right) \frac{\partial}{\partial x^i},$$

where $A = (a_{ij})$ is an $m \times m$ matrix of constants, has flow

$$\exp(\varepsilon \mathbf{v}_A) x = e^{\varepsilon A} x,$$

where $e^{\varepsilon A} = \mathrm{I} + \varepsilon A + \frac{1}{2}\varepsilon^2 A^2 + \cdots$ is the usual matrix exponential.

(c) Consider the group of rotations in the plane

$$\Psi(\varepsilon, (x, y)) = (x \cos \varepsilon - y \sin \varepsilon, x \sin \varepsilon + y \cos \varepsilon).$$

Its infinitesimal generator is a vector field $\mathbf{v} = \xi(x, y)\partial_x + \eta(x, y)\partial_y$, where, according to (1.12),

$$\xi(x, y) = \left.\frac{d}{d\varepsilon}\right|_{\varepsilon=0} (x \cos \varepsilon - y \sin \varepsilon) = -y,$$

$$\eta(x, y) = \left.\frac{d}{d\varepsilon}\right|_{\varepsilon=0} (x \sin \varepsilon + y \cos \varepsilon) = x.$$

Thus $\mathbf{v} = -y\partial_x + x\partial_y$ is the infinitesimal generator, and indeed, the above group transformations agree with the solutions to the system of ordinary differential equations

$$dx/d\varepsilon = -y, \qquad dy/d\varepsilon = x.$$

(d) Finally, consider the local group action

$$\Psi(\varepsilon, (x, y)) = \left(\frac{x}{1 - \varepsilon x}, \frac{y}{1 - \varepsilon x} \right)$$

introduced in Example 1.27(c). Differentiating, as before, we find the infinitesimal generator to be $\mathbf{v} = x^2\partial_x + xy\partial_y$. This demonstrates that a smooth vector field may still generate only a local group action.

The effect of a change of coordinates $y = \psi(x)$ on a vector field $\mathbf{v}$ is determined by its effect on each individual tangent vector $\mathbf{v}|_x$, $x \in M$, as given by (1.6). Thus if $\mathbf{v}$ is a vector field whose expression in the x-coordinates is

$$\mathbf{v} = \sum_{i=1}^{m} \xi^i(x) \frac{\partial}{\partial x^i},$$

and $y = \psi(x)$ is a change of coordinates, then $\mathbf{v}$ has the formula

$$\mathbf{v} = \sum_{j=1}^{m} \sum_{i=1}^{m} \xi^i(\psi^{-1}(y)) \frac{\partial \psi^j}{\partial x^i}(\psi^{-1}(y)) \frac{\partial}{\partial y^j} \tag{1.16}$$

in the y-coordinates.

The next result illustrates our earlier remarks that by suitably choosing local coordinates, we can often simplify the expressions for objects on manifolds, in this case vector fields.

Proposition 1.29. *Suppose* $\mathbf{v}$ *is a vector field not vanishing at a point* $x_0 \in M$: $\mathbf{v}|_{x_0} \neq 0$. *Then there is a local coordinate chart* $y = (y^1, \ldots, y^m)$ *at* x_0 *such that in terms of these coordinates,* $\mathbf{v} = \partial/\partial y^1$.

PROOF. First linearly change coordinates so that $x_0 = 0$ and $\mathbf{v}|_{x_0} = \partial/\partial x^1$. By continuity the coefficient $\xi^1(x)$ of $\partial/\partial x^1$ is positive in a neighbourhood of x_0. Since $\xi^1(x) > 0$, the integral curves of $\mathbf{v}$ cross the hyperplane $\{(0, x^2, \ldots, x^m)\}$ transversally, and hence in a neighbourhood of $x_0 = 0$, each point $x = (x^1, \ldots, x^m)$ can be defined uniquely as the flow of some point $(0, y^2, \ldots, y^m)$ on this hyperplane. Consequently

$$x = \exp(y^1 \mathbf{v})(0, y^2, \ldots, y^m),$$

for y^1 near 0, gives a diffeomorphism from $(x^1, \ldots, x^m)$ to $(y^1, \ldots, y^m)$ which defines the y-coordinates. (Geometrically, we have "straightened out" the integral curves passing through the hyperplane perpendicular to the x^1-axis.) In terms of the y-coordinates, we have by (1.13), for small ε,

$$\exp(\varepsilon \mathbf{v})(y^1, \ldots, y^m) = (y^1 + \varepsilon, y^2, \ldots, y^m),$$

so the flow is just translation in the y^1-direction. Thus every nonvanishing vector field is locally equivalent to the infinitesimal generator of a group of translations. (Of course, the global picture can be very complicated, as the irrational flow on the torus makes clear.) □

Action on Functions

Let $\mathbf{v}$ be a vector field on M and $f: M \to \mathbb{R}$ a smooth function. We are interested in seeing how f changes under the flow generated by $\mathbf{v}$, meaning we look at $f(\exp(\varepsilon \mathbf{v})x)$ as ε varies. In local coordinates, if $\mathbf{v} = \sum \xi^i(x)\partial/\partial x^i$, then using the chain rule and (1.15) we find

$$\begin{aligned} \frac{d}{d\varepsilon} f(\exp(\varepsilon \mathbf{v})x) &= \sum_{i=1}^{m} \xi^i(\exp(\varepsilon \mathbf{v})x) \frac{\partial f}{\partial x^i}(\exp(\varepsilon \mathbf{v})x) \\ &= \mathbf{v}(f)[\exp(\varepsilon \mathbf{v})x]. \end{aligned} \tag{1.17}$$

In particular, at $\varepsilon = 0$,

$$\left.\frac{d}{d\varepsilon}\right|_{\varepsilon=0} f(\exp(\varepsilon \mathbf{v})x) = \sum_{i=1}^{m} \xi^i(x) \frac{\partial f}{\partial x^i}(x) = \mathbf{v}(f)(x).$$

Now the reason underlying our notation for vector fields becomes apparent: the vector field $\mathbf{v}$ acts as a first order partial differential operator on real-valued functions $f(x)$ on M. Furthermore, by Taylor's theorem,

$$f(\exp(\varepsilon \mathbf{v})x) = f(x) + \varepsilon \mathbf{v}(f)(x) + O(\varepsilon^2),$$

so $\mathbf{v}(f)$ gives the *infinitesimal change* in the function f under the flow generated by $\mathbf{v}$. We can continue the process of differentiation and substitution into the Taylor series, obtaining

$$f(\exp(\varepsilon \mathbf{v})x) = f(x) + \varepsilon \mathbf{v}(f)(x) + \frac{\varepsilon^2}{2}\mathbf{v}^2(f)(x) + \cdots + \frac{\varepsilon^k}{k!}\mathbf{v}^k(f)(x) + O(\varepsilon^{k+1}),$$

where $\mathbf{v}^2(f) = \mathbf{v}(\mathbf{v}(f))$, $\mathbf{v}^3(f) = \mathbf{v}(\mathbf{v}^2(f))$, etc. If we assume convergence of the entire Taylor series in ε, then we obtain the *Lie series*

$$f(\exp(\varepsilon \mathbf{v})x) = \sum_{k=0}^{\infty} \frac{\varepsilon^k}{k!}\mathbf{v}^k(f)(x) \tag{1.18}$$

for the action of the flow on f. The same result holds for vector-valued functions $F: M \to \mathbb{R}^n$, $F(x) = (F^1(x), \ldots, F^n(x))$, where we let $\mathbf{v}$ act component-wise on F: $\mathbf{v}(F) = (\mathbf{v}(F^1), \ldots, \mathbf{v}(F^n))$. In particular, if we let F be the coordinate functions x, we obtain (again under assumptions of convergence) a Lie series for the flow itself, given by

$$\exp(\varepsilon \mathbf{v})x = x + \varepsilon\xi(x) + \frac{\varepsilon^2}{2}\mathbf{v}(\xi)(x) + \cdots = \sum_{k=0}^{\infty} \frac{\varepsilon^k}{k!}\mathbf{v}^k(x), \tag{1.19}$$

where $\xi = (\xi^1, \ldots, \xi^m)$, $\mathbf{v}(\xi) = (\mathbf{v}(\xi^1), \ldots, \mathbf{v}(\xi^m))$, etc., providing even further justification for our exponential notation.

According to our new interpretation of the symbols $\partial/\partial x^i$, each tangent vector $\mathbf{v}|_x$ at a point x defines a *derivation* on the space of smooth real-valued functions f defined near x in M. This means that $\mathbf{v}|_x$, when applied to a smooth function f, gives a real number $\mathbf{v}(f) = \mathbf{v}(f)(x)$, and, moreover, this operation determined by $\mathbf{v}$ has the basic derivational properties of

(a) *Linearity*

$$\mathbf{v}(f + g) = \mathbf{v}(f) + \mathbf{v}(g), \tag{1.20}$$

(b) *Leibniz' Rule*

$$\mathbf{v}(f \cdot g) = \mathbf{v}(f) \cdot g + f \cdot \mathbf{v}(g). \tag{1.21}$$

(Here both sides of (1.20), (1.21) are evaluated at the point x.) Conversely, it is not hard to show that every derivation on the space of smooth functions at x is a tangent vector, and in particular is given in local coordinates by $\sum \xi^i \partial/\partial x^i$. (See Exercise 1.12.) This approach is often used to *define* tangent vectors and the tangent bundle in an abstract, coordinate-free manner. Further, if $\mathbf{v}$ is a vector field on M, then $\mathbf{v}(f)$ is a smooth function for any $f: M \to \mathbb{R}$. Thus we can also define vector fields as derivations, i.e. maps satisfying (1.20), (1.21), on the space of smooth functions on M. This point of view is especially useful for defining various operations on vector fields in a coordinate-free manner. (See Warner, [1; Chap. 1] for more details of this construction and the correspondence between tangent vectors and derivations.)

Differentials

Let M and N be smooth manifolds and $F: M \to N$ a smooth map between them. Each parametrized curve $C = \{\phi(\varepsilon): \varepsilon \in I\}$ on M is mapped by F to a parametrized curve $\tilde{C} = F(C) = \{\tilde{\phi}(\varepsilon) = F(\phi(\varepsilon)): \varepsilon \in I\}$ on N. Thus F induces a map from the tangent vector $d\phi/d\varepsilon$ to C at $x = \phi(\varepsilon)$ to the corresponding tangent vector $d\tilde{\phi}/d\varepsilon$ to $\tilde{C}$ at the image point $F(x) = F(\phi(\varepsilon)) = \tilde{\phi}(\varepsilon)$. This induced map is called the *differential* of F, and denoted by

$$dF(\dot{\phi}(\varepsilon)) = \frac{d}{d\varepsilon}\{F(\phi(\varepsilon))\}. \tag{1.22}$$

As every tangent vector $\mathbf{v}|_x \in TM|_x$ is tangent to some curve passing through x, the differential maps the tangent space to M at x to the tangent space to N at $F(x)$:

$$dF: TM|_x \to TN|_{F(x)}.$$

The local coordinate formula for the differential is found using the chain rule in the same manner as the change of variables formula (1.6). If

$$\mathbf{v}|_x = \sum_{i=1}^{m} \xi^i \frac{\partial}{\partial x^i}$$

is a tangent vector at $x \in M$, then

$$dF(\mathbf{v}|_x) = \sum_{j=1}^{n} \left(\sum_{i=1}^{m} \xi^i \frac{\partial F^j}{\partial x^i}(x) \right) \frac{\partial}{\partial y^j} = \sum_{j=1}^{n} \mathbf{v}(F^j(x)) \frac{\partial}{\partial y^j}. \tag{1.23}$$

Note that the differential $dF|_x$ is a linear map from $TM|_x$ to $TN|_{F(x)}$, whose matrix expression in local coordinates is just the Jacobian matrix of F at x.

If we prefer to think of tangent vectors as derivations on the space of smooth functions defined near a point x, then the differential dF has the alternative definition

$$dF(\mathbf{v}|_x)f(y) = \mathbf{v}(f \circ F)(x), \qquad y = F(x), \tag{1.24}$$

for all $\mathbf{v}|_x \in TM|_x$ and all smooth $f: N \to \mathbb{R}$. The equivalence of (1.22) and (1.24) is easily verified using local coordinates.

Example 1.30. Let $M = \mathbb{R}^2$, with coordinates (x, y), and $N = \mathbb{R}$, with coordinate s, and let $F: \mathbb{R}^2 \to \mathbb{R}$ be any map $s = F(x, y)$. Given

$$\mathbf{v}|_{(x,y)} = a \frac{\partial}{\partial x} + b \frac{\partial}{\partial y},$$

then, by (1.23),

$$dF(\mathbf{v}|_{(x,y)}) = \left\{ a\frac{\partial F}{\partial x}(x, y) + b\frac{\partial F}{\partial y}(x, y) \right\} \frac{d}{ds}\bigg|_{F(x,y)}.$$

For example, if $F(x, y) = \alpha x + \beta y$ is a linear projection, then

$$dF(\mathbf{v}|_{(x,y)}) = (a\alpha + b\beta)\frac{d}{ds}\bigg|_{s=\alpha x+\beta y}.$$

Lemma 1.31. *If $F: M \to N$ and $H: N \to P$ are smooth maps between manifolds, then*

$$d(H \circ F) = dH \circ dF, \tag{1.25}$$

where $dF: TM|_x \to TN|_{y=F(x)}$, $dH: TN|_y \to TP|_{z=H(y)}$, and $d(H \circ F): TM|_x \to TP|_{z=H(F(x))}$.

The proof is immediate from either of the two definitions. In local coordinates, (1.25) just says that the Jacobian matrix of the composition of two functions is the product of their respective Jacobian matrices.

It is important to note that if $\mathbf{v}$ is a vector field on M, then in general $dF(\mathbf{v})$ will *not* be a well-defined vector field on N. For one thing, $dF(\mathbf{v})$ may not be defined on all of N; for another, if two points x and $\tilde{x}$ in M are mapped to the same point $y = F(x) = F(\tilde{x})$ in N, there is no guarantee that $dF(\mathbf{v}|_x)$ and $dF(\mathbf{v}|_{\tilde{x}})$ (both of which are in $TN|_y$) are the same. For instance, if $\mathbf{v} = y\partial_x + \partial_y$ and $s = F(x, y) = \alpha x + \beta y$ is the projection of Example 1.30, then

$$dF(\mathbf{v}|_{(x,y)}) = (\alpha y + \beta)\frac{d}{ds}\bigg|_{s=\alpha x+\beta y},$$

which is not a well-defined vector field on $\mathbb{R}$ unless $\alpha = 0$. However, if F is a diffeomorphism onto N, then $dF(\mathbf{v})$ is always a vector field on N. More generally, two vector fields $\mathbf{v}$ on M and $\mathbf{w}$ on N are said to be *F-related* if $dF(\mathbf{v}|_x) = \mathbf{w}|_{F(x)}$ for all $x \in M$. If $\mathbf{v}$ and $\mathbf{w} = dF(\mathbf{v})$ are F-related, then F maps integral curves of $\mathbf{v}$ to integral curves of $\mathbf{w}$, with

$$F(\exp(\varepsilon\mathbf{v})x) = \exp(\varepsilon\, dF(\mathbf{v}))F(x). \tag{1.26}$$

Lie Brackets

The most important operation on vector fields is their Lie bracket or commutator. This is most easily defined in terms of their actions as derivations on functions. Specifically, if $\mathbf{v}$ and $\mathbf{w}$ are vector fields on M, then their *Lie bracket* $[\mathbf{v}, \mathbf{w}]$ is the unique vector field satisfying

$$[\mathbf{v}, \mathbf{w}](f) = \mathbf{v}(\mathbf{w}(f)) - \mathbf{w}(\mathbf{v}(f)) \tag{1.27}$$

for all smooth functions $f: M \to \mathbb{R}$. It is easy to verify that $[\mathbf{v}, \mathbf{w}]$ is indeed a vector field. In local coordinates, if

$$\mathbf{v} = \sum_{i=1}^{m} \xi^i(x) \frac{\partial}{\partial x^i}, \qquad \mathbf{w} = \sum_{i=1}^{m} \eta^i(x) \frac{\partial}{\partial x^i},$$

then

$$[\mathbf{v}, \mathbf{w}] = \sum_{i=1}^{m} \{\mathbf{v}(\eta^i) - \mathbf{w}(\xi^i)\} \frac{\partial}{\partial x^i} = \sum_{i=1}^{m} \sum_{j=1}^{m} \left\{ \xi^j \frac{\partial \eta^i}{\partial x^j} - \eta^j \frac{\partial \xi^i}{\partial x^j} \right\} \frac{\partial}{\partial x^i}. \tag{1.28}$$

(Note that in (1.27) the terms involving second order derivatives of f cancel.) For example, if

$$\mathbf{v} = y \frac{\partial}{\partial x}, \qquad \mathbf{w} = x^2 \frac{\partial}{\partial x} + xy \frac{\partial}{\partial y},$$

then

$$[\mathbf{v}, \mathbf{w}] = \mathbf{v}(x^2) \frac{\partial}{\partial x} + \mathbf{v}(xy) \frac{\partial}{\partial y} - \mathbf{w}(y) \frac{\partial}{\partial x} = xy \frac{\partial}{\partial x} + y^2 \frac{\partial}{\partial y}.$$

Proposition 1.32. *The Lie bracket has the following properties*:

(a) Bilinearity

$$\begin{aligned} [c\mathbf{v} + c'\mathbf{v}', \mathbf{w}] &= c[\mathbf{v}, \mathbf{w}] + c'[\mathbf{v}', \mathbf{w}], \\ [\mathbf{v}, c\mathbf{w} + c'\mathbf{w}'] &= c[\mathbf{v}, \mathbf{w}] + c'[\mathbf{v}, \mathbf{w}'], \end{aligned} \tag{1.29}$$

where c, c' are constants.

(b) Skew-Symmetry

$$[\mathbf{v}, \mathbf{w}] = -[\mathbf{w}, \mathbf{v}]. \tag{1.30}$$

(c) Jacobi Identity

$$[\mathbf{u}, [\mathbf{v}, \mathbf{w}]] + [\mathbf{w}, [\mathbf{u}, \mathbf{v}]] + [\mathbf{v}, [\mathbf{w}, \mathbf{u}]] = 0. \tag{1.31}$$

The proof is left to the reader. (*Hint*: Use (1.27) as your definition—trying to verify the Jacobi identity using the local coordinate formula (1.28) is horrible.)

The first definition (1.27) of the Lie bracket ensures that it is coordinate-free. (This can also be checked from the local coordinate formula (1.28), but is a fairly tedious computation.) More generally, if $F: M \to N$ is any smooth map, and $\mathbf{v}$ and $\mathbf{w}$ are vector fields on M such that $dF(\mathbf{v})$, $dF(\mathbf{w})$ are F-related to well-defined vector fields on N, then their Lie brackets are also F-related:

$$dF([\mathbf{v}, \mathbf{w}]) = [dF(\mathbf{v}), dF(\mathbf{w})]. \tag{1.32}$$

To prove this, given $f: N \to \mathbb{R}$, if $y = F(x) \in N$, then by (1.24),

$$\begin{aligned} dF([\mathbf{v}, \mathbf{w}])f(y) &= [\mathbf{v}, \mathbf{w}]\{f(F(x))\} = \mathbf{v}(\mathbf{w}\{f(F(x))\}) - \mathbf{w}(\mathbf{v}\{f(F(x))\}) \\ &= \mathbf{v}\{dF(\mathbf{w})f(F(x))\} - \mathbf{w}\{dF(\mathbf{v})f(F(x))\} \\ &= dF(\mathbf{v})dF(\mathbf{w})f(y) - dF(\mathbf{w})dF(\mathbf{v})f(y) \\ &= [dF(\mathbf{v}), dF(\mathbf{w})]f(y), \end{aligned}$$

as required.

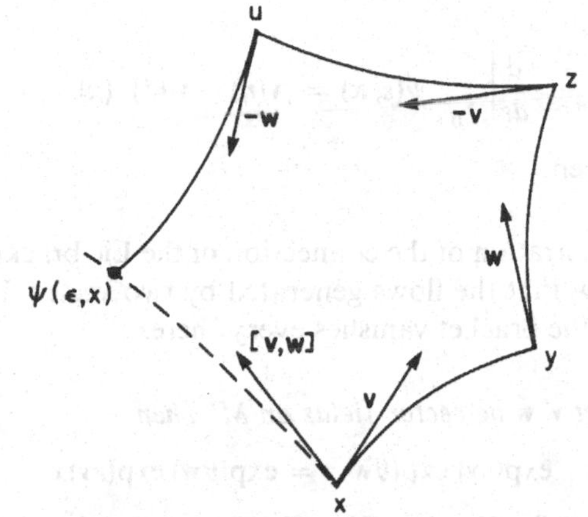

Figure 5. Commutator construction of the Lie bracket.

There is a more geometric characterization of the Lie bracket of two vector fields as the "infinitesimal commutator" of the two one-parameter groups $\exp(\varepsilon \mathbf{v})$ and $\exp(\varepsilon \mathbf{w})$.

Theorem 1.33. *Let* $\mathbf{v}$ *and* $\mathbf{w}$ *be smooth vector fields on a manifold* M. *For each* $x \in M$, *the commutator*

$$\psi(\varepsilon, x) = \exp(-\sqrt{\varepsilon}\mathbf{w})\exp(-\sqrt{\varepsilon}\mathbf{v})\exp(\sqrt{\varepsilon}\mathbf{w})\exp(\sqrt{\varepsilon}\mathbf{v})x$$

defines a smooth curve for sufficiently small $\varepsilon \geqslant 0$. *The Lie bracket* $[\mathbf{v}, \mathbf{w}]|_x$ *is the tangent vector to this curve at the end-point* $\psi(0, x) = x$:

$$[\mathbf{v}, \mathbf{w}]|_x = \frac{d}{d\varepsilon}\bigg|_{\varepsilon=0+} \psi(\varepsilon, x). \tag{1.33}$$

PROOF. Let $x = (x^1, \ldots, x^m)$ be local coordinates, so that

$$\mathbf{v} = \sum_{i=1}^m \xi^i(x)\frac{\partial}{\partial x^i}, \qquad \mathbf{w} = \sum_{i=1}^m \eta^i(x)\frac{\partial}{\partial x^i}.$$

Set $y = \exp(\sqrt{\varepsilon}\mathbf{v})x$, $z = \exp(\sqrt{\varepsilon}\mathbf{w})y$, $u = \exp(-\sqrt{\varepsilon}\mathbf{v})z$, so that $\psi(\varepsilon, x) = \exp(-\sqrt{\varepsilon}\mathbf{w})u$. Then we use the Taylor series expansions (1.18), (1.19) for the action of the flow generated by a vector field repeatedly:

$$\begin{aligned}
\psi(\varepsilon, x) &= u - \sqrt{\varepsilon}\eta(u) + \tfrac{1}{2}\varepsilon \mathbf{w}(\eta)(u) + O(\varepsilon^{3/2}) \\
&= z - \sqrt{\varepsilon}\{\eta(z) + \xi(z)\} + \varepsilon\{\tfrac{1}{2}\mathbf{w}(\eta)(z) + \mathbf{v}(\eta)(z) + \tfrac{1}{2}\mathbf{v}(\xi)(z)\} + O(\varepsilon^{3/2}) \\
&= y - \sqrt{\varepsilon}\xi(y) + \varepsilon\{\mathbf{v}(\eta)(y) - \mathbf{w}(\xi)(y) + \tfrac{1}{2}\mathbf{v}(\xi)(y)\} + O(\varepsilon^{3/2}) \\
&= x + \varepsilon\{\mathbf{v}(\eta)(x) - \mathbf{w}(\xi)(x)\} + O(\varepsilon^{3/2}).
\end{aligned}$$

Therefore

$$\left.\frac{d}{d\varepsilon}\right|_{\varepsilon=0+} \psi(\varepsilon, x) = \{\mathbf{v}(\eta) - \mathbf{w}(\xi)\}(x),$$

and (1.33) is proven. □

As another illustration of the connection of the Lie bracket with the commutator, we show that the flows generated by two vector fields commute if and only if their Lie bracket vanishes everywhere.

Theorem 1.34. *Let* $\mathbf{v}$, $\mathbf{w}$ *be vector fields on* M. *Then*

$$\exp(\varepsilon\mathbf{v})\exp(\theta\mathbf{w})x = \exp(\theta\mathbf{w})\exp(\varepsilon\mathbf{v})x \tag{1.34}$$

for all $x \in M$ *and* $\varepsilon, \theta \in V$, *where* $V \subset R^2$ *is a connected open subset containing* $(0,0)$ *such that both sides of* (1.34) *are defined at all points therein, if and only if*

$$[\mathbf{v}, \mathbf{w}] = 0$$

everywhere.

Proof. Theorem 1.33 immediately shows that if the flows commute, i.e. if (1.34) holds, then the Lie bracket vanishes. Conversely, suppose $[\mathbf{v}, \mathbf{w}] = 0$, and let $x \in M$. If both $\mathbf{v}$ and $\mathbf{w}$ vanish at x, then flows of both vector fields leave x fixed, and hence they obviously commute at x. Otherwise, at least one vector field is not zero at x, say $\mathbf{v}|_x \neq 0$. Using Proposition 1.29, we can choose local coordinates $y = (y^1, \ldots, y^m)$ near x so that $\mathbf{v} = \partial/\partial y^1$ everywhere in these coordinates. Then if $\mathbf{w} = \sum \eta^i(y)\partial/\partial y^i$,

$$0 = [\mathbf{v}, \mathbf{w}] = \sum_{i=1}^{m} \frac{\partial \eta^i}{\partial y^1}\frac{\partial}{\partial y^i}.$$

Therefore each η^i is independent of y^1. The flow generated by $\mathbf{v}$ in these coordinates is just

$$\exp(\varepsilon\mathbf{v})(y^1, \ldots, y^m) = (y^1 + \varepsilon, y^2, \ldots, y^m).$$

The flow generated by $\mathbf{w}$ is a solution of the system of ordinary differential equations

$$\frac{dy^i}{d\theta} = \eta^i(y^2, \ldots, y^m), \qquad i = 1, \ldots, m.$$

Consider the functions

$$y(\theta, \varepsilon) = \exp(\theta\mathbf{w})\exp(\varepsilon\mathbf{v})y = \exp(\theta\mathbf{w})(y^1 + \varepsilon, y^2, \ldots, y^m)$$

and

$$\begin{aligned}\tilde{y}(\theta, \varepsilon) &= \exp(\varepsilon\mathbf{v})\exp(\theta\mathbf{w})y = \exp(\varepsilon\mathbf{v})y(\theta, 0)\\ &= (y^1(\theta, 0) + \varepsilon, y^2(\theta, 0), \ldots, y^m(\theta, 0)).\end{aligned}$$

Since y^1 does not appear on the right-hand side of the differential equations for the flow of $\mathbf{w}$, as functions of θ both y and $\tilde{y}$ are solutions, and both have the same initial conditions

$$y(0, \varepsilon) = (y^1 + \varepsilon, y^2, \ldots, y^m) = \tilde{y}(0, \varepsilon).$$

By uniqueness, $y(\theta, \varepsilon) = \tilde{y}(\theta, \varepsilon)$, which proves (1.34) for θ, ε sufficiently small.

To prove (1.34) in general, consider the following two subsets of the (θ, ε) plane: first V is the connected component of

$$\hat{V} = \{(\theta, \varepsilon)\colon \text{ both sides of (1.34) are defined at } (\theta, \varepsilon)\}$$

containing the origin; second $U = \hat{U} \cap V$, where

$$\hat{U} = \{(\theta, \varepsilon)\colon \text{ both sides of (1.34) are defined and equal at } (\theta, \varepsilon)\}.$$

By what we have just shown, U is open. On the other hand, by continuity, if (1.34) holds at $(\theta_i, \varepsilon_i) \in U$, and $(\theta_i, \varepsilon_i) \to (\theta^*, \varepsilon^*) \in V$, then (1.34) holds at $(\theta^*, \varepsilon^*)$. Thus U is both open and closed as a subset of V, so by connectivity $U = V$. *Warning*: It is not, in general, true that $\hat{U} = \hat{V}$! □

Tangent Spaces and Vector Fields on Submanifolds

Suppose $N \subset M$ is a submanifold of M parametrized by the immersion ϕ: $\tilde{N} \to M$. The tangent space to N at $y \in N$ is, by definition, the image of the tangent space to $\tilde{N}$ at the corresponding point $\tilde{y}$:

$$TN|_y = d\phi(T\tilde{N}|_{\tilde{y}}), \qquad y = \phi(\tilde{y}) \in N.$$

Note that $TN|_y$ is a subspace of $TM|_y$ of the same dimension as N. There is an analogous characterization of the tangent space to an implicitly defined submanifold:

Proposition 1.35. *Let $F\colon M \to \mathbb{R}^n$, $n \leqslant m$, be of maximal rank on $N = \{x\colon F(x) = 0\}$, so $N \subset M$ is an implicitly defined, regular $(m-n)$-dimensional submanifold. Given $y \in N$, the tangent space to N at y is precisely the kernel of the differential of F at y:*

$$TN|_y = \{\mathbf{v} \in TM|_y\colon dF(\mathbf{v}) = 0\}.$$

PROOF. If $\phi(\varepsilon)$ parametrizes a smooth curve $C \subset N$ passing through $y = \phi(\varepsilon_0)$, then $F(\phi(\varepsilon)) = 0$ for all ε. Differentiating with respect to ε, we see that

$$0 = \frac{d}{d\varepsilon} F(\phi(\varepsilon)) = dF(\dot{\phi}(\varepsilon)),$$

hence the tangent vector $\dot{\phi}$ to C is in the kernel of dF. The converse follows by a dimension count, using the fact that dF has rank n at y. □

Example 1.36. Consider the sphere $S^2 = \{x^2 + y^2 + z^2 = 1\}$ in $\mathbb{R}^3$. At each point $p = (x, y, z)$ on the sphere, the tangent space $TS^2|_p$ is given as the kernel of the differential of the defining function $F(x, y, z) = x^2 + y^2 + z^2 - 1$ at p. Thus

$$TS^2|_{(x,y,z)} = \left\{ a\frac{\partial}{\partial x} + b\frac{\partial}{\partial y} + c\frac{\partial}{\partial z} : 2ax + 2by + 2cz = 0 \right\}.$$

Identifying $T\mathbb{R}^3|_p$ with $\mathbb{R}^3$, so that $\mathbf{v}|_p = a\partial_x + b\partial_y + c\partial_z$ becomes the vector (a, b, c), we see that $TS^2|_p$ consists of all vectors $\mathbf{v}|_p$ in $\mathbb{R}^3$ which are orthogonal to the radial vector $p = (x, y, z)$. Thus the tangent space $TS^2|_p$ agrees with the usual geometric tangent plane to S^2 at the point p. (The same argument generalizes to any implicitly defined surface $S = \{F(x, y, z) = 0\}$ in $\mathbb{R}^3$, where dF corresponds to the normal vector ∇F.)

Let N be a submanifold of M. If $\mathbf{v}$ is a vector field on M, then $\mathbf{v}$ restricts to a vector field on N if and only if $\mathbf{v}$ is everywhere tangent to N, meaning that $\mathbf{v}|_y \in TN|_y$ for each $y \in N$. In this case, using the definition of $TN|_y$, we immediately deduce the existence of a corresponding vector field $\tilde{\mathbf{v}}$ on the parametrization space $\tilde{N}$ satisfying $d\phi(\tilde{\mathbf{v}}) = \mathbf{v}$ on N.

Lemma 1.37. *If* $\mathbf{v}$ *and* $\mathbf{w}$ *are tangent to a submanifold* N, *then so is* $[\mathbf{v}, \mathbf{w}]$.

PROOF. Let $\tilde{\mathbf{v}}$ and $\tilde{\mathbf{w}}$ be the corresponding vector fields on $\tilde{N}$. Then by (1.32),

$$d\phi[\tilde{\mathbf{v}}, \tilde{\mathbf{w}}] = [d\phi(\tilde{\mathbf{v}}), d\phi(\tilde{\mathbf{w}})] = [\mathbf{v}, \mathbf{w}].$$

at each point of N. But this says that $[\mathbf{v}, \mathbf{w}]|_y \in TN|_y = d\phi(T\tilde{N}|_{\tilde{y}})$ for each $y \in N$. □

For example, in the case of the sphere S^2, since $z\partial_x - x\partial_z$ and $z\partial_y - y\partial_z$ are both tangent to S^2, so is $[z\partial_x - x\partial_z, z\partial_y - y\partial_z] = y\partial_x - x\partial_y$.

Frobenius' Theorem

We have already seen how each vector field $\mathbf{v}$ on a manifold M determines an integral curve through each point of M, such that $\mathbf{v}$ is tangent to the curve everywhere. Frobenius' theorem deals with the more general case of determining "integral submanifolds" of systems of vector fields, with the property that each vector field is tangent to the submanifold at each point.

Definition 1.38. Let $\mathbf{v}_1, \ldots, \mathbf{v}_r$ be vector fields on a smooth manifold M. An *integral submanifold* of $\{\mathbf{v}_1, \ldots, \mathbf{v}_r\}$ is a submanifold $N \subset M$ whose tangent space $TN|_y$ is spanned by the vectors $\{\mathbf{v}_1|_y, \ldots, \mathbf{v}_r|_y\}$ for each $y \in N$. The system of vector fields $\{\mathbf{v}_1, \ldots, \mathbf{v}_r\}$ is *integrable* if through every point $x_0 \in M$ there passes an integral submanifold.

Note that if N is an integral submanifold of $\{\mathbf{v}_1, \ldots, \mathbf{v}_r\}$, then the dimension of the subspace of $TM|_y$ spanned by $\{\mathbf{v}_1|_y, \ldots, \mathbf{v}_r|_y\}$, which by definition is $TN|_y$, is equal to the dimension of N at each point $y \in N$. This does *not* exclude the possibility that the dimension of the subspace of $TM|_x$ spanned by $\{\mathbf{v}_1|_x, \ldots, \mathbf{v}_r|_x\}$ varies as x ranges over the entire manifold M; this just means that the given set of vector fields can have integral submanifolds of different dimensions.

Lemma 1.37 immediately gives necessary conditions that a system of vector fields be integrable. Namely, if N is an integral submanifold, then each vector field in the collection must be tangent to N at each point. Thus the Lie bracket of any pair of vector fields in the collection must again be tangent to N, and hence in the span of the set of vector fields at each point.

Definition 1.39. A system of vector fields $\{\mathbf{v}_1, \ldots, \mathbf{v}_r\}$ on M is *in involution* if there exist smooth real-valued functions $h_{ij}^k(x)$, $x \in M$, $i, j, k = 1, \ldots, r$, such that for each $i, j = 1, \ldots, r$,

$$[\mathbf{v}_i, \mathbf{v}_j] = \sum_{k=1}^{r} h_{ij}^k \cdot \mathbf{v}_k.$$

Frobenius' theorem, as generalized by Hermann to the case when the integral submanifolds have varying dimensions, states that this necessary condition is also sufficient:

Theorem 1.40. *Let $\mathbf{v}_1, \ldots, \mathbf{v}_r$ be smooth vector fields on M. Then the system $\{\mathbf{v}_1, \ldots, \mathbf{v}_r\}$ is integrable if and only if it is in involution.*

This theorem is *not* true as stated if the system is generated by infinitely many vector fields; see Exercise 1.13. There is, however, a useful generalization provided we make an additional restriction on the system. Let $\mathscr{H}$ be a collection of vector fields which forms a vector space. We say $\mathscr{H}$ is *in involution* if $[\mathbf{v}, \mathbf{w}] \in \mathscr{H}$ whenever $\mathbf{v}$ and $\mathbf{w}$ are in $\mathscr{H}$. In the above finite-dimensional case, $\mathscr{H}$ can be taken to be the set of linear combinations $\sum f_i(x)\mathbf{v}_i$ of the "basis" vector fields $\mathbf{v}_i$, with the f_i being arbitrary smooth real-valued functions on M (in which case $\mathscr{H}$ is called *finitely generated*). Let $\mathscr{H}|_x$ be the subspace of $TM|_x$ spanned by $\mathbf{v}|_x$ for all $\mathbf{v} \in \mathscr{H}$. An *integral submanifold* of $\mathscr{H}$ is a connected submanifold $N \subset M$ such that $TN|_y = \mathscr{H}|_y$ for all $y \in N$. We say that $\mathscr{H}$ is *rank-invariant* if for any vector field $\mathbf{v} \in \mathscr{H}$, the dimension of the subspace $\mathscr{H}|_{\exp(\varepsilon\mathbf{v})x}$ along the flow generated by $\mathbf{v}$ is a constant, independent of ε. (It can, of course, depend on the initial point x.) Note that since the integral curve $\exp(\varepsilon\mathbf{v})x$ of $\mathbf{v}$ emanating from a point x should be contained in an integral submanifold N, rank-invariance is certainly a necessary condition for complete integrability. Rank-invariance follows automatically if $\mathscr{H}$ is finitely generated, or consists of analytic vector fields on an analytic manifold.

Theorem 1.41. *Let $\mathscr{H}$ be a system of vector fields on a manifold M. Then $\mathscr{H}$ is integrable if and only if it is in involution and rank-invariant.*

In essence, the proof proceeds by direct construction of the integral submanifolds. If $x \in N$, then we can realize the integral submanifold through x by examining successive integral curves starting at x:

$$N = \{\exp(\mathbf{v}_1)\exp(\mathbf{v}_2)\cdots\exp(\mathbf{v}_k)x \colon k \geqslant 1, \mathbf{v}_i \in \mathscr{H}\}.$$

The rank invariance will imply that $\mathscr{H}|_y$ for any $y \in N$ has the correct dimension. The details of the proof that N is a submanifold can be found in Hermann, [2]. Borrowing terminology from the more usual constant-rank case, we call the collection of all maximal integral submanifolds of an integrable system of vector fields a *foliation* of the manifold M; the integral submanifolds themselves are also referred to as *leaves* of the foliation.

Example 1.42. Consider the vector fields

$$\mathbf{v} = -y\frac{\partial}{\partial x} + x\frac{\partial}{\partial y}, \qquad \mathbf{w} = 2xz\frac{\partial}{\partial x} + 2yz\frac{\partial}{\partial y} + (z^2 + 1 - x^2 - y^2)\frac{\partial}{\partial z}$$

on $\mathbb{R}^3$. An easy computation proves that $[\mathbf{v}, \mathbf{w}] = 0$, so by Frobenius' theorem $\{\mathbf{v}, \mathbf{w}\}$ is integrable. Given (x, y, z), the subspace of $T\mathbb{R}^3|_{(x,y,z)}$ spanned by $\mathbf{v}|_{(x,y,z)}$ and $\mathbf{w}|_{(x,y,z)}$ is two-dimensional, except on the z-axis $\{x = y = 0\}$ and the circle $\{x^2 + y^2 = 1, z = 0\}$, where it is one-dimensional. It is not difficult to check that both the circle and the z-axis are one-dimensional integral submanifolds of $\{\mathbf{v}, \mathbf{w}\}$. All other integral submanifolds are two-dimensional tori

$$\zeta(x, y, z) = (x^2 + y^2)^{-1/2}(x^2 + y^2 + z^2 + 1) = c,$$

defined for $c > 2$. Indeed,

$$d\zeta(\mathbf{v}) = \mathbf{v}(\zeta) = 0, \qquad d\zeta(\mathbf{w}) = \mathbf{w}(\zeta) = 0,$$

everywhere, so by Proposition 1.35, both $\mathbf{v}$ and $\mathbf{w}$ are tangent to each level set of ζ where $\nabla\zeta \neq 0$. (See Section 2.1 for some general techniques for constructing integral submanifolds.)

An integrable system of vector fields $\{\mathbf{v}_1, \ldots, \mathbf{v}_r\}$ is called *semi-regular* if the dimension of the subspace of $TM|_x$ spanned by $\{\mathbf{v}_1|_x, \ldots, \mathbf{v}_r|_x\}$ does not vary from point to point. In this case all the integral submanifolds have the same dimension. In analogy with the concept of a regular group action, we say that an integrable system of vector fields is *regular* if it is semi-regular, and, in addition, each point x in M has arbitrarily small neighbourhoods U with the property that each maximal integral submanifold intersects U in a pathwise connected subset. Although semi-regularity is a local property, which can be deduced using coordinates, regularity depends on the global structure of the system and is extremely difficult to check without explicitly finding the integral submanifolds. Any semi-regular system can be made regular, however, by restriction to a suitably small open subset of M. For example, the system in Example 1.42 is regular on the open subset $\mathbb{R}^3 \setminus (\{x = y = 0\} \cup \{x^2 + y^2 = 1, z = 0\})$ obtained by deleting the z-axis and the unit circle from $\mathbb{R}^3$.

For semi-regular systems of vector fields, Frobenius' theorem actually gives a means of "flattening out" the integral submanifolds by appropriate choice of local coordinates, just as we did for the integral curves of a single vector field in Proposition 1.29.

Theorem 1.43. *Let $\{\mathbf{v}_1, \ldots, \mathbf{v}_r\}$ be an integrable system of vector fields such that the dimension of the span of $\{\mathbf{v}_1|_x, \ldots, \mathbf{v}_r|_x\}$ in $TM|_x$ is a constant s, independent of $x \in M$. Then for each $x_0 \in M$ there exist* flat local coordinates *$y = (y^1, \ldots, y^m)$ near x_0 such that the integral submanifolds intersect the given coordinate chart in the "slices" $\{y: y^1 = c_1, \ldots, y^{m-s} = c_{m-s}\}$, where $c_1, \ldots, c_{m-s}$ are arbitrary constants. If, in addition, the system is regular, then the coordinate chart can be chosen so that each integral submanifold intersects it in at most one such slice.*

For the system in Example 1.42, near any point (x_0, y_0, z_0) with $z_0 \neq 0$ and not on the z-axis, flat local coordinates are given by $\tilde{x} = x$, $\tilde{y} = y$, $\tilde{z} = \zeta(x, y, z)$. The tangent space to the plane $\{\tilde{z} = \text{constant}\}$ is spanned by the vector fields

$$\frac{\partial}{\partial \tilde{x}} = \frac{\partial}{\partial x} - \frac{x(x^2 + y^2 - z^2 - 1)}{2z(x^2 + y^2)} \frac{\partial}{\partial z},$$

$$\frac{\partial}{\partial \tilde{y}} = \frac{\partial}{\partial y} - \frac{y(x^2 + y^2 - z^2 - 1)}{2z(x^2 + y^2)} \frac{\partial}{\partial z}.$$

Note that $\{\partial/\partial\tilde{x}, \partial/\partial\tilde{y}\}$ and $\{\mathbf{v}, \mathbf{w}\}$ both span the same subspace of $T\mathbb{R}^3$ at each point (x, y, z) with $z(x^2 + y^2) \neq 0$, so we have indeed locally "flattened out" the tori of Example 1.42. A more physically interesting set of flat local coordinates for $\{\mathbf{v}, \mathbf{w}\}$ are provided by the toroidal coordinates (θ, ψ, η), defined by

$$x = \frac{\sinh \eta \cos \psi}{\cosh \eta - \cos \theta}, \qquad y = \frac{\sinh \eta \sin \psi}{\cosh \eta - \cos \theta}, \qquad z = \frac{\sin \theta}{\cosh \eta - \cos \theta},$$

which arise in the theory of separation of variables for Laplace's equation, cf. Moon and Spencer, [1]. The reader can check that the level surfaces $\{\eta = c\}$ are precisely the integral tori for the system $\{\mathbf{v}, \mathbf{w}\}$; in fact $\mathbf{v} = \partial_\psi$, $\mathbf{w} = -2\partial_\theta$ under the change of coordinates!

1.4. Lie Algebras

If G is a Lie group, then there are certain distinguished vector fields on G characterized by their invariance (in a sense to be defined shortly) under the group multiplication. As we shall see, these invariant vector fields form a finite-dimensional vector space, called the Lie algebra of G, which is in a precise sense the "infinitesimal generator" of G. In fact almost all the information in the group G is contained in its Lie algebra. This fundamental observation is the cornerstone of Lie group theory; for example, it enables us to replace complicated nonlinear conditions of invariance under a group action by relatively simple linear infinitesimal conditions. The power of this method cannot be overestimated—indeed almost the entire range of applications of Lie groups to differential equations ultimately rests on this one construction!

We begin with the global Lie group picture, addressing the analogous construction for local Lie groups subsequently. Let G be a Lie group. For any group element $g \in G$, the *right multiplication map*

$$R_g \colon G \to G$$

defined by

$$R_g(h) = h \cdot g$$

is a diffeomorphism, with inverse

$$R_{g^{-1}} = (R_g)^{-1}.$$

A vector field $\mathbf{v}$ on G is called *right-invariant* if

$$dR_g(\mathbf{v}|_h) = \mathbf{v}|_{R_g(h)} = \mathbf{v}|_{hg}$$

for all g and h in G. Note that if $\mathbf{v}$ and $\mathbf{w}$ are right-invariant, so is any linear combination $a\mathbf{v} + b\mathbf{w}$, $a, b \in \mathbb{R}$; hence the set of all right-invariant vector fields forms a vector space.

Definition 1.44. The *Lie algebra* of a Lie group G, traditionally denoted by the corresponding lowercase German letter $\mathfrak{g}$, is the vector space of all right-invariant vector fields on G.

Note that any right-invariant vector field is uniquely determined by its value at the identity because

$$\mathbf{v}|_g = dR_g(\mathbf{v}|_e), \tag{1.35}$$

since $R_g(e) = g$. Conversely, any tangent vector to G at e uniquely determines a right-invariant vector field on G by formula (1.35). Indeed,

$$dR_g(\mathbf{v}|_h) = dR_g(dR_h(\mathbf{v}|_e)) = d(R_g \circ R_h)(\mathbf{v}|_e) = dR_{hg}(\mathbf{v}|_e) = \mathbf{v}|_{hg},$$

proving the right-invariance of $\mathbf{v}$. Therefore we can identify the Lie algebra $\mathfrak{g}$ of G with the tangent space to G at the identity element

$$\mathfrak{g} \simeq TG|_e. \tag{1.36}$$

In particular, g is a finite-dimensional vector space of the same dimension as the underlying Lie group.

In addition to its vector space structure, such a Lie algebra is further equipped with a skew-symmetric bilinear operation, namely the Lie bracket. Indeed, if $\mathbf{v}$ and $\mathbf{w}$ are right-invariant vector fields on G, so is their Lie bracket $[\mathbf{v}, \mathbf{w}]$, since by (1.32)

$$dR_g[\mathbf{v}, \mathbf{w}] = [dR_g(\mathbf{v}), dR_g(\mathbf{w})] = [\mathbf{v}, \mathbf{w}].$$

This motivates the general definition of a Lie algebra.

Definition 1.45. A *Lie algebra* is a vector space $\mathfrak{g}$ together with a bilinear operation

$$[\cdot, \cdot]: \mathfrak{g} \times \mathfrak{g} \to \mathfrak{g},$$

called the *Lie bracket* for $\mathfrak{g}$, satisfying the axioms

(a) *Bilinearity*

$$[c\mathbf{v} + c'\mathbf{v}', \mathbf{w}] = c[\mathbf{v}, \mathbf{w}] + c'[\mathbf{v}', \mathbf{w}], \quad [\mathbf{v}, c\mathbf{w} + c'\mathbf{w}'] = c[\mathbf{v}, \mathbf{w}] + c'[\mathbf{v}, \mathbf{w}'],$$

for constants $c, c' \in \mathbb{R}$,

(b) *Skew-Symmetry*

$$[\mathbf{v}, \mathbf{w}] = -[\mathbf{w}, \mathbf{v}],$$

(c) *Jacobi Identity*

$$[\mathbf{u}, [\mathbf{v}, \mathbf{w}]] + [\mathbf{w}, [\mathbf{u}, \mathbf{v}]] + [\mathbf{v}, [\mathbf{w}, \mathbf{u}]] = 0,$$

for all $\mathbf{u}, \mathbf{v}, \mathbf{v}', \mathbf{w}, \mathbf{w}'$ in $\mathfrak{g}$.

In this book most Lie algebras will be finite-dimensional vector spaces. (An interesting infinite-dimensional Lie algebra is given by the space of all smoth vector fields on a manifold M. However, infinite-dimensional algebras are considerably more difficult to work with.) We begin with some easy examples of Lie algebras.

Example 1.46. If $G = \mathbb{R}$, then there is, up to constant multiple, a single right-invariant vector field, namely $\partial_x = \partial/\partial x$. In fact, given $x, y \in \mathbb{R}$,

$$R_y(x) = x + y,$$

hence

$$dR_y(\partial_x) = \partial_x.$$

Similarly, if $G = \mathbb{R}^+$, then the single independent right-invariant vector field is $x\partial_x$. Finally, for SO(2) the vector field ∂_θ is again the unique independent right-invariant one. Note that the Lie algebras of $\mathbb{R}$, $\mathbb{R}^+$ and SO(2) are all the same, being one-dimensional vector spaces with trivial Lie brackets ($[\mathbf{v}, \mathbf{w}] = 0$ for all $\mathbf{v}, \mathbf{w}$). This shouldn't be surprising, as the reader can easily check from the general definition that there *is* only one one-dimensional Lie algebra, namely $\mathfrak{g} = \mathbb{R}$, with necessarily trivial Lie bracket.

Example 1.47. Here we compute the Lie algebra of the general linear group $\mathrm{GL}(n)$. Note that since $\mathrm{GL}(n)$ is n^2-dimensional, we can identify the Lie algebra $\mathrm{gl}(n) \simeq \mathbb{R}^{n^2}$ with the space of all $n \times n$ matrices. Indeed, coordinates on $\mathrm{GL}(n)$ are provided by the matrix entries x_{ij}, $i, j = 1, \ldots, n$, so the tangent space to $\mathrm{GL}(n)$ at the identity is the set of all vector fields

$$\mathbf{v}_A|_\mathbb{1} = \sum_{i,j} a_{ij} \frac{\partial}{\partial x_{ij}}\bigg|_\mathbb{1},$$

where $A = (a_{ij})$ is an arbitrary $n \times n$ matrix. Now given $Y = (y_{ij}) \in \mathrm{GL}(n)$, the matrix $R_Y(X) = XY$ has entries

$$\sum_{k=1}^{n} x_{ik} y_{kj}.$$

Therefore, according to (1.35), we find

$$\begin{aligned}
\mathbf{v}_A|_Y &= dR_Y(\mathbf{v}_A|_\mathbb{1}) \\
&= \sum_{l,m} \sum_{i,j} a_{ij} \frac{\partial}{\partial x_{ij}} \left(\sum_k x_{lk} y_{km} \right) \frac{\partial}{\partial x_{lm}} \\
&= \sum_{i,j,m} a_{ij} y_{jm} \frac{\partial}{\partial x_{im}},
\end{aligned}$$

or, in terms of $X \in \mathrm{GL}(n)$,

$$\mathbf{v}_A|_X = \sum_{i,j} \left(\sum_k a_{ik} x_{kj} \right) \frac{\partial}{\partial x_{ij}}. \tag{1.37}$$

To compute the Lie bracket:

$$\begin{aligned}
[\mathbf{v}_A, \mathbf{v}_B] &= \sum_{\substack{i,j,k \\ l,m,p}} \left\{ a_{lp} x_{pm} \frac{\partial}{\partial x_{lm}} (b_{ik} x_{kj}) - b_{lp} x_{pm} \frac{\partial}{\partial x_{lm}} (a_{ik} x_{kj}) \right\} \frac{\partial}{\partial x_{ij}} \\
&= \sum_{i,j,k} \left[\sum_l (b_{il} a_{lk} - a_{il} b_{lk}) \right] x_{kj} \frac{\partial}{\partial x_{ij}} \\
&= \mathbf{v}_{[A,B]},
\end{aligned}$$

where $[A, B] \equiv BA - AB$ is the *matrix commutator*. Therefore, *the Lie algebra* $\mathrm{gl}(n)$ *of the general linear group* $\mathrm{GL}(n)$ *is the space of all* $n \times n$ *matrices with the Lie bracket being the matrix commutator.*

One-Parameter Subgroups

Suppose $\mathfrak{g}$ is the Lie algebra of a Lie group G. The next result shows that there is a one-to-one correspondence between one-dimensional subspaces of $\mathfrak{g}$ and (connected) one-parameter subgroups of G.

Proposition 1.48. *Let $\mathbf{v} \neq 0$ be a right-invariant vector field on a Lie group G. Then the flow generated by* $\mathbf{v}$ *through the identity, namely*

$$g_\varepsilon = \exp(\varepsilon\mathbf{v})e \equiv \exp(\varepsilon\mathbf{v}) \tag{1.38}$$

is defined for all $\varepsilon \in \mathbb{R}$ *and forms a one-parameter subgroup of G, with*

$$g_{\varepsilon+\delta} = g_\varepsilon \cdot g_\delta, \qquad g_0 = e, \qquad g_\varepsilon^{-1} = g_{-\varepsilon}, \tag{1.39}$$

isomorphic to either $\mathbb{R}$ *itself or the circle group* SO(2). *Conversely, any connected one-dimensional subgroup of G is generated by such a right-invariant vector field in the above manner.*

Proof. For ε, δ sufficiently small, (1.39) follows from the right-invariance of $\mathbf{v}$ and (1.26):

$$\begin{aligned} g_\delta \cdot g_\varepsilon = R_{g_\varepsilon}(g_\delta) &= R_{g_\varepsilon}[\exp(\delta\mathbf{v})e] \\ &= \exp[\delta \cdot dR_{g_\varepsilon}(\mathbf{v})]R_{g_\varepsilon}(e) \\ &= \exp(\delta\mathbf{v})g_\varepsilon \\ &= \exp(\delta\mathbf{v})\exp(\varepsilon\mathbf{v})e \\ &= \exp[(\delta+\varepsilon)\mathbf{v}]e = g_{\delta+\varepsilon}. \end{aligned}$$

Thus g_ε is at least a local one-parameter subgroup. In particular, $g_0 = e$, and $g_{-\varepsilon} = g_\varepsilon^{-1}$ for ε small. Furthermore, g_ε is defined at least for $-\frac{1}{2}\varepsilon_0 \leqslant \varepsilon \leqslant \frac{1}{2}\varepsilon_0$, for some $\varepsilon_0 > 0$, so we can inductively define

$$g_{m\varepsilon_0+\varepsilon} = g_{m\varepsilon_0} \cdot g_\varepsilon, \qquad -\tfrac{1}{2}\varepsilon_0 \leqslant \varepsilon \leqslant \tfrac{1}{2}\varepsilon_0,$$

for m an integer. The above calculation shows that g_ε is a smooth curve in G satisfying (1.39) for all ε, δ, proving that the flow is globally defined and forms a subgroup. If $g_\varepsilon = g_\delta$ for some $\varepsilon \neq \delta$, then it is not hard to show that $g_{\varepsilon_0} = e$ for some least positive $\varepsilon_0 > 0$, and that g_ε is periodic with period ε_0, i.e. $g_{\varepsilon+\varepsilon_0} = g_\varepsilon$. In this case $\{g_\varepsilon\}$ is isomorphic to SO(2) (take $\theta = 2\pi\varepsilon/\varepsilon_0$). Otherwise $g_\varepsilon \neq g_\delta$ for all $\varepsilon \neq \delta$, and $\{g_\varepsilon\}$ is isomorphic to $\mathbb{R}$.

Conversely, if $H \subset G$ is a one-dimensional subgroup, we let $\mathbf{v}|_e$ be any nonzero tangent vector to H at the identity. Using the isomorphism (1.36) we extend $\mathbf{v}$ to a right-invariant vector field on all of G. Since H is a subgroup it follows that $\mathbf{v}|_h$ is tangent to H at any $h \in H$, and therefore H is the integral curve of $\mathbf{v}$ passing through e. This proves the converse. □

Example 1.49. Suppose $G = \mathrm{GL}(n)$ with Lie algebra gl(n), the space of all $n \times n$ matrices with commutator as the Lie bracket. If $A \in \mathrm{gl}(n)$, then the corresponding right-invariant vector field $\mathbf{v}_A$ on GL(n) has the expression (1.37). The one-parameter subgroup $\exp(\varepsilon\mathbf{v}_A)e$ is found by integrating the system of n^2 ordinary differential equations

$$\frac{dx_{ij}}{d\varepsilon} = \sum_{k=1}^{n} a_{ik}x_{kj}, \qquad x_{ij}(0) = \delta_j^i, \qquad i, j = 1, \ldots, n,$$

involving the matrix entries of A. The solution is just the matrix exponential $X(\varepsilon) = e^{\varepsilon A}$, which is the one-parameter subgroup of $GL(n)$ generated by a matrix A in $gl(n)$.

Example 1.50. Consider the torus T^2 with group multiplication

$$(\theta, \rho) \cdot (\theta', \rho') = (\theta + \theta', \rho + \rho') \bmod 2\pi.$$

Clearly the Lie algebra of T^2 is spanned by the right-invariant fields $\partial/\partial\theta$, $\partial/\partial\rho$ with trivial Lie bracket: $[\partial_\theta, \partial_\rho] = 0$. Let

$$\mathbf{v}_\omega = \partial_\theta + \omega\partial_\rho$$

for some $\omega \in \mathbb{R}$. Then the corresponding one-parameter subgroup is

$$\exp(\varepsilon\mathbf{v}_\omega)(0, 0) = (\varepsilon, \varepsilon\omega) \bmod 2\pi, \qquad \varepsilon \in \mathbb{R},$$

which is precisely the subgroup H_ω discussed on pages 17–18. In particular, if ω is rational, H_ω is a closed, one-parameter subgroup isomorphic to $SO(2)$, while if ω is irrational, H_ω is a dense subgroup isomorphic to $\mathbb{R}$. This shows that it is rather difficult in general to tell the precise character of a one-parameter subgroup just from knowledge of its infinitesimal generator.

Subalgebras

In general a *subalgebra* $\mathfrak{h}$ of a Lie algebra $\mathfrak{g}$ is a vector subspace which is closed under the Lie bracket, so $[\mathbf{v}, \mathbf{w}] \in \mathfrak{h}$ whenever $\mathbf{v}, \mathbf{w} \in \mathfrak{h}$. If H is a Lie subgroup of a Lie group G, any right-invariant vector field $\mathbf{v}$ on H can be extended to a right-invariant vector field on G. (Just set $\mathbf{v}|_g = dR_g(\mathbf{v}|_e)$, $g \in G$.) In this way the Lie algebra $\mathfrak{h}$ of H is realized as a subalgebra of the Lie algebra $\mathfrak{g}$ of G. The correspondence between one-parameter subgroups of a Lie group G and one-dimensional subspaces $\mathfrak{h}$ (subalgebras) of its Lie algebra $\mathfrak{g}$ generalizes to provide a complete one-to-one correspondence between Lie subgroups of G and subalgebras of $\mathfrak{g}$.

Theorem 1.51. *Let G be a Lie group with Lie algebra $\mathfrak{g}$. If $H \subset G$ is a Lie subgroup, its Lie algebra is a subalgebra of $\mathfrak{g}$. Conversely, if $\mathfrak{h}$ is any s-dimensional subalgebra of $\mathfrak{g}$, there is a unique connected s-parameter Lie subgroup H of G with Lie algebra $\mathfrak{h}$.*

The main idea in the proof of this theorem can be outlined as follows. Let $\mathbf{v}_1, \ldots, \mathbf{v}_s$ be a basis of $\mathfrak{h}$, which defines a system of vector fields on G. Since $\mathfrak{h}$ is a subalgebra, each Lie bracket $[\mathbf{v}_i, \mathbf{v}_j]$ is again an element of $\mathfrak{h}$, and hence in the span of $\{\mathbf{v}_1, \ldots, \mathbf{v}_s\}$. Thus $\mathfrak{h}$ defines an involutive system of vector fields on G. Furthermore, it is easily seen that at each point $g \in G$, $\{\mathbf{v}_1|_g, \ldots, \mathbf{v}_s|_g\}$ are linearly independent tangent vectors, so the system is semi-regular. By Frobenius' Theorem 1.40, there is a maximal s-dimensional submanifold of this system passing through e, and this submanifold is the Lie subgroup H

corresponding to $\mathfrak{h}$. It is not too hard to check that H is indeed a subgroup, and we know it is a submanifold. The main technical complication in the proof comes from showing the group operations of multiplication and inversion induced from those of G are smooth in the manifold structure of H. The interested reader can look at Warner, [1; Theorem 3.19] for the complete proof.

Example 1.52. The preceding theorem greatly simplifies the computation of Lie algebras of Lie groups which can be realized as Lie subgroups of the general linear group $\mathrm{GL}(n)$. Namely, if $H \subset \mathrm{GL}(n)$ is a subgroup, then its Lie algebra $\mathfrak{h}$ will be a subalgebra of the Lie algebra $\mathfrak{gl}(n)$ of all $n \times n$ matrices, with Lie bracket being the matrix commutator. Moreover, we can find $\mathfrak{h} \simeq TH|_e$ just by looking at all one-parameter subgroups of $\mathrm{GL}(n)$ which are contained in H:

$$\mathfrak{h} = \{A \in \mathfrak{gl}(n) \colon e^{\varepsilon A} \in H \text{ for } \varepsilon \in \mathbb{R}\}.$$

For example, to find the Lie algebra of the orthogonal group $\mathrm{O}(n)$, we need to find all $n \times n$ matrices A such that

$$(e^{\varepsilon A})(e^{\varepsilon A})^T = \mathrm{I}.$$

Differentiating with respect to ε and setting $\varepsilon = 0$ we find

$$A + A^T = 0.$$

Therefore $\mathfrak{so}(n) = \{A \colon A \text{ is skew-symmetric}\}$ is the Lie algebra of both $\mathrm{O}(n)$ and $\mathrm{SO}(n)$. Lie algebras of other matrix Lie groups are found similarly.

We have seen that there is a general one-to-one correspondence between subalgebras of the Lie algebra of a given Lie group and connected Lie subgroups of the same group. In particular, every subalgebra of $\mathfrak{gl}(n)$ gives rise to a matrix Lie group, i.e. a Lie subgroup of $\mathrm{GL}(n)$. More generally, if $\mathfrak{g}$ is any finite-dimensional (abstract) Lie algebra, the question arises as to whether there is a corresponding Lie group G with the given space $\mathfrak{g}$ as its Lie algebra. The answer to this question is affirmative, and, in fact, reduces to the matrix case by the following important theorem of Ado.

Theorem 1.53. *Let $\mathfrak{g}$ be a finite-dimensional Lie algebra. Then $\mathfrak{g}$ is isomorphic to a subalgebra of $\mathfrak{gl}(n)$ for some n.*

As a direct consequence of Ado's theorem and the latter half of Theorem 1.22 we deduce the fundamental correspondence between Lie groups and Lie algebras.

Theorem 1.54. *Let $\mathfrak{g}$ be a finite-dimensional Lie algebra. Then there exists a unique connected, simply-connected Lie group G^* having $\mathfrak{g}$ as its Lie algebra. Moreover, if G is any other connected Lie group with Lie algebra $\mathfrak{g}$, then $\pi \colon G^* \to G$ is the simply-connected covering group of G.*

Indeed, we need only realize $\mathfrak{g}$ as a subalgebra of $\mathfrak{gl}(n)$ for some n, and take $\tilde{G}$ to be the corresponding Lie subgroup of $GL(n)$. Then G^* will be the simply-connected covering group of $\tilde{G}$, guaranteed by Theorem 1.22.

It is important to emphasize that it is *not* true that every Lie group G is isomorphic to a subgroup of $GL(n)$ for some n. In particular, the simply-connected covering group of $SL(2, \mathbb{R})$ is *not* realizable as a matrix Lie group!

The Exponential Map

The *exponential map* $\exp: \mathfrak{g} \to G$ is obtained by setting $\varepsilon = 1$ in the one-parameter subgroup generated by $\mathbf{v}$:

$$\exp(\mathbf{v}) \equiv \exp(\mathbf{v})e.$$

One readily proves that the differential

$$d\exp: T\mathfrak{g}|_0 \simeq \mathfrak{g} \to TG|_e \simeq \mathfrak{g}$$

of exp at 0 is the identity map. (See Exercise 1.27.) Thus, by the inverse function theorem, exp determines a local diffeomorphism from $\mathfrak{g}$ onto a neighbourhood of the identity element in G. Consequently, every group element g sufficiently close to the identity can be written as an exponential: $g = \exp(\mathbf{v})$ for some $\mathbf{v} \in \mathfrak{g}$. In general, $\exp: \mathfrak{g} \to G$ is globally neither one-to-one nor onto. (See Exercise 1.28.) However, using Proposition 1.24, we can always write any group element g as a finite product of exponentials

$$g = \exp(\mathbf{v}_1)\exp(\mathbf{v}_2)\cdots\exp(\mathbf{v}_k)$$

for some $\mathbf{v}_1, \ldots, \mathbf{v}_k$ in $\mathfrak{g}$. The net effect of this observation is that the proof of the invariance of some object under the entire Lie group reduces to a proof of its invariance just under one-parameter subgroups of G, which in turn will be implied by a form of "infinitesimal invariance" under the corresponding infinitesimal generators in $\mathfrak{g}$. With a little more work, we can actually reduce to just proving "invariance" under a basis $\{\mathbf{v}_1, \ldots, \mathbf{v}_r\}$ of $\mathfrak{g}$, with any group element being expressible in the form

$$g = \exp(\varepsilon^1\mathbf{v}_{i_1})\exp(\varepsilon^2\mathbf{v}_{i_2})\cdots\exp(\varepsilon^k\mathbf{v}_{i_k}) \qquad (1.40)$$

for suitable $\varepsilon^j \in \mathbb{R}$, $1 \leqslant i_j \leqslant r$, $j = 1, \ldots, k$. (See Exercise 1.27.)

Lie Algebras of Local Lie Groups

Turning to the local version we consider a local Lie group $V \subset \mathbb{R}^r$ with multiplication $m(x, y)$. The corresponding right multiplication map $R_y: V \to \mathbb{R}^r$ is $R_y(x) = m(x, y)$. A vector field $\mathbf{v}$ on V is right-invariant if and only if

$$dR_y(\mathbf{v}|_x) = \mathbf{v}|_{R_y(x)} = \mathbf{v}|_{m(x,y)}$$

whenever x, y and $m(x, y)$ are in V. As in the case of global Lie groups, it is easy to check that any right-invariant vector field is determined uniquely by its value at the origin (identity element), $\mathbf{v}|_x = dR_x(\mathbf{v}|_0)$, and hence the Lie algebra $\mathfrak{g}$ for the local Lie group V, determined as the space of right-invariant vector fields on V, is an r-dimensional vector space. In fact, we can determine $\mathfrak{g}$ directly from the formula for the group multiplication.

Proposition 1.55. *Let $V \subset \mathbb{R}^r$ be a local Lie group with multiplication $m(x, y)$, $x, y \in V$. Then the Lie algebra $\mathfrak{g}$ of right-invariant vector fields on V is spanned by the vector fields*

$$\mathbf{v}_k = \sum_{i=1}^{r} \xi_k^i(x) \frac{\partial}{\partial x^i}, \qquad k = 1, \ldots, r,$$

where

$$\xi_k^i(x) = \frac{\partial m^i}{\partial x^k}(0, x). \tag{1.41}$$

Here the m^i's are the components of m, and the $\partial/\partial x^k$ denote derivatives with respect to the first set of r variables in $m(x, y)$, after which the values $x = 0$, $y = x$ are to be substituted.

Proof. Since $R_y(x) = m(x, y)$, we have

$$\mathbf{v}_k|_y = dR_y\left(\sum_{i=1}^{r} \xi_k^i(0) \frac{\partial}{\partial x^i}\right) = \sum_{i,j} \xi_k^i(0) \frac{\partial m^j}{\partial x^i}(0, y) \frac{\partial}{\partial x^j}.$$

Thus it suffices to prove that

$$\xi_k^i(0) = \delta_k^i,$$

i.e.

$$\frac{\partial m^i}{\partial x^k}(0, 0) = \begin{cases} 1, & i = k, \\ 0, & i \neq k. \end{cases}$$

But this follows trivially from the fact that $m(x, 0) = x$ is the identity map. □

Example 1.56. Consider the local Lie group of Example 1.21. The Lie algebra $\mathfrak{g}$ is one-dimensional, spanned by the vector field $\xi(x)\partial_x$, where, by (1.41),

$$\xi(x) = \frac{\partial m}{\partial x}(0, x) = (x - 1)^2.$$

Thus $\mathbf{v} = (x - 1)^2 \partial_x$ is the unique independent right-invariant vector field on V. Note that the local group homomorphism $\chi\colon \mathbb{R} \to V$ of Example 1.23 maps the invariant vector field ∂_t on $\mathbb{R}$ to $-\mathbf{v} = d\chi(\partial_t)$.

Structure Constants

Suppose $\mathfrak{g}$ is any finite-dimensional Lie algebra, so by Theorem 1.54 $\mathfrak{g}$ is the Lie algebra of some Lie group G. If we introduce a basis $\{\mathbf{v}_1, \ldots, \mathbf{v}_r\}$ of $\mathfrak{g}$, then the Lie bracket of any two basis vectors must again lie in $\mathfrak{g}$. Thus there are certain constants c_{ij}^k, $i, j, k = 1, \ldots, r$, called the *structure constants* of $\mathfrak{g}$ such that

$$[\mathbf{v}_i, \mathbf{v}_j] = \sum_{k=1}^{r} c_{ij}^k \mathbf{v}_k, \qquad i, j = 1, \ldots, r. \tag{1.42}$$

Note that since the $\mathbf{v}_i$'s form a basis, if we know the structure constants, then we can recover the Lie algebra $\mathfrak{g}$ just by using (1.42) and the bilinearity of the Lie bracket. The conditions of skew-symmetry and the Jacobi identity place further constraints on the structure constants:

(i) *Skew-symmetry*

$$c_{ij}^k = -c_{ji}^k, \tag{1.43}$$

(ii) *Jacobi identity*

$$\sum_{k=1}^{r} (c_{ij}^k c_{kl}^m + c_{li}^k c_{kj}^m + c_{jl}^k c_{ki}^m) = 0. \tag{1.44}$$

Conversely, it is not difficult to show that any set of constants c_{ij}^k which satisfy (1.43), (1.44) are the structure constants for some Lie algebra $\mathfrak{g}$.

Of course, if we choose a new basis of $\mathfrak{g}$, then in general the structure constants will change. If $\hat{\mathbf{v}}_i = \sum_j a_{ij} \mathbf{v}_j$, then

$$\hat{c}_{ij}^k = \sum_{l,m,n} a_{il} a_{jm} b_{nk} c_{lm}^n, \tag{1.45}$$

where (b_{ij}) is the inverse matrix to (a_{ij}). Thus two sets of structure constants determine the same Lie algebra if and only if they are related by (1.45). Consequently, from Theorem 1.54 we see that there is a one-to-one correspondence between equivalence classes of structure constants c_{ij}^k satisfying (1.43), (1.44) and connected, simply-connected Lie groups G whose Lie algebras have the given structure constants relative to some basis. Thus, in principle, the entire theory of Lie groups reduces to a study of the algebraic equations (1.43), (1.44); however, this is perhaps an excessively simplistic point of view!

Commutator Tables

The most convenient way to display the structure of a given Lie algebra is to write it in tabular form. If $\mathfrak{g}$ is an r-dimensional Lie algebra, and $\mathbf{v}_1, \ldots, \mathbf{v}_r$ form a basis for $\mathfrak{g}$, then the *commutator table* for $\mathfrak{g}$ will be the $r \times r$ table whose (i, j)-th entry expresses the Lie bracket $[\mathbf{v}_i, \mathbf{v}_j]$. Note that the table is always skew-symmetric since $[\mathbf{v}_i, \mathbf{v}_j] = -[\mathbf{v}_j, \mathbf{v}_i]$; in particular, the diagonal entries are all zero. The structure constants can be easily read off the commutator table; namely c_{ij}^k is the coefficient of $\mathbf{v}_k$ in the (i,j)-th entry of the table.

For example, if $\mathfrak{g} = \mathfrak{sl}(2)$, the Lie algebra of the special linear group SL(2), which consists of all 2×2 matrices with trace 0, and we use the basis

$$A_1 = \begin{pmatrix} 0 & 1 \\ 0 & 0 \end{pmatrix}, \qquad A_2 = \begin{pmatrix} \frac{1}{2} & 0 \\ 0 & -\frac{1}{2} \end{pmatrix}, \qquad A_3 = \begin{pmatrix} 0 & 0 \\ 1 & 0 \end{pmatrix},$$

then we obtain the commutator table

	A_1	A_2	A_3
A_1	0	A_1	$-2A_2$
A_2	$-A_1$	0	A_3
A_3	$2A_2$	$-A_3$	0

Thus, for example,

$$[A_1, A_3] = A_3 A_1 - A_1 A_3 = -2A_2,$$

and so on. The structure constants are

$$c_{12}^1 = c_{23}^3 = 1 = -c_{21}^1 = -c_{32}^3, \qquad c_{13}^2 = -2 = -c_{31}^2,$$

with all other c_{jk}^i's being zero.

Infinitesimal Group Actions

Suppose G is a local group of transformations acting on a manifold M via $g \cdot x = \Psi(g, x)$ for $(g, x) \in \mathscr{U} \subset G \times M$. There is then a corresponding "infinitesimal action" of the Lie algebra $\mathfrak{g}$ of G on M. Namely, if $\mathbf{v} \in \mathfrak{g}$ we define $\psi(\mathbf{v})$ to be the vector field on M whose flow coincides with the action of the one-parameter subgroup $\exp(\varepsilon\mathbf{v})$ of G on M. This means that for $x \in M$,

$$\psi(\mathbf{v})|_x = \frac{d}{d\varepsilon}\bigg|_{\varepsilon=0} \Psi(\exp(\varepsilon\mathbf{v}), x) = d\Psi_x(\mathbf{v}|_e), \tag{1.46}$$

where $\Psi_x(g) \equiv \Psi(g, x)$. Note further that since

$$\Psi_x \circ R_g(h) = \Psi(h \cdot g, x) = \Psi(h, g \cdot x) = \Psi_{g \cdot x}(h)$$

wherever defined, we have

$$d\Psi_x(\mathbf{v}|_g) = d\Psi_{g \cdot x}(\mathbf{v}|_e) = \psi(\mathbf{v})|_{g \cdot x}$$

for any $g \in G_x$. It follows from property (1.32) of the Lie bracket that ψ is a Lie algebra homomorphism from $\mathfrak{g}$ to the Lie algebra of vector fields on M:

$$[\psi(\mathbf{v}), \psi(\mathbf{w})] = \psi([\mathbf{v}, \mathbf{w}]), \qquad \mathbf{v}, \mathbf{w} \in \mathfrak{g}. \tag{1.47}$$

Therefore the set of all vector fields $\psi(\mathbf{v})$ corresponding to $\mathbf{v} \in \mathfrak{g}$ forms a Lie algebra of vector fields on M. Conversely, given a finite-dimensional Lie algebra of vector fields on M, there is always a local group of transformations whose infinitesimal action is generated by the given Lie algebra.

Theorem 1.57. *Let $\mathbf{w}_1, \ldots, \mathbf{w}_r$ be vector fields on a manifold M satisfying*

$$[\mathbf{w}_i, \mathbf{w}_j] = \sum_{k=1}^{r} c_{ij}^k \mathbf{w}_k, \qquad i, j = 1, \ldots, r,$$

for certain constants c_{ij}^k. *Then there is a Lie group G whose Lie algebra has the given c_{ij}^k as structure constants relative to some basis $\mathbf{v}_1, \ldots, \mathbf{v}_r$, and a local group action of G on M such that $\psi(\mathbf{v}_i) = \mathbf{w}_i$ for $i = 1, \ldots, r$, where ψ is defined by* (1.46).

Usually we will omit explicit reference to the map ψ and identify the Lie algebra $\mathfrak{g}$ with its image $\psi(\mathfrak{g})$, which forms a Lie algebra of vector fields on M. In this language, we recover $\mathfrak{g}$ from the group transformations by the basic formula

$$\mathbf{v}|_x = \frac{d}{d\varepsilon}\bigg|_{\varepsilon=0} \exp(\varepsilon \mathbf{v})x, \qquad \mathbf{v} \in \mathfrak{g}. \tag{1.48}$$

A vector field $\mathbf{v}$ in $\mathfrak{g}$ is called an *infinitesimal generator* of the group action G. Theorem 1.57 says that if we know infinitesimal generators $\mathbf{w}_1, \ldots, \mathbf{w}_r$, which form a basis for a Lie algebra, then we can always exponentiate to find a local group of transformations whose Lie algebra coincides with the given one.

Example 1.58. Lie proved that up to diffeomorphism there are precisely three finite-dimensional Lie algebras of vector fields on the real line $M = \mathbb{R}$. These are

(a) The algebra spanned by ∂_x: This generates an action of $\mathbb{R}$ on M as a one-parameter group of translations: $x \mapsto x + \varepsilon$.

(b) The two-dimensional Lie algebra spanned by ∂_x and $x\partial_x$, the second vector field generating the group of dilatations $x \mapsto \lambda x$: Note that

$$[\partial_x, x\partial_x] = \partial_x,$$

so this Lie algebra is isomorphic to the 2×2 matrix Lie algebra spanned by

$$\begin{pmatrix} 0 & 1 \\ 0 & 0 \end{pmatrix} \quad \text{and} \quad \begin{pmatrix} 1 & 0 \\ 0 & 0 \end{pmatrix}.$$

This generates the Lie group of all upper triangular matrices of the form

$$A = \begin{pmatrix} \alpha & \beta \\ 0 & 1 \end{pmatrix}, \qquad \alpha > 0.$$

The corresponding action on $\mathbb{R}$ is the group $x \mapsto \alpha x + \beta$ of affine transformations; we leave it to the reader to check that this indeed defines a Lie group action, whose infinitesimal generators agree with the given ones.

(c) The three-dimensional algebra spanned by

$$\mathbf{v}_1 = \partial_x, \qquad \mathbf{v}_2 = x\partial_x, \qquad \mathbf{v}_3 = x^2\partial_x,$$

the third vector field generating the local group of "inversions"

$$x \longmapsto \frac{x}{1 - \varepsilon x}, \qquad |\varepsilon| < \frac{1}{x}.$$

The commutator table for this Lie algebra is as follows:

	$\mathbf{v}_1$	$\mathbf{v}_2$	$\mathbf{v}_3$
$\mathbf{v}_1$	0	$\mathbf{v}_1$	$2\mathbf{v}_2$
$\mathbf{v}_2$	$-\mathbf{v}_1$	0	$\mathbf{v}_3$
$\mathbf{v}_3$	$-2\mathbf{v}_2$	$-\mathbf{v}_3$	0

If we replace $\mathbf{v}_3$ by $-\mathbf{v}_3 = -x^2\partial_x$, then we find the same commutator table as $\mathfrak{sl}(2)$ with basis

$$A_1 = \begin{pmatrix} 0 & 1 \\ 0 & 0 \end{pmatrix}, \qquad A_2 = \begin{pmatrix} \frac{1}{2} & 0 \\ 0 & -\frac{1}{2} \end{pmatrix}, \qquad A_3 = \begin{pmatrix} 0 & 0 \\ 1 & 0 \end{pmatrix}.$$

There is thus a local action of the special linear group SL(2) on the real line with ∂_x, $x\partial_x$ and $-x^2\partial_x$ serving as the infinitesimal generators. It is not difficult to see that this group action is just the *projective group*

$$x \longmapsto \frac{\alpha x + \beta}{\gamma x + \delta}, \qquad \begin{pmatrix} \alpha & \beta \\ \gamma & \delta \end{pmatrix} \in \mathrm{SL}(2),$$

being the real analogue of the complex group of linear fractional transformations.

1.5. Differential Forms

Originally developed as a tool for the multi-dimensional generalization of Stokes' theorem, differential forms play a fundamental role in the topological aspects of differential geometry. Although in this book I have tended to de-emphasize the use of differential forms, there are several occasions, most notably the variational complex of Section 5.4, in which the language of differential forms is especially effective. This section provides a rapid introduction to the theory of differential forms for the reader who is interested in pursuing these more theoretical aspects of the subject. We begin with the basic definition.

Definition 1.59. Let M be a smooth manifold and $TM|_x$ its tangent space at x. The space $\bigwedge_k T^*M|_x$ of *differential k-forms at x* is the set of all k-linear alternating functions

$$\omega: TM|_x \times \cdots \times TM|_x \to \mathbb{R}.$$

Specifically, if we denote the evaluation of ω on the tangent vectors $\mathbf{v}_1, \ldots, \mathbf{v}_k \in TM|_x$ by $\langle \omega; \mathbf{v}_1, \ldots, \mathbf{v}_k \rangle$, then the basic requirements are that for all tangent vectors at x,

$$\langle \omega; \mathbf{v}_1, \ldots, c\mathbf{v}_i + c'\mathbf{v}_i', \ldots, \mathbf{v}_k \rangle = c\langle \omega; \mathbf{v}_1, \ldots, \mathbf{v}_i, \ldots, \mathbf{v}_k \rangle + c'\langle \omega; \mathbf{v}_1, \ldots, \mathbf{v}_i', \ldots, \mathbf{v}_k \rangle$$

for $c, c' \in \mathbb{R}$, $1 \leqslant i \leqslant k$, and

$$\langle \omega; \mathbf{v}_{\pi 1}, \ldots, \mathbf{v}_{\pi k} \rangle = (-1)^\pi \langle \omega; \mathbf{v}_1, \ldots, \mathbf{v}_k \rangle$$

for every permutation π of the integers $\{1, \ldots, k\}$, with $(-1)^\pi$ denoting the sign of π. The space $\bigwedge_k T^*M|_x$ is, in fact, a vector space under the obvious operations of addition and scalar multiplication. A 0-form at x is, by convention, just a real number, while the space $T^*M|_x = \bigwedge_1 T^*M|_x$ of one-forms, called the *cotangent space* to M at x, is the space of linear functions on $TM|_x$, i.e. the dual vector space to the tangent space at x. A *smooth differential k-form* ω on M (or *k-form* for short) is a collection of smoothly varying alternating k-linear maps $\omega|_x \in \bigwedge_k T^*M|_x$ for each $x \in M$, where we require that for all smooth vector fields $\mathbf{v}_1, \ldots, \mathbf{v}_k$,

$$\langle \omega; \mathbf{v}_1, \ldots, \mathbf{v}_k \rangle(x) \equiv \langle \omega|_x; \mathbf{v}_1|_x, \ldots, \mathbf{v}_k|_x \rangle$$

is a smooth, real-valued function of x. In particular, a 0-form is just a smooth real-valued function $f: M \to \mathbb{R}$.

If $(x^1, \ldots, x^m)$ are local coordinates, then $TM|_x$ has basis $\{\partial/\partial x^1, \ldots, \partial/\partial x^m\}$. The dual cotangent space has a dual basis, which is traditionally denoted $\{dx^1, \ldots, dx^m\}$; thus $\langle dx^i; \partial/\partial x^j \rangle = \delta_j^i$ for all i, j, where δ_j^i is 1 for $i = j$ and 0 otherwise. A differential one-form ω thereby has the local coordinate expression

$$\omega = h_1(x)\, dx^1 + \cdots + h_m(x)\, dx^m,$$

where each coefficient function $h_j(x)$ is smooth. Note that for any vector field $\mathbf{v} = \sum \xi^i(x)\partial/\partial x^i$,

$$\langle \omega; \mathbf{v} \rangle = \sum_{i=1}^m h_i(x)\xi^i(x)$$

is a smooth function. Of particular importance are the one-forms given by the differentials of real-valued functions:

$$df = \sum_{i=1}^m \frac{\partial f}{\partial x^i} dx^i, \quad \text{with} \quad \langle df; \mathbf{v} \rangle = \mathbf{v}(f).$$

To proceed to higher order differential forms, we note that, given a collection of differential one-forms $\omega_1, \ldots, \omega_k$, we can form a differential k-form

$\omega_1 \wedge \cdots \wedge \omega_k$, called the *wedge product*, using the determinantal formula

$$\langle \omega_1 \wedge \cdots \wedge \omega_k; \mathbf{v}_1, \ldots, \mathbf{v}_k \rangle = \det(\langle \omega_i; \mathbf{v}_j \rangle), \tag{1.49}$$

the right-hand side being the determinant of a $k \times k$ matrix with indicated (i, j) entry. Note that, by the usual properties of determinants, the wedge product itself is both multi-linear and alternating

$$\omega_1 \wedge \cdots \wedge (c\omega_i + c'\omega_i') \wedge \cdots \wedge \omega_k = c(\omega_1 \wedge \cdots \wedge \omega_i \wedge \cdots \wedge \omega_k)$$
$$+ c'(\omega_1 \wedge \cdots \wedge \omega_i' \wedge \cdots \wedge \omega_k),$$
$$\omega_{\pi 1} \wedge \cdots \wedge \omega_{\pi k} = (-1)^{\pi} \omega_1 \wedge \cdots \wedge \omega_k.$$

It is not hard to see that in local coordinates, $\bigwedge_k T^*M|_x$ is spanned by the basis k-forms

$$dx^I \equiv dx^{i_1} \wedge \cdots \wedge dx^{i_k},$$

where I ranges over all strictly increasing multi-indices $1 \leqslant i_1 < i_2 < \cdots < i_k \leqslant m$. Thus $\bigwedge_k T^*M|_x$ has dimension $\binom{m}{k}$; in particular, $\bigwedge_k T^*M|_x \simeq \{0\}$ if $k > m$. Any smooth differential k-form on M has the local coordinate expression

$$\omega = \sum_I \alpha_I(x)\, dx^I,$$

where, for each strictly increasing multi-index I, the coefficient α_I is a smooth, real-valued function. For example, a two-form in $\mathbb{R}^3$ takes the form

$$\omega = \alpha(x, y, z)\, dy \wedge dz + \beta(x, y, z)\, dz \wedge dx + \gamma(x, y, z)\, dx \wedge dy, \tag{1.50}$$

using the basis $dy \wedge dz$, $dz \wedge dx = -dx \wedge dz$, and $dx \wedge dy$, attuned to the notation for surface integrals. We have

$$\langle \omega; \xi\partial_x + \eta\partial_y + \zeta\partial_z, \hat{\xi}\partial_x + \hat{\eta}\partial_y + \hat{\zeta}\partial_z \rangle = \alpha(\eta\hat{\zeta} - \hat{\eta}\zeta) + \beta(\zeta\hat{\xi} - \hat{\zeta}\xi) + \gamma(\xi\hat{\eta} - \hat{\xi}\eta).$$

If

$$\omega = \omega_1 \wedge \cdots \wedge \omega_k, \qquad \theta = \theta_1 \wedge \cdots \wedge \theta_l,$$

are "decomposible" forms, their *wedge product* is the form

$$\omega \wedge \theta = \omega_1 \wedge \cdots \wedge \omega_k \wedge \theta_1 \wedge \cdots \wedge \theta_l,$$

with the definition extending bilinearly to more general types of forms:

$$(c\omega + c'\omega') \wedge \theta = c(\omega \wedge \theta) + c'(\omega' \wedge \theta),$$
$$\omega \wedge (c\theta + c'\theta') = c(\omega \wedge \theta) + c'(\omega \wedge \theta'),$$

for $c, c' \in \mathbb{R}$. This wedge product is associative:

$$\omega \wedge (\theta \wedge \zeta) = (\omega \wedge \theta) \wedge \zeta,$$

and *anti*-commutative,

$$\omega \wedge \theta = (-1)^{kl} \theta \wedge \omega$$

for ω a k-form and θ an l-form. For example, the wedge product of (1.50) with a one-form $\theta = \lambda\, dx + \mu\, dy + \nu\, dz$ is the three-form

$$\omega \wedge \theta = (\alpha\lambda + \beta\mu + \gamma\nu)\, dx \wedge dy \wedge dz.$$

Pull-Back and Change of Coordinates

If $F\colon M \to N$ is a smooth map between manifolds, its differential dF maps tangent vectors on M to tangent vectors on N. There is thus an induced linear map F^*, called the *pull-back* or *codifferential* of F, which takes differential k-forms on N back to differential k-forms on M,

$$F^*\colon \textstyle\bigwedge_k T^*N|_{F(x)} \to \textstyle\bigwedge_k T^*M|_x.$$

It is defined so that if $\omega \in \bigwedge_k T^*N|_{F(x)}$,

$$\langle F^*(\omega); \mathbf{v}_1, \ldots, \mathbf{v}_k\rangle = \langle \omega; dF(\mathbf{v}_1), \ldots, dF(\mathbf{v}_k)\rangle$$

for any set of tangent vectors $\mathbf{v}_1, \ldots, \mathbf{v}_k \in TM|_x$. In contrast to the differential, the pull-back *does* take smooth differential forms on N back to smooth differential forms on M. If $x = (x^1, \ldots, x^m)$ are local cordinates on M and $y = (y^1, \ldots, y^n)$ coordinates on N, then

$$F^*(dy^i) = \sum_{j=1}^{m} \frac{\partial y^i}{\partial x^j}\cdot dx^j, \quad \text{where} \quad y = F(x),$$

gives the action of F^* on the basis one-forms. We conclude that in general

$$F^*\left(\sum_I \alpha_I(y)\, dy^I\right) = \sum_{I,J} \alpha_I(F(x)) \frac{\partial y^I}{\partial x^J} dx^J, \tag{1.51}$$

where $\partial y^I/\partial x^J$ stands for the Jacobian determinant $\det(\partial y^{i_\kappa}/\partial x^{j_\nu})$ corresponding to the increasing multi-indices $I = (i_1, \ldots, i_k)$, $J = (j_1, \ldots, j_k)$. In particular, if $y = F(x)$ determines a change of coordinates on M, then (1.51) provides the corresponding change of coordinates for differential k-forms on M. Note also that the pull-back preserves the algebraic operation of wedge product:

$$F^*(\omega \wedge \theta) = F^*(\omega) \wedge F^*(\theta).$$

Interior Products

If ω is a differential k-form and $\mathbf{v}$ a smooth vector field, then we can form a $(k-1)$-form $\mathbf{v} \lrcorner\, \omega$, called the *interior product* of $\mathbf{v}$ with ω, defined so that

$$\langle \mathbf{v} \lrcorner\, \omega; \mathbf{v}_1, \ldots, \mathbf{v}_{k-1}\rangle = \langle \omega; \mathbf{v}, \mathbf{v}_1, \ldots, \mathbf{v}_{k-1}\rangle$$

for every set of vector fields $\mathbf{v}_1, \ldots, \mathbf{v}_{k-1}$. The interior product is clearly bilinear in both its arguments, so it suffices to determine it for basis elements:

$$\frac{\partial}{\partial x^i} \,\lrcorner\, (dx^{j_1} \wedge \cdots \wedge dx^{j_k})$$

$$= \begin{cases} (-1)^{\kappa-1}\, dx^{j_1} \wedge \cdots \wedge dx^{j_{\kappa-1}} \wedge dx^{j_{\kappa+1}} \wedge \cdots \wedge dx^{j_k}, & i = j_\kappa, \\ 0, & i \neq j_\kappa \quad \text{for all } \kappa. \end{cases}$$

For example,

$$\partial_x \,\lrcorner\, dx \wedge dy = dy, \qquad \partial_x \,\lrcorner\, dz \wedge dx = -dz, \qquad \partial_x \,\lrcorner\, dy \wedge dz = 0,$$

so that if ω is as in (1.50),

$$(\xi\partial_x + \eta\partial_y + \zeta\partial_z) \,\lrcorner\, \omega = (\zeta\beta - \eta\gamma)\, dx + (\xi\gamma - \zeta\alpha)\, dy + (\eta\alpha - \xi\beta)\, dz.$$

Note that the interior product acts as an *anti-derivation* on forms, meaning that

$$\mathbf{v} \,\lrcorner\, (\omega \wedge \theta) = (\mathbf{v} \,\lrcorner\, \omega) \wedge \theta + (-1)^k \omega \wedge (\mathbf{v} \,\lrcorner\, \theta) \tag{1.52}$$

whenever ω is a k-form, θ an l-form.

The Differential

Besides the purely algebraic operations of wedge and interior products, there are two important differential operations. The first of these generalizes the concept of the differential of a smooth function (or 0-form) to an arbitrary differential k-form. In local coordinates, if $\omega = \sum \alpha_I(x)\, dx^I$ is a smooth differential k-form on a manifold M, its *differential* or *exterior derivative* is the $(k+1)$-form

$$d\omega = \sum_I d\alpha_I \wedge dx^I = \sum_{I,j} \frac{\partial \alpha_I}{\partial x^j} dx^j \wedge dx^I. \tag{1.53}$$

Proposition 1.60. *The differential d, taking k-forms to $(k+1)$-forms, has the following properties*:

(a) Linearity

$$d(c\omega + c'\omega') = c\, d\omega + c'\, d\omega' \quad \textit{for } c, c' \textit{ constant},$$

(b) Anti-derivation

$$d(\omega \wedge \theta) = (d\omega) \wedge \theta + (-1)^k \omega \wedge (d\theta), \quad \textit{for } \omega \textit{ a } k\textit{-form}, \theta \textit{ an } l\textit{-form}. \tag{1.54}$$

(c) Closure

$$d(d\omega) \equiv 0. \tag{1.55}$$

(d) Commutation with Pull-Back

$$F^*(d\omega) = d(F^*\omega), \tag{1.56}$$

for $F: M \to N$ smooth, ω a k-form on N.

The proofs of these properties are reasonably straightforward. Linearity is obvious and the anti-derivational property is an easy consequence of Leibniz' rule. To check closure, we need only prove $d(df) = 0$ for f a smooth function since we can then use properties (a) and (b) to extend this to the general case (1.53). However,

$$d(df) = \sum_{i,j=1}^{m} \frac{\partial^2 f}{\partial x^i\, \partial x^j} dx^i \wedge dx^j = \sum_{i<j} \left(\frac{\partial^2 f}{\partial x^i\, \partial x^j} - \frac{\partial^2 f}{\partial x^j\, \partial x^i} \right) dx^i \wedge dx^j$$

by the alternating property of the wedge product, so closure just reduces to the equality of mixed partial derivatives. In fact, properties (a), (b), and (c) together with the action of d on functions serve to *uniquely* characterize the differential and so d is in fact independent of the choice of local coordinates. Finally, the proof of (1.56) needs only be done in the case of functions: $F^*(df) = dF^*(f)$, where $F^*(f) = f \circ F$, and then it reduces to the ordinary chain rule. □

If $M = \mathbb{R}^3$, then the differential of a one-form,

$$d(\lambda\, dx + \mu\, dy + \nu\, dz) = (\nu_y - \mu_z)\, dy \wedge dz + (\lambda_z - \nu_x)\, dz \wedge dx + (\mu_x - \lambda_y)\, dx \wedge dy,$$

can be identified with curl of its coefficients: $\nabla \times \boldsymbol{\lambda} \equiv \nabla \times (\lambda, \mu, \nu)$. Similarly, the differential of a two-form

$$d(\alpha\, dy \wedge dz + \beta\, dz \wedge dx + \gamma\, dx \wedge dy) = (\alpha_x + \beta_y + \gamma_z)\, dx \wedge dy \wedge dz$$

can be identified with the divergence $\nabla \cdot \boldsymbol{\alpha} \equiv \nabla \cdot (\alpha, \beta, \gamma)$. The closure property (1.55) of d therefore translates into the familiar vector calculus identities

$$\nabla \times (\nabla f) = 0, \qquad \nabla \cdot (\nabla \times \boldsymbol{\lambda}) = 0.$$

The reader may find it instructive to see which vector calculus identities are implied by the anti-derivational property (1.54) in this case.

The de Rham Complex

Given a manifold M, we let $\bigwedge_k = \bigwedge_k(M)$ denote the space of all smooth differential k-forms on M. The differential d, mapping k-forms to $(k + 1)$-forms, serves to define a "complex"

$$0 \to \mathbb{R} \to \textstyle\bigwedge_0 \xrightarrow{d} \bigwedge_1 \xrightarrow{d} \bigwedge_2 \xrightarrow{d} \cdots \xrightarrow{d} \bigwedge_{m-1} \xrightarrow{d} \bigwedge_m \to 0$$

called the *de Rham complex* of M. In general, a *complex* is defined as a sequence of vector spaces, and linear maps between successive spaces, with the property that the composition of any pair of successive maps is identically 0. In the present case, this last requirement is a restatement of the closure property $d \circ d = 0$, (1.55), of the differential. The initial map $\mathbb{R} \to \bigwedge_0$ takes a constant $c \in \mathbb{R}$ to the constant function (0-form) $f(x) \equiv c$ on M. Note that $dc = 0$ for any constant c.

The definition of a complex requires that the kernel of one of the linear maps contains the image of the preceding map. The complex is *exact* if this containment is, in fact, equality. In the case of the de Rham complex, exactness means that a *closed* differential k-form ω, meaning that $d\omega \equiv 0$, is necessarily an *exact* differential k-form, meaning that there exists a $(k-1)$-form θ with $\omega = d\theta$. (For $k = 0$, it says that a smooth function f is closed, $df = 0$, if and only if it is constant.) Clearly, any exact form is closed, but the converse need not hold. A simple example is when $M = \mathbb{R}^2 \setminus \{0\}$, on which $\omega = (x^2 + y^2)^{-1}(y\,dx - x\,dy)$ is easily seen to be closed, but is not the differential of any smooth, single-valued function defined on all of M. Thus the de Rham complex is *not* in general exact. Remarkably, the extent to which it fails to be exact measures purely topological information about the manifold M. This result, the celebrated de Rham theorem, lies beyond the scope of this book and we refer the interested reader to the books of Warner, [1], and Bott and Tu, [1], for a development of this beautiful theory.

On the local side, for special types of domains in Euclidean space $\mathbb{R}^m$, there is only trivial topology and we *do* have exactness of the de Rham complex. This result, known as the *Poincaré lemma*, will hold for *star-shaped* domains $M \subset \mathbb{R}^m$, where "star-shaped" means that whenever $x \in M$, so is the entire line segment joining x to the origin: $\{\lambda x : 0 \leqslant \lambda \leqslant 1\} \subset M$.

Theorem 1.61. *Let $M \subset \mathbb{R}^m$ be a star-shaped domain. Then the de Rham complex over M is exact.*

Example 1.62. For $M \subset \mathbb{R}^m$, any m-form ω is uniquely determined by a single smooth function f, with $\omega = f(x)\,dx^1 \wedge \cdots \wedge dx^m$ relative to the standard volume form. (This does depend on our choice of coordinates.) Similarly, an $(m-1)$-form ξ is determined by an m-tuple of smooth functions $p = (p_1, \ldots, p_m)$, so that

$$\xi = \sum_{j=1}^{m} (-1)^{j-1} p_j(x)\,dx^{\hat{j}},$$

where

$$dx^{\hat{j}} \equiv dx^1 \wedge \cdots \wedge dx^{j-1} \wedge dx^{j+1} \wedge \cdots \wedge dx^m.$$

The differential $\omega = d\xi$ is then determined as the usual divergence of p:

$$f(x) = \operatorname{div} p(x) = \sum_{j=1}^{m} \partial p_j / \partial x^j.$$

Note that any m-form on $\mathbb{R}^m$ is always closed, $d\omega = 0$, as there are no nonzero $(m+1)$-forms. Exactness of the de Rham complex at the $\bigwedge_m$-stage, then, says that any function f defined on a star-shaped subdomain of $\mathbb{R}^m$ can always be written as a divergence: $f = \operatorname{div} p$ for some p. Similarly, an $(m-1)$-form η is determined by $m(m-1)/2$ functions $q_{jk}(x)$, $j, k = 1, \ldots, m$, with $q_{jk} = -q_{kj}$, so that

$$\eta = \sum_{\substack{j,k=1 \\ j<k}}^{m} (-1)^{j+k-1} q_{jk}(x)\,dx^{\widehat{jk}},$$

where

$$dx^{\widehat{jk}} \equiv dx^1 \wedge \cdots \wedge dx^{j-1} \wedge dx^{j+1} \wedge \cdots \wedge dx^{k-1} \wedge dx^{k+1} \wedge \cdots \wedge dx^m.$$

Note that $d\eta = \zeta$ is equivalent to the condition that p be a "generalized curl" of q:

$$p_j(x) = \sum_{k=1}^{m} \partial q_{jk}/\partial x^k, \qquad j = 1, \ldots, m.$$

Exactness of the de Rham complex at this stage then says that any vector field $p(x)$, defined over a star-shaped domain in $\mathbb{R}^m$, which is divergence-free: $\operatorname{div} p \equiv 0$, is necessarily the generalized curl of some such q.

Lie Derivatives

Let $\mathbf{v}$ be a vector field on a manifold M. We are often interested in how certain geometric objects on M, such as functions, differential forms and other vector fields, vary under the flow $\exp(\varepsilon\mathbf{v})$ induced by $\mathbf{v}$. The *Lie derivative* of such an object will in effect tell us its infinitesimal change when acted on by the flow. (Our standard integration procedures will tell us how to reconstruct the variation under the flow from this infinitesimal version.) For instance, the behaviour of a function under the flow induced by a vector field $\mathbf{v}$ has already been established so $\mathbf{v}(f)$, cf. (1.17), will be the "Lie derivative" of the function f with respect to $\mathbf{v}$.

More generally, let σ be a differential form or vector field defined over M. Given a point $x \in M$, after "time" ε it has moved to $\exp(\varepsilon\mathbf{v})x$ and the goal is to compare the value of σ at $\exp(\varepsilon\mathbf{v})x$ with its original value at x. However, $\sigma|_{\exp(\varepsilon\mathbf{v})x}$ and $\sigma|_x$, as they stand are, strictly speaking, incomparable as they belong to different vector spaces, e.g. $TM|_{\exp(\varepsilon\mathbf{v})x}$ and $TM|_x$ in the case of a vector field. To effect any comparison, we need to "transport" $\sigma|_{\exp(\varepsilon\mathbf{v})x}$ back to x in some natural way, and then make our comparison. For vector fields, this natural transport is the inverse differential

$$\phi_\varepsilon^* \equiv d\exp(-\varepsilon\mathbf{v})\colon TM|_{\exp(\varepsilon\mathbf{v})x} \to TM|_x,$$

whereas for differential forms we use the pull-back map

$$\phi_\varepsilon^* \equiv \exp(\varepsilon\mathbf{v})^*\colon \textstyle\bigwedge_k T^*M|_{\exp(\varepsilon\mathbf{v})x} \to \bigwedge_k T^*M|_x.$$

This allows us to make the general definition of a Lie derivative.

Definition 1.63. Let $\mathbf{v}$ be a vector field on M and σ a vector field or differential form defined on M. The *Lie derivative* of σ with respect to $\mathbf{v}$ is the object whose value at $x \in M$ is

$$\mathbf{v}(\sigma)|_x = \lim_{\varepsilon \to 0} \frac{\phi_\varepsilon^*(\sigma|_{\exp(\varepsilon\mathbf{v})x}) - \sigma|_x}{\varepsilon} = \frac{d}{d\varepsilon}\bigg|_{\varepsilon=0} \phi_\varepsilon^*(\sigma|_{\exp(\varepsilon\mathbf{v})x}). \tag{1.57}$$

(Note that $\mathbf{v}(\sigma)$ is an object of the same type as σ.)

In the case that σ is a vector field, its Lie derivative is a by now familiar object—the Lie bracket!

Proposition 1.64. *Let* $\mathbf{v}$ *and* $\mathbf{w}$ *be smooth vector fields on* M. *The Lie derivative of* $\mathbf{w}$ *with respect to* $\mathbf{v}$ *coincides with the Lie bracket of* $\mathbf{v}$ *and* $\mathbf{w}$:

$$\mathbf{v}(\mathbf{w}) = [\mathbf{v}, \mathbf{w}]. \tag{1.58}$$

PROOF. Let $(x^1, \ldots, x^m)$ be local coordinates, with $\mathbf{v} = \sum \xi^i(x)\partial/\partial x^i$, $\mathbf{w} = \sum \eta^i(x)\partial/\partial x^i$. Expanding in powers of ε, we see that

$$\mathbf{w}|_{\exp(\varepsilon\mathbf{v})x} = \sum_{i=1}^{m} [\eta^i(x) + \varepsilon\mathbf{v}(\eta^i) + O(\varepsilon^2)]\frac{\partial}{\partial x^i},$$

hence, using (1.23) and (1.19),

$$d\exp(-\varepsilon\mathbf{v})[\mathbf{w}|_{\exp(\varepsilon\mathbf{v})x}] = \sum_{i=1}^{m} \{\eta^i(x) + \varepsilon[\mathbf{v}(\eta^i) - \mathbf{w}(\xi^i)] + O(\varepsilon^2)\}\frac{\partial}{\partial x^i}.$$

Substituting into the definition (1.57), we deduce (1.58) from (1.28). □

Turning to differential forms, we find that the Lie derivative can be most easily reconstructed from its basic properties:

(a) *Linearity*

$$\mathbf{v}(c\omega + c'\omega') = c\mathbf{v}(\omega) + c'\mathbf{v}(\omega'), \qquad c, c' \text{ constant}, \tag{1.59}$$

(b) *Derivation*

$$\mathbf{v}(\omega \wedge \theta) = \mathbf{v}(\omega) \wedge \theta + \omega \wedge \mathbf{v}(\theta), \tag{1.60}$$

(c) *Commutation with the Differential*

$$\mathbf{v}(d\omega) = d\mathbf{v}(\omega). \tag{1.61}$$

The commutation property is proved using the analogous property of pull-backs (1.56). The derivational property is proved just like Leibniz' rule. In fact, a Leibniz-type argument extends to Lie derivatives of more general bilinear combinations of geometric objects. Thus we have the useful formula

$$\mathbf{v}(\mathbf{w} \,\lrcorner\, \omega) = [\mathbf{v}, \mathbf{w}] \,\lrcorner\, \omega + \mathbf{w} \,\lrcorner\, \mathbf{v}(\omega), \tag{1.62}$$

for vector fields $\mathbf{v}$ and $\mathbf{w}$ and ω a differential form. (See Exercise 1.35.)

In local coordinates, the Lie derivative of a differential form is determined as follows. If

$$\mathbf{v} = \sum_{i=1}^{m} \xi^i(x)\frac{\partial}{\partial x^i},$$

then

$$\mathbf{v}(dx^i) = d\mathbf{v}(x^i) = d\xi^i = \sum_{j=1}^{m} \frac{\partial \xi^i}{\partial x^j} dx^j.$$

Therefore, we have the general formula

$$\mathbf{v}\left(\sum_I \alpha_I(x)\,dx^I\right) = \sum_I \left\{\mathbf{v}(\alpha_I)\,dx^I + \sum_{\kappa=1}^{k} \alpha_I\,dx^{i_1} \wedge \cdots \wedge d\xi^{i_\kappa} \wedge \cdots \wedge dx^{i_k}\right\}. \tag{1.63}$$

For example, if $M = \mathbb{R}^2$ and

$$\mathbf{v} = \xi(x, y)\partial_x + \eta(x, y)\partial_y,$$

then the Lie derivative of a two-form is

$$\begin{aligned}\mathbf{v}(\gamma(x, y)\,dx \wedge dy) &= \mathbf{v}(\gamma)\,dx \wedge dy + \gamma\,d\xi \wedge dy + \gamma\,dx \wedge d\eta \\ &= \{\xi\gamma_x + \eta\gamma_y + \gamma\xi_x + \gamma\eta_y\}\,dx \wedge dy.\end{aligned}$$

For instance, the Lie derivative of $dx \wedge dy$ with respect to $\mathbf{v} = -y\partial_x + x\partial_y$, the generator of the rotation group, is identically 0 and reflects the fact that rotations in $\mathbb{R}^2$ preserve area. (See Exercise 1.36.) Note that the three properties (1.59–1.61) along with its action on smooth functions serve to define the Lie derivative operation uniquely.

Proposition 1.65. *A differential k-form on M is invariant under the flow of a vector field* $\mathbf{v}$:

$$\omega|_{\exp(\varepsilon\mathbf{v})x} = \exp(-\varepsilon\mathbf{v})^*(\omega|_x),$$

if and only if $\mathbf{v}(\omega) = 0$ *everywhere. (A similar result holds for vector fields.)*

PROOF. Applying $\phi_\varepsilon^* = \exp(\varepsilon\mathbf{v})^*$ to (1.57) and using the basic group property of the flow, we find

$$\exp(\varepsilon\mathbf{v})^*(\mathbf{v}(\omega)|_{\exp(\varepsilon\mathbf{v})x}) = \frac{d}{d\varepsilon}\{\exp(\varepsilon\mathbf{v})^*(\omega|_{\exp(\varepsilon\mathbf{v})x})\} \tag{1.64}$$

for all ε where defined. From this the proposition is easily deduced. □

The most important formula for our purposes is one that relates the Lie derivative and the differential.

Proposition 1.66. *Let* ω *be a differential form and* $\mathbf{v}$ *be a vector field on M. Then*

$$\mathbf{v}(\omega) = d(\mathbf{v} \,\lrcorner\, \omega) + \mathbf{v} \,\lrcorner\, (d\omega). \tag{1.65}$$

PROOF. Define the operator $\mathscr{L}_\mathbf{v}(\omega)$ by the right-hand side of (1.65). Since the Lie derivative is uniquely determined by its action on functions and the properties (1.59–1.61), it suffices to check that $\mathscr{L}_\mathbf{v}$ enjoys the same properties. First

$$\mathscr{L}_\mathbf{v}(f) = \mathbf{v} \,\lrcorner\, df = \langle df; \mathbf{v}\rangle = \mathbf{v}(f),$$

so the action on functions is the same. Linearity of $\mathscr{L}_{\mathbf{v}}$ is clear, while the closure property (1.55) of d immediately proves the commutation property:

$$\mathscr{L}_{\mathbf{v}}(d\omega) = d(\mathbf{v} \rfloor d\omega) = d\mathscr{L}_{\mathbf{v}}(\omega).$$

Finally, if ω is a k-form and θ an l-form, we use (1.52), (1.54) to prove that

$$\begin{aligned}
&\mathscr{L}_{\mathbf{v}}(\omega \wedge \theta) \\
&= d[(\mathbf{v} \rfloor \omega) \wedge \theta + (-1)^k \omega \wedge (\mathbf{v} \rfloor \theta)] + \mathbf{v} \rfloor [(d\omega) \wedge \theta + (-1)^k \omega \wedge (d\theta)] \\
&= d(\mathbf{v} \rfloor \omega) \wedge \theta + (-1)^{k-1}(\mathbf{v} \rfloor \omega) \wedge d\theta + (-1)^k (d\omega) \wedge (\mathbf{v} \rfloor \theta) \\
&\quad + (-1)^{2k} \omega \wedge d(\mathbf{v} \rfloor \theta) + (\mathbf{v} \rfloor d\omega) \wedge \theta + (-1)^{k+1}(d\omega) \wedge (\mathbf{v} \rfloor \theta) \\
&\quad + (-1)^k (\mathbf{v} \rfloor \omega) \wedge (d\theta) + (-1)^{2k} \omega \wedge (\mathbf{v} \rfloor d\theta) \\
&= \mathscr{L}_{\mathbf{v}}(\omega) \wedge \theta + \omega \wedge \mathscr{L}_{\mathbf{v}}(\theta),
\end{aligned}$$

the remaining terms cancelling. □

Homotopy Operators

The key to the proof of the exactness of the de Rham complex (or any other complex for that matter) lies in the construction of suitable *homotopy operators*. By definition, these are linear operators $h: \bigwedge_k \to \bigwedge_{k-1}$, taking differential k-forms to $(k-1)$-forms, and satisfying the basic identity

$$\omega = dh(\omega) + h(d\omega) \tag{1.66}$$

for all k-forms ω. (The case $k = 0$ is slightly different, as explained below.) The discovery of such a set of operators immediately implies exactness of the complex. For if ω is closed, $d\omega = 0$, then (1.66) reduces to $\omega = d\theta$ where $\theta = h(\omega)$, so ω is exact. Thus we need only concentrate on finding these homotopy operators.

Let us look back at the Lie derivative formula (1.65). If we could treat the Lie derivative as an ordinary derivative, then we could integrate both sides of (1.65) and deduce the homotopy formula (1.66). More rigorously, we can integrate the Lie derivative formula (1.64) with respect to ε; using (1.65) and (1.56), we find

$$\begin{aligned}
\exp(\varepsilon\mathbf{v})^*[\omega|_{\exp(\varepsilon\mathbf{v})x}] - \omega|_x &= \int_0^\varepsilon \exp(\tilde\varepsilon\mathbf{v})^*[\mathbf{v}(\omega)|_{\exp(\tilde\varepsilon\mathbf{v})x}]\, d\tilde\varepsilon \\
&= \int_0^\varepsilon \{d[\exp(\tilde\varepsilon\mathbf{v})^*(\mathbf{v} \rfloor \omega|_{\exp(\tilde\varepsilon\mathbf{v})x})] \\
&\qquad + \exp(\tilde\varepsilon\mathbf{v})^*[\mathbf{v} \rfloor d\omega|_{\exp(\tilde\varepsilon\mathbf{v})x}]\}\, d\tilde\varepsilon.
\end{aligned}$$

If we define the operator

$$h_{\mathbf{v}}^{\varepsilon}(\omega)|_x \equiv \int_0^\varepsilon \exp(\tilde\varepsilon\mathbf{v})^*[\mathbf{v} \rfloor \omega|_{\exp(\tilde\varepsilon\mathbf{v})x}]\, d\tilde\varepsilon,$$

then we have the homotopy-like formula

$$\exp(\varepsilon\mathbf{v})^*[\omega|_{\exp(\varepsilon\mathbf{v})x}] - \omega|_x = dh_{\mathbf{v}}^{\varepsilon}(\omega)|_x + h_{\mathbf{v}}^{\varepsilon}(d\omega)|_x, \tag{1.67}$$

which is valid for any manifold M, any differential form ω, any vector field $\mathbf{v}$ and all $\varepsilon \in \mathbb{R}$ such that $\exp(\varepsilon\mathbf{v})x$ is defined.

We are now in a position to prove the Poincaré lemma (Theorem 1.61) by constructing a homotopy operator over the star-shaped domain $M \subset \mathbb{R}^m$. Note that the scaling vector field $\mathbf{v}_0 = \sum x^i \partial/\partial x^i$ has flow $\exp(\varepsilon\mathbf{v})x = e^{\varepsilon}x$, which, for $x \in M$, remains in M for all $\varepsilon \leqslant 0$. If $\omega = \sum \alpha_I(x)\, dx^I$ is a k-form defined on all of M, then for $\varepsilon \leqslant 0$,

$$\exp(\varepsilon\mathbf{v}_0)^*[\omega|_{\exp(\varepsilon\mathbf{v}_0)x}] = \sum_I \alpha_I(e^{\varepsilon}x)e^{k\varepsilon}\, dx^I,$$

since $\exp(\varepsilon\mathbf{v}_0)^*(dx^i) = d(e^{\varepsilon}x^i) = e^{\varepsilon}\, dx^i$. We can write this formula in a simpler manner if we denote ω by $\omega[x]$, whereby

$$\exp(\varepsilon\mathbf{v}_0)^*\omega[x] = \omega[e^{\varepsilon}x] = \sum \alpha_I(e^{\varepsilon}x)d(e^{\varepsilon}x^{i_1}) \wedge \cdots \wedge d(e^{\varepsilon}x^{i_k}).$$

(In other words, we substitute $e^{\varepsilon}x^i$ for each x^i wherever it occurs in ω, including the differentials dx^i.) In this special case, (1.67) with $\mathbf{v} = \mathbf{v}_0$ reads

$$\omega[e^{\varepsilon}x] - \omega[x] = dh_0^{\varepsilon}(\omega) + h_0^{\varepsilon}(d\omega), \tag{1.68}$$

where, for $\varepsilon \leqslant 0$,

$$h_0^{\varepsilon}(\omega) \equiv \int_0^{\varepsilon} (\mathbf{v}_0 \lrcorner\, \omega)[e^{\tilde{\varepsilon}}x]\, d\tilde{\varepsilon} = -\int_{\exp\varepsilon}^{1} (\mathbf{v}_0 \lrcorner\, \omega)[\lambda x]\frac{d\lambda}{\lambda},$$

(using the change of variables $\lambda = e^{\tilde{\varepsilon}}$). Now let $\varepsilon \to -\infty$. If ω is a k-form, and $k > 0$, then $\omega[e^{\varepsilon}x] \to 0$ as $\varepsilon \to -\infty$. Thus (1.68) reduces to the homotopy formula (1.66) with homotopy operator

$$h(\omega) = \int_0^1 (\mathbf{v}_0 \lrcorner\, \omega)[\lambda x]\frac{d\lambda}{\lambda}. \tag{1.69}$$

(Note that in this formula, we first compute the interior product $\mathbf{v}_0 \lrcorner\, \omega$ and then evaluate at λx.) If, however, $k = 0$, so ω is a smooth function $f(x)$, (1.68) reduces to the alternative formula

$$f(x) - f(0) = dh(f) + h(df) = h(df)$$

in the limit as $\varepsilon \to -\infty$, leading to the initial injection $\mathbb{R} \to \bigwedge_0$ in the de Rham complex. We have thus completed the proof of the Poincaré lemma.

Example 1.67. Consider a planar star-shaped domain $M \subset \mathbb{R}^2$. If

$$\omega = \alpha(x, y)\, dx + \beta(x, y)\, dy$$

is any one-form, then

$$\mathbf{v}_0 \lrcorner\, \omega = (x\partial_x + y\partial_y) \lrcorner\, \omega = x\alpha(x, y) + y\beta(x, y).$$

Therefore, the function $h(\omega)$ obtained by applying our homotopy operator (1.69) to ω is

$$h(\omega) = \int_0^1 \{\lambda x \alpha(\lambda x, \lambda y) + \lambda y \beta(\lambda x, \lambda y)\} \frac{d\lambda}{\lambda}$$

$$= \int_0^1 \{x \alpha(\lambda x, \lambda y) + y \beta(\lambda x, \lambda y)\}\, d\lambda.$$

Similarly, applying h to a two-form leads to the one-form

$$h[\gamma(x, y)\, dx \wedge dy] = \int_0^1 \{\lambda^2 x \gamma(\lambda x, \lambda y)\, dy - \lambda^2 y \gamma(\lambda x, \lambda y)\, dx\} \frac{d\lambda}{\lambda}$$

$$= -\left\{\int_0^1 \lambda y \gamma(\lambda x, \lambda y)\, d\lambda\right\} dx + \left\{\int_0^1 \lambda x \gamma(\lambda x, \lambda y)\, d\lambda\right\} dy,$$

the differentials dx and dy not being affected by the λ-integration. In particular, for the above one-form,

$$d\omega = (\beta_x - \alpha_y)\, dx \wedge dy,$$

so the homotopy formula (1.66) reduces to two formulae, for α and β, the first of which is

$$\alpha(x, y) = \frac{\partial}{\partial x} \int_0^1 \{x \alpha(\lambda x, \lambda y) + y \beta(\lambda x, \lambda y)\}\, d\lambda$$

$$- \int_0^1 \lambda y [\beta_x(\lambda x, \lambda y) - \alpha_y(\lambda x, \lambda y)]\, d\lambda.$$

The reader may enjoy directly verifying this latter statement. In particular, if $d\omega = 0$, then $\omega = df$ where $f = h(\omega)$ is as above. (A similar kind of result holds for two-forms.)

Integration and Stokes' Theorem

Although it is not a subject central to the theme of this book, it would be unfair to omit a brief discussion of integration and Stokes' theorem from our introduction to differential forms. Indeed, differential forms arise as "the objects one integrates on manifolds". To define integration, we need to first *orient* the m-dimensional manifold M with a nonvanishing m-form ω defined over all of M. A second nonvanishing m-form $\tilde{\omega}$ defines the same orientation if it is a positive scalar multiple of ω at each point. There are precisely two orientations on such a manifold M. (Not every manifold is orientable, for instance, a Möbius band is not.) In particular, we can orient $\mathbb{R}^m$ (and any open subset thereof) by choosing the volume form $dx^1 \wedge \cdots \wedge dx^m$. A map

$F: \tilde{M} \to M$ between two oriented m-dimensional manifolds is *orientation-preserving* if the pull-back of the orientation form on M determines the same orientation on $\tilde{M}$ as the given one. If M is oriented, then we can cover M by orientation-preserving coordinate charts $\chi_\alpha: U_\alpha \to V_\alpha$ whose overlap functions $\chi_\beta \circ \chi_\alpha^{-1}$ are orientation-preserving diffeomorphisms on $\mathbb{R}^m$.

If M is an oriented m-dimensional manifold, we can define the integral $\int_M \omega$ of any m-form ω on M. In essence, we chop up M into component *oriented* coordinate charts and add up the individual integrals

$$\int_{U_\alpha} \omega = \int_{V_\alpha} (\chi_\alpha^{-1})^* \omega = \int_{V_\alpha} f(x)\, dx^1 \wedge \cdots \wedge dx^m,$$

the latter integral being an ordinary multiple integral over $V_\alpha \subset \mathbb{R}^m$. The change of variables formula for multiple integrals assures us that this definition is coordinate-free. More generally, $\int_M \omega = \int_{\tilde{M}} F^* \omega$ whenever $F: \tilde{M} \to M$ is orientation-preserving.

Stokes' theorem relates integrals of m-forms over a compact m-dimensional manifold M to integrals of $(m-1)$-forms over the boundary ∂M. The simplest manifold with boundary is the upper half space $\mathbb{H}^m \equiv \{(x^1, \ldots, x^m): x^m \geq 0\}$ of $\mathbb{R}^m$, with $\partial \mathbb{H}^m = \{(x^1, \ldots, x^{m-1}, 0)\} \simeq \mathbb{R}^{m-1}$. Any other manifold with boundary is defined using coordinate charts $\chi_\alpha: U_\alpha \to V_\alpha$ where $V_\alpha \subset \mathbb{H}^m$ is open, meaning $V_\alpha = \mathbb{H}^m \cap \tilde{V}_\alpha$ where $\tilde{V}_\alpha \subset \mathbb{R}^m$ is open. The boundary of the chart is $\partial U_\alpha = \chi_\alpha^{-1}[\partial V_\alpha]$, $\partial V_\alpha = V_\alpha \cap \partial \mathbb{H}^m$, and ∂M is the union of all such boundaries of coordinate charts. Thus ∂M is a smooth $(m-1)$-dimensional manifold, without boundary.

The boundary of $\mathbb{H}^m$ is given an "induced" orientation $(-1)^m\, dx^1 \wedge \cdots \wedge dx^{m-1}$ from the volume form $dx^1 \wedge \cdots \wedge dx^m$ determining the orientation of $\mathbb{H}^m$ itself. If M is an oriented manifold with boundary, then ∂M inherits an induced orientation so that any oriented coordinate chart $\chi_\alpha: U_\alpha \to V_\alpha$ on M restricts to an oriented coordinate chart $\partial \chi_\alpha: \partial U_\alpha \to \partial V_\alpha$ on ∂M. With these definitions, we can state the general form of Stokes' theorem.

Theorem 1.68. *Let M be a compact, oriented, m-dimensional manifold with boundary ∂M. Let ω be a smooth $(m-1)$-form defined on M. Then $\int_{\partial M} \omega = \int_M d\omega$.*

Using the identification of the differential d with the usual vector differential operations in $\mathbb{R}^3$, the reader can check that Theorem 1.68 reduces to the usual forms of Stokes' theorem and the Divergence theorem of vector calculus. More generally, there is an intimate connection between the de Rham complex, Stokes' theorem and the underlying topology of M, but this would lead us too far afield to discuss any further, and so we conclude our brief introduction to this subject.

Notes

In this chapter we have only been able to give the briefest of introductions to the vast and important subjects of Lie groups and differentiable manifolds. There are a number of excellent books which can be profitably studied by the reader interested in delving further into these areas, including those by Warner, [1], Boothby, [1], Thirring, [1; Chap. 2] and Miller, [2]. Pontryagin, [1], is useful as a reference for the local Lie group approach to the subject and includes many otherwise hard-to-find proofs of important theorems. Many other works could be mentioned as well.

Historically, the two subjects of differentiable manifolds and Lie groups have been closely intertwined throughout their development, each inspiring further work in the other. Lie himself, though, drew his original motivation from the spectacular success of Galois' group theory applied to the solution of polynomial equations and sought to erect a similar theory for the solution of differential equations using *his* theory of continuous groups. Although Lie fell sort of this goal (the more refined Picard–Vessiot theory being the correct "Galois theory of differential equations"—see Pommaret, [2]), his seminal influence in all aspects of the subject continues to this day. Remarkably, Lie arrived at the fundamental concept of a (local) Lie group through his research into the analysis of systems of partial differential equations (predating Frobenius), leading to the concept of a "function group," which we will encounter under the name "Poisson structure" in Chap. 6, and then, finally, to Lie groups. The interested reader is well advised to look up Hawkins' fascinating and illuminating historical essays, [1], [2], [3], on the early history of Lie groups.

In Lie's time, all Lie groups were local groups and arose concretely as groups of transformations on some Euclidean space. The global, abstract approach was quite slow in maturing, and the first modern definition of a manifold with coordinate charts appears in Cartan, [2]. (Cartan himself played a fundamental role in the history of Lie groups; his definition of manifold was inspired by Weyl's book, [1], on Riemann surfaces as well as closely related ideas of Schreier, [1].) The passage from the local Lie group to the present-day definition using manifold theory was also accomplished by Cartan, [2]. Cartan also introduced the concept of the simply-connected covering group of a Lie group and noted, in [3], that the simply-connected covering group of $SL(2, \mathbb{R})$ is not a subgroup of any matrix group $GL(n)$. (Interestingly, there is a realization of this global group as an open subset of $\mathbb{R}^3$ due to Bargmann, [1].) A more accessible example of a Lie group which cannot be realized as a group of matrices can be found in Birkhoff, [1].

Lie's fundamental tool in his theory was the infinitesimal form of a Lie group, now called the Lie algebra. In its local version, the correspondence between a Lie group and the right- (or left-)invariant vector fields forming its Lie algebra is known as the first fundamental theorem of Lie. The reconstruction of a local Lie group from its Lie algebra is known as Lie's second fundamental theorem; a proof not relying on Ado's theorem can be found in Pontryagin, [1; Theorem 89]. The construction of a global Lie group from its Lie algebra, though, is due to Cartan, [3]; see also Pontryagin, [1; Theorem 96]. The proof based on the contemporaneous theorem of Ado [1] (see also Jacobson, [1; Chap. 6]), outlined here is more recent. Lie's third fundamental theorem states that the structure constants determine the Lie algebra, and hence the Lie group. The complete proof of the general correspondence between subgroups of a Lie group and subalgebras of its Lie algebra can be found in Warner, [1; Theorems 3.19 and 3.28]. Theorem 1.19 on closed subgroups of Lie groups is due to Cartan, [2]; see Warner, [1; Theorem 3.42] for the proof. Theorem 1.57 on the reconstruction of a transformation group from its infinitesimal generators dates back to Lie; see Pontryagin, [1; Theorem 98] for a proof. The definition used here of a regular group of transformations is based on the monograph of Palais, [1], and is further developed in Chapter 3.

While vector fields have their origins in the study of mathematical physics, the modern geometrical formulation owes much to the work of Poincaré, [1], whose influence, like Lie's, pervades the entire subject. The notation for a vector field employed here and throughout modern differential geometry, however, comes from Lie's notation for the infinitesimal generators of a group of transformations. Flows of vector fields arise, naturally enough, in fluid mechanics; see Wilczynski, [1], for an early connection between their physical and group-theoretic interpretations.

Frobenius' Theorem 1.43 originally appears as a theorem on the nature of the solutions to certain systems of homogeneous, first order, linear partial differential equations; see Frobenius, [1], and the discussion of invariants in Section 2.1. Its translation into a theorem a differential geometry first appears in Chevalley's influential book, [1], on Lie groups. (In this book, most of the modern definitions and theorems in this subject are assembled together for the first time.) A proof of the semi-regular version of Frobenius' theorem can be found in Narasimhan, [1], and Warner, [1; Theorem 1.64], and, using a modern method due to Weinstein, in Abraham and Marsden, [1; p. 93]. The extension of this result to systems of vector fields of varying rank is due to Hermann, [1], [2]. This result has subsequently been generalized much further—see Sussmann, [1]—but much work, especially on the structure of the singular sets, remains to be done. In these and other references, the terms "distribution" or "differential system" have been applied to what we have simply called a system of vector fields. (The former is especially confusing as it appears in a completely unrelated context in functional analysis.) Furthermore, our term "integrable" is more commonly referred to as "completely integrable", but this latter term has very different connotations in the study of Hamiltonian systems, which motivates our choice of the former.

Differential forms have their origins in the work of Grassmann and the attempts to find a multi-dimensional generalization of Stokes' theorem. In the hands of Poincaré and Cartan they became a powerful tool for the study of differential geometry, topology and differential equations. Already in Poincaré, [1], we find the basic concepts of wedge product (p. 25), interior product (p. 33), the differential and the multi-dimensional form of Stokes' theorem (p. 10) as well as the lemma bearing his name. See also Cartan, [1], for further developments and applications to differential equations. The concept of the Lie derivative, though, while in essence due to Poincaré (see also Cartan, [1; p. 82]), was first formally defined by Schouten and his coworkers; see Schouten and Struik, [1; p. 142]. This last reference also contains the basic homotopy formula proof of the Poincaré lemma for the first time. Finally, the connections with topology stemming from de Rham's theorem can be found in Warner, [1; Chap. 5], and Bott and Tu, [1].

Exercises

1.1. *Real projective m-space* is defined to be the set of all lines through the origin in $\mathbb{R}^{m+1}$. Specifically, we define an equivalence relation on $\mathbb{R}^{m+1}\backslash\{0\}$ by setting $x \sim y$ if and only if $x = \lambda y$ for some nonzero scalar λ. Then $\mathbb{RP}^m$ is the set of equivalence classes in $\mathbb{R}^{m+1}\backslash\{0\}$.
   (a) Prove that $\mathbb{RP}^m$ is a manifold of dimension m by exhibiting coordinate charts.
   (b) Prove that $\mathbb{RP}^1 \simeq S^1$ are the same smooth manifolds, and exhibit a diffeomorphism.
   (c) Let S^m be the unit sphere in $\mathbb{R}^{m+1}$. Prove that the map $F: S^m \to \mathbb{RP}^m$ which associates to $x \in S^m$ its equivalence class in $\mathbb{RP}^m$ is a smooth covering map. What is the inverse image $F^{-1}\{z\}$ of a point $z \in \mathbb{RP}^m$?

*1.2. *Grassmann Manifolds*: Let $0 < m < n$.
   (a) Prove that the space $\mathrm{GL}(m, n)$ of all $m \times n$ matrices of maximal rank is an analytic manifold of dimension $m \cdot n$.
   (b) Let $\mathrm{Grass}(m, n)$ denote the set of all m-dimensional subspaces of $\mathbb{R}^n$. Show that $\mathrm{Grass}(m, n)$ can be given the structure of an analytic manifold of dimension $m(n - m)$. (*Hint*: To any basis of such an m-dimensional subspace, associate the matrix in $\mathrm{GL}(m, n)$ whose rows are the basis. Show that the basis can be chosen so that this matrix has the same m columns as the identity matrix; the remaining entries will give local coordinates for $\mathrm{Grass}(m, n)$.)
   (c) Let $F: \mathrm{GL}(m, n) \to \mathrm{Grass}(m, n)$ be the map assigns to a matrix A the subspace of $\mathbb{R}^n$ spanned by its rows. Prove that F is an analytic map between manifolds.
   (d) Prove that $\mathrm{Grass}(m, n)$ and $\mathrm{Grass}(n - m, n)$ are diffeomorphic manifolds. In particular, $\mathrm{Grass}(1, n) \simeq \mathrm{Grass}(n - 1, n) \simeq \mathbb{RP}^{n-1}$.

1.3. Let $\phi(t) = ((\sqrt{2} + \cos 2t) \cos 3t,\ (\sqrt{2} + \cos 2t) \sin 3t, \sin 2t)$ for $0 \leqslant t \leqslant 2\pi$. Prove that the image of ϕ is a regular closed curve in $\mathbb{R}^3$—a "trefoil knot".

1.4. Let

$$\phi(u, v) = \left(2 \cos u + v \sin\frac{u}{2} \cos u, 2 \sin u + v \sin\frac{u}{2} \sin u, v \cos\frac{u}{2}\right)$$

for $0 \leqslant u < 2\pi$, $-1 < v < 1$. Prove that ϕ is a regular immersion, whose image is a Möbius band in $\mathbb{R}^3$.

1.5. Prove that if $N \subset M$ is a compact submanifold, then N is a regular submanifold. (Boothby, [1; page 79].)

1.6. Prove that the m-dimensional sphere S^m is simply connected if $m \geqslant 2$. What about real projective space $\mathbb{RP}^m$? (See Exercise 1.1.)

1.7. Let $M = \mathbb{R}^2 \backslash \{0\}$. Prove that $(x, y) \mapsto (e^x \cos y, e^x \sin y)$ defines a covering map from $\mathbb{R}^2$ onto M, hence $\mathbb{R}^2$ is the simply-connected cover of $\mathbb{R}^2 \backslash \{0\}$.

1.8. Let $M = \mathbb{R}^2 \backslash \{0\}$. Prove that $\Psi(\varepsilon, (r, \theta)) = (re^{-\varepsilon} + (1 - e^{-\varepsilon}), \theta + \varepsilon)$, $\varepsilon \in \mathbb{R}$, written in polar coordinates, determines a one-parameter group of transformations. What is its infinitesimal generator? Prove that every orbit is a regular submanifold of M, but the group action is *not* regular.

1.9. Consider the system of vector fields

$$\mathbf{v}_1 = x\partial_y - y\partial_x + z\partial_w - w\partial_z, \qquad \mathbf{v}_2 = z\partial_x - x\partial_z + w\partial_y - y\partial_w,$$

on the unit sphere $S^3 \subset \mathbb{R}^4$.
(a) Prove that $\{\mathbf{v}_1, \mathbf{v}_2\}$ form an integrable system. What are the integral submanifolds in S^3?
(b) Let $\pi: S^3 \backslash \{(0, 0, 0, 1)\} \to \mathbb{R}^3$ be stereographic projection (as in Example 1.3). What are the vector fields $d\pi(\mathbf{v}_1)$ and $d\pi(\mathbf{v}_2)$ on $\mathbb{R}^3$? What are their integral submanifolds?

1.10. Is it possible to construct a system of three vector fields $\mathbf{u}$, $\mathbf{v}$ and $\mathbf{w}$ on $\mathbb{R}^3$ such that $[\mathbf{u}, \mathbf{v}] = 0 = [\mathbf{u}, \mathbf{w}]$, but $[\mathbf{v}, \mathbf{w}] \neq 0$? Is it possible to construct an integrable system with the above commutation relations? If so, what would the integral submanifolds of such a system look like?

1.11. Prove that the vector field

$$\mathbf{v} = (-y - 2z(x^2 + y^2))\partial_x + x\partial_y + x(x^2 + y^2 - z^2 - 1)\partial_z$$

does not form a regular system on $\mathbb{R}^3$. Prove that any integral curve of $\mathbf{v}$ lies in one of the tori looked at in Example 1.42. Prove that the flow generated by $\mathbf{v}$, when restricted to one of the above tori, is isomorphic to either the rational or irrational flow on the torus, depending on the size of the torus.

1.12. Suppose $\mathbf{v}$ is a smooth linear map on the space of smooth functions defined near a point $x \in M$ which satisfies (1.20–1.21). Prove that $\mathbf{v}$ is a tangent vector to M at x. (Warner, [1; p. 12].)

*1.13. Let $M = \mathbb{R}^2$ and consider the system of vector fields $\mathscr{H}$ spanned by $\mathbf{v}_0 = \partial_x$ and all vector fields of the form $f(x)\partial_y$ where $f: \mathbb{R} \to \mathbb{R}$ is any smooth function such that all derivatives $f^{(n)}(0) = 0$ vanish for all $n = 0, 1, 2, \ldots$.
(a) Prove that $\mathscr{H}$ is involutive.
(b) Prove that $\mathscr{H}$ has *no* integral submanifold passing through any point $(0, y)$ on the y-axis.
(c) How do you reconcile this with Frobenius' Theorem 1.40 or 1.41? (Nagano, [1].)

*1.14. Let $\{\mathbf{v}_1, \ldots, \mathbf{v}_r\}$ be a finite, involutive system of vector fields on a manifold M. Prove that the system is always rank-invariant. (Thus Theorem 1.40 is a special case of Theorem 1.41). (Hermann, [1].)

1.15. Prove that the set of all nonsingular upper triangular matrices forms a Lie group $\mathrm{T}(n)$. What is its Lie algebra?

1.16. Consider the $2n \times 2n$ matrix

$$\mathbf{J} = \begin{pmatrix} 0 & \mathbf{I} \\ -\mathbf{I} & 0 \end{pmatrix},$$

where each $\mathbf{I}$ is an $n \times n$ identity matrix. The symplectic group $\mathrm{Sp}(n)$ is defined to be the set of all $2n \times 2n$ matrices A such that $A^T \mathbf{J} A = \mathbf{J}$. Prove that $\mathrm{Sp}(n)$ is a Lie group and compute its dimension. What is its Lie algebra?

1.17. Prove that if $H \subset G$ is a connected one-parameter subgroup of a Lie group G then H is isomorphic to either $\mathrm{SO}(2)$ or $\mathbb{R}$.

1.18. Prove that if G and H are Lie groups, then their Cartesian product $G \times H$ is also a Lie group.

1.19. Let G and H be Lie groups and suppose G acts (globally) on H as a group of transformations, via $h \mapsto g \cdot h$, $g \in G$, $h \in H$, with $g \cdot (h_1 \cdot h_2) = (g \cdot h_1) \cdot (g \cdot h_2)$. Define the *semi-direct product* of G and H, denoted $G \ltimes H$, to be the Lie group whose manifold structure is just the Cartesian product $G \times H$, but whose group multiplication is given by

$$(g, h) \cdot (\tilde{g}, \tilde{h}) = (g \cdot \tilde{g}, h \cdot (g \cdot \tilde{h})).$$

(a) Prove that $G \ltimes H$ is a Lie group.
(b) How is the Lie algebra of $G \ltimes H$ related to those of G and H?
(c) Prove that the Euclidean group $\mathrm{E}(m)$, consisting of all the translations and rotations of $\mathbb{R}^m$, is a semi-direct product of the rotation group $\mathrm{SO}(m)$ with the vector group $\mathbb{R}^m$, $\mathrm{SO}(m)$ acting on $\mathbb{R}^m$ as a group of rotations. (See also Exercise 1.29.)

1.20. Let $V = \{(x, y): |x| < 1\} \subset \mathbb{R}^2$ and define the map $m: V \times V \to \mathbb{R}^2$ by

$$m(x, y; z, w) = (xz + x + z, xw + w + y(z+1)^{-1}), \qquad (x, y), (z, w) \in V.$$

Prove that m determines a multiplication map making V into a local Lie group by constructing an inverse map $i: V_0 \to V$ on a suitable subdomain V_0. What is the Lie algebra of V?

1.21. Prove that every two-dimensional Lie algebra is either (a) abelian (all brackets are 0) or (b) isomorphic to the Lie algebra with basis $\{\mathbf{v}, \mathbf{w}\}$ satisfying $[\mathbf{v}, \mathbf{w}] = \mathbf{w}$. Find a 2×2 matrix representation of the Lie algebra in part (b). Find the corresponding simply-connected Lie group. Construct a local group isomorphism from the local Lie group of Exercise 1.20 to this global Lie group. (Jacobson, [1; p. 11].)

1.22. Prove that $\mathbb{R}^3$ forms a Lie algebra with Lie bracket determined by the vector cross product: $[\mathbf{v}, \mathbf{w}] = \mathbf{v} \times \mathbf{w}$, $\mathbf{v}, \mathbf{w} \in \mathbb{R}^3$. What are the structure constants for this Lie algebra with respect to the standard basis of $\mathbb{R}^3$? Prove that this Lie algebra is isomorphic to $\mathfrak{so}(3)$, the Lie algebra of the three-dimensional rotation group. Show that the isomorphism can be constructed so that a given vector $\mathbf{v} \in \mathbb{R}^3$ corresponds to the infinitesimal generator of the one-parameter group of right-handed rotations about the axis in the direction of $\mathbf{v}$.

*1.23. Prove that every complex Lie group contains a two-dimensional subgroup. Is the same true for real Lie groups? (Cohen, [1; p. 50].)

*1.24. A subgroup H of a group G is called *normal* if for every $h \in H$ and every $g \in G$, $g^{-1}hg \in H$. Let G/H denote the set of equivalence classes of G, where g and $\hat{g}$ are equivalent if and only if $g = \hat{g}h$ for some $h \in H$.

(a) Prove that if $H \subset G$ is normal, then G/H can be given the structure of a group in a natural way.

(b) Prove that a Lie subgroup H of a Lie group G is normal if and only if its Lie algebra $\mathfrak{h} \subset \mathfrak{g}$ has the property that $[\mathbf{v}, \mathbf{w}] \in \mathfrak{h}$ whenever $\mathbf{v} \in \mathfrak{g}$ and $\mathbf{w} \in \mathfrak{h}$.

(c) Prove that if $H \subset G$ is a normal Lie subgroup, the *quotient group* G/H is a Lie group with Lie algebra $\mathfrak{g}/\mathfrak{h}$. Explain.

(d) Find all normal subgroups of the two-dimensional Lie groups of Exercise 1.21.

(e) Does SO(3) have any normal subgroups?

*1.25. Let G be a Lie group. The *commutator subgroup* H is defined to be the subgroup generated by the elements $ghg^{-1}h^{-1}$ for $g, h \in G$.

(a) Prove that H is a Lie subgroup of G, and that the Lie algebra of H is the *derived subalgebra* of $\mathfrak{g}$, given by $\mathfrak{h} = \{[\mathbf{v}, \mathbf{w}]: \mathbf{v}, \mathbf{w} \in \mathfrak{g}\}$.

(b) Prove that the commutator subgroup of SO(3) is SO(3) itself. What about SO(m)?

(c) What are the commutator subgroups of the two-dimensional Lie groups discussed in Exercise 1.21?

(d) Is every element in H of the form $ghg^{-1}h^{-1}$?

1.26. Prove Proposition 1.24. (*Hint*: Show that $\bigcup U^k$ is both open and closed in G.) (Warner, [1; p. 93].)

1.27. Let G be a Lie group with Lie algebra $\mathfrak{g}$.

(a) Prove that the exponential map $\exp: \mathfrak{g} \to G$ is a local diffeomorphism from a neighbourhood of $0 \in \mathfrak{g}$ to a neighbourhood of the identity in G.

(b) Prove the "normal coordinate" formula (1.40). (Warner, [1; pp. 103, 109].)

*1.28. Let SL(2) denote the Lie group of 2×2 real matrices of determinant $+1$, and $\mathfrak{sl}(2)$ its Lie algebra.

(a) Let $A \in \mathfrak{sl}(2)$. Prove that

$$\exp(A) = \begin{cases} (\cosh \delta)\mathrm{I} + (\delta^{-1} \sinh \delta)A, & \delta = \sqrt{-\det A}, \quad \det A < 0, \\ (\cos \delta)\mathrm{I} + (\delta^{-1} \sin \delta)A, & \delta = \sqrt{\det A}, \quad \det A > 0. \end{cases}$$

What about the case $\det A = 0$?

(b) Consider the matrix $M = \begin{pmatrix} \lambda & 0 \\ 0 & \lambda^{-1} \end{pmatrix} \in \mathrm{SL}(2)$, where $\lambda \neq 0$. Prove that M lies on exactly one one-parameter subgroup of SL(2) if $\lambda > 0$, on infinitely many one-parameter subgroups if $\lambda = -1$, and on *no* one-parameter subgroup if $-1 \neq \lambda < 0$. (This shows that the exponential map $\exp: \mathfrak{g} \to G$ is, in general, neither one-to-one nor onto!) (Helgason, [1; p. 126].)

*1.29. A diffeomorphism $\psi: \mathbb{R}^m \to \mathbb{R}^m$ is called an *isometry* if it preserves distance, i.e. $|d\psi(\mathbf{v})| = |\mathbf{v}|$ for all $\mathbf{v} \in T\mathbb{R}^m|_x$, $x \in \mathbb{R}^m$, where $|\cdot|$ is the usual Euclidean metric $\sum (dx^i)^2$, i.e.

$$|\mathbf{v}|^2 = \sum (\xi^i)^2, \qquad \mathbf{v} = \sum \xi^i \partial/\partial x^i.$$

(a) Prove that a vector field $\mathbf{v} = \sum \xi^i(x)\partial/\partial x^i$ on $\mathbb{R}^m$ generates a one-parameter group of isometries if and only if its coefficient functions satisfy the system of partial differential equations

$$\frac{\partial \xi^i}{\partial x^j} + \frac{\partial \xi^j}{\partial x^i} = 0, \quad i \neq j, \qquad \frac{\partial \xi^i}{\partial x^i} = 0, \quad i = 1, \dots, m.$$

(b) Prove that the (connected) group of isometries of $\mathbb{R}^m$, called the Euclidean group $E(m)$, is generated by translations and rotations, and hence is an $m(m+1)/2$-dimensional Lie group.

(c) What if $|\cdot|$ is replaced by some non-Euclidean metric? For example, consider the Lorentz metric $(dx^1)^2 + (dx^2)^2 + (dx^3)^2 - (dx^4)^2$ on $\mathbb{R}^4$. (Eisenhart, [1; Chap. 6].)

*1.30. A diffeomorphism $\psi: \mathbb{R}^m \to \mathbb{R}^m$ is called a *conformal transformation* if $|d\psi(\mathbf{v})| = \lambda(x)|\mathbf{v}|$ for all $\mathbf{v} \in T\mathbb{R}^m|_x$, $x \in \mathbb{R}^m$, where λ is some scalar-valued function of x, and $|\mathbf{v}|$ is as in Exercise 1.29.

(a) Prove that a vector field $\mathbf{v} = \sum \xi^i(x)\partial/\partial x^i$ on $\mathbb{R}^m$ generates a one-parameter group of conformal transformations if and only if it satisfies

$$\frac{\partial \xi^i}{\partial x^j} + \frac{\partial \xi^j}{\partial x^i} = 0, \quad i \neq j, \qquad \frac{\partial \xi^i}{\partial x^i} = \mu(x), \quad i = 1, \dots, m, \tag{$*$}$$

for some undetermined function $\mu(x)$.

(b) Prove that if $m \geqslant 3$ then the conformal group of $\mathbb{R}^m$ is an $(m+1) \times (m+2)/2$-dimensional Lie group. Find its infinitesimal generators. (*Hint*: Prove that $(*)$ implies that *all* third order derivatives of the coefficient functions are identically zero.) What about $m = 2$?

(c) Prove that the inversion $I(x) = x/|x|^2$, $0 \neq x \in \mathbb{R}^m$ is a conformal transformation on $\mathbb{R}^m \backslash \{0\}$.

(d) Prove that for $m \geqslant 3$, the group of conformal transformations is generated by the groups of translations and rotations, the scaling group $x \mapsto \lambda x$, $\lambda > 0$, and the inversion of part (c).

(e) Discuss the case of the conformal group for the Lorentz metric in $\mathbb{R}^4$. (See Exercise 1.29(c).) (Eisenhart, [1; Chap. 6].)

*1.31. Let $\pi: S^m \backslash \{(0, \dots, 0, 1)\} \to \mathbb{R}^m$ be stereographic projection from the unit sphere in $\mathbb{R}^{m+1}$. Prove that if $A \in SO(m+1)$ is any rotation of S^m, then A induces a conformal transformation $\pi \circ A \circ \pi^{-1}$ of $\mathbb{R}^m$. Which of the conformal transformations constructed in Exercise 1.30 correspond to rotations of S^m?

1.32. Let G be a local group of transformations acting on a smooth manifold M. For each $x \in M$, the *isotropy group* is defined to be $G^x = \{g \in G: g \cdot x = x\}$. Prove that G^x is a (local) subgroup of G with Lie algebra $\mathfrak{g}^x = \{\mathbf{v} \in \mathfrak{g}: \mathbf{v}|_x = 0\}$. Find the isotropy subgroups and subalgebras of the rotation group $SO(3)$ acting on $\mathbb{R}^3$. Suppose $y = g \cdot x$. How is the isotropy subgroup G^y related to G^x?

1.33. In most treatments of Lie groups, the Lie algebra is defined as the space of *left*-invariant vector fields on the Lie group rather than the right-invariant vector fields employed here. In this problem we compare these two approaches.
(a) Define what is meant by a left-invariant vector field on a Lie group G. Prove that the space of all left-invariant vector fields on G forms a Lie algebra, denoted $\mathfrak{g}_L$, which we can identify with $TG|_e$.
(b) Using subscripts L and R to denote the two Lie algebras, prove that $[\mathbf{v}, \mathbf{w}]_L = -[\mathbf{v}, \mathbf{w}]_R$ where $\mathbf{v}, \mathbf{w}$ are identified with their values at e.
(c) Let G act on a manifold M as in Definition 1.25. If $\mathbf{v} \in \mathfrak{g}_L$ or $\mathfrak{g}_R$, we can define the corresponding infinitesimal generator $\psi(\mathbf{v})$ on M. Show that while (1.47) holds for right-invariant vector fields it is *false* for left-invariant vector fields. What happens to the Lie bracket formula (1.47) in this case? On the other hand, prove that if $\mathbf{v}$ is a left-invariant vector field and $\Psi_g(x) = \Psi(g, x)$, then

$$d\Psi_g(\psi(\mathbf{v})|_x) = \psi(\mathbf{v})|_{g\cdot x},$$

i.e. the infinitesimal generators of the action of G on M behave naturally with respect to the group transformations. Show that this is *false* for right-invariant vector fields.
(d) How does this all change if we let G act on M on the *right*, i.e. set $x \cdot g = \Psi(x, g)$ with $x \cdot (g \cdot h) = (x \cdot g) \cdot h$?
(Marsden, Ratiu and Weinstein, [1].)

1.34. Let $\boldsymbol{\alpha} = (\alpha, \beta, \gamma)$ be a vector field on $\mathbb{R}^3$ with $\nabla \cdot \boldsymbol{\alpha} = 0$. Use the homotopy operator (1.69) to construct a vector field $\boldsymbol{\lambda}$ with $\nabla \times \boldsymbol{\lambda} = \boldsymbol{\alpha}$. Similarly, if $\nabla \times \boldsymbol{\lambda} = \mathbf{0}$, find a function f with $\boldsymbol{\lambda} = \nabla f$.

1.35. (a) Let $\mathbf{v}$ and $\mathbf{w}$ be vector fields, ω a one-form. Prove that

$$\mathbf{v}\langle\omega; \mathbf{w}\rangle = \langle\mathbf{v}(\omega); \mathbf{w}\rangle + \langle\omega; [\mathbf{v}, \mathbf{w}]\rangle.$$

(b) More generally, if ω is a k-form prove that

$$\mathbf{v}\langle\omega; \mathbf{w}_1, \ldots, \mathbf{w}_k\rangle = \langle\mathbf{v}(\omega); \mathbf{w}_1, \ldots, \mathbf{w}_k\rangle + \sum_{i=1}^{k} \langle\omega; \mathbf{w}_1, \ldots, [\mathbf{v}, \mathbf{w}_i], \ldots, \mathbf{w}_k\rangle.$$

(c) Deduce that

$$\mathbf{v}(\mathbf{w} \,\lrcorner\, \omega) = \mathbf{w} \,\lrcorner\, \mathbf{v}(\omega) + [\mathbf{v}, \mathbf{w}] \,\lrcorner\, \omega.$$

1.36. Let $\omega = dx^1 \wedge \cdots \wedge dx^m$ be the volume m-form on $\mathbb{R}^m$ and let $\mathbf{v} = \sum \xi^i(x)\partial/\partial x^i$ be a vector field.
(a) Prove that the Lie derivative of the volume form is $\mathbf{v}(\omega) = \operatorname{div} \xi \cdot \omega$, where $\operatorname{div} \xi = \sum \partial\xi^i/\partial x^i$ is the ordinary divergence.
(b) Prove that the flow $\phi_\varepsilon = \exp(\varepsilon\mathbf{v})$ generated by $\mathbf{v}$ preserves volume, meaning $\operatorname{Vol}(\phi_\varepsilon[S]) = \operatorname{Vol}(S)$ for any $S \subset \mathbb{R}^m$ such that $\phi_\varepsilon(x)$ is defined for all $x \in S$, if and only if $\operatorname{div} \xi = 0$ everywhere.

1.37. Let $\partial_i = \partial/\partial x^i$, and dx^i, $i = 1, \ldots, m$, be the standard bases for $T\mathbb{R}^m$ and $T^*\mathbb{R}^m$ respectively. Let ω be an r-form on $\mathbb{R}^m$. Prove the following formulae:

$$\partial_k \,\lrcorner\, (dx^l \wedge \omega) = -dx^l \wedge (\partial_k \,\lrcorner\, \omega), \quad \text{whenever} \quad k \neq l,$$

$$\partial_k \,\lrcorner\, (dx^k \wedge \omega) = \omega - dx^k \wedge (\partial_k \,\lrcorner\, \omega),$$

$$\sum_{k=1}^{m} \partial_k \,\lrcorner\, (dx^k \wedge \omega) = (m - r)\omega.$$

CHAPTER 2

Symmetry Groups of Differential Equations

The symmetry group of a system of differential equations is the largest local group of transformations acting on the independent and dependent variables of the system with the property that it transform solutions of the system to other solutions. The main goal of this chapter is to determine a useful, systematic, computational method that will explicitly determine the symmetry group of any given system of differential equations. We restrict our attention to connected local Lie groups of symmetries, leaving aside problems involving discrete symmetries such as reflections, in order to take full advantage of the infinitesimal techniques developed in the preceding chapter. Before pressing on to the case of differential equations, it is vital that we deal adequately with the simpler situation presented by symmetry groups of systems of algebraic equations, and this is done in the first section. Section 2.2 investigates the precise definition of a symmetry group of a system of differential equations, which requires knowledge of how the group elements actually transform the solutions. The corresponding infinitesimal method rests on the important concept of "prolonging" a group action to the spaces of derivatives of the dependent variables represented in the system. The key "prolongation formula" for an infinitesimal generator of a group of transformations, given in Theorem 2.36, then provides the basis for the systematic determination of symmetry groups of differential equations. Applications to physically important partial differential equations, including the heat equation, Burgers' equation, the Korteweg-de Vries equation and Euler's equations for ideal fluid flow are presented in Section 2.4.

In the case of ordinary differential equations, Lie showed how knowledge of a one-parameter symmetry group allows us to reduce the order of the equation by one. In particular, a first order equation with a known one-parameter symmetry group can be integrated by a single quadrature. The situation is more delicate in the case of higher dimensional symmetry groups; it is not in general possible to reduce the order of an equation invariant under an r-parameter symmetry group by r using only quadratures. We will discuss in detail how the theory proceeds for multi-parameter symmetry groups of higher order equations and systems of ordinary differential equations.

The last section of this chapter deals with some more technical mathematical issues, and may safely be omitted by an application-oriented reader at first. The basic converse to the theorem on existence of symmetry groups says when one can conclude that every (continuous) symmetry group has been obtained by the above methods. Besides the algebraic maximal rank condition, an additional existence result known as "local solvability" is required. In the case of analytic systems, these questions are related to the problem of existence of noncharacteristic directions for the system, relative to which the Cauchy–Kovalevskaya existence theorem is applicable. Such systems are designated as "normal systems", but there do exist "abnormal systems", of which several examples are presented. The correct understanding of these matters will be crucial to the formulation and proof of Noether's theorems relating symmetry groups and conservation laws to be presented in Chapter 5.

2.1. Symmetries of Algebraic Equations

Before considering symmetry groups of differential equations, it is essential that we deal properly with the conceptually simpler case of symmetry groups of systems of algebraic equations. By a "system of algebraic equations" we mean a system of equations

$$F_\nu(x) = 0, \qquad \nu = 1, \ldots, l,$$

in which $F_1(x), \ldots, F_l(x)$ are smooth real-valued functions defined for x in some manifold M. (Note that the adjective "algebraic" is only used to distinguish this case from the case of systems of differential equations; it does *not* mean that the F_ν must be polynomials—just any differentiable functions.) A *solution* is a point $x \in M$ such that $F_\nu(x) = 0$ for $\nu = 1, \ldots, l$. A *symmetry group* of the system will be a local group of transformations G acting on M with the property that G transforms solutions of the system to other solutions. In other words, if x is a solution, g a group element and $g \cdot x$ is defined, then we require that $g \cdot x$ also be a solution. In this section we will be primarily concerned with finding easily verifiable conditions that a given group of transformations be a symmetry group of such a system.

Invariant Subsets

More generally, we can look at symmetry groups of arbitrary subsets of the given manifold.

Definition 2.1. Let G be a local group of transformations acting on a manifold M. A subset $\mathscr{S} \subset M$ is called *G-invariant*, and G is called a *symmetry group* of $\mathscr{S}$, if whenever $x \in \mathscr{S}$, and $g \in G$ is such that $g \cdot x$ is defined, then $g \cdot x \in \mathscr{S}$.

Example 2.2. Let $M = \mathbb{R}^2$.

(a) If G_c is the one-parameter group of translations

$$(x, y) \mapsto (x + c\varepsilon, y + \varepsilon), \qquad \varepsilon \in \mathbb{R},$$

where c is some fixed constant, then the lines $x = cy + d$ are easily seen to be G_c-invariant, being precisely the orbits of G_c. It can also be readily seen that any invariant subset of $\mathbb{R}^2$ is just the union of some collection of such lines. For example, the strip $\{(x, y): k_1 < x - cy < k_2\}$ is G_c-invariant.

(b) As a second elementary example, let G^α be the one-parameter group of scale transformations

$$(x, y) \mapsto (\lambda x, \lambda^\alpha y), \qquad \lambda > 0,$$

where α is a constant. The origin $(0, 0)$ is a G^α-invariant subset, as are the positive and negative x- and y-axes, e.g. $\{(x, 0): x > 0\}$. Also, the axes themselves, being unions of invariant subsets, are also invariant. Thus the subvariety $\{(x, y): xy = 0\}$ consisting of both coordinate axes is invariant. Other invariant sets are of the form $y = k|x|^\alpha$ for $x > 0$ or $x < 0$, and unions of these orbits of G^α.

In most of our applications, the set $\mathscr{S}$ will be the set of solutions or *subvariety* determined by the common zeros of a collection of smooth functions $F = (F_1, \ldots, F_l)$,

$$\mathscr{S} = \mathscr{S}_F = \{x: F_\nu(x) = 0, \nu = 1, \ldots, l\}.$$

If $\mathscr{S}_1$ and $\mathscr{S}_2$ are G-invariant sets, so are $\mathscr{S}_1 \cup \mathscr{S}_2$ and $\mathscr{S}_1 \cap \mathscr{S}_2$.

Invariant Functions

Besides looking at the symmetries of the solution set of a system of algebraic equations, we can look at the symmetries of the function $F(x)$ which defines them.

Definition 2.3. Let G be a local group of transformations acting on a manifold M. A function $F: M \to N$, where N is another manifold, is called a *G-invariant function* if for all $x \in M$ and all $g \in G$ such that $g \cdot x$ is defined,

$$F(g \cdot x) = F(x).$$

A real-valued G-invariant function $\zeta: M \to \mathbb{R}$ is simply called an *invariant* of G. Note that $F: M \to \mathbb{R}^l$ is G-invariant if and only if each component F_ν of $F = (F_1, \ldots, F_l)$ is an invariant of G.

Example 2.4. (a) Let G_c be the group of translations in the plane presented in Example 2.2(a). Then the function

$$\zeta(x, y) = x - cy$$

is an invariant of G_c since

$$\zeta(x + c\varepsilon, y + \varepsilon) = \zeta(x, y)$$

for all ε. In fact, it is not difficult to see that *every* invariant of this translation group is of the form $\tilde{\zeta}(x, y) = f(x - cy)$, where f is a smooth function of the single variable $x - cy$.

(b) For the scaling group

$$G^1: (x, y) \mapsto (\lambda x, \lambda y), \qquad \lambda > 0,$$

the function

$$\zeta(x, y) = x/y$$

is an invariant defined on the upper and lower half planes $\{y \neq 0\}$. Other invariants include the angular coordinate $\theta = \tan^{-1}(y/x)$ which is smooth on $\mathbb{R}^2 \setminus \{(x, y): x \leqslant 0\}$, say, but not globally single-valued, and the function

$$\tilde{\zeta}(x, y) = xy/(x^2 + y^2),$$

which is smooth everywhere except at the origin. There is, in this case, no smooth, nonconstant, globally defined invariant of G^1. Similar remarks apply to the more general scaling groups G^α of Example 2.2(b) when $\alpha > 0$.

If $F: M \to \mathbb{R}^l$ is a G-invariant function, then clearly every level set of F is a G-invariant subset of M. However, it is *not* true that if the set of zeros of a smooth function, $\{x: F(x) = 0\}$, is an invariant subset of M, then the function itself is invariant. For instance, as we saw in the previous example $\{(x, y): xy = 0\}$ is an invariant subset of the scaling group G^1. However, $F(x, y) = xy$ is not an invariant function for this group since

$$F(\lambda x, \lambda y) = \lambda^2 xy \neq F(x, y)$$

for $\lambda \neq 1$. However, if *every* level set of F is invariant, then F is an invariant function.

Proposition 2.5. *If G acts on M, and $F: M \to \mathbb{R}^l$ is a smooth function, then F is a G-invariant function if and only if every level set $\{F(x) = c\}$, $c \in \mathbb{R}^l$, is a G-invariant subset of M.*

The proof of this result is left to the reader. Thus in Example 2.4(a), the lines $x = cy + d$ are just the level sets of the G_c-invariant function $\zeta(x, y) = x - cy$ and hence are automatically G_c-invariant subsets. Alternatively, the G_c-invariance of ζ follows from the fact that each level set is G_c-invariant, indeed an orbit of G_c.

Another way of looking at the preceding observations is that the "symmetry group" of the solution set $\mathscr{S}_F = \{F(x) = 0\}$ of some system of algebraic equations is, in general, larger than the "symmetry group" of the function F determining it. Here "symmetry group" means, somewhat imprecisely, the largest group of transformations leaving the subvariety or function invariant. For algebraic equations, such a group will not usually be finite dimensional, but the idea underlying these remarks should be clear. The importance of widening our concept of symmetry to those of the solution set, rather than the defining functions, will become evident when we treat symmetry groups of differential equations, and will lead to a much wider variety of symmetry groups.

Infinitesimal Invariance

The great power of Lie group theory lies in the crucial observation that one can replace the complicated, nonlinear conditions for the invariance of a subset or function under the group transformations themselves by an equivalent linear condition of infinitesimal invariance under the corresponding infinitesimal generators of the group action. This infinitesimal criterion will be readily verifiable in practice, and will thereby provide the key to the explicit determination of the symmetry groups of systems of differential equations. Its importance cannot be overemphasized. We begin with the simpler case of an invariant function. Here the infinitesimal criterion for invariance follows directly from the basic formula describing how functions change under the flow generated by a vector field.

Proposition 2.6. *Let G be a connected group of transformations acting on the manifold M. A smooth real-valued function $\zeta\colon M \to \mathbb{R}$ is an invariant function for G if and only if*

$$\mathbf{v}(\zeta) = 0 \quad \textit{for all} \quad x \in M, \tag{2.1}$$

and every infinitesimal generator $\mathbf{v}$ of G.

PROOF. According to (1.17), if $x \in M$,

$$\frac{d}{d\varepsilon}\zeta(\exp(\varepsilon\mathbf{v})x) = \mathbf{v}(\zeta)[\exp(\varepsilon\mathbf{v})x]$$

whenever $\exp(\varepsilon\mathbf{v})x$ is defined. Setting $\varepsilon = 0$ proves the necessity of (2.1). Conversely, if (2.1) holds everywhere, then

$$\frac{d}{d\varepsilon}\zeta(\exp(\varepsilon\mathbf{v})x) = 0$$

where defined, hence $\zeta(\exp(\varepsilon\mathbf{v})x)$ is a constant for the connected, local one-parameter subgroup $\exp(\varepsilon\mathbf{v})$ of $G_x = \{g \in G\colon g \cdot x \text{ is defined}\}$. But by (1.40), every element of G_x can be written as a finite product of exponentials of infinitesimal generators $\mathbf{v}_i$ of G, hence $\zeta(g \cdot x) = \zeta(x)$ for all $g \in G_x$. □

If $\mathbf{v}_1, \ldots, \mathbf{v}_r$ form a basis for $\mathfrak{g}$, the Lie algebra of infinitesimal generators of G, then Proposition 2.6 says that $\zeta(x)$ is an invariant if and only if $\mathbf{v}_k(\zeta) = 0$ for $k = 1, \ldots, r$. In local coordinates,

$$\mathbf{v}_k = \sum_{i=1}^{m} \xi_k^i(x) \frac{\partial}{\partial x^i},$$

so ζ must be a solution to the homogeneous system of linear, first order partial differential equatons

$$\mathbf{v}_k(\zeta) = \sum_{i=1}^{m} \xi_k^i(x) \frac{\partial \zeta}{\partial x^i} = 0, \qquad k = 1, \ldots, r. \tag{2.2}$$

Example 2.7. For the translation group G_c of Example 2.4(a) the infinitesimal generator is $\mathbf{v} = c\partial_x + \partial_y$. Then

$$\mathbf{v}(x - cy) = (c\partial_x + \partial_y)(x - cy) = c - c = 0,$$

so the infinitesimal criterion is satisfied. A similar computation verifies the infinitesimal criterion (2.1) for the invariants of the scale group G^α, whose infinitesimal generator is $x\partial_x + \alpha y \partial_y$.

For the case of the solution set of a system of algebraic equations $F(x) = 0$, the infinitesimal criterion of invariance requires additional conditions to be placed on the defining functions F, namely the maximal rank condition of Definition 1.7. (If F happens to be a G-invariant function, then by Proposition 2.5 this maximal rank condition can be dropped, but in general it is essential.)

Theorem 2.8. *Let G be a connected local Lie group of transformations acting on the m-dimensional manifold M. Let $F\colon M \to \mathbb{R}^l$, $l \leq m$, define a system of algebraic equations*

$$F_\nu(x) = 0, \qquad \nu = 1, \ldots, l,$$

and assume that the system is of maximal rank, *meaning that the Jacobian matrix $(\partial F_\nu / \partial x^k)$ is of rank l at every solution x of the system. Then G is a symmetry group of the system if and only if*

$$\mathbf{v}[F_\nu(x)] = 0, \qquad \nu = 1, \ldots, l, \quad \textit{whenever} \quad F(x) = 0, \tag{2.3}$$

for every infinitesimal generator $\mathbf{v}$ of G. (Note especially that (2.3) is required to hold only for solutions x of the system.)

Proof. The necessity of (2.3) follows from differentiating the identity

$$F(\exp(\varepsilon \mathbf{v})x) = 0,$$

in which x is a solution, and $\mathbf{v}$ is an infinitesimal generator of G, with respect to ε and setting $\varepsilon = 0$.

To prove sufficiency, let x_0 be a solution to the system. Using the maximal rank condition, we can choose local coordinates $y = (y^1, \ldots, y^m)$ such that $x_0 = 0$ and F has the simple form $F(y) = (y^1, \ldots, y^l)$, cf. Theorem 1.8. Let

$$\mathbf{v} = \xi^1(y)\frac{\partial}{\partial y^1} + \cdots + \xi^m(y)\frac{\partial}{\partial y^m}$$

be any infinitesimal generator of G, expressed in the new coordinates. Condition (2.3) means that

$$\mathbf{v}(y^\nu) = \xi^\nu(y) = 0, \qquad \nu = 1, \ldots, l, \tag{2.4}$$

whenever $y^1 = y^2 = \cdots = y^l = 0$. Now the flow $\phi(\varepsilon) = \exp(\varepsilon\mathbf{v})\cdot x_0$ of $\mathbf{v}$ through $x_0 = 0$ satisfies the system of ordinary differential equations

$$\frac{d\phi^i}{d\varepsilon} = \xi^i(\phi(\varepsilon)), \qquad \phi^i(0) = 0, \quad i = 1, \ldots, m.$$

By (2.4) and the uniqueness of solutions to this initial-value problem, we conclude that $\phi^\nu(\varepsilon) = 0$ for $\nu = 1, \ldots, l$, and ε sufficiently small. We have thus shown that if x_0 is a solution to $F(x) = 0$, $\mathbf{v}$ is an infinitesimal generator of G, and ε is sufficiently small, then $\exp(\varepsilon\mathbf{v})x_0$ is again a solution to the system. Since the solution set $\mathcal{S}_F = \{x : F(x) = 0\}$ is closed, the group property (1.13) and continuity of $\exp(\varepsilon\mathbf{v})$ allows us to draw the same conclusion for all $g = \exp(\varepsilon\mathbf{v})$ in the connected one-parameter subgroup of G_{x_0} generated by $\mathbf{v}$. Another application of (1.40), similar to that in the proof of Proposition 2.6, completes the proof of the theorem in general. □

Example 2.9. Let $G = \mathrm{SO}(2)$ be the rotation group in the plane, with infinitesimal generator $\mathbf{v} = -y\partial_x + x\partial_y$. The unit circle $S^1 = \{x^2 + y^2 = 1\}$ is an invariant subset for SO(2) as it is the solution set of the invariant function $\zeta(x, y) = x^2 + y^2 - 1$; indeed

$$\mathbf{v}(\zeta) = -2xy + 2xy = 0$$

everywhere, so (2.3) is verified on the unit circle itself. The maximal rank condition does hold for ζ since its gradient $\nabla\zeta = (2x, 2y)$ does not vanish on S^1, but, as remarked before the theorem, since ζ is already an invariant function, we don't really need to check this.

As a less trivial example, consider the function

$$F(x, y) = x^4 + x^2y^2 + y^2 - 1.$$

We have

$$\mathbf{v}(F) = -4x^3y - 2xy^3 + 2x^3y + 2xy = -2xy(x^2 + 1)^{-1}F(x, y),$$

hence $\mathbf{v}(F) = 0$ whenever $F = 0$. Moreover,

$$\nabla F = (4x^3 + 2xy^2, 2x^2y + 2y)$$

vanishes only when $x = y = 0$, which is not a solution to $F(x, y) = 0$, hence the maximal rank condition is verified. We conclude that the solution set $\{(x,y): x^4 + x^2y^2 + y^2 = 1\}$ is a rotationally-invariant subset of $\mathbb{R}^2$. Indeed, we can factor F as

$$x^4 + x^2y^2 + y^2 - 1 = (x^2 + 1)(x^2 + y^2 - 1),$$

hence the solution set is just the unit circle. Note that $F(x, y)$ is *not* an SO(2)-invariant function in this case; in fact, the other level sets of F are *not* rotationally invariant.

Finally, to appreciate the importance of the maximal rank condition, consider the function

$$H(x, y) = y^2 - 2y + 1.$$

The solution set $\{H(x, y) = 0\}$ is just the horizontal line $\{y = 1\}$ which is certainly not rotationally invariant. However,

$$\mathbf{v}(H) = 2xy - 2x = 2x(y - 1) = 0$$

whenever $H(x, y) = 0$, so the infinitesimal condition (2.3) does hold in this case. The problem is that $\nabla H = (0, 2y - 2)$ vanishes everywhere on the solution set, so that the maximal rank condition fails to hold.

The maximal rank condition needed to apply our infinitesimal symmetry criterion will play a key role in the development of the theory, both for algebraic and differential equations. We will subsequently need several elementary consequences of this condition, which we state here for ease of reference. The proofs are outlined in Exercise 2.5.

Proposition 2.10. *Let $F: M \to \mathbb{R}^l$ be of maximal rank on the subvariety $\mathscr{S}_F = \{x: F(x) = 0\}$. Then a real-valued function $f: M \to \mathbb{R}$ vanishes on $\mathscr{S}_F$ if and only if there exist smooth functions $Q_1(x), \ldots, Q_l(x)$ such that*

$$f(x) = Q_1(x)F_1(x) + \cdots + Q_l(x)F_l(x), \tag{2.5}$$

for all $x \in M$.

Again, the maximal rank condition is essential. For example, suppose $F(x, y) = y^2 - 2y + 1$. Then the function $f(x, y) = y - 1$ vanishes for all solutions of $F(x, y) = 0$, namely $\mathscr{S}_F = \{y = 1\}$, but there is no smooth function $Q(x, y)$ such that $f(x, y) = Q(x, y)F(x, y)$.

Proposition 2.10 says that we can replace the infinitesimal criterion (2.3) for invariance by the equivalent condition

$$\mathbf{v}(F_\nu) = \sum_{\mu=1}^{l} Q_{\nu\mu}(x)F_\mu(x), \qquad \nu = 1, \ldots, l, \qquad x \in M, \tag{2.6}$$

for functions $Q_{\nu\mu}: M \to \mathbb{R}$, $\mu, \nu = 1, \ldots, l$, to be determined. This was indeed how we proved invariance in the second case in Example 2.9, with $Q(x, y) =$

$-2xy/(x^2+1)$. *Both* (2.3) and (2.6) are useful conditions for checking invariance, and will both be employed in various examples.

The functions $Q_\nu(x)$ in (2.5) are not in general uniquely determined. For example, let.

$$F_1(x, y, z) = x, \qquad F_2(x, y, z) = y,$$

so the solution set $\mathcal{S} = \{F_1 = F_2 = 0\}$ is the z-axis in $\mathbb{R}^3$. The function

$$f(x, y, z) = xz + y^2$$

vanishes on $\mathcal{S}$, and indeed can be written both as

$$f = zF_1 + yF_2, \quad \text{or} \quad f = (z - y)F_1 + (x + y)F_2.$$

In general, if

$$f(x) = \sum_\nu Q_\nu(x)F_\nu(x) = \sum_\nu \tilde{Q}_\nu(x)F_\nu(x),$$

then the differences $R_\nu(x) = Q_\nu(x) - \tilde{Q}_\nu(x)$ satisfy the homogeneous system

$$\sum_{\nu=1}^{l} R_\nu(x)F_\nu(x) = 0 \tag{2.7}$$

for all $x \in M$. The folowing provides a useful necessary condition for such functions.

Proposition 2.11. *Let $F\colon M \to \mathbb{R}^l$ be of maximal rank on $\mathcal{S}_F = \{F(x) = 0\}$. Suppose $R_1(x), \ldots, R_l(x)$ are real-valued functions satisfying* (2.7) *for all $x \in M$. Then $R_\nu(x) = 0$ for all $x \in \mathcal{S}_F$. Equivalently, there exist functions $S_\nu^\mu(x)$, for $\nu, \mu = 1, \ldots, l$, such that*

$$R_\nu(x) = \sum_{\mu=1}^{l} S_\nu^\mu(x)F_\mu(x), \qquad x \in M. \tag{2.8}$$

Moreover, the S_ν^μ can be chosen to be skew-symmetric in their indices:

$$S_\nu^\mu(x) = -S_\mu^\nu(x),$$

in which case (2.8) *is necessary and sufficient for* (2.7) *to hold everywhere.*

Local Invariance

It is also useful to introduce the concept of a locally-invariant function or subset for a group of transformations. In this case we only require invariance for group transformations sufficiently near the identity.

Definition 2.12. Let G be a local group of transformations, acting on the manifold M. A subset $\mathcal{S} \subset M$ is called *locally G-invariant* if for every $x \in \mathcal{S}$ there is a neighbourhood $\tilde{G}_x \subset G_x$ of the identity in G such that $g \cdot x \in \mathcal{S}$ for all $g \in \tilde{G}_x$. A smooth function $F\colon U \to N$, where U is some open subset of M, is called *locally G-invariant* if for each $x \in U$ there is a neighbourhood $\tilde{G}_x \subset G_x$ of e in G such that $F(g \cdot x) = F(x)$ for all $g \in \tilde{G}_x$. F is called *globally G-invariant* (even though it is only defined on an open subset of M) if $F(g \cdot x) = F(x)$ for all $x \in U$, $g \in G$ such that $g \cdot x \in U$ also.

Example 2.13. Let G be the group of horizontal translations

$$(x, y) \mapsto (x + \varepsilon, y)$$

in $\mathbb{R}^2$. Then the line segment

$$\{(x, y)\colon y = 0, \ -1 < x < 1\}$$

is locally G-invariant, but not G-invariant.

Similarly, the function

$$\zeta(x, y) = \begin{cases} 0, & y \leqslant 0, \quad \text{or} \quad y > 0 \quad \text{and} \quad x > 0, \\ e^{-1/y}, & y > 0 \quad \text{and} \quad x < 0, \end{cases}$$

is smooth and locally G-invariant on $U = \mathbb{R}^2 \setminus \{(0, y)\colon y \geqslant 0\}$, since

$$\zeta(x + \varepsilon, y) = \zeta(x, y)$$

for $|\varepsilon| < |x|$; ζ is clearly not globally G-invariant.

Proposition 2.14. *Let $N \subset M$ be a submanifold of M. Then N is locally G-invariant if and only if for each $x \in N$, $\mathfrak{g}|_x \subset TN|_x$. In other words, N is locally G-invariant if and only if the infinitesimal generators $\mathbf{v}$ of G are everywhere tangent to N.*

The proof is left to the reader; see Exercise 2.1.

Invariants and Functional Dependence

Often we are interested in determining precisely "how many" invariants a given group of transformations has. To make this problem precise, we first note that if $\zeta^1(x), \ldots, \zeta^k(x)$ are invariants (either local or global) of a group of transformations, and $F(z^1, \ldots, z^k)$ is any smooth function, then $\zeta(x) = F(\zeta^1(x), \ldots, \zeta^k(x))$ will also be an invariant (of the same sort). Such an invariant adds no new knowledge to the given problem, and is termed "functionally dependent" on the preceding invariants $\zeta^1, \ldots, \zeta^k$. In practice, we need only classify functionally independent invariants of a group action, the other invariants all being obtained by relations of the above form.

Definition 2.15. Let $\zeta^1(x), \ldots, \zeta^k(x)$ be smooth, real-valued functions defined on a manifold M. Then

(a) $\zeta^1, \ldots, \zeta^k$ are called *functionally dependent* if for each $x \in M$ there is a neighbourhood U of x and a smooth real-valued function $F(z^1, \ldots, z^k)$, not identically zero on any open subset of $\mathbb{R}^k$, such that

$$F(\zeta^1(x), \ldots, \zeta^k(x)) = 0 \tag{2.9}$$

for all $x \in U$.

(b) $\zeta^1, \ldots, \zeta^k$ are called *functionally independent* if they are not functionally dependent when restricted to any open subset $U \subset M$; in other words, if $F(z^1, \ldots, z^k)$ is such that (2.9) hold for all x in some open $U \subset M$, then $F(z^1, \ldots, z^k) \equiv 0$ for all z in some open subset of $\mathbb{R}^k$ (which is contained in the image of U).

For example, the functions x/y and $xy/(x^2 + y^2)$ are functionally dependent on $\{(x, y): y \neq 0\}$ since

$$\frac{xy}{x^2 + y^2} = \frac{x/y}{1 + (x/y)^2} = f\left(\frac{x}{y}\right)$$

there. On the other hand, x/y and $x + y$ are functionally independent where defined, since if $F(x + y, x/y) \equiv 0$ for (x, y) in any open subset of $\mathbb{R}^2$, then, by the inverse function theorem, the image set contains an open subset of $\mathbb{R}^2$ on which $F = 0$.

Note that functional dependence and functional independence do not exhaust the range of possibilities except in the case of analytic functions, where the vanishing of (2.9) in some open set implies its vanishing everywhere. For example, the smooth functions

$$\eta(x, y) = x, \qquad \zeta(x, y) = \begin{cases} x, & y \leqslant 0, \\ x + e^{-1/y}, & y > 0, \end{cases}$$

are dependent on the lower half plane $\{y < 0\}$, independent on the upper half plane $\{y > 0\}$, but *neither* on the entire (x, y)-plane. Finally, we note that $\zeta^1, \ldots, \zeta^k$ may be locally functionally dependent, but there may be no nonzero function $F(z^1, \ldots, z^k)$ such that (2.9) holds for all x in M. For instance, the image $\{(\zeta^1(x), \ldots, \zeta^k(x)): x \in M\}$ may be dense in some open subset of $\mathbb{R}^k$, so (2.9) would only hold with $F \equiv 0$ there.

The classical necessary and sufficient condition that $\zeta^1(x), \ldots, \zeta^k(x)$ be functionally dependent is that their $k \times m$ Jacobian matrix $(\partial\zeta^i/\partial x^j)$ be of rank $\leqslant k - 1$ everywhere. (See the notes at the end of this chapter regarding the proof of this result.)

Theorem 2.16. *Let* $\zeta = (\zeta^1, \ldots, \zeta^k)$ *be a smooth function from* M *to* $\mathbb{R}^k$. *Then* $\zeta^1(x), \ldots, \zeta^k(x)$ *are functionally dependent if and only if* $d\zeta|_x$ *has rank strictly less than* k *for all* $x \in M$.

The basic theorem regarding number of independent invariants of a group of transformations is the following.

Theorem 2.17. *Let G act semi-regularly on the m-dimensional manifold M with s-dimensional orbits. If $x_0 \in M$, then there exist precisely $m - s$ functionally independent local invariants $\zeta^1(x), \ldots, \zeta^{m-s}(x)$ defined in a neighbourhood of x_0. Moreover, any other local invariant of the group action defined near x_0 is of the form*

$$\zeta(x) = F(\zeta^1(x), \ldots, \zeta^{m-s}(x)) \tag{2.10}$$

for some smooth function F. If the action of G is regular, then the invariants can be taken to be globally invariant in a neighbourhood of x_0.

PROOF. Using Frobenius' Theorem 1.43, we can find flat local coordinates $y = \psi(x)$ near x_0 for the system of vector fields $\mathfrak{g}$ spanned by the infinitesimal generators of G, such that the orbits of G are the slices $\{y^1 = c_1, \ldots, y^{m-s} = c_{m-s}\}$. Then the new coordinates $y^1 = \zeta^1(x), \ldots, y^{m-s} = \zeta^{m-s}(x)$ themselves are local invariants for G, being constant on each slice. Moreover, any other invariant of G must also be constant on these slices, and hence a function of $y^1, \ldots, y^{m-s}$ only. Finally, if G acts regularly, we can choose our flat coordinate chart such that each orbit intersects it in at most one slice. In this case, $y^1, \ldots, y^{m-s}$ actually form global invariants. □

In classical terminology, the invariants constructed in this theorem are called a *complete set of functionally independent invariants.* We have shown that once we have found such a complete set, any other invariant of G can be expressed as a function of these invariants. There is an analogous result for invariant subvarieties.

Proposition 2.18. *Let G act semi-regularly on M and let $\zeta^1(x), \ldots, \zeta^{m-s}(x)$ be a complete set of functionally independent invariants defined on an open subset $W \subset M$. If a subvariety $\mathscr{S}_F = \{x: F(x) = 0\}$ is G-invariant, then for each solution $x_0 \in \mathscr{S}_F$ there is a neighbourhood $\tilde{W} \subset W$ of x_0, and an "equivalent" G-invariant function $\tilde{F}(x) = \tilde{F}(\zeta^1(x), \ldots, \zeta^{m-s}(x))$ whose solution set coincides with that of F in $\tilde{W}$:*

$$\mathscr{S}_F \cap \tilde{W} = \mathscr{S}_{\tilde{F}} \cap \tilde{W} = \{x \in \tilde{W}: \tilde{F}(\zeta^1(x), \ldots, \zeta^{m-s}(x)) = 0\}.$$

PROOF. Note first that we can complete the set of invariants $y^1 = \zeta^1(x), \ldots, y^{m-s} = \zeta^{m-s}(x)$ to be flat local coordinates $y = (y^1, \ldots, y^m)$ for G near x_0. In fact, the remaining coordinates $\hat{y} = (y^{m-s+1}, \ldots, y^m)$ can be chosen from among the given coordinates $(x^1, \ldots, x^m)$ so $\hat{y} = \hat{x} = (x^{i_1}, \ldots, x^{i_s})$. For example, if $\partial(\zeta^1, \ldots, \zeta^{m-s})/\partial(x^1, \ldots, x^{m-s}) \neq 0$ at x_0, then we can set $\hat{x} = (x^{m-s+1}, \ldots, x^m)$. Thus the change of coordinates is of the form $y = \psi(x) = (\zeta(x), \hat{x})$, in which $\zeta(x)$ denotes the invariants and $\hat{x}$ are called *parametric variables.* We write $F(x) = F^*(y) = F^*(\zeta(x), \hat{x})$ in terms of these coordinates, so $F^* = F \circ \psi^{-1}$. Set

$$\tilde{F}(\zeta(x)) = F^*(\zeta(x), \hat{x}_0),$$

where $\hat{x}_0$ is the value of the parametric variables $\hat{x}$ at x_0. Since $\mathcal{S}_F$ is G-invariant, and the orbits of G in these coordinates are the common level sets (or slices) $\{\zeta(x) = c\}$ of the invariants, we find $F^*(\zeta(x), \hat{x}) = 0$ if and only if $F^*(\zeta(x), \hat{x}_0) = 0$ since both points lie in the same slice. □

Note that unless F itself is G-invariant, the corresponding $\tilde{F}$ will not be the same function; only their solution sets coincide. For instance, in the case presented in Example 2.9, $F(x, y) = x^4 + x^2y^2 + y^2 - 1$ has the same solution set as the SO(2)-invariant function $\tilde{F}(x, y) = x^2 + y^2 - 1$ even though they clearly disagree elsewhere.

Methods for Constructing Invariants

It remains to show how one finds the invariants of a given group action. First suppose G is a one-parameter group of transformations acting on M, with infinitesimal generator

$$\mathbf{v} = \xi^1(x)\frac{\partial}{\partial x^1} + \cdots + \xi^m(x)\frac{\partial}{\partial x^m}$$

expressed in some given local coordinates. A local invariant $\zeta(x)$ of G is a solution of the linear, homogeneous first order partial differential equation

$$\mathbf{v}(\zeta) = \xi^1(x)\frac{\partial \zeta}{\partial x^1} + \cdots + \xi^m(x)\frac{\partial \zeta}{\partial x^m} = 0. \tag{2.11}$$

Theorem 2.17 says that if $\mathbf{v}|_x \neq 0$, then there exist $m - 1$ functionally independent invariants, hence $m - 1$ functionally independent solutions of the partial differential equation (2.11) in a neighbourhood of x_0.

The classical theory of such equations shows that the general solution of (2.11) can be found by integrating the corresponding *characteristic system* of ordinary differential equations, which is

$$\frac{dx^1}{\xi^1(x)} = \frac{dx^2}{\xi^2(x)} = \cdots = \frac{dx^m}{\xi^m(x)}. \tag{2.12}$$

The general solution of (2.12) can be written in the form

$$\zeta^1(x^1, \ldots, x^m) = c_1, \ldots, \zeta^{m-1}(x^1, \ldots, x^m) = c_{m-1},$$

in which $c_1, \ldots, c_{m-1}$ are the constants of integration, and the $\zeta^i(x)$ are functions independent of the c_j's. It is then easily seen that the functions $\zeta^1, \ldots, \zeta^{m-1}$ are the required functionally independent solutions to (2.11). Any other invariant, i.e. any other solution of (2.11), will necessarily be a function of $\zeta^1, \ldots, \zeta^{m-1}$. We illustrate this technique with a couple of examples.

Example 2.19. (a) Consider the rotation group SO(2), which has infinitesimal generator $\mathbf{v} = -y\partial_x + x\partial_y$. The corresponding characteristic system is

$$\frac{dx}{-y} = \frac{dy}{x}.$$

This first order ordinary differential equation is easily solved; the solutions are $x^2 + y^2 = c$ for c an arbitrary constant. Thus, $\zeta(x, y) = x^2 + y^2$, or any function thereof, is the single independent invariant of the rotation group.

(b) Consider the vector field

$$\mathbf{v} = -y\frac{\partial}{\partial x} + x\frac{\partial}{\partial y} + (1 + z^2)\frac{\partial}{\partial z}$$

defined on $\mathbb{R}^3$. Note that $\mathbf{v}$ never vanishes, so we can find two independent invariants of the one-parameter group generated by $\mathbf{v}$, in a neighbourhood of any point in $\mathbb{R}^3$. The characteristic system in this case is

$$\frac{dx}{-y} = \frac{dy}{x} = \frac{dz}{1 + z^2}.$$

The first of these two equations was solved in part (a), so one of the invariants is the radius $r = \sqrt{x^2 + y^2}$. To find the other invariant, note that r is a constant for all solutions of the characteristic system, so we can replace x by $\sqrt{r^2 - y^2}$ before integrating. This leads to the equation

$$\frac{dy}{\sqrt{r^2 - y^2}} = \frac{dz}{1 + z^2},$$

which has solution

$$\arcsin\frac{y}{r} = \arctan z + k$$

for k an arbitrary constant. Thus

$$\arctan z - \arcsin\frac{y}{r} = \arctan z - \arctan\frac{y}{x}$$

is a second independent invariant for $\mathbf{v}$. A slightly simpler expression comes by taking the tangent of this invariant, which is $(xz - y)/(yz + x)$, so

$$r = \sqrt{x^2 + y^2} \quad \text{and} \quad \zeta = \frac{xz - y}{yz + x}$$

provide a complete set of functionally independent invariants (provided $yz \neq -x$). As usual, any function of r and ζ is also an invariant, so, for instance

$$\tilde{\zeta} = \frac{r}{\sqrt{1 + \zeta^2}} = \frac{x + yz}{\sqrt{1 + z^2}}$$

is also an invariant, which in conjunction with r forms yet another pair of independent invariants. (This iterative technique of using knowledge of some invariants to simplify the computation of the remaining invariants is extremely useful for solving characteristic systems in general.)

The computation of independent invariants for r-parameter groups of transformations when $r > 1$ can get very complicated. If $\mathbf{v}_k = \sum \xi_k^i(x)\partial/\partial x^i$, $k = 1, \ldots, r$, form a basis for the infinitesimal generators, then the invariants are found by solving the system of homogeneous, linear, first order partial differential equations

$$\mathbf{v}_k(\zeta) = \sum_{i=1}^{m} \xi_k^i(x)\frac{\partial \zeta}{\partial x^i} = 0, \qquad k = 1, \ldots, r.$$

In other words, each invariant ζ must be a *joint invariant* of all the vector fields $\mathbf{v}_1, \ldots, \mathbf{v}_k$. One way to proceed is to first compute the invariants of one of the vector fields, say $\mathbf{v}_1$. Since any joint invariant ζ must in particular be an invariant of $\mathbf{v}_1$, we can write ζ as some function of the computed invariants of $\mathbf{v}_1$. Thus, we should re-express the remaining vector fields $\mathbf{v}_2, \ldots, \mathbf{v}_r$ using the invariants of $\mathbf{v}_1$ as coordinates, and then find joint invariants of these "new" $r - 1$ vector fields. The procedure then works inductively, leading eventually to the joint invariants of all the vector fields expressed in terms of the joint invariants of the first $r - 1$ of them. The process will become clearer in an example.

Example 2.20. Consider the vector fields

$$\mathbf{v} = -y\frac{\partial}{\partial x} + x\frac{\partial}{\partial y}, \qquad \mathbf{w} = 2xz\frac{\partial}{\partial x} + 2yz\frac{\partial}{\partial y} + (z^2 + 1 - x^2 - y^2)\frac{\partial}{\partial z}$$

on $\mathbb{R}^3$. These were considered in Example 1.42, where it was shown that they generate a two-parameter abelian group of transformations on $\mathbb{R}^3$, which is regular on $M = \mathbb{R}^3 \setminus (\{x = y = 0\} \cup \{x^2 + y^2 = 1, z = 0\})$. An invariant $\zeta(x, y, z)$ is a solution to the pair of equations $\mathbf{v}(\zeta) = 0 = \mathbf{w}(\zeta)$. First note that independent invariants of $\mathbf{v}$ are just $r = \sqrt{x^2 + y^2}$ and z. We now re-express $\mathbf{w}$ in terms of r and z,

$$\mathbf{w} = 2rz\frac{\partial}{\partial r} + (z^2 + 1 - r^2)\frac{\partial}{\partial z}.$$

Since ζ must be a function of the invariants r, z of $\mathbf{v}$, it must be a solution to the differential equation

$$\mathbf{w}(\zeta) = 2rz\frac{\partial \zeta}{\partial r} + (z^2 + 1 - r^2)\frac{\partial \zeta}{\partial z} = 0.$$

The characteristic system here is

$$\frac{dr}{2rz} = \frac{dz}{z^2 + 1 - r^2}.$$

Solving this ordinary differential equation, we find that

$$\zeta = \frac{z^2 + r^2 + 1}{r} = \frac{x^2 + y^2 + z^2 + 1}{\sqrt{x^2 + y^2}}$$

is the single independent invariant of this group. (This result was given in Example 1.42, but without the details of the intervening calculation.)

2.2. Groups and Differential Equations

Suppose we are considering a system $\mathscr{S}$ of differential equations involving p independent variables $x = (x^1, \ldots, x^p)$, and q dependent variables $u = (u^1, \ldots, u^q)$. The solutions of the system will be of the form $u = f(x)$, or, in components, $u^\alpha = f^\alpha(x^1, \ldots, x^p)$, $\alpha = 1, \ldots, q$.† Let $X = \mathbb{R}^p$, with coordinates $x = (x^1, \ldots, x^p)$, be the space representing the independent variables, and let $U = \mathbb{R}^q$, with coordinates $u = (u^1, \ldots, u^q)$, represent the dependent variables. A symmetry group of the system $\mathscr{S}$ will be a local group of transformations, G, acting on some open subset $M \subset X \times U$ in such a way that "G transforms solutions of $\mathscr{S}$ to other solutions of $\mathscr{S}$". Note that we are allowing arbitrary nonlinear transformations of both the independent and dependent variables in our definition of symmetry.

To proceed rigorously, we must explain exactly how a given transformation g in the Lie group G transforms a function $u = f(x)$. We begin by identifying the function $u = f(x)$ with its graph

$$\Gamma_f = \{(x, f(x)) : x \in \Omega\} \subset X \times U,$$

where $\Omega \subset X$ is the domain of definition of f. Note that Γ_f is a certain p-dimensional submanifold of $X \times U$. If $\Gamma_f \subset M_g$, the domain of definition of the group transformation g, then the transform of Γ_f by g is just

$$g \cdot \Gamma_f = \{(\tilde{x}, \tilde{u}) = g \cdot (x, u) : (x, u) \in \Gamma_f\}.$$

The set $g \cdot \Gamma_f$ is not necessarily the graph of another single-valued function $\tilde{u} = \tilde{f}(\tilde{x})$. However, since G acts smoothly and the identity element of G leaves Γ_f unchanged, by suitably shrinking the domain of definition Ω of f we ensure that for elements g near the identity, the transform $g \cdot \Gamma_f = \Gamma_{\tilde{f}}$ is the graph of some single-valued smooth function $\tilde{u} = \tilde{f}(\tilde{x})$. We write $\tilde{f} = g \cdot f$ and call the function $\tilde{f}$ the *transform* of f by g.

Example 2.21. Let $p = 1$, $q = 1$, so $X = \mathbb{R}$, with a single independent variable x, and $U = \mathbb{R}$ with a single dependent variable u. (We are thus in the situation of a single ordinary differential equation involving a single function $u = f(x)$.)

† We will consistently employ Latin subscripts or superscripts to refer to the independent variables and Greek subscripts or superscripts to refer to the dependent variables.

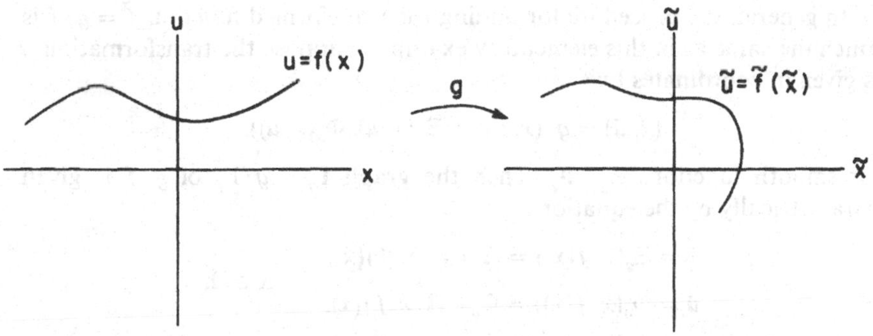

Figure 6. Action of a group transformation on a function.

Let $G = SO(2)$ be the rotation group acting on $X \times U \simeq \mathbb{R}^2$. The transformations in G are given by

$$(\tilde{x}, \tilde{u}) = \theta \cdot (x, u) = (x \cos \theta - u \sin \theta, x \sin \theta + u \cos \theta). \qquad (2.13)$$

Suppose $u = f(x)$ is a function, whose graph is a subset $\Gamma_f \subset X \times U$. The group $SO(2)$ acts on f by rotating its graph. Clearly, if the angle θ is sufficiently large, the rotated graph $\theta \cdot \Gamma_f$ will no longer be the graph of a single-valued function. However, if $f(x)$ is defined on a finite interval $a \leqslant x \leqslant b$, and $|\theta|$ is not too large, then $\theta \cdot \Gamma_f$ will be the graph of a well-defined function $\tilde{u} = \tilde{f}(\tilde{x})$, with $\Gamma_{\tilde{f}} = \theta \cdot \Gamma_f$.

As a specific example, consider the linear function

$$u = f(x) = ax + b.$$

The graph of f is a straight line, so its rotation through angle θ will be another straight line, which, as long as it is not vertical, will be the graph of another linear function $\theta \cdot f = \tilde{f}$, the transform of f by the rotation through angle θ. To find the precise formula for $\theta \cdot f$, note that by (2.13) a point $(x, u) = (x, ax + b)$ on the graph of f is rotated to the point

$$(\tilde{x}, \tilde{u}) = (x \cos \theta - (ax + b) \sin \theta, x \sin \theta + (ax + b) \cos \theta).$$

In order to find $\tilde{u} = \tilde{f}(\tilde{x})$, we must eliminate x from this pair of equations; this is possible provided $\cot \theta \neq a$ (in particular, for θ sufficiently near 0), so that the graph is not vertical. We find

$$x = \frac{\tilde{x} + b \sin \theta}{\cos \theta - a \sin \theta},$$

hence $\theta \cdot f = \tilde{f}$ is given by

$$\tilde{u} = \tilde{f}(\tilde{x}) = \frac{\sin \theta + a \cos \theta}{\cos \theta - a \sin \theta} \tilde{x} + \frac{b}{\cos \theta - a \sin \theta},$$

which, as we noticed earlier, is again a linear function.

In general, the procedure for finding the transformed function $\tilde{f} = g \cdot f$ is much the same as in this elementary example. Suppose the transformation g is given in coordinates by

$$(\tilde{x}, \tilde{u}) = g \cdot (x, u) = (\Xi_g(x, u), \Phi_g(x, u)),$$

for smooth functions Ξ_g, Φ_g. Then the graph $\Gamma_{\tilde{f}} = g \cdot \Gamma_f$ of $g \cdot f$ is given parametrically by the equations

$$\begin{aligned} \tilde{x} &= \Xi_g(x, f(x)) = \Xi_g \circ (\mathbb{1} \times f)(x), \\ \tilde{u} &= \Phi_g(x, f(x)) = \Phi_g \circ (\mathbb{1} \times f)(x), \end{aligned} \qquad x \in \Omega.$$

Here $\mathbb{1}$ denotes the identity function of X, so $\mathbb{1}(x) = x$, and $\times$ is the Cartesian product of functions. To find $\tilde{f} = g \cdot f$ explicitly, we must eliminate x from these two systems of equations. Since for $g = e$, $\Xi_e \circ (\mathbb{1} \times f) = \mathbb{1}$, we know that, provided g is sufficiently near the identity, the Jacobian matrix of $\Xi_g \circ (\mathbb{1} \times f)$ is nonsingular and hence by the inverse function theorem we can locally solve for x:

$$x = [\Xi_g \circ (\mathbb{1} \times f)]^{-1}(\tilde{x}).$$

Substitution into the second system yields the required equation for the transform $g \cdot f$:

$$g \cdot f = [\Phi_g \circ (\mathbb{1} \times f)] \circ [\Xi_g \circ (\mathbb{1} \times f)]^{-1}, \tag{2.14}$$

which holds whenever the second factor is invertible. This general formula is slightly complicated, but this was to be expected from our experience with just linear functions and the rotation group.

Example 2.22. Consider the special case in which the group G transforms just the independent variables x. Thus the transformations in G take the special form

$$(\tilde{x}, \tilde{u}) = g \cdot (x, u) = (\Xi_g(x), u),$$

in which Ξ_g is, in fact, a diffeomorphism of X with $\Xi_g^{-1} = \Xi_{g^{-1}}$ where defined. If $\Gamma_f = \{(x, f(x))\}$ is the graph of a smooth function, then its transform $g \cdot \Gamma_f = \{g \cdot (x, f(x))\}$ is *always* the graph of a smooth function. Indeed

$$(\tilde{x}, \tilde{u}) = g \cdot (x, f(x)) = (\Xi_g(x), f(x)).$$

Thus we can easily eliminate x by inverting Ξ_g, with

$$\tilde{u} = \tilde{f}(\tilde{x}) = f(\Xi_g^{-1}(\tilde{x})) = f(\Xi_{g^{-1}}(\tilde{x})).$$

For example, if G is a group of translations

$$(x, u) \mapsto (x + \varepsilon a, u), \qquad \varepsilon \in \mathbb{R},$$

for $a \in X$ fixed, then the transform of the function $u = f(x)$ is the translate

$$\tilde{u} = \tilde{f}(\tilde{x}) = f(\tilde{x} - \varepsilon a)$$

of f.

The same sort of result holds in the more general case of a *projectable* or *fiber-preserving* group of transformations, in which the action on the independent variables does not depend on the dependent variables:

$$g\cdot(x, u) = (\Xi_g(x), \Phi_g(x, u)).$$

For example, the one-parameter group

$$g_\varepsilon\colon (x, t, u) \mapsto (x + 2\varepsilon t, t, e^{-\varepsilon x - \varepsilon^2 t}u), \qquad \varepsilon \in \mathbb{R},$$

arises as a symmetry group of the heat equation. (See Example 2.41.) If $u = f(x, t)$ is any functon, then its transform by g_ε is

$$\tilde{u} = e^{-\varepsilon x - \varepsilon^2 t}\cdot u = e^{-\varepsilon x - \varepsilon^2 t}\cdot f(x, t),$$

which must now be written in terms of $(\tilde{x}, \tilde{t}) = g_\varepsilon\cdot(x, t) = (x + 2\varepsilon t, t)$. Therefore

$$\begin{aligned}\tilde{u} &= e^{-\varepsilon(\tilde{x} - 2\varepsilon\tilde{t}) - \varepsilon^2\tilde{t}}\cdot f(\tilde{x} - 2\varepsilon\tilde{t}, \tilde{t})\\ &= e^{-\varepsilon\tilde{x} + \varepsilon^2\tilde{t}}\cdot f(\tilde{x} - 2\varepsilon\tilde{t}, \tilde{t})\end{aligned}$$

is the transformed function in this particular case. (Note the disparity with the expressions for the group transformations themselves. The reader is advised to do several examples to gain familiarity with how this works in practice.)

We can now give a rigorous definition of the concept of a symmetry group of a system of differential equations.

Definition 2.23. Let $\mathscr{S}$ be a system of differential equations. A *symmetry group* of the system $\mathscr{S}$ is a local group of transformations G acting on an open subset M of the space of independent and dependent variables for the system with the property that whenever $u = f(x)$ is a solution of $\mathscr{S}$, and whenever $g\cdot f$ is defined for $g \in G$, then $u = g\cdot f(x)$ is also a solution of the system. (By *solution* we mean any smooth solution $u = f(x)$ defined on any subdomain $\Omega \subset X$.)

For example, in the case of the ordinary differential equation $u_{xx} = 0$, the rotation group SO(2) considered in Example 2.21 is obviously a symmetry group, since the solutions are all linear functions and SO(2) takes any linear function to another linear function. Another easy example is given by the heat equation $u_t = u_{xx}$. Here the group of translations

$$(x, t, u) \mapsto (x + \varepsilon a, t + \varepsilon b, u), \qquad \varepsilon \in \mathbb{R},$$

is a symmetry group since $u = f(x - \varepsilon a, t - \varepsilon b)$ is a solution to the heat equation whenever $u = f(x, t)$ is. The reader might enjoy checking that the group presented at the end of Example 2.22 is also a symmetry group of the heat equation, meaning that

$$u = e^{-\varepsilon x + \varepsilon^2 t} f(x - 2\varepsilon t, t)$$

is a solution of the heat equation whenever $u = f(x, t)$ is.

One of the obvious advantages of knowing a symmetry group of a system of differential equations is that we can construct new solutions of the system from known ones. Namely, if we know $u = f(x)$ is a solution, then according to the definition, $\tilde{u} = g \cdot f(\tilde{x})$ is also a solution for any group element g, so we have the possibility of constructing whole families of solutions just by transforming a known solution by all possible group elements. For example, in the case of the above symmetry group of the heat equaton, starting with the trivial constant solution $u = c$, we deduce the existence of a two-parameter family of exponential solutions

$$u(x, t) = ce^{-\varepsilon x + \varepsilon^2 t}.$$

We could further subject these to the translation group, but in this case no new solutions are obtained. The reader might try seeing what happens to other known solutions of the heat equation, e.g. the fundamental solution, under the above group actions.

The primary goal of this chapter is to establish a workable criterion that can be readily checked to determine whether a given group of transformations is or is not a symmetry group of a given system of differential equations. This criterion will be infinitesimal, in direct analogy with the criterion of Theorem 2.8 for systems of algebraic equations. In fact, once we have developed the appropriate geometrical setting for studying systems of differential equations, we will be able to directly invoke Theorem 2.8 to establish an infinitesimal criterion of invariance. Once we have this criterion in hand, not only will we be able to simply check whether a given group is a symmetry group of our system of differential equations, we will actually be able to compute the most general symmetry group of the system through a series of fairly routine calculations.

2.3. Prolongation

Before implementing our program of finding symmetries of differential equations by employing analogues of the infinitesimal methods for algebraic equations discussed in Section 2.1, we need to replace the somewhat nebulous notion of a "system of differential equations" by a concrete geometric object determined by the vanishing of certain functions. To do this we need to "prolong" the basic space $X \times U$ representing the independent and dependent variables under consideration to a space which also represents the various partial derivatives occurring in the system. This construction is a greatly simplified version of the theory of jet bundles occurring in the differential-geometric theory of partial differential equations. So as to avoid the introduction of too much extraneous machinery, we work exclusively in Euclidean space here. (See Section 3.5 for a generalization.)

Given a smooth real-valued function $f(x) = f(x^1, \ldots, x^p)$ of p independent variables, there are

$$p_k \equiv \binom{p+k-1}{k}$$

different k-th order partial derivatives of f. We employ the multi-index notation

$$\partial_J f(x) = \frac{\partial^k f(x)}{\partial x^{j_1} \partial x^{j_2} \cdots \partial x^{j_k}}$$

for these derivatives. In this notation, $J = (j_1, \ldots, j_k)$ is an *unordered* k-tuple of integers, with entries $1 \leqslant j_\kappa \leqslant p$ indicating which derivatives are being taken. The *order* of such a multi-index, which we denote by $\#J \equiv k$, indicates how may derivatives are being taken. More generally, if $f: X \to U$ is a smooth function from $X \simeq \mathbb{R}^p$ to $U \simeq \mathbb{R}^q$, so $u = f(x) = (f^1(x), \ldots, f^q(x))$, there are $q \cdot p_k$ numbers $u_J^\alpha = \partial_J f^\alpha(x)$ needed to represent all the different k-th order derivatives of the components of f at a point x. We let $U_k \equiv \mathbb{R}^{q \cdot p_k}$ be the Euclidean space of this dimension, endowed with coordinates u_J^α corresponding to $\alpha = 1, \ldots, q$, and all multi-indices $J = (j_1, \ldots, j_k)$ of order k, designed so as to represent the above derivatives. Furthermore, set $U^{(n)} = U \times U_1 \times \cdots \times U_n$ to be the Cartesian product space, whose coordinates represent all the derivatives of functions $u = f(x)$ of all orders from 0 to n. Note that $U^{(n)}$ is a Euclidean space of dimension

$$q + qp_1 + \cdots + qp_n = q\binom{p+n}{n} \equiv qp^{(n)}.$$

A typical point in $U^{(n)}$ will be denoted by $u^{(n)}$, so $u^{(n)}$ has $q \cdot p^{(n)}$ different components u_J^α where $\alpha = 1, \ldots, q$, and J runs over all unordered multi-indices $J = (j_1, \ldots, j_k)$ with $1 \leqslant j_k \leqslant p$ and $0 \leqslant k \leqslant n$. (By convention, for $k = 0$ there is just one such multi-index, denoted by 0, and u_0^α just refers to the component u^α of u itself.)

Example 2.24. Consider the case $p = 2$, $q = 1$. Then $X \simeq \mathbb{R}^2$ has coordinates $(x^1, x^2) = (x, y)$, and $U \simeq \mathbb{R}$ has the single coordinate u. The space U_1 is isomorphic to $\mathbb{R}^2$ with coordinates (u_x, u_y) since these represent all the first order partial derivatives of u with respect to x and y. Similarly, $U_2 \simeq \mathbb{R}^3$ has coordinates (u_{xx}, u_{xy}, u_{yy}) representing the second order partial derivatives of u, and, in general, $U_k \simeq \mathbb{R}^{k+1}$, since there are $k + 1$ distinct k-th order partial derivatives of u, namely $\partial^k u/\partial x^i \partial y^{k-i}$, $i = 0, \ldots, k$. Finally, the space $U^{(2)} = U \times U_1 \times U_2 \simeq \mathbb{R}^6$, with coordinates $u^{(2)} = (u; u_x, u_y; u_{xx}, u_{xy}, u_{yy})$, represents all derivatives of u with respect to x and y of order at most 2.

Given a smooth function $u = f(x)$, so $f: X \to U$, there is an induced function $u^{(n)} = \mathrm{pr}^{(n)} f(x)$, called the n-th *prolongation* of f, which is defined by the equations

$$u_J^\alpha = \partial_J f^\alpha(x).$$

Thus $\mathrm{pr}^{(n)} f$ is a function from X to the space $U^{(n)}$, and for each x in X, $\mathrm{pr}^{(n)} f(x)$ is a vector whose $q \cdot p^{(n)}$ entries represent the values of f and all its derivatives up to order n at the point x. For example, in the case $p = 2$, $q = 1$ discussed above, given $u = f(x, y)$, the second prolongation $u^{(2)} = \mathrm{pr}^{(2)} f(x, y)$ is given by

$$(u; u_x, u_y; u_{xx}, u_{xy}, u_{yy}) = \left(f; \frac{\partial f}{\partial x}, \frac{\partial f}{\partial y}; \frac{\partial^2 f}{\partial x^2}, \frac{\partial^2 f}{\partial x\, \partial y}, \frac{\partial^2 f}{\partial y^2}\right), \tag{2.15}$$

all evaluated at (x, y). (Another way of looking at the n-th prolongation $\mathrm{pr}^{(n)} f(x)$ is that it represents the Taylor polynomial of degree n for f at the point x, since the derivatives of order $\leqslant n$ determine the Taylor polynomial and vice versa.)

The total space $X \times U^{(n)}$, whose coordinates represent the independent variables, the dependent variables *and* the derivatives of the dependent variables up to order n is called the n-th order *jet space* of the underlying space $X \times U$. (The n-th prolongation $\mathrm{pr}^{(n)} f(x)$ is also known as the *n-jet* of f, but we will stick to the more suggestive term "prolongation".) Often we are not interested in differential equations defined over all of $X \times U$, but only on some open subset $M \subset X \times U$. In this case, we define the n-jet space

$$M^{(n)} \equiv M \times U_1 \times \cdots \times U_n$$

of M. If $u = f(x)$ is a function whose graph lies in M, the n-th prolongation $\mathrm{pr}^{(n)} f(x)$ is a function whose graph lies in the n-jet space $M^{(n)}$.

Systems of Differential Equations

A system $\mathscr{S}$ of n-th order differential equations in p independent and q dependent variables is given as a system of equations

$$\Delta_\nu(x, u^{(n)}) = 0, \qquad \nu = 1, \ldots, l,$$

involving $x = (x^1, \ldots, x^p)$, $u = (u^1, \ldots, u^q)$ and the derivatives of u with respect to x up to order n. The functions $\Delta(x, u^{(n)}) = (\Delta_1(x, u^{(n)}), \ldots, \Delta_l(x, u^{(n)}))$ will be assumed to be smooth in their arguments, so Δ can be viewed as a smooth map from the jet space $X \times U^{(n)}$ to some l-dimensional Euclidean space,

$$\Delta: X \times U^{(n)} \to \mathbb{R}^l.$$

The differential equations themselves tell where the given map Δ vanishes on $X \times U^{(n)}$, and thus determine a subvariety

$$\mathscr{S}_\Delta = \{(x, u^{(n)}): \Delta(x, u^{(n)}) = 0\} \subset X \times U^{(n)}$$

of the total jet space. We can identify the system of differential equations with its corresponding subvariety, thereby realizing the "abstract" relations among the various derivatives of u determined by the system as some concrete, geometrical subset $\mathscr{S}_\Delta$ of the jet space $X \times U^{(n)}$. We will use the same symbol "Δ" as shorthand for both the system of differential equations $\Delta(x, u^{(n)}) = 0$ and the map $\Delta: X \times U^{(n)} \to \mathbb{R}^l$ which determines it. This should not be the cause of any confusion.

From this point of view, a smooth *solution* of the given system of differential equations is a smooth function $u = f(x)$ such that

$$\Delta_\nu(x, \mathrm{pr}^{(n)} f(x)) = 0, \qquad \nu = 1, \ldots, l,$$

whenever x lies in the domain of f. This is just a restatement of the fact that the derivatives $\partial_J f^\alpha(x)$ of f must satisfy the algebraic constraints imposed by the system of differential equations. This condition is equivalent to the statement that the graph of the prolongation $\mathrm{pr}^{(n)} f(x)$ must lie entirely within the subvariety $\mathscr{S}_\Delta$ determined by the system:

$$\Gamma_f^{(n)} \equiv \{(x, \mathrm{pr}^{(n)} f(x))\} \subset \mathscr{S}_\Delta = \{\Delta(x, u^{(n)}) = 0\}.$$

We can thus take an n-th order system of differential equations to *be* a subvariety $\mathscr{S}_\Delta$ in the n-jet space $X \times U^{(n)}$ and a solution to *be* a function $u = f(x)$ such that the graph of the n-th prolongation $\mathrm{pr}^{(n)} f$ is contained in the subvariety $\mathscr{S}_\Delta$. So far we have not done anything but reformulate the basic problem of finding solutions of systems of differential equations in a more geometrical form, ideally suited to our investigation into symmetry groups thereof. It is perhaps worthwhile pausing at this point to consider a simple example.

Example 2.25. Consider the case of Laplace's equation in the plane

$$u_{xx} + u_{yy} = 0. \tag{2.16}$$

Here $p = 2$ since there are two independent variables x and y, and $q = 1$ since there is one dependent variable u. Also $n = 2$ since the equation is second order, so we are in the situation described in Example 2.24. In terms of the coordinates $(x, y; u; u_x, u_y; u_{xx}, u_{xy}, u_{yy})$ of $X \times U^{(2)}$, (2.16) defines a linear subvariety (a "hyperplane") there, and this is the set $\mathscr{S}_\Delta$ for Laplace's equation. A solution $u = f(x, y)$ must satisfy

$$\frac{\partial^2 f}{\partial x^2} + \frac{\partial^2 f}{\partial y^2} = 0$$

for all (x, y). This is clearly the same as requiring that the graph of the second prolongation $\mathrm{pr}^{(2)} f$ lie in $\mathscr{S}_\Delta$. For instance, if

$$f(x, y) = x^3 - 3xy^2,$$

then (using (2.15))

$$\mathrm{pr}^{(2)} f(x, y) = (x^3 - 3xy^2; 3x^2 - 3y^2, -6xy; 6x, -6y, -6x),$$

which lies in $\mathcal{S}_\Delta$ since the fourth and sixth entries add up to 0: $6x + (-6x) = 0$.

Prolongation of Group Actions

Now suppose G is a local group of transformations acting on an open subset $M \subset X \times U$ of the space of independent and dependent variables. There is an induced local action of G on the n-jet space $M^{(n)}$, called the n-th *prolongation* of G (or, more correctly, the n-th prolongation of the action of G on M) and denoted $\mathrm{pr}^{(n)} G$. This prolongation is defined so that it transforms the derivatives of functions $u = f(x)$ into the corresponding derivatives of the transformed function $\tilde{u} = \tilde{f}(\tilde{x})$. More rigorously, suppose $(x_0, u_0^{(n)})$ is a given point in $M^{(n)}$. Choose any smooth function $u = f(x)$ defined in a neighbourhood of x_0, whose graph lies in M, and has the given derivatives at x_0:

$$u_0^{(n)} = \mathrm{pr}^{(n)} f(x_0), \quad \text{i.e.} \quad u_{J0}^\alpha = \partial_J f^\alpha(x_0).$$

For example, f might be the n-th order Taylor polynomial at x_0 corresponding to the given values $u_0^{(n)}$:

$$f^\alpha(x) = \sum_J \frac{u_{J0}^\alpha}{\tilde{J}!}(x - x_0)^J, \qquad \alpha = 1, \ldots, q. \tag{2.17}$$

(Here the sum is over all multi-indices $J = (j_1, \ldots, j_k)$ with $0 \leqslant k \leqslant n$; also

$$(x - x_0)^J = (x^{j_1} - x_0^{j_1})(x^{j_2} - x_0^{j_2}) \cdots (x^{j_k} - x_0^{j_k}).$$

Further, given J, set $\tilde{J} = (\tilde{j}_1, \ldots, \tilde{j}_p)$, where $\tilde{j}_i$ equals the number of j_κ's which equal i. For instance, if $J = (1, 1, 1, 2, 4, 4)$, $p = 4$, $k = 6$, then $\tilde{J} = (3, 1, 0, 2)$. With this notation $\tilde{J}! \equiv \tilde{j}_1! \tilde{j}_2! \cdots \tilde{j}_p!$.)

If g is an element of G sufficiently near the identity, the transformed function $g \cdot f$ as given by (2.14) is defined in a neighbourhood of the corresponding point $(\tilde{x}_0, \tilde{u}_0) = g \cdot (x_0, u_0)$, with $u_0 = f(x_0)$ being the zeroth order components of $u_0^{(n)}$. We then determine the action of the prolonged group transformation $\mathrm{pr}^{(n)} g$ on the point $(x_0, u_0^{(n)})$ by evaluating the derivatives of the transformed function $g \cdot f$ at $\tilde{x}_0$; explicitly

$$\mathrm{pr}^{(n)} g \cdot (x_0, u_0^{(n)}) = (\tilde{x}_0, \tilde{u}_0^{(n)}),$$

where

$$\tilde{u}_0^{(n)} \equiv \mathrm{pr}^{(n)}(g \cdot f)(\tilde{x}_0). \tag{2.18}$$

It is a relatively straight-forward exercise to check, using the chain rule, that this definition of $\mathrm{pr}^{(n)} g \cdot (x_0, u_0^{(n)})$ depends only on the derivatives of f at x_0 up to order n, i.e. on $(x_0, u_0^{(n)})$ itself, and hence is independent of the choice of representative function f for $(x_0, u_0^{(n)})$. Thus the prolonged group action is well defined. Again, put more succinctly, to define the action of $\mathrm{pr}^{(n)} g$ on a point in $M^{(n)}$, choose a function whose derivatives agree with the given values; transform the function according to (2.14), and re-evaluate the derivatives.

Example 2.26. Let $p = q = 1$, so $X \times U \simeq \mathbb{R}^2$, and consider the action of the rotation group SO(2) as discussed in Example 2.21. We calculate here the first prolongation $\mathrm{pr}^{(1)}\,\mathrm{SO}(2)$. Note first that $X \times U^{(1)} \simeq \mathbb{R}^3$, with coordinates (x, u, u_x). Given a function $u = f(x)$, the first prolongation is

$$\mathrm{pr}^{(1)} f(x) = (f(x), f'(x)).$$

Now given a point $(x^0, u^0, u_x^0) \in X \times U^{(1)}$, and a rotation in SO(2) characterized by the angle θ, we wish to find the corresponding transformed point

$$\mathrm{pr}^{(1)}\, \theta \cdot (x^0, u^0, u_x^0) = (\tilde{x}^0, \tilde{u}^0, \tilde{u}_x^0)$$

(provided it exists). Choose the linear Taylor polynomial

$$f(x) = u^0 + u_x^0(x - x^0) = u_x^0 \cdot x + (u^0 - u_x^0 x^0)$$

as a representative function, noting that

$$f(x^0) = u^0, \qquad f'(x^0) = u_x^0,$$

as required. According to the calculations of Example 2.21, the transform of f by a rotation through angle θ is the linear function

$$\tilde{f}(\tilde{x}) = \theta \cdot f(\tilde{x}) = \frac{\sin\theta + u_x^0 \cos\theta}{\cos\theta - u_x^0 \sin\theta}\,\tilde{x} + \frac{u^0 - u_x^0 x^0}{\cos\theta - u_x^0 \sin\theta},$$

which is well defined provided $u_x^0 \neq \cot\theta$. Then

$$\tilde{x}^0 = x^0 \cos\theta - u^0 \sin\theta,$$

hence

$$\tilde{u}^0 = \tilde{f}(\tilde{x}^0) = x^0 \sin\theta + u^0 \cos\theta,$$

as we already knew. As for the first order derivative, we find

$$\tilde{u}_x^0 = \tilde{f}'(\tilde{x}^0) = \frac{\sin\theta + u_x^0 \cos\theta}{\cos\theta - u_x^0 \sin\theta}.$$

Therefore, dropping the 0-superscripts, we find that the prolonged action $\mathrm{pr}^{(1)}\,\mathrm{SO}(2)$ on $X \times U^{(1)}$ is given by

$$\mathrm{pr}^{(1)}\, \theta \cdot (x, u, u_x) = \left(x\cos\theta - u\sin\theta,\; x\sin\theta + u\cos\theta,\; \frac{\sin\theta + u_x\cos\theta}{\cos\theta - u_x\sin\theta} \right), \tag{2.19}$$

which is defined for $|\theta| < |\operatorname{arccot} u_x|$. Note that even though SO(2) is a linear, globally defined group of transformations, its first prolongation is both nonlinear and only locally defined. From this relatively simple example the reader can appreciate the complexity of the operation of prolonging a group of transformations!

The reader will note that in the above example, the first prolongation $\mathrm{pr}^{(1)} G$ acts on the original variables (x, u) exactly the same way that G itself does; only the action on the derivative u_x provides new information. This remark holds in general. Namely, given the n-th prolongation $\mathrm{pr}^{(n)} G$ acting on the variables $(x, u^{(n)})$, if we restrict our attention to just the derivatives up to order $k \leqslant n$, so just look at the variables $(x, u^{(k)})$, then the action of $\mathrm{pr}^{(n)} G$ there agrees with the earlier prolongation $\mathrm{pr}^{(k)} G$. In particular, for $k = 0$, $\mathrm{pr}^{(0)} G$ agrees with G itself, acting on $M^{(0)} = M$. This result can be stated more precisely by defining a natural projection $\pi_k^n \colon M^{(n)} \to M^{(k)}$, where $\pi_k^n(x, u^{(n)}) = (x, u^{(k)})$, $u^{(k)}$ just consisting of the components u_J^α, $\# J \leqslant k$ of $u^{(n)}$ itself. For example, if $p = 2$, $q = 1$,

$$\pi_1^2(x, y; u; u_x, u_y; u_{xx}, u_{xy}, u_{yy}) = (x, y; u; u_x, u_y),$$

while

$$\pi_0^2(x, y; u; u_x, u_y; u_{xx}, u_{xy}, u_{yy}) = (x, y; u).$$

We then have

$$\pi_k^n \circ \mathrm{pr}^{(n)} g = \mathrm{pr}^{(k)} g, \qquad n \geqslant k, \tag{2.20}$$

for any group element $g \in G$. Another way of looking at this remark is that if we already know the k-th order prolonged group action $\mathrm{pr}^{(k)} G$, then to compute the n-th order prolongation $\mathrm{pr}^{(n)} G$ we need only find how the derivatives u_J^α of orders $k < \# J \leqslant n$ transform, since the action on k-th and lower order derivatives is already determined.

Invariance of Differential Equations

Suppose we are given an n-th order system of differential equations, or, equivalently, a subvariety $\mathcal{S}_\Delta$ of the jet space $M^{(n)} \subset X \times U^{(n)}$. A symmetry group of this system was defined to be a local group of transformations G acting on $M \subset X \times U$ which transforms solutions of the system to other solutions. We will establish the connection between this symmetry condition and the geometric condition that the corresponding subvariety $\mathcal{S}_\Delta$ be invariant under the prolonged group action $\mathrm{pr}^{(n)} G$. This observation will effectively reduce the problem of determining symmetry groups of differential equations to the more tractable problem of determining when some subvariety (in this case $\mathcal{S}_\Delta$) is invariant under some local group of transformations (in this case the prolonged group $\mathrm{pr}^{(n)} G$). In this way all the tools developed in Section 2.1 for symmetries of algebraic equations are at our disposal for the study of symmetries of differential equations. This alone should demonstrate the effectiveness of our geometric reformulation of the notion of differential equation which has been developed in this section.

Theorem 2.27. *Let M be an open subset of $X \times U$ and suppose $\Delta(x, u^{(n)}) = 0$ is an n-th order system of differential equations defined over M, with corresponding subvariety $\mathcal{S}_\Delta \subset M^{(n)}$. Suppose G is a local group of transformations acting on M whose prolongation leaves $\mathcal{S}_\Delta$ invariant, meaning that when-*

ever $(x, u^{(n)}) \in \mathscr{S}_\Delta$, *we have* $\mathrm{pr}^{(n)} g \cdot (x, u^{(n)}) \in \mathscr{S}_\Delta$ *for all* $g \in G$ *such that this is defined. Then G is a symmetry group of the system of differential equations in the sense of Definition* 2.23.

PROOF. The proof just consists of untangling the various definitions. Suppose $u = f(x)$ is a local solution to $\Delta(x, u^{(n)}) = 0$. This means that the graph

$$\Gamma_f^{(n)} = \{(x, \mathrm{pr}^{(n)} f(x))\}$$

of the prolongation $\mathrm{pr}^{(n)} f$ lies entirely within $\mathscr{S}_\Delta$. If $g \in G$ is such that the transformed function $g \cdot f$ is well defined, the graph of its prolongation, namely $\Gamma_{g \cdot f}^{(n)}$, is the same as the transform of the graph of $\mathrm{pr}^{(n)} f$ by the prolonged group transformation $\mathrm{pr}^{(n)} g$:

$$\Gamma_{g \cdot f}^{(n)} = \mathrm{pr}^{(n)} g(\Gamma_f^{(n)}).$$

(This is just a restatement of the basic formula (2.18) defining the prolonged group action.) Now, since $\mathscr{S}_\Delta$ is invariant under $\mathrm{pr}^{(n)} g$, the graph of $\mathrm{pr}^{(n)}(g \cdot f)$ again lies entirely in $\mathscr{S}_\Delta$. But this is just another way of saying that the transformed function $g \cdot f$ is a solution to the system Δ. □

Later (Theorem 2.71) we will present a converse to this result, subject to some additional hypotheses on the system itself.

Prolongation of Vector Fields

As with the group transformations themselves, we can also define the prolongation of the corresponding infinitesimal generators. Indeed, these will just be the infinitesimal generators of the prolonged group action.

Definition 2.28. Let $M \subset X \times U$ be open and suppose $\mathbf{v}$ is a vector field on M, with corresponding (local) one-parameter group $\exp(\varepsilon \mathbf{v})$. The n-th *prolongation* of $\mathbf{v}$, denoted $\mathrm{pr}^{(n)} \mathbf{v}$, will be a vector field on the n-jet space $M^{(n)}$, and is defined to be the infinitesimal generator of the corresponding prolonged one-parameter group $\mathrm{pr}^{(n)}[\exp(\varepsilon \mathbf{v})]$. In other words,

$$\mathrm{pr}^{(n)} \mathbf{v}|_{(x, u^{(n)})} = \frac{d}{d\varepsilon}\bigg|_{\varepsilon=0} \mathrm{pr}^{(n)} [\exp(\varepsilon \mathbf{v})](x, u^{(n)}) \tag{2.21}$$

for any $(x, u^{(n)}) \in M^{(n)}$.

Note that since the coordinates $(x, u^{(n)})$ on $M^{(n)}$ consists of the independent variables $(x^1, \ldots, x^p)$ and all derivatives u_J^α of the dependent variables up to order n, a vector field on $M^{(n)}$ will in general take the form

$$\mathbf{v}^* = \sum_{i=1}^{p} \xi^i \frac{\partial}{\partial x^i} + \sum_{\alpha=1}^{q} \sum_J \phi_\alpha^J \frac{\partial}{\partial u_J^\alpha},$$

the latter sum ranging over all multi-indices J of orders $0 \leqslant \#J \leqslant n$; the coefficient functions ξ^i, ϕ_α^J could depend on all the variables $(x, u^{(n)})$. In the case $\mathbf{v}^*$ is the prolongation $\mathrm{pr}^{(n)}\,\mathbf{v}$ of a vector field

$$\mathbf{v} = \sum_{i=1}^{p} \xi^i(x, u)\frac{\partial}{\partial x^i} + \sum_{\alpha=1}^{q} \phi_\alpha(x, u)\frac{\partial}{\partial u^\alpha},$$

the coefficients ξ^i, ϕ_α^J of $\mathbf{v}^* = \mathrm{pr}^{(n)}\,\mathbf{v}$ will be determined by the coefficients ξ^i, ϕ_α of $\mathbf{v}$ itself. According to (2.20), the prolonged group action $\mathrm{pr}^{(n)}[\exp(\varepsilon\mathbf{v})]$, when restricted to just the zeroth order variables x, u of $M^{(0)} = M$, agrees with the ordinary group action $\exp(\varepsilon\mathbf{v})$ on M. Therefore the coefficients ξ^i and $\phi_\alpha^0 = \phi_\alpha$ of $\mathbf{v}^* = \mathrm{pr}^{(n)}\,\mathbf{v}$ must agree with the corresponding coefficients ξ^i, ϕ_α of $\mathbf{v}$ itself. Thus

$$\mathrm{pr}^{(n)}\,\mathbf{v} = \sum_{i=1}^{p} \xi^i\frac{\partial}{\partial x^i} + \sum_{\alpha=1}^{q}\sum_{J} \phi_\alpha^J\frac{\partial}{\partial u_J^\alpha}, \tag{2.22}$$

where ξ^i, $\phi_\alpha = \phi_\alpha^0$ come directly from $\mathbf{v}$. Moreover, if $\#J = k$, the coefficient ϕ_α^J of $\partial/\partial u_J^\alpha$ will only depend on k-th and lower order derivatives of u, $\phi_\alpha^J = \phi_\alpha^J(x, u^{(k)})$, since, again by (2.20), the corresponding group transformations of k-th order derivatives only involve k-th and lower order derivatives. This can be stated formally, using the projection maps in (2.20), as

$$d\pi_k^n(\mathrm{pr}^{(n)}\,\mathbf{v}) = \mathrm{pr}^{(k)}\,\mathbf{v}, \qquad n \geqslant k, \tag{2.23}$$

where $\mathrm{pr}^{(0)}\,\mathbf{v} = \mathbf{v}$ for $k = 0$. This indicates the possibility of recursively constructing the various prolongations of a given vector field. Our principal remaining task, then, is to find a general formula for the coefficients ϕ_α^J of the prolongation of a vector field. Before tackling this question, however, we look at a simple example and then draw some general conclusions on the computation of symmetry groups of differential equations.

Example 2.29. Consider the rotation group $\mathrm{SO}(2)$ acting on $X \times U \simeq \mathbb{R}^2$ as discussed in Examples 2.21 and 2.26. The corresponding infinitesimal generator is

$$\mathbf{v} = -u\frac{\partial}{\partial x} + x\frac{\partial}{\partial u},$$

with

$$\exp(\varepsilon\mathbf{v})(x, u) = (x\cos\varepsilon - u\sin\varepsilon,\ x\sin\varepsilon + u\cos\varepsilon)$$

being the rotation through angle ε. The first prolongation takes the form

$$\mathrm{pr}^{(1)}[\exp(\varepsilon\mathbf{v})](x, u, u_x) = \left(x\cos\varepsilon - u\sin\varepsilon,\ x\sin\varepsilon + u\cos\varepsilon,\ \frac{\sin\varepsilon + u_x\cos\varepsilon}{\cos\varepsilon - u_x\sin\varepsilon}\right).$$

According to (2.21), the first prolongation of $\mathbf{v}$ is obtained by differentiating these expressions with respect to ε and setting $\varepsilon = 0$. An easy computation shows that

$$\mathrm{pr}^{(1)}\mathbf{v} = -u\frac{\partial}{\partial x} + x\frac{\partial}{\partial u} + (1+u_x^2)\frac{\partial}{\partial u_x}. \tag{2.24}$$

Note that in accordance with (2.23) the first two terms in $\mathrm{pr}^{(1)}\mathbf{v}$ agree with those in $\mathbf{v}$ itself.

Infinitesimal Invariance

Combining Theorems 2.27 and 2.8, we immediately deduce the important infinitesimal condition for a group G to be a symmetry group of a given system of differential equations. Of course, to apply the latter theorem, we need a corresponding maximal rank condition for the system of differential equations.

Definition 2.30. Let

$$\Delta_\nu(x, u^{(n)}) = 0, \qquad \nu = 1, \ldots, l,$$

be a system of differential equations. The system is said to be of *maximal rank* if the $l \times (p + qp^{(n)})$ Jacobian matrix

$$\mathrm{J}_\Delta(x, u^{(n)}) = \left(\frac{\partial \Delta_\nu}{\partial x^i}, \frac{\partial \Delta_\nu}{\partial u_J^\alpha}\right)$$

of Δ with respect to all the variables $(x, u^{(n)})$ is of rank l whenever $\Delta(x, u^{(n)}) = 0$.

Thus, for instance, Laplace's equation

$$\Delta = u_{xx} + u_{yy} = 0$$

is of maximal rank, since the Jacobian matrix with respect to all the variables $(x, y; u; u_x, u_y; u_{xx}, u_{xy}, u_{yy})$ in $X \times U^{(2)}$ (cf. Example 2.25) is

$$\mathrm{J}_\Delta = (0, 0; 0; 0, 0; 1, 0, 1),$$

which is clearly of rank 1 everywhere. However, the rather silly equivalent equation

$$\tilde{\Delta} = (u_{xx} + u_{yy})^2 = 0$$

is *not* of maximal rank, since

$$\mathrm{J}_{\tilde{\Delta}} = (0, 0; 0; 0, 0; 2(u_{xx} + u_{yy}), 0, 2(u_{xx} + u_{yy}))$$

vanishes whenever $(u_{xx} + u_{yy})^2 = 0$.

The maximal rank condition is not much of a restriction, since according to Lemma 1.12 if the subvariety $\mathscr{S}_\Delta = \{\Delta(x, u^{(n)}) = 0\}$ is a regular submanifold of $M^{(n)}$, then there is an (algebraically) equivalent system of differential equations $\tilde{\Delta}(x, u^{(n)}) = 0$ such that

$$\mathscr{S}_\Delta = \mathscr{S}_{\tilde{\Delta}} = \{\tilde{\Delta}(x, u^{(n)}) = 0\},$$

and $\tilde{\Delta}$ is of maximal rank.

Theorem 2.31. *Suppose*

$$\Delta_\nu(x, u^{(n)}) = 0, \qquad \nu = 1, \ldots, l,$$

is a system of differential equations of maximal rank defined over $M \subset X \times U$. If G is a local group of transformations acting on M, and

$$\mathrm{pr}^{(n)}\, \mathbf{v}[\Delta_\nu(x, u^{(n)})] = 0, \quad \nu = 1, \ldots, l, \quad \textit{whenever} \quad \Delta(x, u^{(n)}) = 0, \tag{2.25}$$

for every infinitesimal generator $\mathbf{v}$ of G, then G is a symmetry group of the system.

The proof, as remarked above, is immediate from Theorems 2.8 and 2.27. Again, as for Theorem 2.27, it will be shown in Section 2.6 that, provided the system Δ satisfy certain additional "local solvability" conditions, (2.25) is in fact both necessary and sufficient for G to be a symmetry group. In this case, *all* (connected) symmetry groups can be systematically determined through an analysis of the infinitesimal criterion (2.25), as will be seen in numerous examples in Section 2.4. As a consequence of the maximal rank condition and Proposition 2.10, we see that we can replace (2.25) by the equivalent condition that there exist functions $Q_{\nu\mu}(x, u^{(n)})$ such that

$$\mathrm{pr}^{(n)}\, \mathbf{v}[\Delta_\nu(x, u^{(n)})] = \sum_{\mu=1}^{l} Q_{\nu\mu}(x, u^{(n)})\Delta_\mu(x, u^{(n)}) \tag{2.26}$$

holds identically in $(x, u^{(n)}) \in M^{(n)}$. Both (2.25) and (2.26) are useful when analyzing the infinitesimal criterion of invariance.

Example 2.32. Let $G = \mathrm{SO}(2)$ acting on $X \times U = \mathbb{R}^2$ as in Examples 2.29, 2.26 and 2.21. Consider the first order ordinary differential equation

$$\Delta(x, u, u_x) = (u - x)u_x + u + x = 0. \tag{2.27}$$

Note that the Jacobian matrix referred to in Definition 2.30 is

$$\mathrm{J}_\Delta = \left(\frac{\partial \Delta}{\partial x}, \frac{\partial \Delta}{\partial u}, \frac{\partial \Delta}{\partial u_x}\right) = (1 - u_x, 1 + u_x, u - x),$$

which is of rank 1 everywhere. Applying the infinitesimal generator of $\mathrm{pr}^{(1)}\, \mathrm{SO}(2)$, as calculated in (2.24), to (2.27), we find

$$\begin{aligned} \mathrm{pr}^{(1)}\, \mathbf{v}(\Delta) &= -u\frac{\partial \Delta}{\partial x} + x\frac{\partial \Delta}{\partial u} + (1 + u_x^2)\frac{\partial \Delta}{\partial u_x} \\ &= -u(1 - u_x) + x(1 + u_x) + (1 + u_x^2)(u - x) \\ &= u_x[(u - x)u_x + u + x] \\ &= u_x\Delta. \end{aligned}$$

Therefore $\mathrm{pr}^{(1)}\,\mathbf{v}(\Delta) = 0$ whenever $\Delta = 0$, and the infinitesimal criterion of invariance (2.25) is verified. We conclude that the rotation group SO(2) transforms solutions of (2.27) to other solutions. More geometrically, if $u = f(x)$ is a solution, and we rotate the graph of f by any angle θ, the resulting function is again a solution. Indeed, changing to polar coordinates

$$x = r\cos\theta, \qquad u = r\sin\theta,$$

(2.27) becomes

$$\frac{dr}{d\theta} = r.$$

The solutions are thus (pieces of) logarithmic spirals $r = ce^{\theta}$ for c constant. Obviously, rotating any one of these spirals produces another spiral of the same form, so SO(2) is indeed a symmetry group. (The choice of polar coordinates, and the fact that the equation could be readily solved in these coordinates, is, as we shall see, no accident.)

The Prolongation Formula

In light of Theorem 2.31, which connects symmetry groups of a system of differential equations with the infinitesimal criterion of invariance of the system under the prolonged infinitesimal generators of the group, the principal task remaining for us is to find an explicit formula for the prolongation of a vector field. Despite the daunting complexity of the prolonged group action, as determined by (2.18), the prolonged vector fields have a relatively simple, easily computable expression.

Before tackling the general case, it is helpful to illustrate the basic method in a couple of simpler situations. We first investigate the prolongation of a one-parameter group of transformations which acts solely on the independent variables in our system of differential equations. In other words, consider the vector field

$$\mathbf{v} = \sum_{i=1}^{p} \xi^i(x)\frac{\partial}{\partial x^i}$$

on the space $M \subset X \times U$. The group transformations $g_\varepsilon = \exp(\varepsilon\mathbf{v})$ are therefore of the form

$$(\tilde{x}, \tilde{u}) = g_\varepsilon \cdot (x, u) = (\Xi_\varepsilon(x), u)$$

discussed in Example 2.22, in which the components $\tilde{x}^i = \Xi^i_\varepsilon(x)$ satisfy

$$\left.\frac{d\Xi^i_\varepsilon(x)}{d\varepsilon}\right|_{\varepsilon=0} = \xi^i(x), \tag{2.28}$$

cf. (1.48). For simplicity, we consider the case of a single dependent variable $u \in \mathbb{R}$, although the discussion readily generalizes to several dependent variables.

The first jet space $M^{(1)}$ has coordinates $(x, u^{(1)}) = (x^i, u, u_j)$, where $u_j \equiv \partial u/\partial x^j$. The prolonged group action is found as follows: if $(x, u^{(1)})$ is any point in $M^{(1)}$, and $u = f(x)$ is any function with $u_j = \partial f/\partial x^j$, $j = 1, \ldots, p$, then $\mathrm{pr}^{(1)}\, g_\varepsilon \cdot (x, u^{(1)}) = (\tilde{x}, \tilde{u}^{(1)})$, where $\tilde{x} = \Xi_\varepsilon(x)$, $\tilde{u} = u$, and $\tilde{u}_j$ are the derivatives of the transformed function $\tilde{f}_\varepsilon = g_\varepsilon \cdot f$, which, according to (2.14), is given by

$$\tilde{u} = \tilde{f}_\varepsilon(\tilde{x}) = f[\Xi_\varepsilon^{-1}(\tilde{x})] = f[\Xi_{-\varepsilon}(\tilde{x})].$$

(Here we have used the fact that $g_\varepsilon^{-1} = g_{-\varepsilon}$ wherever defined.) Thus

$$\tilde{u}_j = \frac{\partial \tilde{f}_\varepsilon}{\partial \tilde{x}^j}(\tilde{x}) = \sum_{k=1}^{p} \frac{\partial f}{\partial x^k}(\Xi_{-\varepsilon}(\tilde{x})) \cdot \frac{\partial \Xi^k_{-\varepsilon}}{\partial \tilde{x}^j}(\tilde{x}).$$

But $\Xi_{-\varepsilon}(\tilde{x}) = x$, hence

$$\tilde{u}_j = \sum_{k=1}^{p} \frac{\partial \Xi^k_{-\varepsilon}}{\partial \tilde{x}^j}(\Xi_\varepsilon(x)) u_k \tag{2.29}$$

gives the explicit formula for the prolonged group action on the first order derivatives.

To find the infinitesimal generator of $\mathrm{pr}^{(1)}\, g_\varepsilon$, we must differentiate the formulas for the prolonged transformations with respect to ε and set $\varepsilon = 0$. Thus

$$\mathrm{pr}^{(1)}\, \mathbf{v} = \sum_{i=1}^{p} \xi^i(x) \frac{\partial}{\partial x^i} + \sum_{j=1}^{p} \phi^j(x, u^{(1)}) \frac{\partial}{\partial u_j}, \tag{2.30}$$

where $\xi^i(x)$ is as before (since $\mathrm{pr}^{(1)}\, g_\varepsilon$ transforms x and u just as g_ε does) and, by (2.21),

$$\phi^j(x, u^{(1)}) = \frac{d}{d\varepsilon}\bigg|_{\varepsilon=0} \sum_{k=1}^{p} \frac{\partial \Xi^k_{-\varepsilon}}{\partial \tilde{x}^j}(\Xi_\varepsilon(x)) u_k.$$

Since all the functions are smooth, we can interchange the order of differentiation, and so obtain two types of terms multiplying u_k: first, those of the form

$$\frac{\partial}{\partial \tilde{x}^j}\left[\frac{d\Xi^k_{-\varepsilon}}{d\varepsilon}\right](\Xi_\varepsilon(x))\bigg|_{\varepsilon=0} = \frac{\partial}{\partial x^j}\left[\frac{d\Xi^k_{-\varepsilon}}{d\varepsilon}\bigg|_{\varepsilon=0}\right](x) = -\frac{\partial \xi^k}{\partial x^j}(x),$$

where we have used (2.28) and the fact that at $\varepsilon = 0$, $\Xi_0(x) = x$ is the identity; second, those involving two x-derivatives of $\Xi_{-\varepsilon}$:

$$\sum_l \frac{\partial^2 \Xi^k_{-\varepsilon}}{\partial \tilde{x}^j \partial \tilde{x}^l}(\Xi_{-\varepsilon}(x)) \frac{d\Xi^l_\varepsilon}{d\varepsilon}(x)\bigg|_{\varepsilon=0} = 0,$$

which vanish since $\Xi_0(x)$ is the identity, hence at $\varepsilon = 0$ all second order x-derivatives of Ξ_ε vanish. Therefore,

$$\phi^j(x, u, u_x) = -\sum_{k=1}^{p} \frac{\partial \xi^k}{\partial x^j} \cdot u_k \tag{2.31}$$

provides the basic prolongation formula for $\mathrm{pr}^{(1)}\, \mathbf{v}$ in (2.30).

Example 2.33. Let $p = 2$, $q = 1$, and consider the vector field

$$\mathbf{v} = \xi(x, y)\frac{\partial}{\partial x} + \eta(x, y)\frac{\partial}{\partial y}$$

on $X \simeq \mathbb{R}^2$ with coordinates (x, y). According to (2.31), the first prolongation of $\mathbf{v}$ is the vector field

$$\mathrm{pr}^{(1)}\,\mathbf{v} = \mathbf{v} + \phi^x\frac{\partial}{\partial u_x} + \phi^y\frac{\partial}{\partial u_y}, \tag{2.32}$$

where

$$\phi^x = -\frac{\partial \xi}{\partial x}u_x - \frac{\partial \eta}{\partial x}u_y, \qquad \phi^y = -\frac{\partial \xi}{\partial y}u_x - \frac{\partial \eta}{\partial y}u_y.$$

For example, in the case of the rotation group

$$(x, y, u) \mapsto (x\cos\varepsilon - y\sin\varepsilon,\ x\sin\varepsilon + y\cos\varepsilon,\ u)$$

on X, the infinitesimal generator is

$$\mathbf{v} = -y\frac{\partial}{\partial x} + x\frac{\partial}{\partial y}.$$

Here $\xi = -y$, $\eta = x$, and hence $\mathbf{v}$ has first prolongation

$$\mathrm{pr}^{(1)}\,\mathbf{v} = -y\frac{\partial}{\partial x} + x\frac{\partial}{\partial y} - u_y\frac{\partial}{\partial u_x} + u_x\frac{\partial}{\partial u_y}.$$

(The first prolongation of the rotation group,

$$(x, y, u, u_x, u_y) \mapsto (x\cos\varepsilon - y\sin\varepsilon,\ x\sin\varepsilon + y\cos\varepsilon,\ u,$$
$$u_x\cos\varepsilon - u_y\sin\varepsilon,\ u_x\sin\varepsilon + u_y\cos\varepsilon),$$

can be reconstructed either by integrating $\mathrm{pr}^{(1)}\,\mathbf{v}$, as in (1.7), or directly from formula (2.29).)

It is useful to consider one other special case before proceeding to the general prolongation formula. Again we stick to one dependent variable u, but now look at groups that transform only the dependent variable:

$$(\tilde{x}, \tilde{u}) = g_\varepsilon \cdot (x, u) = (x, \Phi_\varepsilon(x, u)).$$

This has infinitesimal generator $\mathbf{v} = \phi(x, u)\partial_u$, where

$$\phi(x, u) = \left.\frac{d}{d\varepsilon}\right|_{\varepsilon=0} \Phi_\varepsilon(x, u).$$

If $u = f(x)$ is a function, the transformed function $\tilde{f}_\varepsilon = g_\varepsilon \cdot f$ is, according to (2.14), just

$$\tilde{u} = \tilde{f}_\varepsilon(x) = \Phi_\varepsilon(x, f(x)). \tag{2.33}$$

To find the prolonged group action, we differentiate:

$$\tilde{u}_j = \frac{\partial \tilde{f}_\varepsilon}{\partial x^j}(x) = \frac{\partial}{\partial x^j}\{\Phi_\varepsilon(x, f(x))\} = \frac{\partial \Phi_\varepsilon}{\partial x^j}(x, f(x)) + \frac{\partial f}{\partial x^j}(x)\frac{\partial \Phi_\varepsilon}{\partial u}(x, f(x)),$$

hence $\mathrm{pr}^{(1)} g_\varepsilon \cdot (x, u^{(1)}) = (x, \tilde{u}^{(1)})$, where

$$\tilde{u}_j = \frac{\partial \Phi_\varepsilon}{\partial x^j} + u_j \frac{\partial \Phi_\varepsilon}{\partial u}. \tag{2.34}$$

The infinitesimal generator

$$\mathrm{pr}^{(1)}\, \mathbf{v} = \mathbf{v} + \sum_{j=1}^{p} \phi^j(x, u^{(1)}) \frac{\partial}{\partial u_j}$$

of $\mathrm{pr}^{(1)} g_\varepsilon$ is obtained from (2.34) by differentiating with respect to ε and setting $\varepsilon = 0$, just as in our previous computations. Thus,

$$\phi^j(x, u^{(1)}) = \frac{d}{d\varepsilon}\bigg|_{\varepsilon=0} \tilde{u}_j = \frac{\partial \phi}{\partial x^j} + u_j \frac{\partial \phi}{\partial u}. \tag{2.35}$$

This gives the prolongation formula in this special case. For example, if $p = 2$, with independent variables x, y,

$$\mathrm{pr}^{(1)}\left[xu^2 \frac{\partial}{\partial u} \right] = xu^2 \frac{\partial}{\partial u} + (u^2 + 2xuu_x)\frac{\partial}{\partial u_x} + 2xuu_y \frac{\partial}{\partial u_y}.$$

Higher order prolongations of either (2.31) or (2.35) are found by further differentiating the relevant group prolongation formula. To give a general version of this, and in preparation for the general form of the prolongation theorem, we need to introduce the concept of a total derivative.

Total Derivatives

The preceding formulae (2.35) for the prolongation of a vector field of the form $\phi(x, u)\partial_u$ can be "simplified" by making the following observation. If $u = f(x)$ is any function, then $\phi^j(x, u^{(1)})$, when evaluated on f and its first order derivatives, is just the derivative of $\phi(x, f(x))$ with respect to x:

$$\phi^j(x, \mathrm{pr}^{(1)} f(x)) = \frac{\partial}{\partial x^j}[\phi(x, f(x))].$$

(Indeed, this is essentially how the ϕ^j were found.) In other words, $\phi^j(x, u^{(1)})$ is obtained from $\phi(x, u)$ by differentiating it with respect to x^j, while treating u as a function of x. The resulting derivative is called the *total derivative* of ϕ with respect to x^j, and denoted

$$\phi^j(x, u^{(1)}) = D_j\phi(x, u) = \frac{\partial \phi}{\partial x^j} + u_j \frac{\partial \phi}{\partial u}.$$

(The term "total" derivative is to distinguish $D_j\phi$ from the "partial" derivative $\partial\phi/\partial x^j$.) The definition of total derivative extends naturally to functions depending on $x = (x^1, \ldots, x^p)$, $u = (u^1, \ldots, u^q)$ and derivatives u_J^α of u.

Definition 2.34. Let $P(x, u^{(n)})$ be a smooth function of x, u and derivatives of u up to order n, defined on an open subset $M^{(n)} \subset X \times U^{(n)}$. The *total derivative* of P with respect to x^i is the unique smooth function $D_iP(x, u^{(n+1)})$ defined on $M^{(n+1)}$ and depending on derivatives of u up to order $n+1$, with the property that if $u = f(x)$ is any smooth function

$$D_iP(x, \mathrm{pr}^{(n+1)} f(x)) = \frac{\partial}{\partial x^i}\{P(x, \mathrm{pr}^{(n)} f(x))\}.$$

In other words, D_iP is obtained from P by differentiating P with respect to x^i while treating all the u^α's and their derivatives as functions of x.

Proposition 2.35. *Given $P(x, u^{(n)})$, the i-th total derivative of P has the general form*

$$D_iP = \frac{\partial P}{\partial x^i} + \sum_{\alpha=1}^{q}\sum_J u_{J,i}^\alpha \frac{\partial P}{\partial u_J^\alpha}, \tag{2.36}$$

where, for $J = (j_1, \ldots, j_k)$,

$$u_{J,i}^\alpha = \frac{\partial u_J^\alpha}{\partial x^i} = \frac{\partial^{k+1} u^\alpha}{\partial x^i \partial x^{j_1} \cdots \partial x^{j_k}}. \tag{2.37}$$

In (2.36) *the sum is over all J's of order $0 \leqslant \#J \leqslant n$, where n is the highest order derivative appearing in P.*

The proof is a straightforward applicaton of the chain rule. For example, in the case $X = \mathbb{R}^2$, with coordinates (x, y), and $U = \mathbb{R}$, there are two total derivatives D_x, D_y, with

$$D_xP = \frac{\partial P}{\partial x} + u_x\frac{\partial P}{\partial u} + u_{xx}\frac{\partial P}{\partial u_x} + u_{xy}\frac{\partial P}{\partial u_y} + u_{xxx}\frac{\partial P}{\partial u_{xx}} + \cdots,$$

$$D_yP = \frac{\partial P}{\partial y} + u_y\frac{\partial P}{\partial u} + u_{xy}\frac{\partial P}{\partial u_x} + u_{yy}\frac{\partial P}{\partial u_y} + u_{xxy}\frac{\partial P}{\partial u_{xx}} + \cdots.$$

Thus, if $P = xuu_{xy}$, then

$$D_xP = uu_{xy} + xu_xu_{xy} + xuu_{xxy}, \qquad D_yP = xu_yu_{xy} + xuu_{xyy}.$$

Higher order total derivatives are defined in analogy with our notation for higher order partial derivatives. Explicitly, if $J = (j_1, \ldots, j_k)$ is a k-th order multi-index, with $1 \leqslant j_\kappa \leqslant p$ for each κ, then the J-th total derivative is denoted

$$D_J = D_{j_1}D_{j_2}\cdots D_{j_k}.$$

(The explicit expressions for $D_J P$ in terms of the partial derivatives of P with respect to u_J^α rapidly become unmanageable.) Note that, as with partial derivatives, the order of differentiation for total derivatives of smooth functions is immaterial. Thus, for the above example,

$$D_x D_y P = D_y D_x P = u_y u_{xy} + u u_{xyy} + x(u_{xy}^2 + u_x u_{xyy} + u_y u_{xxy} + u u_{xxyy}).$$

The General Prolongation Formula

Theorem 2.36. *Let*

$$\mathbf{v} = \sum_{i=1}^{p} \xi^i(x, u) \frac{\partial}{\partial x^i} + \sum_{\alpha=1}^{q} \phi_\alpha(x, u) \frac{\partial}{\partial u^\alpha}$$

be a vector field defined on an open subset $M \subset X \times U$. The n-th prolongation of $\mathbf{v}$ is the vector field

$$\mathrm{pr}^{(n)} \mathbf{v} = \mathbf{v} + \sum_{\alpha=1}^{q} \sum_{J} \phi_\alpha^J(x, u^{(n)}) \frac{\partial}{\partial u_J^\alpha} \tag{2.38}$$

defined on the corresponding jet space $M^{(n)} \subset X \times U^{(n)}$, the second summation being over all (unordered) multi-indices $J = (j_1, \ldots, j_k)$, with $1 \leqslant j_\kappa \leqslant p$, $1 \leqslant k \leqslant n$. The coefficient functions ϕ_α^J of $\mathrm{pr}^{(n)} \mathbf{v}$ are given by the following formula:

$$\phi_\alpha^J(x, u^{(n)}) = D_J\left(\phi_\alpha - \sum_{i=1}^{p} \xi^i u_i^\alpha\right) + \sum_{i=1}^{p} \xi^i u_{J,i}^\alpha, \tag{2.39}$$

where $u_i^\alpha = \partial u^\alpha / \partial x^i$, and $u_{J,i}^\alpha = \partial u_J^\alpha / \partial x^i$, cf. (2.37).

Proof. We first prove the formula for first order derivatives, so $n = 1$ to begin with. Let $g_\varepsilon = \exp(\varepsilon \mathbf{v})$ be the corresponding one-parameter group, with transformations having the formula

$$(\tilde{x}, \tilde{u}) = g_\varepsilon \cdot (x, u) = (\Xi_\varepsilon(x, u), \Phi_\varepsilon(x, u)),$$

wherever defined. Note that

$$\begin{aligned} \xi^i(x, u) &= \left.\frac{d}{d\varepsilon}\right|_{\varepsilon=0} \Xi_\varepsilon^i(x, u), \qquad i = 1, \ldots, p, \\ \phi_\alpha(x, u) &= \left.\frac{d}{d\varepsilon}\right|_{\varepsilon=0} \Phi_\varepsilon^\alpha(x, u), \qquad \alpha = 1, \ldots, q, \end{aligned} \tag{2.40}$$

where Ξ_ε^i, Φ_ε^α are the components of Ξ_ε, Φ_ε. Given $(x, u^{(1)}) \in M^{(1)}$, let $u = f(x)$ be any representative function, so that $u^{(1)} = \mathrm{pr}^{(1)} f(x)$, or, explicitly,

$$u^\alpha = f^\alpha(x), \qquad u_i^\alpha = \partial f^\alpha(x)/\partial x^i.$$

According to (2.14), for ε sufficiently small, the transform of f by the group element g_ε is well defined (at least if the domain of definition of f is a suitably small neighbourhood of x), and is given by

$$\tilde{u} = \tilde{f}_\varepsilon(\tilde{x}) = (g_\varepsilon \cdot f)(\tilde{x}) = [\Phi_\varepsilon \circ (\mathbb{1} \times f)] \circ [\Xi_\varepsilon \circ (\mathbb{1} \times f)]^{-1}(\tilde{x}).$$

Using the chain rule, the Jacobian matrix $J\tilde{f}_\varepsilon(x) = (\partial \tilde{f}_\varepsilon^\alpha / \partial \tilde{x}^i)$ is then

$$J\tilde{f}_\varepsilon(\tilde{x}) = J[\Phi_\varepsilon \circ (\mathbb{1} \times f)](x) \cdot \{J[\Xi_\varepsilon \circ (\mathbb{1} \times f)](x)\}^{-1} \tag{2.41}$$

(provided the inverse is defined), since

$$x = [\Xi_\varepsilon \circ (\mathbb{1} \times f)]^{-1}(\tilde{x}).$$

Writing out the matrix entries of $J\tilde{f}_\varepsilon(\tilde{x})$ thus provides explicit formulae for the first prolongation $\mathrm{pr}^{(1)} g_\varepsilon$.

To find the infinitesimal generator $\mathrm{pr}^{(1)} \mathbf{v}$, we must differentiate (2.41) with respect to ε and set $\varepsilon = 0$. Recall first that if $M(\varepsilon)$ is any invertible square matrix of functions of ε, then

$$\frac{d}{d\varepsilon}[M(\varepsilon)^{-1}] = -M(\varepsilon)^{-1} \frac{dM(\varepsilon)}{d\varepsilon} M(\varepsilon)^{-1}.$$

Also note that since $\varepsilon = 0$ corresponds to the identity transformation,

$$\Xi_0(x, f(x)) = x, \qquad \Phi_0(x, f(x)) = f(x), \tag{2.42}$$

so if I denotes the $p \times p$ identity matrix,

$$J[\Xi_0 \circ (\mathbb{1} \times f)](x) = \mathrm{I}, \qquad J(\Phi_0 \circ (\mathbb{1} \times f)](x) = Jf(x).$$

Now, differentiating (2.41) and setting $\varepsilon = 0$, we find, using Leibniz' rule,

$$\begin{aligned} \frac{d}{d\varepsilon}\bigg|_{\varepsilon=0} J\tilde{f}_\varepsilon(\tilde{x}) &= \frac{d}{d\varepsilon}\bigg|_{\varepsilon=0} J[\Phi_\varepsilon \circ (\mathbb{1} \times f)](x) - Jf(x) \cdot \frac{d}{d\varepsilon}\bigg|_{\varepsilon=0} J[\Xi_\varepsilon \circ (\mathbb{1} \times f)](x) \\ &= J[\phi \circ (\mathbb{1} \times f)](x) - Jf(x) \cdot J[\xi \circ (\mathbb{1} \times f)](x). \end{aligned}$$

In the second equality, $\xi = (\xi^1, \ldots, \xi^p)^T$, and $\phi = (\phi_1, \ldots, \phi_q)^T$ are column vectors and we have used (2.40). The matrix entries of this last formula will give the coefficient functions ϕ_α^k of $\partial/\partial u_k^\alpha$ in $\mathrm{pr}^{(1)} \mathbf{v}$. Namely, the (α, k)-th entry is

$$\phi_\alpha^k(x, \mathrm{pr}^{(1)} f(x)) = \frac{\partial}{\partial x^k}[\phi_\alpha(x, f(x))] - \sum_{i=1}^p \frac{\partial f^\alpha}{\partial x^i} \cdot \frac{\partial}{\partial x^k}[\xi^i(x, f(x))].$$

Thus, by the definition of total derivative,

$$\begin{aligned} \phi_\alpha^k(x, u^{(1)}) &= D_k[\phi_\alpha(x, u)] - \sum_{i=1}^p D_k[\xi^i(x, u)] u_i^\alpha \\ &= D_k\left[\phi_\alpha - \sum_{i=1}^p \xi^i u_i^\alpha\right] + \sum_{i=1}^p \xi^i u_{ki}^\alpha, \end{aligned} \tag{2.43}$$

where $u_{ki}^\alpha = \partial^2 u^\alpha / \partial x^k \partial x^i$. This proves (2.39) in the case $n = 1$.

To prove the theorem in general, we proceed by induction. The key remark is that the $(n+1)$-st jet space $M^{(n+1)}$ can be viewed as a subspace of the first jet space $(M^{(n)})^{(1)}$ of the n-th jet space. This is because each $(n+1)$-st order derivative u_J^α can be viewed as a first order derivative of an n-th order derivative. (This can be done in general in several ways.) It is instructive to look at an illustrative example. For $p = 2$, $q = 1$, the first jet space $M^{(1)}$ has coordinates $(x, y; u; u_x, u_y)$. If we view (u_x, u_y) as *new* dependent variables, say $u_x = v$, $u_y = w$, then $M^{(1)}$ looks just like an open subset of $X \times \tilde{U}$, where X is still two-dimensional, but now $\tilde{U}$ has three dependent variables u, v and w. Thus the first jet space of $M^{(1)}$, i.e. $(M^{(1)})^{(1)}$, will be an open subset of $X \times \tilde{U}^{(1)}$, with coordinates $(x, y; u; v, w; u_x, u_y, v_x, v_y, w_x, w_y)$. Now remembering that $v = u_x$ and $w = u_y$, we see that $M^{(2)} \subset (M^{(1)})^{(1)}$ is the subspace defined by the relations

$$v = u_x, \qquad w = u_y, \qquad v_y = w_x,$$

in $X \times \tilde{U}^{(1)}$, determined by the superfluous variables u_x, u_y in $(M^{(1)})^{(1)}$ and the equality of mixed second order partial derivatives of u.

With this point of view, the inductive procedure for determining $\mathrm{pr}^{(n)}\mathbf{v}$ from $\mathrm{pr}^{(n-1)}\mathbf{v}$ is as follows; we regard $\mathrm{pr}^{(n-1)}\mathbf{v}$ as a vector field on $M^{(n-1)}$ (of a certain special type) and so by our first order prolongation formula can prolong it to $(M^{(n-1)})^{(1)}$. We then restrict the resulting vector field to the subspace $M^{(n)}$, and this will determine the n-th prolongation $\mathrm{pr}^{(n)}\mathbf{v}$. (Of course, we must check that the restriction is possible, but this will follow from the explicit formula.) Now the new "n-th order" coordinates in $(M^{(n-1)})^{(1)}$ are given by $u_{J,k}^\alpha = \partial u_J^\alpha/\partial x^k$, where $J = (j_1, \ldots, j_{n-1})$, $1 \leqslant k \leqslant p$, and $1 \leqslant \alpha \leqslant q$. According to (2.43), the coefficient of $\partial/\partial u_{J,k}^\alpha$ in the first prolongation of $\mathrm{pr}^{(n-1)}\mathbf{v}$ is therefore

$$\phi_\alpha^{J,k} = D_k\phi_\alpha^J - \sum_{i=1}^p D_k\xi^i \cdot u_{J,i}^\alpha. \tag{2.44}$$

(As we will see, (2.44) provides a useful recursion relation for the coefficient functions of $\mathrm{pr}^{(n)}\mathbf{v}$.) It now suffices to check that the formula (2.39) solves the recursion relation (2.44) in closed form. By induction, we find

$$\begin{aligned}
\phi_\alpha^{J,k} &= D_k\left\{D_J\left(\phi_\alpha - \sum_{i=1}^p \xi^i u_i^\alpha\right) + \sum_{i=1}^p \xi^i u_{J,i}^\alpha\right\} - \sum_{i=1}^p D_k\xi^i \cdot u_{J,i}^\alpha \\
&= D_k D_J\left(\phi_\alpha - \sum_{i=1}^p \xi^i u_i^\alpha\right) + \sum_{i=1}^p (D_k\xi^i \cdot u_{J,i}^\alpha + \xi^i u_{J,ik}^\alpha) - \sum_{i=1}^p D_k\xi^i \cdot u_{J,i}^\alpha \\
&= D_k D_J\left(\phi_\alpha - \sum_{i=1}^p \xi^i u_i^\alpha\right) + \sum_{i=1}^p \xi^i u_{J,ik}^\alpha,
\end{aligned}$$

where $u_{J,ik}^\alpha = \partial^2 u_J^\alpha/\partial x^i \partial x^k$. Thus $\phi_\alpha^{J,k}$ is of the form (2.39), and the induction step is completed. □

Example 2.37. Let's repeat the case of the rotation group SO(2) acting on $X \times U \simeq \mathbb{R} \times \mathbb{R}$ with infinitesimal generator

$$\mathbf{v} = -u\frac{\partial}{\partial x} + x\frac{\partial}{\partial u};$$

see Examples 2.26 and 2.29. In this case $\phi = x$, $\xi = -u$, so the first prolongation

$$\mathrm{pr}^{(1)}\,\mathbf{v} = \mathbf{v} + \phi^x\frac{\partial}{\partial u_x}$$

is given by

$$\phi^x = D_x(\phi - \xi u_x) + \xi u_{xx} = D_x(x + uu_x) - uu_{xx} = 1 + u_x^2.$$

Thus we recover the result of (2.24). The coefficient function ϕ^{xx} of $\partial/\partial u_{xx}$ in $\mathrm{pr}^{(2)}\,\mathbf{v}$ is found using either (2.39)

$$\phi^{xx} = D_x^2(\phi - \xi u_x) + \xi u_{xxx} = D_x^2(x + uu_x) - uu_{xxx} = 3u_x u_{xx},$$

or the recursion formula (2.44)

$$\phi^{xx} = D_x\phi^x - u_{xx}D_x\xi = D_x(1 + u_x^2) + u_x u_{xx} = 3u_x u_{xx}.$$

Thus the infinitesimal generator of the second prolongation $\mathrm{pr}^{(2)}\,\mathrm{SO}(2)$ acting on $X \times U^{(2)}$ is

$$\mathrm{pr}^{(2)}\,\mathbf{v} = -u\frac{\partial}{\partial x} + x\frac{\partial}{\partial u} + (1 + u_x^2)\frac{\partial}{\partial u_x} + 3u_x u_{xx}\frac{\partial}{\partial u_{xx}}.$$

(The derivation of this formula directly from the action $\mathrm{pr}^{(2)}\,\mathrm{SO}(2)$ is, needless to say, considerably more complicated.)

Using the infinitesimal criterion of invariance of Theorem 2.31, we immediately deduce that the ordinary differential equation $u_{xx} = 0$ has SO(2) as a symmetry group, since

$$\mathrm{pr}^{(2)}\,\mathbf{v}(u_{xx}) = 3u_x u_{xx} = 0$$

whenever $u_{xx} = 0$. This is just a restatement of the geometric fact that rotations take straight lines to straight lines. For another geometric illustration, consider the function

$$\kappa(x, u^{(2)}) = u_{xx}(1 + u_x^2)^{-3/2}.$$

An easy computation shows that

$$\mathrm{pr}^{(2)}\,\mathbf{v}(\kappa) = 0$$

for all u_x, u_{xx}, hence by Proposition 2.6, κ is an invariant of $\mathrm{pr}^{(2)}\,\mathrm{SO}(2)$:

$$\kappa(\mathrm{pr}^{(2)}\,\theta\cdot(x, u^{(2)})) = \kappa(x, u^{(2)})$$

for any rotation θ. But κ is just the curvature of the curve determined by the graph of $u = f(x)$, so we've just reproved the fact that the curvature of a curve is invariant under rotations. (This is a special case of the theory of differential invariants—see Section 2.5 for further results of this type.)

Example 2.38. Consider the special case $p = 2$, $q = 1$ in the prolongation formula, so we are looking at a partial differential equation involving a function $u = f(x, t)$. A general vector field on $X \times U \simeq \mathbb{R}^2 \times \mathbb{R}$ takes the form

$$\mathbf{v} = \xi(x, t, u)\frac{\partial}{\partial x} + \tau(x, t, u)\frac{\partial}{\partial t} + \phi(x, t, u)\frac{\partial}{\partial u}.$$

The first prolongation of $\mathbf{v}$ is the vector field

$$\mathrm{pr}^{(1)}\,\mathbf{v} = \mathbf{v} + \phi^x \frac{\partial}{\partial u_x} + \phi^t \frac{\partial}{\partial u_t},$$

where, using (2.39),

$$\begin{aligned} \phi^x &= D_x(\phi - \xi u_x - \tau u_t) + \xi u_{xx} + \tau u_{xt} \\ &= D_x\phi - u_x D_x\xi - u_t D_x\tau \\ &= \phi_x + (\phi_u - \xi_x)u_x - \tau_x u_t - \xi_u u_x^2 - \tau_u u_x u_t, \\ \phi^t &= D_t(\phi - \xi u_x - \tau u_t) + \xi u_{xt} + \tau u_{tt} \\ &= D_t\phi - u_x D_t\xi - u_t D_t\tau \\ &= \phi_t - \xi_t u_x + (\phi_u - \tau_t)u_t - \xi_u u_x u_t - \tau_u u_t^2, \end{aligned} \tag{2.45}$$

the subscripts on ϕ, ξ, τ denoting partial derivatives. Similarly,

$$\mathrm{pr}^{(2)}\,\mathbf{v} = \mathrm{pr}^{(1)}\,\mathbf{v} + \phi^{xx}\frac{\partial}{\partial u_{xx}} + \phi^{xt}\frac{\partial}{\partial u_{xt}} + \phi^{tt}\frac{\partial}{\partial u_{tt}},$$

where, for instance,

$$\begin{aligned} \phi^{xx} &= D_x^2(\phi - \xi u_x - \tau u_t) + \xi u_{xxx} + \tau u_{xxt} \\ &= D_x^2\phi - u_x D_x^2\xi - u_t D_x^2\tau - 2u_{xx}D_x\xi - 2u_{xt}D_x\tau \\ &= \phi_{xx} + (2\phi_{xu} - \xi_{xx})u_x - \tau_{xx}u_t + (\phi_{uu} - 2\xi_{xu})u_x^2 - 2\tau_{xu}u_x u_t \\ &\quad - \xi_{uu}u_x^3 - \tau_{uu}u_x^2 u_t + (\phi_u - 2\xi_x)u_{xx} - 2\tau_x u_{xt} - 3\xi_u u_x u_{xx} \\ &\quad - \tau_u u_t u_{xx} - 2\tau_u u_x u_{xt}. \end{aligned} \tag{2.46}$$

These formulae will be used in the following section to compute symmetry groups of some well-known evolution equations.

Properties of Prolonged Vector Fields

Theorem 2.39. *Suppose* $\mathbf{v}$ *and* $\mathbf{w}$ *are smooth vector fields on* $M \subset X \times U$. *Then their prolongations have the properties*

$$\mathrm{pr}^{(n)}(c\mathbf{v} + c'\mathbf{w}) = c \cdot \mathrm{pr}^{(n)}\,\mathbf{v} + c' \cdot \mathrm{pr}^{(n)}\,\mathbf{w},$$

for c, c' *constant, and*

$$\mathrm{pr}^{(n)}[\mathbf{v}, \mathbf{w}] = [\mathrm{pr}^{(n)}\,\mathbf{v}, \mathrm{pr}^{(n)}\,\mathbf{w}]. \tag{2.47}$$

Proof. The linearity is left to the reader. The Lie bracket property can be proved by direct computation using (2.38), (2.39). However, it is easier to proceed as follows. Note first that if g, h are group elements of some transformation group, then

$$\mathrm{pr}^{(n)}(g \cdot h) = \mathrm{pr}^{(n)}\,g \cdot \mathrm{pr}^{(n)}\,h,$$

and, if we use the fact that M is a subset of some Euclidean space,

$$\mathrm{pr}^{(n)}(g + h) = \mathrm{pr}^{(n)}\,g + \mathrm{pr}^{(n)}\,h,$$

where $(g + h) \cdot x = g \cdot x + h \cdot x$ by definition. Let $\mathbb{1}$ denote the identity map of M, so $\mathbb{1}^{(n)} = \mathrm{pr}^{(n)}\,\mathbb{1}$ is the identity map of $M^{(n)}$. Using the characterization of the Lie bracket in Theorem 1.33,

$$
\begin{aligned}
&[\mathrm{pr}^{(n)}\,\mathbf{v}, \mathrm{pr}^{(n)}\,\mathbf{w}] \\
&\quad = \lim_{\varepsilon \to 0+} \frac{\mathrm{pr}^{(n)} \exp(-\sqrt{\varepsilon}\mathbf{w}) \exp(-\sqrt{\varepsilon}\mathbf{v}) \exp(\sqrt{\varepsilon}\mathbf{w}) \exp(\sqrt{\varepsilon}\mathbf{v}) - \mathbb{1}^{(n)}}{\varepsilon} \\
&\quad = \mathrm{pr}^{(n)} \left\{ \lim_{\varepsilon \to 0+} \frac{\exp(-\sqrt{\varepsilon}\mathbf{w}) \exp(-\sqrt{\varepsilon}\mathbf{v}) \exp(\sqrt{\varepsilon}\mathbf{w}) \exp(\sqrt{\varepsilon}\mathbf{v}) - \mathbb{1}}{\varepsilon} \right\} \\
&\quad = \mathrm{pr}^{(n)}[\mathbf{v}, \mathbf{w}].
\end{aligned}
$$

□

Corollary 2.40. *Let* Δ *be a system of differential equations of maximal rank defined over* $M \subset X \times U$. *The set of all infinitesimal symmetries of this system forms a Lie algebra of vector fields on* M. *Moreover, if this Lie algebra is finite-dimensional, the (connected component of the) symmetry group of the system is a local Lie group of transformations acting on* M.

Characteristics of Symmetries

Finally, we note an equivalent, computationally useful way of writing down the general prolongation formula (2.39). Given $\mathbf{v}$ as above, set

$$Q_\alpha(x, u^{(1)}) = \phi_\alpha(x, u) - \sum_{i=1}^{p} \xi^i(x, u) u_i^\alpha, \qquad \alpha = 1, \ldots, q; \tag{2.48}$$

the q-tuple $Q(x, u^{(1)}) = (Q_1, \ldots, Q_q)$ is referred to as the *characteristic* of the vector field $\mathbf{v}$. With this definition, (2.39) takes the form

$$\phi_\alpha^J = D_J Q_\alpha + \sum_{i=1}^{p} \xi^i u_{J,i}^\alpha. \tag{2.49}$$

Substituting into (2.38) and rearranging terms, we find

$$\mathrm{pr}^{(n)}\,\mathbf{v} = \sum_{\alpha=1}^{q} \sum_J D_J Q_\alpha \frac{\partial}{\partial u_J^\alpha} + \sum_{i=1}^{p} \xi^i \left\{ \frac{\partial}{\partial x^i} + \sum_{\alpha=1}^{q} \sum_J u_{J,i}^\alpha \frac{\partial}{\partial u_J^\alpha} \right\}.$$

The terms in brackets we recognize to be just the total derivative operators, as given by (2.36), hence

$$\mathrm{pr}^{(n)}\,\mathbf{v} = \mathrm{pr}^{(n)}\,\mathbf{v}_Q + \sum_{i=1}^{p} \xi^i D_i, \tag{2.50}$$

where, by definition

$$\mathbf{v}_Q = \sum_{\alpha=1}^{q} Q_\alpha(x, u^{(1)}) \frac{\partial}{\partial u^\alpha}, \qquad \mathrm{pr}^{(n)}\,\mathbf{v}_Q = \sum_{\alpha=1}^{q} \sum_J D_J Q_\alpha \frac{\partial}{\partial u_J^\alpha}. \tag{2.51}$$

In all the above formulae, the summations extend over all multi-indices J of order $0 \leqslant \# J \leqslant n$. Of course, the two terms on the right-hand side of (2.50) are just formal algebraic expressions since they each involve $(n+1)$-st order derivative of the u's. Only when they are combined together do the terms involving the $(n+1)$-st order derivatives cancel and we have a genuine vector field on the jet space $M^{(n)}$. The importance of (2.50) will become manifest once we discuss generalized symmetries in Chapter 5.

2.4. Calculation of Symmetry Groups

Theorem 2.31, when coupled with the prolongation formulae (2.38), (2.39) provides an effective computational procedure for finding the most general (connected) symmetry group of almost any system of partial differential equations of interest. In this procedure, one lets the coefficients $\xi^i(x, u)$, $\phi_\alpha(x, u)$ of the infinitesimal generator $\mathbf{v}$ of a hypothetical one-parameter symmetry group of the system be unknown functions of x and u. The coefficients ϕ_α^J of the prolonged infinitesimal generator $\mathrm{pr}^{(n)}\,\mathbf{v}$ will be certain explicit expressions involving the partial derivatives of the coefficients ξ^i and ϕ_α with respect to both x and u. The infinitesimal criterion of invariance (2.25) will thus involve x, u and the derivatives of u with respect to x, as well as $\xi^i(x, u)$, $\phi_\alpha(x, u)$ and their partial derivatives with respect to x and u. After eliminating any dependencies among the derivatives of the u's caused by the system itself (since (2.25) need only hold on solutions of the system), we can then equate the coefficients of the remaining unconstrained partial derivatives of u to zero. This will result in a large number of elementary partial differential equations for the coefficient functions ξ^i, ϕ_α of the infinitesimal generator, called the *determining equations* for the symmetry group of the given system.

In most instances, these determining equations can be solved by elementary methods, and the general solution will determine the most general infinitesimal symmetry of the system. Corollary 2.40 assures us that the resulting system of infinitesimal generators forms a Lie algebra of symmetries; the general symmetry group itself can then be found by exponentiating the given vector fields. The process will become clearer in the following examples.

Example 2.41. *The Heat Equation.* Consider the equation for the conduction of heat in a one-dimensional rod

$$u_t = u_{xx}, \tag{2.52}$$

the thermal diffusivity having been normalized to unity. Here there are two independent variables x and t, and one dependent variable u, so $p = 2$ and $q = 1$ in our notation. The heat equation is of second order, $n = 2$, and can be identified with the linear subvariety in $X \times U^{(2)}$ determined by the vanishing of $\Delta(x, t, u^{(2)}) = u_t - u_{xx}$.

Let

$$\mathbf{v} = \xi(x, t, u)\frac{\partial}{\partial x} + \tau(x, t, u)\frac{\partial}{\partial t} + \phi(x, t, u)\frac{\partial}{\partial u} \tag{2.53}$$

be a vector field on $X \times U$. We wish to determine all possible coefficient functions ξ, τ and ϕ so that the corresponding one-parameter group $\exp(\varepsilon\mathbf{v})$ is a symmetry group of the heat equation. According to Theorem 2.31, we need to know the second prolongation

$$\mathrm{pr}^{(2)}\,\mathbf{v} = \mathbf{v} + \phi^x\frac{\partial}{\partial u_x} + \phi^t\frac{\partial}{\partial u_t} + \phi^{xx}\frac{\partial}{\partial u_{xx}} + \phi^{xt}\frac{\partial}{\partial u_{xt}} + \phi^{tt}\frac{\partial}{\partial u_{tt}}$$

of $\mathbf{v}$, whose coefficients were calculated in Example 2.38. Applying $\mathrm{pr}^{(2)}\,\mathbf{v}$ to (2.52), we find the infinitesimal criterion (2.25) to be

$$\phi^t = \phi^{xx}, \tag{2.54}$$

which must be satisfied whenever $u_t = u_{xx}$. Substituting the general formulae (2.45), (2.46) into (2.54), replacing u_t by u_{xx} whenever it occurs, and equating the coefficients of the various monomials in the first and second order partial derivatives of u, we find the determining equations for the symmetry group of the heat equation to be the following:

Monomial	Coefficient	
$u_x u_{xt}$	$0 = -2\tau_u$	(a)
u_{xt}	$0 = -2\tau_x$	(b)
u_{xx}^2	$-\tau_u = -\tau_u$	(c)
$u_x^2 u_{xx}$	$0 = -\tau_{uu}$	(d)
$u_x u_{xx}$	$-\xi_u = -2\tau_{xu} - 3\xi_u$	(e)
u_{xx}	$\phi_u - \tau_t = -\tau_{xx} + \phi_u - 2\xi_x$	(f)
u_x^3	$0 = -\xi_{uu}$	(g)
u_x^2	$0 = \phi_{uu} - 2\xi_{xu}$	(h)
u_x	$-\xi_t = 2\phi_{xu} - \xi_{xx}$	(j)
1	$\phi_t = \phi_{xx}$	(k)

(As usual, subscripts indicate derivatives.) The solution of the determining equations is elementary. First, (a) and (b) require that τ be just a function of t. Then (e) shows that ξ doesn't depend on u, and (f) requires $\tau_t = 2\xi_x$, so $\xi(x, t) = \frac{1}{2}\tau_t x + \sigma(t)$, where σ is some function of t only. Next, by (h), ϕ is linear in u, so

$$\phi(x, t, u) = \beta(x, t)u + \alpha(x, t)$$

for certain functions α and β. According to (j), $\xi_t = -2\beta_x$, so β is at most quadratic in x, with

$$\beta = -\tfrac{1}{8}\tau_{tt}x^2 - \tfrac{1}{2}\sigma_t x + \rho(t).$$

Finally, the last equation (k) requires that both α and β be solutions of the heat equation,

$$\alpha_t = \alpha_{xx}, \qquad \beta_t = \beta_{xx}.$$

Using the previous form of β, we find

$$\tau_{ttt} = 0, \qquad \sigma_{tt} = 0, \qquad \rho_t = -\tfrac{1}{4}\tau_{tt}.$$

Thus τ is quadratic in t, σ is linear in t, and we can read off the formulae for ξ and ϕ directly from those of ρ, σ and τ. Since we have now satisfied all the determining equations, we conclude that the most general infinitesimal symmetry of the heat equation has coefficient functions of the form

$$\begin{aligned} \xi &= c_1 + c_4 x + 2c_5 t + 4c_6 xt, \\ \tau &= c_2 + 2c_4 t + 4c_6 t^2, \\ \phi &= (c_3 - c_5 x - 2c_6 t - c_6 x^2)u + \alpha(x, t), \end{aligned}$$

where $c_1, \ldots, c_6$ are arbitrary constants and $\alpha(x, t)$ an arbitrary solution of the heat equation. Thus the Lie algebra of infinitesimal symmetries of the heat equation is spanned by the six vector fields

$$\begin{aligned} \mathbf{v}_1 &= \partial_x, \\ \mathbf{v}_2 &= \partial_t, \\ \mathbf{v}_3 &= u\partial_u, \\ \mathbf{v}_4 &= x\partial_x + 2t\partial_t, \\ \mathbf{v}_5 &= 2t\partial_x - xu\partial_u, \\ \mathbf{v}_6 &= 4tx\partial_x + 4t^2\partial_t - (x^2 + 2t)u\partial_u, \end{aligned} \tag{2.55}$$

and the infinite-dimensional subalgebra

$$\mathbf{v}_\alpha = \alpha(x, t)\partial_u,$$

where α is an arbitrary solution of the heat equation. The commutation relations between these vector fields is given by the following table, the entry in row i and column j representing $[\mathbf{v}_i, \mathbf{v}_j]$:

	$\mathbf{v}_1$	$\mathbf{v}_2$	$\mathbf{v}_3$	$\mathbf{v}_4$	$\mathbf{v}_5$	$\mathbf{v}_6$	$\mathbf{v}_\alpha$
$\mathbf{v}_1$	0	0	0	$\mathbf{v}_1$	$-\mathbf{v}_3$	$2\mathbf{v}_5$	$\mathbf{v}_{\alpha_x}$
$\mathbf{v}_2$	0	0	0	$2\mathbf{v}_2$	$2\mathbf{v}_1$	$4\mathbf{v}_4 - 2\mathbf{v}_3$	$\mathbf{v}_{\alpha_t}$
$\mathbf{v}_3$	0	0	0	0	0	0	$-\mathbf{v}_\alpha$
$\mathbf{v}_4$	$-\mathbf{v}_1$	$-2\mathbf{v}_2$	0	0	$\mathbf{v}_5$	$2\mathbf{v}_6$	$\mathbf{v}_{\alpha'}$
$\mathbf{v}_5$	$\mathbf{v}_3$	$-2\mathbf{v}_1$	0	$-\mathbf{v}_5$	0	0	$\mathbf{v}_{\alpha''}$
$\mathbf{v}_6$	$-2\mathbf{v}_5$	$2\mathbf{v}_3 - 4\mathbf{v}_4$	0	$-2\mathbf{v}_6$	0	0	$\mathbf{v}_{\alpha'''}$
$\mathbf{v}_\alpha$	$-\mathbf{v}_{\alpha_x}$	$-\mathbf{v}_{\alpha_t}$	$\mathbf{v}_\alpha$	$-\mathbf{v}_{\alpha'}$	$-\mathbf{v}_{\alpha''}$	$-\mathbf{v}_{\alpha'''}$	0

where

$$\alpha' = x\alpha_x + 2t\alpha_t, \qquad \alpha'' = 2t\alpha_x + x\alpha,$$
$$\alpha''' = 4tx\alpha_x + 4t^2\alpha_t + (x^2 + 2t)\alpha.$$

Note that since Corollary 2.40 assures us that the totality of infinitesimal symmetries must be a Lie algebra, we can conclude that if $\alpha(x, t)$ is any solution of the heat equation, so are α_x, α_t, and α', α'' and α''' as given above.

The one-parameter groups G_i generated by the $\mathbf{v}_i$ are given in the following table. The entries give the transformed point $\exp(\varepsilon\mathbf{v}_i)(x, t, u) = (\tilde{x}, \tilde{t}, \tilde{u})$:

$$
\begin{aligned}
G_1&: \quad (x + \varepsilon, t, u),\\
G_2&: \quad (x, t + \varepsilon, u),\\
G_3&: \quad (x, t, e^\varepsilon u),\\
G_4&: \quad (e^\varepsilon x, e^{2\varepsilon} t, u),\\
G_5&: \quad (x + 2\varepsilon t, t, u\cdot\exp(-\varepsilon x - \varepsilon^2 t)),\\
G_6&: \quad \left(\frac{x}{1 - 4\varepsilon t}, \frac{t}{1 - 4\varepsilon t}, u\sqrt{1 - 4\varepsilon t}\exp\left\{\frac{-\varepsilon x^2}{1 - 4\varepsilon t}\right\}\right),\\
G_\alpha&: \quad (x, t, u + \varepsilon\alpha(x, t)).
\end{aligned}
\tag{2.56}
$$

Since each group G_i is a symmetry group, (2.14) implies that if $u = f(x, t)$ is a solution of the heat equation, so are the functions

$$
\begin{aligned}
u^{(1)} &= f(x - \varepsilon, t),\\
u^{(2)} &= f(x, t - \varepsilon),\\
u^{(3)} &= e^\varepsilon f(x, t),\\
u^{(4)} &= f(e^{-\varepsilon} x, e^{-2\varepsilon} t),\\
u^{(5)} &= e^{-\varepsilon x + \varepsilon^2 t} f(x - 2\varepsilon t, t),\\
u^{(6)} &= \frac{1}{\sqrt{1 + 4\varepsilon t}} \exp\left\{\frac{-\varepsilon x^2}{1 + 4\varepsilon t}\right\} f\left(\frac{x}{1 + 4\varepsilon t}, \frac{t}{1 + 4\varepsilon t}\right),\\
u^{(\alpha)} &= f(x, t) + \varepsilon\alpha(x, t),
\end{aligned}
$$

where ε is any real number and $\alpha(x, t)$ any other solution to the heat equation. (See Example 2.22 for a detailed discussion of how these expressions are derived from the group transformations.)

The symmetry groups G_3 and G_α thus reflect the linearity of the heat equation; we can add solutions and multiply them by constants. The groups G_1 and G_2 demonstrate the time- and space-invariance of the equation, reflecting the fact that the heat equation has constant coefficients. The well-known scaling symmetry turns up in G_4, while G_5 represents a kind of Galilean boost to a moving coordinate frame. The last group G_6 is a genuinely local group of transformations. Its appearance is far from obvious from basic physical principles, but it has the following nice consequence. If we let $u = c$ be just a constant solution, then we immediately conclude that the function

$$u = \frac{c}{\sqrt{1 + 4\varepsilon t}} \exp\left\{\frac{-\varepsilon x^2}{1 + 4\varepsilon t}\right\}$$

is a solution. In particular, if we set $c = \sqrt{\varepsilon/\pi}$ we obtain the fundamental solution to the heat equation at the point $(x_0, t_0) = (0, -1/4\varepsilon)$. To obtain the fundamental solution

$$u = \frac{1}{\sqrt{4\pi t}} \exp\left\{\frac{-x^2}{4t}\right\}$$

we need to translate this solution in t using the group G_2 (with ε replaced by $-1/4\varepsilon$).

The most general one-parameter group of symmetries is obtained by considering a general linear combination $c_1 \mathbf{v}_1 + \cdots + c_6 \mathbf{v}_6 + \mathbf{v}_\alpha$ of the given vector fields; the explicit formulae for the group transformations are very complicated. Alternatively, we can use (1.40), and represent an arbitrary group transformation g as the composition of transformations in the various one-parameter subgroups $G_1, \ldots, G_6, G_\alpha$. In particular, if g is near the identity, it can be represented uniquely in the form

$$g = \exp(\mathbf{v}_\alpha) \cdot \exp(\varepsilon_6 \mathbf{v}_6) \cdot \ldots \cdot \exp(\varepsilon_1 \mathbf{v}_1).$$

Thus the most general solution obtainable from a given solution $u = f(x, t)$ by group transformations is of the form

$$u = \frac{1}{\sqrt{1 + 4\varepsilon_6 t}} \exp\left\{\varepsilon_3 - \frac{\varepsilon_5 x + \varepsilon_6 x^2 - \varepsilon_5^2 t}{1 + 4\varepsilon_6 t}\right\}$$
$$\times f\left(\frac{e^{-\varepsilon_4}(x - 2\varepsilon_5 t)}{1 + 4\varepsilon_6 t} - \varepsilon_1, \frac{e^{-2\varepsilon_4} t}{1 + 4\varepsilon_6 t} - \varepsilon_2\right) + \alpha(x, t),$$

where $\varepsilon_1, \ldots, \varepsilon_6$ are real constants and α an arbitrary solution to the heat equation.

Example 2.42. *Burgers' Equation.* A nonlinear equation closely allied with the heat equation is Burgers' equation, which, for symmetry group purposes, is convenient to take in "potential form"

$$u_t = u_{xx} + u_x^2. \tag{2.57}$$

Note that if we differentiate this with respect to x and substitute $v = u_x$, we derive the more usual form

$$v_t = v_{xx} + 2vv_x \tag{2.58}$$

of Burgers' equation; it represents the simplest wave equation combining both dissipative and nonlinear effects, and therefore appears in a wide variety of physical applications.

The symmetry group of (2.57) will again the generated by vector fields of the form (2.53). Applying the second prolongation $\mathrm{pr}^{(2)}\,\mathbf{v}$ to (2.57), we find that ξ, τ, ϕ must satisfy the symmetry conditions

$$\phi^t = \phi^{xx} + 2u_x\phi^x, \tag{2.59}$$

where the coefficients ϕ^t, ϕ^x, ϕ^{xx} of $\mathrm{pr}^{(2)}\,\mathbf{v}$ were determined in Example 2.38, and we are allowed to substitute $u_{xx} + u_x^2$ for u_t, wherever it occurs in (2.59). We could, at this juncture, write out (2.59) in full detail and equate coefficients of the various first and second order derivatives of u to get the full determining equations, as was done in the previous example. In practice, however, it is far more expedient to tackle the solution of the symmetry equations in stages, first extracting information from the higher order derivatives appearing in them, and then using this information to simplify the prolongation formulae at the lower order stages. Working this way, "from the top down", is extremely efficient, and, even more to the point, well-nigh the only course available for higher order systems of equations, for which the full system of determining equations would take many pages to write down in full detail.

In the present case, using (2.45), (2.46) and keeping in mind that u_t has been replaced by $u_{xx} + u_x^2$, we find that the coefficients of $u_x u_{xt}$ and u_{xt} require that $\tau_u = \tau_x = 0$, so τ is a function of t only. (Note that this already simplifies the formulae for ϕ^x and ϕ^{xx} quite a bit.) The coefficient of $u_x u_{xx}$ implies that ξ doesn't depend on u, while from that of u_{xx} we find that $\tau_t = 2\xi_x$, so $\xi(x, t) = \frac{1}{2}\tau_t x + \sigma(t)$. The coefficient of u_x^2 is

$$\phi_u - \tau_t = \phi_{uu} + 2\phi_u - 2\xi_x,$$

hence

$$\phi = \alpha(x, t)e^{-u} + \beta(x, t).$$

The coefficient of u_x requires

$$\xi_t = -2\phi_{xu} - 2\phi_x = -2\beta_x,$$

hence $\beta = -\frac{1}{8}\tau_{tt}x^2 - \frac{1}{2}\sigma_t x + \rho(t)$. The remaining terms not involving any derivatives of u are just

$$\phi_t = \phi_{xx},$$

This implies that

$$\xi = c_1 + c_4 x + 2c_5 t + 4c_6 xt,$$
$$\tau = c_2 + 2c_4 t + 4c_6 t^2,$$
$$\phi = \alpha(x, t)e^{-u} + c_3 - c_5 x - 2c_6 t - c_6 x^2,$$

where $c_1, \ldots, c_6$ are arbitrary constants and $\alpha(x, t)$ is an arbitrary solution to the *heat equation*: $\alpha_t = \alpha_{xx}$. The symmetry algebra is thus generated by

$$\begin{aligned}
\mathbf{v}_1 &= \partial_x, \\
\mathbf{v}_2 &= \partial_t, \\
\mathbf{v}_3 &= \partial_u, \\
\mathbf{v}_4 &= x\partial_x + 2t\partial_t, \\
\mathbf{v}_5 &= 2t\partial_x - x\partial_u, \\
\mathbf{v}_6 &= 4tx\partial_x + 4t^2\partial_t - (x^2 + 2t)\partial_u,
\end{aligned} \tag{2.60}$$

and

$$\mathbf{v}_\alpha = \alpha(x, t)e^{-u}\partial_u,$$

where α is any solution to the heat equation.

Note the remarkable similarity between the symmetry algebra for Burgers' equation and that derived previously for the heat equation! Indeed, if we replace u by $w = e^u$, then $\mathbf{v}_1, \ldots, \mathbf{v}_\alpha$ are changed over to the corresponding vector fields (2.55) with w replacing u. Indeed, if we set $w = e^u$ in Burgers' equation, we find

$$w_t = u_t e^u, \qquad w_{xx} = (u_{xx} + u_x^2)e^u,$$

hence w satisfies the heat equation

$$w_t = w_{xx}!$$

We have rediscovered the famous Hopf–Cole transformation reducing solutions of Burgers' equation to positive solutions of the heat equation. (For the usual form (2.58) of Burgers' equation, this takes the form

$$v = (\log w)_x = w_x/w.$$

It is much more difficult to deduce this transformation from the symmetry properties of (2.58), which, as the reader may check, has only a five-parameter symmetry group.) Since we've reduced (2.57) to the heat equation, there is no further need to discuss symmetry properties here.

Example 2.43. *The Wave Equation.* Consider the wave equation

$$u_{tt} - u_{xx} - u_{yy} = 0 \tag{2.61}$$

in two spatial dimensions. A typical vector field on the space of independent and dependent variables takes the form

$$\mathbf{v} = \xi \frac{\partial}{\partial x} + \eta \frac{\partial}{\partial y} + \tau \frac{\partial}{\partial \tau} + \phi \frac{\partial}{\partial u},$$

where ξ, η, τ, ϕ depend on x, y, t, u. In this example, it is easier to work with the infinitesimal criterion of invariance in the form (2.26), which, in the present case, takes the form

$$\phi^{tt} - \phi^{xx} - \phi^{yy} = Q \cdot (u_{tt} - u_{xx} - u_{yy}) \tag{2.62}$$

in which $Q(x, y, t, u^{(2)})$ can depend on up to second order derivatives of u. The coefficient functions ϕ^{tt}, ϕ^{xx}, ϕ^{yy} of $\mathrm{pr}^{(2)}\,\mathbf{v}$ are determined by expressions similar to those in (2.46) but with extra terms involving the y-derivatives thrown in; for example,

$$\begin{aligned}\phi^{tt} &= D_t^2(\phi - \xi u_x - \eta u_y - \tau u_t) + \xi u_{xtt} + \eta u_{ytt} + \tau u_{ttt}\\ &= D_t^2\phi - u_x D_t^2\xi - u_y D_t^2\eta - u_t D_t^2\tau - 2u_{xt} D_t\xi - 2u_{yt} D_t\eta - 2u_{tt} D_t\tau,\end{aligned}$$

etc.

To solve (2.62), we look first at the terms involving the mixed second order partial derivatives of u, namely u_{xy}, u_{xt} and u_{yt}, each of which occurs linearly on the left-hand side. This requires that ξ, η and τ do not depend on u, and, moreover

$$\xi_y + \eta_x = 0, \qquad \xi_t - \tau_x = 0, \qquad \eta_t - \tau_y = 0. \tag{2.63}$$

The coefficients of the remaining second order derivatives of u yield the relations

$$\phi_u - 2\tau_t = \phi_u - 2\xi_x = \phi_u - 2\eta_y = Q,$$

hence

$$\tau_t = \xi_x = \eta_y. \tag{2.64}$$

The equations (2.63), (2.64) are the equations for an infinitesimal conformal transformation on $\mathbb{R}^3$ with Lorentz metric $dt^2 - dx^2 - dy^2$, cf. Exercise 1.30. It is not difficult to show that ξ, η, τ are quadratic polynomials of x, y, t of the form

$$\begin{aligned}\xi &= c_1 + c_4 x - c_5 y + c_6 t + c_8(x^2 - y^2 + t^2) + 2c_9 xy + 2c_{10} xt,\\ \eta &= c_2 + c_5 x + c_4 y + c_7 t + 2c_8 xy + c_9(-x^2 + y^2 + t^2) + 2c_{10} yt,\\ \tau &= c_3 + c_6 x + c_7 y + c_4 t + 2c_8 xt + 2c_9 yt + c_{10}(x^2 + y^2 + t^2),\end{aligned}$$

where $c_1, \ldots, c_{10}$ are constants. For instance, we find

$$\xi_{xxx} = \eta_{xxy} = -\xi_{xyy} = -\tau_{yyt} = -\eta_{ytt} = -\xi_{xtt} = -\tau_{xxt} = -\xi_{xxx},$$

hence all these third order derivatives vanish; similar arguments prove that *all* third order derivatives of ξ, η and τ are zero, and the structure of the resulting quadratic polynomials follows easily from (2.63), (2.64).

Next the coefficient of u_x^2 (or u_y^2 or u_t^2) in (2.62) says $\phi_{uu} = 0$, so

$$\phi(x, y, t, u) = \beta(x, y, t)u + \alpha(x, y, t).$$

Finally, the coefficients of the linear terms in the first order derivatives of u, and the terms without u in them at all yield the relations

$$2\beta_x = \xi_{xx} + \xi_{yy} - \xi_{tt},$$
$$2\beta_y = \eta_{xx} + \eta_{yy} - \eta_{tt},$$
$$2\beta_t = \tau_{tt} - \tau_{xx} - \tau_{yy},$$
$$\alpha_{tt} - \alpha_{xx} - \alpha_{yy} = 0.$$

Thus α is any solution of the wave equation, and

$$\beta = c_{11} - c_8 x - c_9 y - c_{10} t.$$

This gives the most general solution of the determining equations of the symmetry group of the wave equation. We have thus reproved the well-known result that the infinitesimal symmetry group of the wave equation is spanned by the ten vector fields

$$\partial_x, \quad \partial_y, \quad \partial_t, \qquad \text{translations,}$$

$$\mathbf{r}_{xy} = -y\partial_x + x\partial_y, \quad \mathbf{r}_{xt} = t\partial_x + x\partial_t, \quad \mathbf{r}_{yt} = t\partial_y + y\partial_t, \qquad \text{hyperbolic rotations,}$$

$$\mathbf{d} = x\partial_x + y\partial_y + t\partial_t, \qquad \text{dilatation,} \tag{2.65}$$

$$\left.\begin{aligned} \mathbf{i}_x &= (x^2 - y^2 + t^2)\partial_x + 2xy\partial_y + 2xt\partial_t - xu\partial_u, \\ \mathbf{i}_y &= 2xy\partial_x + (y^2 - x^2 + t^2)\partial_y + 2yt\partial_t - yu\partial_u, \\ \mathbf{i}_t &= 2xt\partial_x + 2yt\partial_y + (x^2 + y^2 + t^2)\partial_t - tu\partial_u, \end{aligned}\right\} \qquad \text{inversions,}$$

which generate the conformal algebra for $\mathbb{R}^3$ with the given Lorentz metric, and the additional vector fields

$$u\partial_u, \qquad \mathbf{v}_\alpha = \alpha(x, y, t)\partial_u,$$

for α an arbitrary solution of the wave equation, reflecting the linearity of the equation.

The corresponding group transformations for the translations and dilatation are easily found. Of the rotations, owing to the indefinite character of the underlying metric $dt^2 - dx^2 - dy^2$, only the rotations in the (x, y)-plane are true rotations; the other two are "hyperbolic rotations". For example $\mathbf{r}_{xt}$

generates the group

$$(x, y, t) \mapsto (x \cosh \varepsilon + t \sinh \varepsilon, y, x \sinh \varepsilon + t \cosh \varepsilon).$$

The inversional groups can be constructed, as in Exercise 1.30, from the primary inversion

$$I(x, y, t) = \left(\frac{x}{t^2 - x^2 - y^2}, \frac{y}{t^2 - x^2 - y^2}, \frac{t}{t^2 - x^2 - y^2}\right),$$

which is defined provided (x, y, t) does not lie in the light cone $t^2 = x^2 + y^2$. We find that the group generated by $\mathbf{i}_x$ say, is given by first inverting, then translating the x-direction, and then re-inverting:

$$\exp(\varepsilon \mathbf{i}_x) = I \circ \exp(\varepsilon \partial_x) \circ I.$$

The general formula is

$$(x, y, t) \mapsto \left(\frac{x + \varepsilon(t^2 - x^2 - y^2)}{1 - 2\varepsilon x - \varepsilon^2(t^2 - x^2 - y^2)}, \frac{y}{1 - 2\varepsilon x - \varepsilon^2(t^2 - x^2 - y^2)}, \frac{t}{1 - 2\varepsilon x - \varepsilon^2(t^2 - x^2 - y^2)}\right),$$

which is well defined even for (x, y, t) in the light cone (which is an invariant subvariety). The corresponding transformation of u under $\exp(\varepsilon \mathbf{i}_x)$ is then

$$u \mapsto \sqrt{1 - 2\varepsilon x - \varepsilon^2(t^2 - x^2 - y^2)}\, u.$$

We conclude that if $u = f(x, y, t)$ is a solution to the wave equation, so is

$$\tilde{u} = \frac{1}{\sqrt{1 + 2\varepsilon x - \varepsilon^2(t^2 - x^2 - y^2)}} f\left(\frac{x - \varepsilon(t^2 - x^2 - y^2)}{1 + 2\varepsilon x - \varepsilon^2(t^2 - x^2 - y^2)}, \frac{y}{1 + 2\varepsilon x - \varepsilon^2(t^2 - x^2 - y^2)}, \frac{t}{1 + 2\varepsilon x - \varepsilon^2(t^2 - x^2 - y^2)}\right).$$

Example 2.44. *The Korteweg-de Vries Equation.* As a higher order example, we consider the Korteweg-de Vries equation

$$u_t + u_{xxx} + u u_x = 0, \tag{2.66}$$

which arises in the theory of long waves in shallow water and other physical systems in which both nonlinear and dispersive effects are relevant. A vector field $\mathbf{v} = \xi \partial_x + \tau \partial_t + \phi \partial_u$ generates a one-parameter symmetry group if and only if

$$\phi^t + \phi^{xxx} + u\phi^x + u_x \phi = 0 \tag{2.67}$$

whenever u satisfies (2.66). Here ϕ^t and ϕ^x, the coefficients of the first prolongation of $\mathbf{v}$, are determined by the explicit prolongation formulae (2.45); the coefficient of $\partial/\partial u_{xxx}$ in $\mathrm{pr}^{(3)}\,\mathbf{v}$ is

$$\phi^{xxx} = D_x^3 \phi - u_x D_x^3 \xi - u_t D_x^3 \tau - 3u_{xx} D_x^2 \xi - 3u_{xt} D_x^2 \tau - 3u_{xxx} D_x \xi - 3u_{xxt} D_x \tau.$$

Substituting into (2.67) and replacing u_t by $-u_{xxx} - uu_x$ wherever it occurs, we obtain the determining equations for the symmetry group. To analyze these, we work our way down the order of the derivatives which appear. The coefficient of u_{xxt} is $D_x\tau = 0$, hence τ depends only on t. The coefficient of u_{xx}^2 shows that $\xi_u = 0$. From the coefficient of u_{xxx}, we find $\tau_t = 3\xi_x$ (the ϕ_u-terms cancelling), hence $\xi = \frac{1}{3}\tau_t x + \sigma(t)$. Now the coefficient of u_{xx} reveals that $\phi_{uu} = 0 = \phi_{xu}$, so ϕ is linear in u, the coefficient of u being a function of t alone. The remaining terms in (2.67) are those involving u_x, which give

$$-\xi_t - u(\phi_u - \tau_t) + u(\phi_u - \xi_x) + \phi = 0,$$

and those without any derivatives of u,

$$\phi_t + \phi_{xxx} + u\phi_x = 0.$$

These all have the general solution

$$\xi = c_1 + c_3 t + c_4 x,$$
$$\tau = c_2 + 3c_4 t,$$
$$\phi = c_3 - 2c_4 u,$$

where c_1, c_2, c_3, c_4 are arbitrary constants. Therefore the symmetry algebra of the Korteweg-de Vries equation is spanned by the four vector fields

$$\begin{aligned} \mathbf{v}_1 &= \partial_x, && \text{space translation,} \\ \mathbf{v}_2 &= \partial_t, && \text{time translation,} \\ \mathbf{v}_3 &= t\partial_x + \partial_u, && \text{Galilean boost,} \\ \mathbf{v}_4 &= x\partial_x + 3t\partial_t - 2u\partial_u, && \text{scaling.} \end{aligned} \tag{2.68}$$

Their commutator table is

	$\mathbf{v}_1$	$\mathbf{v}_2$	$\mathbf{v}_3$	$\mathbf{v}_4$
$\mathbf{v}_1$	0	0	0	$\mathbf{v}_1$
$\mathbf{v}_2$	0	0	$\mathbf{v}_1$	$3\mathbf{v}_2$
$\mathbf{v}_3$	0	$-\mathbf{v}_1$	0	$-2\mathbf{v}_3$
$\mathbf{v}_4$	$-\mathbf{v}_1$	$-3\mathbf{v}_2$	$2\mathbf{v}_3$	0

Exponentiation shows that if $u = f(x, t)$ is a solution of the Korteweg-de Vries equation, so are

$$\begin{aligned} u^{(1)} &= f(x - \varepsilon, t), \\ u^{(2)} &= f(x, t - \varepsilon), \\ u^{(3)} &= f(x - \varepsilon t, t) + \varepsilon, \\ u^{(4)} &= e^{-2\varepsilon} f(e^{-\varepsilon}x, e^{-3\varepsilon}t). \end{aligned} \qquad \varepsilon \in \mathbb{R}.$$

These can easily be checked by inspection. (For the reader familiar with the many remarkable "soliton" properties of the Korteweg-de Vries equation, this list of symmetries may seem disappointingly small. Further symmetry properties, reflecting the existence of infinitely many conservation laws and, presumably, the linearization of the inverse scattering method, cf. Newell, [1], will require our development of the theory of generalized symmetries in Chapters 5 and 7.)

Example 2.45. *The Euler Equations.* As a last illustration of the basic method of computing symmetry groups, we consider the system of Euler equations for the motion of an inviscid, incompressible ideal fluid in a three-dimensional domain. Here there are four independent variables, $\mathbf{x} = (x, y, z)$ being spatial coordinates and t the time, together with four dependent variables, the velocity field $\mathbf{u} = (u, v, w)$ and the pressure p. (The density ρ is normalized to be 1.) In vector notation, the system has the form

$$\frac{\partial \mathbf{u}}{\partial t} + \mathbf{u} \cdot \nabla \mathbf{u} = -\nabla p, \qquad \nabla \cdot \mathbf{u} = 0, \tag{2.69}$$

in which the components of the nonlinear terms $\mathbf{u} \cdot \nabla \mathbf{u}$ are

$$(uu_x + vu_y + wu_z, uv_x + vv_y + wv_z, uw_x + vw_y + ww_z).$$

An infinitesimal symmetry of the Euler equations will be a vector field of the form

$$\mathbf{v} = \xi \partial_x + \eta \partial_y + \zeta \partial_z + \tau \partial_t + \phi \partial_u + \psi \partial_v + \chi \partial_w + \pi \partial_p,$$

where $\xi, \eta, \ldots, \pi$ are functions of $\mathbf{x}$, t, $\mathbf{u}$ and p. Applying the first prolongation $\mathrm{pr}^{(1)} \mathbf{v}$ to the Euler equations (2.69), we find the following system of symmetry equations

$$\phi^t + u\phi^x + v\phi^y + w\phi^z + u_x\phi + u_y\psi + u_z\chi = -\pi^x, \tag{2.70a}$$

$$\psi^t + u\psi^x + v\psi^y + w\psi^z + v_x\phi + v_y\psi + v_z\chi = -\pi^y, \tag{2.70b}$$

$$\chi^t + u\chi^x + v\chi^y + w\chi^z + w_x\phi + w_y\psi + w_z\chi = -\pi^z, \tag{2.70c}$$

$$\phi^x + \psi^y + \chi^z = 0, \tag{2.70d}$$

which must be satisfied whenever $\mathbf{u}$ and p satisfy (2.69). Here ϕ^t, ψ^x, etc. are the coefficients of the first order derivatives $\partial/\partial u_t$, $\partial/\partial v_x$, etc. appearing in $\mathrm{pr}^{(1)} \mathbf{v}$; typical expressions for these functions follow from the prolongation formula (2.43), so

$$\phi^t = D_t\phi - u_xD_t\xi - u_yD_t\eta - u_zD_t\zeta - u_tD_t\tau,$$

$$\psi^x = D_x\psi - v_xD_x\xi - v_yD_x\eta - v_zD_x\zeta - v_tD_x\tau,$$

and so on.

Since (2.70) need only hold on solutions of (2.69), we can substitute for p_x, p_y, p_z and w_z wherever they occur in (2.70) using their expressions from the four equations in (2.69). We may then equate all the coefficients of the remaining first order derivatives of $\mathbf{u}$, p in (2.70) and solve the resulting system of determining equations for $\xi, \eta, \ldots, \pi$.

As a first step, let us show that the symmetry is necessarily projectable, meaning that ξ, η, ζ and τ only depend on $\mathbf{x}$ and t. The coefficient of p_t in (2.70a) is

$$\phi_p - \xi_p u_x - \eta_p u_y - \zeta_p u_z - \tau_p u_t = D_x \tau = \tau_x + \tau_u u_x + \tau_v v_x + \tau_w w_x + \tau_p p_x,$$

Therefore $\tau_v = \tau_w = 0$, and, by consideration of the same coefficient in (2.70b), $\tau_u = 0$ also. Furthermore, if we substitute for p_x according to (2.69), we find

$$\phi_p = \tau_x, \qquad \psi_p = \tau_y, \qquad \chi_p = \tau_z, \tag{2.71}$$

$$\xi_p = u\tau_p, \qquad \eta_p = v\tau_p, \qquad \zeta_p = w\tau_p, \tag{2.72}$$

where the equations for ψ_p and χ_p come from similar considerations in (2.70b, c). Next consider the quadratic monomial $v_t v_x$ in (2.70a). This can also arise from the monomials $p_y v_x$, $p_y v_t$ and p_y^2, all of which only appear in π^x. The resulting coefficient is $0 = -\eta_v$. Similarly, the coefficient of $v_t w_x$ in (2.70a) proves that $\eta_w = 0$. Further analysis of quadratic terms in (2.70a, b, c) proves that ξ, η, ζ are independent of u, v, w. Then differentiating (2.72) with respect to u, v and w we find $\tau_p = 0$, hence $\xi_p = \eta_p = \zeta_p = 0$ and the symmetry is projectable.

The next step is to look at the coefficients of u_t, v_t and w_t in (2.70d), keeping in mind that these can also arise from ∇p upon substitution. This implies that

$$\phi_p + \tau_x = \psi_p + \tau_y = \chi_p + \tau_z = 0.$$

Comparison with (2.71) proves that τ depends on t alone, and ϕ, ψ, χ are independent of the pressure. Consider next the coefficients of v_t and v_x in (2.70a), which are

$$\phi_v = -\eta_x, \qquad \phi_v = -\eta_x - \pi_v.$$

Thus $\pi_v = 0$, and, by similar considerations, π does not depend on u or w either. From the coefficients of u_t and w_t we also find that

$$\phi_u = \tau_t - \xi_x + \pi_p, \qquad \phi_w = -\zeta_x,$$

and so on. These all imply that ϕ, ψ, χ have the general form

$$\phi = (\tau_t - \xi_x + \pi_p)u - \eta_x v - \zeta_x w + \hat{\phi},$$

$$\psi = -\xi_y u + (\tau_t - \eta_y + \pi_p)v - \zeta_y w + \hat{\psi},$$

$$\chi = -\xi_z u - \eta_z v + (\tau_t - \zeta_z + \pi_p)w + \hat{\chi},$$

where $\hat{\phi}$, $\hat{\psi}$ and $\hat{\chi}$ depend only on $\mathbf{x}$ and t. The coefficients of the spatial derivatives of $\mathbf{u}$ in (2.70a, b, c) then require

$$\hat{\phi} = \xi_t, \qquad \hat{\psi} = \eta_t, \qquad \hat{\chi} = \zeta_t,$$
$$\xi_x = \eta_y = \zeta_z = \tau_t + \tfrac{1}{2}\pi_p,$$
$$\xi_y + \eta_x = \xi_z + \zeta_x = \eta_z + \zeta_y = 0.$$

In particular, the spatial component $\xi\partial_x + \eta\partial_y + \zeta\partial_z$ of $\mathbf{v}$ generates a (time-dependent) conformal symmetry group of $\mathbb{R}^3$ with the Euclidean metric. The remaining terms in (2.70) involve no derivatives of $\mathbf{u}$ or p. These require ξ, η, ζ to be linear in x, y, z and, furthermore,

$$\xi_{yt} = \xi_{zt} = \eta_{xt} = \eta_{zt} = \zeta_{xt} = \zeta_{yt} = 0,$$
$$\xi_{xt} = \eta_{yt} = \zeta_{zt} = \tau_{tt},$$
$$\xi_{tt} = -\pi_x, \qquad \eta_{tt} = -\pi_y, \qquad \zeta_{tt} = -\pi_z.$$

Therefore

$$\begin{aligned}
\xi &= \delta_t x + c_1 y - c_2 z + \alpha,\\
\eta &= -c_1 x + \delta_t y + c_3 z + \beta,\\
\zeta &= c_2 x - c_3 y + \delta_t z + \gamma,\\
\tau &= 2\delta + c_4 t + c_5,\\
\phi &= -(\delta_t + c_4)u + c_1 v - c_2 w + \alpha_t,\\
\psi &= -c_1 u - (\delta_t + c_4)v + c_3 w + \beta_t,\\
\chi &= c_2 u - c_3 v - (\delta_t + c_4)w + \gamma_t,\\
\pi &= -2(\delta_t + c_4)p - \tfrac{1}{2}\delta_{tt}(x^2 + y^2 + z^2) - \alpha_{tt}x - \beta_{tt}y - \gamma_{tt}z + \theta,
\end{aligned}$$

in which α, β, γ, δ and θ are functions of t, and c_1, c_2, c_3, c_4, c_5 constants. Finally, the divergence-free condition (2.70d) imposes the further restriction that $\delta_{tt} = 0$, so $\delta = c_6 t + c_7$.

We have thus shown that the symmetry group of the Euler equations in three dimensions is generated by the vector fields

$$\left.\begin{aligned}
\mathbf{v}_\alpha &= \alpha\partial_x + \alpha_t\partial_u - \alpha_{tt}x\partial_p,\\
\mathbf{v}_\beta &= \beta\partial_y + \beta_t\partial_v - \beta_{tt}y\partial_p,\\
\mathbf{v}_\gamma &= \gamma\partial_z + \gamma_t\partial_w - \gamma_{tt}z\partial_p,
\end{aligned}\right\} \quad \text{(moving coordinates)}$$

$$\mathbf{v}_0 = \partial_t, \qquad \text{(time translation)}$$

$$\left.\begin{aligned}
\mathbf{d}_1 &= x\partial_x + y\partial_y + z\partial_z + t\partial_t,\\
\mathbf{d}_2 &= t\partial_t - u\partial_u - v\partial_v - w\partial_w - 2p\partial_p,
\end{aligned}\right\} \quad \text{(scaling)} \tag{2.73}$$

$$\left.\begin{aligned}
\mathbf{r}_{xy} &= y\partial_x - x\partial_y + v\partial_u - u\partial_v,\\
\mathbf{r}_{zx} &= x\partial_z - z\partial_x + u\partial_w - w\partial_u,\\
\mathbf{r}_{yz} &= z\partial_y - y\partial_z + w\partial_v - v\partial_w,
\end{aligned}\right\} \quad \text{(rotations)}$$

$$\mathbf{v}_\theta = \theta\partial_p, \qquad \text{(pressure changes)}$$

in which α, β, γ and θ are arbitrary functions of t. The corresponding one-parameter groups of symmetries of the Euler equations are then:

(a) Transformation to an arbitrarily moving coordinate system:

$$G_{\boldsymbol{\alpha}}\colon (\mathbf{x}, t, \mathbf{u}, p) \mapsto (\mathbf{x} + \varepsilon\boldsymbol{\alpha}(t), t, \mathbf{u} + \varepsilon\boldsymbol{\alpha}_t, p - \varepsilon\mathbf{x}\cdot\boldsymbol{\alpha}_{tt} - \tfrac{1}{2}\varepsilon^2\boldsymbol{\alpha}\cdot\boldsymbol{\alpha}_{tt}),$$

where $\boldsymbol{\alpha} = (\alpha, \beta, \gamma)$ and $G_{\boldsymbol{\alpha}}$ is generated by the linear combination $\mathbf{v}_{\boldsymbol{\alpha}} = \mathbf{v}_\alpha + \mathbf{v}_\beta + \mathbf{v}_\gamma$ of the first three vector fields.

(b) Time translations:

$$G_0\colon (\mathbf{x}, t, \mathbf{u}, p) \mapsto (\mathbf{x}, t + \varepsilon, \mathbf{u}, p).$$

(c) Scale transformations:

$$G_1\colon (\mathbf{x}, t, \mathbf{u}, p) \mapsto (\lambda\mathbf{x}, \lambda t, \mathbf{u}, p),$$

$$G_2\colon (\mathbf{x}, t, \mathbf{u}, p) \mapsto (\mathbf{x}, \lambda t, \lambda^{-1}\mathbf{u}, \lambda^{-2}p),$$

where $\lambda = e^\varepsilon$ is a multiplicative group parameter.

(d) The group

$$\mathrm{SO}(3)\colon (\mathbf{x}, t, \mathbf{u}, p) \mapsto (R\mathbf{x}, t, R\mathbf{u}, p)$$

of simultaneous rotations in both space and the velocity field $\mathbf{u}$. Here R is an arbitrary 3×3 orthogonal matrix.

(e) Pressure changes:

$$G_p\colon (\mathbf{x}, t, \mathbf{u}, p) \mapsto (\mathbf{x}, t, \mathbf{u}, p + \varepsilon\theta(t)).$$

The corresponding action on solutions of the Euler equations says that if $\mathbf{u} = \mathbf{f}(\mathbf{x}, t)$, $p = g(\mathbf{x}, t)$ are solutions, so are

$$\begin{aligned}
G_{\boldsymbol{\alpha}}&\colon & \mathbf{u} &= \mathbf{f}(\mathbf{x} - \varepsilon\boldsymbol{\alpha}(t), t) + \varepsilon\boldsymbol{\alpha}_t, & p &= g(\mathbf{x} - \varepsilon\boldsymbol{\alpha}(t), t) - \varepsilon\mathbf{x}\cdot\boldsymbol{\alpha}_{tt} + \tfrac{1}{2}\varepsilon^2\boldsymbol{\alpha}\cdot\boldsymbol{\alpha}_{tt},\\
G_0&\colon & \mathbf{u} &= \mathbf{f}(\mathbf{x}, t - \varepsilon), & p &= g(\mathbf{x}, t - \varepsilon),\\
G_1&\colon & \mathbf{u} &= \mathbf{f}(\lambda\mathbf{x}, \lambda t), & p &= g(\lambda\mathbf{x}, \lambda t),\\
G_2&\colon & \mathbf{u} &= \lambda\mathbf{f}(\mathbf{x}, \lambda t), & p &= \lambda^2 g(\mathbf{x}, \lambda t),\\
\mathrm{SO}(3)&\colon & \mathbf{u} &= R\mathbf{f}(R^{-1}\mathbf{x}, t), & p &= g(R^{-1}\mathbf{x}, t),\\
G_p&\colon & \mathbf{u} &= \mathbf{f}(\mathbf{x}, t), & p &= g(\mathbf{x}, t) + \varepsilon\theta(t).
\end{aligned}$$

(In G_1 and G_2 we have replaced λ by λ^{-1}.) Note that in our change to a moving coordinate system $G_{\boldsymbol{\alpha}}$, we must adjust the pressure according to the induced acceleration $\varepsilon\boldsymbol{\alpha}_{tt}$. The final group G_p results from the fact that the pressure p is only defined up to the addition of an arbitrary function of t. This completes the list of symmetries of the Euler equations.

2.5. Integration of Ordinary Differential Equations

One of the most appealing applications of Lie group theory is to the problem of integrating ordinary differential equations. Lie's fundamental observation was that knowledge of a sufficiently large group of symmetries of a system of ordinary differential equations allows one to integrate the system by quadratures (indefinite integrals) and thereby deduce the general solution. This approach unifies and significantly extends the various special methods introduced for the integration of certain types of first order equations such as homogeneous, separable, exact and so on. Similar results hold for systems of ordinary differential equations. In this section, a comprehensive survey of these methods is presented.

First Order Equations

We begin by considering a single first order ordinary differential equation

$$\frac{du}{dx} = F(x, u). \tag{2.74}$$

It will be shown that if this equation is invariant under a one-parameter group of transformations, then it can be integrated by quadrature. If G is a one-parameter group of transformations on an open subset $M \subset X \times U \simeq \mathbb{R}^2$, let

$$\mathbf{v} = \xi(x, u)\frac{\partial}{\partial x} + \phi(x, u)\frac{\partial}{\partial u}$$

be its infinitesimal generator. The first prolongation of $\mathbf{v}$ is the vector field

$$\mathrm{pr}^{(1)}\,\mathbf{v} = \xi\frac{\partial}{\partial x} + \phi\frac{\partial}{\partial u} + \phi^x\frac{\partial}{\partial u_x}, \tag{2.75}$$

where

$$\phi^x = D_x\phi - u_x D_x\xi = \phi_x + (\phi_u - \xi_x)u_x - \xi_u u_x^2.$$

Thus the infinitesimal condition that G be a symmetry group of (2.74) is

$$\frac{\partial\phi}{\partial x} + \left(\frac{\partial\phi}{\partial u} - \frac{\partial\xi}{\partial x}\right)F - \frac{\partial\xi}{\partial u}F^2 = \xi\frac{\partial F}{\partial x} + \phi\frac{\partial F}{\partial u}, \tag{2.76}$$

and any solution $\xi(x, u)$, $\phi(x, u)$ of the partial differential equation (2.76) generates a one-parameter symmetry group of our ordinary differential equation. Of course, in practice finding solutions of the determining equation (2.76) is usually a much more difficult problem than solving the original ordinary differential equation. However, led on by inspired guess-work, or geometric intuition, we may be able to ascertain a particular solution of (2.76) which will allow us to integrate (2.74). Herein lies the art of Lie's method.

Once we have found a symmetry group G, there are several different methods we can employ to integrate (2.74). Suppose $\mathbf{v}$ is the infinitesimal generator of the symmetry group, and assume that $\mathbf{v}|_{(x_0, u_0)} \neq 0$. (If the vector field $\mathbf{v}$ vanishes at a point (x_0, u_0), then we will expect some kind of singularity for solutions near this point. The behaviour of solutions $u = f(x)$ near such a singularity can be deduced by extrapolation once the equation has been integrated at nearby values of x.) According to Proposition 1.29, we can introduce new coordinates

$$y = \eta(x, u), \qquad w = \zeta(x, u), \tag{2.77}$$

near (x_0, u_0) such that in the (y, w)-coordinates the symmetry vector field has the simple translational form $\mathbf{v} = \partial/\partial w$, with first prolongation

$$\mathrm{pr}^{(1)}\, \mathbf{v} = \mathbf{v} = \partial/\partial w$$

also. Thus in the new coordinate system, in order to be invariant, the differential equation must be independent of w, so (2.74) is equivalent to the elementary equation

$$\frac{dw}{dy} = H(y),$$

for some function H. This equation is trivially integrated by quadrature, with

$$w = \int H(y)\, dy + c$$

for some constant c. Re-substituting the expressions (2.77) for w and y, we obtain a solution $u = f(x)$ of our original system in implicit form.

The change of variables (2.77) is constructed using the methods for finding group invariants presented in Section 2.1. Indeed, (1.16) implies that $\mathbf{v}$ is transformed into the form $\partial/\partial w$ provided η and ζ satisfy the linear partial differential equations

$$\mathbf{v}(\eta) = \xi \frac{\partial \eta}{\partial x} + \phi \frac{\partial \eta}{\partial u} = 0, \tag{2.78a}$$

$$\mathbf{v}(\zeta) = \xi \frac{\partial \zeta}{\partial x} + \phi \frac{\partial \zeta}{\partial u} = 1. \tag{2.78b}$$

The first of these equations just says that $\eta(x, u)$ is an invariant of the group generated by $\mathbf{v}$. We can thus find η by solving the associated characteristic ordinary differential equation

$$\frac{dx}{\xi(x, u)} = \frac{du}{\phi(x, u)}. \tag{2.79}$$

Often the corresponding solution ζ of (2.78b) can be found by inspection. More systematically, we can introduce an auxiliary variable v and note that $\zeta(x, u)$ satisfies (2.78b) if and only if the function $\chi(x, u, v) = v - \zeta(x, u)$ is an invariant of the vector field $\mathbf{w} = \mathbf{v} + \partial_v = \xi \partial_x + \phi \partial_u + \partial_v$. Thus we require

$$\mathbf{w}(\chi) = \xi \frac{\partial \chi}{\partial x} + \phi \frac{\partial \chi}{\partial u} + \frac{\partial \chi}{\partial v} = 0.$$

This we can again solve by the method of characteristics,

$$\frac{dx}{\xi(x, u)} = \frac{du}{\phi(x, u)} = \frac{dv}{1}, \tag{2.80}$$

where we seek a solution of the form $v - \zeta(x, u) = k$, for k an arbitrary constant of integration.

In general, it may be just as difficult to solve (2.79) and (2.80), being again ordinary differential equations, as it was to integrate the original differential equation. In particular, if

$$\phi(x, u)/\xi(x, u) = F(x, u), \tag{2.81}$$

then we automatically have a solution of the determining equation (2.76), so such a vector field $\mathbf{v} = \xi\partial_x + \phi\partial_u$ is always a symmetry of the equation. In this case, finding the invariant $\eta(x, u)$ of the group, i.e. solving (2.79), is exactly the same problem as integrating the original equation, so the method is of no help. Only when the group of symmetries is of a reasonably simple form, so that we can explicitly solve (2.79), (2.80), do we stand any chance of making progress towards the solution of our problem.

Example 2.46. A homogeneous equation is one of the form

$$\frac{du}{dx} = F\left(\frac{u}{x}\right),$$

where F only depends on the ratio of u to x. Such an equation has the group of scaling transformations

$$G: (x, u) \mapsto (\lambda x, \lambda u), \qquad \lambda > 0,$$

as a symmetry group. This can be seen directly from the form of the first prolongation of G,

$$\mathrm{pr}^{(1)}\, G: (x, u, u_x) \mapsto (\lambda x, \lambda u, u_x),$$

which obviously leaves the equation invariant. Alternatively, we can look at the infinitesimal generator

$$\mathbf{v} = x\frac{\partial}{\partial x} + u\frac{\partial}{\partial u},$$

which, according to (2.75), has first prolongation $\mathrm{pr}^{(1)}\, \mathbf{v} = \mathbf{v}$, and use the infinitesimal criterion of invariance.

New coordinates y, w satisfying (2.78) are given by

$$y = \frac{u}{x}, \qquad w = \log x.$$

Employing the chain rule, we find

$$\frac{du}{dx} = \frac{du/dy}{dx/dy} = \frac{x(1 + yw_y)}{xw_y} = \frac{1 + yw_y}{w_y},$$

so the equation takes the form

$$\frac{dw}{dy} = \frac{1}{F(y) - y}.$$

This has the solution

$$w = \int \frac{dy}{F(y) - y} + c,$$

which in turn defines u implicitly as a function of x once we set $w = \log x$, $y = u/x$. For example, if the equation is

$$\frac{du}{dx} = \frac{u^2 + 2xu}{x^2} = \left(\frac{u}{x}\right)^2 + 2\frac{u}{x},$$

so $F(y) = y^2 + 2y$, then, in the coordinates $y = u/x$, $w = \log x$, we have

$$\frac{dw}{dy} = \frac{1}{y^2 + y}.$$

The solution is

$$w = -\log(1 + y^{-1}) + c,$$

or, in terms of the original variables,

$$\log x = -\log\left(1 + \frac{x}{u}\right) + c.$$

This can be solved explicitly for u as a function of x:

$$u = \frac{x^2}{\tilde{c} - x},$$

where $\tilde{c} = e^c$.

Although the answer is of course the same, the above procedure is not quite the usual one learned in a first course in ordinary differential equations. Here the roles of w and y are reversed, with w being the new independent variable. For many first order equations, it is often expedient to adopt this latter strategy. In the present case, we can drop the logarithm and treat x and $y = u/x$ as the new variables. Then

$$\frac{du}{dx} = \frac{d}{dx}(xy) = x\frac{dy}{dx} + y,$$

and we obtain the solution in the form

$$\int \frac{dy}{F(y) - y} = \int \frac{dx}{x} = \log x + c.$$

The equivalence of the methods is clear. Finally, note that in general the origin $u = x = 0$ is a singular point, corresponding to the point where $\mathbf{v}$ vanishes.

Example 2.47. Let G be the rotation group SO(2), whose infinitesimal generator

$$\mathrm{pr}^{(1)}\,\mathbf{v} = -u\frac{\partial}{\partial x} + x\frac{\partial}{\partial u} + (1+u_x^2)\frac{\partial}{\partial u_x}$$

was computed in Example 2.29. It is a straightforward computation to check that any equation of the form

$$\frac{du}{dx} = \frac{u + xH(r)}{x - uH(r)}, \tag{2.82}$$

where $H(r) = H(\sqrt{x^2+u^2})$ is any function of the radius, admits SO(2) as a symmetry group. Polar coordinates r, θ, with $x = r\cos\theta$, $u = r\sin\theta$, are the new coordinates satisfying (2.78) since $\mathbf{v} = \partial/\partial\theta$ in these coordinates. Furthermore,

$$\frac{du}{dx} = \frac{du/dr}{dx/dr} = \frac{\sin\theta + r\theta_r\cos\theta}{\cos\theta - r\theta_r\sin\theta}.$$

Substituting into (2.82) and solving for $d\theta/dr$, we find

$$\frac{d\theta}{dr} = \frac{1}{r}H(r),$$

hence

$$\theta = \int \frac{H(r)}{r}\,dr + c$$

is the general solution. For example, if $H(r) = 1$, we have the equation of Example 2.32.

An alternative method for solving first order equations invariant under a one-parameter group is based on the construction of an integrating factor. We rewrite (2.74) as a total differential equation

$$P(x,u)\,dx + Q(x,u)\,du = 0, \tag{2.83}$$

so $F = -P/Q$. The equation is *exact* provided $\partial P/\partial u = \partial Q/\partial x$, and in this case we can find the solution in implicit form $T(x,u) = c$ by requiring

$$\frac{\partial T}{\partial x} = P, \qquad \frac{\partial T}{\partial u} = Q.$$

(This assumes that the domain M is simply-connected.) If (2.83) is not exact, we must search for an integrating factor $R(x,u)$ such that when we multiply by R we do obtain an exact equation.

Theorem 2.48. *Suppose the equation $P\,dx + Q\,du = 0$ has a one-parameter symmetry group with infinitesimal generator $\mathbf{v} = \xi\partial_x + \phi\partial_u$. Then the function*

$$R(x,u) = \frac{1}{\xi(x,u)P(x,u) + \phi(x,u)Q(x,u)} \tag{2.84}$$

is an integrating factor.

Proof. Using the infinitesimal criterion of invariance (2.76), we find that **v** is a symmetry of (2.83) if and only if

$$\left(\xi\frac{\partial P}{\partial x}+\phi\frac{\partial P}{\partial u}\right)Q-\left(\xi\frac{\partial Q}{\partial x}+\phi\frac{\partial Q}{\partial u}\right)P+\frac{\partial\phi}{\partial x}Q^2-\left(\frac{\partial\phi}{\partial u}-\frac{\partial\xi}{\partial x}\right)PQ-\frac{\partial\xi}{\partial u}P^2=0. \tag{2.85}$$

The condition that R be an integrating factor is

$$\frac{\partial}{\partial u}(RP)=\frac{\partial}{\partial x}(RQ).$$

Substituting the formula for R, this becomes

$$R^2\left\{\phi\left(Q\frac{\partial P}{\partial u}-P\frac{\partial Q}{\partial u}\right)-\frac{\partial\xi}{\partial u}P^2-\frac{\partial\phi}{\partial u}PQ\right\}$$
$$=R^2\left\{\xi\left(P\frac{\partial Q}{\partial x}-Q\frac{\partial P}{\partial x}\right)-\frac{\partial\xi}{\partial x}PQ-\frac{\partial\phi}{\partial x}Q^2\right\}.$$

Comparison with the symmetry condition (2.85) proves the theorem. □

For example, in the case of the rotation group, the equation takes the general form

$$(u+xH(r))\,dx+(uH(r)-x)\,du=0.$$

The integrating factor is then

$$\frac{1}{-u(u+xH)+x(uH-x)}=\frac{-1}{x^2+u^2}.$$

For example, let $H(r)=1$, so we have

$$(u+x)\,dx+(u-x)\,du=0.$$

Multiplying by $(x^2+u^2)^{-1}$ we get an exact equation

$$0=\frac{u+x}{x^2+u^2}\,dx+\frac{u-x}{x^2+u^2}\,du=d\left[\tfrac{1}{2}\log(x^2+u^2)-\arctan\frac{u}{x}\right],$$

hence we re-derive the logarithmic spiral solutions $r=ce^{\theta}$ found in Example 2.32.

Note that if

$$\xi P+\phi Q\equiv 0$$

for all (x,u), then the integrating factor does not exist. This happens precisely in the case (2.81) when the computation of the symmetry group invariants is the same problem as solving the ordinary differential equation itself. In this case, both the invariant method and the integrating factor method fail to provide solutions.

In practice, the integrating factor method is perhaps easier to implement in that we do not need to find the solutions η, ζ to the auxiliary pair of partial differential equations (2.78). However, if one must consider a large number of equations all with the same symmetry group, this slight advantage is nullified by the relative difficulty of finding potentials T for each of the requisite exact differentials.

Higher Order Equations

Symmetry groups can also be used to aid in the solution of higher order ordinary differential equations. The integration method based on the invariants of the group extends straightforwardly. Let

$$\Delta(x, u^{(n)}) = \Delta(x, u, u_x, \ldots, u_n) = 0, \tag{2.86}$$

where $u_n \equiv d^n u/dx^n$, be a single n-th order differential equation involving the single dependent variable u. The basic result in this case is that if we know a one-parameter symmetry group of this equation, then we can reduce the order of the equation by one.

To see this, we first choose coordinates $y = \eta(x, u)$, $w = \zeta(x, u)$ as in (2.78) such that the group transforms into a group of translations with infinitesimal generator $\mathbf{v} = \partial/\partial w$. Employing the chain rule, we can express the derivatives of u with respect to x in terms of y, w and the derivatives of w with respect to y,

$$\frac{d^k u}{dx^k} = \delta_k\left(y, w, \frac{dw}{dy}, \ldots, \frac{d^k w}{dy^k}\right),$$

for certain functions δ_k. Substituting these expressions into our equation, we find an equivalent n-th order equation

$$\tilde{\Delta}(y, w^{(n)}) = \tilde{\Delta}(y, w, w_y, \ldots, w_n) = 0 \tag{2.87}$$

in terms of the new coordinates y and w. Moreover, since the original system (2.86) has the invariance group G, so does the transformed system. In terms of the (y, w)-coordinates, the infinitesimal generator has trivial prolongation

$$\mathrm{pr}^{(n)}\, \mathbf{v} = \mathbf{v} = \partial/\partial w.$$

The infinitesimal criteron of invariance implies

$$\mathrm{pr}^{(n)}\, \mathbf{v}(\tilde{\Delta}) = \frac{\partial \tilde{\Delta}}{\partial w} = 0 \quad \text{whenever} \quad \tilde{\Delta}(y, w^{(n)}) = 0.$$

This means, as in Proposition 2.18, that there is an equivalent equation

$$\hat{\Delta}\left(y, \frac{dw}{dy}, \ldots, \frac{d^n w}{dy^n}\right) = 0$$

which is independent of w, i.e. $\tilde{\Delta}(y, w^{(n)}) = 0$ if and only if $\hat{\Delta}(y, w^{(n)}) = 0$. Now we have accomplished our goal; setting $z = w_y$ we have an $(n-1)$-st order equation for z,

$$\hat{\Delta}(y, z, \ldots, d^{n-1}z/dy^{n-1}) = \hat{\Delta}(y, z^{(n-1)}) = 0, \tag{2.88}$$

whose solutions provide the general solution to our original equation. Namely, if $z = h(y)$ is a solution of (2.88), then $w = \int h(y)\,dy + c$ is a solution of (2.87), and hence, by replacing w and y by their expressions in terms of x and u, implicitly defines a solution of the original equation.

Example 2.49. As an elementary example, consider the case of a second order equation in which x does not occur explicitly,

$$\Delta(u, u_x, u_{xx}) = 0.$$

This equation is clearly invariant under the group of translations in the x-direction, with infinitesimal generator $\partial/\partial x$. In order to change this into the vector field $\partial/\partial w$, corresponding to translations of the dependent variable, it suffices to reverse the roles of dependent and independent variable, so we set $y = u$, $w = x$. Then

$$\frac{du}{dx} = \frac{1}{w_y}, \qquad \frac{d^2u}{dx^2} = -\frac{w_{yy}}{w_y^3},$$

so our equation becomes

$$\Delta\left(y, \frac{1}{w_y}, -\frac{w_{yy}}{w_y^3}\right) = 0,$$

which is a first order equation for $z = w_y$:

$$\hat{\Delta}(y, z, z_y) \equiv \Delta(y, z^{-1}, -z^{-3}z_y) = 0.$$

For example, to solve

$$u_{xx} - 2uu_x = 0,$$

we have the corresponding first order equation

$$-z^{-3}z_y - 2yz^{-1} = 0$$

for $z = dw/dy = (du/dx)^{-1}$. This can easily be solved by separation, with solution

$$z = (y^2 + c)^{-1}.$$

Thus, if $c = c'^2 > 0$, we find

$$w = \int z\,dy = \frac{1}{c'}\arctan\frac{y}{c'} + \tilde{c},$$

or, in terms of x and u,

$$u = c'\tan(c'x + d), \qquad d = -\tilde{c}c'.$$

(For $c < 0$, we have a hyperbolic tangent, for $c = 0$, we get the limiting solution $u = -(x + d)^{-1}$.)

Example 2.50. Consider a homogeneous second order linear equation

$$u_{xx} + p(x)u_x + q(x)u = 0. \tag{2.89}$$

This is clearly invariant under the group of scale transformations

$$(x, u) \mapsto (x, \lambda u),$$

with infinitesimal generator $\mathbf{v} = u\partial_u$. Coordinates (y, w) which straighten out $\mathbf{v}$ are given by $y = x$, $w = \log u$ (provided $u \neq 0$), with $\mathbf{v} = \partial_w$ in these coordinates. We have

$$u = e^w, \qquad u_x = w_x e^w, \qquad u_{xx} = (w_{xx} + w_x^2)e^w,$$

so the equation becomes

$$w_{xx} + w_x^2 + p(x)w_x + q(x) = 0,$$

which is independent of w. We have thus reconstructed the well-known transformation between a linear second order equation and a first order Riccati equation; namely $z = w_x = u_x/u$ changes (2.89) into the Riccati equation

$$z_x = -z^2 - p(x)z - q(x).$$

Differential Invariants

Besides trying to determine the most general symmetry group of a given differential equation, we can turn the whole procedure around and ask the complementary question: What is the most general type of differential equation which admits a given group as a group of symmetries? An answer to this question will not only provide us with a catalogue of large classes of ordinary differential equations which can be integrated by a common method, but also familiarity with the various types of equations which arise from known groups will aid in the recognition of symmetry groups for other equations.

According to Section 2.2, an n-th order ordinary differential equation $\Delta(x, u^{(n)}) = 0$ admits a group G as a symmetry group if and only if the corresponding subvariety $\mathscr{S}_\Delta \subset M^{(n)}$ is invariant under the n-th prolongation $\mathrm{pr}^{(n)} G$. Furthermore, according to Proposition 2.18, there is an equivalent equation $\tilde{\Delta} = 0$ describing the subvariety $\mathscr{S}_\Delta$, where $\tilde{\Delta}$ depends only on the invariants of the group action, which in this case is $\mathrm{pr}^{(n)} G$. The invariants of a prolonged group action play an important role in this procedure, and are known as "differential invariants".

Definition 2.51. Let G be a local group of transformations acting on $M \subset X \times U$. An n-th order *differential invariant* of G is a smooth function $\eta: M^{(n)} \to \mathbb{R}$, depending on x, u and derivatives of u, such that η is an invariant of the prolonged group action $\mathrm{pr}^{(n)} G$:

$$\eta(\mathrm{pr}^{(n)} g \cdot (x, u^{(n)})) = \eta(x, u^{(n)}), \qquad (x, u^{(n)}) \in M^{(n)},$$

for all $g \in G$ such that $\mathrm{pr}^{(n)} g \cdot (x, u^{(n)})$ is defined.

Although the definition makes sense when there are several independent and several dependent variables, we will primarily be interested in the ordinary differential equation case $p = q = 1$.

Example 2.52. Suppose $G = \mathrm{SO}(2)$ is the rotation group acting on $X \times U \simeq \mathbb{R}^2$ with generator $\mathbf{v} = -u\partial_x + x\partial_u$. The first order differential invariants are the ordinary invariants of the first prolongation $\mathrm{pr}^{(1)} \mathrm{SO}(2)$, which has infinitesimal generator

$$\mathrm{pr}^{(1)} \mathbf{v} = -u\frac{\partial}{\partial x} + x\frac{\partial}{\partial u} + (1 + u_x^2)\frac{\partial}{\partial u_x}.$$

If we relabel the variables (x, u, u_x) by (x, y, z), then we are precisely in the situation covered by Example 2.19(b). Translating the result obtained there into the present context, we find that the functions

$$y = \sqrt{x^2 + u^2} \quad \text{and} \quad w = \frac{xu_x - u}{x + uu_x} \tag{2.90}$$

provide a complete set of first order differential invariants for SO(2). For second order invariants, we would also include the curvature invariant κ found in Example 2.37. Any other second order differential invariant must be a function of these three independent invariants.

For higher order differential invariants there is an easy short cut which allows us to construct all differential invariants from knowledge of the lowest order ones.

Proposition 2.53. *Let G be a group of transformations acting on $M \subset X \times U \simeq \mathbb{R}^2$. Suppose $y = \eta(x, u^{(n)})$ and $w = \zeta(x, u^{(n)})$ are n-th order differential invariants of G. Then the derivative*

$$\frac{dw}{dy} = \frac{dw/dx}{dy/dx} \equiv \frac{D_x\zeta}{D_x\eta} \tag{2.91}$$

is an $(n + 1)$-st order differential invariant for G.

PROOF. The proof requires the following formula. Let $\zeta(x, u^{(n)})$ be any smooth function and $\mathbf{v} = \xi\partial_x + \phi\partial_u$ any vector field. Then

$$\mathrm{pr}^{(n+1)} \mathbf{v}(D_x\zeta) = D_x[\mathrm{pr}^{(n)} \mathbf{v}(\zeta)] - D_x\xi \cdot D_x\zeta. \tag{2.92}$$

Using the alternative formulation (2.50) of the prolongation formula, we see that

$$\mathrm{pr}^{(n+1)} \mathbf{v}(D_x\zeta) = \mathrm{pr}^{(n+1)} \mathbf{v}_Q(D_x\zeta) + \xi D_x^2\zeta,$$

while

$$D_x[\mathrm{pr}^{(n)}\, \mathbf{v}(\zeta)] = D_x[\mathrm{pr}^{(n)}\, \mathbf{v}_Q(\zeta)] + D_x(\xi D_x \zeta).$$

Therefore (2.92) reduces to the simpler formula

$$\mathrm{pr}^{(n+1)}\, \mathbf{v}_Q(D_x \zeta) = D_x[\mathrm{pr}^{(n)}\, \mathbf{v}_Q(\zeta)].$$

This latter formula is a special case of a general commutation rule for vector fields and total derivatives—which will be proved in Lemma 5.12. (It is, however, not difficult for the reader to prove directly here.)

Proceeding to the proof of (2.91), let $\mathbf{v}$ be any infinitesimal generator of G. Using (2.92), and the fact that $\mathrm{pr}^{(n+1)}\, \mathbf{v}$ is a derivation,

$$\begin{aligned} \mathrm{pr}^{(n+1)}\, \mathbf{v}\left[\frac{dw}{dy}\right] &= \frac{1}{(D_x\eta)^2}\{\mathrm{pr}^{(n+1)}\, \mathbf{v}(D_x\zeta)\cdot D_x\eta - D_x\zeta\cdot \mathrm{pr}^{(n+1)}\, \mathbf{v}(D_x\eta)\} \\ &= \frac{1}{(D_x\eta)^2}\{D_x[\mathrm{pr}^{(n)}\, \mathbf{v}(\zeta)]\cdot D_x\eta - D_x\xi\cdot D_x\zeta\cdot D_x\eta \\ &\qquad - D_x\zeta\cdot D_x[\mathrm{pr}^{(n)}\, \mathbf{v}(\eta)] + D_x\zeta\cdot D_x\xi\cdot D_x\eta\} \\ &= 0 \end{aligned}$$

since $\mathrm{pr}^{(n)}\, \mathbf{v}(\zeta) = 0 = \mathrm{pr}^{(n)}\, \mathbf{v}(\eta)$ by assumption. Thus dw/dy is infinitesimally invariant under the action of $\mathrm{pr}^{(n+1)}\, G$, and hence by Proposition 2.6 is an invariant. □

Corollary 2.54. *Suppose G is a one-parameter group of transformations acting on $M \subset X \times U \simeq \mathbb{R}^2$. Let $y = \eta(x, u)$ and $w = \zeta(x, u, u_x)$ be a complete set of functionally independent invariants of the first prolongation $\mathrm{pr}^{(1)}\, G$. Then the derivatives*

$$y, w, dw/dy, \ldots, d^{n-1}w/dy^{n-1}$$

provide a complete set of functionally independent invariants for the n-th prolongation $\mathrm{pr}^{(n)}\, G$ for $n \geq 1$.

To check the independence, it suffices to note that the k-th derivative $d^k w/dy^k$ depends explicitly on $u_{k+1} = d^{k+1}u/dx^{k+1}$, and hence is independent of the previous invariants $y, w, \ldots, d^{k-1}w/dy^{k-1}$, which are only functions of $x, u, \ldots, u_k$.

Example 2.55. Return to the second order invariants of the rotation group SO(2) discussed in the previous Example 2.52. It follows from Corollary 2.54 that y, w and the derivative

$$\frac{dw}{dy} = \frac{dw/dx}{dy/dx} = \frac{\sqrt{x^2+u^2}}{(x+uu_x)^3}[(x^2+u^2)u_{xx} - (1+u_x^2)(xu_x - u)]$$

form a complete set of functionally independent invariants for the second prolongation $\mathrm{pr}^{(2)}\, \mathrm{SO}(2)$. Note that this means *any* other second order differential invariant of the rotation group can be written in terms of y, w and dw/dy; for instance, the curvature invariant found previously has expression

$$\kappa = \frac{u_{xx}}{(1+u_x^2)^{3/2}} = \frac{w_y}{(1+w^2)^{3/2}} + \frac{w}{y(1+w^2)^{1/2}},$$

as the reader can check.

Once we know the differential invariants for a group of transformations acting on $M \subset X \times U$, we can determine the structure of all differential equations which admit the given group as a symmetry group. In the case G is a one-parameter group, we thus know all equations which can be integrated using G.

Proposition 2.56. *Let G be a local group of transformations acting on $M \subset X \times U$. Assume $\mathrm{pr}^{(n)} G$ acts semi-regularly on an open subset of $M^{(n)}$, and let $\eta^1(x, u^{(n)}), \ldots, \eta^k(x, u^{(n)})$ be a complete set of functionally independent n-th order differential invariants. An n-th order differential equation $\Delta(x, u^{(n)}) = 0$ admits G as a symmetry group if and only if there is an equivalent equation*

$$\tilde{\Delta}(\eta^1(x, u^{(n)}), \ldots, \eta^k(x, u^{(n)})) = 0$$

involving only the differential invariants of G. In particular, if G is a one-parameter group of transformations, any n-th order differential equation having G as a symmetry group is equivalent to an $(n-1)$-st order equation

$$\tilde{\Delta}(y, w, dw/dy, \ldots, d^{n-1}w/dy^{n-1}) = 0 \tag{2.93}$$

involving the invariants $y = \eta(x, u)$, $w = \zeta(x, u, u_x)$ of $\mathrm{pr}^{(1)} G$ and their derivatives.

The proof is immediate from Proposition 2.18 and Corollary 2.54. □

Example 2.57. For example, we can completely classify all first and second order differential equations admitting the rotation group SO(2) as a symmetry group. Any first order equation invariant under SO(2) is equivalent to an equation involving only the invariants (2.90). Solving for w, we find every such equation takes the form

$$\frac{xu_x - u}{x + uu_x} = H(\sqrt{x^2 + u^2})$$

for some function H. But this is precisely the form (2.82) discussed in Example 2.47 once we solve for u_x. Thus (2.82) is the most general first order ordinary differential equation invariant under the rotation group SO(2).

Similarly, any second order equation invariant under SO(2) is equivalent to one involving y, w and the curvature $\kappa = u_{xx}(1 + u_x^2)^{-3/2}$, i.e.

$$u_{xx} = (1 + u_x^2)^{3/2} H\left(\sqrt{x^2 + u^2}, \frac{xu_x - u}{x + uu_x}\right),$$

where $H(y, w)$ is any function of the first order invariants. This can be integrated once, as in Example 2.47, by setting $r = \sqrt{x^2 + u^2}$, $\theta = \arctan(u/x)$. We find

$$w = r\theta_r, \qquad \kappa = \frac{r\theta_{rr} + r^2\theta_r^3 + 2\theta_r}{(1 + r^2\theta_r^2)^{3/2}},$$

the latter being the expression for the curvature of a curve $\theta = \theta(r)$ expressed in polar coordinates. Thus the equation becomes a first order equation

$$r\frac{dz}{dr} = (1 + r^2z^2)^{3/2}H(r, rz) - (r^2z^3 + 2z)$$

involving only $z = d\theta/dr$, from which we can determine $\theta(r) = \int z(r)\,dr + c$.

The preceding proposition also indicates an alternative method for reducing the order of a differential equation invariant under a one-parameter group by using the differential invariants of the group. Namely, the differential equation $\Delta(x, u^{(n)}) = 0$ must be equivalent to an equation (2.93) involving only the invariants $y, w, \ldots, d^{n-1}w/dy^{n-1}$ of the n-th prolongation of G. But (2.93) is automatically an $(n-1)$-st order equation for w as a function of y, so that merely by re-expressing the original equation in terms of the given list of differential invariants, we have automatically reduced its order by one. *Moreover*, once we know the solution $w = h(y)$ of the reduced equation (2.93), the solution of the original equation is found by integrating the auxiliary first order equation

$$\zeta(x, u, u_x) = h[\eta(x, u)] \tag{2.94}$$

obtained by substituting for y and w their expressions in terms of the original variables x and u. Since (2.94) depends only on the invariants y and w of $\mathrm{pr}^{(1)} G$, it clearly has G as a one-parameter symmetry group and hence can be integrated by the methods for first order equations discussed previously. We have thus, by a completely different method, re-established the basic fact that an ordinary differential equation invariant under a one-parameter group can be reduced in order by one.

Example 2.58. Consider the second order equation

$$x^2u_{xx} + xu_x^2 = uu_x. \tag{2.95}$$

This is invariant under the scaling group $(x, u) \mapsto (\lambda x, \lambda u)$.

Let us first try to integrate (2.95) using the method of differential invariants. We find that the invariants of the second prolonged group action are

$$y = \frac{u}{x}, \qquad w = u_x, \qquad \frac{dw}{dy} = \frac{x^2u_{xx}}{xu_x - u}.$$

The new equation involving w and y is therefore

$$(w - y)\frac{dw}{dy} + w^2 = yw.$$

This has two families of solutions; either $w = y$ or $dw/dy = -w$, the latter integrating to $w = ce^{-y}$ for some constant c. Reverting back to the original variables, we obtain two homogeneous first order equations, as guaranteed by the form of (2.94):

$$\frac{du}{dx} = \frac{u}{x}, \quad \text{or} \quad \frac{du}{dx} = ce^{-u/x}.$$

The first has solutions $u = kx$; the second has implicit solutions

$$\int \frac{dy}{ce^{-y} - y} = \log x + k, \tag{2.96}$$

where $y = u/x$. Here (2.96) is the "general" solution to (2.95), the linear functions being a one-parameter family of singular solutions.

The integration of (2.95) by our earlier method is quite a bit more tricky. As in Example 2.46, we set $y = u/x$, $\tilde{w} = \log x$, so that in terms of y and w the infinitesimal generator is $\partial/\partial\tilde{w}$. We further have

$$u_x = \frac{1 + y\tilde{w}_y}{\tilde{w}_y}, \qquad xu_{xx} = \frac{\tilde{w}_y^2 - \tilde{w}_{yy}}{\tilde{w}_y^3},$$

so the equation takes the form of a first order Riccati equation

$$\frac{dz}{dy} = (y + 1)z^2 + z \tag{2.97}$$

for $z = d\tilde{w}/dy$. The solution proceeds either by using general methods for integrating Riccati equations, or, more expediently, by noticing that it admits a one-parameter symmetry group with generator $\mathbf{w} = (z + yz^2)\partial_z$. Therefore, by Theorem 2.48, $R = (z + yz^2)^{-1}$ is an integrating factor for (2.97). We find

$$T(y, z) = y + \log(y + z^{-1}) = \tilde{c}$$

to be the integral, hence the solutions of (2.97) are given by $z = (ce^{-y} - y)^{-1}$. Recalling the definition of $z = \tilde{w}_y$, we see that we can integrate this latter expression to recover the general solution (2.96) to (2.95). The singular solutions $u = kx$ do not appear in this case since they do not correspond to functions of the form $\tilde{w} = h(y)$. They can be found by choosing alternative coordinates, e.g. $\hat{w} = \log u$ instead of $\tilde{w}$.

One interesting point is that the symmetry group of the Riccati equation (2.97) generated by $\mathbf{w}$ does *not* appear to have a counterpart for the original equation. (In fact, it can be shown that the scaling group is the *only* symmetry group of (2.95); cf. Exercise 2.25a.) Thus, reducing the order of an ordinary differential equation may result in an equation with new symmetries, whereby the order can be yet further reduced!

Multi-parameter Symmetry Groups

If an ordinary differential equation $\Delta(x, u^{(n)}) = 0$ is invariant under an r-parameter group, then intuition tells us that we should be able to reduce the order of the equation by r. In one sense, this somewhat naïve presumption is correct, but the problem may be that we cannot reconstruct the solution of the original n-th order equation from that of the reduced $(n - r)$-th order equation by quadratures alone. More specifically, suppose G is an r-parameter group of transformations acting on $M \subset X \times U$. *Assume*, for simplicity, that the r-th prolongation $\mathrm{pr}^{(r)} G$ acting on $M^{(r)}$ has r-dimensional orbits. (More degenerate cases can be treated analogously, although technical complications may arise.) Since $M^{(r)}$ is $(r + 2)$-dimensional, this means that locally there exist exactly two independent r-th order differential invariants of G, say

$$y = \eta(x, u^{(r)}), \qquad w = \zeta(x, u^{(r)}). \tag{2.98}$$

Note that every further prolongation $\mathrm{pr}^{(n)} G$ also has r-dimensional orbits. (This is because they project down to the orbits of $\mathrm{pr}^{(r)} G$ in $M^{(r)}$, so are at least r-dimensional, but G itself is r-dimensional, so the orbits can never have more than r dimensions; see Exercise 3.17.) Therefore, $\mathrm{pr}^{(n)} G$ has $n - r + 2$ independent differential invariants, which by Proposition 2.53 we can take to be

$$y, w, dw/dy, \ldots, d^{n-r}w/dy^{n-r}.$$

If $\Delta(x, u^{(n)}) = 0$ is invariant under the entire symmetry group G, then by Proposition 2.18 there is an equivalent equation

$$\tilde{\Delta}(y, w, dw/dy, \ldots, d^{n-r}w/dy^{n-r}) = 0 \tag{2.99}$$

involving only the invariants of $\mathrm{pr}^{(n)} G$. In this sense, we have reduced the n-th order system for u as a function of x to an $(n - r)$-th order system for w as a function of y.

The principal problem at this juncture is that it is unclear how we determine the solution $u = f(x)$ of the original system from the general solution $w = h(y)$ of the reduced system (2.99). Using the expressions (2.98) for the invariants y, w, we find that we must solve an auxiliary r-th order equation

$$\zeta(x, u^{(r)}) = h[\eta(x, u^{(r)})] \tag{2.100}$$

to determine u. This auxiliary equation, being expressed in terms of differential invariants, retains G as an r-parameter symmetry group. However, in contrast to the one-parameter situation, there is no assurance that we will be able to integrate (2.100) completely by quadratures, thereby explicitly determining the solution of our original equation. The difficulty in this regard is apparent in the following example.

Example 2.59. Recalling Example 1.58(c), consider the action of SL(2) as the projective group

$$(x, u) \mapsto ((\alpha x + \beta)/(\gamma x + \delta), u)$$

on the line. The infinitesimal generators are

$$\mathbf{v}_1 = \frac{\partial}{\partial x}, \qquad \mathbf{v}_2 = x\frac{\partial}{\partial x}, \qquad \mathbf{v}_3 = x^2\frac{\partial}{\partial x},$$

from which we see that u and its Schwarzian derivative,

$$y = u, \qquad w = 2u_x^{-3}u_{xxx} - 3u_x^{-4}u_{xx}^2$$

form a complete set of functionally independent invariants for the prolongation $\mathrm{pr}^{(3)}\,\mathrm{SL}(2)$.

We conclude that any differential equation $\Delta(x, u^{(n)}) = 0$ which is invariant under the full projective group is equivalent to an $(n-3)$-rd order equation

$$\Delta\left(y, w, \frac{dw}{dy}, \ldots, \frac{d^{n-3}w}{dy^{n-3}}\right) = 0, \tag{2.101}$$

involving only the invariants of $\mathrm{pr}^{(n)}\,\mathrm{SL}(2)$. For instance, since

$$\frac{dw}{dy} = \frac{dw/dx}{u_x} = \frac{2u_x^2u_{xxxx} - 12u_xu_{xx}u_{xxx} + 12u_{xx}^3}{u_x^6},$$

any fourth order equation admitting SL(2) as a symmetry group is equivalent to one of the form

$$2u_x^2u_{xxxx} - 12u_xu_{xx}u_{xxx} + 12u_{xx}^3 = u_x^6H(u, u_x^{-4}(2u_xu_{xxx} - 3u_{xx}^2)).$$

The reduced system (2.101) in this case is the first order equation $dw/dy = H(y, w)$.

However, once we have solved the reduced equations (2.101) for $w = h(y)$, we are left with the task of determining the corresponding solutions $u = f(x)$ by solving the auxiliary equation

$$2u_xu_{xxx} - 3u_{xx}^2 = u_x^4h(u) \tag{2.102}$$

obtained by substituting for y and w. This equation remains invariant under SL(2), so we can use this knowledge to try to integrate it. In particular, it is invariant under the translation subgroup generated by ∂_x, which has invariants $y = u$, $z = u_x$, in terms of which (2.102) reduces to

$$2z\frac{d^2z}{dy^2} - \left(\frac{dz}{dy}\right)^2 = z^2h(y).$$

This latter equation is invariant under the scale group $(y, z) \mapsto (y, \lambda z)$ (reflecting the symmetry of (2.102) under the scale group $(x, u) \mapsto (\lambda^{-1}x, u)$) and hence can be reduced to a first order Riccati equation

$$2\frac{dv}{dy} + v^2 = h(y) \tag{2.103}$$

for $v = (\log z)_y = z_y/z$. However, at this point we are stuck. We have already used the translational and scaling symmetries to reduce (2.102) to a first order equation, but there is no remnant of the inversional symmetries generated by $\mathbf{v}_3$ which can be used to integrate the standard Riccati equation (2.103). Thus the best that can be said is that the solution of n-th order differential equation invariant under the projective group can be found from the general solution of a reduced $(n-3)$-rd order equation by using two quadratures *and* the solution of an auxiliary first order Riccati equation.

This whole example is illustrative of an important point. If we reduce the order of an ordinary differential equation using only a subgroup of the full symmetry group, we may very well lose any additional symmetry properties present in the full group. Only special types of subgroups, namely the normal subgroups presented in Exercise 1.24, will enable us to retain the full symmetry properties under reduction. Before discussing this case, it helps to return to symmetries of algebraic equations once again.

Let G be an r-parameter group acting on $M \subset \mathbb{R}^m$ and let $H \subset G$ be a subgroup. Suppose that $\eta(x) = (\eta^1(x), \ldots, \eta^{m-s}(x))$ form a complete set of functionally independent invariants of H. If H happens to be a *normal* subgroup, meaning that $ghg^{-1} \in H$ whenever $g \in G$, $h \in H$, then there is an induced action of G on the subset $\tilde{M} \subset \mathbb{R}^{m-s}$ determined by these invariants $y = (y^1, \ldots, y^{m-s}) = \eta(x)$:

$$\tilde{g} \cdot y = \tilde{g} \cdot \eta(x) = \eta(g \cdot x), \qquad g \in G, \quad x \in M. \tag{2.104}$$

Note that for any $h \in H$

$$\tilde{g} \cdot \eta(hx) = \eta(g \cdot hx) = \eta(\hat{h} \cdot gx) = \eta(gx) = \tilde{g} \cdot \eta(x),$$

where $\hat{h} = ghg^{-1} \in H$; from this it is easy to see that this action on $\tilde{M}$ is well defined. (In fact, H acts trivially on $\tilde{M}$, so (2.104) actually defines an action of the quotient group G/H; see Exercise 3.11.)

According to Proposition 2.18, any H-invariant subset of M can be written as the zero set $\mathscr{S}_F = \{F(x) = 0\}$ of some H-invariant function $F(x) = \tilde{F}(\eta(x))$. It is not hard to see that, assuming H is a normal subgroup, $\mathscr{S}_F \subset M$ is invariant under the full group G if and only if the reduced subvariety $\mathscr{S}_{\tilde{F}} = \{y: \tilde{F}(y) = 0\} \subset \tilde{M}$ is invariant under the induced action of G on $\tilde{M}$.

For the infinitesimal version, let us introduce s further variables $\hat{x} = (\hat{x}^1, \ldots, \hat{x}^s)$ completing $y = \eta(x)$ to a set of local coordinates $(y, \hat{x})$ on M. Using the infinitesimal criterion of normality from Exercise 1.24(b), we see that each infinitesimal generator of G must be of the form

$$\mathbf{v}_k = \sum_{i=1}^{m-s} \eta_k^i(y) \frac{\partial}{\partial y^i} + \sum_{j=1}^{s} \xi_k^j(y, \hat{x}) \frac{\partial}{\partial \hat{x}^j}, \qquad k = 1, \ldots, r, \tag{2.105}$$

in these coordinates, where each η^i is independent of the parametric variables $\hat{x}$. Thus $\mathbf{v}_k$ reduces to a vector field

$$\tilde{\mathbf{v}}_k = \sum_{i=1}^{m-s} \eta_k^i(y) \frac{\partial}{\partial y^i}, \qquad k = 1, \ldots, r,$$

generating the reduced action of G on $\tilde{M}$. These we can use to check the invariance of the reduced subvariety $\mathscr{S}_{\tilde{F}}$, and hence that of $\mathscr{S}_F$.

Similar results hold for differential equations. Assume, as above, that the r-parameter group G acts on $M \subset X \times U \simeq \mathbb{R}^2$ and suppose $H \subset G$ is an s-parameter subgroup whose prolongation $\mathrm{pr}^{(s)} H$ has s-dimensional orbits in $M^{(s)}$. (As before, degenerate cases can also be treated if required.) Let $y = \eta(x, u^{(s)})$, $w = \zeta(x, u^{(s)})$ be a complete set of functionally independent differential invariants for H on $M^{(s)}$, with corresponding invariants $w^{(m)} = \zeta^{(m)}(x, u^{(s+m)})$ on $M^{(s+m)}$, $m \geqslant 0$. Any n-th order ordinary differential equation admitting H as a symmetry group can be written in the form

$$\Delta(x, u^{(n)}) = \tilde{\Delta}(\eta(x, u^{(s)}), \zeta^{(n-s)}(x, u^{(n)})) = \tilde{\Delta}(y, w^{(n-s)}) = 0,$$

using only the invariants $y, w, \ldots, d^{n-s}w/dy^{n-s}$ of $\mathrm{pr}^{(n)} H$. Moreover, since H is a normal subgroup of G, there is an induced action of G on $\tilde{M} \subset Y \times W \simeq \mathbb{R}^2$, with

$$\begin{aligned} \tilde{g} \cdot (y, w) &= \tilde{g} \cdot (\eta(x, u^{(s)}), \zeta(x, u^{(s)})) \\ &= (\eta(\mathrm{pr}^{(s)} g \cdot (x, u^{(s)})), \zeta(\mathrm{pr}^{(s)} g \cdot (x, u^{(s)}))), \qquad g \in G, \end{aligned} \tag{2.106}$$

cf. (2.104). Similarly, the action of G on $M^{(n)}$ reduces to an action of G on the space $\tilde{M}^{(n-s)}$ determining the derivatives of w with respect to y. It is not too difficult to see that *this* reduced action coincides with the prolongation of the action of G on $\tilde{M}$ defined by (2.106); in other words

$$\mathrm{pr}^{(n-s)} \tilde{g} \cdot (\eta(x, u^{(s)}), \zeta^{(n-s)}(x, u^{(n)})) = (\eta(\mathrm{pr}^{(s)} g \cdot (x, u^{(s)})), \zeta^{(n-s)}(\mathrm{pr}^{(n)} g \cdot (x, u^{(n)}))).$$

(To check this, look at what happens to a representative smooth H-invariant function $u = f(x)$.)

Translating our earlier results for algebraic equations, we deduce the following result on normal subgroups of symmetry groups of ordinary differential equations.

Theorem 2.60. *Let $H \subset G$ be an s-parameter normal subgroup of a Lie group of transformations acting on $M \subset X \times U \simeq \mathbb{R}^2$ such that $\mathrm{pr}^{(s)} H$ has s-dimensional orbits in $M^{(s)}$. Let $\Delta(x, u^{(n)}) = 0$ be an n-th order ordinary differential equation admitting H as a symmetry group, with corresponding reduced equation $\tilde{\Delta}(y, w^{(n-s)}) = 0$ for the invariants $y = \eta(x, u^{(s)})$, $w = \zeta(x, u^{(s)})$ of H. There is an induced action of the quotient group G/H on $\tilde{M} \subset Y \times W$ and Δ admits all of G as a symmetry group if and only if the H-reduced equation $\tilde{\Delta}$ admits the quotient group G/H as a symmetry group.*

An especially important example is the case of a two-parameter symmetry group. Here, owing to the special structure of two-dimensional Lie groups, we can use the preceding theorem to carry out the reduction in order by two using only quadratures.

Theorem 2.61. *Let $\Delta(x, u^{(n)}) = 0$ be an n-th order ordinary differential equation invariant under a two-parameter symmetry group G. Then there is an $(n-2)$-nd order equation $\hat{\Delta}(z, v^{(n-2)}) = 0$ with the property that the general solution to Δ can be found by a pair of quadratures from the general solution to $\hat{\Delta}$.*

PROOF. According to Exercise 1.21, we can find a basis $\{\mathbf{v}, \mathbf{w}\}$ for any two-dimensional Lie algebra $\mathfrak{g}$ with the property

$$[\mathbf{v}, \mathbf{w}] = k\mathbf{v} \tag{2.107}$$

for some constant k. (In fact, k can be taken to be 0 if $\mathfrak{g}$ is abelian and 1 in all other cases.) The one-parameter subgroup H generated by $\mathbf{v}$ is then a normal subgroup of G, with one-parameter quotient group G/H. To effect the reduction of Δ, we begin by determining first order differential invariants $y = \eta(x, u)$, $w = \zeta(x, u, u_x)$ for H using our earlier methods. By Proposition 2.56, our n-th order equation is equivalent to an $(n-1)$st order equation $\tilde{\Delta}(y, w^{(n-1)}) = 0$; moreover, once we know the solution $w = h(y)$ of this latter equation, we can reconstruct the solution to Δ by solving the corresponding first order equation (2.94) using a single quadrature. Since H is normal, the reduced equation $\tilde{\Delta}$ is invariant under the action of G/H on the variables (y, w), and hence we can employ our earlier methods for one-parameter symmetry groups to reduce the order yet again by one. On an infinitesimal level, suppose $(x, y, w) = (x, \eta(x, u), \zeta(x, u, u_x))$ form local coordinates on some subset of $M^{(1)}$. (If x happens to be one of the invariants, we can replace it by u or some combination $\gamma(x, u)$.) As in (2.105), normality, as expressed by (2.107), implies that the vector field $\mathbf{w}$ has first prolongation

$$\operatorname{pr}^{(1)} \mathbf{w} = \alpha(x, y, w)\partial_x + \beta(y, w)\partial_y + \psi(y, w)\partial_w$$

in terms of these coordinates, and hence reduces to a vector field

$$\tilde{\mathbf{w}} = \beta(y, w)\partial_y + \psi(y, w)\partial_w$$

on the space $\tilde{M}$, generating the quotient group action of G/H. Theorem 2.60 assures us that $\tilde{\mathbf{w}}$ remains a symmetry of the preliminary reduced system $\tilde{\Delta}$, and hence we can reduce $\tilde{\Delta}$ in order by one using either the method of differential invariants or that of changing coordinates to straighten out $\tilde{\mathbf{w}}$, leading to a differential equation $\hat{\Delta}(z, v^{(n-2)}) = 0$ of order $n-2$. This completes the reduction procedure, and the proof of the theorem. □

Example 2.62. Consider a second order differential equation of the form

$$x^2 u_{xx} = H(xu_x - u), \tag{2.108}$$

where H is a given function. This equation admits the two-parameter symmetry group

$$(x, u) \mapsto (\lambda x, u + \varepsilon x), \qquad \varepsilon \in \mathbb{R}, \quad \lambda \in \mathbb{R}^+,$$

with infinitesimal generators $\mathbf{v} = x\partial_u$ and $\mathbf{w} = x\partial_x$. Note that $[\mathbf{v}, \mathbf{w}] = -\mathbf{v}$, so the generators are in the correct order to take advantage of Theorem 2.61. According to the basic procedure, we need to first determine the invariants of $\mathbf{v}$, which are x and $w = xu_x - u$, in terms of which (2.108) reduces to the first order equation

$$x\frac{dw}{dx} = H(w).$$

This latter equation is separable, with implicit solution

$$\int \frac{dw}{H(w)} = \log x + c,$$

reflecting the fact that it remains invariant under the reduced group $(x, w) \mapsto (\lambda x, w)$ determined by $\tilde{\mathbf{w}} = x\partial_x$, as guaranteed by the method. From this solution, rewritten in explicit form $w = h(x)$, we reconstruct the general solution to (2.108) by solving the linear equation.

$$xu_x - u = h(x).$$

The integrating factor is $1/x^2$, as can be determined directly from the form of the underlying symmetry group generated by $\mathbf{v}$, and hence we find

$$u = x\left(\int x^{-2}h(x)\,dx + k\right)$$

is the general solution to (2.108).

It is instructive to see what would have happened if, unheeded by the general procedure, we had tried to integrate (2.108) by considering the two one-parameter groups in the reverse order. In this case, the invariants for $\mathbf{w}$ are $y = u$, $z = xu_x$, whence $z_y = xu_x^{-1}u_{xx} + 1$, and the equation reduces to

$$z\left(\frac{dz}{dy} - 1\right) = H(z - y). \qquad (2.109)$$

However, at this stage there is no symmetry property of (2.109) which reflects the symmetry of (2.108) under the group generated by $\mathbf{v}$. This shows that it *is* important to do our reduction procedure in the right order, otherwise we may not end up with the solution. Reversing the procedure, we are left with the intriguing possibility of being able to integrate an $(n-1)$-st order equation by first changing it into an n-th order equation with several symmetries, whose order can then be reduced substantially. For instance, we can solve (2.109) by first substituting $y = u$, $z = xu_x$, which changes it to (2.108), and then integrating the latter equation. This point will be investigated in greater detail in the exercises at the end of the chapter.

Solvable Groups

Turning to yet higher-dimensional symmetry groups, we find, as evidenced by the example of the projective group, that for $r \geqslant 3$, invariance of an n-th order equation under an r-parameter group will not in general imply that we can find the general solution by quadratures from the solution of the corresponding reduced $(n-r)$-th order equation. The problem is that there is not in general a sufficient supply of normal subgroups to ensure the continued applicability of Theorem 2.60 and the reduction procedure for one-parameter groups at each stage. This motivates the following definition of those groups which can be used to fully reduce or "solve" an equation to the extent promised by their dimensionality.

Definition 2.63. Let G be a Lie group with Lie algebra $\mathfrak{g}$. Then G is *solvable* if there exists a chain of Lie subgroups

$$\{e\} = G^{(0)} \subset G^{(1)} \subset G^{(2)} \subset \cdots \subset G^{(r-1)} \subset G^{(r)} = G$$

such that for each $k = 1, \ldots, r$, $G^{(k)}$ is a k-dimensional subgroup of G and $G^{(k-1)}$ is a normal subgroup of $G^{(k)}$. Equivalently, there exists a chain of subalgebras

$$\{0\} = \mathfrak{g}^{(0)} \subset \mathfrak{g}^{(1)} \subset \mathfrak{g}^{(2)} \subset \cdots \subset \mathfrak{g}^{(r-1)} \subset \mathfrak{g}^{(r)} = \mathfrak{g}, \tag{2.110}$$

such that for each k, $\dim \mathfrak{g}^{(k)} = k$ and $\mathfrak{g}^{(k-1)}$ is a normal subalgebra of $\mathfrak{g}^{(k)}$:

$$[\mathfrak{g}^{(k-1)}, \mathfrak{g}^{(k)}] \subset \mathfrak{g}^{(k-1)}.$$

The requirement for solvability is equivalent to the existence of a basis $\{\mathbf{v}_1, \ldots, \mathbf{v}_r\}$ of $\mathfrak{g}$ such that

$$[\mathbf{v}_i, \mathbf{v}_j] = \sum_{k=1}^{j-1} c_{ij}^k \mathbf{v}_k \quad \text{whenever} \quad i < j.$$

Note that any abelian Lie algebra, i.e. one for which the Lie bracket is always zero, is trivially solvable. Any two-dimensional Lie algebra is solvable, since using the basis (2.107), we can set $\mathfrak{g}^{(1)}$ to be the one-dimensional subalgebra generated by $\mathbf{v}$ to produce the chain $\{0\} = \mathfrak{g}^{(0)} \subset \mathfrak{g}^{(1)} \subset \mathfrak{g}^{(2)} = \mathfrak{g}$. The simplest example of a nonsolvable Lie algebra is the three-dimensional algebra $\mathfrak{sl}(2)$.

Theorem 2.64. *Let $\Delta(x, u^{(n)}) = 0$ be an n-th order ordinary differential equation. If Δ admits a solvable r-parameter group of symmetries G such that for $1 \leqslant k \leqslant r$ the orbits of $\mathrm{pr}^{(k)} G^{(k)}$ have dimension k, then the general solution of Δ can be found by quadratures from the general solution of an $(n-r)$-th order differential equation $\tilde{\Delta}(y, w^{(n-r)}) = 0$. In particular, if Δ admits an n-parameter solvable group of symmetries, then (subject to the above technical restrictions) the general solution to Δ can be found by quadratures alone.*

PROOF. The proof proceeds by induction along the chain of subalgebras (2.110) guaranteed by the solvability of G. At the k-th stage, we have used the invariance of Δ under the k-dimensional subalgebra $\mathfrak{g}^{(k)}$ to reduce it to an $(n-k)$-th order equation

$$\tilde{\Delta}^{(k)}(y, w^{(n-k)}) = 0,$$

in which $y, w, dw/dy, \ldots, d^{n-k}w/dy^{n-k}$ form a complete set of functionally independent differential invariants for the n-th prolongation $\mathrm{pr}^{(n)} G^{(k)}$; in particular, $y = \eta(x, u^{(k)})$, $w = \zeta(x, u^{(k)})$ form a complete set of invariants of the k-th prolongation of $G^{(k)}$. We also can reconstruct the general solution $u = f(x)$ from the general solution $w = h(y)$ of $\tilde{\Delta}^{(k)}$ by a series of quadratures.

To pass to the $(k+1)$-st stage, consider a generator $\mathbf{v}_{k+1}$ of $\mathfrak{g}^{(k+1)}$ which does not lie in $\mathfrak{g}^{(k)}$. Since $\mathfrak{g}^{(k)}$ is a normal subalgebra of $\mathfrak{g}^{(k+1)}$, (2.105) says that $\mathrm{pr}^{(k)} \mathbf{v}_{k+1}$ takes the form

$$\begin{aligned}\mathrm{pr}^{(k)} \mathbf{v}_{k+1} &= \mathrm{pr}^{(k-2)} \mathbf{v}_{k+1} + \alpha(y, w)\frac{\partial}{\partial y} + \psi(y, w)\frac{\partial}{\partial w}\\ &\equiv \mathrm{pr}^{(k-2)} \mathbf{v}_{k+1} + \tilde{\mathbf{v}}_{k+1},\end{aligned}$$

in which $\mathrm{pr}^{(k-2)} \mathbf{v}_{k+1}$ depends on the noninvariant coordinates $x, u, \ldots, u_{k-2}$ needed to complete y, w to a coordinate system on $M^{(k)}$.

Theorem 2.60 says that the original equaton Δ is invariant under all of $\mathfrak{g}^{(k+1)}$ if and only if the reduced equation $\tilde{\Delta}^{(k)}$ is invariant under the reduced vector field $\tilde{\mathbf{v}}_{k+1}$, which allows us to implement our reduction procedure for $\tilde{\Delta}^{(k)}$ using the vector field $\tilde{\mathbf{v}}_{k+1}$. Namely, we set

$$\hat{y} = \hat{\eta}(y, w), \qquad \hat{w} = \hat{\zeta}(y, w, w_y)$$

to be independent invariants of the first prolongation $\mathrm{pr}^{(1)} \tilde{\mathbf{v}}_{k+1}$. Then $\hat{y}, \hat{w}, d\hat{w}/d\hat{y}, \ldots, d^{n-k-1}\hat{w}/d\hat{y}^{n-k-1}$ form a complete set of invariants for the $(n-k)$-th prolongation $\mathrm{pr}^{(n-k)} \tilde{\mathbf{v}}_{k+1}$. Since $\tilde{\Delta}^{(k)}$ determines an invariant subvariety of this group, there is an equivalent equation

$$\hat{\Delta}^{(k+1)}(\hat{y}, \hat{w}^{(n-k-1)}) = 0$$

depending only on the invariants of $\mathrm{pr}^{(n-k)} \tilde{\mathbf{v}}_{k+1}$. Moreover, to reconstruct the solutions to $\tilde{\Delta}^{(k)}$ from those, $\hat{w} = \hat{h}(\hat{y})$, to $\hat{\Delta}^{(k+1)}$, we need only solve the first order equation

$$\hat{\zeta}(y, w, w_y) = \hat{h}[\hat{\eta}(y, w)].$$

This is invariant under the one-parameter group generated by $\tilde{\mathbf{v}}_{k+1}$, and hence can be integrated by quadrature. This completes the induction step, and thus proves the theorem. □

Example 2.65. Consider the third order equation

$$u_x^5 u_{xxx} = 3u_x^4 u_{xx}^2 + u_{xx}^3. \tag{2.111}$$

There is a three-parameter group of symmetries, generated by the vector fields

$$\mathbf{v}_1 = \partial_u, \qquad \mathbf{v}_2 = \partial_x, \qquad \mathbf{v}_3 = u\partial_x,$$

which is solvable since

$$[\mathbf{v}_1, \mathbf{v}_2] = 0, \qquad [\mathbf{v}_1, \mathbf{v}_3] = \mathbf{v}_2, \qquad [\mathbf{v}_2, \mathbf{v}_3] = 0.$$

Thus (2.111) can be solved by quadratures. We proceed to implement the reduction procedure given in the proof of Theorem 2.64. First, for $\mathfrak{g}^{(1)}$ generated by $\mathbf{v}_1$, we have invariants x, $v = u_x$, in terms of which (2.111) reduces to

$$v^5 v_{xx} = 3v^4 v_x^2 + v_x^3. \tag{2.112}$$

The second vector field $\mathbf{v}_2$ maintains its form $\tilde{\mathbf{v}}_2 = \partial_x$ when written using the invariants of $\mathfrak{g}^{(1)}$, so to reduce (2.112) for $\mathfrak{g}^{(2)} = \operatorname{Span}\{\mathbf{v}_1, \mathbf{v}_2\}$ we need the invariants

$$y = v, \qquad w = v_x, \qquad w_y = v_{xx}/v_x,$$

of $\operatorname{pr}^{(2)} \tilde{\mathbf{v}}_2$. In terms of these, (2.112) reduces to the first order equation

$$y^5 w_y = 3y^4 w + w^2. \tag{2.113}$$

This last Riccati equaton should retain one further symmetry corresponding to the vector field $\mathbf{v}_3$. Indeed, in terms of x, $y = u_x$, $w = u_{xx}$,

$$\operatorname{pr}^{(2)} \mathbf{v}_3 = u\partial_x - y^2\partial_y - 3yw\partial_w,$$

and the reduced vector field

$$\tilde{\mathbf{v}}_3 = -y^2\partial_y - 3yw\partial_w$$

is a symmetry of (2.113). We can thus integrate (2.113) by setting $t = -1/y$, $z = w/y^3$ (in terms of which $\tilde{\mathbf{v}}_3 = -\partial_t$), so (2.113) becomes

$$\frac{dz}{dt} = z^2.$$

Thus $z = 1/(c - t)$, or, in terms of the invariants of $\tilde{\mathbf{v}}_2$,

$$w = \frac{y^4}{cy + 1}.$$

To find v, we need to solve the autonomous equation

$$\frac{dv}{dx} = \frac{v^4}{cv + 1},$$

(autonomy being guaranteed by invariance under $\mathbf{v}_2$). We find the implicit solution

$$6(x - \tilde{c})v^3 + 3cv + 2 = 6(x - \tilde{c})u_x^3 + 3cu_x + 2 = 0.$$

Solving for u_x, we are left with one final quadrature to produce the general solution to the original equation (2.111).

One interesting thing to note is that although the equation (2.113) is invariant under a reduced vector field corresponding to the symmetry $\mathbf{v}_3$ of (2.111), there is *no* corresponding symmetry of the intermediate reduced equation (2.112). Indeed,

$$\operatorname{pr}^{(3)} \mathbf{v}_3 = u\frac{\partial}{\partial x} - v^2\frac{\partial}{\partial v} - 3vv_x\frac{\partial}{\partial v_x} - (4vv_{xx} + 3v_x^2)\frac{\partial}{\partial v_{xx}}$$

in terms of the invariants x, $v = u_x$ of $\mathbf{v}_1$, but this vector field *cannot* be reduced to one which does not depend on u. As a consequence of this observation, we see that in the general reduction procedure, it is important to wait until we have the invariants for $\mathfrak{g}^{(k)}$ before trying to reduce the next vector field $\mathbf{v}_{k+1}$; one cannot expect $\mathbf{v}_{k+1}$ to naturally reduce relative to an earlier subalgebra $\mathfrak{g}^{(j)}$ if $j < k$!

Systems of Ordinary Differential Equations

Knowledge of a group of symmetries of a system of first order ordinary differential equations has much the same consequences as knowledge of a similar group of symmetries of a single higher order equation. If we know a one-parameter symmetry group, then we can find the solution by quadrature from the solution to a first order system with one fewer equation in it. Similarly, knowledge of an r-parameter solvable group of symmetries allows us to reduce the number of equations by r. These results clearly extend to higher order systems as well, so that invariance of an n-th order system under a one-parameter group, say, allows us to reduce the order of *one* of the equations in the system by one. However, a higher order system can always be replaced by an equivalent first order system, so we are justified in restricting our attention to the latter case.

Theorem 2.66. *Let*

$$\frac{du^\nu}{dx} = F_\nu(x, u), \qquad \nu = 1, \dots, q, \tag{2.114}$$

be a first order system of q ordinary differential equations. Suppose G is a one-parameter group of symmetries of the system. Then there is a change of variables $(y, w) = \psi(x, u)$ under which the system takes the form

$$\frac{dw^\nu}{dy} = H_\nu(y, w^1, \dots, w^{q-1}), \qquad \nu = 1, \dots, q. \tag{2.115}$$

Thus the system reduces to a system of $q - 1$ ordinary differential equations for $w^1, \dots, w^{q-1}$ together with the quadrature

$$w^q(y) = \int H_q(y, w^1(y), \dots, w^{q-1}(y))\, dy + c.$$

Proof. Let $\mathbf{v}$ be the infinitesimal generator of G. Assuming $\mathbf{v}|_{(x,u)} \neq 0$, we can locally find new coordinates $y = \eta(x, u)$, $w^\nu = \zeta^\nu(x, u)$, $\nu = 1, \ldots, q$, such that $\mathbf{v} = \partial/\partial w^q$ in these coordinates. In fact,

$$\eta(x, u), \zeta^1(x, u), \ldots, \zeta^{q-1}(x, u),$$

will be a complete set of functionally independent invariants of G, so

$$\mathbf{v}(\eta) = \mathbf{v}(\zeta^\nu) = 0, \qquad \nu = 1, \ldots, q-1,$$

while $\zeta^q(x, u)$ satisfies

$$\mathbf{v}(\zeta^q) = 1.$$

It is then a simple matter to check that the equivalent first order system for $w^1, \ldots, w^q$ is invariant under the translation group generated by $\mathbf{v} = \partial/\partial w^q$ if and only if the right-hand sides are all independent of w^q, i.e. it is of the form (2.115). □

Example 2.67. Consider an autonomous system of two equations

$$\frac{du}{dx} = F(u, v), \qquad \frac{dv}{dx} = H(u, v).$$

Clearly $\mathbf{v} = \partial/\partial x$ generates a one-parameter symmetry group, so we can reduce this to a single first order equation plus a quadrature. The new coordinates are $y = u$, $w = v$ and $z = x$, in which we are viewing w and z as functions of y. Then

$$\frac{du}{dx} = \frac{1}{dz/dy}, \qquad \frac{dv}{dx} = \frac{dw/dy}{dz/dy},$$

so we have the equivalent system

$$\frac{dw}{dy} = \frac{H(y, w)}{F(y, w)}, \qquad \frac{dz}{dy} = \frac{1}{F(y, w)}.$$

We thus are left with a single first order equation for $w = w(y)$; the corresponding value of $z = z(y)$ is determined by a quadrature:

$$z = \int \frac{dy}{F(y, w)} + c.$$

If we revert to our original variables x, u, v we see that we just have the equation

$$\frac{dv}{du} = \frac{H(u, v)}{F(u, v)}$$

for the phase plane trajectories of the system, the precise motion along these trajectories being then determined by quadrature:

$$x = \int \frac{du}{F(u, v(u))} + c.$$

Theorem 2.68. *Suppose $du/dx = F(x, u)$ is a system of q first order, ordinary differential equations, and suppose G is an r-parameter solvable group of symmetries, acting regularly with r-dimensional orbits. Then the solutions $u = f(x)$ can be found by quadrature from the solutions of a reduced system $dw/dy = H(y, w)$ of $q - r$ first order equations. In particular, if the original system is invariant under a q-parameter solvable group, its general solution can be found by quadratures alone.*

The proof is left to the reader.

Example 2.69. Any linear, two-dimensional system

$$u_t = \alpha(t)u + \beta(t)v,$$
$$v_t = \gamma(t)u + \delta(t)v,$$

is invariant under the one-parameter group of scale transformations $(t, u, v) \mapsto (t, \lambda u, \lambda v)$ with infinitesimal generator $\mathbf{v} = u\partial_u + v\partial_v$, and hence can be reduced to a single first order equation by the method of Theorem 2.66. We set $w = \log u$, $z = v/u$, which straightens out $\mathbf{v} = \partial_w$. These new variables satisfy the transformed system

$$w_t = \alpha(t) + \beta(t)z,$$
$$z_t = \gamma(t) + (\delta(t) - \alpha(t))z - \beta(t)z^2,$$

so if we can solve the Riccati equation for z, we can find w (and hence u and v) by quadrature.

However, if the original system possesses some additional symmetry property, it may be unwise to carry out this preliminary reduction, as the resulting Riccati equation may no longer be invariant under some "reduced" symmetry group. For example, the system

$$u_t = -u + (t + 1)v,$$
$$v_t = u - tv$$

has an additional one-parameter symmetry group with generator $\mathbf{w} = t\partial_u + \partial_v$, as the reader may verify, but the associated Riccati equation

$$z_t = 1 + (1 - t)z - (1 + t)z^2$$

has no obvious symmetry property. The problem is that the vector fields $\mathbf{v}$ and $\mathbf{w}$ generate a solvable, two-dimensional Lie group, but have the commutation relation $[\mathbf{v}, \mathbf{w}] = -\mathbf{w}$, so we should be reducing first with respect to $\mathbf{w}$. To implement the reduction procedure of Theorem 2.68, we need to first straighten out $\mathbf{w} = \partial_{\tilde{w}}$ by choosing coordinates

$$\tilde{w} = v, \qquad \tilde{z} = u - tv.$$

The scaling group still has generator $\mathbf{v} = \tilde{w}\partial_{\tilde{w}} + \tilde{z}\partial_{\tilde{z}}$ in these variables. To straighten its $\tilde{z}$-component we further set $\hat{z} = \log \tilde{z} = \log(u - tv)$, in terms of which

$$\mathbf{w} = \partial_{\tilde{w}}, \qquad \mathbf{v} = \tilde{w}\partial_{\tilde{w}} + \partial_{\hat{z}}.$$

The system now takes the form

$$\frac{d\tilde{w}}{dt} = e^{\hat{z}}, \qquad \frac{d\hat{z}}{dt} = -t - 1,$$

which, as guaranteed by Theorem 2.68, can be integrated by quadratures. We find

$$\hat{z}(t) = -\tfrac{1}{2}(t+1)^2 + \tilde{c},$$

$$\tilde{w}(t) = c \operatorname{erf}[(t+1)/\sqrt{2}] + k,$$

where $\tilde{c} = \log(c\sqrt{2/\pi})$, and

$$\operatorname{erf}(y) = \frac{2}{\sqrt{\pi}} \int_0^y e^{-x^2}\, dx$$

is the standard error function. Thus the general solution to the original system is

$$u(t) = \sqrt{\frac{2}{\pi}}\, c e^{-(t+1)^2/2} + ct \operatorname{erf}\left(\frac{t+1}{\sqrt{2}}\right) + kt, \qquad v(t) = c \operatorname{erf}\left(\frac{t+1}{\sqrt{2}}\right) + k,$$

where c and k are arbitrary constants.

2.6. Nondegeneracy Conditions for Differential Equations

Often one is interested in classifying all the symmetries of a system of differential equations, and so it is important to know when the infinitesimal methods developed in Section 2.3 can construct the most general connected symmetry group of the given system. For this to be the case, it will be necessary to impose an additional nondegeneracy condition, known as "local solvability", beyond the maximal rank condition of Definition 2.30. This relatively unfamiliar and somewhat technical condition requires that the system have solutions for "arbitrary initial data". In this section we discuss this concept and some of its consequences in detail.

Local Solvability

In order to motivate the definition of local solvability, let's see why, in contrast to the case of systems of algebraic equations, the infinitesimal criterion (2.25) is *not* in general a necessary condition for a Lie group G to be a symmetry group of a system of differential equations of maximal rank. For a system of algebraic equations $F(x) = 0$, to each point x_0 on the subvariety $\mathscr{S}_F = \{x: F(x) = 0\}$ there is, tautologously, a solution to the system; namely, x_0 itself! In contrast, if $\Delta(x, u^{(n)}) = 0$ is a system of differential equations, and $(x_0, u_0^{(n)})$ a point on the corresponding subvariety $\mathscr{S}_\Delta = \{(x, u^{(n)})\colon \Delta(x, u^{(n)}) = 0\}$, there is in general no guarantee that there exists a solution $u = f(x)$ of the system which has these particular values for its derivatives at x_0, i.e. $u_0^{(n)} = \mathrm{pr}^{(n)} f(x_0)$. Therefore if G, a local group of transformations, is a symmetry group of the system of differential equations, in the sense that it transforms solutions to solutions, there is no assurance that G will leave the entire subvariety $\mathscr{S}_\Delta$ invariant. We can only conclude that those points $(x_0, u_0^{(n)})$ in $\mathscr{S}_\Delta$ for which there does exist such a solution are transformed into other such points in $\mathscr{S}_\Delta$ under group transformations. Therefore, to prove the necessity of the infinitesimal criterion of invariance, we need to assume that every point in $\mathscr{S}_\Delta$ has a corresponding solution.

Definition 2.70. A system of n-th order differential equations $\Delta(x, u^{(n)}) = 0$ is *locally solvable* at the point

$$(x_0, u_0^{(n)}) \in \mathscr{S}_\Delta = \{(x, u^{(n)})\colon \Delta(x, u^{(n)}) = 0\}$$

if there exists a smooth solution $u = f(x)$ of the system, defined for x in a neighbourhood of x_0, which has the prescribed "initial conditions" $u_0^{(n)} = \mathrm{pr}^{(n)} f(x_0)$. The system is *locally solvable* if it is locally solvable at every point of $\mathscr{S}_\Delta$. A system of differential equations is *nondegenerate* if at every point $(x_0, u_0^{(n)}) \in \mathscr{S}_\Delta$ it is both locally solvable and of maximal rank.

For a system of ordinary differential equations, this condition of local solvability coincides with the usual initial value problem, with $(x_0, u_0^{(n)})$ corresponding to the usual initial data in this case. For instance, for a single second order equation

$$u_{xx} = F(x, u, u_x), \tag{2.116}$$

the initial data for local solvability consist of four numbers $(x^0, u^0, u_x^0, u_{xx}^0)$ subject only to the condition that they satisfy the equation, i.e.

$$u_{xx}^0 = F(x^0, u^0, u_x^0).$$

We are then required to find a solution $u = f(x)$, defined for x near x^0, such that

$$u^0 = f(x^0), \qquad u_x^0 = f'(x^0), \qquad u_{xx}^0 = f''(x^0).$$

Clearly the first two of these conditions form the usual initial value problem for (2.116), and we are thus assured of the existence of a solution $u = f(x)$ satisfying these two conditions. (Indeed, we only need to assume that F is continuous.) The third condition $u_{xx}^0 = f''(x^0)$ is then given to us "for free" since f is a solution, even at x^0, so

$$f''(x^0) = F(x^0, f(x^0), f'(x^0)) = F(x^0, u^0, u_x^0) = u_{xx}^0.$$

Similar reasoning shows that nonsingular systems of ordinary differential equations are always locally solvable.

For systems of partial differential equations, the problem of local solvability is of a completely different character than the more usual existence problems, e.g. Cauchy problems or boundary value problems. In the case of local existence, the initial data is only being prescribed at a single point x_0, whereas one ordinarily requires the specification of the data along an entire submanifold of the space of independent variables. For example, in the case of the wave equation

$$u_{tt} - u_{xx} = 0,$$

the question of local solvability becomes that of determining whether for every set of initial values

$$(x^0, t^0; u^0; u_x^0, u_t^0; u_{xx}^0, u_{xt}^0, u_{tt}^0),$$

subject only to the condition $u_{tt}^0 = u_{xx}^0$, there exists a solution $u = f(x, t)$ of the wave equation in a neighbourhood of (x^0, t^0) with

$$u^0 = f(x^0, t^0), \qquad u_x^0 = \frac{\partial f}{\partial x}(x^0, t^0), \qquad u_t^0 = \frac{\partial f}{\partial t}(x^0, t^0),$$

$$u_{xx}^0 = \frac{\partial^2 f}{\partial x^2}(x^0, t^0), \qquad u_{xt}^0 = \frac{\partial^2 f}{\partial x \partial t}(x^0, t^0), \qquad u_{tt}^0 = \frac{\partial^2 f}{\partial t^2}(x^0, t^0).$$

Clearly in this case the answer is yes, since by design $u_{xx}^0 = u_{tt}^0$, so we can take f to be the polynomial solution

$$\begin{aligned} f(x, t) = u^0 + u_x^0(x - x^0) + u_t^0(t - t^0) + \tfrac{1}{2}u_{xx}^0[(x - x^0)^2 + (t - t^0)^2] \\ + u_{xt}^0(x - x^0)(t - t^0), \end{aligned}$$

hence the wave equation is locally solvable. (Note that there is no question of uniqueness for the solutions to the local existence problem—even in this simple example no such result is valid.) The reader should contrast this problem with the usual Cauchy problem, in which the initial data is specified along the entire x-axis:

$$u(x, 0) = g(x), \qquad \frac{\partial u}{\partial t}(x, 0) = h(x).$$

There are two principal reasons why a system of partial differential equations might fail to be locally solvable. The first is that the system may have integrability conditions obtained by cross-differentiating the various equations. For example, the over-determined system

$$u_x = yu, \qquad u_y = 0, \tag{2.117}$$

is not locally solvable since at any point (x_0, y_0) there is no smooth solution $u(x, y)$ with "initial conditions"

$$u^0 = u(x_0, y_0) = 1, \qquad u_x^0 = u_x(x_0, y_0) = y_0, \qquad u_y^0 = u_y(x_0, y_0) = 0,$$

values which algebraically satisfy the two equations. Indeed, cross-differentiation shows that

$$0 = u_{xy} = (yu)_y = yu_y + u,$$

hence $u(x, y) \equiv 0$ is the only solution.

Another interesting example, which actually arises from a variational problem (cf. Section 4.1), is the second order system

$$\begin{aligned} u_{xx} + v_{xy} + v_x &= 0, \\ u_{xy} + v_{yy} - u_x &= 0. \end{aligned} \tag{2.118}$$

As it stands, (2.118) is not locally solvable since there is an additional relationship among second order derivatives, namely

$$u_{xx} + v_{xy} = 0,$$

obtained by differentiating the first equation with respect to y, the second with respect to x and subtracting. This in turn implies $v_x = 0$ and $u_{xx} = 0$, so any assignation of initial values

$$(x^0, y^0; u^0, v^0; u_x^0, u_y^0, v_x^0, v_y^0; u_{xx}^0, u_{xy}^0, u_{yy}^0, v_{xx}^0, v_{xy}^0, v_{yy}^0)$$

which satisfies (2.118), but which does not have $v_x^0 = v_{xx}^0 = v_{xy}^0 = u_{xx}^0 = 0$, has no local solution pertaining to it.

The second source of systems which are not locally solvable are certain smooth, but not analytic, systems of differential equations which have no solutions. The original example of such a system was discovered by Lewy, [1], who showed that there exist smooth functions $h(x, y, z)$ such that the first order system

$$\begin{aligned} u_x - v_y + 2yu_z + 2xv_z &= h(x, y, z), \\ u_y + v_x - 2xu_z + 2yv_z &= 0, \end{aligned}$$

has no smooth (or even C^1) solutions on any open subset of $\mathbb{R}^3$. A related example is given by Nirenberg, [1; p. 8], who constructs a function $h(x, y)$ such that the homogeneous linear system

$$\begin{aligned} u_x - h(x, y)v_y &= 0, \\ v_x + h(x, y)u_y &= 0, \end{aligned} \tag{2.119}$$

has only constant solutions in a neighbourhood of the origin.

As we will see, for analytic systems, the Cauchy-Kovalevskaya theorem provides the key to the proof of local solvability. For C^∞ systems, the question is much more delicate, owing to the Lewy-type phenomena, and very few general results are known. Before investigating the analytic case in more detail, we apply the local solvability criterion to the infinitesimal condition for group invariance of a system of differential equations.

Invariance Criteria

Theorem 2.71. *Let $\Delta(x, u^{(n)}) = 0$ be a nondegenerate system of differential equations. A connected local group of transformations G acting on an open subset $M \subset X \times U$ is a symmetry group of the system if and only if*

$$\mathrm{pr}^{(n)}\, \mathbf{v}[\Delta_\nu(x, u^{(n)})] = 0, \quad \nu = 1, \ldots, l, \quad \textit{whenever} \quad \Delta(x, u^{(n)}) = 0, \qquad (2.120)$$

for every infinitesimal generator $\mathbf{v}$ *of G.*

Proof. We already know that (2.120) is sufficient for G to be a symmetry group, so we need only prove the necessity of this condition. In light of the algebraic counterpart of this result in Theorem 2.8, it suffices to prove that the subvariety $\mathscr{S}_\Delta = \{\Delta(x, u^{(n)}) = 0\}$ is an invariant subset of the prolonged group action $\mathrm{pr}^{(n)} G$ whenever G transforms solutions of the system to other solutions. Let $(x_0, u_0^{(n)}) \in \mathscr{S}_\Delta$. Using the local solvability, let $u = f(x)$ be a solution of the system defined in a neighbourhood of x_0 such that $u_0^{(n)} = \mathrm{pr}^{(n)} f(x_0)$. If g is a group element such that $\mathrm{pr}^{(n)} g \cdot (x_0, u_0^{(n)})$ is defined, then by appropriately shrinking the domain of definition of f, we can ensure that the transformed function $\tilde{f} = g \cdot f$ is a well-defined function in a neighbourhood of $\tilde{x}_0$, where $(\tilde{x}_0, \tilde{u}_0) = g \cdot (x_0, u_0)$. Since G is a symmetry group, $u = \tilde{f}(x)$ is also a solution to the system. Moreover, by the definition of the prolonged group action, (2.18),

$$\mathrm{pr}^{(n)} g \cdot (x_0, u_0^{(n)}) = (\tilde{x}_0, \mathrm{pr}^{(n)}(g \cdot f)(\tilde{x}_0)) = (\tilde{x}_0, \tilde{u}_0^{(n)}),$$

hence the transformed point $(\tilde{x}_0, \tilde{u}_0^{(n)})$ must again lie in $\mathscr{S}_\Delta$. This proves that $\mathscr{S}_\Delta$ is an invariant subset of G, and the theorem follows. □

In order to appreciate the necessity of the local solvability condition in Theorem 2.71, we discuss a couple of examples. First consider the Nirenberg system (2.119). Since the only solutions near the origin are constants, the translational group $(x, y, u, v) \mapsto (x, y + \varepsilon, u, v)$ is a symmetry group. However, the infinitesimal criterion (2.120) does not hold; applying $\mathrm{pr}^{(1)} \mathbf{v} = \partial_y$ to the first equation we get $h_y v_y$, which is not zero as an algebraic consequence of the system. This group is also, for the same reason, a symmetry group of the over-determined system (2.117). However,

$$\mathrm{pr}^{(1)}\, \mathbf{v}(u_x - yu) = -u,$$

which does not vanish as an algebraic consequence of (2.117), and again the infinitesimal criterion (2.120) does not apply. However, we can get (2.120) to be both necessary and sufficient for G to be a symmetry group if we only require that it hold at points of local solvability:

Theorem 2.72. *Let $\Delta(x, u^{(n)}) = 0$ be a system of differential equations of maximal rank. A local group of transformations G is a symmetry group of the system if and only if for every point $(x_0, u_0^{(n)}) \in \mathscr{S}_\Delta$ at which the system is locally solvable, we have*

$$\mathrm{pr}^{(n)}\, \mathbf{v}(\Delta_\nu)(x_0, u_0^{(n)}) = 0, \qquad \nu = 1, \ldots, l,$$

for all infinitesimal generators $\mathbf{v}$ of G.

The proof is immediate. □

The Cauchy-Kovalevskaya Theorem

For analytic systems of partial differential equations, the Cauchy-Kovalevskaya theorem plays a pivotal role in the existence theory. Besides being the principal general existence result for solutions of such systems, this theorem also provides the key to the general theory of characteristics, which underlies any serious investigation of the behaviour of solutions of systems of partial differential equations. As we will see, the Cauchy-Kovalevskaya theorem also gives a proof of the local solvability of most analytic systems of differential equations.

In its original form, the Cauchy-Kovalevskaya theorem treats the Cauchy problem on the initial hyperplane $\{t = t_0\}$ for a system in *Kovalevskaya form*

$$u_{nt}^\alpha \equiv \frac{\partial^n u^\alpha}{\partial t^n} = \Gamma_\alpha(y, t, \widetilde{u^{(n)}}), \qquad \alpha = 1, \ldots, q. \tag{2.121}$$

Here $(y, t) = (y^1, \ldots, y^{p-1}, t)$ are the independent variables, and $\widetilde{u^{(n)}}$ denotes all partial derivatives of u with respect to both y and t up to order n *except* the derivatives u_{nt}^β which appear on the left-hand side of (2.121). The *Cauchy data* for this system is given by

$$\frac{\partial^k u^\alpha}{\partial t^k}(y, t_0) = h_k^\alpha(y), \qquad \alpha = 1, \ldots, q, \quad k = 0, \ldots, n-1, \tag{2.122}$$

where the h_k^α are analytic functions on the hyperplane $\{t = t_0\}$ for y in a neighbourhood of a point $y_0 \in \mathbb{R}^{p-1}$.

Theorem 2.73. *Suppose the functions Γ_α in the Kovalevskaya system* (2.121) *are analytic in their arguments, and the Cauchy data $h_k^\alpha(y)$ in* (2.122) *are also analytic functions for y near y_0. Then there exists a unique analytic solution $u = f(y, t)$ for the Cauchy problem* (2.121), (2.122) *defined for (y, t) in some neighbourhood of the point (y_0, t_0).*

This theorem immediately proves the local solvability of the Kovalevskaya system (2.121).

Corollary 2.74. *If Δ is an analytic system in Kovalevskaya form* (2.121), *then Δ is locally solvable.*

Proof. Note that for the local solvability problem for (2.121) we can prescribe the lower order t-derivatives $\widetilde{u_0^{(n)}}$ at the initial point (y_0, t_0) in an arbitrary manner, the remaining n-th order derivatives, namely $u^\alpha_{nt,0}$, are then determined by the requirement that $(y_0, t_0, u_0^{(n)})$ be a solution to Δ. Given $(y_0, t_0, u_0^{(n)})$, choose analytic functions $h^\alpha_k(y)$, $k = 0, \ldots, n-1$, $\alpha = 1, \ldots, q$, such that the y-derivatives

$$\partial_J h^\alpha_k(y_0) = \partial^i h^\alpha_k(y_0)/\partial y^{j_1} \cdots \partial y^{j_i}, \qquad 0 \leqslant i \leqslant n-k,$$

agree with the corresponding prescribed values $u^\alpha_{kt,J0}$, where $u^\alpha_{kt,J} \equiv D_J(u^\alpha_{kt})$. (Again, $h^\alpha_k(y)$ could be an appropriate Taylor polynomial.) The corresponding solution $u = f(y, t)$ to the Cauchy problem (2.121), (2.122) ensured by the Cauchy-Kovalevskaya theorem then solves the local existence problem for Δ. Indeed

$$\partial_J \partial^k_t f^\alpha(y_0, t_0) = \partial_J h^\alpha_k(y_0) = u^\alpha_{kt,J0}$$

for $0 \leqslant k \leqslant n-1$, $\#J \leqslant n-k$, while the n-th order derivatives $\partial^n_t f^\alpha(y_0, t_0)$ and $u^\alpha_{nt,0}$ agree because both satisfy the given equations (2.121) at (y_0, t_0) with the same values of $\widetilde{u_0^{(n)}}$. □

More generally, we can admit different order t-derivatives on the left-hand side, whereby a system will be in *general Kovalevskaya form* if

$$\partial^{n_\alpha} u^\alpha/\partial t^{n_\alpha} = \Gamma_\alpha(t, y, \widetilde{u^{(n)}}), \tag{2.123}$$

in which $n = \max\{n_1, \ldots, n_q\}$, and $\widetilde{u^{(n)}}$ denotes all derivatives of each u^β up to order n_β except the particular derivatives $\partial^{n_\beta} u^\beta/\partial t^{n_\beta}$ appearing on the left-hand side. The Cauchy problem (2.122) is the same except that for each α, the index k runs from 0 to $n_\alpha - 1$. All the results of this section, including Corollary 2.74, remain valid for these more general Kovalevskaya forms; the proofs are only slightly more complicated, and are left for the reader to fill in the details.

Characteristics

The range of applicability of the Cauchy-Kovalevskaya existence theorem, and hence the local solvability theorem of Corollary 2.74, can be greatly extended by allowing the possibility of transforming a given system of analytic differential equatons into a system in Kovalevskaya form (2.121) (or (2.123)) by a change of independent variables. To begin with, the system

$$\Delta_\nu(x, u^{(n)}) = 0, \qquad \nu = 1, \ldots, q,$$

must have the same number of equations as unknowns (dependent variables) to stand any chance of being transformed into a system of Kovalevskaya form, and we restrict our attention here to such systems.

It turns out that it suffices to consider changes of variable of the simple form

$$t = \psi(x), \qquad y = (y^1, \ldots, y^{p-1}) = (x^1, \ldots, x^{i-1}, x^{i+1}, \ldots, x^p), \quad (2.124)$$

in which ψ is a smooth, real-valued function with nonzero gradient, $\nabla\psi(x_0) \neq 0$ at the point x_0 under investigation, and i is chosen so that $\partial\psi(x_0)/\partial x^i \neq 0$ so the change of variables (2.124) is locally invertible. Note that the initial hyperplane $\{t = t_0\}$ in the (y, t) coordinate comes from the level set $S = \{x \colon \psi(x) = t_0\}$ in the original coordinates, so the Cauchy problem in the x-coordinates consists of prescribing initial data on the hypersurface S. Under the change of variables (2.124), there is a corresponding system

$$\tilde{\Delta}_\nu(y, t, u^{(n)}) = 0, \qquad \nu = 1, \ldots, q, \quad (2.125)$$

involving y, t and derivatives of u with respect to y and t up to order n obtained by re-expressing the x-derivatives of u in terms of the y and t derivatives. We can apply the Cauchy-Kovalevskaya theorem to the transformed system (2.125) *provided* we can solve it for the n-th order t-derivatives u^α_{nt} in terms of y, t and the remaining derivatives $\widetilde{u^{(n)}}$. By the implicit function theorem, this is possible in a neighbourhood of a point $(y_0, t_0, u_0^{(n)})$ provided the $q \times q$ matrix M with entries

$$\mathsf{M}_{\alpha\nu} = \partial\tilde{\Delta}_\nu(y_0, t_0, u_0^{(n)})/\partial u^\alpha_{nt}, \qquad \alpha, \nu = 1, \ldots, q,$$

is nonsingular: $\det \mathsf{M} \neq 0$.

Let us see what this matrix M looks like. If u^α_J is any n-th order x-derivative of u, then by the chain rule,

$$u^\alpha_J = \frac{\partial^n u^\alpha}{\partial x^{j_1} \cdots \partial x^{j_n}} = \frac{\partial\psi}{\partial x^{j_1}} \cdot \frac{\partial\psi}{\partial x^{j_2}} \cdot \cdots \cdot \frac{\partial\psi}{\partial x^{j_n}} \cdot \frac{\partial^n u^\alpha}{\partial t^n} + \cdots \equiv (\nabla\psi)_J u^\alpha_{nt} + \cdots,$$

where the omitted terms involve various n-th and lower order derivatives of u^α with respect to y and t, except the key derivative u^α_{nt}. Therefore, if we form the $q \times q$ matrix $\mathsf{M}(\omega) = \mathsf{M}_\Delta(\omega; x_0, u_0^{(n)})$ whose entries are the homogeneous polynomials

$$\mathsf{M}_{\alpha\nu}(\omega) = \sum_{\#J=n} \frac{\partial\Delta_\nu}{\partial u^\alpha_J}(x_0, u_0^{(n)}) \cdot \omega_J, \qquad \alpha, \nu = 1, \ldots, q, \quad (2.126)$$

of degree n depending on $\omega = (\omega_1, \ldots, \omega_p)$, with $\omega_J \equiv \omega_{j_1}\omega_{j_2} \cdots \omega_{j_n}$, then the above matrix is obtained by evaluating $\mathsf{M}(\omega)$ at $\omega = \nabla\psi(x_0)$.

Definition 2.75. Let Δ be an n-th order system of differential equations having the same number of equations as unknowns. Given a point $(x_0, u_0^{(n)}) \in \mathcal{S}_\Delta$, form the $q \times q$ matrix of polynomials (2.126). A nonzero p-tuple ω is said to define a *noncharacteristic direction* (respectively *characteristic direction*) to Δ at $(x_0, u_0^{(n)})$ if $\mathsf{M}(\omega)$ is nonsingular (respectively singular). A hypersurface $S = \{\psi(x) = c\}$, $\nabla\psi \neq 0$, is called *noncharacteristic* at $(x_0, u_0^{(n)})$ if $\omega = \nabla\psi(x_0)$ determines a noncharacteristic direction there.

In particular, if the highest order derivatives in the system $\Delta(x, u^{(n)}) = 0$ occur linearly with coefficients only depending on x, then the matrix $\mathsf{M}(\omega)$ determining the characteristic directions depends only on x_0, so we can omit reference to the particular solution $u_0^{(n)}$ and refer unambiguously to a characteristic or noncharacteristic direction at x_0 itself. This is the case occurring most frequently in physical systems.

Our earlier considerations show that we can apply the Cauchy-Kovalevskaya theorem to the Cauchy problem provided the initial data lies on a noncharacteristic hypersurface.

Theorem 2.76. *If $\Delta(x, u^{(n)}) = 0$ is an analytic system of differential equations and S is a noncharacteristic, analytic hypersurface for Δ at $(x_0, u_0^{(n)})$, then there exists a local analytic solution to the Cauchy problem*

$$\Delta(x, u^{(n)}) = 0,$$

$$\frac{\partial^k u}{\partial n^k} = h_k(x), \qquad x \in S, \quad k = 0, \ldots, n-1,$$

in a neighbourhood of x_0. Here the h_k are analytic functions on S, and $\partial/\partial n$ denotes the normal derivative for S.

Example 2.77. (a) In the case of the one-dimensional wave equation

$$u_{tt} - c^2 u_{xx} = H(x, t, u, u_x, u_t),$$

a direction $\omega = (\tau, \xi)$ is characteristic if and only if

$$\tau^2 - c^2\xi^2 = 0.$$

For c constant, we recover the familiar characteristic curves

$$\psi(x, t) = x \pm ct = k.$$

Any curve not tangent to these lines can be used for valid Cauchy data.

(b) The equations of linear isotropic elasticity are known as *Navier's equations*. In two dimensions they take the form

$$\begin{aligned} (2\mu + \lambda)u_{xx} + \mu u_{yy} + (\mu + \lambda)v_{xy} &= 0, \\ (\mu + \lambda)u_{xy} + \mu v_{xx} + (2\mu + \lambda)v_{yy} &= 0, \end{aligned} \tag{2.127}$$

where λ and μ are constants known as the *Lamé moduli*. The 2×2 matrix $\mathsf{M}(\xi, \eta) = \mathsf{M}(\omega)$ determining the characteristics has the form

$$\mathsf{M}(\xi, \eta) = \begin{pmatrix} (2\mu + \lambda)\xi^2 + \mu\eta^2 & (\mu + \lambda)\xi\eta \\ (\mu + \lambda)\xi\eta & \mu\xi^2 + (2\mu + \lambda)\eta^2 \end{pmatrix}.$$

Then $\omega = (\xi, \eta)$ is characteristic if and only if

$$\det \mathsf{M}(\xi, \eta) = (2\mu + \lambda)\mu(\xi^2 + \eta^2)^2 = 0.$$

Thus unless $\mu = 0$ or $2\mu + \lambda = 0$, in which case *every* direction is characteristic, there are no real characteristic directions to Navier's equations. Note that the case $\mu = 0, \lambda = 1$ yields the leading order terms in the not locally-solvable system (2.118), hence this latter system has every direction characteristic.

Normal Systems

Corollary 2.74 will provide an immediate solution to the local solvability problem for an analytic system provided we can find at least one noncharacteristic direction to the system at the point $(x_0, u_0^{(n)})$ of interest. As Example 2.77(b) makes clear, not every system of partial differential equations satisfies this basic requirement, so we need to distinguish those systems which do.

Definition 2.78. A system of q differential equations $\Delta(x, u^{(n)}) = 0$ in q dependent variables $u = (u^1, \ldots, u^q)$ is *normal* at the point $(x_0, u_0^{(n)}) \in \mathscr{S}_\Delta$ if there exists at least one noncharacteristic direction ω for Δ there. The system is *normal* if it is normal at each point of $\mathscr{S}_\Delta$.

Theorem 2.79. *A system of differential equations is normal at $(x_0, u_0^{(n)})$ if and only if there is a change of variables $(y, t) = \chi(x)$ transforming it into a system in Kovalevskaya form near $(y_0, t_0) = \chi(x_0)$.*

Corollary 2.80. *If a system of differential equations is both analytic and normal at $(x_0, u_0^{(n)})$ then it is locally solvable at $(x_0, u_0^{(n)})$.*

We just change variables and invoke Corollary 2.74 for the resulting Kovalevskaya system. Later we will see that Corollary 2.80 admits a converse!

Prolongation of Differential Equations

Definition 2.81. Let

$$\Delta_\nu(x, u^{(n)}) = 0, \qquad \nu = 1, \ldots, l,$$

be an n-th order system of differential equations defined by the vanishing of a smooth function $\Delta: M^{(n)} \to \mathbb{R}^l$. The k-th *prolongation* of this system is the $(n + k)$-th order system of differential equations

$$\Delta^{(k)}(x, u^{(n+k)}) = 0$$

obtained by differentiating the equations in Δ in all possible ways up to order k. In other words, $\Delta^{(k)}$ consists of the $\binom{p+k-1}{k}\cdot l$ equations

$$D_J\Delta_\nu(x, u^{(n+k)}) = 0,$$

where $\nu = 1, \ldots, l$, and J runs over all multi-indices of orders $0 \leqslant \#J \leqslant k$.

For example, the first prolongation of the heat equation

$$u_t = u_{xx}$$

is the third order system

$$u_t = u_{xx}, \qquad u_{xt} = u_{xxx}, \qquad u_{tt} = u_{xxt}.$$

The second prolongation appends the additional fourth order equations

$$u_{xxt} = u_{xxxx}, \qquad u_{xtt} = u_{xxxt}, \qquad u_{ttt} = u_{xxtt},$$

and so on.

Proposition 2.82. *If $u = f(x)$ is a smooth solution of a system $\Delta(x, u^{(n)}) = 0$, then it is also a solution to every prolongation of the system $\Delta^{(k)}(x, u^{(n+k)}) = 0$, $k = 0, 1, 2, \ldots$.*

Definition 2.83. A system of differential equatons is called *totally nondegenerate* if it and all its prolongations are both of maximal rank and locally solvable.

As we will see in a moment, any analytic system in Kovalevskaya form, and hence any normal analytic system, is always totally nondegenerate. Surprisingly, in the case of analytic systems with the same number of equations as unknowns, these are the only totally nondegenerate systems; if an analytic system is not normal, some prolongation of it is either not of maximal rank or not locally solvable. The C^∞ case is more complicated, owing to the appearance of the Lewy phenomenon of nonexistence.

Theorem 2.84. *An analytic system of differential equations*

$$\Delta_\nu(x, u^{(n)}) = 0, \qquad \nu = 1, \ldots, q,$$

involving the same number of equations as dependent variables $u^1, \ldots, u^q$, is totally nondegenerate if and only if it is normal.

PROOF. The Cauchy-Kovalevskaya theorem immediately proves that any normal system is totally nondegenerate. Indeed, by choosing a noncharacteristic direction, we can assume that the system is in Kovalevskaya form

(2.121). The k-th prolongation of such a system takes the form

$$u^\alpha_{(n+l)t,J} = D_t^l D_J\{\Gamma_\alpha(y, t, \widetilde{u^{(n)}})\}, \tag{2.128}$$

where $J = (j_1, \ldots, j_i)$ runs over all multi-indices with $1 \leqslant j_\kappa \leqslant p-1$, $l + i \leqslant k$. Also D_J denotes the corresponding i-th order total derivative with respect to $y = (y^1, \ldots, y^{p-1})$, and $u^\alpha_{(n+l)t,J} = D_J[u^\alpha_{(n+l)t}]$. The right-hand side of (2.128) depends on derivatives $u^\beta_{mt,K}$ where $m < n + l$. We can therefore inductively solve for the derivatives $u^\beta_{mt,K}$, $m \geqslant n$, in terms of y, t and derivatives $u^\gamma_{jt,L}$ with $j < n$. Thus (2.128) is equivalent to a system of the form

$$u^\alpha_{(n+l)t,J} = \Gamma^{J,l}_\alpha(y, t, \widetilde{u^{(n+k)}}), \tag{2.129}$$

in which $l + \#J \leqslant k$ and $\widetilde{u^{(n+k)}}$ denotes all derivatives of u up to order $n + k$ except those involving n or more t-derivatives. The maximal rank condition for $\Delta^{(k)}$ follows easily since the submatrix of the full Jacobian matrix for (2.129), cf. Definition 2.30, corresponding to all the partial derivatives

$$\frac{\partial}{\partial u^\beta_{mt,K}}[u^\alpha_{(n+l)t,J} - \Gamma^{J,l}_\alpha], \qquad m \geqslant n,$$

is the identity matrix.

The local solvability of (2.129) follows from the Cauchy-Kovalevskaya theorem. We can specify the derivatives $\widetilde{u_0^{(n+k)}}$ at a point y_0, t_0 arbitrarily; the values of the remaining derivatives in $u_0^{(n+k)}$ will then be determined by the prolonged system itself. Let $h^\alpha_m(y)$, $m = 0, \ldots, n-1$, $\alpha = 1, \ldots, q$, be analytic functions taking the prescribed values

$$\partial_J h^\alpha_m(y_0) = u^\alpha_{mt,J0}, \qquad \#J \leqslant n + k - m,$$

at (y_0, t_0). Let $u = f(y, t)$ be the analytic solution to the resulting Cauchy problem given by the Cauchy–Kovalevskaya theorem. Then

$$\partial_J \partial_t^m f(y_0, t_0) = u^\alpha_{mt,J0}$$

for $\#J + m \leqslant n + k$: for $m < n$ this follows from the definition of h^α_m, while for $m \geqslant n$ this follows since both $\mathrm{pr}^{(n+k)} f(y_0, t_0)$ and $(y_0, t_0, u_0^{(n+k)})$ satisfy the k-th prolongation of Δ at this point. Thus $u = f(y, t)$ gives the solution to the local solvability problem for $\Delta^{(k)}$ at $(y_0, t_0, u_0^{(n+k)})$.

The proof of the converse in Theorem 2.84 rests on a beautiful result due to Finzi.

Lemma 2.85. *Suppose*

$$\Delta_\nu(x, u^{(n)}) = 0, \qquad \nu = 1, \ldots, q,$$

is an n-th order system of differential equations. Then Δ has no noncharacteristic directions at $(x_0, u_0^{(n)})$ if and only if there exist homogeneous k-th order differential operators

$$\mathscr{D}_\nu = \sum_{\#J=k} P^J_\nu D_J, \qquad \nu = 1, \ldots, q,$$

not all zero at $(x_0, u_0^{(n)})$, such that at $(x_0, u_0^{(n)})$ the combination

$$\sum_{\nu=1}^{q} \mathscr{D}_\nu \Delta_\nu \equiv Q(x_0, u_0^{(n+k-1)}) \tag{2.130}$$

depends only on derivatives of u of order at most $n + k - 1$.

Moreover, if there are no noncharacteristic directions for Δ for all $(x, u^{(n)})$ in some relatively open subset $\mathscr{S}_\Delta \cap V$, V open in $M^{(n)}$, then the differential operators $\mathscr{D}_\nu$ depend smoothly on $(x, u^{(n)})$. In this case, (2.130) holds for all $(x, u^{(n+k)}) \in M^{(n+k)}$ which project to $(x, u^{(n)}) = \pi_n^{n+k}(x, u^{(n+k)}) \in \mathscr{S}_\Delta \cap V$.

The point of the lemma is the following. Ordinarily, if Δ is an n-th order system of differential equations and $\mathscr{D}_1, \ldots, \mathscr{D}_q$ are k-th order differential operators, one would expect the linear combination $\sum \mathscr{D}_\nu \Delta_\nu$ to depend on $(n + k)$-th order derivatives of the u's. However, in the case Δ has only characteristic directions, one can find certain nontrivial k-th order differential operators $\mathscr{D}_\nu$ such that the combination $\sum \mathscr{D}_\nu \Delta_\nu$ depends on *only* $(n + k - 1)$-st and lower order derivatives, and hence the condition $\sum \mathscr{D}_\nu \Delta_\nu = 0$, which must hold for all solutions, provides an additional integrability condition on $(n + k - 1)$-st order derivatives of the u's which is not directly deduced from the $(k - 1)$-st order prolongaton $\Delta^{(k-1)}$. Conversely, if a system has some nontrivial integrability conditions, Finzi's lemma implies that there cannot be any noncharacteristic directions for the system. We can now appreciate why the system (2.118) failed to have noncharacteristic directions: it is for the same reason that it is not locally solvable! The further ramifications of this result will be explored after we discuss the proof.

PROOF OF LEMMA 2.85. According to Definition 2.75, the system Δ has only characteristic directions at a point if and only if the associated $q \times q$ matrix $\mathsf{M}(\omega)$ of n-th degree polynomials in $\omega = (\omega_1, \ldots, \omega_p)$ is singular for all values of ω:

$$\det \mathsf{M}(\omega) \equiv 0, \qquad \omega \in \mathbb{R}^p.$$

A relatively easy result from linear algebra (see Exercise 2.32) says that this is true if and only if there exists a row vector $\sigma(\omega) = (\sigma^1(\omega), \ldots, \sigma^q(\omega)) \not\equiv 0$ of homogeneous polynomials in ω such that

$$\sigma(\omega) \cdot \mathsf{M}(\omega) \equiv 0 \tag{2.131}$$

for all ω. In our case, suppose

$$\sigma^\nu(\omega) = \sum_{\#J=k} P_\nu^J \omega_J.$$

Then the coefficients P_ν^J of the σ_ν will serve as the coefficients of the operators $\mathscr{D}_\nu$ in (2.130). Indeed, it can easily be seen that if $\#J = k$,

$$D_J[\Delta_\nu(x, u^{(n)})] = \sum_{\alpha=1}^{q} \sum_{\#K=n} \frac{\partial \Delta_\nu}{\partial u_K^\alpha} u_{J,K}^\alpha + \cdots,$$

where $u^\alpha_{J,K}$ denotes the $(n+k)$-th order derivative $D_J(u^\alpha_K)$, and the omitted terms all depend on derivatives of orders at most $n+k-1$. Thus

$$\sum_{\nu=1}^{q} \mathscr{D}_\nu \Delta_\nu = \sum_{\nu=1}^{q} \sum_{\alpha=1}^{q} \sum_{\#J=k} \sum_{\#K=n} P^J_\nu \frac{\partial \Delta_\nu}{\partial u^\alpha_K} u^\alpha_{J,K} + Q(x, u^{(n+k-1)}) \tag{2.132}$$

for some well-defined Q depending on at most $(n+k-1)$-st order derivatives of u. On the other hand, the α-th entry of the product (2.131) of σ and M is, by (2.126),

$$\sum_{\nu=1}^{q} \sum_{\#J=k} \sum_{\#K=n} P^\nu_J \frac{\partial \Delta_\nu}{\partial u^\alpha_K} \omega_J \omega_K \equiv 0$$

at $(x_0, u_0^{(n)})$. Since $u^\alpha_{J,K}$ is also completely symmetric in the indices in J, K, we conclude that at $(x_0, u_0^{(n)})$, the leading summation in (2.132) vanishes, and hence (2.130) holds. The smooth dependence of the differential operators $\mathscr{D}_\nu$ on $(x, u^{(n)})$ if there are no non-characteristic directions in any open subset of $\mathscr{S}_\Delta$ follows because if $\mathsf{M}(\omega) = \mathsf{M}(\omega; x, u^{(n)})$ depends smoothly on the parameters $(x, u^{(n)})$, the polynomials $\sigma(\omega) = \sigma(\omega; x, u^{(n)})$ can also be chosen to depend smoothly on the same parameters.

To prove the converse, it suffices to note that (2.130) can never occur for a system in Kovalevskaya form. Indeed any combination $\sum \mathscr{D}_\nu \Delta_\nu$ with k-th order operators $\mathscr{D}_\nu$ not all zero will always depend on $(n+k)$-th order derivatives, namely the k-th order derivatives of the u^α_{nt}. Thus if $\Delta(x, u^{(n)}) = 0$ has a noncharacteristic direction at $(x_0, u_0^{(n)})$, we can choose coordinates so that the system is in Kovalevskaya form, and hence (2.130) does not hold. □

Suppose a system of differential equations Δ satisfies the hypotheses of Lemma 2.85, so it is not normal at the point $(x_0, u_0^{(n)})$. There are then integrability conditions of the form (2.130) in which some linear combination of equations in the k-th prolongation $\Delta^{(k)}$ depends on at most $(n+k-1)$-st order derivatives. At this stage, two distinct possibilities arise.

(a) The integrability condition $\sum \mathscr{D}_\nu \Delta_\nu = 0$ vanishes at $(x_0, u_0^{(n)})$ by virtue of the algebraic relations among the $(n+k-1)$-st and lower order derivatives already established by $\Delta^{(k-1)}$, or
(b) The integrability condition $\sum \mathscr{D}_\nu \Delta_\nu = 0$ is genuine, not being an algebraic consequence of $\Delta^{(k-1)}$, and introduces a further relation among $(n+k-1)$-st and lower order derivatives.

We formalize this dichotomy into a definition of under-determined and over-determined systems, respectively.

Definition 2.86. Let Δ be an n-th order system of differential equations. Let $(x_0, u_0^{(n)})$ be initial values satisfying the system.

(a) Δ is *over-determined* at $(x_0, u_0^{(n)})$ if for some $k \geqslant 0$ there exist homogeneous k-th order differential operators $\mathscr{D}_1, \ldots, \mathscr{D}_q$, not all zero, such that the linear combination $\sum \mathscr{D}_\nu \Delta_\nu = Q$ of equations in $\Delta^{(k)}$, at the point $(x_0, u_0^{(n)})$, depends only on derivatives of u of order at most $n + k - 1$, *and* the linear combination Q does not vanish as an algebraic consequence of $\Delta^{(k-1)}$.

(b) Δ is *under-determined* at $(x_0, u_0^{(n)})$ if (i) there exists at least one set of homogeneous k-th order operators $\mathscr{D}_1, \ldots, \mathscr{D}_q$, not all zero, with $\sum \mathscr{D}_\nu \Delta_\nu = Q$ depending on at most $(n + k - 1)$-st order derivatives at the point x_0, and (ii) whenever $\mathscr{D}_1, \ldots, \mathscr{D}_q$ satisfy the conditions in part (i), the resulting Q vanishes as an algebraic consequence of the previous prolongation $\Delta^{(k-1)}$.

More succinctly, an over-determined system is one in which there are nontrivial integrability conditions. In this case, some prolongation $\Delta^{(k-1)}$ is *not* locally solvable since we can find a point $(x_0, u_0^{(n+k-1)}) \in \mathscr{S}_{\Delta^{(k-1)}}$ which does not satisfy the new integrability condition introduced by $\Delta^{(k)}$. On the other hand, an under-determined system is one in which the equations in some prolongation $\Delta^{(k)}$ are algebraically dependent, so the maximal rank condition cannot hold. In either case, the system is not totally nondegenerate. The third type of system, the normal systems, are then in a very definite sense precisely determined, and hence (in the analytic case) are the only totally nondegenerate systems; all others are either under- or over-determined. This completes the proof of Theorem 2.84. □

Example 2.87. (a) Consider the second order system (2.118). As it stands the system is over-determined since

$$D_y(u_{xx} + v_{xy} + v_x) - D_x(u_{xy} + v_{yy} - u_x) = v_{xy} + u_{xx},$$

which depends on second order derivatives, but does not vanish as an algebraic consequence of (2.118). On the other hand, if we omit the lower order terms, the system

$$u_{xx} + v_{xy} = 0, \qquad u_{xy} + v_{yy} = 0,$$

which corresponds to Navier's equations (2.127) when $\mu = 0$, $\lambda = 1$, is under-determined, since the combination

$$D_y(u_{xx} + v_{xy}) - D_x(u_{xy} + v_{yy}) \equiv 0$$

vanishes identically. In this latter case the general solution

$$u(x, y) = \phi_y(x, y) + cx, \qquad v(x, y) = -\phi_x(x, y),$$

depends on an arbitrary function $\phi(x, y)$. As a matter of fact, this holds for every under-determined system Δ—there is at least one arbitrary function depending on *all* the independent variables in the form of the general solution. In this case the Cauchy problem does *not* uniquely determine the solution, whereas in the over-determined case there does not, in general, exist a solution to the Cauchy problem. Thus for analytic systems, the normal systems are again precisely determined, here from the viewpoint of the Cauchy problem; over- or under-determined systems are characterized by their lack of existence or uniqueness respectively.

NOTES

The system of partial differential equations for the invariants of a local group of transformations pre-dates Lie's work, having arisen in the problem of Pfaff. Its integration was studied by Jacobi, Mayer, Darboux, Lie and, finally, Frobenius, [1], who proved the general result on the existence of functionally independent solutions. See Forsyth, [1; Vol. 1], or Carathéodory, [1], for a discussion of the classical approaches to this problem. The connection with the corresponding characteristic system of ordinary differential equations is also classical; Kamke, [1; vol. 2, § D4] gives a treatment closest in spirit to that given here, along with other methods of integration—see also Ince, [1; § 2.7].

The concepts of functional independence and dependence are classical, but, surprisingly, most standard proofs of the basic Theorem 2.16 are remarkably deficient, usually assuming that the rank of the differential $d\zeta$ is constant. A modern proof of this result, not requiring extra hypotheses, appears in Narasimhan, [1; Theorem 1.4.14]. An alternative proof can be based on a theorem of A. B. Brown, [1] (see also Milnor, [1; p. 11]) that states that the set $\{\zeta(x): x \in M, \text{rank } d\zeta|_x < k\}$ of *critical values* of a smooth map $\zeta: M \to \mathbb{R}^k$ contains no open subset of $\mathbb{R}^k$, together with a theorem of Whitney (see Kahn, [1; Theorem 1.5]) that states that any closed subset $K \subset \mathbb{R}^k$ can be given as the set of zeros, $K = \{z: F(z) = 0\}$, of some smooth function $F: \mathbb{R}^k \to \mathbb{R}$. To prove Theorem 2.16, then, assuming rank $d\zeta < k$ everywhere, we set $K = \zeta[\bar{U}]$, where $U \subset M$ is any open set with compact closure, and choose F as in Whitney's theorem. Brown's theorem says that F does not vanish on any open subset of $\mathbb{R}^k$, and hence satisfies the requirements of Definition 2.15 for functional dependence. (This direct proof does not, to my knowledge, appear in the literature!)

In the case of analytic systems, the maximal rank condition for Theorem 2.8 can be relaxed to hold only "almost everywhere" on the subvariety $\mathscr{S}_F$, allowing the possibility of singularities. This result, which is not hard to prove, does not, however, seem to generalize to the C^∞ case. A similar generalization for Theorems 2.31 and 2.72 can thus also be proved for analytic systems of differential equations, allowing some singularities in the subvariety $\mathscr{S}_\Delta$.

Lie originally formulated his theory of continuous groups expressly for the study of differential equations, but was well aware of the applicability of his powerful infinitesimal method to the study of invariants and algebraic equations. See Lie, [4], for the algebraic and geometric side of his work. Historical accounts of Lie's work and influence appear in Hawkins, [1], [2], [3], and Wussing, [1; § III.3]. Most of Lie's work on ordinary differential equations appears in his collected papers; the book [5] does not really do justice to the full extent of his discoveries. The key to Lie's approach to integrating higher order ordinary differential equations was his complete classification (up to change of variable) of all transformation groups on the complex plane $\mathbb{C}^2$. Using this he was able to exhaustively list all possible reductions in order for a single ordinary differential equation; see Lie, [3], for these results, along with many explicit examples of interest. Lie's results in [3] include the results of Section 2.5 on multi-parameter symmetry groups of higher order ordinary differential equations; all of the other treatments in the earlier literature, including Cohen, [1], Ince, [1; Chap. 3], Markus, [1], and Ovsiannikov, [3; § 8], only do the case of one- and two-parameter groups. See Krause and Michel, [1], for more details on the kinds of symmetry groups admitted by ordinary differential equations. Theorem 2.68 on solvable symmetry groups of first order systems of ordinary differential equatons, though, is due to Bianchi, [1; § 167]; see also Eisenhart, [2; § 36]. This result clearly includes the corresponding Theorem 2.64 on higher order equations, but I was unable to find an explicit statement of the latter result in the literature; see Bluman and Kumei, [2], for further developments.

Most of Lie's work on symmetry groups of partial differential equations was concerned with linear systems of first order equations, which, by the method of characteristics, are essentially equivalent to systems of ordinary differential equations. However, in [2] and [6], Lie did look into symmetries of higher order partial differential equations. In [2; Part 1], Lie computes the symmetry groups of a number of second order partial differential equations in two independent variables, including the heat equation whose symmetry group appears at the end of § 13. This group was recomputed by Appell, [1], and, in the higher dimensional case, Goff, [1]. Lie's work on higher order partial differential equations, however, was not developed at all by other researchers, one possible reason being that, in contrast to the case of ordinary differential equations, knowledge of the symmetry group of a system of partial differential equations did *not* aid one in determining the general solution to the system. (One intriguing possibility, though, is the "group splitting method" of Vessiot, [1]; see Ovsiannikov, [3, § 26], for a modern presentation.) The only other early work on symmetries of partial differential equations of which I am aware is the work of Bateman, [1], Cunningham, [1], and Carmichael, [1], on the symmetries of the wave equation and Maxwell's equations. Apart from this, and despite the availability of Noether's theorem after 1918, work on the theory and applications of symmetry groups of partial differential equations came to a complete standstill; it was not until the appearance of Birkhoff's book, [2], on hydrodynamics, that group methods in the study of the important partial differential equations of mathematical physics began to revive. Under the leadership of Ovsiannikov, [1], [2], in the late 1950's and 1960's, the Soviet school made great progress in the study of symmetry groups of many of these systems. Interest in the methods in the West grew through the works of Bluman and Cole, [1], [2], and the books of Ames, [1], resulting in a great surge of research activity in these areas in the past 15 years. See Holm, [1], for an extensive list of early references up to 1976.

The basic method for computing symmetry groups, using the prolongation formula for their infinitesimal generators, dates back to Lie. Indeed, the recursive form (2.44) of the prolongation formula appears in Lie, [2; § 11], [3; § 1]; see also Eisenhart, [2; equation (28.12)]. The explicit formula (2.39), though, first appears in Olver, [2]. The alternative form (2.50) using the characteristic of such a vector field will be discussed in more detail in Chapter 5; Seshadri and Na, [1; § 3.2(e)] use it as an alternative method for computing ordinary symmetries. In this book, I have chosen to adopt an extremely simplified version of the theory of jets, due in its modern form to Ehresmann, [1], [2], which serves to clearly delineate the geometric foundations of the prolongation theory. A readable account of the more abstract, differential-geometric approach to jets can be found in Golubitsky and Guillemin, [1]; see also Section 3.5. Needless to say, all the results stated here have many alternative restatements and reformulations, using more and more technical and abstract mathematical machinery, a pointless exercise enjoyed by a number of researchers. The net result, of course, is always the same no matter how one tries to dress it up; the unfortunate reader of these versions comes away thoroughly confused, learning nothing of the ease and efficacy of applying this theory to concrete problems. I hope that this book has, for the most part, avoided such pitfalls, and that the use of local coordinates and illustrative examples will genuinely educate the reader interested in applications.

By now, the literature on examples of explicit computations of symmetry groups of specific systems of differential equations has grown too voluminous to attempt to list here. The reader can find references in the book of Ovsiannikov, [3], as well as an extensive, but by no means complete, bibliography in Steinberg, [2]. The calculation of the symmetry group of the Euler equations is due to Buchnev, [1]. The derivation of the Hopf–Cole transformation (actually originally due to Forsyth, [1; Vol. 6, p. 102]—see also Whitham, [2; Chap. 4]) using group-theoretic methods can be found in Kumei and Bluman, [1], along with generalizations. The actual computations for finding the symmetry group of a given system of differential equations are quite mechanical, and are thus amenable to implementation on a computer using a symbolic manipulation program. Several versions have been developed, including Rosenau and Schwarzmeier, [1], Steinberg, [1], Rosencrans, [2], Schwarz, [1], and Champagne, Hereman and Winternitz, [1]. The latter reference contains a complete survey (as of 1991) of available packages, including a discussion of their strengths and weaknesses. (A more up-to-date survey by W. Hereman will appear in *Euromath Bull.* **2** in summer, 1993.)

There are several alternative approaches to the theory of symmetry groups of differential equations worth mentioning. For linear equations, Kalnins, Miller, Boyer and others (see Miller, [2], [3]) have emphasized the use of differential operators rather than vector fields to determine symmetries. The relation between their method and Lie's is made clear in Section 5.2. Ames, [1], proposed a method based on the group transformations themselves, circumventing the introduction of the infinitesimal generators, but this seems to have limited applicability. Poloszny and Rubel, [1], and Rubel, [1], use a method based on the theory of "motions" of a differential equation. Seshadri and Na, [1], make the point that one can considerably simplify the computation of symmetries if one imposes, *a priori*, restrictions on the form of the group, e.g., that it be projectable or a scaling group. An approach based on differential forms was proposed by Harrison and Estabrook, [1], and developed by Edelen, [1], who also describes symbolic manipulation programs based on this method; see also Gragert, Kersten and Martini, [1]. The resuts are the same as the present approach, but their method suffers the drawback of having to first re-express a system of partial differential equations as the integrability conditions for a set of differential forms before one can proceed to the computation of symmetries. Nevertheless, the method can be useful, especially for constructing Bäcklund transformations. An extension of the present infinitesimal method to free boundary problems can be found in Benjamin and Olver, [1].

The more technical matters raised in the final secton of this chapter have only recently been seen to be of importance for symmetry group methods. The connection between existence of solutions and the theory of characteristics dates back to the work of Kovalevskaya, [1]. (The present development of this theory most closely parallels the presentation in Petrovskii, [1].) Bourlet, [1], was the first to demonstrate the existence of systems which could not be solved using the Cauchy–Kovalevskaya theorem in any direction, but did not pursue the matter. Subsequently a number of researchers in the last century, including Delassus, [1], and Riquier, [1], developed quite elaborate existence theorems for systems of partial differential equations generalizing the Cauchy–Kovalevskaya theorem. However, it was not until Finzi, [1], proved the important Lemma 2.85 (see also Hadamard, [1; §25a]) that the true connections between solvability and integrability conditions became evident. The consequent definitions of over- and under-determined systems proposed here are new; see also Olver, [11]. Normality is a more classical concept; see also Vinogradov, [4], for a more technical version of this definition. Although there are definite connections between our definitions and the Spencer, Goldschmidt, *et al.* theory of over-determined systems of partial differential equations, the present terminology is *more* precise. Comparing with the definitions in Pommaret, [1, §V.6.6], (which are for linear systems only) we find Pommaret's underdetermined systems to always have fewer equations than unknowns, whereas his over-determined systems include *both* the under- and over-determined systems of Definition 2.86. These issues are also closely related to questions on the "degree of determinancy" of a system of partial differential equations, which arise in relativity, and were discussed, but never fully resolved, by Cartan and Einstein, [1]. The question of local solvability of systems of partial differential equations is closely connected with the general Riquier existence theory, see Ritt, [1; Chap. 8] for a discussion. Nirenberg, [1; p. 15] proves the local solvability of fairly general types of elliptic systems. Nonsolvability due to integrability conditions was recognized in the last century; the Lewy type of nonsolvable C^∞ systems is much more recent. See Lewy, [1], and Nirenberg, [1; p. 8] for examples. Applications of these results to symmetry group theory have appeared previously in Olver, [2], [7], [11], and Vinogradov, [5].

Exercises

2.1. Let G be a local group of transformations acting on the manifold M.
 (a) Prove that a subset $\mathcal{S} \subset M$ is G-invariant if and only if $\mathcal{S} = \bigcup \mathcal{O}$ is a union of orbits of G.
 (b) Prove Proposition 2.14.
 (c) Prove that a function $F: M \to \mathbb{R}^l$ is G-invariant if and only if F is constant on the orbits of G.
 (d) Prove that the only invariants of the irrational flow on the torus are the constant functions.

2.2. Let G be the one-parameter group of transformations of $\mathbb{R}^3$ generated by the vector field $\mathbf{v}$ of Exercise 1.11. Prove that G has only one independent global invariant.

2.3. Let G act on the manifold M, and let $H \subset G$ be a subgroup. Prove that if $\mathcal{S} \subset M$ is a (locally) H-invariant subset, and $g \in G$ is defined on all of $\mathcal{S}$, then $g \cdot \mathcal{S} = \{g \cdot x \colon x \in \mathcal{S}\}$ is (locally) invariant under the conjugate subgroup $gHg^{-1} = \{ghg^{-1} \colon h \in H\}$.

2.4. A system of submanifolds of M is called G-invariant if the group elements g map one submanifold to another submanifold in the system. For example, the set of parallel lines $\{y = kx + b\}$, k fixed, is invariant under *any* translation group of $\mathbb{R}^2$. Prove that the level sets of a function $F: M \to \mathbb{R}^l$ are invariant under the transformation group G if and only if $\mathbf{v}(F) = H(F)$ for every infinitesimal generator $\mathbf{v}$ of G, where H, depending on $\mathbf{v}$, is some function defined on the range of F. (Eisenhart, [2; p. 82].)

2.5. (a) Prove the local version of Proposition 2.10.
 (b) Prove the global version using a partition of unity—see Kahn [1; Theorem 1.4].
 (c) Prove Proposition 2.11. (*Hint*: Use Theorem 1.8.)
 (d) Prove that if $R_i(x)$, $i = 1, \ldots, p$, are smooth, then

$$\sum_{i=1}^{p} R_i(x)(x^i - c_i) = \sum_{i=1}^{p} a_i x^i + b$$

is an affine function of x if and only if $R_i(x) = a_i + S_i(x)$, where

$$\sum_{i=1}^{p} S_i(x)(x^i - c_i) \equiv 0,$$

or, equivalently,

$$S_i(x) = \sum_{j=1}^{p} S_{ij}(x)(x^j - c_j)$$

where $S_{ij} = -S_{ji}$.

2.6. Let $X = \mathbb{R}$, $U = \mathbb{R}$ and consider the one-parameter group

$$g_\varepsilon\colon (x, u) \mapsto (x \cos(r\varepsilon) - u \sin(r\varepsilon), x \sin(r\varepsilon) + u \cos(r\varepsilon)),$$

where $r^2 = x^2 + u^2$. Let $u = f(x)$ be a function defined for all $x \in \mathbb{R}$. Prove that for *any* $\varepsilon \neq 0$, the transformed function $\tilde{u} = g_\varepsilon \cdot f(x)$ is *not* a globally defined function for $\tilde{x} \in \mathbb{R}$. How does this affect our construction of the prolonged group action?

2.7. Find the determining equations for the symmetry group of the nonlinear wave equation $u_t = uu_x$ and determine some particular symmetry groups. Do your results change if the coefficient of u_x is replaced by $f(u)$ for some function f? (See also Example 5.7.)

2.8. The Fokker-Planck equation is

$$u_t = u_{xx} + (xu)_x = u_{xx} + xu_x + u.$$

Find the symmetry group and interpret geometrically. Use the group transformations to determine some particular solutions to this equation. (Bluman and Cole, [2; § 2.10].)

2.9. Find the symmetry group of the telegraph equation $u_{tt} = u_{xx} + u$. Compare this group with that of the equivalent first order system $u_t + u_x = v$, $v_t - v_x = u$.

2.10. Groups of higher order equations and their equivalent first order systems are not always comparable. For instance, compute the symmetry group of the two-dimensional wave equation $u_{tt} = u_{xx}$, and compare this with the symmetry group of the equivalent system $u_t = v$, $u_x = w$, $v_t = w_x$, $v_x = w_t$. What about the two-dimensional Laplace equation? (Olver, [2], Ibragimov, [1; § 17.1].)

*2.11. Prove that the symmetry group (2.65) of the two-dimensional wave equation (omitting the trivial linear symmetries $u \mapsto \lambda u + \alpha(x, t)$) is locally isomorphic to the group SO(3, 2) of linear isometries $z \mapsto Rz$ of $\mathbb{R}^5$ with metric $(dz^1)^2 + (dz^2)^2 + (dz^3)^2 - (dz^4)^2 - (dz^5)^2$. (See Exercise 1.29.) (Miller, [3; p. 223].)

2.12. Find the symmetry group of the m-dimensional heat equation $u_t = \Delta u$, $x \in \mathbb{R}^m$. How does it compare with the one-dimensional case? (Goff, [1].)

2.13. Discuss the symmetry group of the Helmholtz equation $\Delta u + \lambda u = 0$ for λ a fixed constant, $x \in \mathbb{R}^3$. (Miller, [3; § 3.1].)

2.14. Discuss the symmetry group of the biharmonic equation $\Delta^2 u = \Delta(\Delta u) = 0$, $x \in \mathbb{R}^m$. How is it related to the symmetry group for Laplace's equation?

*2.15. Prove that the symmetry group for the Navier-Stokes equations

$$\frac{\partial \mathbf{u}}{\partial t} + \mathbf{u} \cdot \nabla \mathbf{u} = -\nabla p + \nu \Delta \mathbf{u}, \qquad \nabla \cdot \mathbf{u} = 0,$$

where $\mathbf{u} \in \mathbb{R}^2$ or $\mathbb{R}^3$, ν is the viscosity, is the same as that of the corresponding system of Euler equations ($\nu = 0$). (Buchnev, [1], Lloyd, [1].)

*2.16. (a) Maxwell's equations for the electric field $E \in \mathbb{R}^3$ and the magnetic field $B \in \mathbb{R}^3$ take the vector form

$$E_t = \nabla \times B, \qquad B_t = -\nabla \times E, \qquad \nabla \cdot E = 0, \qquad \nabla \cdot B = 0.$$

Discuss the symmetries of this system.

(b) An equivalent formulation is obtained by introducing the vector potential A with $B = \nabla \times A$, and noting that $\nabla \times (A_t + E) = 0$, hence there exists a scalar potential ϕ satisfying $A_t + E = \nabla\phi$. The resulting system is

$$\frac{\partial^2 A}{\partial t^2} + \nabla \times (\nabla \times A) = \nabla \frac{\partial \phi}{\partial t}, \qquad \nabla \cdot \frac{\partial A}{\partial t} = \Delta\phi.$$

How does the symmetry group of this latter system compare with the previous form of Maxwell's equations? (Ovsiannikov, [3; p. 394], Fushchich and Nikitin, [1], [2], and Pohjanpelto, [1].)

*2.17. Perform a symmetry analysis of Navier's equations (2.127) of linear isotropic elasticity. Discuss the difference between the two- and three-dimensional cases. Do your results depend on the values of the Lamé moduli λ and μ? (Olver, [9].)

2.18. *Group Classification.* Often a system of differential equations arising from a physical problem will involve some arbitrary functions whose precise forms depend on the specific physical system under consideration. For example, the general equation of nonlinear heat conduction takes the form

$$u_t = D_x(K(u)u_x), \tag{$*$}$$

where $K(u)$ depends on the particular type of conductor being modelled. There are often good physical motivations for studying those equations in which the form of the arbitrary functions provides a larger symmetry group than would otherwise be applicable. The problem of determining such functions is known as the *group classification problem.* Perform a group classification on the nonlinear heat conduction equation by proving:

(a) If K is arbitrary (i.e. not any of the following special forms), then ($*$) has a three-parameter symmetry group.

(b) If $K(u) = (au + b)^m$ for $m \neq -\frac{4}{3}$, $a \neq 0$, the symmetry group is four-dimensional.

(c) For $K(u) = ce^{\alpha u}$, there is a four-parameter group.

(d) For $K(u) = (au + b)^{-4/3}$, $a \neq 0$, there is a five-parameter group.

(e) For $K(u)$ constant the group is infinite-dimensional.

(Ovsiannikov, [3; pp. 68–73], Lisle, [1]; see also Lie, [2].)

2.19. Consider a first order homogeneous linear partial differential equation

$$\sum_{i=1}^{p} \xi^i(x)\frac{\partial u}{\partial x^i} = 0, \tag{$*$}$$

and let $\mathbf{v} = \sum \xi^i(x)\partial_i$ be the corresponding vector field.

(a) Show that $\mathbf{w} = \sum \eta^i(x)\partial_i$ generates a one-parameter symmetry group if and only if $[\mathbf{v}, \mathbf{w}] = \gamma\mathbf{v}$ for some scalar-valued function $\gamma(x)$.

(b) Suppose $p = 2$. Show that if $\mathbf{w}$ generates a nontrivial symmetry group, meaning $\mathbf{w} \neq \lambda\mathbf{v}$ for some function $\lambda(x)$, then we can find the general solution to ($*$) by quadrature (provided we know the invariants of $\mathbf{w}$).

(c) What about $p \geqslant 3$?

(Lie, [5; p. 434].)

2.20. Prove that the system $u_x = 0$, $u_y + xu_z = 0$ is not locally solvable. Prove that the group generated by $\mathbf{v} = x\partial_z$ is a symmetry group, but $\mathbf{v}$ does not satisfy the infinitesimal criterion (2.120).

*2.21. Suppose the differential equation $P(x, u^{(n)}) = 0$, $x \in \mathbb{R}^p$, $u \in \mathbb{R}$, admits an infinite-dimensional symmetry group with generators $\rho(x)\partial_u$ where ρ is an arbitrary solution to a linear differential equation $\Delta[\rho] = 0$. Prove that P is equivalent to a nonhomogeneous version of the same equation: $\Delta[u] = f(x)$. (Kumei and Bluman, [1].)

*2.22. (a) Prove that a differential equation $P(x, u^{(n)}) = 0$ is equivalent to a *linear* differential equation $\Delta[\tilde{u}] = f(\tilde{x})$ under a change of variables $x = \Xi(\tilde{x}, \tilde{u})$, $u = \Phi(\tilde{x}, \tilde{u})$ if and only if it admits an infinite-dimensional abelian symmetry group with generators of the form

$$\mathbf{v} = \rho(\Xi(\tilde{x}, \tilde{u}), \Phi(\tilde{x}, \tilde{u}))\left\{\frac{\partial \Xi}{\partial \tilde{u}}\frac{\partial}{\partial \tilde{x}} + \frac{\partial \Phi}{\partial \tilde{u}}\frac{\partial}{\partial \tilde{u}}\right\},$$

where $\rho(x, u)$ is an arbitrary solution to a linear differential equation. (*Hint*: Change variables and use the previous exercise.)

(b) Discuss our derivation of the Hopf–Cole transformation in Example 2.42 in light of this result.

(c) Apply this technique to linearize the Thomas equation

$$u_{xt} + \alpha u_t + \beta u_x + \gamma u_x u_t = 0,$$

which arises in the study of chemical exchange processes.

(d) Apply this technique to the potential form $u_t = u_x^{-2} u_{xx}$ of the nonlinear diffusion equation $v_t = D_x(v^{-2} v_x)$ of importance in porous media flow and solid state physics.

(Kumei and Bluman, [1], Whitham [2; p. 95], Rosen, [1], Fokas and Yortsos, [1], Bluman and Kumei, [1].)

*2.23. Two evolution equations, $u_t = P(x, u^{(n)})$ and $v_s = Q(y, v^{(m)})$, are said to be *related* if there exists a change of variables

$$t = T(s, y), \qquad x = \Xi(s, y), \qquad u = \Phi(s, y, v),$$

changing one into an equation equivalent to the other.

(a) Prove that if $u_t = P$ is related to $u_s = Q$, then

$$\mathbf{v} = \frac{\partial T}{\partial s}\frac{\partial}{\partial t} + \frac{\partial \Xi}{\partial s}\frac{\partial}{\partial x} + \frac{\partial \Phi}{\partial s}\frac{\partial}{\partial u}$$

is a symmetry of $u_t = P$.

(b) Prove that if

$$\mathbf{v} = \tau(t)\frac{\partial}{\partial t} + \xi(t, x)\frac{\partial}{\partial x} + \phi(t, x, u)\frac{\partial}{\partial u} \qquad (*)$$

is a symmetry of $u_t = P$, then there is a related evolution equation $v_s = Q$ with $\mathbf{v} = \partial_s$ in the new coordinates. (In fact, for a large class of evolution equations, (*) is the most general symmetry, so we have a one-to-one correspondence between related evolution equations and symmetries.)

(c) Find a transformation relating the Korteweg–de Vries equation $u_t = u_{xxx} + uu_x$ and the equation $v_s = v_{yyy} + vv_y + 1$. (Kalnins and Miller, [2].)

2.24. Let $\alpha \in \mathbb{R}$. Find the most general first order ordinary differential equation invariant under the scaling group $(x, u) \mapsto (\lambda x, \lambda^\alpha u)$, $\lambda > 0$. How are these equations solved by quadrature?

2.25. (a) Prove that the scaling group $(x, u) \mapsto (\lambda x, \lambda u)$ is the only continuous symmetry group of equation (2.95). (Hint: First change coordinates to straighten out the scaling generator.)

(b) Prove that the second order equation $u_{xx} = xu + \tan(u_x)$ has *no* continuous symmetry groups! (Cohen, [1; p. 206].)

2.26. (a) Prove that the symmetry group of the equation $u_{xx} = 0$ is eight-dimensional, generated by

$$\partial_x, \quad x\partial_x, \quad u\partial_x, \quad xu\partial_x + u^2\partial_u,$$
$$\partial_u, \quad x\partial_u, \quad u\partial_u, \quad x^2\partial_x + xu\partial_u.$$

Prove that the corresponding group is the projective group in the plane, namely,

$$(x, u) \mapsto \left(\frac{ax + bu + c}{\alpha x + \beta u + \gamma}, \frac{dx + eu + f}{\alpha x + \beta u + \gamma}\right),$$

where $\det \begin{vmatrix} a & b & c \\ d & e & f \\ \alpha & \beta & \gamma \end{vmatrix} \neq 0$. Interpret these transformations geometrically.

(b) For $n \geq 3$, prove that the symmetry group of $d^n u/dx^n = 0$ is $(n + 4)$-dimensional.

(Lie, [3], Markus, [1].)

**2.27. (a) Prove that a second order ordinary differential equation admits a symmetry group of dimension at most 8. Moreover, if the equation has an eight-parameter symmetry group, then it can be transformed into the elementary equation $u_{xx} = 0$. Show, in particular, that a homogeneous linear second order ordinary differential equation has an eight-parameter symmetry group, and find the explicit transformation that reduces it to $u_{xx} = 0$. (Note that you will have to know a basis for the solution space to the equation, so this method is of no use for solving the equation!)

(b) Prove that, for $n \geq 3$, an n-th order ordinary differential equation has at most an $(n + 4)$-parameter symmetry group. (Lie, [5].)

Interestingly, although the corresponding maximal dimension for symmetry groups of second order systems of ordinary differential equations is known, a significant open problem is to determine the maximal dimension of the symmetry group of a higher order system of ordinary differential equations. Bounds are known, cf. González–Gascon and González–López, [1], but there are no known systems that achieve the bounds.

2.28. Prove that SL(2) is not a solvable Lie group. How about SO(3)?

*2.29. Consider the ordinary differential equation Δ: $u_x^2 - 4u = 0$.

(a) Prove that Δ is of maximal rank everywhere.

(b) Prove that all prolongations $\Delta^{(m)}$ are of maximal rank provided $u \neq 0$. However, $u_{xxx}^2 = 0$ is a combination of the equations in $\Delta^{(5)}$, but $u_{xxx} = 0$ is *not*. Discuss.

(Ritt, [1; p. 79].)

2.30. Suppose we know the general solution to the first order ordinary differential equation $u_x = F_0(x, u)$. This knowledge implies that we can, using the methods of Section 2.5, integrate any first order ordinary differential equation invariant under the one-parameter group generated by the vector field $\mathbf{v}_0 = \partial_x + F_0(x, u)\partial_u$. Determine the most general first order equation $u_x = F_1(x, u)$ admitting $\mathbf{v}_0$ as a symmetry. This method can then be iterated to give a "tower" of first order ordinary differential equations $u_x = F_n(x, u)$, $n \geqslant 0$, the n-th equation providing a symmetry, and hence an integration method, for the $(n+1)$-st equation. Investigate this method for the translation, scaling and rotation groups. Describe the tower that starts with a first order linear ordinary differential equation. (Beyer, [1].)

*2.31. *Equations Invariant under "Nonlocal Symmetries"*. Lest the reader think that all methods for integrating ordinary differential equations reduce to the invariance of the equation under some symmetry group of the type presented here, we offer the following cautionary problems.

(a) An *exponential vector field* is a formal expression of the form

$$\mathbf{v}^* = e^{\int P(x,u)\,dx}\left(\xi(x, u)\frac{\partial}{\partial x} + \phi(x, u)\frac{\partial}{\partial u}\right),$$

where $\int P(x, u)\,dx$ is, formally, the integral of the function $P(x, u)$, once we choose a function $u = f(x)$. Thus

$$D_x\left[\int P(x, u)\,dx\right] = P(x, u), \qquad D_x^2\left[\int P(x, u)\,dx\right] = D_x P,$$

and so on. Substituting $\mathbf{v}^*$ into the prolongation formula (2.50) (with ξ replaced by $e^{\int P\,dx}\xi$, ϕ by $e^{\int P\,dx}\phi$), prove that

$$\mathrm{pr}^{(n)}\,\mathbf{v}^* = e^{\int P(x,u)\,dx}\cdot\mathbf{v}^{(n)},$$

where $\mathbf{v}^{(n)}$ is an ordinary vector field on $M^{(n)}$. For example, if $\mathbf{v}^* = e^{\int u\,dx}\,\partial_u$, then

$$\mathrm{pr}^{(n)}\,\mathbf{v}^* = \sum_{k=0}^{n} D_x^k[e^{\int u\,dx}]\frac{\partial}{\partial u_k}$$

$$= e^{\int u\,dx}[\partial_u + u\partial_{u_x} + (u_x + u^2)\partial_{u_{xx}} + \cdots].$$

(b) Prove that one can choose differential invariants for an exponential vector field of the form

$$y = \eta(x, u), \qquad w = \zeta(x, u, u_x), \qquad w_n = d^n w/dy^n, \qquad n = 1, 2, \ldots,$$

just as for an ordinary vector field. What are the third order differential invariants of $e^{\int u\,dx}\partial_u$?

(c) Prove that an ordinary differential equation $\Delta(x, u^{(n)}) = 0$ which is invariant under an exponential vector field: $\mathrm{pr}^{(n)}\,\mathbf{v}^*(\Delta) = 0$ whenever $\Delta = 0$, can be reduced in order by one. Use this to reduce the equation

$$u_{xx} - uu_x = H(x, u_x - \tfrac{1}{2}u^2)$$

to a first order ordinary differential equation.

(d) Conversely, prove that if Δ can be reduced in order by one by setting $v = \gamma(x, u, u_x)$, then Δ is invariant under the exponential vector field

$$\mathbf{v}^* = \exp\left[-\int \frac{\partial\gamma/\partial u}{\partial\gamma/\partial u_x}\,dx\right]\frac{\partial}{\partial u}.$$

(e) How might these symmetries arise in practice? Consider the "wrong" reduction procedure used in Example 2.62 to obtain (2.109). We would like to say that the other symmetry $\mathbf{v}$ of (2.108) remains a symmetry of (2.109). However, $\mathrm{pr}^{(1)}\,\mathbf{v} = x\partial_u + \partial_{u_x} = x(\partial_y + \partial_z)$ is not a well-defined vector field in the (y, z)-coordinates. Prove that $\mathrm{pr}^{(1)}\,\mathbf{v}$ is an *exponential* vector field in these coordinates. (*Hint*: Show $x = \exp(\int z^{-1}\,dy)$.) Moreover, it remains a symmetry of (2.109). Use this informaton to complete the integration of (2.108).

*2.32. (a) Prove that the second order equation

$$u_{xx} = D_x[(x + x^2)e^u]$$

has a trivial symmetry group.

(b) Show that, nevertheless, the equation can be explicitly solved by quadrature.

(c) Does the equation have any exponential symmetries? (See Exercise 2.31.) (Ibragimov, [2].)

2.33. Let $\mathsf{M}(\omega)$ be a $q \times q$ matrix of homogeneous n-th order polynomials in ω. Prove that $\det \mathsf{M}(\omega) = 0$ for all ω if and only if there is a vector $\sigma(\omega)$ of homogeneous polynomials such that $\mathsf{M}(\omega)\sigma(\omega) = 0$ for all ω. What is the minimal degree of the polynomials required for σ? Generalize to the case when the polynomials in M are homogeneous, but not all of the same degree. (Finzi, [1].)

2.34. Is a system of evolution equations always normal?

2.35. Let $\Delta_\nu(x, u^{(n)})$, $\nu = 1, \ldots, l$, be a totally nondegenerate system of differential equations. Prove that a function $Q(x, u^{(m)}) = 0$ vanishes for all solutions $u = f(x)$ to Δ if and only if there exist differential operators $\mathscr{D}_\nu = \sum_J P_\nu^J(x, u^{(m)})D_J$, $\nu = 1, \ldots, l$, such that $Q = \sum_\nu \mathscr{D}_\nu\Delta_\nu$ for all functions $u = f(x)$. (*Hint*: Use Proposition 2.10.)

CHAPTER 3

Group-Invariant Solutions

When one is confronted with a complicated system of partial differential equations arising from some physically important problem, the discovery of any explicit solution whatsoever is of great interest. Explicit solutions can be used as models for physical experiments, as benchmarks for testing numerical methods, etc., and often reflect the asymptotic or dominant behaviour of more general types of solutions. The methods used to find group-invariant solutions, generalizing the well-known techniques for finding similarity solutions, provide a systematic computational method for determining large classes of special solutions. These group-invariant solutions are characterized by their invariance under some symmetry group of the system of partial differential equations; the more symmetrical the solution, the easier it is to construct. The fundamental theorem on group-invariant solutions roughly states that the solutions which are invariant under a given r-parameter symmetry group of the system can all be found by solving a system of differential equations involving r fewer independent variables than the original system. In particular, if the number of parameters is one less than the number of independent variables in the physical system, $r = p - 1$, then all the corresponding group-invariant solutions can be found by solving a system of *ordinary differential equations*. In this way, one reduces an intractable set of partial differential equations to a simpler set of ordinary differential equations which one might stand a chance of solving explicitly. In practical applications, these group-invariant solutions can, in most instances, be effectively found and, often, are the only explicit solutions which are known.

This chapter is organized so that the applications-oriented reader can immediately learn the practical implementation of the method of constructing group-invariant solutions without having to delve into the theoretical foundations needed to justify the method. Section 3.1 outlines the method, based on the construction of invariants for the given group action. This is illustrated in Section 3.2 by a number of interesting examples, including the heat equation, the Korteweg–de Vries equation and the Euler equations of ideal fluid flow. Further examples are indicated in the exercises at the end of the chapter as well as the cited references. The third section deals with the problem of classifying group-invariant solutions. Since there are almost always an infinite number of different symmetry groups one might employ to find invariant solutions, a means of determining which groups give fundamentally different types of invariant solutions is essential for gaining a complete understanding of the solutions which might be available. Since any transformation in the full symmetry group will take a solution to another solution, we need only find invariant solutions which are not related by a transformation in the full symmetry group. This classification problem can be solved by looking at the adjoint representation of the symmetry group on its Lie algebra, and includes an analogous classification of the different subgroups of the full symmetry group.

The remaining two sections of this chapter are devoted to a rigorous presentation of the theoretical basis of the group-invariant solution method and can safely be omitted if one is only interested in applying these techniques. A rigorous, global geometrical setting for these results is provided by the quotient manifold of a manifold under some regular group of transformations. Each point on the quotient manifold will correspond to an orbit of the group, so the quotient manifold has, essentially, s fewer dimensions where s is the dimension of the orbits of the group. Group-invariant objects on the original manifold will have natural counterparts on the quotient manifold which serve to completely characterize them. In particular, a system of partial differential equations which is invariant under the given transformation group will have a corresponding reduced system of differential equations on the quotient manifold, the number of independent variables having thereby been reduced by r. Solutions of the reduced system will correspond to group-invariant solutions of the original system. The one complicating detail in this method is that even when the original manifold is an open subset of some Euclidean space, the quotient manifold is not in any natural way an open subset of a "reduced" Euclidean space, so our earlier construction of jet spaces and symmetry groups is not immediately applicable. At this point there are two routes available. The more concrete avenue of attack would be to restrict to suitably smaller open subsets of Euclidean space, thereby forcing the quotient manifold to also be the subset of some Euclidean space through a choice of new independent and dependent variables on it. However, in this approach, constructions become very unpleasantly coordinate-dependent and lose much of their innate simplicity. The more abstract approach, and the one adopted here, is to generalize our construction of jet spaces, prolongations and differential equations to arbitrary smooth manifolds. This is done by "completing" the ordinary jet spaces so as to include "functions" determined by arbitrary p-dimensional submanifolds, which may be multiply-valued or have infinite derivatives. Although this method requires a fair amount of abstraction and mathematical sophistication just to state the definitions, the principal results on group-invariant solutions retain their strong geometrical flavour and, as far as the proofs are concerned, become practically trivial. The more technically complicated local coordinate picture is then straightforwardly derived from this abstract reformulation of the reduction procedure.

3.1. Construction of Group-Invariant Solutions

Consider a system of partial differential equations Δ defined over an open subset $M \subset X \times U \simeq \mathbb{R}^p \times \mathbb{R}^q$ of the space of independent and dependent variables. Let G be a local group of transformations acting on M. Roughly, a solution $u = f(x)$ of the system is said to be G-invariant if it remains unchanged by all the group transformations in G, meaning that for each $g \in G$, the functions f and (provided it is defined) $g \cdot f$ agree on their common domains of definition. For example, the fundamental solution $u = \log(x^2 + y^2)$ for the two-dimensional Laplace equation $u_{xx} + u_{yy} = 0$ is invariant under the one-parameter rotation group $SO(2)$: $(x, y, u) \mapsto (x \cos \theta - y \sin \theta, x \sin \theta + y \cos \theta, u)$, acting on the independent variables x, y. More rigorously, we can define a *G-invariant solution* of a system of partial differential equations as a solution $u = f(x)$ whose graph $\Gamma_f \equiv \{(x, f(x))\} \subset M$ is a locally G-invariant subset of M; see Definition 2.12.

If G is a symmetry group of a system of partial differential equations Δ, then, under some additional regularity assumptions on the action of G, we can find all the G-invariant solutions to Δ by solving a reduced system of differential equations, denoted by Δ/G, which will involve fewer independent variables than the original system Δ. To see how this reduction is effected, we begin by making the simplifying assumption that G act *projectably* on M. This means that the transformations in G all take the form $(\tilde{x}, \tilde{u}) = g \cdot (x, u) = (\Xi_g(x), \Phi_g(x, u))$, i.e. the changes in the independent variables x do not depend on the dependent variables u. (More general nonprojectable group actions will be treated subsequently, but the basic technique is the same.) There is then a projected group action $\tilde{x} = g \cdot x = \Xi_g(x)$ on an open subset $\Omega \subset X$. We make the regularity assumption that both the action of G on M and the projected action of G on Ω are *regular* in the sense of Definition 1.26, and that the orbits of both of these actions have the same dimension s, where s is strictly less than p, the number of independent variables in the system. (The case $s = p$ is fairly trivial, while if $s > p$, no G-invariant functions exist. Usually s will be the same as the dimension of G itself, but this need not be the case.) Under these assumptions, Theorem 2.17 implies that locally there exist $p - s$ functionally independent invariants $y^1 = \eta^1(x), \ldots, y^{p-s} = \eta^{p-s}(x)$ of the projected group action on $\Omega \subset X$. Each of these functions is also an invariant of the full group action on M, and, furthermore, we can find q additional invariants of the action of G on M, of the form $v^1 = \zeta^1(x, u), \ldots, v^q = \zeta^q(x, u)$, which, together with the η's provide a complete set of $p + q - s$ functionally independent invariants for G on M. We write this complete collection of invariants concisely as

$$y = \eta(x), \qquad v = \zeta(x, u). \tag{3.1}$$

In the construction of the reduced system of differential equations for the G-invariant solutions to Δ, the y's will play the role of the new independent variables, and the v's the role of the new dependent variables. Note in particular that there are s fewer independent variables $y^1, \ldots, y^{p-s}$ which will appear in this reduced system, where s is the dimension of the orbits of G.

There is now a one-to-one correspondence between G-invariant functions $u = f(x)$ on M and arbitrary functions $v = h(y)$ involving the new variables. To explain this correspondence, we begin by invoking the implicit function theorem to solve the system $y = \eta(x)$ for $p - s$ of the independent variables, say $\tilde{x} = (x^{i_1}, \ldots, x^{i_{p-s}})$, in terms of the new variables $y^1, \ldots, y^{p-s}$ and the remaining s old independent variables, denoted as $\hat{x} = (x^{j_1}, \ldots, x^{j_s})$. Thus we have the solution

$$\tilde{x} = \gamma(\hat{x}, y) \tag{3.2}$$

for some well-defined function γ. The first $p - s$ of the old independent variables $\tilde{x}$ are known as *principal variables*, and the remaining s of these variables $\hat{x}$ are the *parametric variables*, as they will, in fact, enter parametrically into all the subsequent formulae. The precise manner in which one splits the variables x into principal and parametric variables is restricted only by the requirement that the $(p - s) \times (p - s)$ submatrix $(\partial\eta^j/\partial\tilde{x}^i)$ of the full Jacobian matrix $\partial\eta/\partial x$ is invertible, so that the implicit function theorem is applicable; otherwise, the choice is entirely arbitrary. We need to make a further *transversality* assumption on the action of G on M, cf. (3.35), that allows us to solve the other system of invariants $v = \zeta(x, u)$ for all of the dependent variables $u^1, \ldots, u^q$ in terms of $x^1, \ldots, x^p$, and $v^1, \ldots, v^q$, and hence in terms of the new variables y, v and the parametric variables $\hat{x}$:

$$u = \tilde{\delta}(x, v) = \tilde{\delta}(\hat{x}, \gamma(\hat{x}, y), v) \equiv \delta(\hat{x}, y, v) \tag{3.3}$$

near any point $(x_0, u_0) \in M$.

If $v = h(y)$ is any smooth function, then (3.3) coupled with (3.1) produces a corresponding G-invariant function on M, of the form

$$u = f(x) = \delta(\hat{x}, \eta(x), h(\eta(x))). \tag{3.4}$$

Conversely, if $u = f(x)$ is any G-invariant function on M, then it is not too difficult to see that there necessarily exists a function $v = h(y)$ such that f and the corresponding function (3.4) locally agree. Thus, we have seen how G-invariance of functions serves to decrease the number of variables upon which they depend.

We are now interested in finding all the G-invariant solutions to some system of partial differential equations

$$\Delta_\nu(x, u^{(n)}) = 0, \qquad \nu = 1, \ldots, l.$$

In other words, we want to know when a function of the form (3.4) corresponding to a function $v = h(y)$ is a solution to Δ. This will impose certain constraints on the function h; these are found by computing the formulae for the derivatives of a function of the form (3.4) with respect to x in terms of the derivatives of $v = h(y)$ with respect y, and then substituting these into the system of differential equations Δ. Thus we need to know how the derivatives of functions $v = h(y)$ are related to the derivatives of the corresponding G-invariant function $u = f(x)$. However, this is an easy application of the chain rule. Differentiating (3.4) with respect to x leads to a system of equations of the form

$$\frac{\partial u}{\partial x} = \frac{\partial}{\partial x}[\delta(\hat{x}, y, v)] = \frac{\partial \delta}{\partial \hat{x}} + \frac{\partial \delta}{\partial y}\frac{\partial \eta}{\partial x} + \frac{\partial \delta}{\partial v}\frac{\partial v}{\partial y}\frac{\partial \eta}{\partial x},$$

since $y = \eta(x)$. Here, $\partial u/\partial x$, etc., denote Jacobian matrices of first order derivatives of the indicated variables. Moreover, using (3.2), we can rewrite $\partial\eta/\partial x$ in terms of y and the parametric variables $\hat{x}$. Thus we obtain an equation of the form

$$\partial u/\partial x = \delta_1(\hat{x}, y, v, \partial v/\partial y)$$

expressing the first order derivatives of any G-invariant function u with respect to x in terms of y, v, the first order derivatives of v with respect to y, *plus* the parametric variables $\hat{x}$. Continuing to differentiate using the chain rule, and substituting according to (3.2) whenever necessary, we are led to general formulae

$$u^{(n)} = \delta^{(n)}(\hat{x}, y, v^{(n)}),$$

for all the derivatives of such a u with respect to x up to order n in terms of y, v, the derivatives of v with respect to y up to order n, and the ubiquitous parametric variables $\hat{x}$. At this point, it is worth considering a specific example.

Example 3.1. Consider the one-parameter scaling group

$$(x, t, u) \mapsto (\lambda x, \lambda t, u), \qquad \lambda \in \mathbb{R}^+,$$

acting on $X \times U \simeq \mathbb{R}^3$. On the upper half space $M \equiv \{t > 0\}$, the action is regular, with global independent invariants $y = x/t$ and $v = u$. If we treat v as a function of y, we can compute formulae for the derivatives of u with respect to x and t in terms of y, v and the derivatives of v with respect to y, along with a single parametric variable, which we designate to be t, so that x will be the corresponding principal variable. We find, using the chain rule, that if

$u = v = v(y) = v(x/t)$, then

$$u_x = t^{-1}v_y, \qquad u_t = -t^{-2}xv_y = -t^{-1}yv_y.$$

Further differentiations yield the second order formulae

$$u_{xx} = t^{-2}v_{yy}, \qquad u_{xt} = -t^{-2}(yv_{yy} + v_y), \qquad u_{tt} = t^{-2}(y^2v_{yy} + 2yv_y), \tag{3.5}$$

and so on.

Once the relevant formulae relating derivatives of u with respect to x to those of v with respect to y have been determined, the reduced system of differential equations for the G-invariant solutions to the system Δ is found by substituting these expressions into the system wherever they occur. In general, this leads to a system of equations of the form

$$\tilde{\Delta}_\nu(\hat{x}, y, v^{(n)}) = 0, \qquad \nu = 1, \ldots, l,$$

still involving the parametric variables $\hat{x}$. If G is actually a symmetry group for Δ, this resulting system will in fact always be *equivalent* to a system of equations, denoted

$$(\Delta/G)_\nu(y, v^{(n)}) = 0, \qquad \nu = 1, \ldots, l,$$

which are independent of the parametric variables, and thus constitute a genuine system of differential equations for v as a function of y. This is the reduced system Δ/G for the G-invariant solutions to the system Δ. Every solution $v = h(y)$ of Δ/G will correspond, via (3.4), to a G-invariant solution to Δ, and, moreover, every G-invariant solution can be constructed in this manner.

Example 3.2. The one-dimensional wave equation $u_{tt} - u_{xx} = 0$ is invariant under the scaling group presented in Example 3.1. To construct the corresponding scale-invariant solutions, we need only substitute the derivative formulae (3.5) into the wave equation, and solve the resulting ordinary differential equation. Upon substituting, we find the equation

$$t^{-2}(y^2v_{yy} + 2yv_y - v_{yy}) = 0.$$

As promised by the general theory, this equation is equivalent to an equation

$$(y^2 - 1)v_{yy} + 2yv_y = 0$$

in which the parametric variable t no longer appears. This latter ordinary differential equation is the reduced equation for the scale-invariant solutions to the wave equation. It is easily integrated, with general solution

$$v = c \log|(y - 1)/(y + 1)| + c',$$

where c, c' are arbitrary constants. Replacing the variables y, v in the solution by their expressions in terms of x, t, u, we deduce the general scale-invariant solution to the wave equation (for the particular scaling symmetry group under consideration) to be

$$u = c \log|(x - t)/(x + t)| + c'.$$

For the reader's convenience, we summarize the basic computational procedures for finding group-invariant solutions of a given system of partial differential equations from the beginning. We list the steps in order, starting with the computation of the symmetry group.

(I) Find all the infinitesimal generators **v** of symmetry groups of the system using the basic prolongation methods from Chapter 2, specifically the infinitesimal criterion (2.25).

(II) Decide on the "degree of symmetry" s of the invariant solutions. Here $1 \leqslant s \leqslant p$ will correspond to the dimension of the orbits of some subgroup of the full symmetry group. The reduced systems of differential equations for the invariant solutions will depend on $p - s$ independent variables. Thus to reduce the system of partial differential equations to a system of ordinary differential equations, we need to choose $s = p - 1$. In general, the smaller s is the more invariant solutions there will be, but the harder the reduced system Δ/G will be to solve explicitly.

(III) Find all s-dimensional subgroups G of the full symmetry group found in part I. This is equivalent (Theorem 1.51) to finding all s-dimensional subalgebras of the full Lie algebra of infinitestimal symmetries **v**. To each such subgroup or subalgebra there will correspond a set of group-invariant solutions reflecting the symmetries inherent in G itself. The problem of classifying subalgebras of a given Lie algebra will be explored in detail in Section 3.3. (In principle, an s-dimensional subgroup G may have orbits of dimension smaller than s, and, as we remarked earlier, it is the dimension of the orbits which matters. In practice, however, this mode of degeneracy rarely occurs, so we can content ourselves with fixing the dimension of the subgroup.)

(IV) Fixing the symmetry group G, we construct a complete set of functionally independent invariants, as in Section 2.1, which we divide into two classes

$$\begin{aligned} y^1 &= \eta^1(x, u), \ldots, y^{p-s} = \eta^{p-s}(x, u), \\ v^1 &= \zeta^1(x, u), \ldots, v^q = \zeta^q(x, u), \end{aligned} \tag{3.6}$$

corresponding to the new independent and dependent variables respectively. If G acts projectably, the choice of independent and dependent variables is prescribed by requiring the η^i's to be independent of u; in the more general case, there is quite a bit of freedom in this choice, and different choices will lead to seemingly different reduced systems, all of which are related by some form of "hodograph" transformation.

(V) Provided G acts transversally, (cf. Proposition 3.37) we can solve (3.6) for $p - s$ of the x's, which we denote by $\tilde{x}$, and all of the u's in terms of y, v and the remaining s parametric variables $\hat{x}$.

$$\tilde{x} = \gamma(\hat{x}, y, v), \qquad u = \delta(\hat{x}, y, v). \tag{3.7}$$

Furthermore, considering v as a function of y we can use (3.6), (3.7) and the chain rule to differentiate and thereby find expressions for the x-derivatives of any G-invariant u in terms of y, v, y-derivatives of v and the parametric variables $\hat{x}$:

$$u^{(n)} = \delta^{(n)}(\hat{x}, y, v^{(n)}). \tag{3.8}$$

(VI) Substitute the expressions (3.7), (3.8) into the system $\Delta(x, u^{(n)}) = 0$. The resulting system of equations will always be equivalent to a system of differential equations for $v = h(y)$ independent of the parametric variables $\hat{x}$:

$$\Delta/G(y, v^{(n)}) = 0. \tag{3.9}$$

At this stage we have constructed the reduced system of differential equations for the G-invariant solutions.

(VII) Solve the reduced system (3.9). For each solution $v = h(y)$ of Δ/G there corresponds a G-invariant solution $u = f(x)$ of the original system, which is given implicitly by the relation

$$\zeta(x, u) = h[\eta(x, u)]. \tag{3.10}$$

Repeating steps IV through VII for each symmetry group G determined in step III will yield a complete set of group-invariant solutions for our systems.

3.2. Examples of Group-Invariant Solutions

Before attempting to prove that the basic procedure for constructing group-invariant solutions outlined in the preceding section works, we will illustrate the method with some systematic examples, constructing group-invariant solutions of the Korteweg–de Vries, heat and Euler equations. These will lead naturally into the problem of how to classify group-invariant solutions in such a way as to find "all" such solutions with a minimum of computational difficulty. Before addressing this question, however, we begin with our examples.

Example 3.3. *The Heat Equation.* The symmetry group of the heat equation

$$u_t = u_{xx}$$

was computed in Example 2.41; it consisted of a six-parameter group of symmetries particular to the heat equation itself plus an infinite-dimensional subgroup stemming from the linearity of the equation. For each one-parameter subgroup of the full symmetry group there will be a corresponding class of group-invariant solutions which will be determined from a reduced ordinary differential equation, whose form will in general depend on the particular subgroup under investigation.

(a) *Travelling Wave Solutions.* In general, travelling wave solutions to a partial differential equation arise as special group-invariant solutions in which the group under consideration is a translation group on the space of independent variables. In the present example, consider the translation group

$$(x, t, u) \mapsto (x + c\varepsilon, t + \varepsilon, u), \qquad \varepsilon \in \mathbb{R},$$

generated by $\partial_t + c\partial_x$, in which c is a fixed constant, which will determine the speed of the waves. Global invariants of this group are

$$y = x - ct, \qquad v = u, \tag{3.11}$$

so that a group-invariant solution $v = h(y)$ takes the familiar form $u = h(x - ct)$ determining a wave of unchanging profile moving at the constant velocity c. Solving for the derivatives of u with respect to x and t in terms of those of v with respect to y we find

$$u_t = -cv_y, \qquad u_x = v_y, \qquad u_{xx} = v_{yy},$$

and so on. Substituting these expressions into the heat equation, we find the reduced ordinary differential equation for the travelling wave solutions to be

$$-cv_y = v_{yy}.$$

The general solution of this linear, constant coefficient equation is

$$v(y) = k\,e^{-cy} + l$$

for k, l arbitrary constants. Substituting back according to (3.11), we find the most general travelling wave solution to the heat equation to be an exponential of the form

$$u(x, t) = k\,e^{-c(x-ct)} + l.$$

(b) *Scale-Invariant (Similarity) Solutions.* There are two one-parameter groups of scaling symmetries of the heat equation, and we consider a linear combination

$$x\partial_x + 2t\partial_t + 2au\partial_u, \qquad a \in \mathbb{R},$$

of their infinitesimal generators, which corresponds to the scaling group

$$(x, t, u) \mapsto (\lambda x, \lambda^2 t, \lambda^{2a} u), \qquad \lambda \in \mathbb{R}^+.$$

On the half space $\{(x, t, u): t > 0\}$, global invariants of this one-parameter group are provided by the functions

$$y = x/\sqrt{t}, \qquad v = t^{-a}u.$$

Solving for the derivatives of u in terms of those of v, we find

$$u = t^a v,$$

$$u_x = t^{a-1/2}v_y, \qquad u_{xx} = t^{a-1}v_{yy},$$

$$u_t = -\tfrac{1}{2}xt^{a-3/2}v_y + at^{a-1}v = t^{a-1}(-\tfrac{1}{2}yv_y + av).$$

Here we are treating t as the parametric variable, and we have succeeded in expressing the relevant derivatives of u with respect to x and t in terms of y, v, the derivatives of v with respect to y, and the parametric variable t as in (3.8).

Substituting these expressions into the heat equation, we find

$$t^{a-1}v_{yy} = t^{a-1}(-\tfrac{1}{2}yv_y + av).$$

As guaranteed by the general theory, this equation is equivalent to one in which the parametric variable t does not occur, namely

$$v_{yy} + \tfrac{1}{2}yv_y - av = 0,$$

which forms the reduced equation for the scale-invariant solutions. The solutions of this linear ordinary differential equation can be written in terms of parabolic cylinder functions. Indeed, if we set

$$w = v\exp(\tfrac{1}{8}y^2),$$

then w satisfies a scaled form of Weber's differential equation,

$$w_{yy} = [(a + \tfrac{1}{4}) + \tfrac{1}{16}y^2]w.$$

The general solution of this equation is

$$w(y) = k\mathrm{U}\left(2a + \tfrac{1}{2}, \frac{y}{\sqrt{2}}\right) + \tilde{k}\mathrm{V}\left(2a + \tfrac{1}{2}, \frac{y}{\sqrt{2}}\right),$$

where $\mathrm{U}(b, z)$, $\mathrm{V}(b, z)$ are parabolic cylinder functions, cf. Abramowitz and Stegun, [1; § 19.1]. Thus the general scale-invariant solution to the heat equation takes the form

$$u(x, t) = t^a e^{-x^2/8t}\left\{k\mathrm{U}\left(2a + \tfrac{1}{2}, \frac{x}{\sqrt{2t}}\right) + \tilde{k}\mathrm{V}\left(2a + \tfrac{1}{2}, \frac{x}{\sqrt{2t}}\right)\right\}.$$

Particular values of a lead to special scale-invariant solutions which are expressible in terms of elementary functions. For instance, if $a = 0$, we obtain the probability solution

$$u(x, t) = k^* \operatorname{erf}(x/\sqrt{2t}) + \tilde{k}^*,$$

where erf is the error function. Since $\mathrm{U}(-n - \tfrac{1}{2}, z) = e^{-z^2/4}\,\mathrm{He}_n(z)$ where He_n is the n-th Hermite polynomial, if $a = -(n + 1)/2$ we obtain the solutions

$$u(x, t) = t^{-(n+1)/2}\, e^{-x^2/4t}\, \mathrm{He}_n(x/\sqrt{2t}),$$

which include the source solution ($n = 0$). Similarly, the relation $\mathrm{V}(n + \tfrac{1}{2}, z) = \sqrt{2/\pi}e^{z^2/4}\,\mathrm{He}_n^*(z)$, where $\mathrm{He}_n^*(z) = (-i)^n\,\mathrm{He}_n(iz)$, leads to the heat polynomials (see Widder, [1])

$$x, \quad x^2 + 2t, \quad x^3 + 6xt, \quad \text{etc.},$$

as special scale-invariant solutions.

(c) *Galilean-Invariant Solutions.* The one-parameter group of Galilean boosts, generated by $\mathbf{v}_5 = 2t\partial_x - xu\partial_u$ has global invariants $y = t$, $v = u \exp(x^2/4t)$ on the upper half space $\{t > 0\}$. We find

$$u_t = \left(v_y + \frac{x^2}{4t^2}v\right)e^{-x^2/4t}, \qquad u_{xx} = \left(\frac{x^2}{4t^2} - \frac{1}{2t}\right)v\, e^{-x^2/4t}.$$

Therefore, for the heat equation the reduced equation for Galilean-invariant solutions is a *first order* ordinary differential equation $2yv_y + v = 0$, despite the fact that the heat equation was a second order partial differential equation. The solution is $v(y) = k/\sqrt{y}$. Hence the most general Galilean-invariant solution is a scalar multiple of the source solution,

$$u(x, t) = \frac{k}{\sqrt{t}}e^{-x^2/4t},$$

which we earlier found as a scale-invariant solution. Thus a given solution may be invariant under more than one subgroup of the full symmetry group.

We can clearly extend this list of group-invariant solutions by considering further one-parameter subgroups obtained from more general linear combinations of the infinitesimal generators of the full symmetry group. At the moment, however, without some means of classifying these solutions, it is somewhat pointless to continue. Once we have determined the correct classification procedure, we will return to this question and find (in a sense) the most general group-invariant solutions to the heat equation. See Example 3.17.

Example 3.4. The symmetry group of the Korteweg–de Vries equation

$$u_t + u_{xxx} + uu_x = 0$$

was computed in Example 2.44. Let us look at particular group-invariant solutions.

(a) *Travelling Wave Solutions.* Here the group is the same translational group already looked at in the previous example. In terms of the invariants $y = x - ct$, $v = u$, the reduced equation is

$$v_{yyy} + vv_y - cv_y = 0.$$

This can be immediately integrated once,

$$v_{yy} + \tfrac{1}{2}v^2 - cv = k,$$

and a second integration is performed after multiplying by v_y:

$$\tfrac{1}{2}v_y^2 = -\tfrac{1}{6}v^3 + \tfrac{1}{2}cv^2 + kv + l, \tag{3.12}$$

where k and l are arbitrary constants. The general solution can be written in terms of elliptic functions, $u = \wp(x - ct + \delta)$, δ being an arbitrary phase shift. If $u \to 0$ sufficiently rapidly as $|x| \to \infty$, then $k = l = 0$ in (3.12). This equation has real solutions

$$v = 3c\,\mathrm{sech}^2[\tfrac{1}{2}\sqrt{c}y + \delta],$$

provided the wave speed c is positive. These produce the celebrated "one soliton" solutions

$$u(x, t) = 3c\,\mathrm{sech}^2[\tfrac{1}{2}\sqrt{c}(x - ct) + \delta]$$

to the Korteweg–de Vries equation. (If $c = 0$, we also obtain the singular stationary solution $u = -12(x + \delta)^{-2}$.) More generally, if we only require u to be bounded, we obtain the periodic "cnoidal wave" solutions

$$u(x, t) = a\,\mathrm{cn}^2[\lambda(x - ct) + \delta] + m,$$

where cn is the Jacobi elliptic function of modulus $k = \sqrt{(r_3 - r_2)/(r_3 - r_1)}$, $a = r_3 - r_2$, $\lambda = \sqrt{(r_3 - r_1)/6}$, $m = r_2$ and $r_1 < r_2 < r_3$ are the roots of the cubic polynomial on the right-hand side of (3.12).

(b) *Galilean-Invariant Solutions.* Next look at the one-parameter group of Galilean boosts generated by $t\partial_x + \partial_u$. Here, for $t > 0$, $y = t$ and $v = tu - x$ are independent invariants, from which we calculate

$$u = y^{-1}(x + v), \qquad u_x = y^{-1}, \qquad u_{xxx} = 0, \qquad u_t = y^{-2}(yv_y - v - x),$$

where x is the parametric variable. The reduced equation is simply $dv/dy = 0$, so the general Galilean-invariant solution is $u = (x + \delta)/t$ for δ an arbitrary constant.

A more interesting class of solutions with Galilean-like invariance can be found by adding a time translational component to this group. The generator $t\partial_x + a\partial_t + \partial_u$, $a \neq 0$, has global invariants

$$y = x - \tfrac{1}{2}bt^2, \qquad v = u - bt,$$

where $b = 1/a$. We have

$$u = v + bt, \qquad u_x = v_y, \qquad u_{xxx} = v_{yyy}, \qquad u_t = -btv_y + b,$$

so the reduced equation is

$$v_{yyy} + vv_y + b = 0.$$

This integrates once, leading to a second-order equation

$$v_{yy} + \tfrac{1}{2}v^2 + by + c = 0$$

known as the *first Painlevé transcendent.* Its solutions $v = h(y)$ are meromorphic in the entire complex plane, but are essentially new functions not expressible in terms of standard special functions. The corresponding solutions of the Korteweg–de Vries equation take the form

$$u(x, t) = h(x - \tfrac{1}{2}bt^2) + bt.$$

(c) *Scale-Invariant Solutions.* Finally consider the group of scaling symmetries

$$(x, t, u) \mapsto (\lambda x, \lambda^3 t, \lambda^{-2}u).$$

Invariants on the half space $\{t > 0\}$ are

$$y = t^{-1/3}x, \qquad v = t^{2/3}u.$$

We find

$$u_x = t^{-1}v_y, \qquad u_{xxx} = t^{-5/3}v_{yyy}, \qquad u_t = -\tfrac{1}{3}t^{-5/3}(yv_y + 2v),$$

so that the reduced equation is

$$v_{yyy} + vv_y - \tfrac{1}{3}yv_y - \tfrac{2}{3}v = 0.$$

It is by no means obvious how to solve this third order ordinary differential equation directly. However, motivated by a transformation discovered by Miura, [1], for the Korteweg–de Vries equation itself (see Exercise 5.11), let us set

$$v = \frac{dw}{dy} - \tfrac{1}{6}w^2.$$

The equation for w is

$$\begin{aligned} 0 &= w_{yyyy} - \tfrac{1}{3}ww_{yyy} - \tfrac{1}{3}ww_y^2 - \tfrac{1}{6}w^2w_{yy} + \tfrac{1}{18}w^3w_y - \tfrac{1}{3}yw_{yy} + \tfrac{1}{9}yww_y - \tfrac{2}{3}w_y + \tfrac{1}{9}w^2 \\ &= (D_y - \tfrac{1}{3}w)(w_{yyy} - \tfrac{1}{6}w^2w_y - \tfrac{1}{3}yw_y - \tfrac{1}{3}w). \end{aligned}$$

Therefore every solution to the "modified" third-order equation

$$w_{yyy} - \tfrac{1}{6}w^2w_y - \tfrac{1}{3}yw_y - \tfrac{1}{3}w = 0$$

gives rise to a scale-invariant solution of the Korteweg–de Vries equation by the above transformation. The latter equation can be integrated once:

$$w_{yy} = \tfrac{1}{18}w^3 + \tfrac{1}{3}yw + k$$

for some constant k. This equation is the *second Painlevé transcendent*, which shares similar properties to the first. See Ince, [1], for an extensive discussion of these equations.

Example 3.5. For the Euler equations of three-dimensional incompressible ideal fluid flow,

$$\begin{aligned} \mathbf{u}_t + \mathbf{u}\cdot\nabla\mathbf{u} &= -\nabla p, \\ \nabla\cdot\mathbf{u} &= 0, \end{aligned} \tag{3.13}$$

there are four independent variables: $\mathbf{x} = (x, y, z)$ and t, so we can discuss solutions which are invariant under one-, two- and three-parameter subgroups of the full symmetry group, which was determined in Example 2.45. Here we look at a couple of such subgroups, leading to solutions of physical or mathematical interest. In all cases, the group will contain a one-parameter subgroup of either uniform, time-independent translations in a fixed direction, which we may as well take to be the z-axis, or rotations around a fixed axis, again taken as the z-axis.

(a) *Translationally-Invariant Solutions.* For solutions invariant under the translation group generated by ∂_z, the three-dimensional Euler equations (3.13) reduce to their two-dimensional counterparts, which have the same form but with $\mathbf{u} = (u, v)$, p depending only on $\mathbf{x} = (x, y)$, t, together with an equation

$$w_t + uw_x + vw_y = 0 \qquad (3.14)$$

for the vertical component of the velocity, which can be integrated by solving the characteristic equation $dt = dx/u = dy/v$. Of course, the two-dimensional Euler equations are still far too difficult to solve explicitly, so we look for solutions invariant under a second one-parameter group.

For the time-dependent translational group G_β generated by $\mathbf{v}_\beta = \beta\partial_y + \beta_t\partial_v - \beta_{tt}y\partial_p$, $\beta(t) \neq 0$, invariants are given by

$$t, \qquad x, \qquad \tilde{u} = u, \qquad \tilde{v} = v - (\beta_t/\beta)y, \qquad \tilde{p} = p + \tfrac{1}{2}(\beta_{tt}/\beta)y^2,$$

with t, x being independent variables. The reduced system

$$\tilde{u}_t + \tilde{u}\tilde{u}_x = -\tilde{p}_x, \qquad \tilde{v}_t + \tilde{u}\tilde{v}_x + (\beta_t/\beta)\tilde{v} = 0, \qquad \tilde{u}_x + (\beta_t/\beta) = 0, \quad (3.15)$$

is readily solved:

$$\tilde{u} = [-\beta_t x + \sigma_t]/\beta, \qquad \tilde{v} = h(\beta x - \sigma)/\beta,$$
$$\tilde{p} = [(\tfrac{1}{2}\beta\beta_{tt} - \beta_t^2)x^2 + (2\beta_t\sigma_t - \beta\sigma_{tt})x + \tau]/\beta^2,$$

where $\sigma(t)$, $\tau(t)$ are arbitrary functions of t, and h is an arbitrary function of the single invariant $\beta(t)x - \sigma(t)$ for the second equation in (3.15). Thus we obtain the G_β-invariant solution

$$u = [-\beta_t x + \sigma_t]/\beta, \qquad v = [\beta_t y + h(\beta x - \sigma)]/\beta,$$
$$p = [(\tfrac{1}{2}\beta\beta_{tt} - \beta_t^2)x^2 - \tfrac{1}{2}\beta\beta_{tt}y^2 + (2\beta_t\sigma_t - \beta\sigma_{tt})x + \tau]/\beta^2,$$

of the two-dimensional Euler equations. In particular, if $\beta(t) \equiv 1$, we can further solve (3.14) explicitly, with

$$u = \sigma_t, \qquad v = h(x - \sigma(t)), \qquad w = H(x - \sigma(t), y - th(x - \sigma(t))),$$
$$p = -\sigma_{tt}x + \tau(t),$$

being the three-dimensional solution invariant under the group of translations in the y- and z-directions. (Here $\sigma(t)$, $\tau(t)$, $h(\xi)$, $H(\xi, \eta)$ are arbitrary smooth functions.)

Although this determines all solutions of the two-dimensional Euler equations which are invariant under the group generated by $\mathbf{v}_\beta$, it is instructive to see what happens if we try to determine the more specialized solutions which are invariant under the two-parameter group generated by $\mathbf{v}_\beta$ and $\mathbf{v}_\alpha = \alpha\partial_x + \alpha_t\partial_u - \alpha_{tt}x\partial_p$. Invariants are

$$\tilde{u} = u - (\alpha_t/\alpha)x, \qquad \tilde{v} = v - (\beta_t/\beta)y, \qquad \tilde{p} = p + (\alpha_{tt}/2\alpha)x^2 + (\beta_{tt}/2\beta)y^2,$$

which are functions of the sole remaining independent variable t. The reduced system of ordinary differential equation is

$$\tilde{u}_t + (\alpha_t/\alpha)\tilde{u} = 0 = \tilde{v}_t + (\beta_t/\beta)\tilde{v}, \qquad (3.16a)$$

plus the divergence-free condition

$$(\alpha_t/\alpha) + (\beta_t/\beta) = 0. \tag{3.16b}$$

An important point here is that *unless* $\alpha(t) = k/\beta(t)$ for some constant k, (3.16b) is inconsistent and there are *no* solutions to the reduced equations. In other words, there is no guarantee in general that the reduced system of differential equations for some symmetry group be consistent, and hence no guarantee that any such solutions exist.

(b) *Rotationally-Invariant Solutions.* For the group of rotations about the z-axis, generated by $-y\partial_x + x\partial_y - v\partial_u + u\partial_v$, invariants are provided by $r = \sqrt{x^2 + y^2}$, z, t, p and the cylindrical components of the velocity $\hat{u} = u\cos\theta + v\sin\theta$, $\hat{v} = -u\sin\theta + v\cos\theta$, $\hat{w} = w$. The reduced equations are

$$\begin{aligned} \hat{u}_t + \hat{u}\hat{u}_r + \hat{w}\hat{u}_z - r^{-1}\hat{v}^2 &= -p_r, \\ \hat{v}_t + \hat{u}\hat{v}_r + \hat{w}\hat{v}_z + r^{-1}\hat{u}\hat{v} &= 0, \\ \hat{w}_t + \hat{u}\hat{w}_r + \hat{w}\hat{w}_z &= -p_z, \\ (r\hat{u})_r + (r\hat{w})_z &= 0, \end{aligned} \tag{3.17}$$

cf. Berker, [1]. If we further assume translational invariance under ∂_z, so $\hat{u}$, $\hat{v}$, $\hat{w}$, p are independent of z, then we can solve (3.17) explicitly:

$$\hat{u} = \sigma_t/r, \qquad \hat{v} = r^{-1}h[\tfrac{1}{2}r^2 - \sigma(t)], \qquad \hat{w} = \tilde{h}[\tfrac{1}{2}r^2 - \sigma(t)],$$

$$p = -\sigma_{tt}\log r - \tfrac{1}{2}r^{-2}\sigma_t^2 + \int_0^r s^{-3}h[\tfrac{1}{2}s^2 - \sigma(t)]^2\,ds + \tau(t),$$

where $\sigma(t)$, $\tau(t)$ and $h(\xi)$, $\tilde{h}(\xi)$ are arbitrary functions. These are the most general solutions depending only on t and the cylindrical radius r.

Are there solutions which are completely rotationally-invariant, i.e. have the full SO(3) invariance group? Although SO(3) acts projectably on $\mathbb{R}^3 \times \mathbb{R}^3$ via $(\mathbf{x}, \mathbf{u}) \mapsto (R\mathbf{x}, R\mathbf{u})$, $R \in \mathrm{SO}(3)$, and regularly with three-dimensional orbits on an open subset of $\mathbb{R}^3 \times \mathbb{R}^3$, the projected group action $\mathbf{x} \mapsto R\mathbf{x}$ on $\mathbb{R}^3$ has only two-dimensional orbits. In this case the transversality conditions (3.33) are violated and we are unable to construct a reduced system $\Delta/\mathrm{SO}(3)$. Another way to see this is to look at the invariants for SO(3), which are t, $|\mathbf{x}|$, $\mathbf{x}\cdot\mathbf{u}$, $|\mathbf{u}|$, p, and note that there are one too many independent and one too few dependent variables to carry through the reduction procedure. Thus no SO(3)-invariant solutions can be constructed by this technique.

As a final example, we look directly at the Euler equations in cylindrical coordinates (3.17). This system has a number of symmetry groups, most of which come from symmetry groups for the full Euler equations (3.13). There is, however, one additional symmetry generator, $\mathbf{v}^* = r^{-2}(\hat{v}^{-1}\partial_{\hat{v}} - \partial_p)$, which does not come from a symmetry of (3.13)! Thus, reducing a system Δ by a known symmetry group G may lead to a system Δ/G with *additional* symmetry properties not shared by the original system. Let us look for solutions invariant under the one-parameter group generated by $\partial_t - \mathbf{v}^*$, which has invariants r, z, $\hat{u}$, $\hat{w}$, $\omega = \tfrac{1}{2}r^2\hat{v}^2 + t$, $q = p - r^{-2}t$. These satisfy the reduced system

$$\hat{u}\hat{u}_r + \hat{w}\hat{u}_z - 2r^{-3}\omega = -q_r, \qquad \hat{u}\omega_r + \hat{w}\omega_z = 1,$$
$$\hat{u}\hat{w}_r + \hat{w}\hat{w}_z = -q_z, \qquad (r\hat{u})_r + (r\hat{w})_z = 0. \tag{3.18}$$

This system is still too complicated to solve in general; however, following Kapitanskii, [1], we look for solutions with the ansatz

$$\hat{u} = \hat{u}(r), \qquad \omega = \omega(r), \qquad \hat{w} = \xi(r)z + \eta(r).$$

The first and last equations in (3.18) imply

$$\xi = -\hat{u}_r - r^{-1}\hat{u}, \qquad q_r = -\hat{u}\hat{u}_r + 2r^{-3}\omega.$$

Differentiating the third equation with respect to r, we find the compatibility condition

$$0 = -q_{rz} = (\hat{u}\xi_r + \xi^2)_r z + (\hat{u}\eta_r + \xi\eta)_r,$$

hence $\hat{u}\xi_r + \xi^2 = k$, $\hat{u}\eta_r + \xi\eta = l$ are constant. Using the above formula for ξ, we find that $\hat{u}$ must satisfy the ordinary differential equation

$$\hat{u}\hat{u}_{rr} - \hat{u}_r^2 - r^{-1}\hat{u}\hat{u}_r - 2r^{-2}\hat{u}^2 + k = 0.$$

This equation admits a two-parameter group of symmetries, generated by $\mathbf{w} = r\partial_r + u\partial_u$, $\tilde{\mathbf{w}} = r^{-1}\partial_r - r^{-2}\hat{u}\partial_{\hat{u}}$, and hence can be integrated using the methods of Section 2.5. For $k < 0$ we have

$$\hat{u}(r) = ar^{-1}\cosh(br^2 + \delta),$$

a, b, δ arbitrary constants, hence

$$\omega(r) = -\frac{\arctan \exp(br^2 + \delta)}{2ab^2}, \qquad \xi(r) = -2ab \sinh(br^2 + \delta),$$

$$q(r, z) = -\tfrac{1}{2}kz^2 - lz - \tfrac{1}{2}[\hat{u}(r)]^2 + \int_{r_0}^{r} 2s^{-3}\,\omega(s)\,ds,$$

and, from the last equation of (3.18),

$$\eta(r) = -2b \sinh(br^2 + \delta),$$

and $l = -4ab^2$. Thus the general such solution is

$$\hat{u} = \hat{u}(r), \quad \hat{v} = r^{-1}\sqrt{2\omega(r) - 2t}, \quad \hat{w} = \xi(r)z + \eta(r), \quad p = tr^{-2} + q(r, z),$$

with $\hat{u}$, ω, ξ, η, q being as above. Kapitanskii notes that since $\hat{v}$ is given by a square root, these solutions can be arranged to provide solutions of the Euler equations on cylindrical domains which blow up in finite time ($|\nabla \mathbf{u}| \to \infty$) even though the normal component of u on the boundary is smooth for all t. The reason, of course, is that the singularity $\omega(r) = t$ can be arranged to cross the boundary without affecting the normal component of $\mathbf{u}$. (Similar sorts of behaviour can be arranged for the simpler translationally-invariant solutions.) This observation, therefore, does not answer the outstanding problem of whether smooth solutions to the three-dimensional Euler equations can develop singularities after a finite time.

3.3. Classification of Group-Invariant Solutions

In general, to each s-parameter subgroup H of the full symmetry group G of a system of differential equations in $p > s$ independent variables, there will correspond a family of group-invariant solutions. Since there are almost always an infinite number of such subgroups, it is not usually feasible to list all possible group-invariant solutions to the system. We need an effective, systematic means of classifying these solutions, leading to an "optimal system" of group-invariant solutions from which every other such solution can be derived. Since elements $g \in G$ not in the subgroup H will transform an H-invariant solution to some other group-invariant solution, only those solutions not so related need be listed in our optimal system. The basic result is the following:

Proposition 3.6. *Let G be the symmetry group of a system of differential equations Δ and let $H \subset G$ be an s-parameter subgroup. If $u = f(x)$ is an H-invariant solution to Δ and $g \in G$ is any other group element, then the transformed function $u = \tilde{f}(x) = g \cdot f(x)$ is a $\tilde{H}$-invariant solution, where $\tilde{H} = gHg^{-1}$ is the conjugate subgroup to H under g.*

The proof follows directly from Exercise 2.3 using the graph Γ_f as the invariant subset. As a consequence of this result, the problem of classifying group-invariant solutions reduces to the problem of classifying subgroups of the full symmetry group G under conjugation, We thus need to study the conjugacy map $h \mapsto ghg^{-1}$ on a Lie group in detail, after which we will return to our original classification problem.

The Adjoint Representation

Let G be a Lie group. For each $g \in G$, group conjugation $K_g(h) \equiv ghg^{-1}$, $h \in G$, determines a diffeomorphism on G. Moreover, $K_g \circ K_{g'} = K_{gg'}$, $K_e = \mathbb{1}_G$, so K_g determines a global group action of G on itself, with each conjugacy map K_g being a group homomorphism: $K_g(hh') = K_g(h)K_g(h')$, etc. The differential $dK_g\colon TG|_h \to TG|_{K_g(h)}$ is readily seen to preserve the right-invariance of vector fields, and hence determines a linear map on the Lie algebra of G, called the *adjoint representation*:

$$\operatorname{Ad} g(\mathbf{v}) \equiv dK_g(\mathbf{v}), \qquad \mathbf{v} \in \mathfrak{g}. \tag{3.19}$$

Note that the adjoint representation gives a global linear action of G on $\mathfrak{g}$:

$$\operatorname{Ad}(g \cdot g') = \operatorname{Ad} g \circ \operatorname{Ad} g', \qquad \operatorname{Ad} e = \mathbb{1}.$$

If $\mathbf{v} \in \mathfrak{g}$ generates the one-parameter subgroup $H = \{\exp(\varepsilon\mathbf{v})\colon \varepsilon \in \mathbb{R}\}$, then by (1.22) $\operatorname{Ad} g(\mathbf{v})$ is easily seen to generate the conjugate one-parameter subgroup $K_g(H) = gHg^{-1}$. This remark readily generalizes to higher dimensional subgroups using the fact that they are completely determined by their one-parameter subgroups.

Proposition 3.7. *Let H and $\tilde{H}$ be connected, s-dimensional Lie subgroups of the Lie group G with corresponding Lie subalgebras $\mathfrak{h}$ and $\tilde{\mathfrak{h}}$ of the Lie algebra $\mathfrak{g}$ of G. Then $\tilde{H} = gHg^{-1}$ are conjugate subgroups if and only if $\tilde{\mathfrak{h}} = \operatorname{Ad} g(\mathfrak{h})$ are conjugate subalgebras.*

The adjoint representation of a Lie group on its Lie algebra is often most easily reconstructed from its infinitesimal generators. If $\mathbf{v}$ generates the one-parameter subgroup $\{\exp(\varepsilon \mathbf{v})\}$, then we let $\operatorname{ad} \mathbf{v}$ be the vector field on $\mathfrak{g}$ generating the corresponding one-parameter group of adjoint transformations:

$$\operatorname{ad} \mathbf{v}|_{\mathbf{w}} \equiv \left.\frac{d}{d\varepsilon}\right|_{\varepsilon=0} \operatorname{Ad}(\exp(\varepsilon \mathbf{v}))\mathbf{w}, \qquad \mathbf{w} \in \mathfrak{g}. \tag{3.20}$$

A fundamental fact is that the infinitesimal adjoint action agrees (up to sign) with the Lie bracket on $\mathfrak{g}$:

Proposition 3.8. *Let G be a Lie group with Lie algebra $\mathfrak{g}$. For each $\mathbf{v} \in \mathfrak{g}$, the adjoint vector $\operatorname{ad} \mathbf{v}$ at $\mathbf{w} \in \mathfrak{g}$ is*

$$\operatorname{ad} \mathbf{v}|_{\mathbf{w}} = [\mathbf{w}, \mathbf{v}] = -[\mathbf{v}, \mathbf{w}], \tag{3.21}$$

where we are using the identification of $T\mathfrak{g}|_{\mathbf{w}}$ with $\mathfrak{g}$ itself since $\mathfrak{g}$ is a vector space.

PROOF. We identify $\mathfrak{g} \simeq TG|_e$. Using (3.20), the definition (3.19) of the adjoint representation, and the right-invariance of $\mathbf{w}$, we find

$$\begin{aligned} \operatorname{ad} \mathbf{v}|_{\mathbf{w}} &= \lim_{\varepsilon \to 0} \frac{dK_{\exp(\varepsilon \mathbf{v})}[\mathbf{w}|_e] - \mathbf{w}|_e}{\varepsilon} \\ &= \lim_{\varepsilon \to 0} \frac{d\exp(\varepsilon \mathbf{v})[\mathbf{w}|_{\exp(-\varepsilon \mathbf{v})}] - \mathbf{w}|_e}{\varepsilon}. \end{aligned}$$

If we replace ε by $-\varepsilon$, this last expression is the same as the definition (1.57) of the Lie derivative of $\mathbf{w}$ with respect to $\mathbf{v}$, so (3.21) follows from Proposition 1.64. □

Note. In most references, the adjoint map $\operatorname{ad} \mathbf{v}|_{\mathbf{w}}$ has the other sign $+[\mathbf{v}, \mathbf{w}]$. The reason is our choice in Chapter 1 of right-invariant vector fields to define the Lie algebra, rather than the more traditional left-invariant vector fields. (The reasons for this choice were discussed in Exercise 1.33.) In this book, we will consistently use (3.21) for the infinitesimal adjoint action.

In the case $G \subset \mathrm{GL}(n)$ is a matrix Lie group with Lie algebra $\mathfrak{g} \subset \mathfrak{gl}(n)$, the above formulae are particularly easy to verify. Since $K_A(B) = ABA^{-1}$, where $A, B \in G$ are $n \times n$ matrices, the adjoint map is also given by conjugation

$$\operatorname{Ad} A(X) = AXA^{-1}, \qquad A \in G, \quad X \in \mathfrak{g}.$$

Letting $A = e^{\varepsilon Y}$, $Y \in \mathfrak{g}$, and differentiating with respect to ε, we find

$$\operatorname{ad} Y|_X = YX - XY = [X, Y],$$

agreeing with the commutator bracket on $\mathrm{gl}(n)$.

Example 3.9. Let $G = \mathrm{SO}(3)$ be the group of rotations in $\mathbb{R}^3$. The Lie algebra $\mathfrak{so}(3)$ is spanned by the matrices

$$A^x = \begin{bmatrix} 0 & 0 & 0 \\ 0 & 0 & -1 \\ 0 & 1 & 0 \end{bmatrix}, \qquad A^y = \begin{bmatrix} 0 & 0 & 1 \\ 0 & 0 & 0 \\ -1 & 0 & 0 \end{bmatrix}, \qquad A^z = \begin{bmatrix} 0 & -1 & 0 \\ 1 & 0 & 0 \\ 0 & 0 & 0 \end{bmatrix},$$

generating the one-parameter subgroups

$$R^x_\theta = \begin{bmatrix} 1 & 0 & 0 \\ 0 & \cos\theta & -\sin\theta \\ 0 & \sin\theta & \cos\theta \end{bmatrix}, \qquad R^y_\theta = \begin{bmatrix} \cos\theta & 0 & \sin\theta \\ 0 & 1 & 0 \\ -\sin\theta & 0 & \cos\theta \end{bmatrix},$$

$$R^z_\theta = \begin{bmatrix} \cos\theta & -\sin\theta & 0 \\ \sin\theta & \cos\theta & 0 \\ 0 & 0 & 1 \end{bmatrix},$$

of counterclockwise rotations about the coordinate axes. The adjoint action of, say, R^x_θ on the generator A^y can be found by differentiating the product $R^x_\theta R^y_\varepsilon R^x_{-\theta}$ with respect to ε and setting $\varepsilon = 0$. We find

$$\operatorname{Ad} R^x_\theta(A^y) = \begin{bmatrix} 0 & -\sin\theta & \cos\theta \\ \sin\theta & 0 & 0 \\ -\cos\theta & 0 & 0 \end{bmatrix} = \cos\theta \cdot A^y + \sin\theta \cdot A^z,$$

and similarly,

$$\operatorname{Ad} R^x_\theta(A^x) = A^x, \qquad \operatorname{Ad} R^x_\theta(A^z) = -\sin\theta \cdot A^y + \cos\theta \cdot A^z.$$

Thus the adjoint action of the subgroup R^x_θ of rotations around the x-axis in physical space is the same as the group of rotations around the A^x-axis in the Lie algebra space $\mathfrak{so}(3)$. Similar remarks apply to the other subgroups, so if $R \in \mathrm{SO}(3)$ is any rotation matrix relative to the given (x, y, z)-coordinates in $\mathbb{R}^3$, its adjoint map $\operatorname{Ad} R$ acting on $\mathfrak{so}(3) \simeq \mathbb{R}^3$ has the same matrix representation R relative to the induced basis $\{A^x, A^y, A^z\}$. (The fact that the adjoint representation of $\mathrm{SO}(3)$ agrees with its natural physical representation is accidental and will *not* hold for other matrix Lie groups.) Finally, the infinitesimal generators of the adjoint action are found by differentiation; for example,

$$\operatorname{ad} A^x|_{A^y} = \frac{d}{d\theta}\bigg|_{\theta=0} \operatorname{Ad}(R^x_\theta)A^y = A^z,$$

which agrees with the commutator

$$[A^y, A^x] = A^x A^y - A^y A^x = A^z.$$

Conversely, if we know the infinitesimal adjoint action ad $\mathfrak{g}$ of a Lie algebra $\mathfrak{g}$ on itself, we can reconstruct the adjoint representation Ad G of the underlying Lie group, either by integrating the system of linear ordinary differential equations

$$\frac{d\mathbf{w}}{d\varepsilon} = \text{ad}\, \mathbf{v}|_{\mathbf{w}}, \qquad \mathbf{w}(0) = \mathbf{w}_0, \tag{3.22}$$

with solution

$$\mathbf{w}(\varepsilon) = \text{Ad}(\exp(\varepsilon \mathbf{v}))\mathbf{w}_0,$$

or, perhaps more simply, by summing the Lie series (cf. (1.19))

$$\begin{aligned} \text{Ad}(\exp(\varepsilon \mathbf{v}))\mathbf{w}_0 &= \sum_{n=0}^{\infty} \frac{\varepsilon^n}{n!}(\text{ad}\, \mathbf{v})^n(\mathbf{w}_0) \\ &= \mathbf{w}_0 - \varepsilon[\mathbf{v}, \mathbf{w}_0] + \frac{\varepsilon^2}{2}[\mathbf{v}, [\mathbf{v}, \mathbf{w}_0]] - \cdots. \end{aligned} \tag{3.23}$$

(The convergence of (3.23) follows since (3.22) is a linear system of ordinary differential equations, for which (3.23) is the corresponding matrix exponential.)

Example 3.10. The Lie algebra spanned by $\mathbf{v}_1 = \partial_x$, $\mathbf{v}_2 = \partial_t$, $\mathbf{v}_3 = t\partial_x + \partial_u$, $\mathbf{v}_4 = x\partial_x + 3t\partial_t - 2u\partial_u$ generates the symmetry group of the Korteweg–de Vries equation. To compute the adjoint representation, we use the Lie series (3.23) in conjunction with the commutator table in Example 2.44. For instance

$$\begin{aligned} \text{Ad}(\exp(\varepsilon \mathbf{v}_2))\mathbf{v}_4 &= \mathbf{v}_4 - \varepsilon[\mathbf{v}_2, \mathbf{v}_4] + \tfrac{1}{2}\varepsilon^2[\mathbf{v}_2, [\mathbf{v}_2, \mathbf{v}_4]] - \cdots \\ &= \mathbf{v}_4 - 3\varepsilon \mathbf{v}_2. \end{aligned}$$

In this manner, we construct the table

Ad	$\mathbf{v}_1$	$\mathbf{v}_2$	$\mathbf{v}_3$	$\mathbf{v}_4$
$\mathbf{v}_1$	$\mathbf{v}_1$	$\mathbf{v}_2$	$\mathbf{v}_3$	$\mathbf{v}_4 - \varepsilon \mathbf{v}_1$
$\mathbf{v}_2$	$\mathbf{v}_1$	$\mathbf{v}_2$	$\mathbf{v}_3 - \varepsilon \mathbf{v}_1$	$\mathbf{v}_4 - 3\varepsilon \mathbf{v}_2$
$\mathbf{v}_3$	$\mathbf{v}_1$	$\mathbf{v}_2 + \varepsilon \mathbf{v}_1$	$\mathbf{v}_3$	$\mathbf{v}_4 + 2\varepsilon \mathbf{v}_3$
$\mathbf{v}_4$	$e^{\varepsilon}\mathbf{v}_1$	$e^{3\varepsilon}\mathbf{v}_2$	$e^{-2\varepsilon}\mathbf{v}_3$	$\mathbf{v}_4$

(3.24)

with the (i, j)-th entry indicating $\text{Ad}(\exp(\varepsilon \mathbf{v}_i))\mathbf{v}_j$.

Classification of Subgroups and Subalgebras

Definition 3.11. Let G be a Lie group. An *optimal system* of s-parameter subgroups is a list of conjugacy inequivalent s-parameter subgroups with the property that any other subgroup is conjugate to precisely one subgroup in the list. Similarly, a list of s-parameter subalgebras forms an *optimal system* if every s-parameter subalgebra of $\mathfrak{g}$ is equivalent to a unique member of the list under some element of the adjoint representation: $\tilde{\mathfrak{h}} = \operatorname{Ad} g(\mathfrak{h})$, $g \in G$.

Proposition 3.7 says that the problem of finding an optimal system of subgroups is equivalent to that of finding an optimal system of subalgebras, and so we concentrate on the latter. Unfortunately, this problem can still be quite complicated, and, for once, infinitesimal techniques do not seem to be overly useful.

For one-dimensional subalgebras, this classification problem is essentially the same as the problem of classifying the orbits of the adjoint representation, since each one-dimensional subalgebra is determined by a nonzero vector in $\mathfrak{g}$. Although some sophisticated techniques are available for Lie algebras with additional structure, in essence this problem is attacked by the naïve approach of taking a general element $\mathbf{v}$ in $\mathfrak{g}$ and subjecting it to various adjoint transformations so as to "simplify" it as much as possible. We treat a couple of illustrative examples.

Example 3.12. Consider the symmetry algebra $\mathfrak{g}$ of the Korteweg–de Vries equation, whose adjoint representation was determined in Example 3.10. Given a nonzero vector

$$\mathbf{v} = a_1\mathbf{v}_1 + a_2\mathbf{v}_2 + a_3\mathbf{v}_3 + a_4\mathbf{v}_4,$$

our task is to simplify as many of the coefficients a_i as possible through judicious applications of adjoint maps to $\mathbf{v}$.

Suppose first that $a_4 \neq 0$. Scaling $\mathbf{v}$ if necessary, we can assume that $a_4 = 1$. Referring to table (3.24), if we act on such a $\mathbf{v}$ by $\operatorname{Ad}(\exp(-\frac{1}{2}a_3\mathbf{v}_3))$, we can make the coefficient of $\mathbf{v}_3$ vanish:

$$\mathbf{v}' = \operatorname{Ad}(\exp(-\tfrac{1}{2}a_3\mathbf{v}_3))\mathbf{v} = a_1'\mathbf{v}_1 + a_2'\mathbf{v}_2 + \mathbf{v}_4$$

for certain scalars a_1', a_2' depending on a_1, a_2, a_3. Next we act on $\mathbf{v}'$ by $\operatorname{Ad}(\exp(\frac{1}{3}a_2'\mathbf{v}_2))$ to cancel the coefficient of $\mathbf{v}_2$, leading to $\mathbf{v}'' = a_1'\mathbf{v}_1 + \mathbf{v}_4$, and finally by $\operatorname{Ad}(\exp(a_1'\mathbf{v}_1))$ to cancel the remaining coefficient, so that $\mathbf{v}$ is equivalent to $\mathbf{v}_4$ under the adjoint representation. In other words, every one-dimensional subalgebra generated by a $\mathbf{v}$ with $a_4 \neq 0$ is equivalent to the subalgebra spanned by $\mathbf{v}_4$.

The remaining one-dimensional subalgebras are spanned by vectors of the above form with $a_4 = 0$. If $a_3 \neq 0$, we scale to make $a_3 = 1$, and then act on $\mathbf{v}$ by $\operatorname{Ad}(\exp(a_1\mathbf{v}_2))$, so that $\mathbf{v}$ is equivalent to $\mathbf{v}' = a_2'\mathbf{v}_2 + \mathbf{v}_3$ for some a_2'. We can further act on $\mathbf{v}'$ by the group generated by $\mathbf{v}_4$; this has the net effect of scaling the coefficients of $\mathbf{v}_2$ and $\mathbf{v}_3$:

$$\mathbf{v}'' = \mathrm{Ad}(\exp(\varepsilon\mathbf{v}_4))\mathbf{v}' = a_2' e^{3\varepsilon}\mathbf{v}_2 + e^{-2\varepsilon}\mathbf{v}_3.$$

This is a scalar multiple of $\mathbf{v}''' = a_2' e^{5\varepsilon}\mathbf{v}_2 + \mathbf{v}_3$, so, depending on the sign of a_2', we can make the coefficient of $\mathbf{v}_2$ either $+1$, -1 or 0. Thus any one-dimensional subalgebra spanned by $\mathbf{v}$ with $a_4 = 0$, $a_3 \neq 0$ is equivalent to one spanned by either $\mathbf{v}_3 + \mathbf{v}_2$, $\mathbf{v}_3 - \mathbf{v}_2$ or $\mathbf{v}_3$. The remaining cases, $a_3 = a_4 = 0$, are similarly seen to be equivalent either to $\mathbf{v}_2$ $(a_2 \neq 0)$ or to $\mathbf{v}_1$ $(a_2 = a_3 = a_4 = 0)$. The reader can check that no further simplifications are possible.

Recapitulating, we have found an optimal system of one-dimensional subalgebras to be those spanned by

$$\begin{aligned}
&\text{(a)} && \mathbf{v}_4 = x\partial_x + 3t\partial_t - 2u\partial_u, \\
&\text{(b}_1\text{)} && \mathbf{v}_3 + \mathbf{v}_2 = t\partial_x + \partial_t + \partial_u, \\
&\text{(b}_2\text{)} && \mathbf{v}_3 - \mathbf{v}_2 = t\partial_x - \partial_t + \partial_u, \\
&\text{(b}_3\text{)} && \mathbf{v}_3 = t\partial_x + \partial_u, \\
&\text{(c)} && \mathbf{v}_2 = \partial_t, \\
&\text{(d)} && \mathbf{v}_1 = \partial_x,
\end{aligned} \tag{3.25}$$

This list can be reduced slightly if we admit the discrete symmetry $(x, t, u) \mapsto (-x, -t, u)$, not in the connected component of the identity of the full symmetry group, which maps $\mathbf{v}_3 - \mathbf{v}_2$ to $\mathbf{v}_3 + \mathbf{v}_2$, thereby reducing the number of inequivalent subalgebras to five.

Example 3.13. Consider the six-dimensional symmetry algebra $\mathfrak{g}$ of the heat equation (2.55), which is generated by the vector fields

$$\mathbf{v}_1 = \partial_x, \qquad \mathbf{v}_2 = \partial_t, \qquad \mathbf{v}_3 = u\partial_u, \qquad \mathbf{v}_4 = x\partial_x + 2t\partial_t,$$
$$\mathbf{v}_5 = 2t\partial_x - xu\partial_u, \qquad \mathbf{v}_6 = 4tx\partial_x + 4t^2\partial_t - (x^2 + 2t)u\partial_u.$$

(For the moment we are ignoring the trivial infinite-dimensional subalgebras coming from the linearity of the heat equation.) From the commutator table for this algebra, we obtain the following table:

Ad	$\mathbf{v}_1$	$\mathbf{v}_2$	$\mathbf{v}_3$
$\mathbf{v}_1$	$\mathbf{v}_1$	$\mathbf{v}_2$	$\mathbf{v}_3$
$\mathbf{v}_2$	$\mathbf{v}_1$	$\mathbf{v}_2$	$\mathbf{v}_3$
$\mathbf{v}_3$	$\mathbf{v}_1$	$\mathbf{v}_2$	$\mathbf{v}_3$
$\mathbf{v}_4$	$e^{\varepsilon}\mathbf{v}_1$	$e^{2\varepsilon}\mathbf{v}_2$	$\mathbf{v}_3$
$\mathbf{v}_5$	$\mathbf{v}_1 - \varepsilon\mathbf{v}_3$	$\mathbf{v}_2 + 2\varepsilon\mathbf{v}_1 - \varepsilon^2\mathbf{v}_3$	$\mathbf{v}_3$
$\mathbf{v}_6$	$\mathbf{v}_1 + 2\varepsilon\mathbf{v}_5$	$\mathbf{v}_2 - 2\varepsilon\mathbf{v}_3 + 4\varepsilon\mathbf{v}_4 + 4\varepsilon^2\mathbf{v}_6$	$\mathbf{v}_3$

Ad	$\mathbf{v}_4$	$\mathbf{v}_5$	$\mathbf{v}_6$
$\mathbf{v}_1$	$\mathbf{v}_4 - \varepsilon\mathbf{v}_1$	$\mathbf{v}_5 + \varepsilon\mathbf{v}_3$	$\mathbf{v}_6 - 2\varepsilon\mathbf{v}_5 - \varepsilon^2\mathbf{v}_3$
$\mathbf{v}_2$	$\mathbf{v}_4 - 2\varepsilon\mathbf{v}_2$	$\mathbf{v}_5 - 2\varepsilon\mathbf{v}_1$	$\mathbf{v}_6 - 4\varepsilon\mathbf{v}_4 + 2\varepsilon\mathbf{v}_3 + 4\varepsilon^2\mathbf{v}_2$
$\mathbf{v}_3$	$\mathbf{v}_4$	$\mathbf{v}_5$	$\mathbf{v}_6$
$\mathbf{v}_4$	$\mathbf{v}_4$	$e^{-\varepsilon}\mathbf{v}_5$	$e^{-2\varepsilon}\mathbf{v}_6$
$\mathbf{v}_5$	$\mathbf{v}_4 + \varepsilon\mathbf{v}_5$	$\mathbf{v}_5$	$\mathbf{v}_6$
$\mathbf{v}_6$	$\mathbf{v}_4 + 2\varepsilon\mathbf{v}_6$	$\mathbf{v}_5$	$\mathbf{v}_6$

where the (i, j)-th entry gives $\mathrm{Ad}(\exp(\varepsilon\mathbf{v}_i))\mathbf{v}_j$.

Let $\mathbf{v} = a_1\mathbf{v}_1 + a_2\mathbf{v}_2 + \cdots + a_6\mathbf{v}_6$ be an element of $\mathfrak{g}$, which we shall try to simplify using suitable adjoint maps. A key observation here is that the function $\eta(\mathbf{v}) = (a_4)^2 - 4a_2a_6$ is an invariant of the full adjoint action: $\eta(\mathrm{Ad}\, g(\mathbf{v})) = \eta(\mathbf{v})$, $\mathbf{v} \in \mathfrak{g}$, $g \in G$. The detection of such an invariant is important since it places restrictions on how far we can expect to simplify $\mathbf{v}$. For example, if $\eta(\mathbf{v}) \neq 0$, then we cannot simultaneously make a_2, a_4 and a_6 all zero through adjoint maps; if $\eta(\mathbf{v}) < 0$ we cannot make either a_2 or a_6 zero!

To begin the classification process, we concentrate on the a_2, a_4, a_6 coefficients of $\mathbf{v}$. If $\mathbf{v}$ is as above, then

$$\tilde{\mathbf{v}} = \sum_{i=1}^{6} \tilde{a}_i\mathbf{v}_i = \mathrm{Ad}(\exp(\alpha\mathbf{v}_6)) \circ \mathrm{Ad}(\exp(\beta\mathbf{v}_2))\mathbf{v}$$

has coefficients

$$\begin{aligned}
\tilde{a}_2 &= a_2 - 2\beta a_4 + 4\beta^2 a_6,\\
\tilde{a}_4 &= 4\alpha a_2 + (1 - 8\alpha\beta)a_4 - 4\beta(1 - 4\alpha\beta)a_6,\\
\tilde{a}_6 &= 4\alpha^2 a_2 + 2\alpha(1 - 4\alpha\beta)a_4 + (1 - 4\alpha\beta)^2 a_6.
\end{aligned} \tag{3.26}$$

There are now three cases, depending on the sign of the invariant η:

Case 1. If $\eta(\mathbf{v}) > 0$, then we choose β to be either real root of the quadratic equation $4a_6\beta^2 - 2a_4\beta + a_2 = 0$, and $\alpha = a_6/(8\beta a_6 - 2a_4)$ (which is always well defined). Then $\tilde{a}_2 = \tilde{a}_6 = 0$, while $\tilde{a}_4 = \sqrt{\eta(\mathbf{v})} \neq 0$, so $\mathbf{v}$ is equivalent to a multiple of $\tilde{\mathbf{v}} = \mathbf{v}_4 + \tilde{a}_1\mathbf{v}_1 + \tilde{a}_3\mathbf{v}_3 + \tilde{a}_5\mathbf{v}_5$. Acting further by adjoint maps generated respectively by $\mathbf{v}_5$ and $\mathbf{v}_1$ we can arrange that the coefficients of $\mathbf{v}_5$ and $\mathbf{v}_1$ in $\tilde{\mathbf{v}}$ vanish. Therefore, every element with $\eta(\mathbf{v}) > 0$ is equivalent to a multiple of $\mathbf{v}_4 + a\mathbf{v}_3$ for some $a \in \mathbb{R}$. No further simplifications are possible.

Case 2. If $\eta(\mathbf{v}) < 0$, set $\beta = 0$, $\alpha = -a_4/4a_2$ to make $\tilde{a}_4 = 0$. Acting on $\mathbf{v}$ by the group generated by $\mathbf{v}_4$, we can make the coefficients of $\mathbf{v}_2$ and $\mathbf{v}_6$ agree, so $\mathbf{v}$ is equivalent to a scalar multiple of $\tilde{\mathbf{v}} = (\mathbf{v}_2 + \mathbf{v}_6) + \tilde{a}_1\mathbf{v}_1 + \tilde{a}_3\mathbf{v}_3 + \tilde{a}_5\mathbf{v}_5$. Further use of the groups generated by $\mathbf{v}_1$ and $\mathbf{v}_5$ show that $\tilde{\mathbf{v}}$ is equivalent to a scalar multiple of $\mathbf{v}_2 + \mathbf{v}_6 + a\mathbf{v}_3$ for some $a \in \mathbb{R}$.

Case 3. If $\eta(\mathbf{v}) = 0$, there are two subcases. If not all of the coefficients a_2, a_4, a_6 vanish, then we can choose α and β in (3.26) so that $\tilde{a}_2 \neq 0$, but $\tilde{a}_4 = \tilde{a}_6 = 0$, so $\mathbf{v}$ is equivalent to a multiple of $\tilde{\mathbf{v}} = \mathbf{v}_2 + \tilde{a}_1\mathbf{v}_1 + \tilde{a}_3\mathbf{v}_3 + \tilde{a}_5\mathbf{v}_5$. Suppose $\tilde{a}_5 \neq 0$. Then we can make the coefficients of $\mathbf{v}_1$ and $\mathbf{v}_3$ zero using the groups generated by $\mathbf{v}_1$ and $\mathbf{v}_2$, while the group generated by $\mathbf{v}_4$ independently scales the coefficients of $\mathbf{v}_2$ and $\mathbf{v}_5$. Thus such a $\mathbf{v}$ is equivalent to a multiple of either $\mathbf{v}_2 + \mathbf{v}_5$ or $\mathbf{v}_2 - \mathbf{v}_5$. If, on the other hand, $\tilde{a}_5 = 0$, then the group generated by $\mathbf{v}_5$ can be used to reduce $\mathbf{v}$ to a vector of the form $\mathbf{v}_2 + a\mathbf{v}_3$, $a \in \mathbb{R}$.

The last remaining case occurs when $a_2 = a_4 = a_6 = 0$, for which our earlier simplifications were unnecessary. If $a_1 \neq 0$, then using groups generated by $\mathbf{v}_5$ and $\mathbf{v}_6$ we can arrange $\mathbf{v}$ to become a multiple of $\mathbf{v}_1$. If $a_1 = 0$, but $a_5 \neq 0$, we can first act by any map $\mathrm{Ad}(\exp(\varepsilon\mathbf{v}_2))$ to get a nonzero coefficient in front of $\mathbf{v}_1$, reducing to the previous case. The only remaining vectors are the multiples of $\mathbf{v}_3$, on which the adjoint representation acts trivially.

In summary, an optimal system of one-dimensional subalgebras of the heat algebra is provided by those generated by

$$\begin{array}{llll}
\text{(a)} & \mathbf{v}_4 + a\mathbf{v}_3, & \eta > 0, & a \in \mathbb{R}, \\
\text{(b)} & \mathbf{v}_2 + \mathbf{v}_6 + a\mathbf{v}_3, & \eta < 0, & a \in \mathbb{R}, \\
\text{(c1)} & \mathbf{v}_2 - \mathbf{v}_5, & \eta = 0, & \\
\text{(c2)} & \mathbf{v}_2 + \mathbf{v}_5, & \eta = 0, & \\
\text{(d)} & \mathbf{v}_2 + a\mathbf{v}_3, & \eta = 0, & a \in \mathbb{R}, \\
\text{(e)} & \mathbf{v}_1, & \eta = 0, & \\
\text{(f)} & \mathbf{v}_3, & \eta = 0. &
\end{array} \tag{3.27}$$

Again, the discrete symmetry $(x, t, u) \mapsto (-x, t, u)$ will map $\mathbf{v}_2 - \mathbf{v}_5$ to $\mathbf{v}_2 + \mathbf{v}_5$, and the list is reduced by one.

Inclusion of the additional infinite-dimensional symmetry algebra $\{\mathbf{v}_\alpha = \alpha(x, t)\partial_u\}$, α a solution to the heat equation, does not essentially alter this classification. If $\mathbf{v} + \mathbf{v}_\alpha$ is in this larger algebra, with $\mathbf{v} \neq 0$ in the above six-dimensional heat algebra, then we can always find $\mathbf{v}_\beta = \beta(x, t)\partial_u$ such that $\mathrm{Ad}(\exp(\mathbf{v}_\beta))(\mathbf{v} + \mathbf{v}_\alpha) = \mathbf{v}$. For instance, if $\mathbf{v} = \mathbf{v}_1 = \partial_x$, then

$$\beta(x, t) = -\int_0^x \alpha(y, t)\, dy - \int_0^t \alpha_x(0, s)\, ds$$

will do. (The reader should check that β is a solution to the heat equation.) The only remaining vectors not equivalent to ones in the six-dimensional algebra are thus those of the form $\mathbf{v}_\alpha$ only. We will not attempt to classify these vectors as they do *not* lead to group-invariant solutions to the heat equation.

Once we have classified one-dimensional subalgebras of a Lie algebra, we can go on to find optimal systems of higher dimensional subalgebras. Lack of space precludes us from pursuing this interesting problem any further here, so we refer the reader to Ovsiannikov, [3; § 14.8], for some of the techniques available.

Classification of Group-Invariant Solutions

Definition 3.14. An *optimal system* of s-parameter group-invariant solutions to a system of differential equations is a collection of solutions $u = f(x)$ with the following properties:

(i) Each solution in the list is invariant under some s-parameter symmetry group of the system of differential equations.
(ii) If $u = \tilde{f}(x)$ is any other solution invariant under an s-parameter symmetry group, then there is a further symmetry g of the system which maps $\tilde{f}$ to a solution $f = g \cdot \tilde{f}$ on the list.

Proposition 3.15. *Let G be the full symmetry group of a system of partial differential equations Δ. Let $\{H_\alpha\}$ be an optimal system of s-parameter subgroups of G. Then the collecton of all H_α-invariant solutions, for H_α in the optimal system, forms an optimal system of s-parameter group-invariant solutions to Δ.*

The proof is immediate from Proposition 3.6. Moreover, our earlier classification of subalgebras is now directly applicable to the classification of group-invariant solutions.

Example 3.16. For the Korteweg–de Vries equation, we've already done all the work to provide a complete list of invariant solutions in our earlier treatment of Example 3.4. Indeed, according to our optimal system of one-dimensional subalgebras (3.25) of the full symmetry algebra, we need only find group-invariant solutions for the one-parameter subgroups generated by: (a) $\mathbf{v}_4$—scaling; (b) $\mathbf{v}_3 + \mathbf{v}_2$—modified Galilean boosts; (c) $\mathbf{v}_3$—Galilean boosts; (d) $\mathbf{v}_2$—time translations; and (e) $\mathbf{v}_1$—space translations. All of these except the last were determined in Example 3.4, to which we refer to reader. The space translationally-invariant solutions are all constant, and hence trivially appear among the other solutions. Any other group-invariant solution of the Korteweg–de Vries equation can thus be found by transforming one of the solutions of Example 3.4 by an appropriate group element.

For example, the travelling wave solutions, which correspond to the symmetry group generated by $\mathbf{v}_2 + c\mathbf{v}_1 = \partial_t + c\partial_x$ can be recovered from the stationary solutions $u = f(x)$, invariant under the group generated by $\mathbf{v}_2 = \partial_t$. Referring to table (3.24), we see that

$$\operatorname{Ad}(\exp(c\mathbf{v}_3))\mathbf{v}_2 = \mathbf{v}_2 + c\mathbf{v}_1,$$

where $\mathbf{v}_3 = t\partial_x + \partial_u$ generates the one-parameter Galilean symmetry group for the Korteweg–de Vries equation. According to Proposition 3.6, if $u = f(x)$ is any stationary solution, then $\tilde{f} = \exp(c\mathbf{v}_3)f$ will be a travelling wave solution with velocity c. Using the formulas in Example 2.64, we see that

$$\tilde{f}(x, t) = f(x - ct) + c,$$

where $f(x)$ is any elliptic functon satisfying

$$f''' + ff' = 0.$$

In particular, if

$$u = f_0(x) = 3c\,\mathrm{sech}^2(\tfrac{1}{2}\sqrt{c}x + \delta) - c,$$

which, as the reader can check, is a stationary solution to the Korteweg–de Vries equation for any $c > 0$, we recover the one-soliton solution with velocity c.

Example 3.17. Finally, we look at the classification of the group-invariant solutions to the heat equation $u_t = u_{xx}$. The construction of the group-invariant solutions for each of the one-dimensional subgroups in the optimal system (3.27) proceeds in the same fashion as in Example 3.3, and we merely list the results.

(a)
$$\mathbf{v}_4 + a\mathbf{v}_3 = x\partial_x + 2t\partial_t + 2au\partial_u.$$

Invariants are $y = x/\sqrt{t}$, $v = t^{-a}u$; the reduced equation is $v_{yy} + \frac{1}{2}yv_y - av = 0$, and the invariant solutions are our earlier parabolic cylinder solutions

$$u(x, t) = t^a e^{-x^2/8t}\left\{k\mathrm{U}\left(2a + \tfrac{1}{2}, \frac{x}{\sqrt{2t}}\right) + \tilde{k}\mathrm{V}\left(2a + \tfrac{1}{2}, \frac{x}{\sqrt{2t}}\right)\right\}.$$

(b)
$$\mathbf{v}_2 + \mathbf{v}_6 + a\mathbf{v}_3 = 4tx\partial_x + (4t^2 + 1)\partial_t - (x^2 + 2t - a)u\partial_u.$$

Invariants are

$$y = (4t^2 + 1)^{-1/2}x, \qquad v = (4t^2 + 1)^{1/4}u \cdot \exp\left\{(4t^2 + 1)^{-1}tx^2 + \frac{a}{2}\arctan(2t)\right\}.$$

The reduced equation is

$$v_{yy} + (a + y^2)v = 0.$$

Invariant solutions are expressed in terms of parabolic cylinder functions with imaginary arguments (Abramowitz and Stegun, [1; § 19.17])

$$u(x, t) = (4t^2 + 1)^{-1/4}\left\{k\mathrm{W}\left(-\frac{a}{2}, \frac{x}{\sqrt{8t^2 + 2}}\right) + \tilde{k}\mathrm{W}\left(-\frac{a}{2}, \frac{-x}{\sqrt{8t^2 + 2}}\right)\right\}\exp\left\{\frac{-tx^2}{4t^2 + 1} - \frac{a}{2}\arctan(2t)\right\}.$$

(c)
$$\mathbf{v}_2 - \mathbf{v}_5 = \partial_t - 2t\partial_x + xu\partial_u.$$

Invariants are:

$$y = x + t^2, \qquad v = u\exp(-xt - \tfrac{2}{3}t^3).$$

The reduced equation is Airy's equation

$$v_{yy} = yv.$$

Solutions are written in terms of Airy functions

$$u(x, t) = \{k \operatorname{Ai}(x + t^2) + \tilde{k} \operatorname{Bi}(x + t^2)\} \exp(xt + \tfrac{2}{3}t^3).$$

The corresponding invariant solutions for $\mathbf{v}_2 + \mathbf{v}_5$ are obtained by replacing x by $-x$.

(d) $$\mathbf{v}_2 + a\mathbf{v}_3 = \partial_t + au\partial_u.$$

Invariants are x, $v = e^{-at}u$, and the reduced equation $v_{xx} = av$ leads to the solutions

$$u(x, t) = \begin{cases} k e^{at} \cosh(\sqrt{a}x + \delta), & a > 0, \\ kx + \tilde{k}, & a = 0, \\ k e^{at} \cos(\sqrt{-a}x + \delta), & a < 0. \end{cases}$$

For the two remaining subalgebras, that generated by $\mathbf{v}_1$ has only constants for its invariant solutions, which already appear in (d), and that generated by $\mathbf{v}_3$ has no invariant solutions. Thus the above solutions constitute an optimal system of group-invariant solutions to the heat equation whereby any other group-invariant solution can be found by transforming one of these solutions by a suitable group element—see page 120.

In our previous encounter with this problem, Example 3.3, we determined group invariant solutions for a couple of subgroups not appearing in the optimal system (3.27). By the general theory, these solutions can be derived from the above solutions by suitable group transformations. For example, since

$$\operatorname{Ad}[\exp(-\tfrac{1}{2}c\mathbf{v}_5)](\mathbf{v}_2 + c\mathbf{v}_1) = \mathbf{v}_2 + \tfrac{1}{4}c^2\mathbf{v}_3,$$

we could have found the travelling wave solutions by transforming the solutions invariant under $\mathbf{v}_2 + a\mathbf{v}_3$, $a = c^2/4$, by the Galilean boost $\exp(\frac{1}{2}c\mathbf{v}_5)$, and indeed $u = k e^{\sqrt{a}x + at} + \tilde{k} e^{-\sqrt{a}x + at}$ gets changed into the travelling wave solution $u = k + \tilde{k} e^{-c(x - ct)}$ when $a = c^2/4$.

3.4. Quotient Manifolds

In order to provide a rigorous formulation of the basic method for finding group-invariant solutions of systems of differential equations outlined in Section 3.1, we need to gain a better understanding of the geometry underlying these constructions. The concept of the quotient manifold of a smooth manifold under a regular group of transformations will provide the natural setting for all group-invariant objects. Ultimately, we will see how the reduced system of differential equations for the group-invariant solutions naturally lives on the quotient manifold. We begin by discussing this quotient manifold in general.

Let G be a local group of transformations acting on a smooth manifold M. There is an induced equivalence relation among the points of M, with x being equivalent to y if they lie in the same orbit of G. Let M/G denote the set of equivalence classes, or, equivalently, the set of orbits of G. The projection $\pi: M \to M/G$ associates to each x in M its equivalence class $\pi(x) \in M/G$, which can be identified with the orbit of G passing through x. In particular, $\pi(g \cdot x) = \pi(x)$ for any $g \in G$ such that $g \cdot x$ is defined. Conversely, given a point $w \in M/G$, $\pi^{-1}\{w\}$ will be the orbit determined by w, realized as a subset of M. The quotient space M/G has a natural topology obtained by requiring that the projection $\pi[U]$ of an open subset $U \subset M$ is open in M/G.

In general, the quotient space M/G will be an extremely complicated topological space with no readily comprehensible structure. However, if we further require G to act *regularly* on M, then we can endow M/G with the structure of a smooth manifold. If M is an m-dimensional manifold and G has s-dimensional orbits, then the quotient manifold M/G will be of dimension $m - s$.† Thus the quotient manifold construction has the effect of reducing the dimension by s, the dimension of the orbits of G.

Once we have constructed the quotient manifold, the general philosophy is that any object on M which is invariant under the action of G will have a natural counterpart on the lower-dimensional quotient manifold M/G whose properties completely characterize the original object on M. As a first example, consider a G-invariant function $F: M \to \mathbb{R}^l$. Since $F(g \cdot x) = F(x)$ whenever $g \cdot x$ is defined, F is constant along the orbits of G. Therefore there is a well-defined function $\tilde{F} = F/G: M/G \to \mathbb{R}^l$ such that $\tilde{F}(\pi(x)) = F(x)$ whenever $x \in M$. Conversely, if $\tilde{F}: M/G \to \mathbb{R}^l$ then the function $F: M \to \mathbb{R}^l$ given by $F(x) = \tilde{F}(\pi(x))$, $x \in M$ is clearly a G-invariant function on M. There is thus a one-to-one correspondence between G-invariant functions on M and arbitrary functions on M/G. Note further that in any local coordinate chart, the functions defined on M/G depend on s fewer variables than their counterparts on M merely because we have reduced the dimension of the underlying manifold by s. Thus projection to the quotient manifold has the net effect of reducing the number of degrees of freedom by s, the dimension of the orbits of the group action.

Theorem 3.18. *Let M be a smooth m-dimensional manifold. Suppose G is a local group of transformations which acts regularly on M with s-dimensional orbits. Then there exists a smooth $(m - s)$-dimensional manifold, called the* quotient manifold *of M by G and denoted M/G, together with a projection $\pi: M \to M/G$, which satisfy the following properties.*

(a) *The projection π is a smooth map between manifolds.*
(b) *The points x and y lie in the same orbit of G in M if and only if $\pi(x) = \pi(y)$.*

† It may, however, not be a Hausdorff manifold; see the subsequent discussion.

(c) *If* $\mathfrak{g}$ *denotes the Lie algebra of infinitesimal generators of the action of* G, *then the linear map*

$$d\pi\colon TM|_x \to T(M/G)|_{\pi(x)}$$

is onto, with kernel $\mathfrak{g}|_x = \{\mathbf{v}|_x\colon \mathbf{v} \in \mathfrak{g}\}$.

PROOF. As above, M/G is simply the set of all orbits of G on M. Coordinate charts on M/G are constructed using the flat local coordinate charts on M provided by Frobenius' Theorem 1.43 using the regularity of G. The local coordinates $y_\alpha = (y_\alpha^1, \ldots, y_\alpha^m)$ on such a chart U_α are such that each orbit intersects U_α in at most one slice $\mathcal{O} \cap U_\alpha = \{y_\alpha^1 = c_\alpha^1, \ldots, y_\alpha^{m-s} = c_\alpha^{m-s}\}$, the constants $c_\alpha^1, \ldots, c_\alpha^{m-s}$ uniquely determining the orbit $\mathcal{O}$. The corresponding coordinate chart V_α on M/G is defined as the set of all orbits with nonempty intersection with U_α, so

$$V_\alpha = \{w \in M/G\colon \pi^{-1}\{w\} \cap U_\alpha \neq \varnothing\}.$$

Local coordinates on V_α are determined by the slice coordinates $y_\alpha^1, \ldots, y_\alpha^{m-s}$; in other words the coordinate map $\tilde{\chi}_\alpha\colon V_\alpha \to \mathbb{R}^{m-s}$ is defined so that $\tilde{\chi}_\alpha(w) = (c_\alpha^1, \ldots, c_\alpha^{m-s})$ when $\pi^{-1}\{w\}$ intersects U_α in the slice determined by $y_\alpha^1 = c_\alpha^1, \ldots, y_\alpha^{m-s} = c_\alpha^{m-s}$. Clearly the projection $\pi\colon M \to M/G$ is smooth in these coordinates since $\pi(y_\alpha^1, \ldots, y_\alpha^m) = (y_\alpha^1, \ldots, y_\alpha^{m-s})$ for $y_\alpha \in U_\alpha$. Furthermore,

$$d\pi\left[\frac{\partial}{\partial y_\alpha^i}\right] = \begin{cases} \partial/\partial y_\alpha^i, & i = 1, \ldots, m - s, \\ 0, & i = m - s + 1, \ldots, m, \end{cases}$$

so $d\pi\colon TM|_{y_\alpha} \to T(M/G)|_{\pi(y_\alpha)}$ is onto, with kernel spanned by $\partial/\partial y_\alpha^{m-s+1}, \ldots, \partial/\partial y_\alpha^m$, which is the same as the span of the infinitesimal generators $\mathfrak{g}$ of G at y_α.

The only remaining point is to prove that the overlap functions $\tilde{\chi}_\beta \circ \tilde{\chi}_\alpha^{-1}$ are smooth on the intersection $V_\alpha \cap V_\beta$ of two local coordinate charts on M/G. This is more or less clear if the corresponding flat coordinate charts U_α and U_β are sufficiently small, and intersect on M, but this latter possibility need not occur. However, a fairly straightforward argument based on the connectivity of the orbits of G can be applied here, and the result holds; the details are given in Palais, [1]. □

In order to see a little more clearly what the local coordinates on the quotient manifold mean, consider some general local coordinates $x = (x^1, \ldots, x^m)$ on M. Theorem 2.17 shows that by possibly shrinking the coordinate chart, we can find a complete set of functionally independent invariants $\eta^1(x), \ldots, \eta^{m-s}(x)$, with each orbit intersecting the chart in at most one connected component, which is a level set $\{\eta^1(x) = c_1, \ldots, \eta^{m-s}(x) = c_{m-s}\}$. The constants $c_1, \ldots, c_{m-s}$ uniquely determine the orbit, then, and hence can be chosen as new local coordinates on the quotient manifold M/G, which agree with a set of flat coordinates used in the proof of the theorem. Thus *local coordinates on the quotient manifold* M/G *are provided by a*

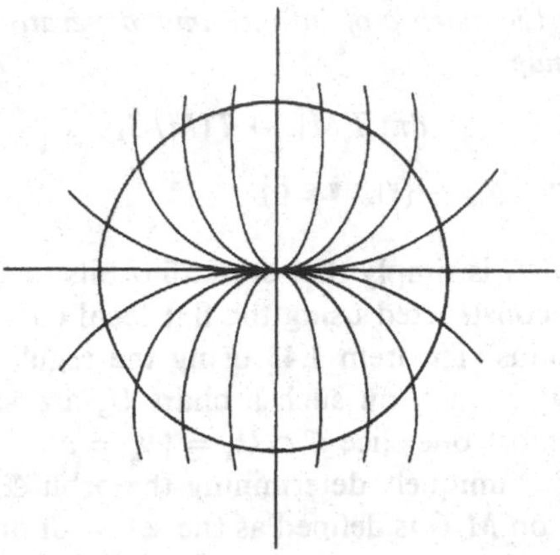

Figure 7. Quotient manifold for $\mathbb{R}^2/G^2$.

complete set of functionally independent invariants for the group action:

$$y^1 = \eta^1(x), \ldots, y^{m-s} = \eta^{m-s}(x).$$

Example 3.19. Consider the group of scale transformations

$$G^2\colon (x, y) \mapsto (\lambda x, \lambda^2 y), \qquad \lambda > 0.$$

The action is regular on $M = \mathbb{R}^2 \backslash \{0\}$, the orbits being semi-parabolas $y = kx^2$ for $x > 0$ or $x < 0$ and the positive and negative y-axis. Since each orbit is uniquely determined by its point of intersection with the unit circle $S^1 = \{x^2 + y^2 = 1\}$, we can identify M/G^2 with S^1. A local coordinate on M/G^2 is provided by the group invariant y/x^2 for $x > 0$ or $x < 0$, or by x^2/y if $y > 0$ or $y < 0$, giving four overlapping coordinate charts on M/G. (A better choice of coordinate on M/G is perhaps given by the multiply-valued "angular" invariant $\theta = \arctan(y/x^2)$.) Clearly, no global coordinate chart valid for all nonzero (x, y) can be found in this case.

The same construction works for any of the two-dimensional scaling groups $G^\alpha\colon (x, y) \mapsto (\lambda x, \lambda^\alpha y)$ provided $\alpha > 0$, the "angular" invariant $\theta = \arctan(y/x^\alpha)$ providing the identification $M/G^\alpha \simeq S^1$. (The case $\alpha < 0$ is discussed in Exercise 3.14.)

As remarked earlier, one technical difficulty that can arise is that the quotient manifold M/G may not satisfy the Hausdorff separation property, and so we are naturally led to consider a more general notion of manifold than is usual. In other words, although M/G will always satisfy the requirements (a) and (b) in Definition 1.1, there may exist distinct points y and $\tilde{y}$ in M/G which cannot be "separated" by open neighbourhoods, i.e. if U is *any* neighbourhood of y and $\tilde{U}$ *any* neighbourhood of $\tilde{y}$, then $U \cap \tilde{U} \neq \varnothing$. One can develop the

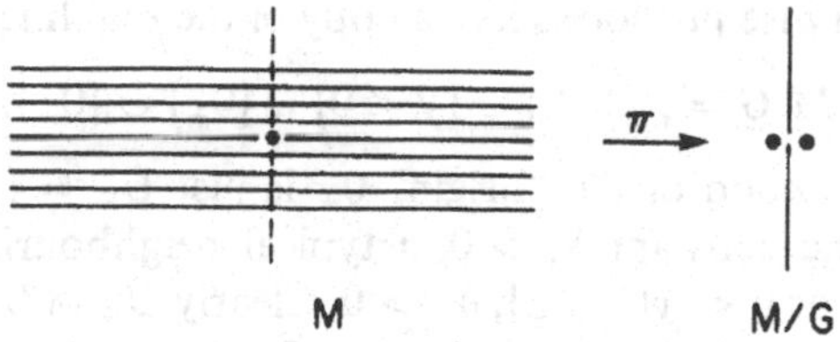

Figure 8. Non-Hausdorff quotient manifold.

entire theory of manifolds omitting the Hausdorff axiom, in which case, as shown by Palais, [1], the quotient manifold construction keeps one in the same "category". Alternatively, the approach most often taken in practice is to remove non-Hausdorff "singularities" in M/G by restricting attention to a suitably small open submanifold $\tilde{M}$ of the original manifold M with $\tilde{M}/G \subset M/G$ an open, Hausdorff submanifold. For instance, $\tilde{M}$ might be a coordinate chart on which we construct a complete set of functionally independent invariants, in which case $\tilde{M}/G$ will have global coordinates provided by these invariants, and so can be realized as an open subset of the Euclidean space $\mathbb{R}^{m-s}$.

Example 3.20. Consider the vector field $\mathbf{v} = (x^2 + y^2)\partial_x$ on $\mathbb{R}^2$. The one-parameter group G generated by $\mathbf{v}$ takes the form $(\tilde{x}, \tilde{y}) = \exp(\varepsilon\mathbf{v})(x, y)$, where $\tilde{y} = y$ and

$$\tilde{x} = \begin{cases} y\tan(\varepsilon y + \arctan(x/y)), & y \neq 0, \\ x/(1 - \varepsilon x), & y = 0. \end{cases}$$

The orbits of G consist of

(a) The origin $(0, 0)$,
(b) The horizontal lines $\{y = c\}$ with $c \neq 0$,
(c) The positive x-axis $\{y = 0, x > 0\}$,
(d) The negative x-axis $\{y = 0, x < 0\}$.

Thus G acts regularly on $M = \mathbb{R}^2\backslash\{0\}$. The quotient manifold M/G is one-dimensional and looks like a copy of the real line but with two "infinitely close" origins! Indeed, the single invariant of G is the coordinate y, each orbit not on the x-axis being uniquely determined by its vertical displacement. Thus the point in M/G determined by the horizontal lines (b) are coordinatized by $\pi(x, y) = y$, $x \neq 0$, so we can identify this part of M/G with the positive and negative real axes, corresponding to the images of the upper and lower half-planes respectively under the projection $\pi\colon M \to M/G$. However, there must be *two* points in M/G corresponding to the positive and negative x-axes in $\mathbb{R}^2$; we denote these by 0_+ and 0_- respectively, which can be viewed as two distinct, but infinitely close, origins of the quotient manifold

M/G, which otherwise just looks like a copy of the real line:

$$M/G = \{y > 0\} \cup \{y < 0\} \cup \{0_+\} \cup \{0_-\}.$$

A basic neighbourhood of the "origin" 0_+ is just $U_+ = \{y: y = 0_+ \text{ or } 0 < |y| < \delta_+\}$ for some constant $\delta_+ > 0$, a typical neighbourhood of 0_- being $U_- = \{y: y = 0_- \text{ or } 0 < |y| < \delta_-\}$, $\delta_- > 0$. Clearly $U_+ \cap U_- \neq \varnothing$ no matter what δ_+ and δ_- are, so the points 0_+ and 0_- do not satisfy the Hausdorff separation property. (A more physically relevant example of a non-Hausdorff quotient space is provided by the scaling groups G^α of Example 3.19 when $\alpha < 0$; see Exercise 3.14.)

Proposition 3.21. *Let G act regularly on the manifold M with s-dimensional orbits.*

(a) *A smooth function* $F: M \to \mathbb{R}^l$ *is G-invariant if and only if there is a smooth function* $\tilde{F} = F/G: M/G \to \mathbb{R}^l$ *such that* $\tilde{F}(x) = F[\pi(x)]$ *for all* $x \in M$.
(b) *A smooth n-dimensional submanifold* $N \subset M$ *is G-invariant if and only if there is a smooth* $(n - s)$*-dimensional submanifold* $\tilde{N} = N/G \subset M/G$ *such that* $\tilde{N} = \pi[N]$ *and hence* $N = \pi^{-1}[\tilde{N}]$.
(c) *A subvariety* $\mathcal{S}_F = \{x: F(x) = 0\}$ *defined by a smooth function* $F: M \to \mathbb{R}^l$ *is G-invariant if and only if there is a smooth subvariety* $\mathcal{S}_{\tilde{F}} = \{y: \tilde{F}(y) = 0\}$ *defined by* $\tilde{F}: M/G \to \mathbb{R}^l$ *such that* $\mathcal{S}_{\tilde{F}} = \pi[\mathcal{S}_F]$. *(In this case, it is not necessarily true that* $\tilde{F} = F \circ \pi$ *unless F itself happens to be G-invariant.)*

This proposition is just a global restatement of Theorem 2.17 and Proposition 2.18, and we leave the details of the construction to the reader.

Dimensional Analysis

In the case of scaling groups, the preceding constructions provide an easy proof of the so-called Pi theorem, which forms the foundation of the method of dimensional analysis. In any physical problem, there are certain fundamental physical quantities, such as length, time, mass, etc., which can all be scaled independently of each other. Let $z^1, \ldots, z^r$ denote these quantities, so the group under consideration transforms according to

$$(z^1, \ldots, z^r) \mapsto (\lambda_1 z^1, \ldots, \lambda_r z^r),$$

where the scaling factors $\lambda = (\lambda_1, \ldots, \lambda_r)$ are arbitrary positive real numbers. Thus the underlying group is just the Cartesian products of r copies of the multiplicative group $\mathbb{R}^+$ of positive real numbers. There also exist certain derived physical quantities, such as velocity, force, fluid density and so on, which also scale under a rescaling of the fundamental physical units. Calling these quantities $x = (x^1, \ldots, x^m)$ and, assuming dimensional homogeneity, the action of our scaling group on the derived quantities takes the form

$$\lambda \cdot (x^1, \ldots, x^m) = (\lambda_1^{\alpha_{11}} \lambda_2^{\alpha_{21}} \cdots \lambda_r^{\alpha_{r1}} x^1, \ldots, \lambda_1^{\alpha_{1m}} \lambda_2^{\alpha_{2m}} \cdots \lambda_r^{\alpha_{rm}} x^m), \tag{3.28}$$

in which the exponents α_{ij}, $i = 1, \ldots, r$, $j = 1, \ldots, m$, are prescribed by the physical dependence of the quantities x^j on the fundamental units y^i. For example, if y^1 denotes length, y^2 time and y^3 mass, then changes in velocity v, being the ratio of length over time, and fluid density ρ, the ratio of mass to volume (or length cubed), are given by

$$\lambda \cdot v = \lambda_1 \lambda_2^{-1} v, \qquad \lambda \cdot \rho = \lambda_1^{-3} \lambda_3 \rho.$$

If a derived quantity remains unchanged under the given scalings, then it is called *dimensionless*. The first part of the Pi theorem tells how many independent dimensionless quantities exist, this being determined by the number of independent invariants of the underlying group action.

In general, certain functional relations of the form $F(x^1, \ldots, x^m) = 0$ among the derived quantities are posited. For instance, for waves in deep water, the velocity v might be determined as a function of wave length l and the gravitational acceleration g. (In this simple model, we ignore surface tension and other effects.) Such a relation is called *unit-free* if it is unchanged under a rescaling of the fundamental quantities. Such unit-free relations are often of great physical significance. The second part of the Pi theorem states that any such unit-free relation can be re-expressed solely in terms of the dimensionless combinations of the derived physical quantities. For instance, in our example, if $\lambda = (\lambda_1, \lambda_2, \lambda_3)$ represent the scalings in length, time and mass respectively, then

$$\lambda \cdot (v, l, g) = (\lambda_1 \lambda_2^{-1} v, \lambda_1 l, \lambda_1 \lambda_2^{-2} g).$$

Obviously, the only dimensionless quantity here is the *Froude number* v^2/lg, or powers thereof. Thus any unit-free relation determining wave speed as a function of wave length and gravitational acceleration must take the form

$$v = c\sqrt{lg},$$

in which only the constant c remains to be determined. We now state the Pi theorem in general.

Theorem 3.22. *Let $z^1, \ldots, z^r$ be fundamental physical quantities which scale independently according to $z^i \mapsto \lambda_i z^i$. Let $x^1, \ldots, x^m$ be the derived quantities scaling according to* (3.28) *for some $r \times m$ matrix of constants $A = (\alpha_{ij})$. Let s be the rank of A. Then there exist $m - s$ independent dimensionless "power products"*

$$\pi^k = (x^1)^{\beta_{1k}} (x^2)^{\beta_{2k}} \cdots (x^m)^{\beta_{mk}}, \qquad k = 1, \ldots, m - s, \tag{3.29}$$

with the property that any other dimensionless quantity can be written as a function of $\pi^1, \ldots, \pi^{m-s}$. If $F(x^1, \ldots, x^m) = 0$ is any unit-free relation among the given derived quantities, then there is an equivalent relation $\tilde{F}(x^1, \ldots, x^m) = 0$ which can be expressed solely in terms of the above dimensionless power products:

$$\tilde{F} = \tilde{F}(\pi^1, \ldots, \pi^{m-s}) = 0.$$

Proof. Consider the positive octant of $\mathbb{R}^m$, $M = \{x = (x^1, \ldots, x^m): x^i > 0, i = 1, \ldots, m\}$. If $G = \mathbb{R}^+ \times \cdots \times \mathbb{R}^+$ is the r-fold Cartesian product of the multiplicative group $\mathbb{R}^+$, then (3.28) determines a global action of G on M. The infinitesimal generators of this action are found by differentiating (3.28) with respect to λ_i and setting $\lambda_1 = \cdots = \lambda_r = 1$. We find

$$\mathbf{v}_i = \alpha_{i1} x^1 \frac{\partial}{\partial x^1} + \alpha_{i2} x^2 \frac{\partial}{\partial x^2} + \cdots + \alpha_{im} x^m \frac{\partial}{\partial x^m}$$

to be the generator corresponding to the i-th copy of $\mathbb{R}^+$ in G. The dimension of the span of $\mathbf{v}_1, \ldots, \mathbf{v}_r$ at $x \in M$ is clearly the same as the rank of the matrix $A = (\alpha_{ij})$, namely s, hence G has s-dimensional orbits. Global invariants for G on the entire octant M are given by power products of the form (3.29) provided $\mathbf{v}_i(\pi^k) = 0$ for $i = 1, \ldots, r$. This holds if and only if the exponents β_{jk} in (3.29) satisfy the linear system

$$\sum_{j=1}^m \alpha_{ij}\beta_{jk} = 0, \qquad i = 1, \ldots, r. \tag{3.30}$$

There are $m - s$ linearly independent solutions to this system leading to $m - s$ functionally independent power products. Moreover, the power products uniquely determine the orbits of G on M. Indeed, if $\pi^k(x) = \pi^k(\tilde{x})$ for all k, set $x^j = e^{t_j}\tilde{x}^j$ for $j = 1, \ldots, m$. The exponents t_j satisfy the linear system $\sum_j t_j \beta_{jk} = 0$ for $k = 1, \ldots, r$. Since we have constructed a basis for the null space of A, this is true if and only if there exist real numbers $s_1, \ldots, s_r$ such that $t_j = \sum_i s_i \alpha_{ij}$. But then $x = \lambda \cdot \tilde{x}$ where $\lambda_i = e^{s_i}$, and hence x and $\tilde{x}$ lie in the same orbit of G. Since each orbit is thus a common level set of the global invariants $\pi^1, \ldots, \pi^{m-s}$, the action of G is automatically regular and $\pi^1, \ldots, \pi^{m-s}$ provide global coordinates on the quotient manifold G/M, which can be identified with the positive octant of $\mathbb{R}^{m-s}$. The second part of the theorem now follows immediately from part (c) of Proposition 3.21. □

Example 3.23. Assume that the resistance D of an object immersed in a fluid is determined by a unit-free function of fluid density ρ, fluid velocity v, object diameter d, and fluid viscosity μ. Letting $\lambda_1, \lambda_2, \lambda_3$ be the scaling parameters of length, time and mass respectively, we obtain the consequent scaling of the relevant derived quantities

$$\lambda \cdot (\rho, v, d, \mu, D) = (\lambda_1^{-3}\lambda_3\rho, \lambda_1\lambda_2^{-1}v, \lambda_1 d, \lambda_1^{-1}\lambda_2^{-1}\lambda_3\mu, \lambda_1\lambda_2^{-2}\lambda_3 D).$$

(For instance, D is written in units of length $\times$ mass/(time)^2, etc.) The matrix A in this case takes the form

$$A = \begin{bmatrix} -3 & 1 & 1 & -1 & 1 \\ 0 & -1 & 0 & -1 & -2 \\ 1 & 0 & 0 & 1 & 1 \end{bmatrix},$$

which is of rank 3. There are thus $5 - 3 = 2$ independent dimensionless power products. To determine them, according to (3.30) we need to analyze the null space of A, which is spanned by the column vectors $(1, 1, 1, -1, 0)^T$, $(-1, -2, -2, 0, 1)^T$. These correspond to the independent power products

$$\pi_1 = R = \frac{\rho v d}{\mu}, \qquad \pi_2 = K = \frac{D}{\rho v^2 d^2},$$

the first of which is the *Reynolds number* for the flow. According to the Pi theorem, any unit-free relation between our five quantities must be of the form $F(R, K) = 0$, or, upon solving for K, the resistance is given by $D = \rho v^2 d^2 h(R)$, where h is a function whose form remains to be determined.

3.5. Group-Invariant Prolongations and Reduction

In the basic program to rigorously implement the general procedure for constructing group-invariant solutions to differential equations, there is one primary hurdle that must be overcome. In general, if the system of differential equations Δ is defined over an open subset $M \subset X \times U$ of the space of independent and dependent variables upon which the symmetry group G acts regularly, then the reduced system of differential equations Δ/G for the G-invariant solutions will naturally live on the quotient manifold M/G. The difficulty is that although M/G has the structure of a smooth manifold, it will not in general be an open subset of any Euclidean space and so our earlier construction of jet spaces and prolonged group actions is no longer applicable. In practice, however, we work in local coordinate charts, and so we can make the more modest assumption that there are $p + q - s$ functionally-independent invariants on M, $\eta^1(x, u), \ldots, \eta^{p+q-s}(x, u)$, determining global cooordinates on M/G, which can therefore be viewed as an open subset of the Euclidean space $\mathbb{R}^{p+q-s}$. Here s denotes the dimension of the orbits of G. At this point, a second difficulty arises in that there is in general no natural way of distinguishing which of the invariants η^j will be the new independent variables and which will be the new dependent variables. If G acts projectably, then, as we have seen, there are precisely $p - s$ invariants which depend only on x, which can be designated as the new independent variables, the remaining q invariants becoming the new dependent variables. In the general case, however, there is no way of determining new independent and dependent variables in a consistent manner, and one is forced to make an arbitrary choice among the given invariants, assigning $p - s$ of them the role of independent variables $y = (y^1, \ldots, y^{p-s})$, and the remaining q the role of dependent variables $v = (v^1, \ldots, v^q)$. In this way, we have forced M/G to be a subset of the Euclidean space $Y \times V \simeq \mathbb{R}^{p-s} \times \mathbb{R}^q$, and hence can determine an explicit form for the reduced system Δ/G by regarding v as a function of y. Already, the roles of independent and dependent variables are starting to blur. Different assignations of invariants will lead to seemingly different expressions for the reduced system, but—and this must be emphasized—these will all be equivalent under an interchange of the roles of independent and dependent variables reminiscent of the hodograph transformation of fluid dynamics.

So far, this would be all right for our purposes, were it not for yet another complication. Once we have selected the new independent variables y and dependent variables v for our reduced system Δ/G, there is no guarantee that a given function $v = h(y)$ will correspond to a smooth, single-valued function $u = f(x)$; vice versa, there exists the possibility of G-invariant functions $u = f(x)$ not corresponding to smooth functions of the form $v = h(y)$ relative to the given choice of independent and dependent variables. The problem in both instances is that a function on one of the spaces may give rise to a "function" with infinite derivatives or multiple values on the other space, and these are excluded by our perhaps artificial division of the coordinates into independent and dependent variables. This point is perhaps made more clear through the use of an illustrative example.

Example 3.24. Suppose $p = 2$, $q = 1$ so $M = X \times U$ has coordinates (x, t, u). Consider the one-parameter group of translations $G: (x, t, u) \mapsto (x, t + \varepsilon, u)$. Suppose instead of choosing the natural invariants x and u to coordinatize $M/G = \mathbb{R}^2$ we were to choose the invariants $y = x + u$ and $v = u$, with y the new independent and v the new dependent variable. Any function $v = h(y)$ on $M/G = Y \times V$ will determine a two-dimensional G-invariant submanifold of M, given by the equation $u = h(x + u)$, but unless $h'(y) \neq 1$, this equation will not determine u explicitly as a function of x (and t). For example, the function $v = y$ in M/G corresponds to the vertical plane $\{x = 0\}$, which is certainly not the graph of a function $u = f(x, t)$. On the other hand, the G-invariant function $u = -x$ reduces to the vertical line $y = 0$, which is not the graph of a function of the form $v = h(y)$. Although this example is somewhat artificial, the phenomena may be unavoidable in more complicated situations.

The principal reason for all these technical complications is our attachment to the distinction between independent and dependent variables, a stance which becomes increasingly untenable in light of the above reasoning. If we abandon this prejudice, the general construction of the reduced system for group-invariant solutions becomes very natural. Once the basic coordinate-free construction has been made, the technicalities involved in introducing particular independent and dependent variables, both on M and M/G, can be handled with a minimum of difficulty. We therefore commence this section with a coordinate-free reformulation of our basic jet space construction, this time valid for arbitrary manifolds, not just open subsets of Euclidean space.

Extended Jet Bundles

Let us begin by looking a bit closer at our earlier constructions of the jet space $M^{(n)}$ for $M \subset X \times U \simeq \mathbb{R}^p \times \mathbb{R}^q$. Each point $(x_0, u_0^{(n)}) \in M^{(n)}$ is determined by the derivatives of a smooth function $u = f(x)$ whose graph passes through the base point $z_0 = (x_0, u_0) \in M$, with $u_0^{(n)} = \mathrm{pr}^{(n)} f(x_0)$. Two such functions are said to be n-th *order equivalent* at z_0 if they determine the same point in $M^{(n)}|_{z_0} \equiv \{(x, u^{(n)}): (x, u) = z_0\}$; in other words, f and $\tilde{f}$ are n-th order equivalent at (x_0, u_0) if their derivatives up to order n agree:

$$\partial_J f^\alpha(x_0) = \partial_J \tilde{f}^\alpha(x_0), \qquad \alpha = 1, \ldots, q, \quad 0 \leqslant \#J \leqslant n.$$

From this point of view, the jet space $M^{(n)}|_{z_0}$ can be regarded as the set of n-th order equivalence classes on the space of all smooth functions $u = f(x)$ whose graphs pass through $z_0 = (x_0, u_0)$.

Thus the important object is not the function f but rather its graph $\Gamma_f = \{(x, f(x))\}$, which is a p-dimensional submanifold of M. However, not every p-dimensional submanifold of M is the graph of a smooth function, so not every such submanifold passing through the point $z_0 \in M$ will determine a point in $M^{(n)}|_{z_0}$. The goal here is to "extend" the jet space $M^{(n)}|_{z_0}$ to include those submanifolds with "vertical tangents". The implicit function theorem tells us which submanifolds are the graphs of smooth functions.

Proposition 3.25. *A p-dimensional submanifold $\Gamma \subset M \subset X \times U$ is the graph of a smooth function $u = f(x)$ if and only if Γ satisfies the properties of being*

(a) Transverse. *For each $z_0 = (x_0, u_0) \in \Gamma$, Γ intersects the vertical space $U_{z_0} = \{(x_0, u): u \in U\}$ transversally, meaning $T\Gamma|_{z_0} \cap TU_{z_0}|_{z_0} = \{0\}$.*
(b) Single-Valued. *Γ intersects each vertical space U_{z_0}, $z_0 \in M$, in at most one point.*

Of course, if we change coordinates on M, the requisite vertical planes will change, so a submanifold which is the graph of a function in one coordinate system may not be in another. For instance, the parabola $u = x^2$ is the graph of a function when $x \in \mathbb{R}$ is the independent and $u \in \mathbb{R}$ the dependent variable, but if we let $y = u$ be the new independent and $v = x$ the new dependent variable, then the parabola fails the transversality condition at the origin and the single-valuedness condition for all $y > 0$. However, if we have a sufficiently small p-dimensional submanifold Γ, we can always arrange local coordinates so that Γ is the graph of a function.

Once we allow arbitrary changes in the independent and dependent variables, it is senseless to exclude certain p-dimensional submanifolds merely because they happen to violate the transversality or single-valuedness conditions in some given set of coordinates. From this standpoint, the role of functions $u = f(x)$ is now played by *arbitrary* p-dimensional submanifolds $\Gamma \subset M$. At this point, we see that we have freed ourselves entirely from our dependence on Euclidean coordinates (x, u) and the definitions to follow will make sense for arbitrary $(p + q)$-dimensional manifolds M.

Definition 3.26. Let Γ and $\tilde{\Gamma}$ be regular p-dimensional submanifolds of the smooth manifold M. We say that Γ and $\tilde{\Gamma}$ have n-th *order contact* at a common point $z_0 \in \Gamma \cap \tilde{\Gamma}$ if and only if there exists a local coordinate chart W containing $z_0 = (x_0, u_0)$ with coordinates $(x, u) = (x^1, \ldots, x^p, u^1, \ldots, u^q)$, such that $\Gamma \cap W$ and $\tilde{\Gamma} \cap W$ coincide with the graphs of smooth functions $u = f(x)$ and $u = \tilde{f}(x)$ which are n-th order equivalent at z_0: $\mathrm{pr}^{(n)} f(x_0) = \mathrm{pr}^{(n)} \tilde{f}(x_0)$.

It is not difficult to see that the property of having n-th order contact at z_0 is independent of the choice of local coordinates at z_0, provided only that Γ and $\tilde{\Gamma}$ are both transverse to the vertical space U_{z_0} and hence are locally the graphs of smooth functions. Clearly n-th order contact determines an equivalence relation on the set of p-dimensional submanifolds passing through a point.

Definition 3.27. Let M be a smooth manifold and p a fixed integer with $0 < p < \dim M$. The *extended jet space* $M_*^{(n)}|_z$, is defined as the set of equivalence classes of the set of all p-dimensional submanifolds passing through z under the equivalence relation of n-th order contact. The *extended jet bundle* is the union of all these spaces: $M_*^{(n)} = \bigcup_{z \in M} M_*^{(n)}|_z$.

If $\Gamma \subset M$ is any p-dimensional submanifold, and $z \in \Gamma$, the n-th *prolongation* $\mathrm{pr}^{(n)} \Gamma|_z \in M_*^{(n)}|_z$ is the equivalence class determined by Γ. If Γ and $\tilde{\Gamma}$ have n-th order contact at z_0, they certainly have k-th order contact for any $k < n$, so there is a natural projection $\pi_k^n\colon M_*^{(n)} \to M_*^{(k)}$, $\pi_k^n(\mathrm{pr}^{(n)} \Gamma) = \mathrm{pr}^{(k)} \Gamma$. In particular, $M_*^{(0)} \simeq M$. The next result makes precise in what sense the extended jet space is the "completion" of the ordinary jet space (in the same way that projective space is the "completion" of Euclidean space).

Theorem 3.28. *If M is a smooth $(p + q)$-dimensional manifold, then the extended jet bundle $M_*^{(n)}$ determined by p-dimensional submanifolds is a smooth $p + q\binom{p+n}{n}$-dimensional manifold. If $\Gamma \subset M$ is any regular p-dimensional submanifold, its prolongaton $\mathrm{pr}^{(n)} \Gamma$ is a regular p-dimensional submanifold of $M_*^{(n)}$. If $\tilde{M} \subset M$ is a local coordinate chart, which determines a local choice of independent and dependent variables (x, u), then the subspace*

$$\tilde{M}^{(n)}|_z \equiv \{\mathrm{pr}^{(n)} \Gamma|_z : z \in \Gamma,\ T\Gamma|_z \cap TU_z|_z = \{0\}\}$$

determined by the transverse submanifolds Γ passing through z, is an open dense subset of the extended jet space $M_*^{(n)}|_z$. Moreover, the union of all such subspaces, $\tilde{M}^{(n)}$, is isomorphic to the ordinary Euclidean jet space: $\tilde{M}^{(n)} \simeq \tilde{M} \times U_1 \times \cdots \times U_n = \{(x, u^{(n)}) : (x, u) \in \tilde{M}\}$. If $\Gamma \subset \tilde{M}$ coincides with the graph of a smooth function $u = f(x)$, then under the above identification its prolongation $\mathrm{pr}^{(n)} \Gamma \subset \tilde{M}^{(n)} \subset \tilde{M}_*^{(n)}$ coincides with the graph of the prolongation of f: $\mathrm{pr}^{(n)} \Gamma = \{(x, \mathrm{pr}^{(n)} f(x))\}$.

Thus, except for the singular subvariety $\mathscr{V}^{(n)}|_z \equiv \tilde{M}_*^{(n)}|_z \setminus \tilde{M}^{(n)}|_z$, consisting of the prolongations of nontransverse submanifolds, the exended jet space looks just like the ordinary jet space discussed in Chapter 2. The proof of this theorem is not difficult; an illustrative example should indicate how one would fill in the details in general.

Example 3.29. Let $M \subset \mathbb{R}^2$ be open and let $p = 1$, so we are considering one-dimensional submanifolds (curves) in M. Two curves determine the same point in $M_*^{(n)}|_{z_0}$ if and only if in some local coordinates near $z_0 = (x_0, u_0)$ they are given as the graphs of functions $u = f(x)$, $u = \tilde{f}(x)$ with the same derivatives up to order n at x_0:

$$u_0 = f(x_0) = \tilde{f}(x_0),\ f'(x_0) = \tilde{f}'(x_0),\ \ldots,\ f^{(n)}(x_0) = \tilde{f}^{(n)}(x_0).$$

In the case $n = 1$, then, the curves Γ and $\tilde{\Gamma}$ have first order contact at z_0 if and only if they are tangent at z_0. Thus $M_*^{(1)}|_{z_0}$ is given by the set of all tangent lines to curves passing through z_0. Since every such line is determined by the angle θ it makes with the horizontal, varying from $\theta = 0$ to $\theta = \pi$, we can identify $M_*^{(1)}|_{z_0}$ with the circle S^1, where the "angular" coordinate satisfies $0 \leqslant \theta < \pi$. Topologically, then, $M^{(1)} \simeq M \times S^1$. Choosing coordinates (x, u) on M, the Euclidean jet space $M^{(1)}|_{z_0}$ is the subset of $M_*^{(1)}|_{z_0}$ given by those curves whose tangent line is not vertical, i.e. $\theta \neq \pi/2$. We can identify this subset with the usual jet space $\{(x, u, u_x)\}$ by setting $u_x = \tan\theta$.

Turning to the case $n = 2$, we see that two curves Γ and $\tilde{\Gamma}$ have second order contact at z_0 if and only if they osculate at z_0, i.e. have the same tangent *and* curvature there. Thus $M_*^{(2)}|_{z_0}$ can be identified with the set of all circles of positive radius passing through z_0, including the degenerate straight lines. I claim that this space is topologically equivalent to a Möbius band! The natural projection $\pi_1^2\colon M_*^{(2)}|_{z_0} \to M_*^{(1)}|_{z_0} \simeq S^1$ associates the common tangent line to any pair of osculating curves, so the inverse image of a point $\theta \in S^1$ (i.e. a tangent line through z_0) is isomorphic to $\mathbb{R}$, the additional coordinate being the signed curvature of the curve with the given tangent direction. Thus, locally $M_*^{(2)}|_{z_0}$ looks like the Cartesian product of a piece of S^1 with $\mathbb{R}$. However, if we fix the curvature, but let θ increase from 0 to π, in essence we

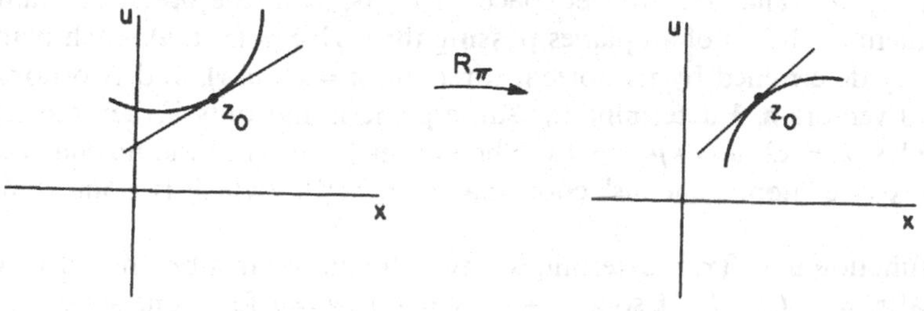

Figure 9. Extended jet space M_*^2 for $M \subset \mathbb{R}^2$.

are rotating the given curve through an angle of π. The result is a curve with the same tangent, but whose curvature has changed sign! Thus as the circle S^1 is traversed, the copy of $\mathbb{R}$ over each point twists once, and we have a Möbius band. In local coordinates (x, u), the open dense subset $M^{(2)}|_{z_0} \subset M_*^{(2)}|_{z_0}$ is obtained by cutting the band along the line $\theta = \pi/2$, the result being isomorphic to the two-dimensional plane $U_1 \times U_2 = \{(u_x, u_{xx})\}$.

Further results on the structure of the extended jet space are given in Exercise 3.16 and Olver, [2].

Differential Equations

Definition 3.30. Let M be a smooth manifold with extended jet bundle $M_*^{(n)}$ determined by p-dimensional submanifolds. A *system of differential equations* over M is determined by a closed subvariety $\mathscr{S}_\Delta^* \subset M_*^{(n)}$. A *solution* to the system is a p-dimensional submanifold Γ whose prolongation lies entirely within the subvariety: $\text{pr}^{(n)}\,\Gamma \subset \mathscr{S}_\Delta^*$.

If we choose a local coordinate chart $\tilde{M} \subset M$, and concentrate on the subset $\tilde{M}^{(n)} \subset \tilde{M}_*^{(n)}$, then we reduce to our previous concept of a system of differential equations: $\mathscr{S}_\Delta \equiv \mathscr{S}_\Delta^* \cap \tilde{M}^{(n)} = \{(x, u^{(n)}): \Delta(x, u^{(n)}) = 0\}$ for some set of smooth functions $\Delta: \tilde{M}^{(n)} \to \mathbb{R}^l$. A transverse, single-valued submanifold Γ, which thus is the graph of a smooth function $u = f(x)$, is a solution in the above sense if and only if the corresponding function f is a solution in the traditional sense: $\Delta(x, \text{pr}^{(n)} f(x)) = 0$. In addition we are allowing the possibility of both multiply-valued solutions and solutions with vertical tangents (infinite derivatives) provided they are in some sense "limits" of classical solutions. Any "traditional" system of differential equations defined over an open subset M of Euclidean space $X \times U$ can always be made into such an "extended" system by taking the closure of its subvariety in $M_*^{(n)}$: $\mathscr{S}_\Delta^* = \overline{\mathscr{S}_\Delta}$.

Example 3.31. Consider the nonlinear wave equation $u_t + uu_x = 0$. Here the underlying space is $M \simeq \mathbb{R}^2 \times \mathbb{R}$, with coordinates (x, t, u), and the equation determines a subvariety in the first jet space $M^{(1)} \simeq \mathbb{R}^5$ with coordinates $(x, t; u; u_x, u_t)$. The extended jet space $M_*^{(1)}|_z$ is, as in the previous example, equivalent to the set of all planes passing through $z = (x, t, u)$. Each plane is uniquely determined by its normal direction $\mathbf{n} = (\lambda, \mu, \nu)$, and two nonzero normal vectors $\mathbf{n}, \tilde{\mathbf{n}}$ determine the same plane if and only if they are scalar multiples, $\tilde{\lambda} = \kappa\lambda$, $\tilde{\mu} = \kappa\mu$, $\tilde{\nu} = \kappa\nu$. The entries $[\lambda, \mu, \nu]$ of the normal vector thus provide "homogeneous" coordinates on $M_*^{(1)}|_z$ (which is isomorphic to $\mathbb{RP}^2$).

A function $u = f(x, t)$ determines a two-dimensional submanifold Γ_f with normal $\mathbf{n} = (-f_x, -f_t, 1)$, so $\lambda = -f_x$, $\mu = -f_t$, $\nu = 1$ form one set of homogeneous coordinates for $\text{pr}^{(1)}\,\Gamma_f$. A more general submanifold $\Gamma =$

$\{F(x, t, u) = 0\}$ has normal $\mathbf{n} = \nabla F$, hence $\mathrm{pr}^{(1)}\,\Gamma = \{(x, t; u; [F_x, F_t, F_u])\}$ in our coordinates. In particular, Γ and Γ_f have the same tangent plane if and only if their homogeneous coordinates are equivalent: $F_x = -\kappa f_x$, $F_t = -\kappa f_t$, $F_u = \kappa$, from which we deduce the familiar formulae $u_x = -F_x/F_u$, $u_t = -F_t/F_u$. Consequently, $M^{(1)}|_z$ is the open subset of $M_*^{(1)}|_z$ where the third homogeneous coordinate ν does not vanish, and in this case $u_x = -\lambda/\nu$, $u_t = -\mu/\nu$.

If we substitute these expressions into the equation, we find an explicit formula for the extended subvariety

$$\mathscr{S}_\Delta^* = \{(x, t; u; [\lambda, \mu, \nu]): \lambda + u\mu = 0\}.$$

A solution is then a two-dimensional submanifold $\Gamma = \{F(x, t, u) = 0\}$ with $\mathrm{pr}^{(1)}\,\Gamma \subset \mathscr{S}_\Delta^*$, meaning $\partial_t F + u\partial_x F = 0$ ($\nabla F \neq 0$). This equation can now be solved directly by the characteristic methods of Section 2.1, leading to $F = F(x - tu, u)$ for the general solution. Alternatively, we can use a "hodograph" coordinate change and choose new independent variables t and u and new dependent variable x. Note that $x_t = -\mu/\lambda$, $x_u = -\nu/\lambda$, so in the $(t, u; x; x_t, x_u)$ coordinates on $M_*^{(1)}$, the equation becomes $x_t = u$, with elementary solution $x = tu + h(u)$, h an arbitrary function of u. Note that although this choice of coordinates leads to globally defined solutions, in the original coordinates $(x, t; u)$ solutions can become multiply-valued, leading to the familiar phenomena of wave breaking. (In our present interpretation, these multiply-valued functions *remain* solutions, while for physical applications, one would replace them by shock solutions.)

Group Actions

If $g: M \to M$ is any diffeomorphism and $\Gamma \subset M$ is a p-dimensional submanifold, then $g \cdot \Gamma = \{g \cdot x: x \in \Gamma\}$ is also a p-dimensional submanifold. Moreover, g preserves the equivalence relation of n-th order contact, so there is an induced diffeomorphism $\mathrm{pr}^{(n)}\,g$ of the extended jet space $M_*^{(n)}$:

$$\mathrm{pr}^{(n)}\,g(\mathrm{pr}^{(n)}\,\Gamma|_z) \equiv \mathrm{pr}^{(n)}\,(g \cdot \Gamma)|_{g \cdot z}, \qquad z \in \Gamma. \tag{3.31}$$

Thus for any local group of transformations G acting on M, there is an induced action $\mathrm{pr}^{(n)}\,G$, the n-th *prolongation* of G, on $M_*^{(n)}$. In any local coordinate chart $\tilde{M} \subset M$, this action agrees with our earlier notion of prolongation on the corresponding Euclidean jet space $\tilde{M}^{(n)} \subset \tilde{M}_*^{(n)}$. Note especially that since *any* p-dimensional submanifold, transverse or not, is now being regarded as the graph of a "function", we no longer have to worry about domains of definition of the prolonged group action; if g is defined on M_g, then $\mathrm{pr}^{(n)}\,g$ is defined on all of $M_{g*}^{(n)}$. In particular, if G is a global group of transformations, its prolongation to $M_*^{(n)}$ is *still* a global group of transformations. (Compare this with Example 2.26.)

If $\mathbf{v}$ is a vector field on M, its prolongation $\mathrm{pr}^{(n)}\,\mathbf{v}$ is the vector field on $M_*^{(n)}$ which generates the prolongation $\mathrm{pr}^{(n)}\,[\exp(\varepsilon\mathbf{v})]$ of the one-parameter group generated by $\mathbf{v}$. Since this agrees with the usual prolongation on any coordinate chart $\tilde{M} \subset M$, we immediately conclude that the formula for $\mathrm{pr}^{(n)}\,\mathbf{v}$ is the same as that given in Theorem 2.36 on the subspace $\tilde{M}^{(n)} \subset \tilde{M}_*^{(n)}$. (Notice that by this remark, we conclude the invariance of (2.38) under arbitrary changes of independent and dependent variables!)

A locally solvable system of differential equations $\mathscr{S}_\Delta^* \subset M_*^{(n)}$ is invariant under the group action of G if and only if $\mathrm{pr}^{(n)}\,G$ preserves $\mathscr{S}_\Delta^*$, i.e. $\mathrm{pr}^{(n)}\,g[\mathscr{S}_\Delta^*] \subset \mathscr{S}_\Delta^*$. The corresponding infinitesimal criterion is that $\mathrm{pr}^{(n)}\,\mathbf{v}$ is tangent to $\mathscr{S}_\Delta^*$ whenever $\mathbf{v}$ is an infinitesimal generator of G. In local coordinates on $\tilde{M} \subset M$, this reduces to our usual infinitesimal criterion of invariance (2.25), which is both necessary and sufficient provided $\mathscr{S}_\Delta = \mathscr{S}_\Delta^* \cap \tilde{M}^{(n)}$ is both locally solvable and of maximal rank, and the complete subvariety $\mathscr{S}_\Delta^* \subset \tilde{M}_*^{(n)}$ is just the closure of $\mathscr{S}_\Delta$. (Otherwise we would have to check invariance in other coordinate systems.) Thus the theory of symmetry groups of systems of differential equations on extended jet bundles does not differ in any essential aspect from our previous theory of symmetry groups of differential equations, and, in fact, reduces to it as soon as local coordinates are introduced on M.

The Invariant Jet Space

The real key that unlocks the geometrical insight behind the construction of group-invariant solutions is the determination of the structure of that subset of the jet space traced out by their prolongations. Suppose G acts on a smooth manifold M, on which some system of differential equations $\mathscr{S}_\Delta^*$ is defined. The G-invariant solutions to the system will be certain p-dimensional submanifolds $\Gamma \subset M$, corresponding to the graphs of functions in local coordinate charts, which are locally invariant under the action of G. In general, these G-invariant submanifolds will not fill up the entire jet space $M_*^{(n)}$, but only a certain subspace $I_*^{(n)} = I_*^{(n)}(G)$ called the *invariant space* of G. It is defined as

$$I_*^{(n)}|_{z_0} \equiv \{z_0^{(n)} \in M_*^{(n)}|_{z_0}\colon \text{there exists a locally } G\text{-invariant } p\text{-dimensional submanifold } \Gamma \text{ passing through } z_0 \text{ with prolongation } z_0^{(n)} = \mathrm{pr}^{(n)}\,\Gamma|_{z_0}\}.$$

In most cases of practical interest, M is an open subset of some fixed Euclidean space, with ordinary jet space $M^{(n)} \subset M_*^{(n)}$. There is a corresponding invariant space $I^{(n)} = I_*^{(n)} \cap M^{(n)}$, which is determined by the prolongations of G-invariant functions $u = f(x)$:

$$I^{(n)}|_{x_0} = \{(x_0, u_0^{(n)}) \in M^{(n)}\colon \text{there exists a locally } G\text{-invariant function defined in a neighbourhood of } x_0 \text{ such that } u_0^{(n)} = \mathrm{pr}^{(n)}\,f(x_0)\}.$$

For practical purposes, the space $I^{(n)}$ is the easiest to work with, while in the theoretical proofs, its extension $I_*^{(n)}$ comes into the forefront.

Example 3.32. Consider the case $p = 2$, $q = 1$, so X has coordinates (x, t) and U the single dependent variable u. Let G be the translation group $(x, t; u) \mapsto (x + \varepsilon, t; u)$, with infinitesimal generator $\partial/\partial x$. A function $u = f(x, t)$ is G-invariant if and only if f is independent of x. Thus

$$I^{(1)} = \{(x, t; u; u_x, u_t): u_x = 0\},$$

since at each point $u_x = \partial f/\partial x$ vanishes, while $u_t = \partial f/\partial t$ can be specified arbitrarily. Similarly,

$$I^{(2)} = \{(x, t; u; u_x, u_t; u_{xx}, u_{xt}, u_{tt}): u_x = u_{xx} = u_{xt} = 0\},$$

and so on. As a useful exercise, at this point the reader should determine $I^{(1)}$ and $I^{(2)}$ in the case $G = \mathrm{SO}(2)$ is the rotation group with infinitesimal generator $-t\partial_x + x\partial_t$.

In Theorem 3.38 we will give an explicit characterization of the invariant space. However, we can already prove most of the important properties of this space even without the explicit formulae.

Proposition 3.33. *Let M be a smooth manifold, and G a local group of transformations acting on M. Then the invariant jet space $I_*^{(n)} \subset M_*^{(n)}$ corresponding to G is invariant under the action of $\mathrm{pr}^{(n)}\, G$ on $M_*^{(n)}$:*

$$\mathrm{pr}^{(n)}\, g[I_*^{(n)}] \subset I_*^{(n)}, \qquad g \in G.$$

PROOF. Let $z_0^{(n)}$ be a point in $I_*^{(n)}|_{z_0}$, so that by definition there exists a locally G-invariant p-dimensional submanifold Γ passing through z_0 with $\mathrm{pr}^{(n)}\, \Gamma|_{z_0} = z_0^{(n)}$. If g is any element of G such that $\tilde{z}_0 = g \cdot z_0$ is defined, then the transformed submanifold $\tilde{\Gamma} = g \cdot \Gamma = \{g \cdot z : z \in \Gamma, g \cdot z \text{ is defined}\}$ is also locally G-invariant. (Why?) Thus, by (3.31),

$$\mathrm{pr}^{(n)}\, g(z_0^{(n)}) = \mathrm{pr}^{(n)}\, g \cdot [\mathrm{pr}^{(n)}\, \Gamma|_{z_0}] = \mathrm{pr}^{(n)}\, \tilde{\Gamma}|_{\tilde{z}_0},$$

which, being the prolongation of a locally G-invariant submanifold, lies in $I_*^{(n)}|_{\tilde{z}_0}$. This completes the proof. (The same proof clearly works for the ordinary invariant space $I^{(n)} \subset M^{(n)}$.) □

Connection with the Quotient Manifold

Since the invariant jet space $I_*^{(n)}$ for a group action is itself invariant under the prolonged group action $\mathrm{pr}^{(n)}\, G$, we can define a quotient space $I_*^{(n)}/\mathrm{pr}^{(n)}\, G$ by contracting the orbits of $\mathrm{pr}^{(n)}\, G$ in $I_*^{(n)}$ to points. In the case G acts regularly on the underlying manifold M, this "prolonged quotient manifold" can be identified with the n-jet space of the corresponding quotient manifold M/G.

This result, which becomes elementary to both state and prove in the language of extended jet bundles (but is considerably more complicated if we stick to ordinary jet spaces, as will be seen subsequently) immediately leads to the reduced system of differential equations for G-invariant solutions:

Proposition 3.34. *Let G be a local group of transformations acting regularly on the $(p+q)$-dimensional manifold M with s-dimensional orbits, $s \leqslant p$, and let M/G be the corresponding $(p+q-s)$-dimensional quotient manifold. Let $M_*^{(n)}$ be the extended n-jet space generated by p-dimensional submanifolds of M, and $I_*^{(n)} \subset M_*^{(n)}$ the corresponding invariant space generated by the G-invariant p-dimensional submanifolds. Then there is a natural projection $\pi^{(n)}: I_*^{(n)} \to (M/G)_*^{(n)}$ onto the extended n-jet space corresponding to $(p-s)$-dimensional submanifolds of M/G with the following properties:*

(a) *If $z \in M$ has image $\pi(z) = w \in M/G$, where $\pi: M \to M/G$ is the natural projection, then*

$$\pi^{(n)}: I_*^{(n)}|_z \to (M/G)_*^{(n)}|_w$$

is a diffeomorphism.

(b) *If $\Gamma \subset M$ is any G-invariant p-dimensional submanifold, with image $\Gamma/G = \pi[\Gamma] \subset M/G$, then*

$$\pi^{(n)}[\mathrm{pr}^{(n)}\,\Gamma|_z] = \mathrm{pr}^{(n)}\,(\Gamma/G)|_w \tag{3.32}$$

for any $z \in \Gamma$ with image $w = \pi(z) \in \Gamma/G$.

(c) *Two points $z^{(n)}$ and $\tilde{z}^{(n)}$ in $I_*^{(n)}$ have the same image in $(M/G)_*^{(n)}$ under $\pi^{(n)}$ if and only if they lie in the same orbit of $\mathrm{pr}^{(n)}\,G$. Thus*

$$I_*^{(n)}/\mathrm{pr}^{(n)}\,G \simeq (M/G)_*^{(n)}$$

with $\pi^{(n)}$ coinciding with the natural projection.

Proof. Almost all of these properties follow directly from the correspondence between G-invariant p-dimensional submanifolds of M and general $(p-s)$-dimensional submanifolds of M/G described in Proposition 3.21, and the following lemma.

Lemma 3.35. *Let Γ and $\tilde{\Gamma}$ be locally G-invariant submanifolds of M with images Γ/G and $\tilde{\Gamma}/G$ in M/G. Then Γ and $\tilde{\Gamma}$ have n-th order contact at $z_0 \in M$ if and only if Γ/G and $\tilde{\Gamma}/G$ have n-th order contact at $w_0 = \pi(z_0) \in M/G$.*

Proof. Choose flat local coordinates $(t, y, v) = (t^1, \ldots, t^s, y^1, \ldots, y^{p-s}, v^1, \ldots, v^q)$ near $z_0 = (t_0, y_0, v_0)$, the orbits of G being the slices $\{y = c, v = \tilde{c}\}$, such that Γ and $\tilde{\Gamma}$ are the graphs of functions $v = f(y, t)$, $v = \tilde{f}(y, t)$ respectively. The G-invariance of Γ and $\tilde{\Gamma}$ implies that f and $\tilde{f}$ are independent of t, and, moreover, in the corresponding local coordinates (y, v) on M/G, Γ/G and $\tilde{\Gamma}/G$ have the same respective formulae $v = f(y)$, $v = \tilde{f}(y)$. The lemma is thus trivial: n-th order contact of Γ and $\tilde{\Gamma}$ means that the n-th order derivatives of f and $\tilde{f}$ with respect to both y and t agree at y_0, t_0. But the t-derivatives are all identically zero, so this is clearly equivalent to the requirement that just the n-th order derivatives of f and $\tilde{f}$ with respect to y agree at y_0, which is the same as Γ/G and $\tilde{\Gamma}/G$ having n-th order contact. □

To prove Proposition 3.34, we define the map $\pi^{(n)}$ using (3.32), the lemma assuring us that it is well defined. Part (a) follows from the correspondence between G-invariant submanifolds of M and their images in M/G. To prove part (c), let Γ and $\tilde{\Gamma}$ be locally G-invariant submanifolds representing $z^{(n)} = \mathrm{pr}^{(n)}\,\Gamma|_z$ and $\tilde{z}^{(n)} = \mathrm{pr}^{(n)}\,\tilde{\Gamma}|_{\tilde{z}}$. The images $\pi^{(n)}(z^{(n)}) = \mathrm{pr}^{(n)}(\Gamma/G)|_{\pi(z)}$ and $\pi^{(n)}(\tilde{z}^{(n)}) = \mathrm{pr}^{(n)}(\tilde{\Gamma}/G)_{\pi(\tilde{z})}$ are the same if and only if Γ/G and $\tilde{\Gamma}/G$ have n-th order contact at $w = \pi(z) = \pi(\tilde{z})$. We conclude that z and $\tilde{z}$ lie in the same orbit of G in M, so by Proposition 1.24 there exist elements $g_1, \ldots, g_k \in G$ such that $\tilde{z} = g_1 \cdot g_2 \cdot \ldots \cdot g_k \cdot z$. Let $\Gamma^* = g_1 \cdot g_2 \cdot \ldots \cdot g_k \cdot \Gamma$. Then Γ^* passes through $\tilde{z}$, and has the same projection $\Gamma^*/G = \Gamma/G$ as Γ. By Lemma 3.35, Γ^* and $\tilde{\Gamma}$ have n-th order contact at $\tilde{z}$. Therefore

$$\begin{aligned}\tilde{z}^{(n)} &= \mathrm{pr}^{(n)}\,\tilde{\Gamma}|_{\tilde{z}} = \mathrm{pr}^{(n)}\,\Gamma^*|_{\tilde{z}} = \mathrm{pr}^{(n)}[g_1 \cdot g_2 \cdot \ldots \cdot g_k \cdot \Gamma]|_{\tilde{z}}\\ &= \mathrm{pr}^{(n)}\,g_1 \cdot \mathrm{pr}^{(n)}\,g_2 \cdot \ldots \cdot \mathrm{pr}^{(n)}\,g_k(\mathrm{pr}^{(n)}\,\Gamma|_z) = \mathrm{pr}^{(n)}\,g_1 \cdot \ldots \cdot \mathrm{pr}^{(n)}\,g_k(z^{(n)}),\end{aligned}$$

so $\tilde{z}^{(n)}$ and $z^{(n)}$ lie in the same orbit of $\mathrm{pr}^{(n)}\,G$. □

The Reduced Equation

Let $\mathcal{S}_\Delta^* \subset M_*^{(n)}$ correspond to a system Δ of partial differential equations admitting the symmetry group G. If Γ is the graph of a G-invariant solution to Δ, then not only is its prolongation $\mathrm{pr}^{(n)}\,\Gamma$ a submanifold of $\mathcal{S}_\Delta^*$, it also necessarily lies in the invariant space $I_*^{(n)}$. This suggests that the determination of such solutions is accomplished through the analysis of the intersection $\mathcal{S}_\Delta^* \cap I_*^{(n)}$ of these two subvarieties, whose invariance under $\mathrm{pr}^{(n)}\,G$ is guaranteed by Proposition 3.33. Using the projection of Proposition 3.34, we will then arrive at the *reduced system* $\mathcal{S}_{\Delta/G}^* = \pi^{(n)}[\mathcal{S}_\Delta^* \cap I_*^{(n)}]$. It is now easy to state and prove the fundamental theorem on the construction of group-invariant solutions.

Theorem 3.36. *Let G be a symmetry group of a system of differential equations $\mathcal{S}_\Delta^* \subset M_*^{(n)}$. A p-dimensional submanifold $\Gamma \subset M$ is a G-invariant solution if and only if the corresponding $(p - s)$-dimensional submanifold $\Gamma/G \subset M/G$ is a solution to the reduced system $\mathcal{S}_{\Delta/G}^* = \pi^{(n)}[\mathcal{S}_\Delta^* \cap I_*^{(n)}] \subset (M/G)_*^{(n)}$.*

Proof. If Γ is such a solution, its prolongation $\mathrm{pr}^{(n)}\,\Gamma$ lies in the intersection $\mathcal{S}_\Delta^* \cap I_*^{(n)}$. Then by Proposition 3.34, $\pi^{(n)}[\mathrm{pr}^{(n)}\,\Gamma] = \mathrm{pr}^{(n)}(\Gamma/G)$ lies in $\mathcal{S}_{\Delta/G}^*$, hence Γ/G is a solution to the reduced system. To prove the converse, note first that by Proposition 3.33, $\mathscr{S}_\Delta^* \cap I_*^{(n)}$ is $\mathrm{pr}^{(n)}$ G-invariant, hence if $z^{(n)} \in I_*^{(n)}$ has projection $\pi^{(n)}(z^{(n)}) \in \mathscr{S}_{\Delta/G}^*$, then $z^{(n)} \in \mathscr{S}_\Delta^* \cap I_*^{(n)}$. (Indeed, we can find $\tilde{z}^{(n)} \in \mathscr{S}_\Delta^* \cap I_*^{(n)}$ lying in the same orbit of $\mathrm{pr}^{(n)}\, G$.) Therefore, if Γ/G is a solution to the reduced system, and $\Gamma = \pi^{-1}(\Gamma/G)$ the corresponding G-invariant submanifold of M, then by (3.32) $\mathrm{pr}^{(n)}\,\Gamma \subset \mathscr{S}_\Delta^* \cap I_*^{(n)}$ since $\pi^{(n)}[\mathrm{pr}^{(n)}\,\Gamma] = \mathrm{pr}^{(n)}(\Gamma/G) \subset \mathscr{S}_{\Delta/G}^*$. Thus Γ is a solution. □

Local Coordinates

Theorem 3.36 does provide the rigorous justification of the general method for constructing group-invariant solutions. Its almost trivial proof is a good illustration of the power of mathematical abstraction for simplifying and simultaneously generalizing seemingly complicated constructions. On the other hand, from a more practical standpoint its slick presentation is rather disconcerting, so we need to bring the abstract jet space constructions back down to earth, which means re-introducing local coordinates. We thus let $(x, u) = (x^1, \ldots, x^p, u^1, \ldots, u^q)$ be local coordinates on M, which we can now regard as an open subset of the Euclidean space $X \times U$, with jet space $M^{(n)} \subset M_*^{(n)}$.

If G is a local group of transformations acting on M, the invariant space $I^{(n)} \subset M^{(n)}$ differs from the extended invariant space $I_*^{(n)} \subset M_*^{(n)}$ just by the images of nontransversal G-invariant submanifolds Γ. In particular, $I^{(0)} \subset M$ consists of all points $z_0 = (x_0, u_0)$ such that there is at least one locally G-invariant function $u = f(x)$ whose graph passes through z_0. Note that while $I_*^{(0)} = M$, provided only that s, the dimension of the orbits of G, does not exceed p, the same cannot be said of $I^{(0)}$. For example, in the case $G = \mathrm{SO}(2)$ acting as the group of rotations on $X \times U \simeq \mathbb{R}^2$, the locally G-invariant functions are $u = \pm\sqrt{c^2 - x^2}$, whose graphs are circular arcs. No such graphs pass through the points on the x-axis, so $I^{(0)} = \{(x, u)\colon u \neq 0\}$ is *strictly* contained in $M = X \times U$. In general, outside $I^{(0)}$ there are no G-invariant functions at all, so we may as well restrict attention to $I^{(0)}$ itself and assume from now on that $M = I^{(0)}$, meaning that through each point of M there passes the graph of some G-invariant function $u = f(x)$. There is a simple explicit characterization of this requirement in the case of a regular group action.

Proposition 3.37. *Let G act regularly on $M \subset X \times U$. Then $z_0 \in M$ lies in $I^{(0)}$ if and only if the orbit of G through z_0 is transverse to the vertical space U_{z_0}, in which case G is said to act* transversally *at z_0.*

Proof. The necessity of transversality of G at z_0 is clear, since if Γ is locally G-invariant, Γ contains a relatively open subset $W \cap \mathscr{O}$ of the orbit $\mathscr{O}$ passing through z_0, so transversality of Γ implies transversality of $\mathscr{O}$. To prove sufficiency, note that by definition, G acts transversally at z_0 if and only if

$$\mathfrak{g}|_{z_0} \cap TU_{z_0}|_{z_0} = \{0\}, \tag{3.33}$$

since $\mathfrak{g}|_{z_0}$ is the tangent space to the orbit $\mathcal{O}$ through z_0. Let $w_0 = \pi(z_0)$ be the image point in M/G. Theorem 3.18 implies that $d\pi[TU_{z_0}|_{z_0}] \equiv TU^*|_{w_0}$ is a q-dimensional subspace of $T(M/G)|_{w_0}$. Let $\tilde{\Gamma}$ be any $(p-s)$-dimensional submanifold transverse to this subspace, meaning $T\tilde{\Gamma}|_{w_0} \cap TU^*|_{w_0} = \{0\}$. Then $\Gamma = \pi^{-1}(\tilde{\Gamma})$ is easily seen to be a G-invariant p-dimensional submanifold of M passing through z_0, transverse to U_{z_0}, and hence $z_0 \in I^{(0)}$. □

In local coordinates, if $\mathfrak{g}$ is spanned by the vector fields

$$\mathbf{v}_k = \sum_i \xi^i_k(x, u)\partial_{x^i} + \sum_\alpha \phi^k_\alpha(x, u)\partial_{u^\alpha}, \qquad k = 1, \ldots, r,$$

then (3.33) is equivalent to the condition that the rank of the $p \times r$ matrix with entries $\xi^i_k(x, u)$ be exactly s, the dimension of the orbit, at $z_0 = (x_0, u_0)$:

$$\operatorname{rank}(\xi^i_k(x_0, u_0)) = \operatorname{rank}(\xi^i_k(x_0, u_0), \phi^k_\alpha(x_0, u_0)) = s. \tag{3.34}$$

For example, the action of SO(2) on $\mathbb{R}^2$ is generated by $-u\partial_x + x\partial_u$. We have $s = \operatorname{rank}(-u, x) = 1$ provided $(x, u) \neq (0, 0)$, while $\operatorname{rank}(-u) = 1$ except when $u = 0$. Thus SO(2) acts transversally everywhere except on the x-axis, which agrees with our earlier computation.

This clears up the connection between transversality and the existence of G-invariant functions, at least on the local level. The question of existence of *globally* defined G-invariant functions is considerably more delicate, and does not follow even if the local transversality condition holds everywhere. See Exercise 3.15 for an example.

From now on, we assume G acts transversally everywhere, so $I^{(0)} = M$. Then the invariant space $I^{(n)}$ can be described explicitly using the infinitesimal generators of G.

Theorem 3.38. *Let G act regularly and transversally on $M \subset X \times U$. Let*

$$\mathbf{v}_k = \sum_{i=1}^p \xi^i_k(x, u)\frac{\partial}{\partial x^i} + \sum_{\alpha=1}^q \phi^k_\alpha(x, u)\frac{\partial}{\partial u^\alpha}, \qquad k = 1, \ldots, r,$$

be a basis for the infinitesimal generators. Then the n-th invariant space $I^{(n)} \subset M^{(n)}$ is determined by the equations

$$I^{(n)} = \{(x, u^{(n)}): D_J Q^k_\alpha(x, u^{(n)}) = 0, k = 1, \ldots, r, \alpha = 1, \ldots, q, \#J \leqslant n-1\},$$

where $Q^k_\alpha = \phi^k_\alpha - \sum_i \xi^i_k u^\alpha_i$ are the characteristics of the vector fields $\mathbf{v}_k$. (See (2.48).)

Proof. We outline the proof for $n = 1$, whereby $I^{(1)}$ is the common vanishing set of the characteristics $Q^k_\alpha(x, u^{(1)})$, leaving the extension to general n to the reader. If $u = f(x)$ is a G-invariant function, then, for $\alpha = 1, \ldots, q$,

$$0 = \mathbf{v}_k(u^\alpha - f^\alpha(x)) = \phi_\alpha^k - \sum_{i=1}^{p} \xi_k^i \frac{\partial f^\alpha}{\partial x^i}$$

must vanish whenever $u = f(x)$. The right-hand side is thus $Q_\alpha^k(x, \mathrm{pr}^{(1)} f(x))$. Since every point in $I^{(1)}$ is determined by the first prolongation of such a G-invariant function, we conclude that $I^{(1)}$ is contained in the set where $Q_\alpha^k = 0$ for all α, k. This part holds for any group action whatsoever.

To prove the converse, we have to use the restrictions on the group action. The easiest way to proceed is to introduce flat local coordinates $(y, v) = (y^1, \ldots, y^p, v^1, \ldots, v^q)$ where the orbits of G are the slices $\{y^1 = c_1, \ldots, y^{p-s} = c_{p-s}, v^1 = \hat{c}_1, \ldots, v^q = \hat{c}_q\}$, and hence at each point (y_0, v_0), the space of infinitesimal generators $\mathfrak{g}|_{(y_0, v_0)}$ is spanned by the tangent vectors $\partial/\partial y^{p-s+1}, \ldots, \partial/\partial y^p$. In this case, the vanishing of all the characteristics is equivalent to the conditions $\partial v^\alpha/\partial y^k = 0$ for all $\alpha = 1, \ldots, q$, $k = p - s + 1, \ldots, p$. To any point $(y_0, v_0^{(1)}) \in M^{(1)}$ satisfying these equations, it is easy to associate a function $v = h(y)$ with $v_0^{(1)} = \mathrm{pr}^{(1)} h(y_0)$. (For instance, h can be constant!) Therefore the reverse inclusion is valid, proving the theorem.

There are, however, two technical points to be dealt with in this coordinate change. The first is that changing coordinates does not alter the characteristics, or, more precisely, does not change their common vanishing set. This follows from the general formula of Exercise 3.21 for the behaviour of characteristics under a change of variables. The other point is that the graph of $u = f(x)$, when re-expressed in the (y, v)-coordinates, may fail to be transverse to the vertical v-space, and hence not be the graph of a well-defined function $v = h(y)$. This, however, is easily rectified by "skewing" the v coordinates through a linear change of variables $\tilde{y} = y + Lv$, $\tilde{v} = v$, for some constant $p \times q$ matrix L. The $(\tilde{y}, \tilde{v})$-coordinates are still flat, and L can always be determined so that the graph of $u = f(x)$ is once again transverse to the vertical space. □

Example 3.39. In the case of the rotation group SO(2), the infinitesimal generator is $\mathbf{v} = -u\partial_x + x\partial_u$, with characteristic $Q = x + uu_x$. The first invariant space at (x, u) with $u \neq 0$ is thus $I^{(1)} = \{(x, u, u_x): x + uu_x = 0\}$. Note that even if $u = 0$, Q does not vanish unless $x = 0$, so $I^{(1)}$ is still described by the vanishing of the characteristic except at the origin $x = u = 0$. There, however, $I^{(1)}$ is still empty, but $Q \equiv 0$ for all u_x, so the regularity of the group action is essential for the validity of Theorem 3.38. Higher order invariant spaces are constructed by differentiating, so for $u \neq 0$,

$$I^{(2)} = \{(x, u, u_x, u_{xx}): x + uu_x = 0,\ 1 + uu_{xx} + u_x^2 = 0\},$$

and so on.

To proceed to the quotient manifold, we make the further assumption that there exist $p + q - s$ globally defined, functionally independent invariants for

G on M, which we partition into new independent variables $y^i = \eta^i(x, u)$, $i = 1, \ldots, p - s$, and new dependent variables $v^\alpha = \zeta^\alpha(x, u)$, $\alpha = 1, \ldots, q$. (This can always be arranged by shrinking the domain M still further.) These provide global coordinates on the quotient manifold M/G, which we can therefore regard as an open subset of the $(p + q - s)$-dimensional Euclidean space $Y \times V$.

As we saw in Proposition 3.34, the projection $\pi^{(n)}$ provides a diffeomorphism between the full invariant space $I_*^{(n)}|_{z_0}$ at a point $z_0 \in M$ and the full extended jet space $(M/G)_*^{(n)}|_{w_0}$ at the image point $w_0 = \pi(z_0) \in M/G$. However, with the introduction of local coordinates on both M and M/G, we must impose transversality requirements on the relevant submanifolds to ensure that they locally look like the graphs of smooth functions. As a result, the basic correspondence between the invariant space $I^{(n)}|_{z_0}$ and the usual jet space $(M/G)^{(n)}|_{w_0}$ loses much of the innate simplicity of the extended version in Proposition 3.34.

Example 3.24 illustrated how graphs of smooth G-invariant functions might project down to nontransverse submanifolds of M/G, or, vice versa, smooth functions on M/G might correspond to nontransverse G-invariant submanifolds of M. To maintain the basic correspondence, then, we must avoid these pathological cases and concentrate on those G-invariant functions on M which correspond to smooth functions on M/G and conversely. More specifically, the invariant space $I^{(n)}|_{z_0}$ traced out by the prolongations of G-invariant functions $u = f(x)$ differs from the extended invariant space $I_*^{(n)}|_{z_0}$ only by those points in the "vertical" subvariety $\mathscr{V}^{(n)}|_{z_0} = M_*^{(n)}|_{z_0} \setminus M^{(n)}|_{z_0}$. Let $(\widetilde{\mathscr{V}/G})^{(n)}|_{w_0} \equiv \pi^{(n)}[\mathscr{V}^{(n)}|_{z_0} \cap I_*^{(n)}|_{z_0}]$ be its image in $(M/G)_*^{(n)}$; a point therein represents the prolongation of a submanifold of M/G passing through w_0 which does *not* correspond to a graph of a smooth G-invariant function passing through z_0. The remainder of the jet space,

$$(\widetilde{M/G})^{(n)}|_{w_0} = (M/G)^{(n)}|_{w_0} \setminus (\widetilde{\mathscr{V}/G})^{(n)}|_{w_0}$$

represents the prolongations of "nice" functions $v = h(y)$ which correspond to locally G-invariant functions $u = f(x)$ near z_0.

Conversely, the usual jet space $(M/G)^{(n)}|_{w_0}$ differs from the extended jet space $(M/G)_*^{(n)}|_{w_0}$ by the vertical subvariety $(\mathscr{V}/G)^{(n)}|_{w_0} \equiv (M/G)_*^{(n)}|_{w_0} \setminus (M/G)^{(n)}|_{w_0}$. Let $\tilde{\mathscr{V}}^{(n)}|_{z_0}$ be its pre-image in $I_*^{(n)}|_{z_0}$, so $\pi^{(n)}[\tilde{\mathscr{V}}^{(n)}|_{z_0}] = (\mathscr{V}/G)^{(n)}|_{w_0}$. A point thereof represents the prolongation of a G-invariant submanifold passing through z_0 which does *not* correspond to the graph of a smooth function $v = h(y)$ on M/G. The remainder of the invariant space

$$\tilde{I}^{(n)}|_{z_0} = I^{(n)}|_{z_0} \setminus \tilde{\mathscr{V}}^{(n)}|_{z_0}$$

contains the prolongations of all "nice" G-invariant functions $u = f(x)$ which correspond to explicit functions $v = h(y)$ on M/G. It is on these "nice" prolongation spaces that a correspondence similar to that of Proposition 3.34 holds.

Proposition 3.40. *The projection* $\pi^{(n)}: I_*^{(n)}|_{z_0} \to (M/G)_*^{(n)}|_{w_0}$ *induces a diffeomorphism* $\tilde{\pi}^{(n)}: \tilde{I}^{(n)}|_{z_0} \to \widetilde{(M/G)}^{(n)}|_{w_0}$ *for* $w_0 = \pi(z_0)$.

So much for geometry; how does this all work out explicitly in the given local coordinates? Functional independence of the invariants η, ζ requires that the Jacobian matrix

$$\mathrm{J} \equiv \begin{pmatrix} \partial\eta^i/\partial x^j & \partial\eta^i/\partial u^\beta \\ \partial\zeta^\alpha/\partial x^j & \partial\zeta^\alpha/\partial u^\beta \end{pmatrix}$$

have rank $p + q - s$ everywhere. The transversality condition (3.34), when coupled with the infinitesimal criterion for invariance of η^i, ζ^α, requires that the last q columns of J have rank q everywhere:

$$\operatorname{rank}(\partial\eta^i/\partial u^\beta, \partial\zeta^\alpha/\partial u^\beta)^T = q. \tag{3.35}$$

By the implicit function theorem, we can then locally solve for all q dependent variables $u^1, \ldots, u^q$ along with $p - s$ of the independent variables, say $\tilde{x} = (x^{i_1}, \ldots, x^{i_{p-s}})$, in terms of $y = (y^1, \ldots, y^{p-s})$, $v = (v^1, \ldots, v^q)$ and the remaining s independent variables $\hat{x} = (x^{j_1}, \ldots, x^{j_s})$:

$$\tilde{x} = \gamma(\hat{x}, y, v), \qquad u = \delta(\hat{x}, y, v). \tag{3.36}$$

For each fixed value y_0, v_0 of the reduced variables, (3.36) determines an orbit of G in M parametrized by the "parametric variables" $\hat{x}$.

If $v = h(y)$ is a function whose graph lies in M/G, then the corresponding G-invariant p-dimensional submanifold of M is determined by the equations

$$\zeta(x, u) = h[\eta(x, u)], \tag{3.37}$$

obtained by replacing y and v by their expressions as invariants on M. This submanifold of M will be the graph of a function $u = f(x)$ if and only if we can solve (3.37) for u as a function of x, which requires that the $q \times q$ matrix $\partial\zeta/\partial u - (\partial h/\partial y)\cdot(\partial\eta/\partial u)$ be nonsingular. (As in Section 3.1, the derivative symbols denote Jacobian matrices.) Since we can identify h with v, it makes sense to write *this* transversality condition as

$$\det\left(\frac{\partial\zeta}{\partial u} - \frac{\partial v}{\partial y}\frac{\partial\eta}{\partial u}\right) \neq 0. \tag{3.38}$$

The contrary case when this determinant vanishes will correspond to G-invariant submanifolds of M which are *not* transverse to the vertical space U_z, and, hence, determine the singular subvariety $\widetilde{(\mathscr{V}/G)}^{(n)}|_w$ which we must avoid!

From (3.37), we can differentiate to find the expressions for the derivatives of u with respect to x in terms of those of v with respect to y. By the chain rule,

$$\frac{\partial\zeta}{\partial x} + \frac{\partial\zeta}{\partial u}\frac{\partial u}{\partial x} = \frac{\partial v}{\partial y}\left(\frac{\partial\eta}{\partial x} + \frac{\partial\eta}{\partial u}\frac{\partial u}{\partial x}\right), \tag{3.39}$$

each derivative again representing a Jacobian matrix of the appropriate size. This can be rewritten in the form

$$\left(\frac{\partial \zeta}{\partial u} - \frac{\partial v}{\partial y}\frac{\partial \eta}{\partial u}\right)\frac{\partial u}{\partial x} = \frac{\partial v}{\partial y}\frac{\partial \eta}{\partial x} - \frac{\partial \zeta}{\partial x}, \tag{3.40}$$

whereby our transversality condition (3.38) permits us to solve explicitly for $\partial u/\partial x$,

$$\frac{\partial u}{\partial x} = \left(\frac{\partial \zeta}{\partial u} - \frac{\partial v}{\partial y}\frac{\partial \eta}{\partial u}\right)^{-1}\left(\frac{\partial v}{\partial y}\frac{\partial \eta}{\partial x} - \frac{\partial \zeta}{\partial x}\right),$$

as a function of x, u and $\partial v/\partial y$. The first $p + q$ variables can in turn be replaced by their expressions (3.36) leading to the formula

$$\frac{\partial u}{\partial x} = \delta_1\left(\hat{x}, y, v, \frac{\partial v}{\partial y}\right),$$

for the first order x-derivatives of a G-invariant function $u = f(x)$ in terms of the first order derivatives of its representative $v = h(y)$.

Higher order derivatives are treated by further differentiating (3.39). If we introduce the *total Jacobian matrices* $\mathrm{D}_x\eta$, $\mathrm{D}_x\zeta$ with entries $D_i\eta^j$, $D_i\zeta^\alpha$ respectively, then (3.39) has the simpler form

$$\mathrm{D}_x\zeta = \frac{\partial v}{\partial y}\cdot \mathrm{D}_x\eta.$$

Differentiating with respect to x, we find, with self-evident notation,

$$\mathrm{D}_x^2\zeta = \frac{\partial v}{\partial y}\mathrm{D}_x^2\eta + \frac{\partial^2 v}{\partial y^2}(\mathrm{D}_x\eta)^2, \tag{3.41}$$

where

$$\mathrm{D}_x^2\zeta = \frac{\partial^2 \zeta}{\partial x^2} + 2\frac{\partial^2 \zeta}{\partial x \partial u}\frac{\partial u}{\partial x} + \frac{\partial^2 \zeta}{\partial u^2}\left(\frac{\partial u}{\partial x}\right)^2 + \frac{\partial \zeta}{\partial u}\frac{\partial^2 u}{\partial x^2}.$$

(We leave it to the reader to fill in the appropriate indices.) If we group the terms involving the second order derivatives of u together, we get an expression of the form

$$\left(\frac{\partial \zeta}{\partial u} - \frac{\partial v}{\partial y}\frac{\partial \eta}{\partial u}\right)\frac{\partial^2 u}{\partial x^2} = \tilde{\delta}_2\left(x, u, \frac{\partial u}{\partial x}, \frac{\partial v}{\partial y}, \frac{\partial^2 v}{\partial y^2}\right).$$

Again, (3.38) allows us to invert the matrix on the left-hand side, leading to an expression for $\partial^2 u/\partial x^2$ in terms of x, u, $\partial u/\partial x$ and v, $\partial v/\partial y$. The first collection of variables can be replaced by their appropriate expressions in terms of y, v, $\partial v/\partial y$ *and* the parametric variables $\hat{x}$, so

$$\frac{\partial^2 u}{\partial x^2} = \delta_2(\hat{x}, y, v^{(2)}),$$

for some well-determined function δ_2. The n-th order case is very similar, replacing (3.41) we have a formula of the form

$$\mathrm{D}_x^n\zeta = \frac{\partial v}{\partial y}\mathrm{D}_x^n\eta + \cdots, \tag{3.42}$$

where the omitted terms depend on lower order total derivatives of η, as well as y-derivatives of v up to order n. Furthermore

$$\mathrm{D}_x^n\zeta = \frac{\partial \zeta}{\partial u}\frac{\partial^n u}{\partial x^n} + \cdots,$$

the omitted terms depending on $(n-1)$-st and lower order derivatives of u. Thus (3.42) is of the form

$$\left(\frac{\partial \zeta}{\partial u} - \frac{\partial v}{\partial y}\frac{\partial \eta}{\partial u}\right)\frac{\partial^n u}{\partial x^n} = \tilde{\delta}_n\left(x, u, \ldots, \frac{\partial^{n-1} u}{\partial x^{n-1}}, \frac{\partial v}{\partial y}, \ldots, \frac{\partial^n v}{\partial y^n}\right).$$

As before, we can invert the matrix on the left, and substitute for all the derivatives of u which appear on the right their formerly computed expressions, leading to an explicit formula

$$u^{(n)} = \delta^{(n)}(\hat{x}, y, v^{(n)}) \tag{3.43}$$

for the n-th order derivatives of u with respect to x in terms of y, v and the derivatives of v with respect to y up to order n, plus the ubiquitous parametric variables $\hat{x}$.

These formulae serve to parametrize the invariant space, so that if $\pi(z) = w$, then

$$\tilde{I}^{(n)}|_z = \{(x, u^{(n)}): u^{(n)} = \delta^{(n)}(\hat{x}, y, v^{(n)}) \text{ for some } (y, v^{(n)}) \in \widetilde{(M/G)}^{(n)}|_w\},$$

so that for each fixed $(y, v^{(n)})$ satisfying (3.38), the corresponding orbit of $\mathrm{pr}^{(n)}\, G$ in $\tilde{I}^{(n)}$ is parametrized by $\hat{x}$. Moreover, the projection $\pi^{(n)}(x, u^{(n)})$ of such a point is simply the point $(y, v^{(n)}) \in (M/G)^{(n)}$ obtained by omitting the variables $\hat{x}$.

Finally, let $\Delta(x, u^{(n)}) = 0$ be a system of partial differential equations on M which determines a subvariety $\mathscr{S}_\Delta \subset M^{(n)}$. The G-invariant solutions of this system (provided they exist) will have prolongations in the intersection $\mathscr{S}_\Delta \cap I^{(n)}$. If we further require each such G-invariant solution $u = f(x)$ to correspond to a smooth function $v = h(y)$ on the quotient manifold, we must further restrict to the subspace $\tilde{I}^{(n)}$ and look at $\mathscr{S}_\Delta \cap \tilde{I}^{(n)}$. Since $\tilde{I}^{(n)}$ is parametrized by $\hat{x}$, y, $v^{(n)}$, as given by (3.43), we can find this intersection by re-expressing Δ in terms of $\hat{x}$, y, $v^{(n)}$, so $\tilde{\Delta}(\hat{x}, y, v^{(n)}) = \Delta(x, u^{(n)})$ whenever (3.43) holds, whereby

$$\mathscr{S}_\Delta \cap \tilde{I}^{(n)} = \{(\hat{x}, y, v^{(n)}) \in \tilde{I}^{(n)}: \tilde{\Delta}(\hat{x}, y, v^{(n)}) = 0\}.$$

Furthermore, if G is a symmetry group of Δ, $\mathscr{S}_\Delta \cap \tilde{I}^{(n)}$ is locally invariant under the prolonged group action $\mathrm{pr}^{(n)}\, G$, for which $(y, v^{(n)})$ form a complete set of independent invariants. By Proposition 2.18, there is an equivalent set of equations, which we call Δ/G, which are independent of the parametric variables $\hat{x}$, so

$$\mathcal{S}_\Delta \cap \tilde{I}^{(n)} = \{(\hat{x}, y, v^{(n)}) \in \tilde{I}^{(n)}: \Delta/G(y, v^{(n)}) = 0\}.$$

From the above form of the projection $\pi^{(n)}$, we immediately conclude that the part of the reduced system in $\widetilde{(M/G)}^{(n)}$, namely

$$\mathcal{S}_{\Delta/G} = \pi^{(n)}(\mathcal{S}_\Delta \cap \tilde{I}^{(n)}) = \{(y, v^{(n)}): \Delta/G(y, v^{(n)}) = 0\}, \tag{3.44}$$

is given by the equations Δ/G. Thus we have completely justified our procedure of Section 3.1. Theorem 3.36, when completed with Proposition 3.40, shows that we have proved the following rigorous version of the construction of group-invariant solutions to systems of partial differential equations.

Theorem 3.41. *Let G be a local group of transformations acting regularly and transversally on $M \subset X \times U$ with globally defined independent invariants, whereby $M/G \subset Y \times V$. Let Δ be a system of partial differential equations defined over M for which G is a symmetry group. Then there is a reduced system of differential equations Δ/G over M/G, determined by* (3.44), *with the property that any G-invariant function $u = f(x)$ on M corresponding to a well-defined function $v = h(y)$ on M/G will be a solution to Δ if and only if its representative h is a solution to Δ/G.*

(Suitable changes of coordinates on M/G will lead to all the G-invariant solutions to Δ, even those which might originally be nontransverse for the original choice of coordinates.) This completes our development of the theory and justification of the group reduction procedure.

NOTES

Despite numerous claims that the concept of a group-invariant solution of a system of partial differential equations did not originate in its full generality until the 1950's, Lie, in one of his last papers, [6], actually did introduce the present general method for finding such solutions. Lie was concerned with solutions to systems of partial differential equations invariant under groups of contact transformations, but his results include the local versions of the present reduction theorems. In Section 65 of the above-mentioned paper he proves that the solutions to a partial differential equation in two independent variables, which are invariant under a one-parameter group, can all be found by solving a related ordinary differential equation. The generalization to systems of partial differential equations invariant under multi-parameter groups, i.e. our Theorem 3.41, is stated and proved in Section 76 of the same paper, but, as far as I am aware, has never before been referred to in any of the literature on this subject!

Lie died before he could make any application of his discovery. Much later, A. J. A. Morgan, [1], and Michal, [1], restated the special case of Lie's result for one-parameter symmetry groups. Subsequently, Ovsiannikov, [1], [2], reproved the general case, again unaware of the earlier work of Lie. Prior to these rediscoveries, a number of special instances of group-invariant solutions, especially similarity solutions, appeared sporadically in the literature, but without any indication that they were special cases of a much more general theory. The first such construction of which I am aware is in a paper of Boltzmann, [1]. After the turn of the century, similarity solutions appear extensively in the work of Prandtl and Blasius, and, later, Falkner and Skan, on boundary layers in fluid mechanics; see Birkhoff, [2; Chap. 5], for these and other references, as well as a discussion of the history of the Pi Theorem 3.22 from dimensional analysis. Sedov, [1], gave great emphasis to the applicability of scaling groups of symmetries and the consequential similarity solutions in the theory of dimensional analysis of complicated systems. (A good modern introduction to the use of similarity methods in engineering applications is the book of Seshadri and Na, [1].) It remained for Birkhoff, [2], to champion the use of more general symmetry groups for constructing explicit solutions to partial differential equations, and thereby directly inspire the rediscovery of Lie's method.

Since Ovsiannikov began his extensive investigations, the reduction method for constructing group-invariant solutions to partial differential equations has become the focus of much research activity, first in the Soviet Union, and, subsequently, in Europe and the United States. There is by now a large body of Soviet papers on the symmetry properties and explicit solutions for the equations of fluid mechanics, including the recent work of Kapitanskii, [1], [2], mentioned in the text; see Ovsiannikov, [3; p. 391] for a complete bibliography. (Alternative techniques for constructing explicit solutions in fluid mechanics can be found in Berker, [1].) The appearance of extra symmetries after performing a group reduction noticed by Kapitanskii has also been looked at by Rosen, [2].

Group-invariant solutions have been used to great effect in the description of the asymptotic behaviour of much more general solutions to systems of partial differential equations. The book of Barenblatt, [1], gives a good introduction to the applications to hyperbolic equations. In the same vein, Ablowitz and Kodama, [1], have given a rigorous analysis of the asymptotic behaviour of solutions to the Korteweg–de Vries equation, proving that any solution decaying to 0 at $\pm\infty$ ultimately breaks up into a finite number of distinct solitons (travelling waves) plus a dispersive tail decaying like the second Painlevé transcendent solution described here. (Incidentally, the complete classification of the group-invariant solutions of the Korteweg–de Vries equation appeared first in Kostin, [1].) Related ideas appear in the St.-Venant problem in elasticity—see Ericksen, [1].

The general connection between completely integrable (soliton) equations such as the Korteweg–de Vries equation and ordinary differential equations of Painlevé type using the mechanism of group-invariance was first conjectured by Ablowitz, Ramani and Segur, [1]. Proofs of certain special cases of the general conjecture, which gives a quite useful test for "integrability", were given by Ablowitz, Ramani and Segur, [2], and McLeod and Olver, [1]. Recently this method has been significantly extended by Weiss, Tabor and Carnevale, [1].

The rigorous foundation of the general method for constructing group-invariant solutions based on Palais' monograph, [1], using quotient manifolds first appeared in Olver, [2]. The present treatment is a much simplified version of this theory. (See also Vinogradov, [5].) For more details of the theory of extended jet bundles, see Golubitsky and Guillemin, [1; p. 172ff.], and Olver, [2].

The adjoint representation of a Lie group on its Lie algebra was known to Lie. Its use in classifying group-invariant solutions appears in Ovsiannikov, [2; § 86], and [3; § 20]. The latter reference contains more details on how to perform the classification of subgroups of a Lie group under the adjoint action. The method has received extensive development by Patera, Winternitz and Zassenhaus: see [1] and the references therein for many examples of optimal systems of subgroups for the important Lie groups of mathematical physics. The classification of the symmetry algebra of the heat equation is originally due to Weisner, [1], in his investigation of the connections between Lie groups and special functions. See also Kalnins and Miller, [1], where this classification is applied to the problem of separation of variables.

A generalization of the concept of a group-invariant solution known as a partially-invariant solution was introduced by Ovsiannikov, [2; § 17], [3; Chap. 6]. In essence, a partially-invariant solution is one whose graph, while not fully invariant under the group transformations, gets mapped into a submanifold of dimension strictly less than $p + r$ thereby. Here p is the number of independent variables and r the dimension of the orbits of G. (Note that a general function's graph would get mapped into a $(p + r)$-dimensional manifold under all the group transformations.) In certain cases these, too, can be found explicitly by solving a reduced system of differential equations in fewer independent variables, but the intervening calculations are quite a bit more complicated than in the fully invariant case. The interested reader can refer to the above-mentioned works of Ovsiannikov for a full development of this theory, and the thesis of Ondich, [1], for more recent developments.

A second possible generalization was proposed by Bluman and Cole, [1], and Ames, [1; Vol. 2, § 2.10], and called the "nonclassical method" for group-invariant solutions. Here one requires not that the entire subvariety $\mathscr{S}_\Delta^*$ be $\mathrm{pr}^{(n)} G$-invariant, but only that its intersection with the invariant space, $\mathscr{S}_\Delta^* \cap I_*^{(n)}$, be $\mathrm{pr}^{(n)} G$-invariant. Although this method does lead to reduced equations, it is a little *too* general in that, once we admit the prolongations of the equations into the picture, *every* group of transformations on M satisfies this requirement and, conversely, *every* solution of the system can be obtained in this manner. See Olver and Rosenau, [1], [2]. A direct method for finding explicit solutions to partial differential equations introduced by Clarkson and Kruskal, [1], has proved very effective. Levi and Winternitz, [1], noted its connection with the nonclassical method, while Clarkson and Nucci, [1], showed that it is not quite as general as the nonclassical approach. Galaktionov, [1], gives a further promising generalization called "nonlinear separation." See Olver, [16], for a review of the available methods.

Exercises

3.1. Consider the axially symmetric wave equation $u_{tt} - u_{xx} - (1/x)u_x = 0$.
(a) What is the symmetry group?
(b) Find and classify the group-invariant solutions.
(c) What is the fundamental solution to this equation?

3.2. The BBM equation $u_t + u_x + uu_x - u_{xxt} = 0$ arises as a model equation for the uni-directional propagation of long waves in shallow water.
(a) What is the symmetry group of this equation?
(b) Find group-invariant solutions corresponding to the various one-parameter subgroups found in part (a). (McLeod and Olver, [1].)

3.3. Determine the scale-invariant solutions to Boltzmann's problem $u_t = (uu_x)_x$, the solutions of which represent diffusion of some material in a medium, the rate of diffusion of which is proportional to the concentration of the material. What other types of group-invariant solutions exist? (Dresner, [1; §4.1].)

*3.4. Discuss the group-invariant solutions to the two-dimensional wave equation. Can you classify them?

3.5. Discuss the scale-invariant solutions to the two-dimensional Euler equations of ideal fluid flow. (The reduced equations are *not*, as far as I know, soluble in closed form!)

3.6. Assume that the fluid resistance of an object is determined by the density of the fluid ρ, the velocity of the object v, the object diameter d, and the compressibility $\rho^{-2}\, dp/d\rho$ of the fluid. Let $c^2 = dp/d\rho$ denote the sound speed. Prove that $D = \rho v^2 d^2 f(M)$, where $M = v/c$ is the Mach number of the fluid. ($M < 1$ corrresponds to subsonic motion, $M > 1$ to supersonic motion.) (Birkhoff, [2; p. 92].)

3.7. In 1947, G. I. Taylor determined the energy released from the first atomic explosion in New Mexico by applying a similarity analysis to the photographs of it. In an expanding spherical shock wave, the radius R will depend on time t, energy E released, and the ambient air density ρ_0 and pressure p_0. Assuming dimensional homogeneity, prove that

$$R = \left(\frac{t^2 E}{\rho_0}\right)^{1/5} h\left[p_0\left(\frac{t^6 E^2}{\rho_0^3}\right)^{1/5}\right]$$

for some function $h(\zeta)$. (For t small, the argument ζ of h is small, so we can approximate $R \simeq h_0 t^{2/5} E^{1/5} \rho_0^{-1/5}$, $h_0 = h(0)$; this was the relation used by Taylor.) (G. I. Taylor, [1], [2].)

***3.8.** Find the orbits of the adjoint representation of the Euclidean groups E(2) and E(3). (See Exercise 1.29.)

***3.9.** Prove that every subalgebra of the Korteweg–de Vries symmetry algebra (2.68) is uniquely equivalent to one subalgebra in the optimal system consisting of 0, the one-dimensional subalgebras (3.25), the subalgebras spanned by

$$\{\mathbf{v}_1, \mathbf{v}_4\}, \{\mathbf{v}_2, \mathbf{v}_4\}, \{\mathbf{v}_3, \mathbf{v}_4\}, \{\mathbf{v}_1, \mathbf{v}_3\}, \{\mathbf{v}_1, \mathbf{v}_2 + \mathbf{v}_3\}, \{\mathbf{v}_1, \mathbf{v}_2 - \mathbf{v}_3\},$$
$$\{\mathbf{v}_1, \mathbf{v}_2\}, \{\mathbf{v}_1, \mathbf{v}_3, \mathbf{v}_4\}, \{\mathbf{v}_1, \mathbf{v}_2, \mathbf{v}_4\}, \{\mathbf{v}_1, \mathbf{v}_2, \mathbf{v}_3\},$$

and the full symmetry algebra itself.

3.10. Consider the differential equation $\Delta[u] = u_{xy} = 0$, on $M = X \times U \simeq \mathbb{R}^3$.
(a) Prove that the one-parameter group G of translations in the x-direction, $(x, y, u) \mapsto (x + \varepsilon, y, u)$ is a symmetry group.
(b) Show that the reduced equation Δ/G on M/G is vacuous, so any function on M/G determines a G-invariant solution to Δ.
(c) More generally, prove that if G is a regular symmetry group of a system of differential equations Δ, and its invariant space $\mathcal{I}^{(n)}$ is a subset of the corresponding subvariety $\mathcal{S}_\Delta \subset M^{(n)}$, then *every* function on M/G gives rise to a G-invariant solution to Δ. How might the condition $\mathcal{I}^{(n)} \subset \mathcal{S}_\Delta \subset M^{(n)}$ be checked in practice? (See also Exercise 3.18.)

3.11. (a) Suppose G is a symmetry group of the system Δ and $H \subset G$ is a normal subgroup acting regularly on $M \subset X \times U$. Prove that the reduced system Δ/H is invariant under the quotient group G/H acting on M/H.
(b) Suppose $p = 2$, $q = 1$ and $\Delta(x, u^{(n)}) = 0$ is a single n-th order partial differential equation. Prove that if Δ is invariant under an $(n + 1)$-parameter solvable Lie group, then all group-invariant solutions corresponding to a particular one-parameter subgroup can be found by quadrature.

3.12. Suppose $\mathbf{v}$ is a vector field on the smooth manifold M and $\dot{x} = \xi(x)$ the system of ordinary differential equations describing the flow of $\mathbf{v}$. Suppose G acts regularly on M, and is a symmetry group of this system. Prove that there is an induced vector field $\tilde{\mathbf{v}} = d\pi(\mathbf{v})$ on the quotient manifold M/G whose flow corresponds to that of $\mathbf{v}$ on M. Discuss how this result applies to Theorem 2.66.

3.13. Let G be a Lie group and $H \subset G$ a closed subgroup, which acts on G itself by right translation: $g \mapsto g \cdot h$, $h \in H$. Prove that the quotient space G/H is a smooth manifold. What is G/H in the case $G = \mathrm{SO}(3)$ and $H \simeq \mathrm{SO}(2)$, a one-parameter subgroup of rotations about a fixed axis?

3.14. (a) Consider the scaling group $G: (x, y) \mapsto (\lambda x, \lambda^{-1} y)$ acting regularly on $M = \mathbb{R}^2 \backslash \{0\}$. Prove that the quotient manifold is not Hausdorff and discuss its structure.
(b) Do the same problem for the scaling group of symmetries of the Korteweg–de Vries equation. (Olver, [2].)

3.15. Let $p = 2$, $q = 1$ and consider the one-parameter group $G: (x, y, u) \mapsto (x + \varepsilon, y + \varepsilon u, u)$. Prove that G acts transversally everywhere, but there are no nonconstant globally-defined G-invariant functions.

3.16. The p-Grassmann bundle of an m-dimensional manifold M, $m \geqslant p$, is defined so that over each point $x \in M$, $\mathrm{Grass}(p, M)|_x = \mathrm{Grass}(p, TM|_x)$ is the Grassmann manifold of p-planes in the tangent space $TM|_x$. (See Exercise 1.2.) Prove that this is the same as the first extended jet bundle: $\mathrm{Grass}(p, M) \simeq M_*^{(1)}$. (Olver, [2].)

3.17. Let G be a group of transformations acting on $M \subset X \times U$ with prolongation $\mathrm{pr}^{(n)} G$ acting on $M^{(n)}$.

(a) Prove that the dimension of the orbits of $\mathrm{pr}^{(n)} G$ is greater than or equal to the dimension of the orbits of G. Give an example where the strict inequality holds.

(b) Prove that if G is an r-parameter group, and G has r-dimensional orbits, then the same is true of $\mathrm{pr}^{(n)} G$.

(c) Prove that if G has r-dimensional orbits, then $I^{(n)} = (I^{(1)})^{(n-1)}$, where $(I^{(1)})^{(n-1)}$ denotes the $(n-1)$-st prolongation of the invariant space $I^{(1)} \subset M^{(1)}$ as determined by Definition 2.81. Interpret Corollary 2.54 in light of this result.

*3.18. *Explicit Characterization of the Invariant Space*

(a) Let $p = q = 1$, and let G be a regular one-parameter group of transformations acting on $M \subset X \times U$ with a single global invariant $\zeta(x, u)$. Prove that the invariant space $I^{(n)} \subset M^{(n)}$ is defined by the equations

$$I^{(n)} = \{(x, u^{(n)}): D_x^k \zeta = 0, k = 1, 2, \ldots, n\}.$$

(b) Let $q = 1$ but p be arbitrary. Let G be a regular one-parameter group of transformations acting on M with global invariants $\eta^1(x, u), \ldots, \eta^p(x, u)$. Prove that

$$I^{(1)} = \{(x, u^{(1)}): \mathrm{D}(\eta^1, \ldots, \eta^p)/\mathrm{D}(x^1, \ldots, x^p) = 0\},$$

where the defining equation stands for the $p \times p$ "total Jacobian determinant"

$$\frac{\mathrm{D}(\eta^1, \ldots, \eta^p)}{\mathrm{D}(x^1, \ldots, x^p)} = \det(D_i \eta^j).$$

What if G is an r-parameter group?

(c) Generalize part (b) to give an explicit characterization of the invariant space $I^{(n)}$ in general.

(d) How is this result related to Theorem 3.38?

3.19. Show that the vector field $\mathbf{v} = 2t\partial_x + \partial_t + 8t\partial_u$ is not a symmetry of the equation $u_{tt} = uu_{xx}$, but nevertheless one can use the method of this chapter to find solutions which are invariant under the one-parameter group generated by $\mathbf{v}$. Show that none of these arise among the standard group-invariant solutions. Explain. (See the following exercise.) (Olver and Rosenau, [2].)

*3.20. *The Nonclassical Method for Group-Invariant Solutions.* In Bluman and Cole, [1], the following method is proposed as a generalization of the reduction method for finding group-invariant solutions.

(a) Let Δ be an n-th order system of partial differential equations over M with corresponding subvariety $\mathscr{S}_\Delta^* \subset M_*^{(n)}$. Let G be a regular group of transformations acting on M with invariant space $I_*^{(n)} \subset M_*^{(n)}$. Prove that if the intersection $\mathscr{S}_\Delta^* \cap I_*^{(n)}$ is invariant under $\mathrm{pr}^{(n)} G$, then there is a reduced system of differential equations Δ/G on the quotient manifold M/G such that all the solutions to Δ/G give rise to G-invariant solutions to Δ and conversely. (Note especially that $\mathscr{S}_\Delta^*$ itself does *not* have to be invariant under $\mathrm{pr}^{(n)} G$, so these groups are more general than symmetry groups as defined in Chapter 2.)

(b) Interpret Exercise 3.19 in light of this result.

(c) Let Δ be *any* system of differential equations and G *any* (regular) group of transformations. Prove that a suitable prolongation of $\mathscr{S}_\Delta^* \cap I_*^{(n)}$ (as per Definition 2.81) is always $\mathrm{pr}^{(n)} G$-invariant. Thus one can use *any* group to effect the reduction of part (a). (*Hint*: Use the prolongation formula (2.50) and the characterization of the invariant space in Theorem 3.38.)

(d) Conversely, show that if $u = f(x)$ is any solution to Δ, then there exists a group G leading to f by the reduction method of part (a).

(Olver and Rosenau, [1], [2].)

3.21. Let $\mathbf{v}$ be a vector field on $M \subset X \times U \simeq \mathbb{R}^p \times \mathbb{R}^q$, and let $Q = (Q_1, \ldots, Q_q)$ be the characteristic of $\mathbf{v}$ in the (x, u)-coordinates, cf. (2.48). Let $y = Y(x, u)$, $v = \Psi(x, u)$ be a change of coordinates on M, and let $\tilde{Q} = (\tilde{Q}_1, \ldots, \tilde{Q}_q)$ be the characteristic of $\mathbf{v}$ in the new (y, v)-coordinates on M. Prove that $\tilde{Q}$ is related to Q by the change of variables formula

$$\tilde{Q}_\beta = \sum_{\alpha=1}^{q} Q_\alpha \left(\frac{\partial \Psi^\beta}{\partial u^\alpha} - \sum_{j=1}^{p} \frac{\partial Y^j}{\partial u^\alpha} \frac{\partial v^\beta}{\partial y^j} \right), \qquad \beta = 1, \ldots, q.$$

*3.22. (a) Find an optimal system of all higher-dimensional subalgebras of the six-dimensional heat algebra (2.55).

(b) Show that, in contrast to the one-dimensional case in Example 3.13, the subalgebras in part (a) do *not* form an optimal system for the full infinite-dimensional heat symmetry algebra.

(Svinolupov and Sokolov, [1].)

CHAPTER 4

Symmetry Groups and Conservation Laws

In the study of systems of differential equations, the concept of a conservation law, which is a mathematical formulation of the familiar physical laws of conservation of energy, conservation of momentum and so on, plays an important role in the analysis of basic properties of the solutions. In 1918, Emmy Noether proved the remarkable result that for systems arising from a variational principle, every conservation law of the system comes from a corresponding symmetry property.† For example, invariance of a variational principle under a group of time translations implies the conservation of energy for the solutions of the associated Euler–Lagrange equations, and invariance under a group of spatial translations implies conservation of momentum. This basic principle constitutes the first fundamental result in the study of classical or quantum-mechanical systems with prescribed groups of symmetries. Noether's method is the principal systematic procedure for constructing conservation laws for complicated systems of partial differential equations.

For the applicability of Noether's theorem, one needs some form of variational structure in the system under consideration. The first section of this chapter gives a rudimentary introduction to the relevant aspects of the calculus of variations, of which the construction of the Euler–Lagrange equations characterizing the minimizers of a variational problem is the most important. Beyond this, not many of the results from the calculus of variations will be required, so the student interested in further studying this important field of mathematics would be well advised to consult any of the standard reference books on the subject. Not every symmetry group of a system of Euler–

† There is now an English translation of Noether's paper, [1], available. The reader is *strongly* urged to read this essential work.

Lagrange equations will give rise to a conservation law; one needs the group to satisfy an additional "variational" property of leaving the variational integral in a certain sense invariant. Section 4.2 develops the theory of variational symmetries, illustrated by a number of examples. In the case of a system of ordinary differential equations in variational form, the variational character of a symmetry group *doubles* the effectiveness of the reduction procedure presented in Section 2.5, so that a system of Euler–Lagrange equations which admits a one-parameter group of variational symmetries can be reduced in order by two.

The third section of this chapter is devoted to the systematic development of the theory of conservation laws of systems of differential equations. An important complication here is the existence of trivial conservation laws, which apply to any system of differential equations and in essence provide no new information on the behaviour of solutions to the particular system being considered. Pending some proofs to be given at the end of Chapter 5, we are able to completely characterize such trivial laws. Each nontrivial conservation law is, for normal systems of differential equations, uniquely characterized by a certain function, called its characteristic. Once we have the connection between conservation laws and their characteristics well in hand, the proof of the so-called "classical form" of Noether's theorem is immediate. In the second half of Section 4.4 we apply the constructions embodied in Noether's theorem to determine conservation laws for a number of systems of physical importance. Lack of space, however, precludes us from applying these conservation laws to the direct study of properties of solutions of the systems, which include global existence results, decay estimates, scattering theory, crack and dislocation problems, stability of solutions and so on; these can be found in the references discussed at the end of the chapter.

4.1. The Calculus of Variations

As usual, to keep things as simple as possible, we will work in Euclidean space, with $X = \mathbb{R}^p$, with coordinates $x = (x^1, \ldots, x^p)$ representing the independent variables, and $U = \mathbb{R}^q$, with coordinates $u = (u^1, \ldots, u^q)$ the dependent variables in our problem. (Extension of these local results to variational problems over smooth manifolds is not difficult; however, global results require the introduction of topological tools—see Anderson, [1] or Vinogradov, [1].). Let $\Omega \subset X$ be an open, connected subset with smooth boundary $\partial\Omega$. A *variational problem* consists of finding the extrema (maxima or minima) of a *functional*

$$\mathscr{L}[u] = \int_\Omega L(x, u^{(n)})\, dx$$

in some class of functions $u = f(x)$ defined over Ω. The integrand $L(x, u^{(n)})$, called the *Lagrangian* of the variational problem $\mathscr{L}$, is a smooth function of

x, u and various derivatives of u. The precise specification of the class of functions over which $\mathscr{L}$ is to be extremized will depend both on boundary conditions which might be pertinent to the physical problem, as well as differentiability conditions required of the extremals $u = f(x)$.

As a simple example, the problem of finding a curve of minimum length joining two points (a, b) and (c, d) in the plane can be cast into variational form as follows. Assume that the minimizing curve is given as the graph of a function $u = f(x)$. The length of such a curve is

$$\mathscr{L}[u] = \int_a^c \sqrt{1 + u_x^2}\, dx.$$

The variational problem consists of minimizing $\mathscr{L}$ over the space of differentiable functions $u = f(x)$, say, such that $b = f(a)$ and $d = f(c)$.

The precise degree of smoothness required of the extrema of a given variational problem, the space of functions being extremized over, and the appropriate norm(s) are quite delicate matters in general and quickly lead into advanced topics in nonlinear functional analysis. The complex issues involved are not directly relevant to our immediate area of inquiry, however, and we therefore adopt the admittedly oversimplifying assumption of only considering smooth (C^∞) extremals of a variational problem. Extending our results on symmetry groups and conservation laws to more general types of functions must then be done on a case by case basis.

The Variational Derivative

In finite dimensions, the extrema of a smooth real-valued function $f(x)$, $x \in \mathbb{R}^m$, are determined by looking at the points where the gradient $\nabla f(x)$ vanishes. The gradient itself is found by seeing how f changes under small changes in x:

$$\langle \nabla f(x), y \rangle = \frac{d}{d\varepsilon}\bigg|_{\varepsilon=0} f(x + \varepsilon y),$$

where $\langle x, y \rangle$ is the usual inner product on $\mathbb{R}^m$. For functionals $\mathscr{L}[u]$, the role of the gradient is played by the "variational derivative" of $\mathscr{L}$. To construct this object, we look at how $\mathscr{L}$ changes under small "variations" in u. The inner product $\langle \cdot\, , \cdot \rangle$ on $\mathbb{R}^m$ is replaced by the L^2 inner product

$$\langle f, g \rangle = \int_\Omega f(x) \cdot g(x)\, dx = \int_\Omega \sum_{\alpha=1}^{q} f^\alpha(x) g^\alpha(x)\, dx$$

between vector-valued functions $f, g\colon \mathbb{R}^p \to \mathbb{R}^q$. This motivates the following definition:

Definition 4.1. Let $\mathscr{L}[u]$ be a variational problem. The *variational derivative* of $\mathscr{L}$ is the unique q-tuple

$$\delta\mathscr{L}[u] = (\delta_1\mathscr{L}[u], \ldots, \delta_q\mathscr{L}[u]),$$

with the property that

$$\frac{d}{d\varepsilon}\bigg|_{\varepsilon=0} \mathscr{L}[f + \varepsilon\eta] = \int_\Omega \delta\mathscr{L}[f(x)]\cdot\eta(x)\,dx \tag{4.1}$$

whenever $u = f(x)$ is a smooth function defined on Ω, and $\eta(x) = (\eta^1(x), \ldots, \eta^q(x))$ is a smooth function with compact support in Ω, so that $f + \varepsilon\eta$ still satisfies any boundary conditions that might be imposed on the space of functions over which we are extremizing $\mathscr{L}$. The component $\delta_\alpha\mathscr{L} = \delta\mathscr{L}/\delta u^\alpha$ is the *variational derivative of* $\mathscr{L}$ *with respect to* u^α.

Proposition 4.2. *If* $u = f(x)$ *is an extremal of* $\mathscr{L}[u]$, *then*

$$\delta\mathscr{L}[f(x)] = 0, \qquad x \in \Omega. \tag{4.2}$$

Proof. Since f is an extremal, for any η of compact support in Ω, $f + \varepsilon\eta$ lies in the same function space, so, as a function of ε, $\mathscr{L}[f + \varepsilon\eta]$ must have an extremum at $\varepsilon = 0$. Therefore, by elementary calculus, (4.1) must vanish for all η of compact support in Ω, hence (4.2) must hold everywhere. (The same argument proves the uniqueness of $\delta\mathscr{L}$.) □

The general formula for the variational derivative is not difficult to find. First of all, interchanging the order of differentiation and integration (which is justified under our assumption of smoothness)

$$\begin{aligned}\frac{d}{d\varepsilon}\bigg|_{\varepsilon=0} \mathscr{L}[f + \varepsilon\eta] &= \int_\Omega \frac{d}{d\varepsilon}\bigg|_{\varepsilon=0} L(x, \mathrm{pr}^{(n)}(f + \varepsilon\eta)(x))\,dx \\ &= \int_\Omega \left\{\sum_{\alpha,J} \frac{\partial L}{\partial u_J^\alpha}(x, \mathrm{pr}^{(n)}f(x))\cdot\partial_J\eta^\alpha(x)\right\} dx,\end{aligned}$$

where the u_J^α are as usual the partial derivatives of u^α, and $\partial_J\eta^\alpha$ the corresponding derivatives of η^α. Since η has compact support, we can use the divergence theorem to integrate the latter expression by parts, with the boundary terms on $\partial\Omega$ vanishing. Each partial derivative $\partial/\partial x^j$, when applied to the derivatives $\partial L/\partial u_J^\alpha$ of the Lagrangian, becomes the total derivative D_j since L depends on x through the function $u = f(x)$ also—see Definition 2.34. Therefore

$$\frac{d}{d\varepsilon}\bigg|_{\varepsilon=0} \mathscr{L}[f + \varepsilon\eta] = \int_\Omega \left\{\sum_{\alpha=1}^q \left[\sum_J (-D)_J \frac{\partial L}{\partial u_J^\alpha}(x, \mathrm{pr}^{(n)} f(x))\right]\eta^\alpha(x)\right\} dx,$$

where, for $J = (j_1, \ldots, j_k)$,

$$(-D)_J = (-1)^k D_J = (-D_{j_1})(-D_{j_2})\cdots(-D_{j_k}).$$

The operator appearing in the preceding formula is of key importance in the calculus of variations.

Definition 4.3. For $1 \leqslant \alpha \leqslant q$, the α-th *Euler operator* is given by

$$\mathsf{E}_\alpha = \sum_J (-D)_J \frac{\partial}{\partial u_J^\alpha}, \tag{4.3}$$

the sum extending over all multi-indices $J = (j_1, \ldots, j_k)$ with $1 \leqslant j_\kappa \leqslant p$, $k \geqslant 0$. Note that to apply E_α to any given function $L(x, u^{(n)})$ of u and its derivatives, only finitely many terms in the summation are required, since L depends on only finitely many derivatives u_J^α.

Thus, according to our calculation, the variational derivative of $\mathscr{L}[u] = \int_\Omega L(x, u^{(n)})\, dx$ is found by applying the Euler operator to the Lagrangian: $\delta\mathscr{L}[u] = \mathsf{E}(L)$, where $\mathsf{E}(L) = (\mathsf{E}_1(L), \ldots, \mathsf{E}_q(L))$. Proposition 4.2 provides the classical necessary conditions for smooth extremals of a variational problem.

Theorem 4.4. *If $u = f(x)$ is a smooth extremal of the variational problem $\mathscr{L}[u] = \int_\Omega L(x, u^{(n)})\, dx$, then it must be a solution of the* Euler–Lagrange equations

$$\mathsf{E}_\nu(L) = 0, \qquad \nu = 1, \ldots, q.$$

Of course, not every solution to the Euler–Lagrange equations is an extremal. The other solutions furnish other types of critical points for the functional.

Example 4.5. Let us look at the special case $p = q = 1$, so we are considering a single function $u = f(x)$ of a single independent variable. The Euler operator here takes the form

$$\mathsf{E} = \sum_{j=0}^{\infty} (-D_x)^j \frac{\partial}{\partial u_j} = \frac{\partial}{\partial u} - D_x \frac{\partial}{\partial u_x} + D_x^2 \frac{\partial}{\partial u_{xx}} - \cdots,$$

where D_x is the total derivative with respect to x, and $u_j = d^j u/dx^j$. The Euler–Lagrange equation for an n-th order variational problem

$$\mathscr{L}[u] = \int_a^b L(x, u^{(n)})\, dx$$

takes the form

$$0 = \mathsf{E}(L) = \frac{\partial L}{\partial u} - D_x \frac{\partial L}{\partial u_x} + D_x^2 \frac{\partial L}{\partial u_{xx}} - \cdots + (-1)^n D_x^n \frac{\partial L}{\partial u_n}.$$

It is a $2n$-th order ordinary differential equation provided L satisfies the nondegeneracy condition $\partial^2 L/\partial u_n^2 \neq 0$. In particular, for a first order variational problem, $L = L(x, u, u_x)$, we recover the familiar second order Euler–

Lagrange equation

$$0 = \frac{\partial L}{\partial u} - D_x \frac{\partial L}{\partial u_x} = \frac{\partial L}{\partial u} - \frac{\partial^2 L}{\partial x\, \partial u_x} - u_x \frac{\partial^2 L}{\partial u\, \partial u_x} - u_{xx} \frac{\partial^2 L}{\partial u_x^2}.$$

Thus, for our curve length minimizing problem, the Euler–Lagrange equation is

$$-D_x\left(\frac{u_x}{\sqrt{1+u_x^2}}\right) = -\frac{u_{xx}}{(1+u_x^2)^{3/2}} = 0.$$

The solutions are all straight lines $u = mx + k$, and these are the only smooth candidates for the minimization problem.

Example 4.6. Perhaps the most famous variational problem comes from Dirichlet's principle for Laplace's equation $\Delta u = 0$. Here we set

$$\mathscr{L}[u] = \int_\Omega \tfrac{1}{2}|\nabla u|^2\, dx = \int_\Omega \tfrac{1}{2} \sum_{i=1}^p u_i^2\, dx,$$

where $u_i = \partial u/\partial x^i$ and $\Omega \subset X \simeq \mathbb{R}^p$. The Euler–Lagrange equation is

$$0 = \mathsf{E}(L) = \sum_{i=1}^p (-D_i)\frac{\partial L}{\partial u_i} = -\sum_{i=1}^p D_i(u_i) = -\Delta u,$$

which agrees with Laplace's equation up to sign. Further examples will appear later in this chapter.

Null Lagrangians and Divergences

Occasionally, for a given variational problem the Euler–Lagrange equations vanish identically and so every function is a possible extremal of the problem. For example, if

$$\mathscr{L}[u] = \int_a^b u u_x\, dx,$$

then

$$\delta\mathscr{L} = u_x - D_x(u) \equiv 0$$

for all u. In this case the variational problem is trivial, since by the fundamental theorem of calculus

$$\mathscr{L}[u] = \int_a^b D_x(\tfrac{1}{2}u^2)\, dx = \tfrac{1}{2}u^2\Big|_{x=a}^{b},$$

so any function $u = f(x)$ satisfying the relevant boundary conditions will give the same value for $\mathscr{L}$.

The situation readily generalizes to the case of several independent variables. If $x = (x^1, \ldots, x^p)$ and $P(x, u^{(n)}) = (P_1(x, u^{(n)}), \ldots, P_p(x, u^{(n)}))$ is a p-tuple of smooth functions of x, u and the derivatives of u, we define the *total divergence* of P to be the function

$$\operatorname{Div} P = D_1 P_1 + D_2 P_2 + \cdots + D_p P_p, \tag{4.4}$$

where each D_j is the total derivative with respect to x^j. For instance, if $p = 2$, and $P = (uu_y, uu_x)$, then

$$\operatorname{Div} P = D_x(uu_y) + D_y(uu_x) = 2uu_{xy} + 2u_x u_y.$$

If a Lagrangian $L(x, u^{(n)})$ can be written as a divergence, so $L = \operatorname{Div} P$ for some p-tuple P, then, by the divergence theorem,

$$\mathscr{L}[u] = \int_\Omega L\,dx = \int_{\partial\Omega} P \cdot dS$$

for any function $u = f(x)$ and any bounded domain Ω with smooth boundary $\partial\Omega$. Thus $\mathscr{L}[f]$ depends only on the boundary behaviour of $u = f(x)$, and will be unaffected by the variations η used in the definition of the variational derivative. Therefore the Euler–Lagrange equations for such a functional are identically 0. Remarkably, these are the only such examples of "null Lagrangians".

Theorem 4.7. *A function $L(x, u^{(n)})$ of x, u and the derivatives of u, defined everywhere on $X \times U^{(n)}$, is a* null Lagrangian, *meaning that the Euler-Lagrange equations $\mathrm{E}(L) \equiv 0$ vanish identically for all x, u, if and only if it is a total divergence: $L = \operatorname{Div} P$, for some p-tuple of functions $P = (P_1, \ldots, P_p)$ of x, u and the derivatives of u.*

Proof. The proof that $\mathrm{E}(\operatorname{Div} P) \equiv 0$ follows from the above remarks, or by direct computation; see Section 5.4. To prove the converse, suppose $L(x, u^{(n)})$ is a null Lagrangian, and consider the derivative

$$\frac{d}{d\varepsilon} L(x, \varepsilon u^{(n)}) = \sum_{\alpha, J} u_J^\alpha \frac{\partial L}{\partial u_J^\alpha}(x, \varepsilon u^{(n)}).$$

Each term in this sum can be integrated by parts; for example

$$u_i^\alpha \frac{\partial L}{\partial u_i^\alpha} = D_i(u^\alpha) \frac{\partial L}{\partial u_i^\alpha} = -u^\alpha D_i \frac{\partial L}{\partial u_i^\alpha} + D_i\left(u^\alpha \frac{\partial L}{\partial u_i^\alpha}\right),$$

leading to an expression of the form

$$\begin{aligned}\frac{d}{d\varepsilon} L(x, \varepsilon u^{(n)}) &= \sum_{\alpha=1}^{q} u^\alpha \sum_J (-D)_J \frac{\partial L}{\partial u_J^\alpha}(x, \varepsilon u^{(n)}) + \operatorname{Div} \hat{P}(\varepsilon; x, u^{(2n)}) \\ &= u \cdot \mathrm{E}(L)(x, \varepsilon u^{(2n)}) + \operatorname{Div} \hat{P}(\varepsilon; x, u^{(2n)})\end{aligned}$$

for some p-tuple $\hat{P}$ of functions of x, u and derivatives of u whose precise form is not of importance. (However, see (5.150), (5.151).) Since $\mathsf{E}(L) \equiv 0$, we can integrate with respect to ε,

$$L(x, u^{(n)}) - L(x, 0) = \operatorname{Div} \tilde{P}, \quad \text{where} \quad \tilde{P}(x, u^{(2n)}) = \int_0^1 \hat{P}(\varepsilon; x, u^{(2n)})\, d\varepsilon.$$

Finally, since L is defined on all of $\mathbb{R}^p$, we can always find a p-tuple $p(x)$ of ordinary functions of x such that $\operatorname{div} p(x) = L(x, 0)$, so the theorem holds with $P = \tilde{P} + p$. □

A corollary of Theorem 4.7 is the basic fact that two Lagrangians L and $\tilde{L}$ have the same Euler–Lagrange expression, $\mathsf{E}(L) = \mathsf{E}(\tilde{L})$, if and only if they differ by a divergence,

$$L = \tilde{L} + \operatorname{Div} P.$$

Invariance of the Euler Operator

Since the Euler–Lagrange equations determine the extremals of a variational problem, their solution set should remain unchanged by a change of variables. This suggests that the Euler operator itself should be, more or less, invariant under a change of variables. Here we derive the basic formula expressing this fact.

Note first that if

$$\tilde{x} = \Xi(x, u), \qquad \tilde{u} = \Phi(x, u), \tag{4.5}$$

is any change of variables, there is an induced change of variables

$$\tilde{u}^{(n)} = \Phi^{(n)}(x, u^{(n)})$$

for the derivatives, given by prolongation. Thus, given a function $u = f(x)$, (4.5) implicitly defines the transformed function $\tilde{u} = \tilde{f}(\tilde{x})$ (provided the conditions required by the implicit function theorem are satisfied). Each functional

$$\mathscr{L}[f] = \int_\Omega L(x, \operatorname{pr}^{(n)} f(x))\, dx$$

will be transformed into a new form

$$\tilde{\mathscr{L}}[\tilde{f}] = \int_{\tilde{\Omega}} \tilde{L}(\tilde{x}, \operatorname{pr}^{(n)} \tilde{f}(\tilde{x}))\, d\tilde{x}.$$

In this latter integral, the transformed domain

$$\tilde{\Omega} = \{\tilde{x} = \Xi(x, f(x)) \colon x \in \Omega\}$$

will depend not only on the original domain Ω, but also on the precise function $u = f(x)$ on which $\mathscr{L}$ is being evaluated. The formula for the new Lagrangian follows readily from the change of variables formula for multiple integrals:

$$L(x, \mathrm{pr}^{(n)} f(x)) = \tilde{L}(\tilde{x}, \mathrm{pr}^{(n)} \tilde{f}(\tilde{x})) \det \mathsf{J}(x, \mathrm{pr}^{(1)} f(x)), \tag{4.6}$$

whenever $(\tilde{x}, \tilde{u})$ are given by (4.5), where J is the Jacobian matrix with entries

$$\mathsf{J}^{ij}(x, \mathrm{pr}^{(1)} f(x)) = \frac{\partial}{\partial x^j}[\Xi^i(x, f(x))] = D_j \Xi^i(x, \mathrm{pr}^{(1)} f(x)), \qquad i, j = 1, \ldots, p,$$

corresponding to the function f. Here we are assuming, for simplicity, that the change of variables is orientation-preserving, so $\det \mathsf{J}(x) > 0$.

Theorem 4.8. *Let $L(x, u^{(n)})$ and $\tilde{L}(\tilde{x}, \tilde{u}^{(n)})$ be two Lagrangians related by the change of variables formula* (4.5), (4.6). *Then*

$$\mathsf{E}_{u^\alpha}(L) = \sum_{\beta=1}^{q} \mathsf{F}_{\alpha\beta}(x, u^{(1)}) E_{\tilde{u}^\beta}(\tilde{L}), \qquad \alpha = 1, \ldots, q, \tag{4.7}$$

whenever $(\tilde{x}, \tilde{u}^{(n)})$ and $(x, u^{(n)})$ are so related, where $\mathsf{F}_{\alpha\beta}$ is the determinant of the following $(p+1) \times (p+1)$ matrix

$$\mathsf{F}_{\alpha\beta} = \det \begin{bmatrix} D_1\Xi^1 & \cdots & D_p\Xi^1 & \partial\Xi^1/\partial u^\alpha \\ \vdots & & \vdots & \vdots \\ D_1\Xi^p & \cdots & D_p\Xi^p & \partial\Xi^p/\partial u^\alpha \\ D_1\Phi^\beta & \cdots & D_p\Phi^\beta & \partial\Phi^\beta/\partial u^\alpha \end{bmatrix}. \tag{4.8}$$

PROOF. Let $u = f(x)$ be a given function defined over a domain Ω and let $\tilde{u} = \tilde{f}(\tilde{x})$, $\tilde{x} \in \tilde{\Omega}$, be the corresponding function in the transformed variables, which is usually well defined as long as Ω is sufficiently small. For ε sufficiently small, the perturbations $u_\varepsilon = f(x, \varepsilon) = f(x) + \varepsilon\eta(x)$, η of compact support in Ω, have corresponding expressions $\tilde{u} = \tilde{f}(\tilde{x}, \varepsilon)$ determined implicitly

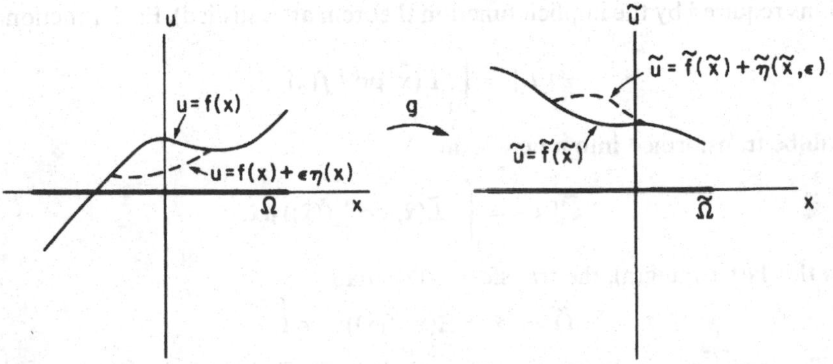

Figure 10. Change of coordinates for the variation of a functional.

by the relations

$$\tilde{x} = \Xi(x, f(x) + \varepsilon\eta(x)), \qquad \tilde{u} = \Phi(x, f(x) + \varepsilon\eta(x)). \tag{4.9}$$

An important point is that since η has compact support in Ω, each $\tilde{f}(\tilde{x}, \varepsilon) = \tilde{f}(\tilde{x}) + \tilde{\eta}(\tilde{x}, \varepsilon)$ is defined over a common domain

$$\tilde{\Omega} = \{\tilde{x} = \Xi(x, f(x)) : x \in \Omega\}$$

independent of ε, and $\tilde{\eta}$ has compact support in $\tilde{\Omega}$. The variational derivative of $\mathscr{L}$ was determined by differentiating $\mathscr{L}[f + \varepsilon\eta]$ with respect to ε at $\varepsilon = 0$; similarly, by a slight generalization of the argument leading to the formula (4.3) for the Euler operator, we find

$$\left.\frac{d}{d\varepsilon}\right|_{\varepsilon=0} \tilde{\mathscr{L}}[\tilde{f}] = \int_{\tilde{\Omega}} \mathsf{E}_{\tilde{u}}(\tilde{L}) \cdot \left.\frac{\partial \tilde{f}}{\partial \varepsilon}\right|_{\varepsilon=0} d\tilde{x}, \tag{4.10}$$

where $\mathsf{E}_{\tilde{u}}(\tilde{L})$ is evaluated at $\tilde{u} = \tilde{f}$. We now need to evaluate $\partial\tilde{f}/\partial\varepsilon$.

Keeping in mind that when variations of $\tilde{\mathscr{L}}$ are computed, the base variables $\tilde{x}$ are not allowed to depend on ε, we find from (4.9) that

$$0 = \sum_{j=1}^{p} D_j \Xi^i \frac{\partial x^j}{\partial \varepsilon} + \sum_{\alpha=1}^{q} \frac{\partial \Xi^i}{\partial u^\alpha} \eta^\alpha,$$

hence, by Cramer's rule,

$$\left.\frac{\partial x^j}{\partial \varepsilon}\right|_{\varepsilon=0} = \frac{-1}{\det \mathsf{J}} \sum_{i=1}^{p} K_{ij} \sum_{\alpha=1}^{q} \frac{\partial \Xi^i}{\partial u^\alpha} \eta^\alpha,$$

where K_{ij} is the (i, j)-th cofactor of the Jacobian matrix $\mathsf{J}(x)$. Therefore,

$$\begin{aligned}
\left.\frac{\partial \tilde{f}^\beta}{\partial \varepsilon}\right|_{\varepsilon=0} &= \sum_{\alpha=1}^{q} \frac{\partial \Phi^\beta}{\partial u^\alpha} \eta^\alpha + \sum_{j=1}^{p} D_j \Phi^\beta \left.\frac{\partial x^j}{\partial \varepsilon}\right|_{\varepsilon=0} \\
&= \frac{1}{\det \mathsf{J}} \sum_{\alpha=1}^{q} \left\{ \frac{\partial \Phi^\beta}{\partial u^\alpha} \det \mathsf{J} - \sum_{i,j=1}^{p} D_j \Phi^\beta \cdot K_{ij} \frac{\partial \Xi^i}{\partial u^\alpha} \right\} \eta^\alpha.
\end{aligned}$$

The reader can recognize the expression in brackets as the column expansion of the determinant (4.8) along the last column, where the intermediary summation $\sum_j D_j \Phi^\beta \cdot K_{ij}$ is the row expansion of the $(i, p + 1)$-st minor along its last row. Thus

$$\left.\frac{\partial \tilde{f}^\beta}{\partial \varepsilon}\right|_{\varepsilon=0} = \frac{1}{\det \mathsf{J}} \sum_{\alpha=1}^{q} \mathsf{F}_{\alpha\beta} \eta^\alpha.$$

Substituting into (4.10) and changing variables, we find

$$\left.\frac{d}{d\varepsilon}\right|_{\varepsilon=0} \tilde{\mathscr{L}}[\tilde{f}] = \int_{\Omega} \left\{ \sum_{\alpha,\beta=1}^{q} \mathsf{F}_{\alpha\beta} \mathsf{E}_{\tilde{u}^\beta}(\tilde{L}) \cdot \eta^\alpha \right\} dx.$$

On the other hand, this must equal

$$\left.\frac{d}{d\varepsilon}\right|_{\varepsilon=0} \mathscr{L}[f + \varepsilon\eta] = \int_{\Omega} \left\{ \sum_{\alpha=1}^{q} \mathsf{E}_{u^\alpha}(L) \eta^\alpha \right\} dx,$$

from which (4.7) follows since η is arbitrary. □

Example 4.9. In the case $p = q = 1$,

$$L(x, u^{(n)}) = \tilde{L}(\tilde{x}, \tilde{u}^{(n)}) D_x \Xi(x, u^{(1)}),$$

and (4.7) simplifies to

$$\mathsf{E}_u(L) = \frac{\partial(\Xi, \Phi)}{\partial(x, u)} \mathsf{E}_{\tilde{u}}(\tilde{L}), \tag{4.11}$$

where we need only the Jacobian determinant

$$\frac{\partial(\Xi, \Phi)}{\partial(x, u)} = \det \begin{pmatrix} \partial\Xi/\partial x & \partial\Xi/\partial u \\ \partial\Phi/\partial x & \partial\Phi/\partial u \end{pmatrix}$$

of partial derivatives of Ξ and Φ. Indeed, the determinant appearing in (4.8),

$$\mathsf{F}_{11}(x) = \det \begin{pmatrix} D_x\Xi & \partial\Xi/\partial u \\ D_x\Phi & \partial\Phi/\partial u \end{pmatrix} = \det \begin{pmatrix} \partial\Xi/\partial x + u_x \partial\Xi/\partial u & \partial\Xi/\partial u \\ \partial\Phi/\partial x + u_x \partial\Phi/\partial u & \partial\Phi/\partial u \end{pmatrix}$$

equals the above determinant by an elementary column operation. For example, if

$$\mathscr{L}[u] = \int_a^b \tfrac{1}{2} u_x^2 \, dx,$$

and we use the hodograph transformation $\tilde{x} = u$, $\tilde{u} = x$, then

$$\tilde{\mathscr{L}}[\tilde{u}] = \int_{\tilde{a}}^{\tilde{b}} \tfrac{1}{2}(\tilde{u}_{\tilde{x}})^{-2} \tilde{u}_{\tilde{x}} \, d\tilde{x} = \int_{\tilde{a}}^{\tilde{b}} (2\tilde{u}_{\tilde{x}})^{-1} \, d\tilde{x},$$

and

$$\mathsf{E}_u(L) = -u_{xx} = -\tilde{u}_{\tilde{x}}^{-3} \tilde{u}_{\tilde{x}\tilde{x}} = -\mathsf{E}_{\tilde{u}}(\tilde{L}),$$

verifying (4.11). However, in general we cannot replace the total derivatives in (4.8) by partial derivatives. (See Exercise 4.15.)

4.2. Variational Symmetries

In order to apply group methods in the calculus of variations, we need to make precise the notion of a symmetry group of a functional

$$\mathscr{L}[u] = \int_{\Omega_0} L(x, u^{(n)}) \, dx. \tag{4.12}$$

The groups considered here will be local groups of transformations G acting on an open subset $M \subset \Omega_0 \times U \subset X \times U$. As discussed in detail in Chapter 2, if $u = f(x)$ is a smooth function defined over a suitably small subdomain $\Omega \subset \Omega_0$, such that the graph of f lies in M, each transformation g in G sufficiently close to the identity will transform f to another smooth function

$\tilde{u} = \tilde{f}(\tilde{x}) = g \cdot f(\tilde{x})$ defined over $\tilde{\Omega} \subset \Omega_0$. (Note that unless G is projectable, $\tilde{\Omega}$ will, in general, not only depend on g, but also on f itself.) A symmetry group G will be one that, roughly speaking, leaves the variational integral $\mathscr{L}$ unchanged for all such f.

Definition 4.10. A local group of transformations G acting on $M \subset \Omega_0 \times U$ is a *variational symmetry group* of the functional (4.12) if whenever Ω is a subdomain with closure $\bar{\Omega} \subset \Omega_0$, $u = f(x)$ is a smooth function defined over Ω whose graph lies in M, and $g \in G$ is such that $\tilde{u} = \tilde{f}(\tilde{x}) = g \cdot f(\tilde{x})$ is a single-valued function defined over $\tilde{\Omega} \subset \Omega_0$, then

$$\int_{\tilde{\Omega}} L(\tilde{x}, \mathrm{pr}^{(n)} \tilde{f}(\tilde{x}))\, d\tilde{x} = \int_{\Omega} L(x, \mathrm{pr}^{(n)} f(x))\, dx. \tag{4.13}$$

Example 4.11. Consider the case when $X = \mathbb{R}$ and we have a first order variational problem

$$\mathscr{L}[u] = \int_a^b L(x, u, u_x)\, dx. \tag{4.14}$$

If L does not depend on x, then the translation group $(x, u) \mapsto (x + \varepsilon, u)$ is a variational symmetry group of $\mathscr{L}$. Indeed, since $\tilde{x} = x + \varepsilon$, $\tilde{u} = u$, if $u = f(x)$ is any function defined over a smaller subinterval $[c, d] \subset (a, b)$, then $\tilde{u} = \tilde{f}(\tilde{x}) = f(\tilde{x} - \varepsilon)$ is defined over $[\tilde{c}, \tilde{d}] = [c + \varepsilon, d + \varepsilon]$, which, for ε sufficiently small, is still a subinterval of (a, b). To verify (4.13), we have

$$\int_{\tilde{c}}^{\tilde{d}} L(\tilde{f}(\tilde{x}), \tilde{f}'(\tilde{x}))\, d\tilde{x} = \int_{\tilde{c}}^{\tilde{d}} L(f(\tilde{x} - \varepsilon), f'(\tilde{x} - \varepsilon))\, d\tilde{x} = \int_c^d L(f(x), f'(x))\, dx,$$

using a change of variables.

Infinitesimal Criterion of Invariance

In accordance with our usual *modus operandi*, we now find the analogous infinitesimal criterion for the invariance of a variational problem under a group of transformations. Again this condition will be necessary and sufficient for a connected group of transformations to be a symmetry group of the variational problem.

Theorem 4.12. *A connected group of transformations G acting on $M \subset \Omega_0 \times U$ is a variational symmetry group of the functional* (4.12) *if and only if*

$$\mathrm{pr}^{(n)} \mathbf{v}(L) + L\, \mathrm{Div}\, \xi = 0 \tag{4.15}$$

for all $(x, u^{(n)}) \in M^{(n)}$ and every infinitesimal generator

$$\mathbf{v} = \sum_{i=1}^{p} \xi^i(x, u) \frac{\partial}{\partial x^i} + \sum_{\alpha=1}^{q} \phi_\alpha(x, u) \frac{\partial}{\partial u^\alpha}$$

of G. In (4.15), Div ξ *denotes the total divergence of the p-tuple* $\xi = (\xi^1, \ldots, \xi^p)$, cf. (4.4).

PROOF. For each $g \in G$, the group transformation

$$(\tilde{x}, \tilde{u}) = g \cdot (x, u) = (\Xi_g(x, u), \Phi_g(x, u))$$

can be regarded as a change of variables, so by the same reasoning as led to (4.6) we can rewrite the symmetry condition (4.13) in the form

$$\int_\Omega L(\tilde{x}, \mathrm{pr}^{(n)}(g \cdot f)(\tilde{x})) \det \mathrm{J}_g(x, \mathrm{pr}^{(1)} f(x))\, dx = \int_\Omega L(x, \mathrm{pr}^{(n)} f(x))\, dx,$$

where the Jacobian matrix has entries

$$\mathrm{J}_g^{ij}(x, u^{(1)}) = D_i \Xi_g^j(x, u^{(1)}).$$

Since this is required to hold for all subdomains Ω and all functions $u = f(x)$, the integrands must agree pointwise:

$$L(\mathrm{pr}^{(n)} g \cdot (x, u^{(n)})) \det \mathrm{J}_g(x, u^{(1)}) = L(x, u^{(n)}) \tag{4.16}$$

for all $(x, u^{(n)}) \in M^{(n)}$. To obtain the infinitesimal version of (4.16) we set $g = g_\varepsilon = \exp(\varepsilon \mathbf{v})$ and differentiate with respect to ε. We need the formula

$$\frac{d}{d\varepsilon}[\det \mathrm{J}_{g_\varepsilon}(x, u^{(1)})] = \mathrm{Div}\, \xi(\mathrm{pr}^{(1)} g_\varepsilon \cdot (x, u^{(1)})) \det \mathrm{J}_{g_\varepsilon}(x, u^{(1)}) \tag{4.17}$$

expressing the fact that the divergence of a vector field measures the rate of change of volume under the induced flow. Indeed, if we replace u by a function $f(x)$, then (4.17) reduces to the identity of Exercise 1.36 for the reduced vector field $\tilde{\mathbf{v}} = \sum_{i=1}^p \xi^i(x, f(x))\, \partial/\partial x^i$.

Using (4.17) and (2.21), the derivative of (4.16) with respect to ε when $g = g_\varepsilon = \exp(\varepsilon \mathbf{v})$ is

$$(\mathrm{pr}^{(n)} \mathbf{v}(L) + L\, \mathrm{Div}\, \xi) \det \mathrm{J}_{g_\varepsilon} = 0, \tag{4.18}$$

the expression in parentheses being evaluated at $(\tilde{x}, \tilde{u}_\varepsilon^{(n)}) = \mathrm{pr}^{(n)} g_\varepsilon \cdot (x, u^{(n)})$. In particular, at $\varepsilon = 0$, g_ε is the identity map and we have proved the necessity of (4.15) for G to be a variational symmetry group. Conversely, if (4.15) holds everywhere, then (4.18) holds for ε sufficiently small. The left-hand side of (4.18), though, is just the derivative of the left-hand side of (4.16) (for $g = g_\varepsilon$) with respect to ε; thus, integrating from 0 to ε we prove (4.16) for g sufficiently near the identity. The usual connectivity arguments complete the proof of (4.16) for all $g \in G$, and hence the theorem. □

Example 4.13. For an easy illustration of Theorem 4.12, we re-derive the result of Example 4.11. The infinitesimal generator of the horizontal translation group is ∂_x, with prolongation $\mathrm{pr}^{(1)} \partial_x = \partial_x$. Also $\xi(x, u) = 1$, so $D_x \xi = 0$. Thus for $\mathscr{L} = \int_a^b L(u, u_x)\, dx$, we find

$$\mathrm{pr}^{(1)} \partial_x(L) + L D_x \xi = 0,$$

trivially, so (4.15) is verified. Another easy example is the arc-length integral $\mathscr{L}_0[u] = \int_a^b \sqrt{1+u_x^2}\,dx$, with $\mathbf{v} = -u\partial_x + x\partial_u$ the generator of the rotation group. We have

$$\mathrm{pr}^{(1)}\,\mathbf{v} = -u\partial_x + x\partial_u + (1+u_x^2)\partial_{u_x},$$

and $\xi = -u$, so

$$\mathrm{pr}^{(1)}\,\mathbf{v}(L) + LD_x\xi = (1+u_x^2)\frac{\partial}{\partial u_x}\sqrt{1+u_x^2} - \sqrt{1+u_x^2}\cdot u_x \equiv 0.$$

Thus Theorem 4.12 implies the geometrically obvious fact that arc-length is unchanged by a rigid rotation.

Symmetries of the Euler–Lagrange Equations

In both of the preceding examples, as the reader may readily check, invariance of the given variational integral under a group of symmetries implies that the associated Euler–Lagrange equations are also invariant under the group. This result holds in general.

Theorem 4.14. *If G is a variational symmetry group of the functional $\mathscr{L}[u] = \int_{\Omega_0} L(x, u^{(n)})\,dx$, then G is a symmetry group of the Euler–Lagrange equations $\mathsf{E}(L) = 0$.*

Intuitively, what is happening is that if $g \in G$ and $u = f(x)$ is an extremal of $\mathscr{L}[u]$, then clearly $\tilde{u} = g \cdot f(\tilde{x})$ (provided it is defined) is an extremal of the transformed variational problem $\tilde{\mathscr{L}}[\tilde{u}]$ coming from $(\tilde{x}, \tilde{u}) = g \cdot (x, u)$. But if G is a variational symmetry group, $\tilde{\mathscr{L}}[\tilde{u}] = \mathscr{L}[\tilde{u}]$, hence $g \cdot f$ is also an extremal of $\mathscr{L}$. The problem is that there are also non-extremal solutions of the Euler–Lagrange equations. One approach would be to use the change of variables formula from Theorem 4.8. Rather than belabour the point here, we refer the reader to Theorem 5.53 for a direct, computational proof.

It is *not* true that every symmetry group of the Euler–Lagrange equations is also a variational symmetry group of the original variational problem! The most common counterexamples are given by groups of scaling transformations.

Example 4.15. *The Wave Equation.* We return to the wave equation $u_{tt} = u_{xx} + u_{yy}$, whose symmetry group was found in Example 2.43. The wave equation is the Euler–Lagrange equation for the variational problem

$$\mathscr{L}[u] = \iiint [\tfrac{1}{2}u_t^2 - \tfrac{1}{2}u_x^2 - \tfrac{1}{2}u_y^2]\,dx\,dy\,dt.$$

Let us find out which of the symmetries listed in (2.65) are variational symmetries of $\mathscr{L}$. We use $L = \frac{1}{2}u_t^2 - \frac{1}{2}u_x^2 - \frac{1}{2}u_y^2$ in the infinitesimal criterion

(4.15). The translations are easily found to be variational symmetries since their prolongations have no effect on the derivatives of u. Next consider a rotation group generator, say $\mathbf{r}_{xy} = -y\partial_x + x\partial_y$. The term Div ξ in (4.15) vanishes. Moreover, $\mathrm{pr}^{(1)}\,\mathbf{r}_{xy} = \mathbf{r}_{xy} - u_y\partial_{u_x} + u_x\partial_{u_y}$, hence (4.15) reads

$$\mathrm{pr}^{(1)}\,\mathbf{r}_{xy}(L) = u_y u_x - u_x u_y = 0,$$

so $\mathscr{L}$ is rotationally invariant under the group generated by $\mathbf{r}_{xy}$. A similar computation shows that $\mathbf{r}_{xt}$ and $\mathbf{r}_{yt}$ also generate variational symmetry groups. Turning to the dilatational subgroup, we have

$$\mathrm{pr}^{(1)}\,\mathbf{d} = x\partial_x + y\partial_y + t\partial_t - u_x\partial_{u_x} - u_y\partial_{u_y} - u_t\partial_{u_t},$$

and, in this case,

$$\mathrm{Div}\,\xi = D_x(x) + D_y(y) + D_t(t) = 3.$$

Therefore

$$\mathrm{pr}^{(1)}\,\mathbf{d}(L) + L\,\mathrm{Div}\,\xi = L,$$

so $\mathbf{d}$ does *not* generate a variational symmetry group of $\mathscr{L}$. However, if we modify the dilatational generator to be

$$\mathbf{m} \equiv \mathbf{d} - \tfrac{1}{2}u\partial_u = x\partial_x + y\partial_y + t\partial_t - \tfrac{1}{2}u\partial_u,$$

then

$$\mathrm{pr}^{(1)}\,\mathbf{m}(L) + L\,\mathrm{Div}\,\xi = -3L + 3L = 0,$$

so $\mathbf{m}$ does generate a variational symmetry group. Finally, consider an inversional group, say that generated by

$$\mathbf{i}_x = (x^2 - y^2 + t^2)\partial_x + 2xy\partial_y + 2xt\partial_t - xu\partial_u.$$

We have

$$\begin{aligned}\mathrm{pr}^{(1)}\,\mathbf{i}_x = \mathbf{i}_x &- (u + 3xu_x + 2yu_y + 2tu_t)\partial_{u_x} \\ &+ (2yu_x - 3xu_y)\partial_{u_y} - (2tu_x + 3xu_t)\partial_{u_t},\end{aligned}$$

and

$$\mathrm{Div}\,\xi = D_x(x^2 - y^2 + t^2) + D_y(2xy) + D_t(2xt) = 6x.$$

Therefore

$$\mathrm{pr}^{(1)}\,\mathbf{i}_x(L) + L\,\mathrm{Div}\,\xi = uu_x - 3x(u_t^2 - u_t^2 - u_y^2) + 6xL = uu_x,$$

and hence $\mathbf{i}_x$ is not a variational symmetry according to Definition 4.10. Neither are the other two one-parameter inversional subgroups of the symmetry group (2.65) nor is any linear combination thereof. Finally, if $\alpha(x, y, t)$ is a solution of the wave equation, with symmetry generator $\mathbf{v}_\alpha = \alpha\partial_u$, we find

$$\mathrm{pr}^{(1)}\,\mathbf{v}_\alpha(L) = \alpha_t u_t - \alpha_x u_x - \alpha_y u_y,$$

so we get a variational symmetry if and only if α is a constant. Thus the variational symmetry group of L is generated by the translations, the "rotations", the scaling group generated by $\mathbf{m}$, and the group generated by ∂_u. Theorem 4.12 assures us that there are no other variational symmetries. (Of course, this could be checked directly by solving (4.15) for the coefficients of the infinitesimal generator $\mathbf{v}$.)

Proposition 4.16. *If* $\mathbf{v}$ *and* $\mathbf{w}$ *are variational symmetries of* $\mathscr{L}[u]$, *then so is their Lie bracket* $[\mathbf{v}, \mathbf{w}]$.

The proof is left to the reader. (See Exercise 4.1.)

Reduction of Order

As we've seen in Section 2.5, knowledge of a one-parameter symmetry group of a single ordinary differential equation allows us to reduce the order of the equation by one. In this section, we will see that knowledge of a one-parameter group of *variational* symmetries for the Euler–Lagrange equation of some variational problem allows us to reduce the order of the equation by *two*! In effect, the variational structure of the differential equation and the symmetry group doubles the power of Lie's integration theory.

The easiest way to see how this happens is to use the invariance of the Euler–Lagrange equations under changes of variable as presented in Theorem 4.8, which allows us to change both independent and dependent variables without affecting the variational nature of the problem. Thus, let $x, u \in \mathbb{R}$, and let $\mathscr{L}[u]$ be an n-th order variational problem with $2n$-th order Euler–Lagrange equations. Suppose $\mathbf{v} = \xi(x, u)\partial_x + \phi(x, u)\partial_u$ is the infinitesimal generator of a one-parameter group of variational symmetries of $\mathscr{L}$. Note that by the definition of variational symmetry, $\mathbf{v}$ will remain a variational symmetry under a change of both independent and dependent variables. As in Section 2.5, we now introduce *particular* new variables $y = \eta(x, u)$, $w = \zeta(x, u)$ so that $\mathbf{v}$ takes the elementary form $\tilde{\mathbf{v}} = \partial/\partial w$ in these new coordinates. Let $\tilde{\mathscr{L}}[w] = \int \tilde{L}(y, w^{(n)})\, dy$ be the corresponding variational problem in the (y, w) variables. According to the above remarks, $\tilde{\mathbf{v}}$ remains a variational symmetry of $\tilde{\mathscr{L}}$, so by the infinitesimal criterion (4.15) we have

$$\mathrm{pr}^{(n)}\, \tilde{\mathbf{v}}(\tilde{L}) = \partial \tilde{L}/\partial w = 0,$$

hence $\tilde{L} = \tilde{L}(y, w_y, w_{yy}, \ldots)$ is independent of w. The Euler–Lagrange equation for $\tilde{\mathscr{L}}$ thus takes the form

$$0 = \mathrm{E}_w(\tilde{L}) = \sum_{j=1}^{n} (-D_y)^j \frac{\partial \tilde{L}}{\partial w_j} = -D_y \left\{ \sum_{j=0}^{n-1} (-D_y)^j \frac{\partial \tilde{L}}{\partial w_{j+1}} \right\}, \qquad (4.19)$$

where $w_j = d^j w/dy^j$. Thus the expression in the brackets is constant, independent of y, and hence a *first integral* of the Euler–Lagrange equations. (This constitutes our first real encounter with Noether's theorem.)

Note further that if we introduce the new dependent variable $v = w_y$, so $v_j = d^j v/dy^j = w_{j+1}$, the expression in brackets can be written as the variational derivative of $\tilde{\mathscr{L}}[v] = \int \hat{L}(y, v^{(n-1)})\, dy$, where

$$\hat{L}(y, v, \ldots, v_{n-1}) = \tilde{L}(y, w_y, \ldots, w_n).$$

Every solution $w = f(y)$ of the original $2n$-th order Euler–Lagrange equation corresponds to a solution $v = h(y)$ of the $(2n - 2)$-nd order equation

$$\mathsf{E}_v(\hat{L})(y, v^{(2n-2)}) = \lambda \tag{4.20}$$

for some constant λ (depending on the initial conditions), where w is recovered by quadrature:

$$w = \int h(y)\, dy + c.$$

Note that we can write (4.20) as a pure Euler–Lagrange equation for

$$\hat{\mathscr{L}}_\lambda[v] = \int [\hat{L}(y, v^{(n-1)}) - \lambda v]\, dy.$$

(Alternatively, λ can be thought of as a Lagrange multiplier, so we are minimizing $\hat{\mathscr{L}}[v]$ subject to the constraint $\int v\, dy = 0$, say; see Courant and Hilbert, [1; p. 218].)

Theorem 4.17. *Let $p = q = 1$. Let $\mathscr{L}[u]$ be an n-th order variational problem with Euler–Lagrange equation of order 2n. Suppose G is a one-parameter group of variational symmetries of $\mathscr{L}$. Then there exists a one-parameter family of variational problems $\hat{\mathscr{L}}_\lambda[v]$ of order $n - 1$, with Euler–Lagrange equations of order $2n - 2$, such that every solution of the Euler–Lagrange equation for $\mathscr{L}[u]$ can be found by quadrature from the solutions to the Euler–Lagrange equation for $\hat{\mathscr{L}}_\lambda[v]$, $\lambda \in \mathbb{R}$.*

Example 4.18. In the case of first order variational problems, as in (4.14), knowledge of a one-parameter group of variational symmetries allows us to integrate the second order Euler–Lagrange equation

$$\mathsf{E}(L) = \frac{\partial L}{\partial u} - D_x \frac{\partial L}{\partial u_x} = 0 \tag{4.21}$$

completely by quadratures. (For a general one-parameter symmetry group of a second order ordinary differential equation, we can only expect to reduce to a first order equation.) Thus if L is independent of u, (4.21) reduces to $D_x(\partial L/\partial u_x) = 0$, hence

$$\frac{\partial L}{\partial u_x}(x, u_x) = \lambda$$

for some constant λ. We can solve this implicit relation for $u_x = F(x, \lambda)$, so the general solution is

$$u = \int F(x, \lambda)\, dx + c.$$

If $L(u, u_x)$ is independent of x, we can reduce to the previous case by using the hodograph change of variables of Example 4.9: $y = u$, $w = x$. A somewhat more direct approach, however, is to note that if we multiply the Euler–Lagrange equation by u_x, we can find a first integral

$$0 = u_x \mathsf{E}(L) = u_x \frac{\partial L}{\partial u} - u_x^2 \frac{\partial^2 L}{\partial u\, \partial u_x} - u_x u_{xx} \frac{\partial^2 L}{\partial u_x^2} = D_x\left(L - u_x \frac{\partial L}{\partial u_x}\right).$$

Thus

$$L(u, u_x) - u_x \frac{\partial L}{\partial u_x}(u, u_x) = \lambda$$

defines $u_x = F(u, \lambda)$ implicitly as a function of u and λ, which we can integrate to recover the solution of the Euler–Lagrange equation:

$$\int \frac{du}{F(u, \lambda)} = x + c.$$

The method can be extended to multi-parameter groups, but, unless they are abelian, we cannot in general expect to reduce the order by two at each stage. (See Exercise 4.11 and the later development of Hamiltonian systems in Chapter 6.) Here we content ourselves with an illustrative example.

Example 4.19. *Kepler's Problem.* We show how the above procedure can be used to immediately integrate the two-dimensional version of Kepler's problem of a mass under a central gravitational force field. The functional is

$$\mathscr{L}[x, y] = \int [\tfrac{1}{2}(x_t^2 + y_t^2) - U(r)]\, dt,$$

in which $(x(t), y(t))$ are the coordinates of the mass, $r^2 = x^2 + y^2$, and U is the potential function; for the three-dimensional gravitational attraction of a mass moving in the (x, y)-plane, $U(r) = -\gamma/r$. The Euler–Lagrange equations are

$$x_{tt} = -\frac{x}{r} U'(r), \qquad y_{tt} = -\frac{y}{r} U'(r).$$

Clearly, $\mathscr{L}$ is invariant under the abelian two-parameter group of time translations and spatial rotations with infinitesimal generators ∂_t and $x\partial_y - y\partial_x$ respectively. Introducing the polar coordinates (r, θ, t), we see that these vector fields become ∂_t and ∂_θ. By analogy with the case of a one-parameter group, this says that we should regard r as the new independent variable and θ and t as the new dependent variables, which should effect a reduction of the second order system to a system solvable by quadratures. Note first that

$$x_t = \frac{1}{t_r}(\cos\theta - r\sin\theta \cdot \theta_r), \qquad y_t = \frac{1}{t_r}(\sin\theta + r\cos\theta \cdot \theta_r),$$

hence in polar coordinates

$$\mathscr{L} = \int \left[\frac{1}{2t_r}(1 + r^2\theta_r^2) - t_r U(r) \right] dr,$$

which is, as expected, independent of both t and θ. The Euler–Lagrange equations can thus be immediately integrated once, leading to

$$\frac{1}{2t_r^2}(1 + r^2\theta_r^2) + U(r) = \lambda, \qquad \frac{r^2\theta_r}{t_r} = \mu,$$

where λ, μ are constants. Note that if we revert back to t as the independent variable, the first equation gives the well-known conservation of energy, while the second equation is just Kepler's second law, $r^2\theta_t = \mu$, that the mass sweeps out equal areas in equal times. Retaining r as the independent variable, however, we can eliminate t_r from these two equations,

$$(2\lambda\mu^{-2}r^4 - 2\mu^{-2}r^4U(r) - r^2)\left(\frac{d\theta}{dr}\right)^2 = 1,$$

hence

$$\theta = \int \frac{dr}{r(2\lambda\mu^{-2}r^2 - 2\mu^{-2}r^2U(r) - 1)^{1/2}} + \theta_0.$$

In particular, if $U(r) = -\gamma r^{-1}$, we can integrate this explicitly,

$$\theta - \theta_0 = \arcsin\left[\frac{p}{\varepsilon}\left(\frac{1}{p} - \frac{1}{r}\right)\right],$$

where

$$\varepsilon^2 = 1 + \frac{2\mu^2\lambda}{\gamma^2}, \qquad p = \frac{\mu^2}{\gamma}.$$

Thus the orbits are conic sections

$$r = \frac{p}{1 - \varepsilon\sin(\theta - \theta_0)}$$

of eccentricity ε. Similarly, we can determine t by a single quadrature:

$$t = \int \frac{r^2\theta_r}{\mu}\,dr + t_0 = \int \frac{r\,dr}{(2\lambda r^2 - 2r^2U(r) - \mu^2)^{1/2}} + t_0,$$

which, in the gravitational case, yields

$$t = \frac{s}{2\lambda} - \frac{\gamma}{(2\lambda)^{3/2}}\log(\sqrt{2\lambda}s + 2\lambda r + \gamma), \qquad s = \sqrt{2\lambda r^2 + 2\gamma r - \mu^2}.$$

We have thus completely solved Kepler's problem by quadratures.

4.3. Conservation Laws

Consider a system of differential equations $\Delta(x, u^{(n)}) = 0$. A *conservation law* is a divergence expression

$$\operatorname{Div} P = 0 \tag{4.22}$$

which vanishes for all solutions $u = f(x)$ of the given system. Here $P = (P_1(x, u^{(n)}), \ldots, P_p(x, u^{(n)}))$ is a p-tuple of smooth functions of x, u and the derivatives of u, and $\operatorname{Div} P = D_1 P_1 + \cdots + D_p P_p$ is its total divergence.

For example, in the case of Laplace's equation, some conservation laws are readily apparent. First of all, the equation itself is a conservation law since

$$\Delta u = \operatorname{Div}(\nabla u) = 0$$

for all solutions u. Multiplying Laplace's equation by $u_i = \partial u/\partial x^i$ yields p further conservation laws.

$$0 = u_i \Delta u = \sum_{j=1}^{p} D_j\left(u_i u_j - \tfrac{1}{2}\delta_i^j \sum_{k=1}^{p} u_k^2\right).$$

Later we will see how to establish yet more conservation laws.

In the case of a system of ordinary differential equations involving a single independent variable $x \in \mathbb{R}$, a conservation law takes the form $D_x P = 0$ for all solutions $u = f(x)$ of the system. This requires that $P(x, u^{(n)})$ be *constant* for all solutions of the system. Thus a conservation law for a system of ordinary differential equations is equivalent to the classical notion of a *first integral* or *constant of the motion* of the system. As we will see, (4.22) is the appropriate generalization of this concept to partial differential equations, and includes familiar concepts of conservation of mass, energy, momentum, etc. arising in physical applications.

In a dynamical problem, one of the independent variables is distinguished as the time t, the remaining variables $x = (x^1, \ldots, x^p)$ being spatial variables. In this case a conservation law takes the form

$$D_t T + \operatorname{Div} X = 0,$$

in which Div is the spatial divergence of X with respect to $x^1, \ldots, x^p$. The *conserved density*, T, and the associated *flux*, $X = (X_1, \ldots, X_p)$, are functions of x, t, u and the derivatives of u with respect to both x and t. In this situation, it is easy to see that for certain types of solutions, the conserved density, when integrated, provides us with a constant of the motion of the system. More specifically, suppose $\Omega \subset \mathbb{R}^p$ is a spatial domain, and $u = f(x, t)$ a solution defined for all $x \in \Omega$, $a \leqslant t \leqslant b$. Consider the functional

$$\mathscr{T}_\Omega[f](t) = \int_\Omega T(x, t, \operatorname{pr}^{(n)} f(x, t))\, dx, \tag{4.23}$$

which, for fixed f, Ω, depends on t alone. The basic conservative property of T states that $\mathscr{T}_\Omega[f]$ depends only on the initial values of f at $t = a$ and the boundary values of f on $\partial\Omega$.

Proposition 4.20. *Suppose T, X are the conserved density and flux for a conservation law of a given system of differential equations. Then for any bounded domain $\Omega \subset \mathbb{R}^p$ with smooth boundary $\partial\Omega$, and any solution $u = f(x, t)$ defined for $x \in \Omega$, $a \leqslant t \leqslant b$, the functional (4.23) satisfies*

$$\mathscr{T}_\Omega[f](t) - \mathscr{T}_\Omega[f](a) = -\int_a^t \int_{\partial\Omega} X(x, \tau, \mathrm{pr}^{(n)} f(x, \tau)) \cdot dS \, d\tau. \quad (4.24)$$

Conversely, if (4.24) holds for all such domains and solutions $u = f(x, t)$, then T, X define a conservation law.

PROOF. By the divergence theorem

$$\frac{d}{dt}\mathscr{T}_\Omega[f](t) = \int_\Omega D_t T(x, t, \mathrm{pr}^{(n+1)} f) \, dx = -\int_{\partial\Omega} X(x, t, \mathrm{pr}^{(n)} f) \cdot dS.$$

Then (4.24) follows upon integration. The converse follows by differentiating (4.24) with respect to t, yielding

$$\int_\Omega \{D_t T(x, t, \mathrm{pr}^{(n+1)} f) + \mathrm{Div}\, X(x, t, \mathrm{pr}^{(n+1)} f)\} \, dx = 0.$$

Since this holds for arbitrary subdomains, the integrand itself must vanish, proving the converse. □

Corollary 4.21. *If $\Omega \subset \mathbb{R}^p$ is bounded, and $u = f(x, t)$ is a solution such that $X(x, t, \mathrm{pr}^{(n)} f(x, t)) \to 0$ as $x \to \partial\Omega$, then $\mathscr{T}_\Omega[f]$ is a constant, independent of t.*

Usually, $X(x, t, 0) \equiv 0$, so one requires that the solution $f(x, t)$ vanish sufficiently rapidly as $x \to \partial\Omega$ (so that there is no flux over $\partial\Omega$), or, if Ω has unbounded components, as $|x| \to \infty$.

Example 4.22. Perhaps the most graphic physical illustration of the relationship between conserved densities and fluxes comes from the equations of compressible, inviscid fluid motion. Let $x \in \mathbb{R}^3$ represent the spatial coordinates, and $u = u(x, t) \in \mathbb{R}^3$ the velocity of a fluid particle at position x and time t. Further let $\rho(x, t)$ be the density, and $p(x, t)$ the pressure; in the particular case of isentropic (constant entropy) flow, pressure $p = P(\rho)$ will depend on density alone. The equation of continuity takes the form

$$\rho_t + \mathrm{Div}(\rho u) = 0,$$

where $\mathrm{Div}(\rho u) = \sum_j \partial(\rho u^j)/\partial x^j$ is the spatial divergence, while momentum balance yields the three equations

$$\frac{\partial u^i}{\partial t} + \sum_{j=1}^{3} u^j \frac{\partial u^i}{\partial x^j} = -\frac{1}{\rho}\frac{\partial p}{\partial x^i}, \qquad i = 1, 2, 3.$$

The equation of continuity is already in the form of a conservation law, with density $T = \rho$ and flux $X = \rho u$. This leads to the integral equation for the conservation of mass

$$\frac{d}{dt}\int_\Omega \rho\, dx = -\int_{\partial\Omega} \rho u\cdot n\, dS.$$

Here $\int_\Omega \rho\, dx$ is clearly the mass of fluid within the domain Ω, while $\rho u\cdot n$, with n the unit normal to $\partial\Omega$, is the instantaneous mass flux of fluid out of a point on the boundary $\partial\Omega$. Thus we see that the net change in mass inside Ω equals the flux of fluid into Ω. In particular, if the normal component of velocity $u\cdot n$ on $\partial\Omega$ vanishes, there is no net change in mass within the domain Ω, and we have a law of conservation of mass:

$$\int_\Omega \rho\, dx = \text{constant}.$$

The momentum balance equations, coupled with the continuity equation, yield three further conservation laws

$$D_t(\rho u^i) + \sum_{j=1}^{3} D_j(\rho u^i u^j + p\delta_i^j) = 0, \qquad i = 1, 2, 3.$$

In integrated form, these are the laws of conservation of linear momentum

$$\frac{d}{dt}\int_\Omega \rho u^i\, dx = -\int_{\partial\Omega} (\rho u^i(u\cdot n) + pn_i)\, dS, \qquad i = 1, 2, 3,$$

(n_i being the i-th component of the normal n). The first term in the boundary integral denotes the transport of momentum ρu^i due to the flow across the surface $\partial\Omega$, while the second term is the net change in momentum due to the pressure across $\partial\Omega$. In this way $X_j = \rho u^i u^j + p\delta_i^j$ does represent the components of momentum flux. Finally, if the flow is isentropic, we can introduce the internal energy $W(\rho) = \int \rho^{-2}P(\rho)\, d\rho$ per unit mass, measuring the work done by the fluid against the pressure. The law of conservation of energy takes the form

$$D_t[\tfrac{1}{2}\rho|u|^2 + \rho W(\rho)] + \operatorname{Div}[(\tfrac{1}{2}\rho|u|^2 + P(\rho) + \rho W(\rho))u] = 0;$$

or, in integrated form,

$$\frac{d}{dt}\int_\Omega [\tfrac{1}{2}\rho|u|^2 + \rho W(\rho)]\, dx = -\int_{\partial\Omega} (\tfrac{1}{2}\rho|u|^2 + P(\rho) + \rho W(\rho))u\cdot n\, dS.$$

Here $\int_\Omega \frac{1}{2}\rho|u|^2\, dx$ is the kinetic energy, while $\int_\Omega \rho W(\rho)\, dx$ the internal (potential) energy of the fluid. The surface integral represents the transport of kinetic and potential energy across $\partial\Omega$ together with the rate of working due to the pressure across the boundary. In particular, if $u\cdot n = 0$ on $\partial\Omega$, both mass and energy are conserved, while momentum is conserved if the pressure $p = 0$ on $\partial\Omega$ also.

Trivial Conservation Laws

There are two distinct ways in which a conservation law could trivially hold. In the *first kind* of triviality, the p-tuple P itself in (4.22) vanishes for all solutions of the given system. This type of triviality is usually easy to eliminate by solving the system and its prolongations $\Delta^{(k)}$ for certain of the variables u_J^α in terms of the remaining variables, and substituting for these distinguished variables wherever they occur. For example, in the case of an evolution equation $u_t = P(x, u^{(n)})$, we can always solve for any time derivative of u, e.g. u_{tt}, u_{xt}, etc., solely in terms of x, u and spatial derivatives of u. As a net result, any dynamical conservation law is equivalent, up to the addition of a trivial conservation law of the first kind, to a conservation law in which the conserved density T depends only on x, t, u and spatial derivatives of u. For evolution equations, this is the usual form of a conservation law.

Example 4.23. Consider the system of first order evolution equations

$$u_t = v_x, \qquad v_t = u_x,$$

which is equivalent to the one-dimensional wave equation $u_{tt} = u_{xx}$. The expression

$$D_t(\tfrac{1}{2}u_t^2 + \tfrac{1}{2}u_x^2) - D_x(u_t u_x) = u_t(u_{tt} - u_{xx}) = 0$$

is clearly a conservation law. According to the above remarks, we can replace the conserved density and flux by ones depending on spatial derivatives, resulting in the equivalent conservation law

$$D_t(\tfrac{1}{2}u_x^2 + \tfrac{1}{2}v_x^2) - D_x(u_x v_x) = 0.$$

These differ by the trivial conservation law

$$D_t(\tfrac{1}{2}u_t^2 - \tfrac{1}{2}v_x^2) + D_x(v_x u_x - u_t u_x) = 0,$$

whose density and flux both vanish on solutions of the system.

A second possible type of triviality occurs when the divergence identity

$$\operatorname{Div} P = 0$$

holds for *all* functions $u = f(x)$, regardless of whether they solve the given system of differential equations. For example, in the case $p = 2$ the identity

$$D_x(u_y) - D_y(u_x) \equiv 0$$

clearly holds for any smooth function $u = f(x, y)$, and hence provides a trivial conservation law of the *second kind* for any partial differential equation involving $u = f(x, y)$. A less obvious example is the identity

$$D_x(u_y v_z - u_z v_y) + D_y(u_z v_x - u_x v_z) + D_z(u_x v_y - u_y v_x) \equiv 0$$

involving Jacobian determinants. Any such p-tuple $P(x, u^{(n)})$, whose divergence vanishes identically, is called a *null divergence*. The conservation law offered by any null divergence does not depend on the particular structure of any given system of differential equations, and we are thus justified in labelling these laws as trivial.

As with the Poincaré lemma, which characterizes the kernel of the ordinary divergence operator (cf. Example 1.62), there is a similar characterization of all null divergences/trivial conservation laws of the second kind.

Theorem 4.24. *Suppose $P = (P_1, \ldots, P_p)$ is a p-tuple of smooth functions depending on $x = (x^1, \ldots, x^p)$, $u = (u^1, \ldots, u^q)$ and derivatives of u, defined on all of the jet space $X \times U^{(n)}$. Then P is a null divergence*: $\operatorname{Div} P \equiv 0$ *if and only if there exist smooth functions Q_{jk}, $j, k = 1, \ldots, p$, depending on x, u and derivatives of u, such that*

$$Q_{jk} = -Q_{kj}, \qquad j, k = 1, \ldots, p, \tag{4.25}$$

and

$$P_j = \sum_{k=1}^{p} D_k Q_{jk}, \qquad j = 1, \ldots, p, \tag{4.26}$$

for all $(x, u^{(n)})$.

In particular, if $p = 3$, Theorem 4.24 says that

$$\operatorname{Div} P = D_1 P_1 + D_2 P_2 + D_3 P_3 \equiv 0$$

if and only if P is a "total curl": $P = \operatorname{Curl} Q$, i.e.

$$P_1 = D_2 Q_3 - D_3 Q_2, \qquad P_2 = D_3 Q_1 - D_1 Q_3, \qquad P_3 = D_1 Q_2 - D_2 Q_1.$$

(Here we identify $Q_{12} = -Q_{21}$ with Q_3, etc.) For our previous example,

$$P = (u_y v_z - u_z v_y, u_z v_x - u_x v_z, u_x v_y - u_y v_x),$$

and the corresponding Q is (uv_x, uv_y, uv_z).

Although for any fixed function $u = f(x)$, Theorem 4.24 reduces to the Poincaré lemma, the fact that the resulting Q_{jk} can be taken to depend just on x, u and derivatives of u for *all* such functions is a considerably more delicate matter. The proof turns out to be rather complicated, and will be deferred until Section 5.4, when we have considerably more algebraic machinery at our disposal.

In general, a *trivial conservation law* will be, by definition, a linear combination of trivial laws of the above two kinds. In other words $\operatorname{Div} P = 0$ is a trivial conservation law of the system if and only if there exist functions Q_{jk} satisfying (4.25) such that (4.26) holds for all solutions of Δ. Two conservation laws P and $\tilde{P}$ are *equivalent* if they differ by a trivial conservation law, so $\tilde{P} = P + R$ where R is trivial. We will only be interested in classifying conservation laws up to equivalence, so by "conservation law" in general we really mean "equivalence class of conservation laws".

Characteristics of Conservation Laws

Consider a conservation law of a totally nondegenerate system of differential equations $\Delta(x, u^{(n)}) = 0$. According to Exercise 2.35, Div P vanishes on all solutions of the system if and only if there exist functions $Q_\nu^J(x, u^{(m)})$ such that

$$\operatorname{Div} P = \sum_{\nu, J} Q_\nu^J D_J \Delta_\nu. \tag{4.27}$$

for all (x, u). Now, each of the terms in (4.27) can be integrated by parts; for example, if $1 \leqslant j \leqslant p$,

$$Q_\nu^j D_j \Delta_\nu = D_j(Q_\nu^j \Delta_\nu) - D_j(Q_\nu^j)\Delta_\nu.$$

In this way, we obtain an equivalent identity

$$\operatorname{Div} P = \operatorname{Div} R + \sum_{\nu=1}^{l} Q_\nu \Delta_\nu \equiv \operatorname{Div} R + Q \cdot \Delta,$$

in which the l-tuple $Q = (Q_1, \ldots, Q_l)$ has entries

$$Q_\nu = \sum_J (-D)_J Q_\nu^J, \tag{4.28}$$

and $R = (R_1, \ldots, R_p)$ (whose precise expression is not required here) depends linearly on the components Δ_ν of the given system of differential equations and their total derivatives. Thus R is a trivial conservation law (of the first kind), and if we replace P by $P - R$, we have an equivalent conservation law of the special form

$$\operatorname{Div} P = Q \cdot \Delta. \tag{4.29}$$

We call (4.29) the *characteristic form* of the conservation law (4.27), and the l-tuple $Q = (Q_1, \ldots, Q_l)$ the *characteristic* of the given conservation law.

In general, unless $l = 1$ the characteristic of a given conservation law is not uniquely determined, this stemming from the fact that the Q_ν in (4.29) are not uniquely determined. Note that if Q and $\tilde{Q}$ both satisfy (4.29) for the same P, then $Q \cdot \Delta = \tilde{Q} \cdot \Delta$. Since Δ is nondegenerate, Proposition 2.11 implies that $Q - \tilde{Q}$ vanishes on all solutions. This motivates the definition of a *trivial characteristic* Q as one which vanishes for all solutions of the system. Two characteristics Q and $\tilde{Q}$ are *equivalent* if they differ by a trivial characteristic, so $Q = \tilde{Q}$ for all solutions $u = f(x)$ to Δ. In general, characteristics are only determined up to equivalence.

Example 4.25. In order to find the characteristic for the conservation law of the wave equation in Example 4.23, we need to rewrite the left-hand side in the form (4.27), which is

$$D_t(\tfrac{1}{2}u_t^2 + \tfrac{1}{2}u_x^2) - D_x(u_x u_t) = u_t D_t(u_t - v_x) + u_t D_x(v_t - u_x).$$

Therefore, according to (4.28), the characteristic is

$$Q = (-D_t(u_t), -D_x(u_t)) = (-u_{tt}, -u_{xt}),$$

and there is an equivalent conservation law in characteristic form, which is found by integrating by parts:

$$D_t(\tfrac{1}{2}u_x^2 - \tfrac{1}{2}u_t^2 + u_t v_x) + D_x(-u_t v_t) = -u_{tt}(u_t - v_x) - u_{xt}(v_t - u_x).$$

It is important to note that replacing the t-derivatives by x-derivatives in this conservation law will, as in Example 4.23, lead to an equivalent conservation law, but that this will *not* in general remain in characteristic form. In the present example, the conserved density is equivalent to $\frac{1}{2}u_x^2 + \frac{1}{2}v_x^2$, the flux to $-u_x v_x$, but the resulting conservation law

$$D_t(\tfrac{1}{2}u_x^2 + \tfrac{1}{2}v_x^2) + D_x(-u_x v_x) = u_x(u_{xt} - v_{xx}) + v_x(v_{xt} - u_{xx})$$

is definitely not in characteristic form. In general, *replacing a conservation law by an equivalent one does not maintain the characteristic form.*

Furthermore, this last conservation law has as its characteristic

$$\tilde{Q} = (-D_x(u_x), -D_x(v_x)) = (-u_{xx}, -v_{xx}),$$

which is *not* the same as our previous characteristic. However, the difference

$$Q - \tilde{Q} = (-(u_{tt} - u_{xx}), -(v_{tt} - v_{xx}))$$

is a trivial characteristic, since it vanishes for all solutions of the system. Thus two equivalent conservation laws can have equivalent, but not identical, characteristics. We finally note that the characteristic forms of the two conservation laws are different, that of the latter law being

$$D_t(\tfrac{1}{2}u_x^2 + \tfrac{1}{2}v_x^2) + D_x(u_x v_x - u_x u_t - v_x v_t) = -u_{xx}(u_t - v_x) - v_{xx}(v_t - u_x).$$

Thus, a *single* (equivalence class of) *conservation laws may have more than one characteristic form.* Finally, we remark that one can add any null divergence to any of the above conservation laws without affecting its validity, or the form(s) of the characteristic.

The preceding example should give the reader a good idea of the algebraic complexity of the general relationship between characteristics and conservation laws. Nevertheless, if we restrict our attention to normal, nondegenerate systems (in particular, normal analytic systems), there is a one-to-one correspondence between equivalence classes of conservation laws and equivalence classes of characteristics, so that each conservation law is uniquely determined by its characteristic and vice versa, provided one keeps the equivalence relations in mind. This result forms the cornerstone for much of the general theory and classification of conservation laws, including Noether's theorem. (Counterexamples in the case of abnormal systems will be discussed in Section 5.3.)

Theorem 4.26. *Let $\Delta(x, u^{(n)}) = 0$ be a normal, totally nondegenerate system of differential equations. Let the p-tuples P and $\tilde{P}$ determine conservation laws with respective characteristics Q and $\tilde{Q}$. Then P and $\tilde{P}$ are equivalent conservation laws if and only if Q and $\tilde{Q}$ are equivalent characteristics.*

Clearly the theorem reduces to proving that a conservation law in characteristic form (4.29) is trivial if and only if its characteristic Q is trivial. Several complications arise because there are two types of triviality for conservation laws which must be treated. The proof itself is quite complicated, and the reader may at first be well advised to skip ahead to Section 4.4 at this point.

As a warm-up exercise for the general proof, we begin with the simple case of a single n-th order ordinary differential equation

$$\Delta(x, u^{(n)}) = \Delta(x, u, u_1, \ldots, u_n) = 0,$$

in which $u_k = d^k u/dx^k$ are the derivatives of the single dependent variable u. A conservation law in characteristic form is

$$D_x P = Q \cdot \Delta,$$

in which $Q(x, u^{(m)})$ is a single function of x, u and the derivatives of u. Note that in this case, the only trivial conservation laws of the second kind are the constants, so the proof will be considerably simplified. First suppose P is a trivial conservation law. Since Δ is nondegenerate,

$$P = \sum_{k=0}^{l} A_k \cdot D_x^k \Delta + c$$

for certain functions A_k, and $c \in \mathbb{R}$. By Leibniz' rule,

$$D_x P = \sum_{k=0}^{l} [D_x A_k \cdot D_x^k \Delta + A_k \cdot D_x^{k+1} \Delta]$$

$$= (D_x A_0) \cdot \Delta + \sum_{k=1}^{l} (A_{k-1} + D_x A_k) \cdot D_x^k \Delta + A_l \cdot D_x^{l+1} \Delta.$$

Equating this to $Q \cdot \Delta$, we find that

$$(D_x A_0 - Q) \cdot \Delta + \sum_{k=1}^{l} (A_{k-1} + D_x A_k) \cdot D_x^k \Delta + A_l \cdot D_x^{l+1} \Delta = 0$$

for all x, u. Now, the prolongations $\Delta^{(l+1)}$ of Δ are assumed to be of maximal rank. According to Proposition 2.11, then, a linear combination of the functions $\Delta, D_x\Delta, \ldots, D_x^{l+1}\Delta$ determining $\Delta^{(l+1)}$ will vanish identically if and only if the coefficients vanish for all solutions of $\Delta^{(l+1)}$. Thus we find

$$A_l = 0, \qquad A_{k-1} + D_x A_k = 0, \qquad k = 1, 2, \ldots, l,$$

and

$$D_x A_0 - Q = 0,$$

whenever $u = f(x)$ is a solution to Δ. An easy induction shows that $A_k = 0$ on solutions to Δ for $k = l, l-1, \ldots, 1, 0$, and hence the final equation requires Q to vanish for all solutions too. This means Q is a trivial characteristic, and hence we've proved that a trivial conservation law necessarily has a trivial characteristic.

In order to prove the converse, we need to solve our equation Δ for the highest order derivative,

$$u_n = \Gamma(x, u, \ldots, u_{n-1}), \tag{4.30}$$

which can be done in a neighbourhood of any point $(x_0, u_0^{(n)})$ at which Δ is normal, which, in the present circumstance means $\partial\Delta(x_0, u_0^{(n)})/\partial u_n \neq 0$. Before continuing, its is important to note that replacing Δ by the algebraically equivalent equation $u_n = \Gamma$ does not affect the structure of the space of conservation laws:

Lemma 4.27. *Suppose Δ and $\tilde{\Delta}$ are two totally nondegenerate systems of partial differential equations which are algebraically equivalent in the sense that their corresponding subvarieties $\mathscr{S}_\Delta$ and $\mathscr{S}_{\tilde{\Delta}}$ in the jet space $M^{(n)}$ coincide:*

$$\mathscr{S}_\Delta = \{(x, u^{(n)}): \Delta(x, u^{(n)}) = 0\} = \mathscr{S}_{\tilde{\Delta}} = \{(x, u^{(n)}): \tilde{\Delta}(x, u^{(n)}) = 0\}.$$

A p-tuple P is then a conservation law for Δ if and only if it is a conservation law for $\tilde{\Delta}$. It is trivial as a conservation law for Δ if and only if it is trivial as a conservation law for $\tilde{\Delta}$. If $\operatorname{Div} P = Q \cdot \Delta$ is in characteristic form for Δ, it is also in characteristic form for $\tilde{\Delta}$: $\operatorname{Div} P = \tilde{Q} \cdot \tilde{\Delta}$. Finally, Q is a trivial characteristic for Δ if and only if $\tilde{Q}$ is a trivial characteristic for $\tilde{\Delta}$.

Proof. The statement that a function $R(x, u^{(n)})$ vanishes for all solutions of Δ is, by local solvability, equivalent to saying that $R(x, u^{(n)}) = 0$ whenever $(x, u^{(n)}) \in \mathscr{S}_\Delta$. Clearly this requirement is independent of the particular functions Δ or $\tilde{\Delta}$ used to characterize the subvariety $\mathscr{S}_\Delta$ (or its prolongations). This trivial observation is sufficient to prove all the statements in the lemma save the last one. The requirement that Q be a trivial characteristic means that the expression $Q \cdot \Delta = R$ vanish to *second order* on some appropriate prolongation $\Delta^{(k)}$ of Δ. (This means that both R and all its partial derivatives $\partial R/\partial x^i$, $\partial R/\partial u_J^\alpha$ vanish on the prolonged subvariety $\mathscr{S}_{\Delta^{(k)}}$.) Again, this geometric condition is clearly independent of the particular functions Δ or $\tilde{\Delta}$ used to characterize $\mathscr{S}_\Delta$, and hence also $\mathscr{S}_{\Delta^{(k)}} = \mathscr{S}_{\tilde{\Delta}^{(k)}}$. □

Returning to our proof of Theorem 4.26 in the ordinary differential equation case, we are trying to show that if Q is a trivial characteristic, then $D_x P = Q \cdot \Delta$ is necessarily a trivial conservation law. By Lemma 4.27, we can assume that Δ has the form (4.30). Moreover, differentiating (4.30) and substituting we can find expressions for higher order derivatives u_{n+k}, $k \geqslant 0$, in terms of $x, u, \ldots, u_{n-1}$. These can be substituted into the conservation law P, leading to an equivalent conservation law $P^*(x, u, \ldots, u_{n-1})$ depending on only $(n-1)$-st and lower order derivatives of u.

Now in the general case, as Example 4.25 made clear, replacing a conservation law by an equivalent one does not necessarily preserve the characteristic form itself, and so we have no reason to expect $D_x P^* = 0$ to be in characteristic form. However, since P^* only depends on $(n-1)$-st and lower order derivatives, the only way n-th and higher order derivatives appear in

$$D_x P^* = \frac{\partial P^*}{\partial x} + u_1 \frac{\partial P^*}{\partial u} + \cdots + u_n \frac{\partial P^*}{\partial u_{n-1}}$$

is in the final term. Thus, by local solvability, $D_x P^* = 0$ on solutions if and only if

$$D_x P^* = Q^*(u_n - \Gamma),$$

where $Q^* = \partial P/\partial u_{n-1}$ is the characteristic which, by the first half of the theorem, is equivalent to the original characteristic Q, and is hence also trivial. Moreover, Q^* only depends on $(n-1)$-st and lower order derivatives of u, so the only way that it can be trivial is if it vanishes identically, $Q^* = \partial P/\partial u_{n-1} \equiv 0$. This implies $D_x P^* \equiv 0$, and hence P^* is a trivial conservation law of the second kind. (In the present case this means P^* is a constant!) Thus P is also trivial, and the theorem is proved in this special case.

The proof of Theorem 4.26 in the general case proceeds along similar lines, although the details, especially in the second part of the proof, become much more complicated. First suppose that P is a trivial conservation law, so that by the nondegeneracy of Δ, there exist functions $A^J_{iv}(x, u^{(m)})$ such that

$$P_i = \sum_{v,J} A^J_{iv} D_J \Delta_v + R_i, \qquad i = 1, \ldots, p, \tag{4.31}$$

where $R = (R_1, \ldots, R_p)$ is a null divergence. In this case

$$\operatorname{Div} P = \sum_{i,v,J} \{D_i A^J_{iv} \cdot D_J \Delta_v + A^J_{iv} D_i D_J \Delta_v\}.$$

Assuming P is in characteristic form, we equate this latter expression to $Q \cdot \Delta$, thereby obtaining a linear combination of the derivatives $D_K \Delta_v$ which vanishes identically in x and u. Again, by the maximal rank condition on the prolongations of Δ, Proposition 2.11 requires that the coefficient of each derivative $D_K \Delta_v$ must vanish whenever $u = f(x)$ is a solution to the system. An easy induction along the same lines as in the ordinary differential equation case shows that each coefficient A^J_{iv} must vanish whenever u is a solution, and, finally, each $Q_v = 0$ whenever $u = f(x)$ is a solution to Δ. Thus a trivial conservation law necessarily has a trivial characteristic and the first half of the theorem is proved.

To prove the converse, we first need to use a change of independent variables $(y, t) = \psi(x)$ which makes the system Δ equivalent to one in Kovalevskaya form

$$u^v_{nt} \equiv \frac{\partial^n u^v}{\partial t^n} = \Gamma_v(y, t, \widetilde{u^{(n)}}), \qquad v = 1, \ldots, q, \tag{4.32}$$

the Γ_v depending on all derivatives u^α_J up to order n except the u^α_{nt}. (See Theorem 2.79. The extension to the more general Kovalevskaya form (2.123) is not difficult, but the notation is more complicated, so this will be left to the reader.) Using Lemma 4.27, it suffices to prove that if $Q = (Q_1, \ldots, Q_q)$

is a trivial characteristic for a system in Kovalevskaya form, then the corresponding conservation law is trivial.

Lemma 4.28. *If Δ is in Kovalevskaya form* (4.32), *and P is a conservation law, then there exists an equivalent conservation law $\bar{P}$ such that*

$$\operatorname{Div} \bar{P} = D_t \bar{T} + \operatorname{Div}_y \bar{Y} = \sum_{\nu=1}^{q} \bar{Q}_\nu(y, t, \widetilde{u^{(m)}})\{u_{nt}^\nu - \Gamma_\nu\}, \tag{4.33}$$

where $\operatorname{Div}_y = D_{y^1} Y_1 + \cdots + D_{y^{p-1}} Y_{p-1}$ denotes the "spatial part" of the total divergence. The new characteristic $\bar{Q} = (\bar{Q}_1, \ldots, \bar{Q}_q)$ depends only on y, t, and derivatives $u_{jt,J}^\alpha$ of orders $j + \#J \leqslant m$ such that $j < n$, which we denote by $\widetilde{u^{(m)}}$; in other words, there are no t-derivatives of u of order $\geqslant n$ occurring in $\bar{Q}$.

Proof. First, as in Section 2.6, each prolongation of a system in Kovalevskaya form (4.32) will provide formulae for the higher-order t-derivatives $u_{kt,K}^\alpha$, $k \geqslant n$, in terms of the derivatives $u_{jt,J}^\beta$, $j < n$, of order strictly less than n in t. If the original conservation law P depends on n-th or higher order t-derivatives of u, we can substitute these formulae and thereby replace P by an equivalent conservation law $\hat{P}$ which is independent of the derivatives $u_{kt,K}^\alpha$, $k \geqslant n$. Hence, the divergence of $\hat{P}$ has the form

$$\begin{aligned} \operatorname{Div} \hat{P} = D_t \hat{T} + \operatorname{Div}_y \hat{Y} &= \sum_{\nu,K} \frac{\partial \hat{T}}{\partial u_{(n-1)t,K}^\nu} u_{nt,K}^\nu + R \\ &= \sum_{\nu,K} \frac{\partial \hat{T}}{\partial u_{(n-1)t,K}^\nu} D_K\{u_{nt}^\nu - \Gamma_\nu\} + \hat{R}, \end{aligned} \tag{4.34}$$

where neither R nor $\hat{R}$ depend on the derivatives $u_{nt,K}^\alpha$. However, to be a conservation law, $\operatorname{Div} \hat{P}$ must vanish on the system (4.32), and this implies that $\hat{R}$ vanishes on (4.32), which is not possible unless $\hat{R} \equiv 0$ vanishes identically. Therefore, (4.34) reduces to an identity

$$\operatorname{Div} \hat{P} = \sum_{\nu,K} Z_\nu^K D_K\{u_{nt}^\nu - \Gamma_\nu\}, \tag{4.35}$$

where the coefficients $Z_\nu^K = \partial \hat{T}/\partial u_{(n-1)t,K}^\nu$ do not depend on the derivatives $u_{nt,K}^\alpha$. Integrating (4.35) by parts, we recover the equivalent conservation law (4.33), with characteristic

$$\bar{Q}_\nu = \sum_K (-D)_K Z_\nu^K. \tag{4.36}$$

Since the multi-indices K in (4.36) only refer to y-derivatives, $\bar{Q}_\nu$ is independent of the n-th order t-derivatives $u_{nt,J}^\alpha$. This proves Lemma 4.28. □

To complete the proof of Theorem 4.26, suppose Q is a trivial characteristic for the conservation law

$$\operatorname{Div} P = Q \cdot \Delta.$$

Replace P by the equivalent conservation law $\bar{P}$ as given by Lemma 4.28. According to the direct half of Theorem 4.26 (which has already been proven!), since P and $\bar{P}$ are equivalent conservation laws, the associated characteristics Q and $\bar{Q}$ are equivalent; hence, $Q - \bar{Q} = 0$ on Δ. But Q already vanishes on Δ, and, according to Lemma 4.28, $\bar{Q}$ does not depend on the derivatives $u^\alpha_{kt,K}$ for $k \geqslant n$. Therefore, the only way that $\bar{Q}$ can vanish on Δ is if it vanishes identically. This means that the conservation law $\bar{P}$ is a trivial conservation law of the second kind (a null divergence); hence P, being equivalent to $\bar{P}$, is also trivial. This completes the proof that, for a system in Kovalevskaya form (and hence any normal system), trivial characteristics necessarily come from trivial conservation laws. □

4.4. Noether's Theorem

The general principle relating symmetry groups and conservation laws was first determined by E. Noether, [1], who stated it in almost complete generality. The version presented in this section is the one most familiar to physicists and engineers, requiring only knowledge of ordinary symmetry group theory as developed in Chapter 2, but is far from the most comprehensive version of Noether's theorem available. We will return to this topic in Section 5.3, where the complete, general form of Noether's theorem, which subsumes the present version, will be proved. Nevertheless, the result here is still of great practical use, and we will illustrate its effectiveness with a number of examples of physical importance.

Theorem 4.29. *Suppose G is a (local) one-parameter group of symmetries of the variational problem $\mathscr{L}[u] = \int L(x, u^{(n)})\, dx$. Let*

$$\mathbf{v} = \sum_{i=1}^{p} \xi^i(x, u)\frac{\partial}{\partial x^i} + \sum_{\alpha=1}^{q} \phi_\alpha(x, u)\frac{\partial}{\partial u^\alpha} \tag{4.37}$$

be the infinitesimal generator of G, and

$$Q_\alpha(x, u) = \phi_\alpha - \sum_{i=1}^{p} \xi^i u^\alpha_i, \qquad u^\alpha_i = \partial u^\alpha/\partial x^i,$$

the corresponding characteristic of $\mathbf{v}$, as in (2.48). Then $Q = (Q_1, \ldots, Q_q)$ is also the characteristic of a conservation law for the Euler–Lagrange equations $\mathsf{E}(L) = 0$; in other words, there is a p-tuple $P(x, u^{(m)}) = (P_1, \ldots, P_p)$ such that

$$\operatorname{Div} P = Q \cdot \mathsf{E}(L) = \sum_{\nu=1}^{q} Q_\nu \mathsf{E}_\nu(L) \tag{4.38}$$

is a conservation law in characteristic form for the Euler–Lagrange equations $\mathsf{E}(L) = 0$.

Proof. We substitute the prolongation formula (2.50) into the infinitesimal invariance criterion (4.15), to find

$$\begin{aligned} 0 &= \mathrm{pr}^{(n)}\,\mathbf{v}(L) + L\,\mathrm{Div}\,\xi \\ &= \mathrm{pr}^{(n)}\,\mathbf{v}_Q(L) + \sum_{i=1}^{p} \xi^i D_i L + L \sum_{i=1}^{p} D_i \xi^i \\ &= \mathrm{pr}^{(n)}\,\mathbf{v}_Q(L) + \mathrm{Div}(L\xi), \end{aligned}$$

where $L\xi$ is the p-tuple with components $(L\xi^1, \ldots, L\xi^p)$. The first term in this equation can be integrated by parts:

$$\begin{aligned} \mathrm{pr}^{(n)}\,\mathbf{v}_Q(L) &= \sum_{\alpha, J} D_J Q_\alpha \frac{\partial L}{\partial u_J^\alpha} \\ &= \sum_{\alpha, J} Q_\alpha \cdot (-D)_J \frac{\partial L}{\partial u_J^\alpha} + \mathrm{Div}\,A \\ &= \sum_{\alpha=1}^{q} Q_\alpha \mathsf{E}_\alpha(L) + \mathrm{Div}\,A, \end{aligned}$$

where $A = (A_1, \ldots, A_p)$ is some p-tuple of functions depending on Q, L and their derivatives whose precise form is not required here. We have proved that

$$\mathrm{pr}^{(n)}\,\mathbf{v}_Q(L) = Q \cdot \mathsf{E}(L) + \mathrm{Div}\,A \tag{4.39}$$

for some A. Therefore,

$$0 = Q \cdot \mathsf{E}(L) + \mathrm{Div}(A + L\xi), \tag{4.40}$$

and (4.38) holds with $P = -(A + L\xi)$. This completes the proof of Noether's theorem. □

From this standpoint, the essence of Noether's theorem is reduced to the integration by parts formula (4.39). To find the explicit expression for the resulting conservation law $P = -(A + L\xi)$, we thus need to find the general formula for A in terms of L and the characteristic Q of the symmetry. The general formula appears in Proposition 5.98; here we look at the case of first order variational problems in detail. (An alternative approach is to construct P directly from the basic formula (4.38) once the characteristic Q is known. This somewhat *ad hoc* tecnique is often useful in practice, when the general formula is rather cumbersome to apply directly.) If $L(x, u^{(1)})$ depends only on first order derivatives, then

$$\mathrm{pr}^{(1)}\,\mathbf{v}_Q(L) = \sum_{\alpha=1}^{q} \left\{ Q_\alpha \frac{\partial L}{\partial u^\alpha} + \sum_{i=1}^{p} D_i Q_\alpha \frac{\partial L}{\partial u_i^\alpha} \right\}.$$

Only the second batch of summands need to be integrated by parts, so we find (4.39) holds with $A_i = \sum_\alpha Q_\alpha\, \partial L / \partial u_i^\alpha$. Thus we have the following version of Noether's theorem for first order variational problems.

Corollary 4.30. *Suppose $\mathscr{L}[u] = \int L(x, u^{(1)})\, dx$ is a first order variational problem, and* $\mathbf{v}$ *as in (4.37) is a variational symmetry. Then*

$$P_i = \sum_{\alpha=1}^{q} \phi_\alpha \frac{\partial L}{\partial u_i^\alpha} + \xi^i L - \sum_{\alpha=1}^{q} \sum_{j=1}^{p} \xi^j u_j^\alpha \frac{\partial L}{\partial u_i^\alpha}, \qquad i = 1, \ldots, p, \tag{4.41}$$

form the components of a conservation law $\operatorname{Div} P = 0$ *for the Euler–Lagrange equations* $\mathsf{E}(L) = 0$.

Example 4.31. Consider a system of n particles moving in $\mathbb{R}^3$ subject to a potential force field. The kinetic energy of this system takes the form

$$K(\dot{\mathbf{x}}) = \tfrac{1}{2} \sum_{\alpha=1}^{n} m_\alpha |\dot{\mathbf{x}}^\alpha|^2,$$

where m_α is the mass and $\mathbf{x}^\alpha = (x^\alpha, y^\alpha, z^\alpha)$ the position of the α-th particle. The potential energy $U(t, x)$ will depend on the specific problem; for instance,

$$U(t, \mathbf{x}) = \sum \gamma_{\alpha\beta} |\mathbf{x}^\alpha - \mathbf{x}^\beta|^{-1}$$

might depend only on the pairwise gravitational interaction between masses, or (if $n = 1$) we may have the central gravitational force of Kepler's problem. Newton's equations of motion

$$m_\alpha \mathbf{x}_{tt}^\alpha = -\nabla_\alpha U \equiv -(U_{x^\alpha}, U_{y^\alpha}, U_{z^\alpha}), \qquad \alpha = 1, \ldots, n,$$

are in variational form, being the Euler–Lagrange equations for the action integral $\int_{-\infty}^{\infty} (K - U)\, dt$.

A vector field

$$\mathbf{v} = \tau(t, \mathbf{x}) \frac{\partial}{\partial t} + \sum_\alpha \xi^\alpha(t, \mathbf{x}) \cdot \frac{\partial}{\partial \mathbf{x}^\alpha} \equiv \tau \frac{\partial}{\partial t} + \sum_\alpha \left(\xi^\alpha \frac{\partial}{\partial x^\alpha} + \eta^\alpha \frac{\partial}{\partial y^\alpha} + \zeta^\alpha \frac{\partial}{\partial z^\alpha} \right)$$

will generate a variational symmetry group if and only if

$$\operatorname{pr}^{(1)} \mathbf{v}(K - U) + (K - U) D_t \tau = 0 \tag{4.42}$$

for all $(t, \mathbf{x})$. Noether's theorem immediately provides a corresponding conservation law or first integral

$$T = \sum_{\alpha=1}^{n} m_\alpha \xi^\alpha \cdot \dot{\mathbf{x}}^\alpha - \tau E = \text{constant}, \tag{4.43}$$

where $E = K + U$ is the total energy of the system. In this example, we investigate what form the potential must take so that certain groups of direct physical interest be variational symmetries, and deduce the form of the corresponding conservation law.

First, the group of time translations has generator $\mathbf{v} = \partial_t$. Since $\operatorname{pr}^{(1)} \mathbf{v} = \mathbf{v}$, (4.42) holds if and only if $\partial U/\partial t = 0$, i.e. U does not depend explicitly on t. The resulting conservation law is just the energy E. Invariance of a physical system under time translations generally implies conservation of energy. Next consider the group of simultaneous translations of all the particles in a fixed direction $\mathbf{a} \in \mathbb{R}^3$. The group $\mathbf{x}^\alpha \mapsto \mathbf{x}^\alpha + \varepsilon\mathbf{a}$ has generator $\mathbf{v} = \sum_\alpha \mathbf{a} \cdot \partial/\partial \mathbf{x}^\alpha$. Again $\mathrm{pr}^{(1)}\,\mathbf{v} = \mathbf{v}$, so (4.42) holds if and only if $\mathbf{v}(U) = 0$ meaning that U is translationally invariant in the given direction. The corresponding first integral is the linear momentum

$$\sum_\alpha m_\alpha \mathbf{a} \cdot \dot{\mathbf{x}}^\alpha = \text{constant}.$$

Again, in most physical systems translational invariance implies conservation of linear momentum. As a last example, consider the group of simultaneous rotations of all the masses about some fixed axis which, for simplicity, we take as the z-axis. The generator of this group is

$$\mathbf{v} = \sum_\alpha \left(x^\alpha \frac{\partial}{\partial y^\alpha} - y^\alpha \frac{\partial}{\partial x^\alpha} \right), \qquad \mathrm{pr}^{(1)}\,\mathbf{v} = \mathbf{v} + \sum_\alpha \left(\dot{x}^\alpha \frac{\partial}{\partial \dot{y}^\alpha} - \dot{y}^\alpha \frac{\partial}{\partial \dot{x}^\alpha} \right).$$

Note that $\mathrm{pr}^{(1)}\,\mathbf{v}(K) = 0$, hence rotations form a variational symmetry group if and only if U is rotationally invariant: $\mathbf{v}(U) = 0$. The conservation law is that of angular momentum

$$\sum_\alpha m_\alpha (x^\alpha \dot{y}^\alpha - y^\alpha \dot{x}^\alpha) = \text{constant}.$$

Again, in general, rotational invariance implies conservation of angular momentum. For example, the n-body problem admits all seven symmetries and thus has conservation of energy, linear and angular momentum, while Kepler's problem only retains energy and angular momentum; the translational invariance no longer holds since one mass has been fixed at the origin.

Example 4.32. *Elastostatics.* In elasticity, conservation laws take on an added importance because they provide nontrivial path-independent integrals, thereby allowing one to investigate singularities such as cracks by integrating appropriate quantities far away from them. Let $x \in \Omega \subset \mathbb{R}^p$ represent the material coordinates of an elastic body in some reference configuration, and $u \in \mathbb{R}^q$ the spatial coordinates representing the deformation, so $u(x)$ is the deformed position of the initial point x. Thus, in physical applications, $p = q = 2$ or 3 for planar or three-dimensional elasticity. In the hyperelastic theory, assuming the absence of body forces, the equilibrium deformations are determined as minima of the energy functional

$$\mathscr{W}[u] = \int_\Omega W(x, u^{(1)})\, dx$$

subject to appropriate boundary conditions on $\partial\Omega$. In most cases W, the *stored energy function*, will depend on material coordinates, deformation, and deformation gradient $\nabla u = (\partial u^\alpha / \partial x^i)$, the last measuring the strain due to the deformation.† The precise form of the stored energy function will depend on

† There do, however, exist theories of "higher grade" materials, allowing dependence of W on higher order derivatives.

the constitutive assumptions governing the type of elastic material of which the body is composed. Nevertheless, certain universal physical constraints will impose certain general restrictions on the form of W. Each of these constraints will appear in the guise of a variational symmetry group of $\mathscr{W}$, and then Noether's theorem will immediately lead to corresponding conservation laws, valid for general elastic materials.

First of all, since W is independent of any external forces, it presumably does not depend on the frame of reference of the observer. This means that W must be invariant under the Euclidean group

$$\mathrm{E}(q): \quad u \mapsto Ru + a, \qquad a \in \mathbb{R}^q, \quad R \in \mathrm{SO}(q),$$

in the spatial variables. Translational invariance implies $W = W(x, \nabla u)$ is independent of u; the corresponding conservation laws are just

$$\sum_{i=1}^{p} D_i(\partial W/\partial u_i^\alpha) = 0, \qquad \alpha = 1, \ldots, q,$$

which are nothing but the Euler–Lagrange equations themselves, expressed in divergence form. The rotational invariance of W:

$$W(x, R\nabla u) = W(x, \nabla u), \qquad R \in \mathrm{SO}(q),$$

leads to conservation laws

$$\sum_{i=1}^{p} D_i \left\{ u^\alpha \frac{\partial W}{\partial u_i^\beta} - u^\beta \frac{\partial W}{\partial u_i^\alpha} \right\} = 0, \qquad \alpha, \beta = 1, \ldots, q,$$

whose characteristics are those of the infinitesimal rotations $u^\alpha \partial_{u^\beta} - u^\beta \partial_{u^\alpha}$.

Further conservation laws can result if we impose additional restrictions on the type of elastic material. For instance, if the body is homogeneous, $W = W(\nabla u)$ does not depend on x. Invariance under the translation group $x \mapsto x + a$, $a \in \mathbb{R}^p$, leads to p further conservation laws.

$$\sum_{i=1}^{p} D_i \left(\sum_{\alpha=1}^{q} u_j^\alpha \frac{\partial W}{\partial u_i^\alpha} - \delta_i^j W \right) = 0,$$

the components of which form Eshelby's celebrated *energy-momentum tensor*. When integrated around the tip of a crack it determines the associated energy-release rate. For a homogeneous, isotropic material, the symmetry group $x \mapsto Qx$, $Q \in \mathrm{SO}(p)$, which requires $W(\nabla u \cdot Q) = W(\nabla u)$, leads to $\frac{1}{2}p(p-1)$ further laws

$$\sum_{i=1}^{p} D_i \left[\sum_{\alpha=1}^{q} (x^j u_k^\alpha - x^k u_j^\alpha) \frac{\partial W}{\partial u_i^\alpha} + (\delta_i^j x^k - \delta_i^k x^j) W \right] = 0$$

corresponding to the infinitesimal generators $x^k \partial/\partial x^j - x^j \partial/\partial x^k$. Further interesting conservation laws can be found by imposing still more restrictions on the nature of the stored energy function W. Restricting to a homogeneous material, if $W(\nabla u)$ is an algebraically homogeneous function of degree n, so

$$W(\lambda \nabla u) = \lambda^n W(\nabla u), \qquad \lambda > 0,$$

for all ∇u, then the scaling group

$$(x, u) \mapsto (\lambda x, \lambda^{(n-p)/n} u), \qquad \lambda > 0,$$

is a variational symmetry group since

$$\int_{\tilde{\Omega}} W(\nabla \tilde{u})\, d\tilde{x} = \int_{\Omega} W(\lambda^{-p/n} \nabla u) \lambda^p \, dx = \int_{\Omega} W(\nabla u)\, dx.$$

(If we just scale x or u individually, we have a symmetry of the Euler–Lagrange equations, but *not* in general a variational symmetry.) The infinitesimal generator of this group is

$$\sum_{i=1}^{p} x^i \frac{\partial}{\partial x^i} + \frac{n-p}{n} \sum_{\alpha=1}^{q} u^\alpha \frac{\partial}{\partial u^\alpha},$$

so the conservation law is

$$\sum_{i=1}^{p} D_i \left\{ \frac{n-p}{n} \sum_{\alpha=1}^{q} u^\alpha \frac{\partial W}{\partial u_i^\alpha} + x^i W - \sum_{j=1}^{p} \sum_{\alpha=1}^{q} x^j u_j^\alpha \frac{\partial W}{\partial u_i^\alpha} \right\} = 0.$$

Now in practice, the algebraic homogeneity assumption on W is rather special. For a general function W, then, a slightly modified form of the above conservation law yields the divergence identity

$$\sum_{i=1}^{p} D_i \left\{ \sum_{\alpha=1}^{q} u^\alpha \frac{\partial W}{\partial u_i^\alpha} + x^i W - \sum_{j=1}^{p} \sum_{\alpha=1}^{q} x^j u_j^\alpha \frac{\partial W}{\partial u_i^\alpha} \right\} = pW,$$

which was used by Knops and Stuart, [1], to prove uniqueness of equilibrium solutions corresponding to homogeneous deformations. (See Exercise 5.35 for a general theorem of this type.) If W is algebraically homogeneous of degree p, then there is a full conformal group of variational symmetries. The infinitesimal generators of the inversional transformations take the form

$$\sum_{j=1}^{p} (x^i x^j - \tfrac{1}{2} \delta_j^i |x|^2) \frac{\partial}{\partial x^j}$$

with corresponding conservation laws

$$\sum_{i=1}^{p} D_i C_i^j \equiv \sum_{i=1}^{p} D_i \left\{ \sum_{k=1}^{p} (x^j x^k - \tfrac{1}{2} \delta_j^k |x|^2) \left(\sum_{\alpha=1}^{q} u_k^\alpha \frac{\partial W}{\partial u_i^\alpha} - \delta_i^k W \right) \right\} = 0.$$

Again, if W is not homogeneous, these turn into divergence identities:

$$\sum_{i=1}^{p} D_i C_i^j = x^j \left[pW - \sum_{\alpha, k} u_k^\alpha \frac{\partial W}{\partial u_k^\alpha} \right].$$

This method of using symmetries of special variational problems to construct useful divergence identities for more general functionals is quite promising. See Pucci and Serrin, [1], and van der Vorst, [1], for further developments and applications.

Divergence Symmetries

A cursory inspection of the proof of Noether's theorem reveals that the hypothesis that the vector field $\mathbf{v}$ generate a group of variational symmetries is overly restrictive for us to deduce the existence of a conservation law. This inspires the following relaxation of the definition of a variational symmetry group.

Definition 4.33. Let $\mathscr{L}[u] = \int L\, dx$ be a functional. A vector field $\mathbf{v}$ on $M \subset X \times U$ is an *infinitesimal divergence symmetry* of $\mathscr{L}$ if there exists a p-tuple $B(x, u^{(m)}) = (B_1, \ldots, B_p)$ of functions of x, u and derivatives of u such that

$$\mathrm{pr}^{(n)}\, \mathbf{v}(L) + L\, \mathrm{Div}\, \xi = \mathrm{Div}\, B \tag{4.44}$$

for all x, u in M.

Compare Theorem 4.12 for the motivation and notation for the "infinitesimal criterion" (4.44). In particular, if $B = 0$ we recover our previous notion of variational symmetry. Each infinitesimal divergence symmetry of a variational problem generates a one-parameter group $g_\varepsilon = \exp(\varepsilon \mathbf{v})$ of transformations on M, but the precise symmetry properties of such groups of *divergence symmetries* is less transparent than for the ordinary groups of variational symmetries. However, we do have the following generalization of Theorem 4.14.

Theorem 4.34. *If $\mathbf{v}$ is an infinitesimal divergence symmetry of a variational problem, then $\mathbf{v}$ generates a symmetry group of the associated Euler–Lagrange equations.*

The proof of this result is deferred until Section 5.3, when a generalization will be developed. In practice, then, to determine divergence symmetries of a given variational problem, one first computes the general symmetry group of the corresponding Euler–Lagrange equations. It is then a fairly straightforward matter to check which linear combination of these symmetries satisfies the additional criterion (4.44) so as to actually be a divergence symmetry. (See also Proposition 5.55.)

The statement of Noether's Theorem 4.29 remains the same if we replace variational symmetry by divergence symmetry in the hypothesis: the characteristic Q of the infinitesimal divergence symmetry remains the characteristic of a conservation law of the Euler–Lagrange equations. The only thing that changes in the proof is the incorporation of the extra term $\mathrm{Div}\, B$ stemming from (4.44) in the formulae so that, for instance, (4.40) is replaced by

$$Q \cdot E(L) + \mathrm{Div}(A + L\xi) = \mathrm{Div}\, B.$$

Thus the conclusion (4.38) holds, with $P = B - A - L\xi$ in this case.

Example 4.35. Let us look at the invariance of the Lagrangian $K - U$ for a system of n masses under Galilean boosts:

$$(t, \mathbf{x}^\alpha) \mapsto (t, \mathbf{x}^\alpha + \varepsilon t\mathbf{a}),$$

where $\mathbf{a} \in \mathbb{R}^3$. The infinitesimal generator of this action has prolongation

$$\mathrm{pr}^{(1)}\,\mathbf{v} = \sum_{\alpha=1}^{n} \left(t\mathbf{a} \cdot \frac{\partial}{\partial \mathbf{x}^\alpha} + \mathbf{a} \cdot \frac{\partial}{\partial \dot{\mathbf{x}}^\alpha} \right);$$

thus

$$\mathrm{pr}^{(1)}\,\mathbf{v}(L) = \sum_{\alpha=1}^{n} m_\alpha \mathbf{a} \cdot \dot{\mathbf{x}}^\alpha - t \sum_{\alpha=1}^{n} \mathbf{a} \cdot \nabla_\alpha U.$$

This never vanishes identically (unless $\mathbf{a} = 0$), so the Galilean boost is never an ordinary variational symmetry. However, the first term in $\mathrm{pr}^{(1)}\,\mathbf{v}(L)$ is a divergence, namely $D_t(\sum m_\alpha \mathbf{a} \cdot \mathbf{x}^\alpha)$, so $\mathbf{v}$ generates a group of divergence symmetries provided U is translationally invariant in the direction of $\mathbf{a}$. The associated first integral is

$$\sum_\alpha m_\alpha \mathbf{a} \cdot \mathbf{x}^\alpha - t \sum_\alpha m_\alpha \mathbf{a} \cdot \dot{\mathbf{x}}^\alpha.$$

The first summation when divided by the total mass $\sum m_\alpha$ determines the position of the centre of mass of the system in the direction $\mathbf{a}$, while the second is just the linear momentum in the same direction. We thus find that if U is translationally invariant in a given direction, not only is the linear momentum in that direction a constant, but the centre of mass in that direction is a linear function of t:

$$\text{Centre of Mass} = t(\text{Momentum})/(\text{Mass}) + c.$$

In particular, if U is invariant under the complete translation group in $\mathbb{R}^3$, the centre of mass of any such system moves linearly in a fixed direction.

Example 4.36. Return to the wave equation in two spatial dimensions considered in Examples 2.43 and 4.15. It has already been shown that, of the full group of symmetries of the wave equation, the translations, rotations and (modified) dilatations are symmetries of the associated variational problem. It is now seen that the inversions, while not variational symmetries in the strict sense, are divergence symmetries. In the case of $\mathbf{i}_x$, we have

$$\mathrm{pr}^{(1)}\,\mathbf{i}_x(L) + L\,\mathrm{Div}\,\xi = uu_x = D_x(\tfrac{1}{2}u^2).$$

There are thus ten conservation laws for the wave equation arising from geometrical symmetry groups—three from translations, three from rotations, one dilatational and, finally, three inversional conservation laws. In the following table, we just list the ten conserved densities, leaving the reader to determine the associated fluxes.

Symmetry	Characteristic	Conserved Density
Translations	u_x	$P_x = u_x u_t$
	u_y	$P_y = u_y u_t$
	u_t	$E = \frac{1}{2}(u_x^2 + u_y^2 + u_t^2)$
Rotations	$xu_y - yu_x$	$A = xP_y - yP_x$
	$xu_t + tu_x$	$M_x = xE + tP_x$
	$yu_t + tu_y$	$M_y = yE + tP_y$
Dilatations	$xu_x + yu_y + tu_t + \frac{1}{2}u$	$D = xP_x + yP_y + \frac{1}{2}uu_t + tE$
Inversions	$(x^2 - y^2 + t^2)u_x + 2xyu_y + 2xtu_t + xu$	$I_x = xD - yA + \frac{1}{2}xuu_t + tM_x$
	$2xyu_x + (y^2 - x^2 + t^2)u_y + 2ytu_t + yu$	$I_y = yD - xA + \frac{1}{2}yuu_t + tM_y$
	$2xtu_x + 2ytu_y + (x^2 + y^2 + t^2)u_t + tu$	$I_t = (x^2 + y^2)E - \frac{1}{2}u^2 + 2tD - t^2$

Consequently, if $u(x, y, t)$ is any global solution to the wave equation decaying sufficiently rapidly as $x^2 + y^2 \to \infty$, then the spatial integrals of each of the above densities is a constant, independent of t. Thus we obtain conservation of energy

$$\mathscr{E} = \iint E \, dx \, dy = \text{constant}$$

and similar statements about linear momenta $\mathscr{P}_x$ and $\mathscr{P}_y$ (the integrals of P_x and P_y) and angular momentum $\mathscr{A}$. The hyperbolic rotations yield the linear dependence of the associated energy moments on t; for instance,

$$-\iint xE \, dx \, dy = \mathscr{P}_x t + \mathscr{C}$$

for some constant $\mathscr{C}$, where $\mathscr{P}_x$ is the constant linear momentum. The dilatational group leads to the useful identity

$$-\frac{d}{dt}\iint \tfrac{1}{2}u^2 \, dx \, dy = \iint (xP_x + yP_y) \, dx \, dy + \mathscr{E}t + \hat{\mathscr{C}},$$

for $\hat{\mathscr{C}}$ constant. The three inversional conservation laws, e.g.

$$\iint [(x^2 + y^2)E - \tfrac{1}{2}u^2) \, dx \, dy = \mathscr{E}t^2 + 2\hat{\mathscr{C}}t + \mathscr{C}^*,$$

while less physically motivated, are of key importance in the development of scattering theory for both linear and nonlinear wave equations.

Finally, there are the symmetry generators $\mathbf{v}_\alpha = \alpha(x, y. t)\partial_u$ stemming from the linearity of the equation. These satisfy

$$\text{pr}^{(1)} \mathbf{v}_\alpha(L) = \alpha_t u_t - \alpha_x u_x - \alpha_y u_y = D_t(\alpha_t u) - D_x(\alpha_x u) - D_y(\alpha_y u),$$

since α is a solution to the wave equation. Thus, except in the special case of constant α, these are not variational symmetries in the sense of Definition

4.10; they do generate divergence symmetries. The corresponding conservation laws are the reciprocity relations

$$D_t(\alpha u_t - \alpha_t u) - D_x(\alpha u_x - \alpha_x u) - D_y(\alpha u_y - \alpha_y u)$$
$$= \alpha(u_{tt} - u_{xx} - u_{yy}) - u(\alpha_{tt} - \alpha_{xx} - \alpha_{yy}) = 0,$$

vanishing whenever α and u both solve the wave equation. In integrated form this law is just Green's formula, as applied to the wave operator. (See Section 5.3 for a general discussion of reciprocity relations.)

NOTES

The calculus of variations has its origins in the work of Euler and the Bernoullis in the eighteenth century, the operator bearing Euler's name first appearing in 1744. However, it was not until the work of Weierstrass and Hilbert in the latter half of the nineteenth century that some semblance of rigor appeared in the subject. The book by Gel'fand and Fomin, [1], gives a reasonable introduction to the calculus of variations, of which we are only using the most elementary ideas here. Conservation laws are of even older origin, although the idea of conservation of energy was not conceptualized until the work of Helmholtz in the 1840's. (See Elkana, [1], for an interesting study of the historical development of this idea.) See Whitham, [2; §6.1], for a more detailed development of the conservation laws of fluid mechanics outlined in Example 4.22.

In this book I have not attempted to present any of the numerous applications of conservation laws to the study of differential equations, but have concentrated just on their systematic derivation using the symmetry group method of Noether. Lax, [2], uses conservation laws (called "entropy-flux pairs" in this context) to prove global existence theorems and determine realistic conditions for shock wave solutions to hyperbolic systems. This is further developed in DiPerna, [1], [2], where extra conservation laws are applied to the decay of shock waves and further existence theorems. Conservation laws have been applied to problems of stability by Benjamin, [1], and Holm, Marsden, Ratiu and Weinstein, [1]. Morawetz, [1] and Strauss, [1], use them in scattering theory. In elasticity, conservation laws (or, rather, their path-independent integral form—see Exercise 4.2) are of key importance in the study of cracks and dislocations; see the papers in Bilby, Miller and Willis, [1]. Knops and Stuart, [1], have used them to prove uniqueness theorems for elastic equilibria. The above is only a small sampling of all the applications which have appeared.

Trivial conservation laws were known for a long time by people in general relativity. Those of the second kind go under the name of "strong conservation laws" since they hold regardless of the underlying field equations; see the review papers of J. G. Fletcher, [1], and Goldberg, [1]. The characteristic form of a conservation law appears in Steudel, [1], but the connection between trivial characteristics and trivial conservation laws of Theorem 4.26 is due to Alonso, [1]. See Vinogradov, [5], and Olver, [11], for related results.

The concept of a variational symmetry, including the basic infinitesimal criterion (4.15), is due to Lie, [7], from his early theory of integral invariants. The first people to notice a connection between symmetries and conservation laws were Jacobi, [1], and later, Schütz, [1]. Engel, [1], developed the correspondence between the conservation of linear and angular momenta and linear motion of the centre of mass with invariance under translational, rotational and Galilean symmetries in the context of classical mechanics. Klein and Hilbert's investigations into Einstein's theory of general relativity inspired Noether to her remarkable paper, [1], in which both the concept of a variational symmetry group and the connection with conservation laws were set down in complete generality. The version of Noether's theorem appearing in this chapter is only a special case of her more general theorem, to be discussed in Section 5.3. The extension of Noether's methods to include divergence symmetries is due to Bessel-Hagen, [1].

Thus by 1922 all the machinery for a detailed, systematic investigation into the symmetry properties and consequent conservation laws of the important equations of mathematical physics was available. Strangely enough, this did not occur until quite recently. One possible explanation is that the constructive infinitesimal methods of Lie for computing symmetry groups were never quite reconciled with the theorem of Noether. In any event, the next significant reference to Noether's paper is in a review article by the physicist Hill, [1], in which the special case of Noether's theorem discussed in this chapter was presented, with implications that this was all Noether had actually proved on the subject. Unfortunately, the next twenty years saw a succession of innumerable papers either re-deriving the basic Noether Theorem 4.29 or purporting to generalize it, while in reality only reproving Noether's original result or special cases thereof. The mathematical physics literature to this day abounds with such papers, and it would be senseless to list them here. (I know of close to 50 such references, but I am certain many more exist!) Some references can be found in the book of Logan, [1], (which again only treats the special form of Noether's theorem for classical symmetry groups) and also other references mentioned below.

The lack of investigation into and appreciation of Noether's theorem has had some interesting consequences. Eshelby's energy-momentum tensor, which has much importance in the study of cracks and dislocations in elastic media, was originally found using *ad hoc* techniques, Eshelby, [1]. It was not related to symmetry properties of the media, as in Example 4.32, until the work of Günther, [1], and Knowles and Sternberg, [1]. An extension to the equations of linear elastodynamics was made by D. C. Fletcher, [1]. Subsequently, Olver, [8], [9], [14], found further undetected symmetries of the equations of linear elasticity, with consequent new conservation laws. Similarly, the important identities of Morawetz, [1], used in scattering theory for the wave equation were initially derived from scratch. Subsequently Strauss, [1], showed how these were related to the conformal invariance of the equation. (The further conservation laws to be found in Chapter 5 have yet to be applied here.) A similar development holds for the work of Baker and Tavel, [1], on conservation laws in optics, and no doubt further examples can be found.

The use of variational symmetry groups to reduce the order of ordinary differential equations which are the Euler–Lagrange equations of some variational problem presented in Theorem 4.17 is not as well known as its Hamiltonian counterpart, Theorem 6.35. A version of Theorem 4.17 for Lagrangians depending on only first order derivatives of the dependent variables is given in Whittaker, [1; p. 55], but I was unable to locate a reference to the full statement of this theorem in the literature.

Exercises

4.1. Let $\mathscr{L}$ be a functional. Prove that if $\mathbf{v}$ and $\mathbf{w}$ generate one-parameter variational symmetry groups of $\mathscr{L}$, then so does their Lie bracket $[\mathbf{v}, \mathbf{w}]$.

4.2. Suppose $p = 2$ and $D_x P + D_y Q = 0$ is a conservation law for a system of differential equations. Prove that if $u(x, y)$ is any solution to the system, the line integral

$$\int_C Q(x, y, u^{(m)})\, dx - P(x, y, u^{(m)})\, dy$$

does not depend on the path C. Generalize to $p > 2$.

4.3. If the case of a mechanical system, such as that in Example 4.31, time-translational invariance implies conservation of energy, space-translational invariance implies conservation of linear momentum (in the given direction) while, as in Example 4.35, Galilean invariance implies linear motion of the centre of mass. Prove that if a system admits laws of conservation of energy and the linear motion of the centre of mass, then it automatically admits the law of conservation of linear momentum as well. (Schütz, [1].)

4.4. The BBM equation $u_t + u_x + uu_x - u_{xxt} = 0$ can be put into variational form by letting $u = v_x$. Find three conservation laws of this equation using Noether's theorem. (Olver, [3].)

4.5. The equation $u_{tt} = u_{xxxx}$ describes the vibrations of a rod. Compute symmetries and conservation laws of this equation using Noether's theorem.

*4.6. Prove that Maxwell's equations, in both the physical form of Exercise 2.16(a) and the potential form of Exercise 2.16(b) are Euler–Lagrange equations. Find the variational principle in each case. Which of the symmetries of Exercise 2.16 lead to conservation laws and what are these laws? (Pohjanpelto, [1], [2].)

*4.7. Find a variational principle for Navier's equations (2.127) of linear elasticity. Discuss symmetries and the associated conservation laws, including triviality, in this instance. Do the same for the abnormal system (2.118). (Olver, [9].)

4.8. The Emden–Fowler equation is

$$\frac{d^2u}{dx^2} + \frac{2}{x}\frac{du}{dx} + u^5 = 0.$$

(a) Determine a variational problem such that the Emden–Fowler equation is the Euler–Lagrange equation thereof. (*Hint*: Multiply by x^2.)
(b) Find a simple variational scaling symmetry and use this to integrate the Emden–Fowler equation.
(Dresner, [1; p. 14], Logan, [1; p. 52], Rosenau, [1].)

4.9. Prove that the damped harmonic oscillator $m\ddot{x} + a\dot{x} + kx = 0$, $m \neq 0$, can be made into the form of an Euler–Lagrange equation by multiplying by $\exp(at/m)$. Prove that the vector field $\mathbf{v} = \partial_t - (ax/2m)\partial_x$ generates a one-parameter group of variational symmetries. Use this to integrate the equation by quadrature. How does this method compare in effort with the usual method of solving linear ordinary differential equations? (Logan, [1; p. 57]; see also Exercise 5.48.)

4.10. Consider an n-th order ordinary differential equation, on $M \subset X \times U \simeq \mathbb{R}^2$,

$$\frac{d^n u}{dx^n} = H(x, u^{(n-1)}).$$

Prove that the first integrals of this equation are the same as the invariants of the one-parameter group generated by

$$\frac{\partial}{\partial x} + u_x\frac{\partial}{\partial u} + u_{xx}\frac{\partial}{\partial u_x} + \cdots + u_{n-1}\frac{\partial}{\partial u_{n-2}} + H(x, u^{(n-1)})\frac{\partial}{\partial u_{n-1}}$$

acting on the jet space $M^{(n-1)}$. Find the solution to $u_{xx} = u$ using this remark. (Cohen [1; pp. 86, 99].)

4.11. Consider a variational problem of the form $\mathscr{L} = \int L(x, u_x^{-1}u_{xx})\,dx$, for $x, u \in \mathbb{R}$.
(a) Prove that the two-parameter group $(x, u) \mapsto (x, au + b)$, $a \neq 0$, is a variational symmetry group.
(b) What is the Euler–Lagrange equation for $\mathscr{L}$?
(c) Show how the Euler–Lagrange equation can be integrated twice using the translational invariance, but that the resulting second order equation is not in general scale-invariant.
(d) Do the same for the scaling symmetry.
(e) Integrate the Euler–Lagrange equation twice by using the two first integrals given by Noether's theorem, but show again that one cannot in general reduce the order any further.
(f) What happens if one uses the methods of Section 2.5 on the equation?

This shows that, whereas a one-parameter variational symmetry group will in general allow one to reduce a system of Euler–Lagrange equations by two, a two-parameter variational symmetry group does *not* in general allow one to reduce the order by four! (This problem will be taken up in a Hamiltonian framework in Chapter 6.)

4.12. Show that if $\mathscr{L}$ is a variational problem depending on a single independent and single dependent variable, and $\mathscr{L}$ is invariant under a two-parameter *abelian* group of symmetries, then one can reduce the order of the corresponding Euler–Lagrange equation by four.

4.13. (a) Suppose $\Delta = \mathsf{E}(L) = 0$ forms the Euler–Lagrange equations of some variational problem, and G is a regular group of variational symmetries (or even divergence symmetries) acting on M. Is the reduced system $\Delta/G = 0$ for the G-invariant solutions of Δ necessarily the Euler–Lagrange equations for some variational problem on the quotient manifold M/G? See Anderson and Fels, Symmetry reduction of variational bicomplexes and the principle of symmetric criticality, *Amer. J. Math.* **119 (1997) 609–670, for details.

(b) Find variational principles for the equations for the group-invariant solutions to the Korteweg–de Vries equation (2.66) found in Example 3.4, using the substitution $u = v_x$ to first put the Korteweg–de Vries equation itself into variational form.

4.14. The heat equation $u_t = u_{xx}$ cannot be put into variational form (except through some artificial tricks—see Exercises 5.36 and 5.37). Prove, however, that the equation for the scale-invariant solutions is equivalent to an Euler–Lagrange equation. (*Hint*: Look for an appropriate function to multiply it by.) Generalize to higher dimensions. (Thus reduction by a symmetry group will usually maintain a variational structure if there is one to begin with, but may also introduce a variational structure where none existed before!)

4.15. Suppose $p = 1$, $q = 2$ and we have a functional

$$\mathscr{L}[u, \tilde{u}] = \int L(x, u, \tilde{u}, u_x, \tilde{u}_x, \ldots)\, dx.$$

Consider the "hodograph" change of variables $y = \tilde{u}$, $v = u$, $\tilde{v} = x$, and let

$$\tilde{\mathscr{L}}[v, \tilde{v}] = \int \tilde{L}(y, v, \tilde{v}, v_y, \tilde{v}_y, \ldots)\, dy$$

be the transformed functional. Prove that the corresponding Euler–Lagrange equations are related by the formula

$$\mathsf{E}_u(L) = \tilde{u}_x \mathsf{E}_v(\tilde{L}), \qquad \mathsf{E}_{\tilde{u}}(L) = -u_x \mathsf{E}_v(\tilde{L}) - \mathsf{E}_{\tilde{v}}(\tilde{L}).$$

4.16. Use Noether's theorem to give an alternative proof of the Reduction Theorem 4.17 that does not directly rely on a change of variables. Apply your result to Exercises 4.8 and 4.9.

CHAPTER 5

Generalized Symmetries

The symmetry groups of differential equations or variational problems considered so far in this book have all been local transformation groups acting "geometrically" on the space of independent and dependent variables. E. Noether was the first to recognize that one could significantly extend the application of symmetry group methods by including derivatives of the relevant dependent variables in the transformations (or, more correctly, their infinitesimal generators). More recently, these "generalized symmetries"† have proved to be of importance in the study of nonlinear wave equations, where it appears that the possession of an infinite number of such symmetries is a characterizing property of "solvable" equations, such as the Korteweg–de Vries equation, which have "soliton" solutions and can be linearized either directly or via inverse scattering.

The first section of this chapter presents the basic theory of generalized vector fields and the associated group transformations, which are now found by solving the Cauchy problem for some associated system of evolution equations. The determination of the generalized symmetries of a system of differential equations is essentially the same as before, although the intervening calculations usually are far more complicated. A second approach to this problem is through the use of a recursion operator, which will generate infinite families of symmetries at once. These are presented in the second

† Some authors have mistakenly attributed the introduction of these symmetries to the work of Lie and Bäcklund, and have given the misleading misnomer of "Lie-Bäcklund transformations". (In particlar, they are *not* the same as true Bäcklund transformations, which do *not* have group properties.) We have chosen the term "generalized symmetry" rather than "Noether transformation" since the latter already has acquired several other meanings in the context of variational problems. A fuller discussion of the curious history of these symmetries appears in the notes at the end of the chapter.

section. For linear systems, recursion operators and symmetries are essentially the same objects, while for nonlinear equations, only very special "solvable" equations appear to have recursion operators.

Many of our earlier applications of geometrical symmetries remain valid for generalized symmetries. In particular, Noether's theorem now provides a complete one-to-one correspondence between one-parameter groups of generalized variational symmetries of some functional and the conservation laws of its associated Euler–Lagrange equations. Thus, one can hope to completely classify conservation laws by constructive symmetry group methods. In particular, the recursion operator interpretation of symmetry groups of linear systems leads at once to infinite families of conservation laws depending on higher order derivatives in very general situations. Recent results have further crystallized the roles of trivial symmetries and conservation laws in the Noether correspondence for totally nondegenerate systems, with the consequence that each nontrivial variational symmetry group gives rise to a nontrivial conservation law, and conversely. Under-determined systems fall under the ambit of Noether's second theorem, which relates infinite-dimensional groups of variational symmetries to dependencies among the Euler–Lagrange equations themselves. All these will be discussed in detail in the third section of this chapter.

Underlying much of our algebraic manipulations involving symmetries, conservation laws, differential operators and the like, a subject best described as the "formal variational calculus", is a certain complex, called the variational complex, doing for the variational calculus what the de Rham complex does for ordinary vector calculus on manifolds. There are three fundamental results which motivate the consideration of this complex: the first is the characterization of the kernel of the Euler operator as the space of total divergences; the second is the characterization, in Theorem 4.24, of the space of null divergences (trivial conservation laws of the second kind) as "total curls"; the third is Helmholtz's version of the inverse problem of the calculus of variations which states when a given set of differential equations forms the Euler–Lagrange equations for some variational problem. All of these results are manifestations of the exactness of the full variational complex at different stages. Although each result could be proved as it stands, the variational complex, whose fundamental role in the geometric theory of the calculus of variations is becoming more and more apparent, provides the unifying theme behind them, and the complete proof of exactness of it is not much more difficult to obtain. Thus we have devoted the last section of this chapter to a self-contained exposition of this complex, together with a much simplified proof of exactness thereof.

5.1. Generalized Symmetries of Differential Equations

Consider a vector field

$$\mathbf{v} = \sum_{i=1}^{p} \xi^i(x, u)\frac{\partial}{\partial x^i} + \sum_{\alpha=1}^{q} \phi_\alpha(x, u)\frac{\partial}{\partial u^\alpha}$$

defined on some open subset M of the space of independent and dependent variables $X \times U$. Provided the coefficient functions ξ^i, ϕ_α depend only on x and u, $\mathbf{v}$ will generate a (local) one-parameter group of transformations $\exp(\varepsilon\mathbf{v})$ acting pointwise on the underlying space M of the type discussed in detail in the previous chapters. A significant generalization of the notion of symmetry group is obtained by relaxing this geometrical assumption, and allowing the coefficient functions ξ^i, ϕ_α to also depend on derivatives of u. In this chapter, we will explore the many consequences of such an extension of the notion of symmetry.

Differential Functions

Before proceeding with the development of the theory of generalized vector fields, it is useful to introduce some notation. Throughout this chapter $M \subset X \times U$ will denote a fixed connected open subset of the space of independent and dependent variables. The prolongations $M^{(n)} \subset X \times U^{(n)}$ are then open subsets of the corresponding jet spaces, with $(x, u^{(n)}) \in M^{(n)}$ if and only if $(x, u) \in M$. We let $\mathcal{A}$ denote the space of smooth functions $P(x, u^{(n)})$ depending on x, u and derivatives of u up to some finite, but unspecified order n, defined for $(x, u^{(n)}) \in M^{(n)}$. The functions in $\mathcal{A}$ are called *differential functions* (in analogy with the differential polynomials of differential algebra). Each differential function is thus a smooth function $P: M^{(n)} \to \mathbb{R}$ for some (finite) n. If $m \geqslant n$, then $P(x, u^{(n)})$ can also be viewed as a function on $M^{(m)}$ since the coordinates $(x, u^{(n)})$ form part of the coordinates $(x, u^{(m)})$ on $M^{(m)}$. If we do not care as to precisely how many derivatives of u that P depends on, we will write $P[u] = P(x, u^{(n)})$ for P, where the square brackets will serve to remind us that P depends on x, u *and* derivatives of u. We further define $\mathcal{A}^l$ to be the vector space of l-tuples of differential functions, $P[u] = (P_1[u], \ldots, P_l[u])$, where each $P_j \in \mathcal{A}$.

Note that $\mathcal{A}$ is an algebra, meaning that we can add differential functions and multiply them together. There are also a number of fundamental differential operators on $\mathcal{A}$ which we have already encountered. Both the partial derivatives $\partial/\partial x^i$ and $\partial/\partial u_J^\alpha$ take a differential function to another differential function, but in general do not preserve the order of derivatives on which they depend. For instance, $P = u_{xxx} + xuu_x$ depends on third order derivatives, but $\partial P/\partial u = xu_x$ only depends on first order derivatives. Similarly, the total derivatives $D_j: \mathscr{A} \to \mathscr{A}$ are linear maps, with $D_j P[u]$ depending on $(n+1)$-st order derivatives when $P[u] = P(x, u^{(n)})$ depends on n-th order derivatives. Two other important operators are the total divergence Div: $\mathscr{A}^p \to \mathscr{A}$ and the Euler operator E: $\mathscr{A} \to \mathscr{A}^q$ defined in the preceding chapter.

Generalized Vector Fields

Definition 5.1. A *generalized vector field* will be a (formal) expression of the form

$$\mathbf{v} = \sum_{i=1}^{p} \xi^i[u] \frac{\partial}{\partial x^i} + \sum_{\alpha=1}^{q} \phi_\alpha[u] \frac{\partial}{\partial u^\alpha} \tag{5.1}$$

in which ξ^i and ϕ_α are smooth differential functions.

Thus, for example,

$$\mathbf{v} = x u_x \frac{\partial}{\partial x} + u_{xx} \frac{\partial}{\partial u}$$

is a generalized vector field in the case $p = q = 1$. For the moment, we will avoid any discussion of the precise meaning of such an object, but work with such generalized vector fields as if they were ordinary vector fields. Thus, in accordance with the prolongation formula of Theorem 2.36, we can define the *prolonged* generalized vector field

$$\mathrm{pr}^{(n)}\, \mathbf{v} = \mathbf{v} + \sum_{\alpha=1}^{q} \sum_{\#J \leq n} \phi_\alpha^J[u] \frac{\partial}{\partial u_J^\alpha},$$

whose coefficients are determined by the formula

$$\phi_\alpha^J = D_J\left(\phi_\alpha - \sum_{i=1}^{p} \xi^i u_i^\alpha \right) + \sum_{i=1}^{p} \xi^i u_{J,i}^\alpha, \tag{5.2}$$

with the same notation as before. Thus, in our previous example,

$$\mathrm{pr}^{(1)}\, \mathbf{v} = x u_x \frac{\partial}{\partial x} + u_{xx} \frac{\partial}{\partial u} + [u_{xxx} - (x u_{xx} + u_x) u_x] \frac{\partial}{\partial u_x},$$

the coefficient of $\partial/\partial u_x$ being computed as

$$D_x(u_{xx} - x u_x^2) + x u_x u_{xx} = D_x(u_{xx}) - D_x(x u_x) u_x.$$

Since all the prolongations of $\mathbf{v}$ have the same general expression for their coefficient functions ϕ_α^J, it is helpful to pass to the "infinite" prolongation, and take care of *all* the derivatives at once. Specifically, given a generalized vector field $\mathbf{v}$, its *infinite prolongation* (or *prolongation* for short) is the formally infinite sum

$$\mathrm{pr}\, \mathbf{v} = \sum_{i=1}^{p} \xi^i \frac{\partial}{\partial x^i} + \sum_{\alpha=1}^{q} \sum_{J} \phi_\alpha^J \frac{\partial}{\partial u_J^\alpha}, \tag{5.3}$$

where each ϕ_α^J is given by (5.2), and the sum in (5.3) now extends over *all* multi-indices $J = (j_1, \ldots, j_k)$ for $k \geqslant 0$, $1 \leqslant j_\kappa \leqslant p$. Note that if $P[u] = P(x, u^{(n)})$ is any differential function, $\text{pr}\,\mathbf{v}(P) = \text{pr}^{(n)}\mathbf{v}(P)$ is again a differential function. In particular, since P depends on only finitely many derivatives of u, only finitely many terms in the sum (5.3) are ever required to compute $\text{pr}\,\mathbf{v}(P)$. Thus questions about the "convergence" of (5.3) never arise.

Whatever the geometrical significance of a generalized vector field (a subject we will explore in depth later in this section) the formal condition that it be an "infinitesimal symmetry" of a system of differential equations is clear.

Definition 5.2. A generalized vector field $\mathbf{v}$ is a *generalized infinitesimal symmetry* of a system of differential equations

$$\Delta_\nu[u] = \Delta_\nu(x, u^{(n)}) = 0, \qquad \nu = 1, \ldots, l,$$

if and only if

$$\text{pr}\,\mathbf{v}[\Delta_\nu] = 0, \qquad \nu = 1, \ldots, l, \tag{5.4}$$

for every smooth solution $u = f(x)$.

This is the direct analogue of the infinitesimal symmetry criterion in Theorems 2.31 and 2.72. According to the latter result, we need to make some nondegeneracy assumptions on the system Δ. Note that by the preceding discussion if the coefficients of $\mathbf{v}$ depend on m-th order derivatives $u^{(m)}$, then the left-hand sides of (5.4) will in general depend on $(m+n)$-th order derivatives. Thus if we are going to require (5.4) to vanish for all solutions of the system, we must impose nondegeneracy conditions not only on the system Δ itself but also on all its prolongations $\Delta^{(k)}$, $k = 0, 1, \ldots$. To avoid always restating this hypothesis, we will assume it throughout this chapter.

Blanket Hypothesis. *Unless stated otherwise, all systems of differential equations are assumed to be totally nondegenerate in the sense of Definition* 2.83; *namely they, and all their prolongations, are of maximal rank and locally solvable.*

In particular, if Δ is a normal, analytic system, as discussed in Section 2.6, then Δ satisfies this hypothesis. In this case (5.4) holds for all solutions if and only if there exist differential operators $\mathscr{D}_{\nu\mu} = \sum P_{\nu\mu}^J D_J$, $P_{\nu\mu}^J \in \mathscr{A}$, such that

$$\text{pr}\,\mathbf{v}(\Delta_\nu) = \sum_{\mu=1}^{l} \mathscr{D}_{\nu\mu}\Delta_\mu \tag{5.5}$$

for all functions $u = f(x)$. (See Exercise 2.33.) Both (5.4) and (5.5) are useful versions of the basic infinitesimal criterion for a generalized symmetry group.

Example 5.3. Consider the heat equation

$$\Delta[u] = u_t - u_{xx} = 0.$$

The generalized vector field $\mathbf{v} = u_x \partial_u$ has prolongation

$$\operatorname{pr} \mathbf{v} = u_x \frac{\partial}{\partial u} + u_{xx} \frac{\partial}{\partial u_x} + u_{xt} \frac{\partial}{\partial u_t} + u_{xxx} \frac{\partial}{\partial u_{xx}} + \cdots.$$

Thus

$$\operatorname{pr} \mathbf{v}(\Delta) = u_{xt} - u_{xxx} = D_x(u_t - u_{xx}) = D_x \Delta,$$

and hence according to (5.5) $\mathbf{v}$ is a generalized symmetry of the heat equation. More generally, any generalized vector field of the form $\mathbf{v} = \mathscr{D}[u]\partial_u$, where $\mathscr{D}$ is any linear, constant-coefficient differential operator, is easily seen to be a generalized symmetry of the heat equation.

Evolutionary Vector Fields

Among all the generalized vector fields, those in which the coefficients $\xi^i[u]$ of the $\partial/\partial x^i$ are zero play a distinguished role.

Definition 5.4. Let $Q[u] = (Q_1[u], \ldots, Q_q[u]) \in \mathscr{A}^q$ be a q-tuple of differential functions. The generalized vector field

$$\mathbf{v}_Q = \sum_{\alpha=1}^{q} Q_\alpha[u] \frac{\partial}{\partial u^\alpha}$$

is called an *evolutionary vector field*, and Q is called its *characteristic*.

Note that according to (5.2), the prolongation of an evolutionary vector field takes a particularly simple form:

$$\operatorname{pr} \mathbf{v}_Q = \sum_{\alpha, J} D_J Q_\alpha \frac{\partial}{\partial u_J^\alpha}. \tag{5.6}$$

Any generalized vector field $\mathbf{v}$ as in (5.1) has an associated *evolutionary representative* $\mathbf{v}_Q$ in which the characteristic Q has entries

$$Q_\alpha = \phi_\alpha - \sum_{i=1}^{p} \xi^i u_i^\alpha, \qquad \alpha = 1, \ldots, q, \tag{5.7}$$

where $u_i^\alpha = \partial u^\alpha / \partial x^i$. (See (2.48).) These two generalized vector fields determine essentially the same symmetry.

Proposition 5.5. *A generalized vector field $\mathbf{v}$ is a symmetry of a system of differential equations if and only if its evolutionary representative $\mathbf{v}_Q$ is.*

Proof. According to the alternative form (2.50) of the prolongation formula,

$$\operatorname{pr} \mathbf{v}[\Delta_\nu] = \operatorname{pr} \mathbf{v}_Q[\Delta_\nu] + \sum_{i=1}^{p} \xi^i D_i \Delta_\nu. \tag{5.8}$$

The second set of terms vanishes on all solutions to Δ, so the proposition follows easily from Definition 5.2. □

For example, the symmetry $u_x \partial_u$ of the heat equation is just the evolutionary representative of the translational symmetry generator $-\partial_x$. Similarly, the Galilean generator $-2t\partial_x + xu\partial_u$ has evolutionary representative $(2tu_x + xu)\partial_u$, which, as the reader can check, is also a symmetry of the heat equation.

We will distinguish between the symmetries discussed in Chapter 2 and the true generalized symmetries here by referring to the former as *geometric symmetries* since they act geometrically on the underlying space $X \times U$. (Another suggestive name in use is *point transformations.*) According to the previous example, every geometric symmetry has an evolutionary representative with characteristic depending on at most first order derivatives. However, not every first order evolutionary symmetry comes from a geometrical group of transformations; the characteristic must be of the specific form (5.7), with ξ^i and ϕ_α depending only on x and u.

Equivalence and Trivial Symmetries

Note that if $\mathbf{v}_Q$ is an evolutionary vector field and the q-tuple Q vanishes on solutions of the system Δ then by (5.6) all the coefficients of the prolongation $\operatorname{pr} \mathbf{v}_Q$ also vanish on all solutions. Therefore $\mathbf{v}_Q$ is automatically a generalized symmetry of the system Δ. Such symmetries are called *trivial*, and we are primarily interested in nontrivial symmetries of the system. A generalized symmetry is *trivial* if its evolutionary form is. Two generalized symmetries $\mathbf{v}$ and $\tilde{\mathbf{v}}$ are called *equivalent* if their difference $\mathbf{v} - \tilde{\mathbf{v}}$ is a trivial symmetry of the system. This induces an equivalence relation on the space of generalized symmetries of the given system; moreover, we will classify symmetries up to equivalence so by a *symmetry* of the system we really mean a whole equivalence class of generalized symmetries, each differing from the other by a trivial symmetry. For example, in the case of the heat equation, the time translation symmetry ∂_t, its evolutionary form $-u_t\partial_u$ and the generalized symmetry $-u_{xx}\partial_u$ are all equivalent, and for all practical purposes determine the self-same symmetry group.

Example 5.6. Let's look at the case of a system of first order ordinary differential equations

$$\frac{du^\alpha}{dt} = P_\alpha(t, u), \qquad \alpha = 1, \ldots, q. \tag{5.9}$$

Suppose we are interested in finding generalized symmetries

$$\mathbf{v} = \tau(t, u, u_t, \ldots)\frac{\partial}{\partial t} + \sum_{\alpha=1}^{q} \phi_\alpha(t, u, u_t, \ldots)\frac{\partial}{\partial u^\alpha}.$$

We simplify the computation by replacing $\mathbf{v}$ by its evolutionary representative

$$\mathbf{v}_Q = \sum_{\alpha=1}^{q} Q_\alpha(t, u, u_t, \dots)\frac{\partial}{\partial u^\alpha}, \quad \text{where} \quad Q_\alpha = \phi_\alpha - \tau u_t^\alpha.$$

Moreover, for solutions $u = f(t)$, the system (5.9) provides expressions for the derivatives du^α/dt solely in terms of u and t. Differentiating (5.9) will similarly lead to expressions for all higher order derivatives $d^k u^\alpha/dt^k$ in terms of just u and t. Under the above notion of equivalence, we are allowed to substitute these expressions into Q, leading to an equivalent vector field of the simple form

$$\mathbf{w} = \sum_{\alpha=1}^{q} \tilde{Q}_\alpha(t, u)\frac{\partial}{\partial u^\alpha}.$$

In other words, for systems of first order ordinary differential equations, any generalized symmetry is always equivalent to a geometric symmetry in which only the dependent variables are transformed.

Computation of Generalized Symmetries

In principle, the computation of generalized symmetries of a given system of differential equations proceeds in the same way as the earlier computations of geometric symmetries, but with the following added features: First we should put the symmetry in evolutionary form $\mathbf{v}_Q$—this has the effect of reducing the number of unknown functions from $p + q$ to just q, while simultaneously simplifying the computation of the prolongation $\text{pr}\,\mathbf{v}_Q$. One must then *a priori* fix the order of derivatives on which the characteristic $Q(x, u^{(m)})$ may depend. The basic trade-off in this regard is that the more derivatives of u that Q depends on, the more possible generalized symmetries there are to be found, but, on the other hand, the more tedious and time-consuming it will be to solve the ensuing symmetry equations. Of course, such an approach cannot hope to find all generalized symmetries (unless one can treat evolutionary vector fields depending on all orders of derivatives $u^{(m)}$ simultaneously) but taking m not too large will often yield important information on the general form of the symmetries. Finally one must deal with the occurrence of trivial symmetries; the easiest way to handle these is to eliminate any superfluous derivatives in Q by substitution using the prolongations of the system, as was done in the preceding example.

Example 5.7. Consider the elementary nonlinear wave equation

$$u_t = uu_x.$$

Suppose $\mathbf{v}_Q = Q[u]\partial_u$ is a generalized symmetry in evolutionary form. Note that we can replace any t-derivatives of u occurring in Q by their corresponding expressions involving only x-derivatives without changing the equivalence class of $\mathbf{v}$. For instance, u_t is replaced by uu_x, u_{xt} by $uu_{xx} + u_x^2$, u_{tt} by $u^2 u_{xx} + 2uu_x^2$ and so on. Thus every symmetry is uniquely equivalent to one with characteristic $Q = Q(x, t, u, u_x, u_{xx}, \ldots)$. The infinitesimal condition (5.4) for invariance is then

$$D_t Q = u D_x Q + u_x Q, \tag{5.10}$$

which must be satisfied for all solutions. To calculate second order symmetries, we require $Q = Q(x, t, u, u_x, u_{xx})$, so (5.10) becomes, upon substituting for u_t according to the equation and simplifying,

$$\frac{\partial Q}{\partial t} - u\frac{\partial Q}{\partial x} + u_x^2 \frac{\partial Q}{\partial u_x} + 3u_x u_{xx} \frac{\partial Q}{\partial u_{xx}} = u_x Q.$$

By the method of characteristics, cf. (2.12), the most general solution of this linear, first order partial differential equation is

$$Q = u_x R\left(x + tu, u, t + \frac{1}{u_x}, \frac{u_{xx}}{u_x^3}\right),$$

where R is an arbitrary function of its arguments. Which of these generalized symmetries correspond to geometric symmetries of the type discussed in Chapter 2? For this to be the case, the characteristic Q must be of the form $Q = \phi - u_x \xi - uu_x \tau$, where ϕ, ξ and τ depend only on x, t and u, and where $\mathbf{v} = \xi \partial_x + \tau \partial_t + \phi \partial_u$ is the corresponding infinitesimal generator. Thus

$$Q = u_x \psi(x + tu, u) + (tu_x + 1)\phi(x + tu, u),$$

for some ψ, and where $-\xi - u\tau = \psi + t\phi$. Thus there is quite a lot of freedom in the forms of ξ and τ; however, if $\xi + u\tau = 0 = \phi$, then the evolutionary form of $\mathbf{v}$ is trivial, $Q = 0$, so every geometric symmetry is equivalent to one in which $\tau = 0$, i.e.

$$\mathbf{v} = -(\psi + t\phi)\partial_x + \phi \partial_u.$$

If we restrict our attention to projectable symmetries, then it can be shown that this subgroup is generated by the following eight vector fields:

$$\begin{aligned}
&\partial_x, && t\partial_x - \partial_u,\\
&\partial_t, && x\partial_t + u^2\partial_u,\\
&x\partial_x + t\partial_t, && xt\partial_x + t^2\partial_t - (x + tu)\partial_u,\\
&x\partial_x + u\partial_u, && x^2\partial_x + xt\partial_t + (x + tu)u\partial_u.
\end{aligned}$$

The preceding example might give the reader an overly optimistic assessment of the computational complexity of the problem of computing generalized symmetries. In practice, given a system of differential equations, the computation of all generalized symmetries of a given order is inherently feasible, but only after a considerable investment of time and computational dexterity on the part of the investigator. The following example, which is still relatively easy, should give a better idea of what is required.

Example 5.8. Here we compute all third order generalized symmetries of Burgers' equation, which we take in potential form

$$u_t = u_{xx} + u_x^2. \tag{5.11}$$

We take our infinitesimal generator in evolutionary form $\mathbf{v} = Q\partial_u$, where we assume Q depends on x, t, u, u_x, u_{xx}, u_{xxx}. The symmetry condition (5.4) is

$$D_t Q = D_x^2 Q + 2u_x D_x Q. \tag{5.12}$$

Since this is only required to hold on solutions, we can substitute for any t derivatives of u therein using (5.11) and its prolongations. Upon analyzing (5.12) in detail, we can read off the coefficients of the various derivatives of u in descending order. The coefficients of the fifth order derivative u_{xxxxx} cancel, so we proceed to terms involving u_{xxxx}. From the only term involving u_{xxxx}^2, we see that Q is affine in u_{xxx},

$$Q = \alpha(t)u_{xxx} + Q'(x, t, u, u_x, u_{xx}),$$

where α depends only on t due to the other terms involving u_{xxxx}.

Proceeding to the terms involving the third order derivative u_{xxx}, we find

$$6\alpha u_{xx}u_{xxx} + \alpha_t u_{xxx} = Q'_{u_{xx}u_{xx}}u_{xxx}^2 + 2Q'_{u_x u_{xx}}u_{xx}u_{xxx} + 2Q'_{uu_{xx}}u_x u_{xxx} + 2Q'_{xu_{xx}}u_{xxx}.$$

Thus Q' is affine in u_{xx}, with

$$Q' = 3\alpha u_x u_{xx} + (\tfrac{1}{2}\alpha_t x + \beta)u_{xx} + Q''(x, t, u, u_x),$$

where $\beta = \beta(t)$ is a function of t alone. The coefficient of u_{xx}^2 in (5.12) now reads

$$6\alpha u_x + \alpha_t x + 2\beta = Q''_{u_x u_x},$$

hence

$$Q = \alpha(u_{xxx} + 3u_x u_{xx} + u_x^3) + (\tfrac{1}{2}\alpha_t x + \beta)(u_{xx} + u_x^2) + A(x, t, u)u_x + B(x, t, u).$$

The only other terms involving u_{xx} are

$$(3\alpha_t u_x + \tfrac{1}{2}\alpha_{tt}x + \beta_t)u_{xx} = (2A_u u_x + 3\alpha_t u_x + 2A_x)u_{xx}.$$

Thus A does not depend on u, and

$$A = \tfrac{1}{8}\alpha_{tt}x^2 + \tfrac{1}{2}\beta_t x + \gamma,$$

where $\gamma = \gamma(t)$ is yet another function of t. The coefficient of u_x^2 now implies

$$B(x, t, u) = \rho(x, t)e^{-u} + \sigma(x, t),$$

with ρ and σ to be determined. The coefficient of u_x reads

$$\tfrac{1}{8}\alpha_{ttt}x^2 + \tfrac{1}{2}\beta_{tt}x + \gamma_t = 2\sigma_x + \tfrac{1}{4}\alpha_{tt},$$

so

$$\sigma(x, t) = \tfrac{1}{48}\alpha_{ttt}x^3 + \tfrac{1}{8}\beta_{tt}x^2 + (\tfrac{1}{2}\gamma_t - \tfrac{1}{8}\alpha_{tt})x + \delta,$$

where $\delta = \delta(t)$. The remaining terms in (5.12), which do not involve derivatives of u, are just

$$\rho_t e^{-u} + \sigma_t = \rho_{xx} e^{-u} + \sigma_{xx}.$$

Thus $\rho(x, t)$ is any solution to the heat equation $\rho_t = \rho_{xx}$, while using the above form of σ, we conclude

$$\alpha_{tttt} = 0, \qquad \beta_{ttt} = 0, \qquad \gamma_{tt} = \tfrac{1}{2}\alpha_{ttt}, \qquad \delta_t = \tfrac{1}{4}\beta_{tt}.$$

Thus α and β are, respectively, cubic and quadratic polynomials in t,

$$\alpha(t) = c_9 t^3 + c_8 t^2 + c_7 t + c_6, \qquad \beta(t) = c_5 t^2 + c_4 t + c_3,$$

where $c_3, \ldots, c_9$ are arbitrary constants, whence

$$\gamma(t) = \tfrac{3}{2}c_9 t^2 + c_2 t + c_1, \qquad \delta(t) = \tfrac{1}{2}c_5 t + c_0,$$

for further constants c_0, c_1, c_2.

Assembling all the information we have obtained, we conclude that every third order generalized symmetry of the potential Burgers' equation has as its characteristic Q a linear, constant-coefficient combination of the following ten "basic" characteristics

$$\begin{aligned}
Q_0 &= 1,\\
Q_1 &= u_x,\\
Q_2 &= tu_x + \tfrac{1}{2}x,\\
Q_3 &= u_{xx} + u_x^2,\\
Q_4 &= t(u_{xx} + u_x^2) + \tfrac{1}{2}xu_x,\\
Q_5 &= t^2(u_{xx} + u_x^2) + txu_x + (\tfrac{1}{2}t + \tfrac{1}{4}x^2),\\
Q_6 &= u_{xxx} + 3u_x u_{xx} + u_x^3,\\
Q_7 &= t(u_{xxx} + 3u_x u_{xx} + u_x^3) + \tfrac{1}{2}x(u_{xx} + u_x^2),\\
Q_8 &= t^2(u_{xxx} + 3u_x u_{xx} + u_x^3) + tx(u_{xx} + u_x^2) + (\tfrac{1}{2}t + \tfrac{1}{4}x^2)u_x,\\
Q_9 &= t^3(u_{xxx} + 3u_x u_{xx} + u_x^3) + \tfrac{3}{2}t^2x(u_{xx} + u_x^2) + (\tfrac{3}{2}t^2 + \tfrac{3}{4}tx^2)u_x\\
&\quad + (\tfrac{3}{4}tx + \tfrac{1}{8}x^3),
\end{aligned} \tag{5.13}$$

plus the infinite family of characteristics

$$Q_\rho = \rho(x, t)e^{-u},$$

where ρ is an arbitrary solution to the heat equation. Of these characteristics, the first six, $Q_0, \ldots, Q_5$, and the characteristics Q_ρ correspond to the geometric symmetries computed in Example 2.42. For example, Q_4 is equivalent to

$$\tilde{Q}_4 = tu_t + \tfrac{1}{2}xu_x,$$

which is the characteristic corresponding to the vector field

$$-\tfrac{1}{2}\mathbf{v}_4 = -\tfrac{1}{2}x\partial_x - t\partial_t$$

generating the scaling group of symmetries.

One could continue in this fashion to compute higher and higher order generalized symmetries, but the computations grow rapidly more and more involved. The reader might try fourth order characteristics $Q = Q(x, t, u, \ldots, u_{xxxx})$ to gain a feeling for this phenomenon. In Section 5.2 we will discover a more systematic means of finding these symmetries.

Group Transformations

What is the group of transformations corresponding to a generalized vector field? If $\mathbf{v}$ is a genuine generalized vector field, its one-parameter group $\exp(\varepsilon\mathbf{v})$ can no longer act geometrically on the underlying domain $M \subset X \times U$ since the coefficients of $\mathbf{v}$ depend on derivatives of u, which are also being transformed. Nor can we define a prolonged group action on any finite jet space $M^{(n)}$ since the coefficients of $\mathrm{pr}^{(n)}\,\mathbf{v}$ will depend on still higher order derivatives of u than appear in $M^{(n)}$. The easiest way to resolve this dilemma is to define an action of the group $\exp(\varepsilon\mathbf{v})$ on a space of smooth functions as follows:† First replace $\mathbf{v}$ by its evolutionary representative $\mathbf{v}_Q$ as above and consider the system of evolution equations

$$\frac{\partial u}{\partial \varepsilon} = Q(x, u^{(m)}), \tag{5.14}$$

where Q is the characteristic of $\mathbf{v}$. The solution (provided it exists) to the Cauchy problem $u(x, 0) = f(x)$ will determine the group action:

$$[\exp(\varepsilon\mathbf{v}_Q)f](x) \equiv u(x, \varepsilon).$$

Here we are forced to assume that the solution to this Cauchy problem is uniquely determined provided the initial data $f(x)$ is chosen in some appropriate space of functions, at least for ε sufficiently small. The resulting flow $\exp(\varepsilon\mathbf{v}_Q)$ will then be on the given function space. Of course, the verification of this hypothesis leads to some very difficult problems on existence and uniqueness of solutions to systems of evolution equations which lie far beyond the scope of this book. Our results are, barring a resolution of these problems, of a somewhat formal nature, but nevertheless will have direct

† See also Exercise 5.8 for an alternative method.

practical applications. Note that our uniqueness assumption implies that $\exp(\varepsilon \mathbf{v}_Q)$ determines a local one-parameter group of transformations on the function space.

Example 5.9. Let $p = 2$, $q = 1$ with coordinates (x, y, u) and consider the translation group G generated by $\mathbf{v} = \partial_x$. The induced action of G on functions $u = f(x, y)$, as defined in Section 2.2, is

$$[\exp(\varepsilon \mathbf{v})f](x, y) = f(x - \varepsilon, y).$$

The evolutionary form of $\mathbf{v}$ is the generalized vector field $\mathbf{v}_Q = -u_x \partial_u$. The associated one-parameter group is determined by solving the Cauchy problem

$$\partial u/\partial \varepsilon = -u_x, \qquad u(x, y, 0) = f(x, y).$$

The solution is

$$[\exp(\varepsilon \mathbf{v}_Q)f](x, y) = u(x, y, \varepsilon) = f(x - \varepsilon, y).$$

Thus $\mathbf{v}$ and $\mathbf{v}_Q$ generate the same action, and in this sense are equivalent vector fields.

Theorem 5.10. *The evolutionary vector field* $\mathbf{v} = \mathbf{v}_Q$ *is a symmetry of the system of differential equations* Δ *if and only if the corresponding group* $\exp(\varepsilon \mathbf{v})$ *transforms solutions of the system to other solutions.*

Remark. This theorem is of course subject to various technical assumptions, namely

(1) Δ is a totally nondegenerate system as in our blanket hypothesis.
(2) The system of evolution equations appropriate to $\mathbf{v}$ is uniquely solvable in some space of functions which includes all the (local) solutions to Δ.
(3) A certain system of linear equations (5.15) appearing in the proof has unique solutions.

PROOF. Let $u_\varepsilon = \exp(\varepsilon \mathbf{v})f$. (N.B.: the ε subscript is *not* a derivative.) If u_ε is a parametrized family of solutions, then

$$0 = \frac{\partial}{\partial \varepsilon}\Delta_\nu(x, u_\varepsilon^{(n)}) = \sum_{\alpha, J} D_J Q_\alpha(x, u_\varepsilon^{(n)}) \frac{\partial \Delta_\nu}{\partial u_J^\alpha}(x, u_\varepsilon^{(n)}) = \operatorname{pr} \mathbf{v}_Q[\Delta_\nu(x, u_\varepsilon^{(n)})].$$

Setting $\varepsilon = 0$ verifies (5.4). Conversely, suppose (5.5) holds. Assume that for ε sufficiently small, the only solution $v = (v^1, \ldots, v^l)$ of the linear system of evolution equations

$$\frac{\partial v^\nu}{\partial \varepsilon} = \sum_\mu \mathscr{D}_{\nu\mu} v^\mu = \sum_{\mu, J} P_{\nu\mu}^J(x, u_\varepsilon^{(m)}(x)) v_J^\mu, \qquad \nu = 1, \ldots, l, \tag{5.15}$$

with zero initial values $v(x, 0) \equiv 0$ is the zero solution $v(x, \varepsilon) \equiv 0$. Then (5.5) and the above computation imply that if $u = f(x)$ is a solution to Δ then

$v^\nu(x, \varepsilon) = \Delta_\nu(x, u_\varepsilon^{(n)})$ satisfies this initial value problem, and hence $\Delta_\nu(x, u_\varepsilon^{(n)}) = 0$ for all ε, proving that u_ε is a solution. □

If $P[u]$ is any differential function, and $u(x, \varepsilon)$ a smooth solution to (5.14), then it is not difficult to see that

$$\frac{d}{d\varepsilon} P[u] = \sum_{\alpha, J} D_J Q_\alpha[u] \frac{\partial P}{\partial u_J^\alpha} = \operatorname{pr} \mathbf{v}_Q(P).$$

In other words, $\operatorname{pr} \mathbf{v}_Q(P)$ determines the infinitesimal change in P under the one-parameter group generated by $\mathbf{v}_Q$:

$$P[\exp(\varepsilon \mathbf{v}_Q)f] = P[f] + \varepsilon \operatorname{pr} \mathbf{v}_Q(P)[f] + O(\varepsilon^2). \tag{5.16}$$

As in (1.18), we can continue to expand in powers of ε, leading to the Lie series

$$P[\exp(\varepsilon \mathbf{v}_Q)f] = \sum_{n=0}^{\infty} \frac{\varepsilon^n}{n!} (\operatorname{pr} \mathbf{v}_Q)^n P[f]. \tag{5.17}$$

(Here $(\operatorname{pr} \mathbf{v}_Q)^2(P) = \operatorname{pr} \mathbf{v}_Q[\operatorname{pr} \mathbf{v}_Q(P)]$, etc.) In particular, if $P[u] = u$, then (5.17) provides a (formal) Lie series solution to the evolutionary system (5.14). (We will not try to analyze the actual convergence of (5.17); see the following example.)

Example 5.11. Let $p = q = 1$ and consider the generalized vector field $\mathbf{v} = u_{xx}\partial_u$. The corresponding one-parameter group will be obtained by solving the Cauchy problem

$$\frac{\partial u}{\partial \varepsilon} = u_{xx}, \qquad u(x, 0) = f(x), \tag{5.18}$$

the solution being $u(x, \varepsilon) = \exp(\varepsilon \mathbf{v})f(x)$. Thus exponentiating the generalized vector field $\mathbf{v} = u_{xx}\partial_u$ is equivalent to solving the heat equation!

Several difficulties will be immediately apparent to any reader familiar with this problem. First, for $\varepsilon < 0$ we are dealing with the "backwards heat equation", which is a classic ill-posed problem and may not even have solutions. Thus we should only expect to have a "semi-group" of transformations generated by $\mathbf{v}$. Secondly, as an example due to Tikhonov makes clear, unless we impose some growth conditions the solution will *not* in general be unique. Furthermore, if $P[u] = u$ in (5.17), we obtain the (formal) series solution

$$u(x, \varepsilon) = f(x) + \varepsilon \frac{\partial^2 f}{\partial x^2} + \frac{\varepsilon^2}{2!} \frac{\partial^4 f}{\partial x^4} + \cdots$$

to (5.18). However, as shown by Kovalevskaya, even if f is analytic, this Lie series for u may not converge. In fact, it will converge only if f is an entire analytic function satisfying the growth condition $|f(x)| \leqslant C \exp(Kx^2)$ for positive constants C, K. (These are the same growth conditions needed to ensure uniqueness of solutions.) This example gives a good indication of some of the difficulties associated with rigorously implementing our exponentiation of generalized vector fields.

Symmetries and Prolongations

The connection between generalized symmetries of systems of differential equations and their prolongations is based on the following important characterization of evolutionary vector fields. It says that except for the trivial translation fields $\partial/\partial x^i$, evolutionary vector fields are uniquely determined by the fact that they commute with the operations of total differentiation.

Lemma 5.12. *If $\mathbf{v}_Q$ is an evolutionary vector field, then*

$$\mathrm{pr}\,\mathbf{v}_Q[D_iP] = D_i[\mathrm{pr}\,\mathbf{v}_Q(P)], \qquad i = 1, \ldots, p, \tag{5.19}$$

for all $P \in \mathscr{A}$. Conversely, given a vector field

$$\mathbf{v}^* = \sum_{k=1}^{p} \xi^k[u]\frac{\partial}{\partial x^k} + \sum_{\alpha=1}^{q}\sum_{J} \phi_\alpha^J[u]\frac{\partial}{\partial u_J^\alpha}$$

for some $\xi^i, \phi_\alpha^J \in \mathscr{A}$, we have $[\mathbf{v}^*, D_i] = 0$ for $i = 1, \ldots, p$, if and only if

$$\mathbf{v}^* = \mathrm{pr}\,\mathbf{v}_Q + \sum_{k=1}^{p} c_k \frac{\partial}{\partial x^k}$$

for some $Q \in \mathscr{A}^q$, $c_1, \ldots, c_p \in \mathbb{R}$.

Proof. Note first the commutation relation

$$\frac{\partial}{\partial u_J^\alpha}(D_iP) = D_i\left(\frac{\partial P}{\partial u_J^\alpha}\right) + \frac{\partial P}{\partial u_{J\setminus i}^\alpha}, \tag{5.20}$$

where $J\setminus i$ is obtained by deleting one i from the multi-index J. (If i does not occur in J, this term is zero by convention.) This implies

$$\begin{aligned}\mathrm{pr}\,\mathbf{v}_Q[D_iP] &= \sum_{\alpha,J} D_JQ_\alpha \frac{\partial}{\partial u_J^\alpha}[D_iP] \\ &= \sum_{\alpha,J} D_JQ_\alpha \cdot D_i\left(\frac{\partial P}{\partial u_J^\alpha}\right) + \sum_{\alpha,J} D_JQ_\alpha \frac{\partial P}{\partial u_{J\setminus i}^\alpha}.\end{aligned}$$

Relabelling $J\setminus i$ as J in the second summation (so J becomes J, i), this is easily seen to equal

$$D_i[\mathrm{pr}\,\mathbf{v}_Q(P)] = \sum_{\alpha,J} D_JQ_\alpha \cdot D_i\left(\frac{\partial P}{\partial u_J^\alpha}\right) + \sum_{\alpha,J} D_iD_JQ_\alpha \cdot \frac{\partial P}{\partial u_J^\alpha}.$$

To prove the converse, we have by (5.20)

$$D_i \cdot \mathbf{v}^* - \mathbf{v}^* \cdot D_i = \sum_{k=1}^{p} D_i\xi^k \frac{\partial}{\partial x^k} + \sum_{J,\alpha} (D_i\phi_\alpha^J - \phi_\alpha^{J,i})\frac{\partial}{\partial u_J^\alpha}.$$

This vanishes if and only if $D_i\xi^k = 0$ for all i, k, and $\phi_\alpha^{J,i} = D_i\phi_\alpha^J$ for all i, J, α. Thus each ξ^k is necessarily a constant and, by induction, $\phi_\alpha^J = D_J Q_\alpha$, where $Q_\alpha = \phi_\alpha^0$ is the coefficient of ∂_{u^α}.

Theorem 5.13. *If $\mathbf{v}_Q$ is a symmetry of the system Δ, then it is also a symmetry of any prolongation $\Delta^{(k)}$.*

PROOF. All the equations in $\Delta^{(k)}$ are of the form $D_J\Delta_\nu = 0$. By the lemma,

$$\operatorname{pr} \mathbf{v}_Q(D_J\Delta_\nu) = D_J(\operatorname{pr} \mathbf{v}_Q(\Delta_\nu)) = 0$$

whenever u is a solution since $\operatorname{pr} \mathbf{v}_Q(\Delta_\nu)$ vanishes on solutions by assumption. □

The Lie Bracket

As with ordinary vector fields, there is a Lie bracket between generalized vector fields, which, owing to the appearance of derivatives of u in their coefficient functions, must arise from the form of their prolongations. As with the usual Lie bracket, the easiest definition is as a commutator, but it can also be related to the corresponding one-parameter groups (see Exercise 5.7).

Definition 5.14. Let $\mathbf{v}$ and $\mathbf{w}$ be generalized vector fields. Their *Lie bracket* $[\mathbf{v}, \mathbf{w}]$ is the unique generalized vector field satisfying

$$\operatorname{pr}[\mathbf{v}, \mathbf{w}](P) = \operatorname{pr} \mathbf{v}[\operatorname{pr} \mathbf{w}(P)] - \operatorname{pr} \mathbf{w}[\operatorname{pr} \mathbf{v}(P)] \tag{5.21}$$

for all differential functions $P \in \mathcal{A}$.

There is a slight complication here in that it is not obvious that the right-hand side of (5.21) really is the prolongation of a generalized vector field. However, this follows from the explicit formulae for the Lie bracket.

Proposition 5.15. (a) *Let $\mathbf{v}_Q$ and $\mathbf{v}_R$ be evolutionary vector fields. Then their Lie bracket $[\mathbf{v}_Q, \mathbf{v}_R] = \mathbf{v}_S$ is also an evolutionary vector field with characteristic*

$$S = \operatorname{pr} \mathbf{v}_Q(R) - \operatorname{pr} \mathbf{v}_R(Q). \tag{5.22}$$

In (5.22), $\operatorname{pr} \mathbf{v}_Q$ *acts component-wise on $R \in \mathcal{A}^q$, with entries* $\operatorname{pr} \mathbf{v}_Q(R_k)$, *and conversely.*

(b) *More generally, if*

$$\mathbf{v} = \sum_i \xi^i[u]\frac{\partial}{\partial x^i} + \sum_\alpha \phi_\alpha[u]\frac{\partial}{\partial u^\alpha}, \qquad \mathbf{w} = \sum_i \eta^i[u]\frac{\partial}{\partial x^i} + \sum_\alpha \psi_\alpha[u]\frac{\partial}{\partial u^\alpha},$$

then

$$[\mathbf{v}, \mathbf{w}] = \sum_{i=1}^p \{\operatorname{pr} \mathbf{v}(\eta^i) - \operatorname{pr} \mathbf{w}(\xi^i)\}\frac{\partial}{\partial x^i} + \sum_{\alpha=1}^q \{\operatorname{pr} \mathbf{v}(\psi_\alpha) - \operatorname{pr} \mathbf{w}(\phi_\alpha)\}\frac{\partial}{\partial u^\alpha}. \tag{5.23}$$

Moreover, if $\mathbf{v}$ has characteristic Q and $\mathbf{w}$ has characteristic R, then $[\mathbf{v}, \mathbf{w}]$ has characteristic S as given by (5.22).

PROOF. In the basic formula (5.21) with $\mathbf{v} = \mathbf{v}_Q$, $\mathbf{w} = \mathbf{v}_R$, the coefficient of $\partial/\partial u^\alpha$ in pr $[\mathbf{v}, \mathbf{w}]$ is clearly given by the α-th component S_α of (5.22). Thus to prove part (a), it suffices to show that [pr $\mathbf{v}_Q$, pr $\mathbf{v}_R$] is an evolutionary vector field, which will necessarily imply that it agrees with pr $\mathbf{v}_S$. This immediately follows from Lemma 5.12: since both pr $\mathbf{v}_Q$ and pr $\mathbf{v}_R$ commute with all total derivatives, the same is true of their commutator, which contains no terms involving any $\partial/\partial x^i$. This proves part (a). Part (b) follows from the prolongation formula (5.8), and is left to the reader. □

For example, if $\mathbf{v} = uu_x\partial_u$, $\mathbf{w} = u_{xx}\partial_u$, then

$$[\mathbf{v}, \mathbf{w}] = (\text{pr}\,\mathbf{v}(u_{xx}) - \text{pr}\,\mathbf{w}(uu_x))\partial_u = 2u_x u_{xx}\partial_u.$$

Proposition 5.16. *The Lie bracket between generalized vector fields has the usual properties of*

(a) Bilinearity:

$$[c\mathbf{u} + c'\mathbf{v}, \mathbf{w}] = c[\mathbf{u}, \mathbf{w}] + c'[\mathbf{v}, \mathbf{w}], \qquad c, c' \in \mathbb{R},$$

(b) Skew-Symmetry:

$$[\mathbf{v}, \mathbf{w}] = -[\mathbf{w}, \mathbf{v}],$$

(c) Jacobi Identity:

$$[\mathbf{u}, [\mathbf{v}, \mathbf{w}]] + [\mathbf{w}, [\mathbf{u}, \mathbf{v}]] + [\mathbf{v}, [\mathbf{w}, \mathbf{u}]] = 0,$$

for any generalized vector fields $\mathbf{u}, \mathbf{v}, \mathbf{w}$.

Indeed, these properties clearly hold when we replace each vector field by its prolongation, and this suffices to prove their validity. The commutator definition (5.21) of the Lie bracket immediately implies:

Proposition 5.17. *The set of generalized symmetries of a nondegenerate system of differential equations forms a Lie algebra.*

Example 5.18. In certain cases, this result can be used to construct new generalized symmetries from known ones. For example, consider the list of symmetries (5.13) of the potential Burgers' equation. Using $\mathbf{v}_i$ to denote the symmetry with characteristic Q_i, we conclude that $[\mathbf{v}_i, \mathbf{v}_j]$ is a symmetry with characteristic pr $\mathbf{v}_i(Q_j)$ − pr $\mathbf{v}_j(Q_i)$ for any i, j. For example,

$$\text{pr}\,\mathbf{v}_6(Q_7) - \text{pr}\,\mathbf{v}_7(Q_6) = -\tfrac{3}{2}(u_{xxxx} + 4u_x u_{xxx} + 3u_{xx}^2 + 6u_x^2 u_{xx} + u_x^4)$$

gives the characteristic Q_{10} of a *new*, fourth order symmetry $\mathbf{v}_{10} \equiv -\tfrac{3}{2}[\mathbf{v}_6, \mathbf{v}_7]$ of Burgers' equation. This process can be repeated indefinitely, so $[\mathbf{v}_7, \mathbf{v}_{10}]$ will be a fifth order symmetry and so on. Thus Burgers' equation has an infinite collection of generalized symmetries depending on progressively higher and higher order derivatives of u. In Fuchssteiner's terminology, $\mathbf{v}_7$ is known as a "master symmetry" for the potential Burger's equation. (See Section 5.2.)

Evolution Equations

Consider a system of evolution equations

$$\frac{\partial u}{\partial t} = P[u] \tag{5.24}$$

in which $P[u] = P(x, u^{(n)}) \in \mathscr{A}^q$ depends on $x \in \mathbb{R}^p$, $u \in \mathbb{R}^q$ and x-derivatives of u only. Substituting according to (5.24) and its derivatives, we see that any evolutionary symmetry must be equivalent to one whose characteristic $Q[u] = Q(x, t, u^{(m)})$ depends only on x, t, u and the x-derivatives of u. On the other hand, (5.24) itself can be thought of as the equations for the flow $\exp(t\mathbf{v}_P)$ of the evolutionary vector field with characteristic P. The symmetry criterion (5.4), which in this case is

$$D_t Q_\nu = \operatorname{pr} \mathbf{v}_Q(P_\nu), \qquad \nu = 1, \ldots, q, \tag{5.25}$$

is readily seen to be equivalent to the following Lie bracket condition on the two generalized vector fields, generalizing the correspondence between symmetries of systems of first order ordinary differential equations and the Lie bracket of the corresponding vector fields.

Proposition 5.19. *An evolutionary vector field $\mathbf{v}_Q$ is a symmetry of the system of evolution equations $u_t = P[u]$ if and only if*

$$\frac{\partial \mathbf{v}_Q}{\partial t} + [\mathbf{v}_P, \mathbf{v}_Q] = 0 \tag{5.26}$$

holds identically in $(x, t, u^{(m)})$. (Here $\partial \mathbf{v}_Q/\partial t$ denotes the evolutionary vector field with characteristic $\partial Q/\partial t$.)

PROOF. Note that according to the prolongation of the system of evolution equations, the derivative $u^\alpha_{J,t} = \partial u^\alpha_J/\partial t$ evolves according to $u^\alpha_{J,t} = D_J P_\alpha[u]$. Using this and the formula for the total derivative, it is easy to see that on solutions

$$D_t Q_\nu = \frac{\partial Q_\nu}{\partial t} + \sum_{\alpha, J} u^\alpha_{J,t} \frac{\partial Q_\nu}{\partial u^\alpha_J} = \frac{\partial Q_\nu}{\partial t} + \operatorname{pr} \mathbf{v}_P(Q_\nu),$$

since Q_ν only depends on x-derivatives of u. Thus (5.25) is equivalent to the equation

$$\frac{\partial Q}{\partial t} + \operatorname{pr} \mathbf{v}_P(Q) = \operatorname{pr} \mathbf{v}_Q(P),$$

which, as there are no more t-derivatives of u present, must hold identically in x, t and u. The equivalence with (5.26) follows easily from the formula (5.22) for the Lie bracket. □

In particular, if $Q[u] = Q(x, u^{(m)})$ does not depend explicitly on t, then (5.26) reduces to the condition that the two vector fields $\mathbf{v}_P$ and $\mathbf{v}_Q$ commute:

$$[\mathbf{v}_P, \mathbf{v}_Q] = 0. \tag{5.27}$$

It is not difficult to show that, under certain existence and uniqueness hypotheses, this condition is equivalent to the condition that the corresponding one-parameter symmetry groups commute:

$$\exp(\varepsilon \mathbf{v}_P) \exp(\tilde{\varepsilon} \mathbf{v}_Q) f = \exp(\tilde{\varepsilon} \mathbf{v}_Q) \exp(\varepsilon \mathbf{v}_P) f \tag{5.28}$$

where defined. (See Exercise 5.7.) Consequently, we have the reciprocity relation that, provided $P, Q \in \mathscr{A}^q$ only depend on x, u and x-derivatives of u, the vector field $\mathbf{v}_Q$ is a generalized symmetry of the system $u_t = P$ if and only if $\mathbf{v}_P$ is a generalized symmetry of $u_t = Q$. In particular, for P as above, the vector field $\mathbf{v}_P$ itself is always a symmetry of $u_t = P[u]$. Indeed, $\mathbf{v}_P$ is equivalent to the evolutionary form of the time translation symmetry group generated by ∂_t, stemming from the "autonomy" of the evolution equation. The reader may try rechecking the symmetry condition (5.26) for the symmetries of Burgers' equation found in Example 5.8.

5.2. Recursion Operators, Master Symmetries and Formal Symmetries

The method of Section 5.1 provides a systematic means of determining all the generalized symmetries of a given order of a system of differential equations, but suffers the drawback that the order of derivatives on which the coefficients of the symmetry depend must be specified in advance. Thus the method cannot simultaneously generate *all* generalized symmetries of the system. In this section we explore a second method for generating symmetries based on the notion of a recursion operator. While this method cannot provide an exhaustive classification of all possible symmetries without further analysis, it does provide a mechanism for generating infinite hierarchies of generalized symmetries, depending on higher order derivatives of u, in one step. Unfortunately, while the verification that a given operator does determine a recursion operator is fairly straightforward, in contrast to the previous method, this technique is not fully constructive. The deduction of the form of the recursion operator (if it exists) requires a certain amount of inspired guesswork, often based on the form of lower order symmetries determined by the earlier method.

Definition 5.20. Let Δ be a system of differential equations. A *recursion operator* for Δ is a linear operator $\mathscr{R}: \mathscr{A}^q \to \mathscr{A}^q$ in the space of q-tuples of differential functions with the property that whenever $\mathbf{v}_Q$ is an evolutionary symmetry of Δ, so is $\mathbf{v}_{\tilde{Q}}$ with $\tilde{Q} = \mathscr{R}Q$.

Thus if we are fortunate enough to know a recursion operator $\mathscr{R}$ for a system of differential equations, we can generate an infinite family of symmetries from any one symmetry $\mathbf{v}_{Q_0}$ merely by applying $\mathscr{R}$ successively to the characteristic Q_0; in other words, each $Q_j = \mathscr{R}^j Q_0$, $j = 0, 1, 2, \ldots$ is the characteristic of a generalized symmetry. Often, but not always, $\mathscr{R}$ will be a $q \times q$ matrix of differential operators.

Example 5.21. As an easy example, we show that the differential operator $\mathscr{R}_1 = D_x$ is a recursion operator for the heat equation $u_t = u_{xx}$. Now $\mathbf{v}_Q$ is a generalized symmetry of the heat equation if and only if $D_t Q = D_x^2 Q$ on all solutions. Then $\tilde{Q} = D_x Q$ is also the characteristic of a symmetry since

$$D_t\tilde{Q} - D_x^2\tilde{Q} = (D_t - D_x^2)D_x Q = D_x(D_t Q - D_x^2 Q) = 0$$

on solutions. Thus, starting with the basic scaling symmetry $u\partial_u$ we can generate a whole hierarchy of generalized symmetries by recursively applying $\mathscr{R}_1$; we find $u_x = \mathscr{R}_1(u)$, $u_{xx} = \mathscr{R}_1^2(u)$, etc. are all characteristics of generalized symmetries of the heat equation. Put another way, the "flow" generated by the heat equation commutes with the "flow" determined by the evolution equations $u_t = \partial^k u/\partial x^k$, $k \geqslant 0$, cf. (5.28).

By the same arguments, the t-derivative D_t is also a recursion operator, but it is trivially related to D_x since $D_t Q = D_x^2 Q$ whenever Q is the characteristic of a symmetry. (In general, $\mathscr{R}^m$ is trivially a recursion operator whenever $\mathscr{R}$ is.) There is, however, a second recursion operator not related to D_x, namely $\mathscr{R}_2 = tD_x + \frac{1}{2}x$. To see this, we find

$$(D_t - D_x^2)(tD_x + \tfrac{1}{2}x)Q = (tD_x + \tfrac{1}{2}x)(D_t Q - D_x^2 Q),$$

so $\tilde{Q} = \mathscr{R}_2 Q$ gives a symmetry whenever Q does. We thus obtain a double infinity of generalized symmetries of the heat equation, by applying $\mathscr{R}_1$ or $\mathscr{R}_2$ successively to $Q_0 = u$. These have characteristics

$$\begin{aligned} &Q_0 = u, \qquad Q_1 = \mathscr{R}_1[u] = u_x, \qquad Q_2 = \mathscr{R}_2[u] = tu_x + \tfrac{1}{2}xu,\\ &Q_3 = \mathscr{R}_1^2[u] = u_{xx}, \qquad Q_4 = \mathscr{R}_2\mathscr{R}_1[u] = tu_{xx} + \tfrac{1}{2}xu_x,\\ &Q_5 = \mathscr{R}_2^2[u] = t^2u_{xx} + txu_x + (\tfrac{1}{2}t + \tfrac{1}{4}x^2)u,\\ &Q_6 = \mathscr{R}_1^3[u] = u_{xxx}, \qquad Q_7 = \mathscr{R}_2\mathscr{R}_1^2[u] = tu_{xxx} + \tfrac{1}{2}xu_{xx}, \quad \text{etc.} \end{aligned} \tag{5.29}$$

(Note that since $\mathscr{R}_1\mathscr{R}_2 = \mathscr{R}_2\mathscr{R}_1 + \frac{1}{2}$, if we are only interested in independent characteristics, it doesn't matter in which order $\mathscr{R}_1$ and $\mathscr{R}_2$ are applied.) The first six of these, $Q_0, \ldots, Q_5$, are (up to sign) the characteristics of the geometric symmetries of the heat equation computed in Example 2.41; the rest of these are genuine generalized symmetries.

The above results for the heat equation actually generalize to arbitrary linear systems of differential equations as follows.

Proposition 5.22. *Let $\Delta[u] = 0$ be a linear system of differential equations, with Δ denoting a linear differential operator. A second linear differential operator $\mathscr{R}: \mathscr{A}^q \to \mathscr{A}^q$ not depending on u or its derivatives is a recursion operator for Δ if and only if $Q = \mathscr{R}[u]$ is the characteristic of a "linear" generalized symmetry to the system.*

In other words, for a linear system every generalized symmetry whose characteristic depends linearly on u and its derivatives determines a recursion operator and conversely. For linear systems, then, the whole theory of (linear) symmetries could be developed using the recursion operators as the fundamental objects. This is the approach favoured by Miller, [3], and his co-workers. Proposition 5.22 provides the link between their approach and the more geometrical Lie theory developed in this book. (The latter has the advantage of simultaneously treating nonlinear systems, which are not covered by the operator method.)

The proof of Proposition 5.22 is easy. If $\mathscr{R}$ is a recursion operator, then $Q = \mathscr{R}[u]$ trivially gives a symmetry since $Q_0 = u$ is the characteristic for the trivial scaling symmetry group $(x, u) \mapsto (x, \lambda u)$ stemming from the linearity of the system. Conversely, if $\mathbf{v}_Q$ is a symmetry, by (5.4) and the linearity of Δ,

$$\operatorname{pr} \mathbf{v}_Q(\Delta[u]) = \Delta[Q] = \Delta\mathscr{R}[u]$$

on all solutions. Total nondegeneracy of Δ implies the existence of a differential operator $\tilde{\mathscr{R}}$ satisfying $\Delta\mathscr{R}[u] = \tilde{\mathscr{R}}\Delta[u]$ for all u, cf. (5.5). It is easily seen that since Δ and $\mathscr{R}$ are independent of u, we can choose $\tilde{\mathscr{R}}$ to also be independent of u, and $\Delta\mathscr{R} = \tilde{\mathscr{R}}\Delta$ identically. Thus if $\tilde{Q} = \mathscr{R}Q$, where Q is the characteristic of a symmetry, so $\Delta[Q] = 0$, then $\Delta[\tilde{Q}] = \tilde{\mathscr{R}}\Delta[Q] = 0$ on solutions, and $\tilde{Q}$ provides another symmetry. □

Example 5.23. For the two-dimensional wave equation, the ten-parameter conformal symmetry group was derived in Example 2.43; the corresponding characteristics are given in Example 4.36. According to Proposition 5.22 there are ten recursion operators, namely

$$D_x, \quad D_y, \quad D_t, \qquad \text{(translations)}$$

$$\mathscr{R}_{xy} = xD_y - yD_x, \quad \mathscr{R}_{xt} = tD_x + xD_t, \quad \mathscr{R}_{yt} = tD_y + yD_t, \qquad \text{(rotations)}$$

$$\mathscr{D} = xD_x + yD_y + tD_t, \qquad \text{(dilatation)}$$

$$\begin{aligned} \mathscr{I}_x &= (x^2 - y^2 + t^2)D_x + 2xyD_y + 2xtD_t + x, \\ \mathscr{I}_y &= 2xyD_x + (y^2 - x^2 + t^2)D_y + 2ytD_t + y, \qquad \text{(inversions)} \\ \mathscr{I}_t &= 2xtD_x + 2ytD_t + (x^2 + y^2 + t^2)D_t + t. \end{aligned} \tag{5.30}$$

Applying successive products of these operators to $Q_0 = u$ leads to vast numbers of generalized symmetries of the wave equation; for example

$$\mathscr{R}_{xy}\mathscr{R}_{xt}[u] = xtu_{xy} - ytu_{xx} + x^2u_{yt} - xyu_{xt} - yu_t,$$

and so on. There are a number of dependencies among the resulting symmetries stemming from relations among the operators, e.g.

$$\mathscr{R}_{xy}D_t - \mathscr{R}_{xt}D_y + \mathscr{R}_{yt}D_x = 0.$$

In his thesis, Delong, [1], proves that there are $(2k + 1)(2k + 2)(2k + 3)/6$ independent k-th order symmetries generated by these recursion operators; for instance, there are 35 independent second order symmetries like the above example. Recently, Shapovalov and Shirokov, [1], proved that every generalized symmetry of the wave equation can be obtained in this way.

Fréchet Derivatives

For nonlinear systems, there is an analogous criterion for a differential operator to be a recursion operator, but to state it we need to introduce the notion of the (formal) Fréchet derivative of a differential function.

Definition 5.24. Let $P[u] = P(x, u^{(n)}) \in \mathscr{A}^r$ be an r-tuple of differential functions. The *Fréchet derivative* of P is the differential operator $\mathsf{D}_P\colon \mathscr{A}^q \to \mathscr{A}^r$ defined so that

$$\mathsf{D}_P(Q) = \frac{d}{d\varepsilon}\bigg|_{\varepsilon=0} P[u + \varepsilon Q[u]] \tag{5.31}$$

for any $Q \in \mathscr{A}^q$.

In other words, to evaluate $\mathsf{D}_P(Q)$ we replace u (and its derivatives) in P by $u + \varepsilon Q$ and differentiate the resulting expression with respect to ε. For example, if $P[u] = u_x u_{xx}$, then

$$\mathsf{D}_P(Q) = \frac{d}{d\varepsilon}\bigg|_{\varepsilon=0} (u_x + \varepsilon D_x Q)(u_{xx} + \varepsilon D_x^2 Q) = u_x D_x^2 Q + u_{xx} D_x Q,$$

so $\mathsf{D}_P = u_x D_x^2 + u_{xx} D_x$. This calculation easily generalizes and shows that the Fréchet derivative of a general r-tuple $P = (P_1, \ldots, P_r)$ is the $q \times r$ matrix differential operator with entries

$$(\mathsf{D}_P)_{\mu\nu} = \sum_J (\partial P_\mu/\partial u_J^\nu) D_J, \qquad \mu = 1, \ldots, r, \quad \nu = 1, \ldots, q, \tag{5.32}$$

the sum being over all multi-indices J. In particular, if $P = \Delta[u]$ is a linear differential polynomial, then $\mathsf{D}_P = \Delta$ is the same as the differential operator determining it. There is an intimate connection between the Fréchet derivative and evolutionary vector fields.

Proposition 5.25. *If $P \in \mathscr{A}^r$ and $Q \in \mathscr{A}^q$, then*

$$\mathsf{D}_P(Q) = \operatorname{pr} \mathbf{v}_Q(P). \tag{5.33}$$

This follows directly from the formulae (5.6), (5.32). Alternatively, one can remark that both sides determine the infinitesimal variation in P under the action of the one-parameter group generated by $\mathbf{v}_Q$, cf. (5.16), and hence must agree. □

The Fréchet derivative is a derivation, meaning that it satisfies a Leibniz rule.

Proposition 5.26. *The Fréchet derivative of the product of two differential functions is*

$$\mathsf{D}_{PQ} = P\mathsf{D}_Q + Q\mathsf{D}_P. \tag{5.34}$$

The proof of this and the following proposition follows directly from the definition (5.31) of the Fréchet derivative.

Proposition 5.27. *Let $P \in \mathscr{A}^s$. Let $\mathscr{D}$ be a constant coefficient $r \times s$ matrix differential operator, so that $\mathscr{D}P \in \mathscr{A}^r$. Then the Fréchet derivative of $\mathscr{D}P$ is the product of two differential operators*

$$\mathsf{D}_{\mathscr{D}P} = \mathscr{D} \cdot \mathsf{D}_P. \tag{5.35}$$

For example, the Fréchet derivative of the total derivative $D_x P$ of a differential function P is $\mathsf{D}_{D_x P} = D_x \cdot \mathsf{D}_P$. Thus, the Fréchet derivative of $P = uu_{xx} + u_x^2 = D_x(uu_x) = D_x^2(\frac{1}{2}u^2)$ is $\mathsf{D}_P = uD_x^2 + 2u_x D_x + u_{xx} = D_x \cdot (uD_x + u_x) = D_x^2 \cdot u$.

Lie Derivatives of Differential Operators

Recall that the Lie derivative of an ordinary geometric object with respect to a vector field, as introduced in Section 1.5, represents the infinitesimal change in the object under the flow induced by the vector field. Analogous concepts exist for generalized vector fields; here we are particularly interested in the Lie derivative of a differential operator with respect to an evolutionary vector field. If $\mathbf{v}_Q$ is an evolutionary vector field, and $\mathscr{D} = \sum P_K[u] D_K$ a differential operator whose coefficients may depend on u, then we define the *Lie derivative* of $\mathscr{D}$ to be the infinitesimal change in $\mathscr{D}$ under the one-parameter group $\exp(t\mathbf{v}_Q)$. This is computed by evaluating the "time" derivative of $\mathscr{D}$:

$$\mathscr{D}_t = \sum_K D_t(P_K) D_K, \tag{5.36}$$

on solutions to the associated evolution equation $u_t = Q$, leading to the formula

$$\operatorname{pr} \mathbf{v}_Q(\mathscr{D}) = \sum_K \operatorname{pr} \mathbf{v}_Q(P_K) D_K. \tag{5.37}$$

For example, if $\mathscr{D} = D_x^3 + 2uD_x + u_x$, and $\mathbf{v}_Q = u_{xx}\partial_u$, then

$$\operatorname{pr} \mathbf{v}_Q(\mathscr{D}) = 2 \operatorname{pr} \mathbf{v}_Q(u)D_x + \operatorname{pr} \mathbf{v}_Q(u_x) = 2u_{xx}D_x + u_{xxx}.$$

The one-parameter group $\exp(t\mathbf{v}_Q)$ in this case is found by solving the heat equation $u_t = u_{xx}$ and $\operatorname{pr} \mathbf{v}_Q(\mathscr{D})$ represents the infinitesimal change in $\mathscr{D}$ when $u(x, t)$ is a solution to the heat equation. Alternatively, we can identify $\operatorname{pr} \mathbf{v}_Q(\mathscr{D})$ with the time derivative $\mathscr{D}_t = 2u_tD_x + u_{xt}$ of $\mathscr{D}$ (note that D_x does not depend on t) evaluated on solutions to the heat equation.

This definition of the Lie derivative extends to matrix differential operators, where now $\operatorname{pr} \mathbf{v}_Q$ acts on the individual entries. A straightforward computation proves that the Lie derivative is a derivation satisfying the following Leibniz rule:

$$\operatorname{pr} \mathbf{v}_Q(\mathscr{D}P) = \operatorname{pr} \mathbf{v}_Q(\mathscr{D})\cdot P + \mathscr{D}[\operatorname{pr} \mathbf{v}_Q(P)]. \tag{5.38}$$

For $q \times q$ matrix differential operators, there is a second type of Lie derivative which arises from a different "tensorial" interpretation of the operator. Here, instead of letting the differential operator act on q-tuples of differential functions, we regard it as a linear transformation on the space of evolutionary vector fields. Specifically, if $\mathscr{R} = (\mathscr{R}^\alpha_\beta)$ is a $q \times q$ matrix differential operator, and $\mathbf{v}_Q$ an evolutionary vector field with characteristic $Q = (Q_1, \ldots, Q_q)^T$ (viewed as a column vector), we define $\mathscr{R}(\mathbf{v}_Q)$ to be the evolutionary vector field with characteristic $\mathscr{R}Q$, so[†]

$$\mathscr{R}(\mathbf{v}_Q) = \mathbf{v}_{\mathscr{R}Q}. \tag{5.39}$$

Note that, although an evolutionary vector field is prescribed by a q-tuple of differential functions, its transformation properties under a change of variables are different (just as the transformation rules for ordinary vector fields are different from those of points).

As before, the Lie derivative of the operator $\mathscr{R}$ with respect to an evolutionary vector field is defined to be its infinitesimal change under the one-parameter group generated by the vector field; however, since the operator acts on vector fields, the formula in this case is different. Indeed, in analogy with the formula for the Lie derivative of an ordinary vector field with respect to another vector field given in Proposition 1.64, the Lie derivative of an evolutionary vector field $\mathbf{v}_P$ with respect to another vector field $\mathbf{v}_Q$ will be provided by the Lie bracket $[\mathbf{v}_Q, \mathbf{v}_P]$ of Definition 5.14. (This can, of course, be justified directly.) To compute the infinitesimal change of the operator $\mathscr{R}$ with respect to a given evolutionary vector field $\mathbf{v}_Q$, we determine the Lie derivative of the vector field $\mathscr{R}\mathbf{v}_P = \mathbf{v}_{\mathscr{R}P}$, where P is an arbitrary q-tuple of differential functions, with respect to $\mathbf{v}_Q$. This is given by the Lie bracket

[†] In the language of tensor analysis, we are interpreting the differential operator $\mathscr{R}$ as a (1, 1)-tensor, since it maps vector fields to vector fields.

$[\mathbf{v}_Q, \mathbf{v}_{\mathscr{R}P}]$, which, according to Proposition 5.15, has characteristic

$$\begin{aligned}\operatorname{pr}\mathbf{v}_Q(\mathscr{R}P) - \operatorname{pr}\mathbf{v}_{\mathscr{R}P}(Q) &= \operatorname{pr}\mathbf{v}_Q(\mathscr{R})P + \mathscr{R}\operatorname{pr}\mathbf{v}_Q(P) - \operatorname{pr}\mathbf{v}_{\mathscr{R}P}(Q)\\ &= \{\operatorname{pr}\mathbf{v}_Q(\mathscr{R}) + \mathscr{R}\mathrm{D}_Q - \mathrm{D}_Q\mathscr{R}\}P\\ &\quad + \mathscr{R}\{\operatorname{pr}\mathbf{v}_Q(P) - \operatorname{pr}\mathbf{v}_P(Q)\},\end{aligned}$$

where we used (5.33) and (5.38). In the final formula, the second set of terms just gives the infinitesimal change in P; hence the first set of terms give the required formula for this alternative type of Lie derivative of the operator $\mathscr{R}$ with respect to $\mathbf{v}_Q$, for which we introduce the following notation and terminology.

Definition 5.28. Let $\mathscr{R}$ be a $q \times q$ matrix differential operator. The (1, 1)-*Lie derivative* of $\mathscr{R}$ with respect to a generalized vector field $\mathbf{v}_Q$ is the differential operator

$$\mathbf{v}_Q[\![\mathscr{R}]\!] = \mathscr{R}_t + [\mathscr{R}, \mathrm{D}_Q] = \operatorname{pr}\mathbf{v}_Q(\mathscr{R}) + [\mathscr{R}, \mathrm{D}_Q]. \tag{5.40}$$

(In (5.40), the time derivative $\mathscr{R}_t$ is evaluated on solutions to the associated evolution equation $u_t = Q$.)

The preceding computation proves the Leibniz rule

$$[\mathbf{v}_Q, \mathbf{v}_{\mathscr{R}P}] = [\mathbf{v}_Q, \mathscr{R}\mathbf{v}_P] = \mathbf{v}_Q[\![\mathscr{R}]\!]\mathbf{v}_P + \mathscr{R}[\mathbf{v}_Q, \mathbf{v}_P], \tag{5.41}$$

for the (1, 1)-Lie derivative.

Finally, we remark that, owing to their invariant definitions, the two kinds of Lie derivatives are unaffected by changes of variables. For instance, if we replace the dependent variable u by $v = \varphi(x, u)$, then the evolutionary vector field with characteristic $Q[u]$ in the u-coordinates has characteristic $\tilde{Q}[v] = \varphi_u \cdot Q[u]$ in the v-coordinates, where φ_u denotes the $q \times q$ Jacobian matrix with entries $\partial\varphi^\alpha/\partial u^\beta$; see Exercise 3.21. Consequently, a $q \times q$ matrix differential operator $\mathscr{R}$ mapping vector fields to vector fields will change to $\tilde{\mathscr{R}} = \varphi_u \cdot \mathscr{R} \cdot \varphi_u^{-1}$ in the v-coordinates. The (1, 1)-Lie derivative is invariant under such transformations, as can also be proved directly, cf. Exercise 5.16.

Criteria for Recursion Operators

Formula (5.33) readily leads to a general characterization of recursion operators.

Theorem 5.29. *Suppose $\Delta[u] = 0$ is a system of differential equations. If $\mathscr{R}\colon \mathscr{A}^q \to \mathscr{A}^q$ is a linear operator such that*

$$\mathrm{D}_\Delta \cdot \mathscr{R} = \tilde{\mathscr{R}} \cdot \mathrm{D}_\Delta \tag{5.42}$$

for all solutions u to Δ, where $\tilde{\mathscr{R}}\colon \mathscr{A}^q \to \mathscr{A}^q$ is a linear differential operator, then $\mathscr{R}$ is a recursion operator for the system.

PROOF. According to (5.33), an evolutionary vector field $\mathbf{v}_Q$ is a symmetry of Δ if and only if

$$\mathsf{D}_\Delta(Q) = \operatorname{pr} \mathbf{v}_Q(\Delta) = 0$$

for all solutions to Δ. If $\mathcal{R}$ satisfies (5.42), and $\tilde{Q} = \mathcal{R}Q$, then

$$\mathsf{D}_\Delta \tilde{Q} = \mathsf{D}_\Delta(\mathcal{R}Q) = \tilde{\mathcal{R}}(\mathsf{D}_\Delta Q) = 0$$

for all solutions. Thus $\tilde{Q}$ is also a symmetry, and the theorem follows. □

For example, suppose $\Delta[u] = u_t - K[u]$ is an evolution equation. Then $\mathsf{D}_\Delta = D_t - \mathsf{D}_K$. If $\mathcal{R}$ is a recursion operator, then it is not hard to see that the operator $\tilde{\mathcal{R}}$ in (5.42) must be the same as $\mathcal{R}$. Therefore, the condition (5.42) in this case reduces to the commutator condition

$$\mathcal{R}_t = [\mathsf{D}_K, \mathcal{R}] \tag{5.43}$$

for a recursion operator of an evolution equation. Note that condition (5.43) is the same as requiring that the (1, 1)-Lie derivative (5.40) of $\mathcal{R}$ with respect to the evolutionary vector field $\mathbf{v}_K$ vanishes:

$$\mathbf{v}_K[\![\mathcal{R}]\!] = 0. \tag{5.44}$$

Therefore, a recursion operator is nothing but an operator (or, rather, a (1, 1)-tensor) which is invariant under the flow of the evolution equation $u_t = K$!

Remark. Condition (5.43) has the form of a *Lax pair*, of fundamental importance to the inverse scattering approach to soliton theories; see Lax, [1], Dickey, [1], and Newell, [1]. However, in most cases, the recursion operator is not the usual spectral operator appearing in the Lax pair, but, rather, the "squared eigenfunction operator".

Example 5.30. Return to the potential Burgers' equation $u_t = K = u_{xx} + u_x^2$ for which we computed generalized symmetries in Examples 5.8 and 5.18. The structure of the resulting characteristics strongly suggests that, like the heat equation, Burgers' equation has two recursion operators. Inspection of Q_0, Q_1, Q_3, Q_6 and Q_{10} leads us to conjecture that $\mathcal{R}_1 = D_x + u_x$ is a recursion operator since $Q_1 = \mathcal{R}_1 Q_0$, $Q_3 = \mathcal{R}_1 Q_1$, etc. To prove this, we note that the Fréchet derivative for right-hand side of Burgers' equation (5.11) is

$$\mathsf{D}_K = D_x^2 + 2u_x D_x.$$

We must verify (5.43). The time derivative of the first recursion operator $\mathcal{R}_1$ on solutions to Burgers' equation is the multiplication operator

$$(\mathcal{R}_1)_t = (D_x + u_x)_t = u_{xt} = u_{xxx} + 2u_x u_{xx}.$$

On the other hand, the commutator is computed using Leibniz' rule for differential operators:

$$\begin{aligned}[\mathrm{D}_K, \mathcal{R}_1] &= (D_x^2 + 2u_x D_x)(D_x + u_x) - (D_x + u_x)(D_x^2 + 2u_x D_x) \\ &= (D_x^3 + 3u_x D_x^2 + 2(u_{xx} + u_x^2)D_x + u_{xxx} + 2u_x u_{xx}) \\ &\quad - (D_x^3 + 3u_x D_x^2 + 2(u_{xx} + u_x^2)D_x) \\ &= u_{xxx} + 2u_x u_{xx}.\end{aligned}$$

Comparing these two verifies (5.43) and proves that $\mathcal{R}_1$ is a recursion operator for the potential Burgers' equation.

There is thus an infinite hierarchy of symmetries, with characteristics $\mathcal{R}_1^k Q_0$, $k = 0, 1, 2, \ldots$. For example, the next characteristic after Q_{10} in the sequence is

$$\begin{aligned}Q_{15} = \mathcal{R}_1 Q_{10} &= u_{xxxxx} + 5u_x u_{xxxx} + 10u_{xx}u_{xxx} + 10u_x^2 u_{xxx} \\ &\quad + 15u_x u_{xx}^2 + 10u_x^3 u_{xx} + u_x^5.\end{aligned}$$

To obtain the characteristics depending on x and t, we require a second recursion operator, which, by inspection, we guess to be

$$\mathcal{R}_2 = t\mathcal{R}_1 + \tfrac{1}{2}x = tD_x + tu_x + \tfrac{1}{2}x.$$

Using the fact that $\mathcal{R}_1$ satisfies (5.43), we find

$$(\mathcal{R}_2)_t = t(\mathcal{R}_1)_t + \mathcal{R}_1 = t[\mathrm{D}_K, \mathcal{R}_1] + \mathcal{R}_1,$$

whereas

$$\begin{aligned}[\mathrm{D}_K, \mathcal{R}_2] &= t[\mathrm{D}_K, \mathcal{R}_1] + [D_x^2 + 2u_x D_x, \tfrac{1}{2}x] \\ &= t[\mathrm{D}_K, \mathcal{R}_1] + (D_x + u_x) = t[\mathrm{D}_K, \mathcal{R}_1] + \mathcal{R}_1,\end{aligned}$$

proving that $\mathcal{R}_2$ is also a recursion operator. There is thus a doubly infinite hierarchy of generalized symmetries of Burgers' equation, with characteristics $\mathcal{R}_2^l \mathcal{R}_1^k Q_0$, $k, l \geq 0$. For instance, $Q_2 = \mathcal{R}_2 Q_0$, $Q_4 = \mathcal{R}_2 \mathcal{R}_1 Q_0$, and so on.

The Korteweg–de Vries Equation

In the case of nonlinear equations, we often have to expand the class of possible recursion operators to include formal "pseudo-differential" operators. As an example, we prove that the operator

$$\mathcal{R} = D_x^2 + \tfrac{2}{3}u + \tfrac{1}{3}u_x D_x^{-1}$$

is a recursion operator for the Korteweg–de Vries equation, in the form

$$u_t = u_{xxx} + uu_x. \tag{5.45}$$

(This is the same as (2.66) under the change of variable $x \mapsto -x$.) The integral operator D_x^{-1} will *only* be defined on those differential functions which are total derivatives, so if $Q = D_x R$, then we set $R = D_x^{-1} Q$.† (Actually, this only defines $D_x^{-1} Q$ up to an additive constant, which we can normalize by requiring $R(0, 0) = 0$.) If $\mathbf{v}_Q$ is a generalized symmetry, then $\mathscr{R}Q$ will only be defined if $Q = D_x R$ for some $R \in \mathscr{A}$. Thus, we could run into difficulties in trying to obtain a full hierarchy of symmetries $\mathscr{R}^k Q$, $k = 0, 1, 2, \ldots$.

Before addressing these difficulties, we first show that formally $\mathscr{R}$ is a recursion operator. The relevant Fréchet derivative is

$$\mathsf{D}_K = D_x^3 + uD_x + u_x,$$

and we will prove that, on solutions of the Korteweg–de Vries equation, (5.43) holds. First,

$$\mathscr{R}_t = \tfrac{2}{3}u_t + \tfrac{1}{3}u_{xt}D_x^{-1} = \tfrac{2}{3}(u_{xxx} + uu_x) + \tfrac{1}{3}(u_{xxxx} + uu_{xx} + u_x^2)D_x^{-1}.$$

Next,

$$\begin{aligned}\mathsf{D}_K\mathscr{R} = D_x^5 + \tfrac{5}{3}uD_x^3 + \tfrac{10}{3}u_xD_x^2 + (3u_{xx} + \tfrac{2}{3}u^2)D_x + \tfrac{5}{3}(u_{xxx} + uu_x)\\ + \tfrac{1}{3}(u_{xxxx} + uu_{xx} + u_x^2)D_x^{-1}.\end{aligned}$$

On the other hand, since $D_x \cdot u = uD_x + u_x$, we have $D_x^{-1} \cdot (uD_x + u_x) = u$. Therefore,

$$\mathscr{R}\mathsf{D}_K = D_x^5 + \tfrac{5}{3}uD_x^3 + \tfrac{10}{3}u_xD_x^2 + (3u_{xx} + \tfrac{2}{3}u^2)D_x + (u_{xxx} + uu_x).$$

Therefore, the required commutator is

$$[\mathsf{D}_K, \mathscr{R}] = \tfrac{2}{3}(u_{xxx} + uu_x) + \tfrac{1}{3}(u_{xxxx} + uu_{xx} + u_x^2)D_x^{-1},$$

which, on comparison with the formula for $\mathscr{R}_t$, verifies (5.43) and proves that $\mathscr{R}$ is a recursion operator for the Korteweg–de Vries equation.

If we start applying $\mathscr{R}$ successively to the translational symmetry $-\partial_x$, with characteristic $Q_0 = u_x$, we first obtain

$$Q_1 = \mathscr{R}Q_0 = u_{xxx} + uu_x,$$

which is equivalent to the characteristic u of the translational symmetry $-\partial_t$. Noting that $Q_1 = D_x(u_{xx} + \frac{1}{2}u^2)$, we find

$$Q_2 = \mathscr{R}Q_1 = u_{xxxxx} + \tfrac{5}{3}uu_{xxx} + \tfrac{10}{3}u_xu_{xx} + \tfrac{5}{6}u^2u_x$$

to be the characteristic of a genuine generalized symmetry, as the reader can check. Similarly

$$\begin{aligned}Q_3 = \mathscr{R}Q_2 = u_{xxxxxxx} + \tfrac{7}{3}uu_{xxxxx} + 7u_xu_{xxxx} + \tfrac{35}{3}u_{xx}u_{xxx} + \tfrac{35}{18}u^2u_{xxx}\\ + \tfrac{70}{9}uu_xu_{xx} + \tfrac{35}{18}u_x^3 + \tfrac{35}{54}u^3u_x\end{aligned}$$

† More generally, one might try defining $D_x^{-1}P = \int_0^x P(x, u^{(n)})\,dx$, but this takes us outside the class of differential functions.

gives a seventh order generalized symmetry. As the following result shows, we can continue this recursive procedure indefinitely, leading to higher and higher order generalized symmetries.

Theorem 5.31. *Let $Q_0 = u_x$. For each $k \geqslant 0$, the differential polynomial $Q_k = \mathscr{R}^k Q_0$ is a total x-derivative, $Q_k = D_x R_k$, and hence we can recursively define $Q_{k+1} = \mathscr{R} Q_k$. Each Q_k is the characteristic of a symmetry of the Korteweg–de Vries equation.*

In fact, by Theorem 5.36, the vector fields $\mathbf{v}_k = \mathbf{v}_{Q_k}$ determine an infinite collection of mutually commuting flows

$$u_t = Q_k[u] = u_{2k+1} + \cdots,$$

called the "higher order Korteweg–de Vries equations". All of the above vector fields are thus symmetries of any one of these remarkable evolution equations.

Proof of Theorem 5.31. We proceed by induction on k, so assume that $Q_k = D_x R_k$ for some $R_k \in \mathscr{A}$. From the form of the recursion operator,

$$Q_{k+1} = D_x^2 Q_k + \tfrac{2}{3} u Q_k + \tfrac{1}{3} u_x D_x^{-1} Q_k = D_x[D_x Q_k + \tfrac{1}{3} u D_x^{-1} Q_k + \tfrac{1}{3} D_x^{-1}(u Q_k)].$$

If we can prove that $uQ_k = D_x S_k$ for some differential polynomial $S_k \in \mathscr{A}$, we will have proved that $Q_{k+1} = D_x R_{k+1}$, where R_{k+1} is the above expression in brackets, which will complete the induction step.

To prove this fact, note first that the formal adjoint of the recursion operator $\mathscr{R}$ is[†]

$$\mathscr{R}^* = D_x^2 + \tfrac{2}{3} u - \tfrac{1}{3} D_x^{-1} \cdot u_x = D_x^{-1} \mathscr{R} D_x.$$

We use this to integrate the expression uQ_k by parts, cf. (5.77), so

$$uQ_k = u \mathscr{R}^k [u_x] = u_x \cdot (\mathscr{R}^*)^k [u] + D_x A_k$$

for some differential function $A_k \in \mathscr{A}$. On the other hand, using a further integration by parts

$$u_x (\mathscr{R}^*)^k [u] = u_x \cdot D_x^{-1} \mathscr{R}^k [u_x] = u_x \cdot D_x^{-1} Q_k = -uQ_k + D_x B_k$$

for some $B_k \in \mathscr{A}$. Substituting into the previous identity, we conclude

$$uQ_k = D_x S_k, \quad \text{where} \quad S_k = \tfrac{1}{2}(A_k + B_k),$$

proving the claim. □

There were two other geometrical symmetry groups of the Korteweg–de Vries equation. The characteristic of the Galilean group $t\partial_x - \partial_u$ is $1 + tu_x$, and

$$3\mathscr{R}(1 + tu_x) = 2u + xu_x + 3t(u_{xxx} + uu_x),$$

[†] See the beginning of the following section.

which is equivalent to the scaling symmetry group. This latter characteristic is not a total derivative, so we cannot re-apply the recursion operator to get a meaningful generalized symmetry. However, the resulting "nonlocal symmetry" is important, as the following subsection demonstrates.

Example 5.32. Another example where the inverse of the total derivative is required is the physical form of Burgers' equation

$$v_t = v_{xx} + vv_x. \tag{5.46}$$

Note that if u satisfies the potential form (5.11), then $v = 2u_x$ satisfies (5.46); u is a "potential" for the physical velocity v. Thus, given a recursion operator $\mathscr{R}$, for the potential Burgers' equation, the transformed operator $\hat{\mathscr{R}} = D_x \mathscr{R} D_x^{-1}$ should be a recursion operator for the physical equation (5.46). (Why?) We conclude that

$$\hat{\mathscr{R}}_1 = D_x + \tfrac{1}{2}v + \tfrac{1}{2}v_x D_x^{-1}, \qquad \hat{\mathscr{R}}_2 = t\hat{\mathscr{R}}_1 + \tfrac{1}{2}x + \tfrac{1}{2}D_x^{-1}, \tag{5.47}$$

are recursion operators for Burgers' equation (5.46), a fact that can be verified directly. The resulting hierarchies of generalized symmetries are directly related to those of the potential Burgers' equation via the "potential" transformation $v = 2u_x$.

Master Symmetries

As in Example 5.18, time-dependent generalized symmetries can often used as an effective alternative to the recursion operator method for generating infinite hierarchies of symmetries of evolution equations. In this guise, they are called "master symmetries." A *master symmetry*, then, is a generalized (or even nonlocal) vector field $\mathbf{w}$ with the property that whenever $\mathbf{v}_Q$ is a generalized symmetry of the evolution equation, so is the Lie bracket $[\mathbf{w}, \mathbf{v}_Q]$. Note that any symmetry of the system satisfies this property, so to be really interesting, the master symmetry should produce *new* symmetries, mapping, say, the n-th member of the hierarchy of symmetries to the $(n + 1)$-st one, as the Burgers' example does.

In the theory of master symmetries, the following extension of the concept of a recursion operator plays a key role. (See (5.40) for notation.)

Definition 5.33. An operator $\mathscr{R}\colon \mathscr{A}^q \to \mathscr{A}^q$ is said to be *hereditary* if it satisfies

$$\mathbf{v}_{\mathscr{R}P}[\![\mathscr{R}]\!] = \mathscr{R} \cdot \mathbf{v}_P[\![\mathscr{R}]\!], \tag{5.48}$$

for all differential functions $P[u] \in \mathscr{A}^q$.

Almost all known recursion operators satisfy the hereditary property, including the recursion operators for the potential and ordinary Burgers' equations, and the Korteweg–de Vries equation. To see this in the case of

$\mathscr{R} = D_x + u_x$, we compute the (1, 1)-Lie derivative

$$\mathbf{v}_P[\![\mathscr{R}]\!] = D_x P + [\mathscr{R}, \mathsf{D}_P].$$

Moreover, by (5.35), $\mathsf{D}_{D_x P} = D_x \mathsf{D}_P$, whereas by (5.34), $\mathsf{D}_{u_x P} = u_x \mathsf{D}_P + P D_x$. Hence (5.48) reduces to

$$\begin{aligned} D_x(D_x + u_x)P &+ [D_x + u_x, (D_x + u_x)\mathsf{D}_P + PD_x] \\ &= (D_x + u_x)\{D_x P + [D_x + u_x, \mathsf{D}_P]\}, \end{aligned}$$

which can be straightforwardly verified. A similar, but more complicated computation, which is left to the reader, proves that the recursion operators for Burgers' equation and the Korteweg–de Vries equation also satisfy the hereditary property.

Proposition 5.34. *Suppose $\mathscr{R}$ is an hereditary operator. Let Q_0 be a differential function such that $\mathbf{v}_{Q_0}[\![\mathscr{R}]\!] = 0$, so that $\mathscr{R}$ is a recursion operator for the evolution equation $u_t = Q_0$. Then $\mathscr{R}$ is also a recursion operator for each of the evolution equations in the hierarchy $u_t = Q_k = \mathscr{R}^k Q_0$, $k = 0, 1, 2, \ldots$.*

PROOF. According to the preceding remarks, to prove $\mathscr{R}$ is a recursion operator for the evolution equation $u_t = Q_k$, we need only prove the vanishing of the Lie derivative

$$\mathbf{v}_{Q_k}[\![\mathscr{R}]\!] = \mathbf{v}_{\mathscr{R}^k Q_0}[\![\mathscr{R}]\!] = \mathscr{R}^k \cdot \mathbf{v}_{Q_0}[\![\mathscr{R}]\!] = 0,$$

the second equality following directly from the hereditary condition (5.48). □

The most common application of Proposition 5.34 is when $\mathscr{R}$ is a hereditary operator which does not depend explicitly on x, and the "seed" function for the hierarchy is $Q_0 = u_x$, the characteristic of the translational group $x \mapsto x - \varepsilon$, with associated evolution equation $u_t = u_x$. It is easy to see that *any* operator $\mathscr{R}$ which does not explicitly depend on x satisfies $\mathbf{v}_{Q_0}[\![\mathscr{R}]\!] = 0$, and so satisfies the basic hypothesis of Proposition 5.34. Indeed, if $\mathscr{R}$ does not depend on x, then, on solutions to $u_t = u_x$, we have $\mathscr{R}_t = \mathscr{R}_x$, the latter being obtained by applying D_x coefficient-wise to $\mathscr{R}$. Moreover, $\mathsf{D}_{u_x} = D_x$, so, in this case the Lie derivative (5.48) reduces to

$$\mathbf{v}_{u_x}[\![\mathscr{R}]\!] = \mathscr{R}_x + [\mathscr{R}, D_x] = 0.$$

Therefore, if $\mathscr{R}$ is an x-independent hereditary operator, Proposition 5.34 proves that $\mathscr{R}$ defines a recursion operator for the hierarchy $Q_k = \mathscr{R}^k(u_x)$. Both the Burgers' and Korteweg–de Vries equations fit into this framework.

Master symmetries arise from applying the recursion operator associated with the hierarchy to an appropriate scaling symmetry. Here "scaling" requires that the recursion operator itself scales under the associated one-parameter group.

Definition 5.35. A generalized vector field $\tilde{\mathbf{v}}_0$ is called a *scaling symmetry* for the operator $\mathscr{R}$ if $\tilde{\mathbf{v}}_0[\![\mathscr{R}]\!] = \lambda\mathscr{R}$ for some constant λ.

Examples are the time-independent parts of the scaling symmetries for the potential Burgers' and Korteweg–de Vries equations—see below. The basic result on master symmetries is the following.

Theorem 5.36. *Suppose $\mathscr{R}$ is an hereditary operator. Let $\mathbf{v}_0 = \mathbf{v}_{Q_0}$ be an evolutionary vector field such that $\mathbf{v}_0[\![\mathscr{R}]\!] = 0$, and let $\tilde{\mathbf{v}} = \mathbf{v}_{\tilde{Q}_0}$ be a scaling symmetry for both the operator $\mathscr{R}$ and the vector field $\mathbf{v}_0$, meaning that $[\tilde{\mathbf{v}}_0, \mathbf{v}_0] = \mu\mathbf{v}_0$ for some constant μ. Then $\mathscr{R}\tilde{\mathbf{v}}_0 = \mathbf{v}_{\mathscr{R}\tilde{Q}_0}$ is a master symmetry of the hierarchy of evolution equations $u_t = Q_k = \mathscr{R}^k(Q_0)$. In fact, if we set $\mathbf{v}_k = \mathbf{v}_{Q_k}$ and $\tilde{\mathbf{v}}_k = \mathbf{v}_{\tilde{Q}_k}$, where $\tilde{Q}_k = \mathscr{R}^k(\tilde{Q}_0)$, then*

$$[\mathbf{v}_j, \mathbf{v}_k] = 0, \qquad [\tilde{\mathbf{v}}_j, \mathbf{v}_k] = (\mu + k\lambda)\mathbf{v}_{j+k}, \qquad [\tilde{\mathbf{v}}_j, \tilde{\mathbf{v}}_k] = \lambda(k - j)\tilde{\mathbf{v}}_{j+k}.$$

PROOF. According to Proposition 5.34, $\mathscr{R}$ is a recursion operator for the hierarchy, hence $\mathbf{v}_j[\![\mathscr{R}]\!] = 0$, $j \geq 0$. The first identity is proved using the Leibniz rule (5.41) and the fact that $\mathscr{R}$ is hereditary:

$$[\mathbf{v}_j, \mathbf{v}_k] = [\mathbf{v}_j, \mathscr{R}^k\mathbf{v}_0] = \mathscr{R}^k[\mathbf{v}_j, \mathbf{v}_0] = \mathscr{R}^{j+k}[\mathbf{v}_0, \mathbf{v}_0] = 0.$$

Similarly, to prove the second identity,

$$\begin{aligned}[\mathbf{v}_j, \tilde{\mathbf{v}}_k] &= [\mathbf{v}_j, \mathscr{R}^k\tilde{\mathbf{v}}_0] = -\mathscr{R}^k[\tilde{\mathbf{v}}_0, \mathscr{R}^j\mathbf{v}_0] = -\mathscr{R}^k\{\tilde{\mathbf{v}}_0[\![\mathscr{R}^j]\!]\mathbf{v}_0 + \mathscr{R}^j[\tilde{\mathbf{v}}_0, \mathbf{v}_0]\} \\ &= (j\lambda + \mu)\mathscr{R}^{j+k}\mathbf{v}_0.\end{aligned}$$

The third identity is left to the reader. □

In the case of the potential Burgers' equation, the scaling symmetry is provided by the time-independent part, $\tilde{Q}_0 = xu_x$, of the geometric scaling symmetry (which was called Q_4 in (5.13)). The basic master symmetry, then, has characteristic $\tilde{Q}_1 = \mathscr{R}(xu_x) = x(u_{xx} + u_x^2) + u_x$, which is (except for the extra term u_x, which doesn't play any role since it commutes with everything) the time-independent part of the symmetry with characteristic $2Q_7$ in (5.13). Theorem 5.36 reconfirms the bracket relations noted earlier.

In the Korteweg–de Vries case, the scaling symmetry has characteristic $\tilde{Q}_0 = xu_x + 2u$, which, again, is the time-independent part of the standard geometric scaling symmetry. The basic master symmetry, then, has characteristic

$$\begin{aligned}\tilde{Q}_1 = \mathscr{R}\tilde{Q}_0 &= (D_x^2 + \tfrac{2}{3}u + \tfrac{1}{3}u_xD_x^{-1})(xu_x + 2u) \\ &= x(u_{xxx} + uu_x) + 4u_{xx} + \tfrac{4}{3}u^2 + \tfrac{1}{3}u_xD_x^{-1}(u),\end{aligned}$$

where we wrote $xu_x + 2u = D_x(xu) + u$ to obtain the final expression. The occurrence of the "integral" $D_x^{-1}(u)$ of u in the resulting formula means that $\tilde{Q}_1$ is not an ordinary differential function. Thus, the master symmetry of the

Korteweg–de Vries equation is not an ordinary generalized symmetry, but rather a "nonlocal vector field," and one must be a bit careful in computations which involve it. It is remarkable that its commutator with any of the higher order Korteweg–de Vries equations is a local evolution equation, a fact that follows from Theorem 5.31 and Theorem 5.36. Despite its importance, we will not pursue the theory of such nonlocal symmetries any further here.

Pseudo-differential Operators

In our discussion of recursion operators, we have already encountered the formal inverse D_x^{-1} of the total derivative operator D_x. By allowing the inverse of D_x, and higher order powers thereof, into the picture, we are naturally led to enlarge our space of differential operators. The net result is a formal, but very useful version of the calculus of pseudo-differential operators in one independent variable.

We begin by recalling the elementary properties of the ring of differential operators in one independent variable x and (just for simplicity) one dependent variable u, most of which have already been used. A *differential operator* is a finite sum

$$\mathscr{D} = \sum_{i=0}^{n} P_i[u] D_x^i, \tag{5.49}$$

where the coefficients $P_i[u]$ are differential functions. The differential operator $\mathscr{D}$ has *order* n provided its leading coefficient is not identically zero: $P_n \not\equiv 0$. A differential operator of order 0 is given by a single differential function, $\mathscr{D} = P_0[u]$ and is referred to as a *multiplication operator*; it is important in what follows to distinguish between differential functions and the multiplication operators they determine.

The (noncommutative) multiplication of differential operators is completely described by the obvious formula

$$D_x^i \cdot D_x^j = D_x^{i+j}, \tag{5.50}$$

valid for $i, j \geqslant 0$, and the elementary Leibniz rule, which begins with the derivational property of D_x:

$$D_x \cdot Q = Q D_x + Q', \qquad Q' = D_x Q. \tag{5.51}$$

(In (5.51), the differential function Q is to be viewed as a multiplication operator.) By induction, we find the general Leibniz rule

$$D_x^n \cdot Q = \sum_{k=0}^{n} \binom{n}{k} Q^{(k)} D_x^{n-k}, \tag{5.52}$$

where $Q^{(k)} = D_x^k Q$. The two rules (5.50) and (5.52) allow us, by linearity, to define the product of any two differential operators. Therefore, in algebraic language, the space of all differential operators forms a noncommutative ring, the identity operator being the multiplication operator determined by the constant function 1.

A differential operator (5.49), then, is a polynomial in the total derivative D_x with differential functions for coefficients. A pseudo-differential operator will be the analogous "Laurent series" (as in complex analysis) obtained by admitting negative powers of D_x.

Definition 5.37. A (formal) *pseudo-differential operator* is a formal infinite series

$$\mathscr{D} = \sum_{i=-\infty}^{n} P_i[u] D_x^i, \tag{5.53}$$

whose coefficients P_i are differential functions. We say that $\mathscr{D}$ has order n provided its leading coefficient is not identically zero: $P_n \not\equiv 0$.

By convention, the zero pseudo-differential operator is said to have order $-\infty$. (The reason for this will appear shortly.) The recursion operator $\mathscr{R} = D_x^2 + \frac{2}{3}u + \frac{1}{3}u_x D_x^{-1}$ for the Korteweg–de Vries equation discussed above provides an example of a pseudo-differential operator of order 2.

An important remark: Given a differential function $P[u]$, while it is clear how to define $\mathscr{D}(P)$ when $\mathscr{D}$ is a differential operator, there is no obvious analogue when $\mathscr{D}$ is a general pseudo-differential operator. We will not even try to make sense of any action of pseudo-differential operators on differential functions here.

Just as we multiply ordinary differential operators, we can define a compatible multiplication process for pseudo-differential operators. We treat D_x^{-1} as the inverse of D_x, so that $D_x^{-1} \cdot D_x = 1 = D_x \cdot D_x^{-1}$; more generally, we continue to allow the product rule (5.50), for arbitrary integers i, j. Now consider the Leibniz rule (5.51). If we multiply (5.51) on both the left and the right by the operator D_x^{-1}, and rearrange terms, we deduce the identity

$$D_x^{-1} \cdot Q = Q D_x^{-1} - D_x^{-1} \cdot Q' D_x^{-1}. \tag{5.54}$$

If we apply identity (5.54) to the second term on the right-hand side (with $Q' = D_x Q$ replacing Q), we deduce

$$D_x^{-1} \cdot Q = Q D_x^{-1} - Q' D_x^{-2} + D_x^{-1} \cdot Q'' D_x^{-2}.$$

Applying (5.54) to the third term on the right-hand side of this latter equation, and continuing, we deduce the general rule

$$D_x^{-1} \cdot Q = Q D_x^{-1} - Q' D_x^{-2} + Q'' D_x^{-2} - \cdots = \sum_{i=0}^{\infty} (-1)^i Q^{(i)} D_x^{-i-1}, \tag{5.55}$$

which allows us to commute the multiplication operator determined by Q with the operator D_x^{-1} and explains why we allow infinite series in our original definition of pseudo-differential operators. An obvious induction allows us to prove a general *Leibniz rule* for commuting multiplication and differentiation operators,

$$D_x^n \cdot Q = \sum_{i=0}^{\infty} \frac{n(n-1)\cdots(n-k+1)}{k!} Q^{(k)} D_x^{n-k}, \tag{5.56}$$

which is valid for *any* integral value of n. (If $n \geqslant 0$, all but finitely many of the terms in (5.56) vanish, and hence this formula does reduce to the differential operator version (5.52)). Extending formulas (5.50) and (5.56) by linearity allows us to express the product of any two pseudo-differential operators as a pseudo-differential operator. For example,

$$(D_x + u^2 D_x^{-1})(D_x^2 + uD_x^{-1}) = D_x^3 + u^2 D_x + u + u_x D_x^{-1} + u^3 D_x^{-2} - u^2 u_x D_x^{-3} + u^2 u_{xx} D_x^{-4} + \cdots .$$

See Exercise 5.19 for an alternative version of the product formula for pseudo-differential operators.

The product of pseudo-differential operators is associative and linear, but not commutative. Indeed, the commutator of two pseudo-differential operators is defined as $[\mathscr{D}, \mathscr{E}] = \mathscr{D} \cdot \mathscr{E} - \mathscr{E} \cdot \mathscr{D}$, and, like the commutator of (scalar) differential operators, obeys the order relation

$$\operatorname{order}[\mathscr{D}, \mathscr{E}] \leqslant \operatorname{order} \mathscr{D} + \operatorname{order} \mathscr{E} - 1. \tag{5.57}$$

The embedding of the ring of differential operators into the larger ring of pseudo-differential operators has many remarkable consequences, due to the additional structure of the latter domain.

Theorem 5.38. *Every nonzero pseudo-differential operator has an inverse.*

Proof. Suppose $\mathscr{D}$ is a pseudo-differential operator of order n as in (5.53), with nonzero leading coefficient P_n. The inverse $\mathscr{E} = \mathscr{D}^{-1}$ will have order $-n$, and take the form

$$\mathscr{E} = \frac{1}{P_n} D_x^{-n} + Q_1 D_x^{-n-1} + Q_2 D_x^{-n-2} + \cdots .$$

Substituting these two expressions into the equation $\mathscr{D} \cdot \mathscr{E} = 1$ leads to a system of equations for the coefficients Q_k of $\mathscr{E}$:

$$P_n Q_1 + nP_n D_x\left(\frac{1}{P_n}\right) + \frac{P_{n-1}}{P_n} = 0,$$

$$P_n Q_2 + nP_n D_x Q_1 + P_{n-1} Q_1 + \binom{n}{2} P_n D_x^2\left(\frac{1}{P_n}\right) + (n-1)P_{n-1} D_x\left(\frac{1}{P_n}\right) + \frac{P_{n-2}}{P_n} = 0,$$

and so on. The first equation can be solved for Q_1; plugging this formula into the second allows us to solve for Q_2 in terms of the P_j, etc., etc.

The resulting pseudo-differential operator $\mathscr{E}$ is, by construction, a right inverse for $\mathscr{D}$. To prove that it is also a left inverse, we multiply the equation $\mathscr{D} \cdot \mathscr{E} = 1$ on the right by $\mathscr{D}$, producing $\mathscr{D} \cdot \mathscr{E} \cdot \mathscr{D} = \mathscr{D}$; hence

$$\mathscr{D} \cdot (\mathscr{E} \cdot \mathscr{D} - 1) = 0.$$

Since $\mathscr{D} \neq 0$, the latter equation proves that $\mathscr{E} \cdot \mathscr{D} = 1$, and hence $\mathscr{E}$ is also a left inverse for $\mathscr{D}$. Here we make use of the easy fact that the space of pseudo-differential operators has no zero divisors, so $\mathscr{D} \cdot \mathscr{F} = 0$ if and only if $\mathscr{D} = 0$ or $\mathscr{F} = 0$, a result which becomes obvious by looking at the leading order term of the product of any two nonzero pseudo-differential operators. □

For example, the inverse of the first order differential operator $\mathscr{D} = D_x + u$ has the form

$$\mathscr{D}^{-1} = D_x^{-1} - uD_x^{-2} + (u_x + u^2)D_x^{-3} - (u_{xx} + 3uu_x + u^3)D_x^{-4} + \cdots.$$

It can be proved that the coefficient of D_n^{-n} in $(D_x + u)^{-1}$ is $(-1)^{n-1}(D_x + u)^{n-2}u$ for $n \geqslant 2$, cf. Exercise 5.20.

Theorem 5.38 proves that the space of pseudo-differential operators forms a skew (noncommutative) field. Yet there is even more structure available: Not only can we take inverses, but also roots.

Theorem 5.39. *Every pseudo-differential operator of order $n > 0$ has an n-th root.*

Proof. The proof is similar to that of Theorem 5.38. Suppose $\mathscr{D}$ is a pseudo-differential operator of the form (5.53), with nonzero leading coefficient P_n. The n-th root $\mathscr{E} = \sqrt[n]{\mathscr{D}}$ will be a first order pseudo-differential operator of the form

$$\mathscr{E} = \sqrt[n]{P_n} D_x + Q_0 + Q_{-1}D_x^{-1} + Q_{-2}D_x^{-2} + \cdots.$$

Substituting into the equation $\mathscr{E}^n = \mathscr{D}$ leads to a system of equations for the coefficients Q_k of $\mathscr{E}$, which, in analogy with the proof of Theorem 5.38, can be recursively solved for the Q_k. □

For example, the square root of the second order operator $\mathscr{D} = D_x^2 + u$ has the form

$$\sqrt{\mathscr{D}} = D_x + \tfrac{1}{2}uD_x^{-1} - \tfrac{1}{4}u_x D_x^{-2} + \tfrac{1}{8}(u_{xx} - u^2)D_x^{-3} + \cdots.$$

Theorem 5.39 allows us to define the fractional powers $\mathscr{D}^{i/n} = (\sqrt[n]{\mathscr{D}})^i$, for all integers i, of any n-th order pseudo-differential operator. We note that, just as with the ordinary powers, the fractional powers of a pseudo-differential oprator commute: $[\mathscr{D}^{i/n}, \mathscr{D}^{j/n}] = 0$. The fractional powers play a key role in Gel'fand and Dikii's approach, [1], [3], to soliton equations such as the Korteweg–de Vries equation.

Formal Symmetries

Consider an n-th order evolution equation

$$u_t = K[u] = K(x, u^{(n)}), \tag{5.58}$$

which does not explicitly depend on t. According to (5.27), a t-independent evolutionary vector field $\mathbf{v}_Q$ determines a symmetry of this equation if and only if the flows commute; this condition has the infinitesimal formulation

$$\operatorname{pr} \mathbf{v}_K(Q) - \operatorname{pr} \mathbf{v}_Q(K) = 0. \tag{5.59}$$

The key observation is that the symmetry condition (5.59) can, in turn, be converted into an equation involving differential operators by taking its Fréchet derivative. (We are effectively "relinearizing" the symmetry condition.)

Lemma 5.40. *Let $u_t = K[u]$ be an evolution equation, and let $Q[u]$ be a differential function. The Fréchet derivative of $Q_t = \operatorname{pr} \mathbf{v}_K(Q) = \mathsf{D}_Q(K)$ has the form*

$$\mathsf{D}_{Q_t} = \mathsf{D}_{\operatorname{pr} \mathbf{v}_K(Q)} = \operatorname{pr} \mathbf{v}_K(\mathsf{D}_Q) + \mathsf{D}_Q \cdot \mathsf{D}_K. \tag{5.60}$$

The proof follows directly from Definition 5.24 of the Fréchet derivative and is left to the reader. Note that, on solutions to $u_t = K$, the first term on the right-hand side of (5.60), which is the (ordinary) Lie derivative of the differential operator D_Q with respect to $\mathbf{v}_K$, can also be written as the time derivative $(\mathsf{D}_Q)_t$ of D_Q. □

Taking the Fréchet derivative of our symmetry condition (5.59) and using (5.60), we deduce the *operator symmetry condition*

$$\operatorname{pr} \mathbf{v}_K(\mathsf{D}_Q) - \operatorname{pr} \mathbf{v}_Q(\mathsf{D}_K) - [\mathsf{D}_K, \mathsf{D}_Q] = 0. \tag{5.61}$$

Condition (5.61) is almost identical to the original symmetry condition (5.59) since the Fréchet derivative of a differential function is zero if and only if the function depends on x only, so that if (5.61) holds, (5.59) is satisfied up to a function of x.

We now concentrate on equations having generalized symmetries of high order, meaning that the order m of the symmetry is much greater that the order n of the evolution equation: $m \gg n$. Note that, according to (5.57), the differential operators in the operator symmetry condition (5.61) have respective orders m, n and $m + n - 1$. Therefore, if $m \gg n$, the dominant (highest order) terms in (5.61) are the first and third, which, on solutions to $u_t = K$, are

$$(\mathsf{D}_Q)_t - [\mathsf{D}_K, \mathsf{D}_Q] = \operatorname{pr} \mathbf{v}_K(\mathsf{D}_Q) + [\mathsf{D}_Q, \mathsf{D}_K].$$

According to Definition 5.28, the latter operator is nothing but the (1, 1)-Lie derivative, cf. (5.40) (but with K playing the role of Q), of the differential operator D_Q with respect to the evolutionary vector field $\mathbf{v}_K$ determined by our evolution equation (5.58). According to (5.61), these terms must vanish modulo a differential operator of order n.

We can generalize this condition by replacing the Fréchet derivative operator D_Q by an arbitrary (pseudo-) differential operator.

Definition 5.41. Let $u_t = K[u]$ be an n-th order evolution equation. A pseudo-differential operator $\mathscr{D}$ of order m is called a *formal symmetry of rank k* if the (1, 1)-Lie derivative $\mathbf{v}_K[\![\mathscr{D}]\!]$ has order at most $n + m - k$; explicitly,

$$\operatorname{order}(\mathscr{D}_t + [\mathscr{D}, \mathrm{D}_K]) \leqslant n + m - k. \tag{5.62}$$

As an immediate consequence of (5.61), we see that genuine symmetries provide formal symmetries whose rank is the same as the order of the symmetry.

Proposition 5.42. *If $Q(x, u^{(m)})$ is the characteristic of an m-th order generalized symmetry of an evolution equation, then its Fréchet derivative D_Q is a formal symmetry of the equation of order m and of rank m.*

Our goal is to determine explicit conditions on our evolution equation (5.58) such that it will possess a formal symmetry operator of a prescribed rank. As we shall see, the higher the rank of the formal symmetry, the more restrictive the conditions imposed on the evolution equation. Eventually, the existence of a formal symmetry of a high enough rank will impose so many conditions on the evolution equation that it will possess a formal symmetry of infinite rank! Indeed, since the condition (5.44) that an operator $\mathscr{R}$ be a recursion operator for the evolution equation $u_t = K$ is that the (1, 1)-Lie derivative of $\mathscr{R}$ vanish, our convention that the zero pseudo-differential operator has order $-\infty$ implies that a recursion operator is the same as a formal symmetry of rank ∞. As we have seen, any evolution equation possessing a recursion operator has an infinite hierarchy of generalized symmetries, and, is, in an appropriate sense, an integrable evolution equation. We therefore propose the following symmetry-based definition of integrability.

Definition 5.43. An evolution equation is called *integrable* if it possesses a nonconstant formal symmetry of rank ∞.

(Note that a constant multiplication operator $\mathscr{D} = c$ is trivially a formal symmetry of order ∞.) For example, it is known that a second order evolution equation is integrable if and only if it has a formal symmetry of rank 5, and a third order evolution equation in which u_{xxx} occurs linearly is integrable if and only if it has a formal symmetry of rank 8. However, there is, as yet, no proof that a general evolution equation having a formal symmetry of sufficiently high rank is integrable in the sense of Definition 5.43, nor are there any realistic estimates of what "sufficiently high" might mean.

Lemma 5.44. *If $\mathscr{D}_1$ and $\mathscr{D}_2$ are formal symmetries of ranks k_1 and k_2, respectively, then their sum $\mathscr{D}_1 + \mathscr{D}_2$ and their product $\mathscr{D}_1 \cdot \mathscr{D}_2$ are formal symmetries of rank (at least) $k = \min\{k_1, k_2\}$.*

Lemma 5.45. *If a pseudo-differential operator $\mathscr{D}$ is a formal symmetry of rank k, its inverse $\mathscr{D}^{-1}$ is a formal symmetry of rank k. If $\mathscr{D}$ has order $m > 0$, then any (fractional) power $\mathscr{D}^{i/m}$ is also a formal symmetry of the same rank k.*

Thus, we can replace a (pseudo-) differential operator $\mathscr{D}$ of positive order $m > 0$ by its m-th root $\sqrt[m]{\mathscr{D}}$, which is a pseudo-differential operator of order 1, without losing the formal symmetry property. In fact, all the formal symmetry operators of a given rank (if any exist) can be completely characterized by a first order formal symmetry of that rank.

Theorem 5.46. *If an evolution equation of order $n \geqslant 2$ possesses a nonconstant formal symmetry of rank k, then it has a first order formal symmetry of rank k. Conversely, if*

$$\mathscr{D} = Q_1 D_x + Q_0 + Q_{-1} D_x^{-1} + \cdots \tag{5.63}$$

is a first order pseudo-differential operator which is a formal symmetry of rank k, then every formal symmetry of rank k has the form

$$\tilde{\mathscr{D}} = c_m \mathscr{D}^m + c_{m-1} \mathscr{D}^{m-1} + \cdots + c_{m-k+2} \mathscr{D}^{m-k+2} + \mathscr{E}, \tag{5.64}$$

where $c_m, \ldots, c_{m-k+2}$ are arbitrary constants, and $\mathscr{E}$ is any pseudo-differential operator of order at most $m - k + 1$.

Proof. The proof relies on a lemma characterizing the leading terms of a formal symmetry of an evolution equation. Note first that, by (5.57), *any* pseudo-differential operator is a formal symmetry of rank 1, so we only get interesting information starting at rank 2.

Lemma 5.47. *Let $\mathscr{D} = P_m[u] D_x^m + \cdots$ be an m-th order pseudo-differential operator which is a formal symmetry of rank $k \geqslant 2$ of the n-th order evolution equation $u_t = K[u]$, where $n \geqslant 2$. Then its leading coefficient is $P_m = c(\partial K/\partial u_n)^{m/n}$ for some constant $c \neq 0$.*

Proof. Define $K_n = \partial K/\partial u_n$. Since $\mathsf{D}_K = K_n D_x^n + \cdots$, the Leibniz rule (5.56) shows that the commutator in the formal symmetry condition (5.62) has leading term

$$[\mathsf{D}_K, \mathscr{D}] = (n K_n D_x P_m - m P_m D_x K_n) D_x^{m+n-1} + \cdots.$$

Since $n \geqslant 2$, the other term $\mathscr{D}_t = \operatorname{pr} \mathbf{v}_K(P_m) D_x^m + \cdots$ is of lower order. Therefore, $\mathscr{D}$ will be a formal symmetry of rank 2 (or more) if and only if $n K_n D_x P_m = m P_m D_x K_n$ or, equivalently, $D_x(K_n^{-m/n} P_m) = 0$. This suffices to prove the lemma. □

To prove the first part of Theorem 5.46, if the formal symmetry has positive order $m > 0$, then, according to Lemma 5.45, its m-th root provides a first order formal symmetry of the same rank. If its order is negative, $m < 0$, then

$\mathcal{D}^{-1/m}$ is the required first order formal symmetry. Finally, if $\mathcal{D}$ has order 0, then according to Lemma 5.47, $\mathcal{D} = c + \hat{\mathcal{D}}$ for some constant c and where $\hat{\mathcal{D}}$ is a pseudo-differential operator of order -1. Since c is trivially a formal symmetry of rank ∞, $\hat{\mathcal{D}}$ is a formal symmetry of rank k, and hence $\hat{\mathcal{D}}^{-1}$ provides the required first order formal symmetry.

To prove the second part, Lemma 5.47 implies that the first order formal symmetry (5.63) has leading term $\mathcal{D} = c\sqrt[n]{K_n}D_x + \cdots$ for some nonzero constant c. Similarly, any m-th order formal symmetry of order $k \geqslant 2$ has leading term $\tilde{\mathcal{D}} = \tilde{c}_m K_n^{m/n} D_x^m + \cdots$. Let $c_m = \tilde{c}_m/c^m$. Then $\hat{\mathcal{D}} = \tilde{\mathcal{D}} - c_m \mathcal{D}^m$ is a pseudo-differential operator of order $\hat{m} < m$; moreover, by the linearity of the Lie derivative, $\hat{\mathcal{D}}$ satisfies order$(\hat{\mathcal{D}}_t + [\hat{\mathcal{D}}, \mathsf{D}_K]) \leqslant n + m - k$, and hence $\hat{\mathcal{D}}$ is a formal symmetry of rank $k - (m - \hat{m})$. The proof now proceeds by an obvious induction on m, the order of the formal symmetry. □

As an illustration of the basic techniques, we discuss the problem of classifying integrable second order evolution equations of the particular form

$$u_t = u_{xx} + P(x, u, u_x). \tag{5.65}$$

(The classification of general second order evolution equations is handled by similar methods, although the calculations are more complicated. See the review paper of Shabat, Mikhailov and Sokolov, [1], for a more comprehensive treatment.) The Fréchet derivative of the right-hand side of (5.65) is

$$\mathsf{D}_K = D_x^2 + P_{u_x} D_x + P_u. \tag{5.66}$$

Consider a first order pseudo-differential operator which defines a formal symmetry of this equation. According to Lemma 5.47, we can, without loss of generality, assume that its leading coefficient is unity, and so the operator has the form

$$\mathcal{D} = D_x + Q_0 + Q_1 D_x^{-1} + Q_2 D_x^{-2} + \cdots, \tag{5.67}$$

where $Q_0, Q_1, \ldots$ are differential functions to be determined so as to satisfy the formal symmetry conditions. In order that $\mathcal{D}$ be a formal symmetry of rank k, the pseudo-differential operator

$$\mathscr{E} = \mathbf{v}_K[\![\mathcal{D}]\!] = \mathcal{D}_t + [\mathcal{D}, \mathsf{D}_K]$$

appearing in (5.62) must have order at most $3 - k$; the fact that $\mathcal{D}$ has leading coefficient 1 implies that it is already a formal symmetry of rank 2, so $\mathscr{E}$ has order at most 1. Requiring that the coefficients of the successive powers of D_x in $\mathscr{E}$ vanish will impose a series of increasingly stringent conditions on the operator, and, eventually, on the equation (5.65) itself, that are necessary for the existence of a formal symmetry of progressively higher and higher orders.

First, the coefficient of D_x in $\mathscr{E}$ is

$$D_x P_{u_x} - 2D_x Q_0.$$

To have a formal symmetry of rank 3, we require

$$Q_0 = \tfrac{1}{2} P_{u_x}. \tag{5.68}$$

(We can ignore the additive constant since, as remarked above, adding a constant to $\mathcal{D}$ does not affect its formal symmetry property.) Next, the coefficient of $D_x^0 = 1$ in $\mathcal{E}$ is

$$D_t Q_0 + D_x P_u - 2D_x Q_1 - P_{u_x} D_x Q_0.$$

In order to have a formal symmetry of rank 4, this quantity must vanish, which, in view of our normalization (5.68) of Q_0, means

$$\begin{aligned} 2D_x Q_1 &= D_t Q_0 + D_x P_u - P_{u_x} D_x Q_0 \\ &= \tfrac{1}{2} P_{u_x u_x} D_x(u_{xx} + P) + \tfrac{1}{2} P_{uu_x}(u_{xx} + P) + D_x P_u - \tfrac{1}{2} P_{u_x} D_x P_{u_x} \qquad (5.69) \\ &= D_x\left[\tfrac{1}{2} P_{u_x}(u_{xx} + P) + P_u - \tfrac{1}{4} P_{u_x}^2\right] + \tfrac{1}{2}(P_{uu_x} - D_x P_{u_x u_x})(u_{xx} + P). \end{aligned}$$

In order that (5.69) be soluble for the differential function Q_1, the right-hand side must lie in the image of the total derivative D_x:

$$\begin{aligned} &(P_{uu_x} - D_x P_{u_x u_x})(u_{xx} + P) \\ &\quad = -P_{u_x u_x u_x} u_{xx}^2 + (P_{uu_x} - P_{xu_x u_x} - u_x P_{uu_x u_x} - P P_{u_x u_x u_x}) u_{xx} \\ &\qquad + (P_{uu_x} - P_{xu_x u_x} - u_x P_{uu_x u_x}) P \in \operatorname{im} D_x. \qquad (5.70) \end{aligned}$$

Applying Theorem 4.7 (or, by inspection), we first see that this expression will certainly not lie in the image of D_x unless the coefficient of u_{xx}^2 vanishes. Therefore, P must be a quadratic polynomial in u_x,

$$P(x, u, u_x) = \alpha(x, u) u_x^2 + \beta(x, u) u_x + \gamma(x, u). \qquad (5.71)$$

Thus, we immediately deduce a strong restriction on the type of evolution equations (5.65) which admit formal symmetries. Any equation (5.65) which admits a formal symmetry of rank 4 is necessarily of the form

$$u_t = u_{xx} + \alpha(x, u) u_x^2 + \beta(x, u) u_x + \gamma(x, u). \qquad (5.72)$$

In particular, according to Proposition 5.42, only these types of equations can possibly admit generalized symmetries of order 4 or more (including admitting recursion operators). Plugging (5.71) into (5.70), and incorporating the terms involving u_{xx} into a total derivative, leads to a rather messy quadratic polynomial in u_x (the cubic terms all cancel) which must lie in the image of D_x. We could continue to analyze the general case directly, but the analysis is fairly complex. However, a simple observation will dramatically simplify our calculations.

Suppose we change variables in our evolution equation, replacing u by $v = \varphi(x, u)$, where φ is a smooth function, with $\varphi_u \neq 0$. (Note that the formal symmetry property, being given by the (1, 1)-Lie derivative, is unaffected by changes of variables.) Then

$$v_t = \varphi_u u_t, \quad v_x = \varphi_u u_x + \varphi_x, \quad v_{xx} = \varphi_u u_{xx} + \varphi_{uu} u_x^2 + 2\varphi_{xu} u_x + \varphi_{xx}. \qquad (5.73)$$

Therefore, given an equation of the form (5.72), the equation for v will take the same form

$$v_t = v_{xx} + \hat{\alpha}(x, v)v_x^2 + \hat{\beta}(x, v)v_x + \hat{\gamma}(x, v),$$

whose coefficients are related to those of the equation (5.72) for u according to

$$\alpha = \frac{\varphi_{uu}}{\varphi_u} + \hat{\alpha}\varphi_u,$$

$$\beta = 2\frac{\varphi_{xu}}{\varphi_u} + 2\hat{\alpha}\varphi_x + \hat{\beta},$$

$$\gamma = \frac{\varphi_{xu} + \hat{\alpha}\varphi_x^2 + \hat{\beta}\varphi_x + \hat{\gamma}}{\varphi_u}.$$

In particular, if we choose φ so that $\varphi_{uu} = \alpha\varphi_u$, we can eliminate the v_x^2 term in (5.73). Therefore, if, in our classification of evolution equations admitting formal symmetries, we are allowed to change variables, we can, without loss of generality (and using u instead of v), assume that the coefficient α in P is zero, so we need only consider evolution equations of the quasi-linear form:

$$u_t = u_{xx} + \beta(x, u)u_x + \gamma(x, u). \tag{5.74}$$

For such equations, the formal symmetry condition (5.70) dramatically simplifies to

$$\beta_u(u_{xx} + \beta u_x + \gamma) = D_x(\beta_u u_x) - \beta_{uu}u_x^2 + (\beta\beta_u - \beta_{xu})u_x + \beta_u\gamma \in \operatorname{im} D_x.$$

The coefficient of u_x^2 must vanish, hence $\beta(x, u) = a(x)u + b(x)$, and we require

$$\begin{aligned} &a^2uu_x + (ab - a_x)u_x + a\gamma(x, u) \\ &\quad = D_x[\tfrac{1}{2}a^2u^2 + (ab - a_x)u] - aa_xu^2 - (a_xb + ab_x - a_{xx})u + a\gamma \in \operatorname{im} D_x. \end{aligned}$$

Therefore,

$$\gamma = a_xu^2 + \frac{a_xb + ab_x - a_{xx}}{a}u + c(x).$$

We can yet further simplify this equation by incorporating the linear change of variables $u \mapsto a(x)u + b(x) - 2(a'(x)/a(x))$. The resulting equation is of the form

$$u_t = u_{xx} + uu_x + h(x). \tag{5.75}$$

The higher rank conditions for formal symmetries of (5.75) are all automatically satisfied; indeed (5.75) is a simple modification of the usual Burgers' equation (5.46) and therefore possesses a recursion operator, cf. Exercise 5.14. Thus, we have proved that every evolution equation of the form (5.65) which admits a formal symmetry of rank 4 (or, as in Proposition 5.42, a generalized symmetry of order 4 or more) is necessarily integrable, admits a recursion operator and generalized symmetries of arbitrarily high order and is, in fact, equivalent to the Burgers'-type equation (5.75). Applying the inverse transformation $u \mapsto \psi(x, u)$ to (5.75) will produce the most general integrable equation of the form (5.65).

A similar analysis will give a complete classification of all integrable second order evolution equations $u_t = Q(x, u, u_x, u_{xx})$. A key simplification comes from allowing a sufficiently wide variety of changes of variables to simplify the analysis as much as possible. The appropriate class consists not solely of the changes of dependent variable $v = \varphi(x, u)$, but, rather, all first order contact transformations. (See Bluman and Kumei, [2], and Ibragimov, [1].) The final result of this analysis, cf. Mikhailov, Shabat and Sokolov, [1], is the following.

Theorem 5.48. *Every second order evolution equation which admits a formal symmetry of rank* 5 or *more is integrable and is equivalent, under a contact transformation, to one of the following*:

$$u_t = u_{xx} + q(x)u,$$

$$u_t = u_{xx} + uu_x + h(x),$$

$$u_t = (u^{-2}u_x + \alpha xu + \beta u)_x,$$

$$u_t = (u^{-2}u_x)_x + 1.$$

5.3. Generalized Symmetries and Conservation Laws

The correspondence between ordinary variational symmetries and conservation laws of systems of Euler–Lagrange equations readily generalizes, a fact recognized even by Noether herself. In fact, once we admit generalized symmetries into the picture, Noether's theorem provides a *one-to-one* correspondence between variational symmetries and conservation laws. In this section we develop this result in the form due to Bessel-Hagen. (See Exercise 5.33 for Noether's original version.) The basic computational results depend on the concept of the adjoint of a differential operator.

Adjoints of Differential Operators

If

$$\mathscr{D} = \sum_J P_J[u]D_J, \qquad P_J \in \mathscr{A},$$

is a differential operator, its (formal) *adjoint* is the differential operator $\mathscr{D}^*$ which satisfies

$$\int_\Omega P \cdot \mathscr{D}Q \, dx = \int_\Omega Q \cdot \mathscr{D}^* P \, dx \tag{5.76}$$

for every pair of differential functions $P, Q \in \mathcal{A}$ which vanish when $u \equiv 0$, every domain $\Omega \subset \mathbb{R}^p$ and every function $u = f(x)$ of compact support in Ω. An easy integration by parts shows that

$$\mathscr{D}^* = \sum_J (-D)_J \cdot P_J,$$

meaning that for any $Q \in \mathcal{A}$,

$$\mathscr{D}^* Q = \sum_J (-D)_J [P_J Q].$$

For example, if

$$\mathscr{D} = D_x^2 + u D_x,$$

then its adjoint is

$$\mathscr{D}^* = (-D_x)^2 + (-D_x)\cdot u = D_x^2 - u D_x - u_x.$$

Similarly, a matrix differential operator $\mathscr{D}: \mathcal{A}^k \to \mathcal{A}^l$ with entries $\mathscr{D}_{\mu\nu}$ has adjoint $\mathscr{D}^*: \mathcal{A}^l \to \mathcal{A}^k$ with entries $\mathscr{D}^*_{\mu\nu} = (\mathscr{D}_{\nu\mu})^*$, the adjoint of the transposed entries of $\mathscr{D}$. Note that $(\mathscr{D}\mathscr{E})^* = \mathscr{E}^*\mathscr{D}^*$ for any operators $\mathscr{D}$, $\mathscr{E}$. An operator $\mathscr{D}$ is *self-adjoint* if $\mathscr{D}^* = \mathscr{D}$; it is *skew-adjoint* if $\mathscr{D}^* = -\mathscr{D}$. For example, $D_x^2 + u$ is self-adjoint, while $D_x^3 + 2uD_x + u_x$ is skew-adjoint. Note that (5.76) is equivalent to the integration by parts formula

$$P \cdot \mathscr{D} Q = Q \cdot \mathscr{D}^* P + \operatorname{Div} A, \tag{5.77}$$

where $A \in \mathcal{A}^p$ is a bilinear expression involving P, Q and their derivatives, with coefficients depending on x, u and derivatives of u. Equivalently

$$\mathsf{E}(P \cdot \mathscr{D} Q) = \mathsf{E}(Q \cdot \mathscr{D}^* P), \tag{5.78}$$

where E is the Euler operator, cf. Theorem 4.7.

Note that if $P \in \mathcal{A}^l$, its Fréchet derivative has adjoint $\mathsf{D}_P^*: \mathcal{A}^l \to \mathcal{A}^q$, which, using (5.32), has entries

$$(\mathsf{D}_P^*)_{\nu\mu} = \sum_J (-D)_J \cdot \frac{\partial P_\mu}{\partial u_J^\nu}, \qquad \mu = 1, \ldots, l, \quad \nu = 1, \ldots, q. \tag{5.79}$$

For example, if $P = u_{xx} + u_x^2$,

$$\mathsf{D}_P = D_x^2 + 2u_x D_x, \qquad \mathsf{D}_P^* = D_x^2 - 2D_x \cdot u_x = D_x^2 - 2u_x D_x - 2u_{xx}.$$

Although (5.79) bears some similarity to the Euler operator, it is in fact a differential operator, not a differential function, and is thus quite different. However, if $P \in \mathcal{A}$,

$$\mathsf{E}(P) = \left(\sum_J (-D)_J \frac{\partial P}{\partial u_J^\nu}\right) = \mathsf{D}_P^*(1),$$

1 denoting the constant differential function. We note finally the important formula for the variational derivative of the product of two functions

$$\mathsf{E}(P \cdot Q) = \mathsf{D}_P^*(Q) + \mathsf{D}_Q^*(P), \qquad P, Q \in \mathcal{A}^l, \tag{5.80}$$

which follows from the Leibniz rule:

$$\mathsf{E}_\nu(P\cdot Q) = \sum_{\mu=1}^{l} \left\{ \sum_J (-D)_J \left[\frac{\partial P_\mu}{\partial u_J^\nu} \cdot Q_\mu \right] + \sum_J (-D)_J \left[\frac{\partial Q_\mu}{\partial u_J^\nu} \cdot P_\mu \right] \right\}.$$

Characteristics of Conservation Laws

Before restricting our attention to Euler–Lagrange equations, we look at conservation laws in general again. Recall that every conservation law of a system of differential equations Δ is equivalent to one in characteristic form

$$\operatorname{Div} P = Q\cdot\Delta = \sum_{\nu=1}^{l} Q_\nu \Delta_\nu. \tag{5.81}$$

Using the notion of a Fréchet derivative, we readily obtain necessary and sufficient conditions for a given l-tuple Q to be the characteristic of a conservation law.

Proposition 5.49. *Let $\Delta = 0$ be a system of differential equations. An l-tuple $Q \in \mathscr{A}^l$ is the characteristic of a conservation law if and only if*

$$\mathsf{D}_\Delta^*(Q) + \mathsf{D}_Q^*(\Delta) = 0 \tag{5.82}$$

for all (x, u).

PROOF. According to Theorem 4.7, $Q\cdot\Delta$ is a total divergence (5.81) if and only if $\mathsf{E}(Q\cdot\Delta) = 0$. Thus (5.82) follows at once from the product rule (5.80). □

In particular, a necessary condition for Q to be the characteristic of a conservation law for Δ is

$$\mathsf{D}_\Delta^*(Q) = 0 \qquad \text{for all solutions to } \Delta, \tag{5.83}$$

since $\mathsf{D}_Q^*(\Delta) = 0$ automatically on solutions. This simplified form of (5.82) can often be used effectively to eliminate many possible l-tuples Q from consideration as characteristics of conservation laws, and thus readily lead to a complete classification of conservation laws for the system.

Example 5.50. Consider Burgers' equation in physical form

$$u_t = u_{xx} + uu_x.$$

If $\tilde{Q}[u] \in \mathscr{A}$ is the characteristic of a conservation law, then we can always replace t-derivatives of u by x-derivatives using the equation, so there is an equivalent characteristic of the form $Q(x, t, u, u_x, \ldots, u_n)$, $u_n = \partial^n u/\partial x^n$. Let us see what (5.83) says about the form of Q. For Burgers' equation,

$$\mathsf{D}_\Delta = D_t - D_x^2 - uD_x - u_x, \quad \text{so} \quad \mathsf{D}_\Delta^* = -D_t - D_x^2 + uD_x.$$

The leading order terms in (5.83) are

$$\mathsf{D}^*_\Delta(Q) = \frac{\partial Q}{\partial u_n}(-u_{n,t} - u_{n+2}) + \cdots = -2\frac{\partial Q}{\partial u_n}u_{n+2} + \cdots,$$

on solutions, the omitted terms depending on $(n+1)$-st and lower order x-derivatives of u. Thus (5.83) implies that $\partial Q/\partial u_n = 0$, so Q actually only depends on $(n-1)$-st and lower order derivatives of u. Proceeding by induction, we conclude that $Q = q(x, t)$ cannot depend on u or its derivatives in any nontrivial way. Moreover,

$$\mathsf{D}^*_\Delta(q) = q_t - q_{xx} + uq_x = 0$$

if and only if q is a constant. Thus the only nontrivial conservation law for Burgers' equation has a constant for its characteristic; the corresponding law is the equation itself:

$$D_t(u) + D_x(-u_x - \tfrac{1}{2}u^2) = 0.$$

Variational Symmetries

As with the geometrical form of Noether's theorem discussed in Chapter 4, the general form of Noether's theorem will only provide a correspondence between conservation laws and *variational symmetries*. These are defined in analogy with the divergence symmetries of Definition 4.33.

Definition 5.51. A generalized vector field

$$\mathbf{v} = \sum_{i=1}^{p} \xi^i \frac{\partial}{\partial x^i} + \sum_{\alpha=1}^{q} \phi_\alpha \frac{\partial}{\partial u^\alpha}$$

is a *variational symmetry* of the functional $\mathscr{L}[u] = \int L(x, u^{(n)})\,dx$ if and only if there exists a p-tuple $B[u] \in \mathscr{A}^p$ of differential functions such that

$$\mathrm{pr}\,\mathbf{v}(L) + L\,\mathrm{Div}\,\xi = \mathrm{Div}\,B \tag{5.84}$$

for all x, u. (Here $\xi = (\xi^1, \ldots, \xi^p)$ is as in (4.15).)

We first show that we can effectively restrict our attention to variational symmetries which are in evolutionary form.

Proposition 5.52. *A generalized vector field* $\mathbf{v}$ *is a variational symmetry of* $\mathscr{L}[u]$ *if and only if its evolutionary representative* $\mathbf{v}_Q$ *is.* (*Note: This statement is false if we omit the divergence term* Div B *in our definition* (5.84).)

PROOF. Using the basic prolongation formula (5.8),

$$\begin{aligned}\mathrm{pr}\,\mathbf{v}(L) + L\,\mathrm{Div}\,\xi &= \mathrm{pr}\,\mathbf{v}_Q(L) + \sum_{i=1}^{p} \xi^i D_i L + L\sum_{i=1}^{p} D_i\xi^i \\ &= \mathrm{pr}\,\mathbf{v}_Q(L) + \sum_{i=1}^{p} D_i(\xi^i L).\end{aligned}$$

Therefore, (5.84) holds if and only if

$$\operatorname{pr} \mathbf{v}_Q(L) = \operatorname{Div} \tilde{B}, \tag{5.85}$$

where $\tilde{B}_i = B_i - L\xi^i$. □

As with ordinary symmetries, every generalized variational symmetry of a variational problem is necessarily a symmetry of the corresponding Euler–Lagrange equations. (The converse of this statement remains *not* true in general.)

Theorem 5.53. *If the generalized vector field* $\mathbf{v}$ *is a variational symmetry of* $\mathscr{L}[u] = \int L(x, u^{(n)})\, dx$, *then* $\mathbf{v}$ *is a generalized symmetry of the Euler–Lagrange equations* $\mathsf{E}(L) = 0$.

The proof is based on the following important commutation formula.

Lemma 5.54. *Suppose* $L \in \mathscr{A}$, $Q \in \mathscr{A}^q$. *Then*

$$\mathsf{E}[\operatorname{pr} \mathbf{v}_Q(L)] = \operatorname{pr} \mathbf{v}_Q[\mathsf{E}(L)] + \mathsf{D}_Q^* \mathsf{E}(L). \tag{5.86}$$

PROOF. According to the integration by parts formula (4.39) and the identity (5.80),

$$\mathsf{E}[\operatorname{pr} \mathbf{v}_Q(L)] = \mathsf{E}[Q \cdot \mathsf{E}(L)] = \mathsf{D}_{\mathsf{E}(L)}^*[Q] + \mathsf{D}_Q^*[\mathsf{E}(L)].$$

We now need the important result that $\Delta = \mathsf{E}(L)$ is an Euler–Lagrange expression if and only if its Fréchet derivative is a self-adjoint differential operator: $\mathsf{D}_\Delta^* = \mathsf{D}_\Delta$. This fundamental theorem, which is the variational analogue of the equality of mixed partial derivatives, and constitutes the solution to the inverse problem of the calculus of variations, will be proved in Section 5.4. (See Theorem 5.92.) Assuming this result, (5.86) follows easily from (5.33) since

$$\mathsf{D}_{\mathsf{E}(L)}^*[Q] = \mathsf{D}_{\mathsf{E}(L)}[Q] = \operatorname{pr} \mathbf{v}_Q[\mathsf{E}(L)]. \qquad \square$$

PROOF OF THEOREM 5.53. By Propositions 5.5 and 5.52 we can replace $\mathbf{v}$ by its evolutionary form $\mathbf{v}_Q$ without affecting the validity of the theorem. If $\mathbf{v}_Q$ is a variational symmetry, (5.85) implies that the left-hand side of (5.86) vanishes. But D_Q^* is a linear differential operator, hence the symmetry condition (5.5) for $\Delta = \mathsf{E}(L)$ holds, completing the proof. □

Thus to find all the variational symmetries of a system of Euler–Lagrange equations, it suffices to use the methods of Sections 5.1 or 5.2 to construct symmetries of the Euler–Lagrange equations and then check which of them satisfy the additional variational requirement (5.84). Actually, we don't need to re-apply $\operatorname{pr} \mathbf{v}$ to the Lagrangian, or even know precisely what the Lagrangian is, since we can use the following intrinsic characterization of a variational symmetry.

Proposition 5.55. *Let $\Delta = 0$ be a system of differential equations whose Fréchet derivative is self-adjoint: $\mathsf{D}_\Delta^* = \mathsf{D}_\Delta$, so Δ is the Euler–Lagrange equations for some variational problem.*† *An evolutionary vector field $\mathbf{v}_Q$ is a variational symmetry thereof if and only if*

$$\operatorname{pr} \mathbf{v}_Q(\Delta) + \mathsf{D}_Q^*(\Delta) = 0 \tag{5.87}$$

for all x, u.

The proof is immediate from the preceding calculations and the solution to the inverse problem in Theorem 5.92. □

Group Transformations

Assuming that the variational symmetry is in evolutionary form, we can deduce that the corresponding group transformations leave the functional itself invariant in the following sense.

Proposition 5.56. *Given the relevant existence and uniqueness results on the Cauchy problem for the associated system of evolution equations, a generalized vector field $\mathbf{v}_Q$ is a variational symmetry of the functional $\mathscr{L}_{\Omega_0}[u] = \int_{\Omega_0} L(x, u^{(n)})\,dx$ if and only if for every subdomain $\Omega \subset \Omega_0$ and every function $u = f(x)$ in the appropriate function space*

$$\mathscr{L}_\Omega[\exp(\varepsilon \mathbf{v}_Q)f] = \mathscr{L}_\Omega[f] + \mathscr{B}_{\partial\Omega}[\varepsilon, f], \tag{5.88}$$

where $\mathscr{B}_{\partial\Omega}$ depends only on the values of $\exp(\varepsilon \mathbf{v}_Q)f$ and its derivatives on the boundary $\partial\Omega$.

Another way of interpreting this result is that a generalized vector field $\mathbf{v}_Q$ is a variational symmetry of a functional $\mathscr{L}$ if and only if $\mathscr{L}$ determines a conservation law for the system of evolution equations $u_t = Q$ prescribing the flow of $\mathbf{v}_Q$.

Proof. Differentiating (5.88) with respect to ε, we find

$$\int_\Omega \operatorname{pr} \mathbf{v}_Q(L)\,dx = \int_{\partial\Omega} B \cdot dS = \int_\Omega (\operatorname{Div} B)\,dx$$

for some $B \in \mathscr{A}^p$ depending on u and its derivatives; both sides of this latter identity are to be evaluated at $u = \exp(\varepsilon \mathbf{v}_Q)f$. Since this holds for an arbitrary subdomain Ω, we conclude the equality of the integrands,

$$\operatorname{pr} \mathbf{v}_Q(L) = \operatorname{Div} B,$$

verifying the infinitesimal criterion (5.85). The converse follows upon integration with respect to ε. □

† This assumes the restriction on the domain M of Theorem 5.92.

Example 5.57. Consider the functional

$$\mathscr{L}[u] = \int_a^b \tfrac{1}{2}u_x^2\,dx, \qquad x, u \in \mathbb{R}.$$

The generalized symmetry $\mathbf{v} = -u_x\partial_u$ is easily seen to be a variational symmetry of $\mathscr{L}$:

$$\operatorname{pr}\mathbf{v}(\tfrac{1}{2}u_x^2) = -u_x u_{xx} = -D_x(\tfrac{1}{2}u_x^2).$$

Indeed $\mathbf{v}$ is just the evolutionary form of the translation field $\tilde{\mathbf{v}} = \partial_x$, and generates the one-parameter group

$$\exp(\varepsilon\mathbf{v})f(x) = f(x-\varepsilon).$$

If $[c, d] \subset (a, b)$ is any subinterval, the boundary contribution in the proof of (5.88) is

$$\mathscr{B}(x, u^{(1)}) = -\tfrac{1}{2}u_x^2\Big|_{x=c}^{d} = \tfrac{1}{2}[f'(c)^2 - f'(d)^2];$$

indeed (5.88) in this case reads

$$\int_c^d \tfrac{1}{2}[f'(x-\varepsilon)]^2\,dx$$

$$= \int_c^d \tfrac{1}{2}[f'(x)]^2\,dx + \int_0^\varepsilon \tfrac{1}{2}\{[f'(c-\tilde{\varepsilon})]^2 - [f'(d-\tilde{\varepsilon})]^2\}\,d\tilde{\varepsilon}.$$

Note especially that we cannot dispense with the boundary contribution in general since the only solution vanishing on the boundary is the trivial solution $u \equiv 0$.

Noether's Theorem

As the reader may have already noticed, in the case that the system of differential equations Δ is the Euler–Lagrange equations for some variational problem, the condition (5.82) for Q to be the characteristic of a conservation law and the condition (5.87) for $\mathbf{v}_Q$ to generate a variational symmetry group coincide. Thus, using Theorem 4.26, we immediately deduce the general form of Noether's theorem.

Theorem 5.58. *A generalized vector field $\mathbf{v}$ determines a variational symmetry group of the functional $\mathscr{L}[u] = \int L\,dx$ if and only if its characteristic $Q \in \mathscr{A}^q$ is the characteristic of a conservation law* $\operatorname{Div} P = 0$ *for the corresponding Euler–Lagrange equations* $\mathrm{E}(L) = 0$. *In particular, if $\mathscr{L}$ is a nondegenerate variational problem, there is a one-to-one correspondence between equivalence classes of nontrivial conservation laws of the Euler–Lagrange equations and equivalence classes of variational symmetries of the functional.*

Note that two variational symmetries are equivalent provided they differ by a trivial symmetry, meaning one whose characteristic vanishes on all solutions of the Euler–Lagrange equations. (However, it is *not* true that a symmetry which happens to be equivalent to a variational symmetry is necessarily variational—see Exercise 5.32.)

Example 5.59. As a first illustration of this result, consider the Kepler problem $\ddot{x} + \mu r^{-3}x = 0$, $\ddot{y} + \mu r^{-3}y = 0$, $\ddot{z} + \mu r^{-3}z = 0$, $r^2 = x^2 + y^2 + z^2$, for a mass moving in a gravitational potential due to a fixed mass at the origin. The associated Lagrangian is $L = \frac{1}{2}(\dot{x}^2 + \dot{y}^2 + \dot{z}^2) - \mu r^{-1}$. We've already seen in Example 4.31 how the conservation laws of energy and angular momenta arise from the variational symmetry groups of time translations and rotations in $\mathbb{R}^3$. Owing to the Newtonian nature of the force field, there are three additional "hidden" generalized variational symmetries of this system, leading to three further independent conservation laws. One such infinitesimal generator is the vector field

$$\mathbf{v}_x = (y\dot{y} + z\dot{z})\partial_x + (\dot{x}y - 2x\dot{y})\partial_y + (\dot{x}z - 2x\dot{z})\partial_z,$$

the other two being obtained by permuting the variables x, y, z. To prove that $\mathbf{v}_x$ is indeed a variational symmetry, we compute

$$\begin{aligned}\mathrm{pr}^{(1)}\,\mathbf{v}_x = \mathbf{v}_x &+ (y\ddot{y} + z\ddot{z} + \dot{y}^2 + \dot{z}^2)\partial_{\dot{x}} + (\ddot{x}y - 2x\ddot{y} - \dot{x}\dot{y})\partial_{\dot{y}} \\ &+ (\ddot{x}z - 2x\ddot{z} - \dot{x}\dot{z})\partial_{\dot{z}},\end{aligned}$$

and hence

$$\begin{aligned}\mathrm{pr}^{(1)}\,\mathbf{v}_x(L) &= (y\dot{y} + z\dot{z})\ddot{x} + (\dot{x}y - 2x\dot{y})\ddot{y} + (\dot{x}z - 2x\dot{z})\ddot{z} \\ &\quad + \mu r^{-3}[(y^2 + z^2)\dot{x} - xy\dot{y} - xz\dot{z}] \\ &= D_t[\dot{x}(y\dot{y} + z\dot{z}) - x(\dot{y}^2 + \dot{z}^2) + \mu r^{-1}x],\end{aligned}$$

verifying (5.85). The corresponding conservation laws are found from the characteristic form (5.81), or, more simply, by noting that $\mathrm{pr}^{(1)}\,\mathbf{v}_x(L)$ itself vanishes on solutions of the Euler–Lagrange equations, so

$$R_x \equiv \dot{x}(y\dot{y} + z\dot{z}) - x(\dot{y}^2 + \dot{z}^2) + \mu r^{-1}x$$

is a first integral of the Kepler problem. Coupled with the other conservation laws R_y and R_z obtained by permuting the variables, we deduce the constancy of the *Runge–Lenz vector*, which can be written as

$$\mathbf{R} \equiv (R_x, R_y, R_z) = \dot{\mathbf{x}} \times \mathbf{A} - \mu\mathbf{x}/|\mathbf{x}| = \dot{\mathbf{x}} \times (\mathbf{x} \times \dot{\mathbf{x}}) - \mu\mathbf{x}/|\mathbf{x}|,$$

where $\mathbf{x} = (x, y, z)$ is the position vector and $\mathbf{A} = \mathbf{x} \times \dot{\mathbf{x}}$ the angular momentum. Physically, $\mathbf{R}$ points along the major axis of the conic section determined by the planetary orbit, its magnitude determining the eccentricity. (See Thirring, [1; p. 147].)

Example 5.60. The sine–Gordon equation $u_{xt} = \sin u$ is the Euler–Lagrange equation for the functional

$$\mathscr{L}[u] = \iint (\tfrac{1}{2}u_x u_t - \cos u)\, dx\, dt.$$

The generalized vector field $\mathbf{v}_1$ with characteristic $Q_1 = u_{xxx} + \frac{1}{2}u_x^3$ is a variational symmetry of $\mathscr{L}$. This can be seen directly, or, slightly easier, by using Proposition 5.55. Note that

$$\mathsf{D}_{Q_1} = D_x^3 + \tfrac{3}{2}u_x^2 D_x, \qquad \mathsf{D}_{Q_1}^* = -D_x^3 - \tfrac{3}{2}u_x^2 D_x - 3u_x u_{xx}.$$

A short calculation shows that

$$\begin{aligned} \operatorname{pr} \mathbf{v}_{Q_1}[u_{xt} - \sin u] &= u_{xxxxt} + \tfrac{3}{2}u_x^2 u_{xxt} + 3u_x u_{xx} u_{xt} - (u_{xxx} + \tfrac{1}{2}u_x^3)\cos u \\ &= -\mathsf{D}_{Q_1}^*[u_{xt} - \sin u], \end{aligned}$$

verifying (5.87). The associated conservation law has characteristic form

$$\begin{aligned} D_t(-\tfrac{1}{2}u_{xx}^2 + \tfrac{1}{8}u_x^4) &+ D_x(u_{xx}u_{xt} - u_{xx}\sin u + \tfrac{1}{2}u_x^2 \cos u) \\ &= (u_{xxx} + \tfrac{1}{2}u_x^3)(u_{xt} - \sin u). \end{aligned}$$

In particular, the conserved density determines a functional

$$\mathscr{T}_1[u] = \int_{-\infty}^{\infty} (\tfrac{1}{8}u_x^4 - \tfrac{1}{2}u_{xx}^2)\, dx,$$

whose value is independent of t whenever $u(x, t)$ is a solution whose derivatives decay rapidly as $|x| \to \infty$.

An even more tedious computation shows that

$$\mathbf{v}_{Q_2} = (u_{xxxxx} + \tfrac{5}{2}u_x^2 u_{xxx} + \tfrac{5}{2}u_x u_{xx}^2 + \tfrac{3}{8}u_x^5)\partial_u$$

is also a variational symmetry, with associated conservation law

$$\mathscr{T}_2 = \int_{-\infty}^{\infty} (\tfrac{1}{2}u_{xxx}^2 - \tfrac{5}{4}u_x^2 u_{xx}^2 + \tfrac{1}{16}u_x^6)\, dx.$$

(See Exercises 5.12 and 5.31 for further results on this equation.)

Self-adjoint Linear Systems

Consider a homogeneous system of linear differential equations $\Delta[u] = 0$ determined by a $q \times q$ matrix of differential operators

$$\Delta_{\mu\nu} = \sum_J a_{\mu\nu}^J(x) D_J, \qquad \mu, \nu = 1, \ldots, q,$$

whose coefficients depend only on x. As is well known, this system is the Euler–Lagrange equations for a variational problem if and only if Δ is self-adjoint: $\Delta^* = \Delta$. In this case, we can take the functional simply to be

$$\mathscr{L}[u] = \tfrac{1}{2}\int u\cdot\Delta[u]\,dx. \tag{5.89}$$

(See also Theorem 5.92.)

Any conservation law for the given self-adjoint linear system can, without loss of generality, be taken in characteristic form $\operatorname{Div} P = Q\cdot\Delta$. By Noether's theorem, the characteristic Q determines a variational symmetry of the corresponding quadratic variational problem. Here we investigate in some detail the cases of *linear* conservation laws, where P is linear in u and its derivatives, and hence Q depends only on x, and quadratic conservation laws, with P being quadratic and Q linear in u and its derivatives. The former case will lead to "reciprocity" relations relating pairs of solutions of the system; the latter will be closely tied to our theory of recursion operators for linear systems developed in the preceding section.

For a linear conservation law, note that $\mathbf{v}_q = \sum q_\alpha(x)\partial_{u^\alpha}$ generates a symmetry group of a linear system if and only if $q(x)$ is a solution itself: $\Delta[q] = 0$. (The group transformations are just $u \mapsto u + \varepsilon q$, reflecting the linearity of Δ.) Also note that the Fréchet derivative in this case is automatically 0, so (5.87) is verified and $\mathbf{v}_q$ is always a variational symmetry. Noether's theorem allows us to conclude the existence of a linear conservation law

$$\operatorname{Div} \hat{P}[u] = q(x)\cdot\Delta[u] \tag{5.90}$$

for any solution $q(x)$ of Δ. Alternatively, we can derive (5.90) directly using our basic integration by parts procedure:

Proposition 5.61. *Let $\Delta[u] = 0$ be a self-adjoint linear system. Then, for any functions $u(x)$, $v(x)$, we have the reciprocity relation*

$$v\cdot\Delta[u] - u\cdot\Delta[v] = \operatorname{Div} P[u, v], \tag{5.91}$$

where $P \in \mathscr{A}^p$ is some bilinear expression involving u and v and their derivatives.

The general formula for P in terms of Δ is quite complicated. In the second order case, however, we can derive a relatively simple expression. It is not difficult to see that any self-adjoint second order matrix differential operator can be written in the particular form

$$\Delta_{\mu\nu} = \sum_{i,j=1}^{p} D_i\cdot a^{ij}_{\mu\nu}(x)D_j + \sum_{i=1}^{p}(b^i_{\mu\nu}(x)\cdot D_i + D_i\cdot b^i_{\mu\nu}(x)) + c_{\mu\nu}(x),$$
$$\mu, \nu = 1, \dots, q,$$

where the coefficients satisfy

$$a^{ij}_{\mu\nu} = a^{ji}_{\nu\mu}, \qquad b^i_{\mu\nu} = -b^i_{\nu\mu}, \qquad c_{\mu\nu} = c_{\nu\mu}.$$

The corresponding variational problem can either be taken in the form (5.89), or, by a simple integration by parts, in first order form

$$\mathscr{L}[u] = \tfrac{1}{2} \int \sum_{\mu,\nu=1}^{q} \left\{ - \sum_{i,j=1}^{p} a_{\mu\nu}^{ij} u_i^\mu u_j^\nu + \sum_{i=1}^{p} b_{\mu\nu}^i (u^\mu u_i^\nu - u_i^\mu u^\nu) + c_{\mu\nu} u^\mu u^\nu \right\} dx. \tag{5.92}$$

If we define the $q \times q$ matrix differential operators $\mathscr{D}^i$, $i = 1, \ldots, p$, with entries

$$\mathscr{D}_{\mu\nu}^i = \sum_{j=1}^{p} a_{\mu\nu}^{ij}(x) D_j + b_{\mu\nu}^i(x),$$

then the reciprocity relation (5.91) holds with

$$P_i = v \cdot \mathscr{D}^i[u] - u \cdot \mathscr{D}^i[v], \qquad i = 1, \ldots, p.$$

Equivalently, we have the integral form

$$\int_{\partial\Omega} (v \cdot \mathscr{D}[u] - u \cdot \mathscr{D}[v]) \cdot dS = \int_{\Omega} (v \cdot \Delta[u] - u \cdot \Delta[v]) \, dx, \tag{5.93}$$

where $v \cdot \mathscr{D}[u] \equiv (v \cdot \mathscr{D}^1[u], \ldots, v \cdot \mathscr{D}^p[u])$.

For instance, in the case of Laplace's equation, (5.93) is the familiar form of Green's formula since $\mathscr{D}[u] = \nabla u$. For Navier's equations

$$\mu \Delta u + (\mu + \lambda) \nabla(\nabla \cdot u) = 0$$

of linear isotropic elasticity, (5.93) is equivalent to the standard Betti reciprocal theorem

$$\int_{\partial\Omega} (u \cdot \sigma[v] - v \cdot \sigma[u]) \, dS = \int_{\Omega} \{u \cdot [\mu \Delta v + (\mu + \lambda) \nabla(\nabla \cdot v)] - v[\mu \Delta u + (\mu + \lambda) \nabla(\nabla \cdot u)]\} \, dx,$$

in which

$$\sigma[u] = \mu(\nabla u + \nabla u^T) + \lambda(\nabla \cdot u) \mathrm{I}$$

is the stress tensor associated with the displacement u. (Here (5.92), which is

$$\mathscr{L}[u] = -\tfrac{1}{2} \int \{\mu \|\nabla u\|^2 + (\mu + \lambda)(\nabla \cdot u)^2\} \, dx$$

is not exactly the same as the usual variational principle derived from the stored energy function, but differs from it only by the null Lagrangian

$$N = \sum_{\substack{i \neq j \\ \alpha \neq \beta}} \mu \frac{\partial(u^\alpha, u^\beta)}{\partial(x^i, x^j)}.$$

Turning to the quadratic conservation laws, the characteristic Q is a linear function of u and its derivatives, hence $Q(x, u^{(m)}) = \mathscr{D}[u]$ for some $q \times q$ matrix of differential operators $\mathscr{D}$ whose coefficients depend only on x.

Noether's theorem implies that Q is the characteristic of a variational symmetry, and hence a symmetry of the Euler–Lagrange equations themselves. Proposition 5.22 implies that $\mathscr{D}$ is a recursion operator for the linear system, so $\Delta\mathscr{D} = \tilde{\mathscr{D}}\Delta$ for some differential operator $\tilde{\mathscr{D}}$. Not every recursion operator gives rise to a variational symmetry, however, but it is easy to characterize those which do.

Proposition 5.62. *A q-tuple $Q = \mathscr{D}[u]$ of linear functions in u and its derivatives forms the characteristic of a conservation law for the linear system $\Delta[u] = 0$ if and only if the product differential operator $\mathscr{D}^* \cdot \Delta$ is skew-adjoint.*

This is an immediate consequence of (5.82) using the fact that the Fréchet derivative of a linear q-tuple $\Delta[u]$ is the same as the differential operator Δ which determines it. In particular, if Δ is self-adjoint, this condition takes the form

$$\Delta \cdot \mathscr{D} = -\mathscr{D}^* \cdot \Delta, \tag{5.94}$$

meaning that the operator $\tilde{\mathscr{D}}$ appearing in the recursion condition $\Delta\mathscr{D} = \tilde{\mathscr{D}}\Delta$ must agree with $-\mathscr{D}^*$. Note that in this case, any *odd* power $\mathscr{D}^{2k+1}$ of $\mathscr{D}$ also satisfies (5.94). We conclude that a self-adjoint linear system with one quadratic conservation law always has an *infinite* hierarchy of such laws

$$\operatorname{Div} P^{(k)} = \mathscr{D}^{2k+1}[u] \cdot \Delta[u]$$

depending on higher and higher order derivatives of u.

Under our nondegeneracy hypothesis, a symmetry operator $\mathscr{D}$ determines a trivial conservation law if and only if it is a multiple of Δ, i.e. $\mathscr{D} = \mathscr{E} \cdot \Delta$ for some differential operator $\mathscr{E}$. The question of how many nontrivial quadratic conservation laws of a given order there are, then, is related to the (complicated) question of how many inequivalent symmetry operators of a given order there are, which was considered in Section 5.2. Note further that if $\mathscr{D}$ is any linear recursion operator, so $\Delta\mathscr{D} = \tilde{\mathscr{D}}\Delta$ for some operator $\tilde{\mathscr{D}}$, then we can always "skew-symmetrize" $\mathscr{D}$ to produce a new recursion operator $\hat{\mathscr{D}} = \frac{1}{2}(\mathscr{D} - \tilde{\mathscr{D}}^*)$ which does satisfy (5.94) and hence does determine a conservation law. To see this, it suffices to take the adjoint of the symmetry condition,

$$\Delta\tilde{\mathscr{D}}^* = (\tilde{\mathscr{D}}\Delta)^* = (\Delta\mathscr{D})^* = \mathscr{D}^*\Delta,$$

using the self-adjointness of Δ, hence

$$\Delta\hat{\mathscr{D}} = \tfrac{1}{2}(\Delta\mathscr{D} - \Delta\tilde{\mathscr{D}}^*) = \tfrac{1}{2}(\tilde{\mathscr{D}}\Delta - \mathscr{D}^*\Delta) = -\hat{\mathscr{D}}^*\Delta.$$

In particular, since for any symmetry operator $\mathscr{D}$, the operator $\tilde{\mathscr{D}}$ has the same leading order terms, we see that there is a one-to-one correspondence between quadratic conservation laws and the skew-adjoint leading terms of recursion operators. For scalar equations, the leading order terms must be of odd order, and for every such term we get a conservation law. If $\mathscr{D}_1, \ldots, \mathscr{D}_k$ are linear first order variational symmetry operators, then the skew-symmetrized product

$$\tfrac{1}{2}[\mathscr{D}_1\mathscr{D}_2\cdots\mathscr{D}_k + (-1)^{k-1}\mathscr{D}_k\mathscr{D}_{k-1}\cdots\mathscr{D}_1] \tag{5.95}$$

gives a k-th (or lower) order variational operator. (For scalar equations, we need only take k odd, and then the operator is k-th order.) In many examples, it appears that every quadratic conservation law can be generated in this way.

Example 5.63. Here we conclude our investigation into the symmetries and conservation laws of the two-dimensional wave equation $u_{tt} = u_{xx} + u_{yy}$ of Examples 2.43, 4.36 and 5.23. Here $\Delta = D_t^2 - D_x^2 - D_y^2$. Of the recursion operators listed in (5.30), the first six, corresponding to translations and rotations, all commute with Δ and are all skew-adjoint and hence all satisfy (5.94). The resulting conservation laws were determined in Example 4.36. For the dilatational operator $\mathscr{D}$, we find $\Delta\mathscr{D} = (\mathscr{D} + 2)\Delta$, but $\mathscr{D}^* = -\mathscr{D} - 3$. Thus $\mathscr{D}$ does not determine a conservation law; however, the modified dilatation operator $\mathscr{M} = \mathscr{D} + \frac{1}{2}$ does satisfy (5.94):

$$\Delta\mathscr{M} = (\mathscr{M} + 2)\Delta, \quad \text{and} \quad \mathscr{M}^* = \mathscr{D}^* + \tfrac{1}{2} = -\mathscr{D} - \tfrac{5}{2} = -\mathscr{M} - 2.$$

See Example 4.36 for the conservation law. Finally, each inversional operator also determines a conservation law, since, for $\mathscr{I}_x$, say,

$$\Delta\mathscr{I}_x = (\mathscr{I}_x + 4x)\Delta, \quad \text{and} \quad \mathscr{I}_x^* = -\mathscr{I}_x - 4x.$$

The corresponding conservation laws were found in Example 4.36.

Higher order quadratic conservation laws are found by looking at "skew-symmetrized" odd order products (5.95) of these basic recursion operators, e.g. $\frac{1}{2}[\mathscr{R}_{xy}\mathscr{M}\mathscr{I}_x + \mathscr{I}_x\mathscr{M}\mathscr{R}_{xy}]$. A partial listing of some of the second order conservation laws and their corresponding symmetry operators is given in the following table. (See also Example 5.65.)

Recursion Operator	Characteristic	Conserved Density
D_x^3	u_{xxx}	$u_{xx}u_{xt}$
$D_x^2D_t$	u_{xxt}	$\frac{1}{2}(u_{xt}^2 + u_{xx}^2 + u_{xy}^2)$
D_t^3	u_{ttt}	$\frac{1}{2}(u_{tt}^2 + u_{xt}^2 + u_{yt}^2)$
$D_x\mathscr{R}_{xy}D_x$	$-yu_{xxx} + xu_{xyy} + u_{xy}$	$u_{xt}(xu_{xy} - yu_{xx})$
$D_x\mathscr{R}_{xy}D_y - \frac{1}{2}D_x^2 - \frac{1}{2}D_y^2$	$-yu_{xxy} + xu_{xyy} - \frac{1}{2}u_{xx} + \frac{1}{2}u_{yy}$	$u_{xx}(yu_{yt} + \frac{1}{2}u_t) - u_{yy}(xu_{xt} +$
$D_x\mathscr{R}_{xt}D_x$	$xu_{xxt} + tu_{xxx} + u_{xt}$	$x\hat{T} + tu_{xx}u_{xt}$
$D_x\mathscr{M}D_x$	$xu_{xxx} + yu_{xxy} + tu_{xxt} + \frac{3}{2}u_{xx}$	$T^* + t\hat{T}$
$\mathscr{R}_{xt}D_t\mathscr{R}_{xt}$	$x^2u_{xtt} + 2xtu_{xxt} + t^2u_{xxx} + xu_{tt} +$ $+ 2tu_{xt} + xu_{xx}$	$(x^2 + t^2)\hat{T} + \frac{1}{2}u_y^2 +$ $+ t(2xu_{xt}u_{tt} - u_yu_{yt})$
$D_x\mathscr{I}_tD_x$	$(x^2 + y^2 + t^2)u_{xxt} + 2xtu_{xxx} +$ $2ytu_{xxy} + 2xu_{xt} + 3tu_{xx}$	$(x^2 + y^2 + t^2)\hat{T} + u_x^2 + 2t$

where

$$\hat{T} = \tfrac{1}{2}(u_{xt}^2 + u_{xx}^2 + u_{xy}^2), \qquad T^* = xu_{xx}u_{xt} + yu_{xx}u_{yt} + \tfrac{1}{2}u$$

Action of Symmetries on Conservation Laws

A second method of generating conservation laws is to apply known symmetry group generators to known conservation laws. Unfortunately, the method is not guaranteed to produce nontrivial laws, but we can determine precisely when it does.

Proposition 5.64. *Let Δ be totally nondegenerate and* $\operatorname{Div} P = 0$ *a conservation law. If $\mathbf{v}_R$ is an evolutionary symmetry of Δ, then the induced p-tuple $\tilde{P} = \operatorname{pr} \mathbf{v}_R(P)$, with entries $\tilde{P}_i = \operatorname{pr} \mathbf{v}_R(P_i)$, is also a conservation law:* $\operatorname{Div} \tilde{P} = 0$. *Moreover, if $\Delta = E(L)$ is a system of Euler–Lagrange equations, P has characteristic Q corresponding to the variational symmetry $\mathbf{v}_Q$, and $\mathbf{v}_R$ is a variational symmetry, then $\tilde{P}$ has characteristic $\tilde{Q}$ corresponding to the Lie bracket $\mathbf{v}_{\tilde{Q}} = [\mathbf{v}_R, \mathbf{v}_Q]$ of the two symmetries.*

PROOF. We assume that the conservation law is in characteristic form (5.81). (Note that if P_0 is a trivial conservation law, so is $\operatorname{pr} \mathbf{v}_R(P_0)$, so this first step is justified.) Applying $\operatorname{pr} \mathbf{v}_R$, we find

$$\operatorname{Div}[\operatorname{pr} \mathbf{v}_R(P)] = \operatorname{pr} \mathbf{v}_R(Q) \cdot \Delta + Q \cdot \operatorname{pr} \mathbf{v}_R(\Delta) \tag{5.96}$$

using (5.19). Since $\operatorname{pr} \mathbf{v}_R(\Delta) = 0$ for solutions of Δ, the right-hand side of (5.96) vanishes on solutions, proving the first part of the theorem. If $\Delta = \mathrm{E}(L)$ and $\mathbf{v}_R$ is variational, then we can use Proposition 5.55 to rewrite the second term in (5.96) and integrate by parts,

$$Q \cdot \operatorname{pr} \mathbf{v}_R(\Delta) = -Q \cdot \mathrm{D}_R^*(\Delta) = -\mathrm{D}_R(Q) \cdot \Delta - \operatorname{Div} B = -\operatorname{pr} \mathbf{v}_Q(R) \cdot \Delta - \operatorname{Div} B$$

for some p-tuple B which depends linearly on Δ and its total derivatives, and hence forms a trivial conservation law of the first kind. Thus, by (5.22),

$$\operatorname{Div}[\operatorname{pr} \mathbf{v}_R(P) + B] = \{\operatorname{pr} \mathbf{v}_R(Q) - \operatorname{pr} \mathbf{v}_Q(R)\} \cdot \Delta = \tilde{Q} \cdot \Delta$$

is the characteristic form of our conservation law and the proof is complete. □

This result is most useful in the case of self-adjoint linear systems. Indeed, if $P \in \mathscr{A}^p$ determines a quadratic conservation law corresponding to the linear characteristic $Q = \mathscr{D}[u]$, and $\mathbf{v}_R$ is a linear symmetry, so $R = \mathscr{E}[u]$ for some differential operator $\mathscr{E}$ satisfying $\Delta \cdot \mathscr{E} = \tilde{\mathscr{E}} \cdot \Delta$, then $\operatorname{pr} \mathbf{v}_R(P)$ yields a conservation law with characteristic $\tilde{Q} = (\mathscr{D} \cdot \mathscr{E} + \tilde{\mathscr{E}}^* \cdot \mathscr{D})[u]$. In particular, if $\mathbf{v}_R$ is a variational symmetry, then $\tilde{Q}$ has characteristic corresponding to the commutator operator $[\mathscr{D}, \mathscr{E}] = \mathscr{D} \cdot \mathscr{E} - \mathscr{E} \cdot \mathscr{D}$.

Example 5.65. For the two-dimensional wave equation, the conserved densities in the table of Example 5.63 are most easily computed using this method. For example, the conservation law with characteristic u_{xxt} can be constructed either by applying the prolongation of the symmetry $\mathbf{v} = \frac{1}{2} u_{xx} \partial_u$ to the energy conservation law with characteristic u_t, or the prolongation of

$\mathbf{w} = \frac{1}{2}u_{xt}\partial_u$ to the momentum conservation law with characteristic u_x. (In the former case, $\Delta D_x^2 = D_x^2 \Delta$, so the new characteristic is indeed $\operatorname{pr}\mathbf{v}(u_t) + \frac{1}{2}(D_x^2)^* u_t = u_{xxt}$.) In the first case, the new density is

$$\operatorname{pr}\mathbf{v}[\tfrac{1}{2}u_t^2 + \tfrac{1}{2}u_x^2 + \tfrac{1}{2}u_y^2] = \tfrac{1}{2}(u_t u_{xxt} + u_x u_{xxx} + u_y u_{xxy}) \equiv T,$$

while in the second it is

$$\operatorname{pr}\mathbf{w}[u_x u_t] = \tfrac{1}{2}(u_t u_{xxt} + u_x u_{xtt}) \equiv \tilde{T}.$$

Since both these densities have the same characteristics, they should be equivalent:

$$T = \tilde{T} + D_x R + D_y S$$

on solutions of the wave equation. In other words, we have the freedom to (a) substitute for derivatives according to the equation and its prolongation, and (b) integrate by parts with respect to x and y (but *not* t); thus $u_t u_{xxt}$ is equivalent to $-u_{xt}^2$, but *not* $u_{xx}u_{tt}$. The reader can verify that T and $\tilde{T}$ are both equivalent to the second order density listed in the above-mentioned table.

As a second example, the conservation law corresponding to the operator $D_x \mathscr{R}_{xt} D_x$ is found by applying the symmetry $\hat{\mathbf{v}} = \frac{1}{2}\mathscr{R}_{xt} D_x[u]\partial_u = \frac{1}{2}(xu_{xt} + tu_{xx})\partial_u$ to the conservation law with characteristic u_x. We find

$$\begin{aligned}\operatorname{pr}\hat{\mathbf{v}}[u_x u_t] &= \tfrac{1}{2}\{D_x(xu_{xt} + tu_{xx})u_t + u_x D_t(xu_{xt} + tu_{xx})\}\\ &= \tfrac{1}{2}(xu_{xxt} + u_{xt} + tu_{xxx})u_t + \tfrac{1}{2}u_x(xu_{xtt} + tu_{xxt} + u_{xx}).\end{aligned}$$

Both lower order terms $u_t u_{xt}$ and $u_x u_{xx}$ are x-derivatives, hence this density is equivalent to the one in the table by a similar integration by parts.

Abnormal Systems and Noether's Second Theorem

The connection between variational symmetries and conservation laws for systems which fail to be totally nondegenerate is less transparent. Although the basic integration by parts formula (4.39) still yields a variational symmetry for each conservation law and vice versa, there is now no guarantee that nontrivial symmetries will give rise to nontrivial conservation laws or the reverse. In the case of analytic systems, we saw that there are two basic types of abnormality possible. Over-determined systems are less well understood in this regard, and the precise relationship between their symmetries and conservation laws has yet to be fully sorted out. Under-determined systems, however, fall under the ambit of Noether's second theorem which is concerned with systems posessing infinite-dimensional groups of variational symmetries. The resulting dependencies among the Euler–Lagrange equations can be re-interpreted as trivial conservation laws determined by nontrivial variational symmetry groups, so the nice one-to-one correspondence of Theorem 5.58 breaks down in the under-determined case.

Theorem 5.66. *The variational problem $\mathscr{L}[u] = \int L\,dx$ admits an infinite-dimensional group of variational symmetries whose characteristics $Q[u; h]$ depend on an arbitrary function $h(x)$ (and its derivatives) if and only if there exist differential operators $\mathscr{D}_1, \ldots, \mathscr{D}_q$, not all zero, such that*

$$\mathscr{D}_1 \mathsf{E}_1(L) + \cdots + \mathscr{D}_q \mathsf{E}_q(L) \equiv 0 \tag{5.97}$$

for all x, u.

PROOF. Assume first that the Euler–Lagrange equations for $\mathscr{L}$ are under-determined, so there is a relation of the form (5.97) among them. Let $h(x)$ be arbitrary. Then an easy integration by parts shows that

$$\begin{aligned} 0 &= h(x)\{\mathscr{D}_1 \mathsf{E}_1(L) + \cdots + \mathscr{D}_q \mathsf{E}_q(L)\} \\ &= \mathscr{D}_1^*[h] \mathsf{E}_1(L) + \cdots + \mathscr{D}_q^*[h] \mathsf{E}_q(L) - \operatorname{Div} P \end{aligned} \tag{5.98}$$

for some p-tuple $P \in \mathscr{A}^p$ depending linearly on $\mathsf{E}(L)$ and its derivatives. If we set $Q_\nu = \mathscr{D}_\nu^*[h]$, $\nu = 1, \ldots, q$, then the above identity is in the form of a conservation law in characteristic form, where Q is the characteristic and $P = P[u; h] \in \mathscr{A}^p$ the conservation law, which is actually trivial (of the first kind). Now we can clearly use (5.98) to prove that for any function $h(x)$, $\mathbf{v}_{Q[u;h]}$ determines a variational symmetry of the functional $\mathscr{L}[u]$.

The proof of the converse is straightforward if $Q_\nu[u; h] = \tilde{\mathscr{D}}_\nu[h]$ are all linear in h and its derivatives, $\tilde{\mathscr{D}}_\nu$ being differential operators whose coefficients can depend on u. Starting with the condition (5.85) that $\mathbf{v}_Q$ be a variational symmetry, we integrate by parts to obtain the corresponding conservation law

$$\operatorname{Div} P = Q \cdot \mathsf{E}(L) = \tilde{\mathscr{D}}_1[h] \mathsf{E}_1(L) + \cdots + \tilde{\mathscr{D}}_q[h] \mathsf{E}_q(L).$$

Further integration by parts, effectively reversing the arguments in (5.98), leads to an identity of the form

$$\operatorname{Div} \tilde{P} = h(x)\{\tilde{\mathscr{D}}_1^* \mathsf{E}_1(L) + \cdots + \tilde{\mathscr{D}}_q^* \mathsf{E}_q(L)\}, \tag{5.99}$$

which holds for an arbitrary function $h(x)$. The proof is completed using the following "formal" version of the du Bois–Reymond lemma of the variational calculus.

Lemma 5.67. *Let $R(x, u^{(n)})$ be a differential function and suppose for every smooth function $h(x)$ there exists $P[u] = P_h[u] \in \mathscr{A}^p$ such that*

$$h(x) R(x, u^{(n)}) = \operatorname{Div} P(x, u^{(m)}).$$

Then $R(x, u^{(n)}) = r(x)$ is a function of x alone.

PROOF. Assume R depends on the n-th and lower order derivatives of u and that $\partial R(x_0, u_0^{(n)})/\partial u_J^\alpha \neq 0$ for some $\#J = n \geq 0$, $(x_0, u_0^{(n)}) \in M^{(n)}$. Choose $h(x)$ such that $\partial_J h(x_0) \neq 0$, but all other derivatives of h of order $\leq n$ vanish at x_0.

A straight-forward calculation shows that

$$\mathsf{E}_\alpha(h\cdot R)(x_0, u_0^{(n)}) = (-1)^n \partial_J h(x_0)\cdot \partial R(x_0, u_0^{(n)})/\partial u_J^\alpha \neq 0.$$

Theorem 4.7 implies that $h\cdot R$ is not a total divergence, contradicting our assumption. An easy induction now proves that R can only depend on x. □

An easy corollary of this result will be quite important for subsequent developments.

Corollary 5.68. *Let $P \in \mathscr{A}^r$ be an r-tuple of differential functions. Then $\int_\Omega P\cdot Q\, dx = 0$ for all $Q \in \mathscr{A}^r$, all $\Omega \subset X$, if and only if $P \equiv 0$ for all x, u.*

PROOF. Using the lemma component-wise, we conclude that $P = p(x)$ depends on x alone. Further, given $1 \leqslant v \leqslant r$, choose $Q_\mu[u] = \delta_\mu^v u^\alpha$ for any $1 \leqslant \alpha \leqslant q$. Then $\mathsf{E}_\alpha(p\cdot Q) = p_v(x) \equiv 0$ by Theorem 4.7, hence $P \equiv 0$ for all x, u. □

Returning to (5.99), we see that

$$\tilde{\mathscr{D}}_1^* \mathsf{E}_1(L) + \cdots + \tilde{\mathscr{D}}_q^* \mathsf{E}_q(L) = r(x)$$

is a function of x alone. If $r \equiv 0$ we're done; otherwise we divide by $r(x)$ and differentiate once more (with respect to any variable x^i) to produce an identity of the required form (5.97). □

More generally, if $h(x)$ appears nonlinearly in $Q[u; h]$, we can nevertheless reduce to the previous case using the following:

Lemma 5.69. *Suppose $Q[u; h]$ is the characteristic of a variational symmetry of $\mathscr{L}$ depending on an arbitrary function $h(x)$. Let $\mathscr{D}_Q = \mathscr{D}_{Q[u;h]}$ denote the Fréchet derivative of Q with respect to h, with entries*

$$\mathscr{D}_Q^v = \sum (\partial Q_v/\partial h_J)\cdot D_J, \qquad v = 1, \ldots, q, \quad (h_J = \partial_J h).$$

Then $Q' = \mathscr{D}_Q[k]$ is the characteristic of a variational symmetry depending linearly on the arbitrary function $k(x)$.

PROOF. By assumption, for any function $h(x)$, there exists a p-tuple $B_h[u] \in \mathscr{A}^p$ such that

$$\operatorname{pr} \mathbf{v}_{Q[u;h]}(L) = \operatorname{Div} B_h.$$

If we replace h by $h + \varepsilon k$ in this identity and differentiate with respect to ε at $\varepsilon = 0$, we obtain

$$\operatorname{pr} \mathbf{v}_{Q'}(L) = \operatorname{Div} B', \qquad B' = \left.\frac{d}{d\varepsilon}\right|_{\varepsilon=0} B_{h+\varepsilon k},$$

which proves the lemma. □

In Theorem 5.66, each of the nontrivial symmetries $Q[u; h]$ (linear in h) gives rise to a trivial conservation law with Q as the characteristic. This remark has a converse also, that says that if a system of Euler–Lagrange equations has a trivial conservation law, which corresponds to a nontrivial variational symmetry, then it is necessarily under-determined, and hence admits an entire infinite-dimensional family of such symmetries depending on an arbitrary function. (See Exercise 5.34.) In applications, these are the "gauge symmetries" of the theory. (In relativity, cf. Goldberg, [1], these "trivial" conservation laws are among the most important identities of the subject. Here, perhaps, our choice of terminology is slightly misleading.)

Example 5.70. *Parametric Variational Problems.* Consider a first order variational problem of the form

$$\mathscr{L}[u, v] = \int L(x, u, v, u_x, v_x)\, dx,$$

with $x \in \mathbb{R}$. Consider the infinite-dimensional symmetry group consisting of arbitrary coordinate changes $x \mapsto \psi(x)$ in the independent variable. Its infinitesimal generators are of the form $\mathbf{v}_h = h(x)\partial_x$ for h an arbitrary function of x. The infinitesimal criterion (4.15) says that this is a variational symmetry group provided

$$h(x)L_x + h'(x)\{-u_x L_{u_x} - v_x L_{v_x} + L\} = 0,$$

subscripts denoting derivatives of L. (Generalizing to divergence symmetries doesn't add anything here.) As both h and h' are arbitrary, L must be independent of x, and of the form $L = u_x \tilde{L}(u, v, v_x/u_x)$. We conclude that we are necessarily dealing with a parametric variational problem

$$\mathscr{L}[u] = \int \tilde{L}\left(u, v, \frac{v_x}{u_x}\right) u_x\, dx = \int \tilde{L}\left(u, v, \frac{dv}{du}\right) du,$$

in which we can treat v, say, as a function of u only.

Noether's second theorem says that there is a dependency between the two original Euler–Lagrange equations

$$\mathsf{E}_u(L) = u_x \tilde{L}_u - D_x\left(\tilde{L} - \frac{v_x}{u_x}\tilde{L}_{v_u}\right) = 0, \qquad \mathsf{E}_v(L) = u_x \tilde{L}_v - D_x \tilde{L}_{v_u} = 0.$$

The evolutionary form of $\mathbf{v}_h$ is $-h(x)(u_x\partial_u + v_x\partial_v)$, so according to (5.97), (5.98) we have the identity

$$u_x \mathsf{E}_u(L) + v_x \mathsf{E}_v(L) = 0.$$

This argument clearly extends to both higher order and higher dimensional problems.

Example 5.71. Consider the variational problem

$$\mathscr{L}[u] = \tfrac{1}{2} \iint (u_x + v_y)^2\, dx\, dy,$$

whose Euler–Lagrange equations,

$$-\mathsf{E}_u(L) = u_{xx} + v_{xy} = 0, \qquad -\mathsf{E}_v(L) = u_{xy} + v_{yy} = 0,$$

were seen to be under-determined in Section 2.6, with $D_y\mathsf{E}_u(L) - D_x\mathsf{E}_v(L) \equiv 0$. The proof of Theorem 5.66 provides the corresponding infinite-dimensional symmetry group, generated by $\mathbf{v}_h = -h_y\partial_u + h_x\partial_v$ for arbitrary $h(x, y)$, with group transformations

$$\exp(\varepsilon\mathbf{v}_h)(u, v) = (u - \varepsilon h_y, v + \varepsilon h_x)$$

obviously leaving $\mathscr{L}$ invariant. Although these groups are certainly nontrivial, the corresponding conservation laws are trivial. For instance, if $h(x, y) = -y$, so $\mathbf{v}_h = \partial_u$ we get the trivial law with components $(u_x + v_y, 0)$, i.e.

$$D_x(u_x + v_y) = u_{xx} + v_{xy}.$$

Admittedly this doesn't look trivial, but if we add in the obviously trivial law (of the first kind) $(y(u_{xy} + v_{yy}), -y(u_{xx} + v_{xy}))$ we obtain an equivalent trivial conservation law of the second kind, since

$$(u_x + v_y) + y(u_{xy} + v_{yy}) = D_y(y(u_x + v_y)),$$
$$-y(u_{xx} + v_{xy}) = -D_x(y(u_x + v_y)).$$

The lesson is that for abnormal systems one must exercise even more care in distinguishing trivial laws from nontrivial laws; here even the characteristics no longer are a foolproof indicator of triviality.

Formal Symmetries and Conservation Laws

There is an intimate connection between formal symmetries and conservation laws of evolution equations that does not appear to bear any obvious relationship to Noether's Theorem. (Indeed, a single evolution equation can never be the Euler–Lagrange equation for a variational problem, a fact that follows immediately from the Helmholtz conditions of Theorem 5.92.) The main observation is that the coefficient of D_x^{-1} in a formal symmetry of the appropriate rank and order will provide a conserved density of the evolution equation.

Definition 5.72. The *residue* of a pseudo-differential operator $\mathscr{D}$ is the coefficient of D_x^{-1}:

$$\operatorname{Res} \sum_{i=-\infty}^{n} P_i D_x^i = P_{-1}. \tag{5.100}$$

Proposition 5.73. *For any pair of pseudo-differential operators, the residue of their commutator is a total x-derivative*: $\operatorname{Res}[\mathscr{D}, \mathscr{E}] = D_x P$ *for some differential function P.*

PROOF. By linearity, it suffices to prove the theorem when $\mathscr{D} = QD_x^n$ and $\mathscr{E} = RD_x^m$ are monomials, with $n \geqslant m$. It is easy to see that, in this case, the residue of their commutator vanishes unless $n \geqslant -m-1 \geqslant 0$, in which case

$$\begin{aligned}\operatorname{Res}[QD_x^n, RD_x^m] &= \binom{n}{-m-1}(PD_x^{n-m+1}Q + (-1)^{n-m+1}QD_x^{n-m+1}P)\\ &= \binom{n}{-m-1}D_x(PD_x^{n-m}Q - (D_xP)D_x^{n-m-1}Q + \cdots \pm (D_x^{n-m}P)Q). \qquad \square\end{aligned}$$

We first show how, given a formal symmetry, one constructs a sequence of conservation laws. Recall that we can, without loss of generality, assume that the formal symmetry is given as a first order pseudo-differential operator, cf. Theorem 5.46.

Theorem 5.74. *If $\mathscr{D}$ is a first order formal symmetry of the n-th order evolution equation $u_t = K$ of rank $k \geqslant n+2$, then the residues of the first $k-n-2$ powers of $\mathscr{D}$,*

$$T_j = \operatorname{Res} \mathscr{D}^j, \quad j = 1, \ldots, k-n-2, \tag{5.101}$$

are conserved densities.

PROOF. Note first that according to Lemma 5.45, each power $\mathscr{R}^j$ is a formal symmetry of rank k also and, therefore, satisfies a formal symmetry condition

$$(\mathscr{D}^j)_t + [\mathscr{D}^j, \mathrm{D}_K] = \mathscr{E}_j, \tag{5.102}$$

where $\mathscr{E}_j$ is a pseudo-differential operator of order at most $n+j-k$. Provided $n+j-k < -1$, the coefficient of D_x^{-1} in $\mathscr{E}_j$ is zero; hence, according to Proposition 5.73, the residue of (5.102) is of the form

$$D_t T_j + D_x X_j = 0,$$

where $D_x X_j$ denotes the residue of the commutator in (5.102). This produces the required conservation law. $\square$

An important point is that there is no guarantee that these conservation laws are nontrivial, and, in the case of Burgers' equation, they are *all* trivial, in accordance with the analysis in Example 5.50.

Example 5.75. Consider the recursion operator $\mathscr{R} = D_x^2 + \frac{2}{3}u + \frac{1}{3}u_x D_x^{-1}$ for the Korteweg–de Vries equation (5.45). As a consequence of Theorem 5.74, the coefficient $\frac{1}{3}u_x$ of D_x^{-1} is a conserved density, but is trivial, being an x-derivative. To get nontrivial conservation laws, we must work with the square root of the recursion operator,

$$\begin{aligned}\sqrt{\mathscr{R}} = D_x &+ \tfrac{1}{3}uD_x^{-1} - \tfrac{1}{18}u^2 D_x^{-3} + \tfrac{1}{9}uu_x D_x^{-4}\\ &+ (-\tfrac{1}{9}uu_{xx} - \tfrac{1}{18}u_x^2 + \tfrac{1}{54}u^3)D_x^{-5} + \cdots,\end{aligned}$$

which, according to Lemma 5.45, is also a formal symmetry of order ∞. The residue of $\sqrt{\mathcal{R}}$ provides the first conserved density, u, of the Korteweg–de Vries equation. Moreover, all the powers $\mathcal{R}^{m/2}$ of this operator are also formal symmetries of order ∞, and hence their residues provide an infinite sequence of conserved densities. For example,

$$\mathcal{R}^{3/2} = D_x^3 + uD_x + u_x + (\tfrac{1}{3}u_{xx} + \tfrac{1}{6}u^2)D_x^{-1} - (\tfrac{1}{18}u_x^2 + \tfrac{1}{54}u^3)D_x^{-2} + \cdots$$

gives the conserved density $u_{xx} + \frac{1}{2}u^2$, which is equivalent to the nontrivial density $\frac{1}{2}u^2$;

$$\mathcal{R}^2 = D_x^4 + \tfrac{4}{3}uD_x^2 + 2u_xD_x + (\tfrac{4}{3}u_{xx} + \tfrac{4}{9}u^2) + (\tfrac{1}{3}u_{xxx} + \tfrac{4}{9}uu_x)D_x^{-1} + \tfrac{1}{9}u_x^2D_x^{-2} + \cdots$$

gives the trivial conserved density $u_{xxx} + \frac{4}{3}uu_x$;

$$\mathcal{R}^{5/2} = D_x^5 + \tfrac{5}{3}uD_x^3 + \tfrac{10}{3}u_xD_x^2 + (\tfrac{10}{3}u_{xx} + \tfrac{5}{6}u^2)D_x + (\tfrac{5}{3}u_{xxx} + \tfrac{5}{3}uu_x) + (\tfrac{1}{3}u_{xxxx} + \tfrac{5}{9}uu_{xx} + \tfrac{5}{18}u_x^2 + \tfrac{5}{54}u^3)D_x^{-1} + \cdots$$

gives the conserved density $u_{xxxx} + \frac{5}{3}uu_{xx} + \frac{5}{6}u_x^2 + \frac{5}{18}u^3$, which is equivalent to a multiple of the next nontrivial conserved density $u_x^2 - \frac{1}{3}u^3$ for the Korteweg–de Vries equation. It can be shown that every integral power of $\mathcal{R}$ has a trivial density as its residue, whereas the half integral powers provide the well-known infinite sequence of (inequivalent) conservation laws of the Korteweg–de Vries equation.

With a little more work, one can produce two further conservation laws from a first order formal symmetry. If

$$\mathcal{D} = Q_1D_x + Q_0 + Q_{-1}D_x^{-1} + \cdots$$

is a formal symmetry of rank at least $n + 1$, then

$$T_{-1} = \frac{1}{Q_1}, \qquad T_0 = \frac{Q_1}{Q_0}, \tag{5.103}$$

are conserved densities; if the formal symmetry has rank n, just T_{-1} is conserved. Therefore, any formal symmetry of rank $k \geqslant n$ of an n-th order evolution equation ($n \geqslant 2$) provides $k - n$ *canonical conserved densities* T_j, $j = -1, \ldots, k - n + 2$. A converse to this result can also be proven: If an evolution equation has $k - n$ canonical conserved densities, then it has a formal symmetry of rank k; see Mikhailov, Shabat and Yamilov, [1]. Indeed, in their approach, the analysis is based primarily on the conditions imposed by the existence of canonical conservation laws rather than the more direct formal symmetry condition. This strategy results in some simplifications in the required calculations.

Just as there is a formal analogue of a symmetry obtained by taking the Fréchet derivative of the basic infinitesimal symmetry condition, there is also the concept of a formal conservation law. Let $T[u]$ be a conserved density which depends only on x, u and x-derivatives of u (so we are excluding explicitly time-dependent conservation laws), and set $R = \mathsf{E}(T)$. We say that the conservation law determined by T has *order* m if R depends on at most m-th order derivatives; note that this definition of order does not change if we replace T by any equivalent conserved density $T + D_x Q$. Applying the Euler operator to the conservation law

$$0 = D_t T + D_x X = \mathrm{pr}\,\mathbf{v}_Q(T) + D_x X,$$

and using (5.86) and Theorem 4.7, we find the equivalent condition

$$\mathrm{pr}\,\mathbf{v}_Q(R) + \mathsf{D}_K^* R = 0. \tag{5.104}$$

We now "linearize" (5.104) by taking its Fréchet derivative, leading via (5.60) to a differential operator identity of the form

$$(\mathsf{D}_R)_t + \mathsf{D}_R \cdot \mathsf{D}_K + \mathsf{D}_K^* \cdot \mathsf{D}_R + \mathscr{E} = 0, \tag{5.105}$$

where $\mathscr{E}$ is a differential operator of order at most n, which depends on the second derivatives of K with respect to the derivatives of u, but plays no role in the subsequent discussion.

Now suppose that the order m of the conservation law is much larger than the order n of the evolution equation, $m \gg n$. Then $\mathscr{E}$ is of much lower order that the other three differential operators appearing in (5.105). Thus, in analogy with the definition of a formal symmetry, we make the following definition of a formal conservation law.

Definition 5.76. A *formal conservation law of rank* k of an n-th order evolution equation $u_t = K$ is a pseudo-differential operator $\mathscr{C}$ of order m which satisfies

$$\mathrm{order}\{\mathscr{C}_t + \mathscr{C} \cdot \mathsf{D}_K + \mathsf{D}_K^* \cdot \mathscr{C}\} \leqslant m + n - k. \tag{5.106}$$

Proposition 5.77. *If T is a conserved density with $R = \mathsf{E}(T)$ of order m, then D_R is a formal conservation law of rank m.*

The most important fact about formal conservation laws is that two of them can be combined to provide a formal symmetry!

Theorem 5.78. *If $\mathscr{C}_1$ and $\mathscr{C}_2$ are formal conservation laws of ranks k_1, k_2, respectively, then $\mathscr{D} = \mathscr{C}_2 \cdot \mathscr{C}_1^{-1}$ is a formal symmetry of rank $k = \min\{k_1, k_2\}$.*

Proof. Let m_i denote the order of $\mathscr{C}_i$, $i = 1, 2$. We have

$$(\mathscr{C}_i)_t + \mathscr{C}_i \cdot \mathsf{D}_K + \mathsf{D}_K^* \cdot \mathscr{C}_i = \mathscr{B}_i, \qquad i = 1, 2,$$

where $\mathscr{B}_i$ is a pseudo-differential operator of order at most $n + m_i - k_i$.

Evaluating the (1, 1)-Lie derivative $\mathbf{v}_K[\![\mathscr{D}]\!]$, we find, on solutions,

$$
\begin{aligned}
(\mathscr{C}_2\cdot\mathscr{C}_1^{-1})_t + [\mathrm{D}_K, \mathscr{C}_2\cdot\mathscr{C}_1^{-1}] &= (\mathscr{C}_2)_t\cdot\mathscr{C}_1^{-1} - \mathscr{C}_2\cdot\mathscr{C}_1^{-1}\cdot(\mathscr{C}_1)_t\cdot\mathscr{C}_1^{-1} \\
&\quad + \mathrm{D}_K\cdot\mathscr{C}_2\cdot\mathscr{C}_1^{-1} - \mathscr{C}_2\cdot\mathscr{C}_1^{-1}\cdot\mathrm{D}_K \\
&= \{(\mathscr{C}_2)_t + \mathscr{C}_2\cdot\mathrm{D}_K + \mathrm{D}_K^*\cdot\mathscr{C}_2\}\cdot\mathscr{C}_1^{-1} \\
&\quad - \mathscr{C}_2\cdot\mathscr{C}_1^{-1}\cdot\{(\mathscr{C}_1)_t + \mathscr{C}_1\cdot\mathrm{D}_K + \mathrm{D}_K^*\cdot\mathscr{C}_1\}\cdot\mathscr{C}_1^{-1} \\
&= \mathscr{B}_2\cdot\mathscr{C}_1^{-1} - \mathscr{C}_2\cdot\mathscr{C}_1^{-1}\cdot\mathscr{B}_1\cdot\mathscr{C}_1^{-1}.
\end{aligned}
$$

Now, $\mathscr{D} = \mathscr{C}_2\cdot\mathscr{C}_1^{-1}$ has order $m_2 - m_1$, whereas the two operators in the final equality have respective orders at most $n + m_2 - m_1 - k_2$, $n + m_2 - m_1 - k_1$. Therefore, order $\mathbf{v}_K[\![\mathscr{D}]\!] \leqslant n + m_2 - m_1 - k$, which proves that $\mathscr{D}$ is a formal symmetry of rank k (at least).

Corollary 5.79. *An evolution equation having two (ordinary) conservation laws of orders k_1, k_2 has a formal symmetry of rank $k \geqslant \min\{k_1, k_2\}$.*

Consequently, assuming our earlier conjecture on the integrability of equations having high order formal symmetries, we deduce that any evolution equation with two conservation laws of sufficiently high order is necessarily integrable. For example, any second order equation possessing two fifth or higher order conservation laws is integrable and equivalent, via a contact transformation, to one of the equations listed in Theorem 5.48. (Actually, since the first two equations in the list do not have any higher order conservation laws, the equation must be equivalent to either the third or fourth equation in that list.)

5.4. The Variational Complex

As mentioned in the introduction to this chapter, the variational complex draws its inspiration from three principal results that have formed the basis of much of our work on symmetries, conservation laws, differential operators and so on. First is the characterization of the kernel of the Euler operator as the space of total divergences given in Theorem 4.7; second is the characterization of all null divergences as total curls given in Theorem 4.24; third is the characterization of Euler–Lagrange equations by the self-adjointness of the their Fréchet derivatives—see the proof of Lemma 5.54. The two latter results especially are not easy to prove, as the reader may have discovered, but, when restated in a more natural differential form language, can be recovered through the construction of suitable homotopy operators similar to those used in the proof of the Poincaré lemma in Section 1.5. (The reader is well advised to become thoroughly familiar with the concepts of ordinary differential forms on manifolds, as developed in Section 1.5, before attempting to explore the more complicated types of forms to be treated here.) Although in this book we will only require the above three special instances of the full variational complex, we have chosen to include it in its entirety because (a) the proofs are not any more difficult in the general case, and (b) a familiarity with this complex will provide the reader with an excellent preparation for further reading and research into recent work on the geometric theory of the calculus of variations on manifolds.

The variational complex naturally splits in two halves. In the first half, the relevant differential forms are expressions involving the differentials dx^i of the independent variables, but whose coefficients are now differential functions. Replacing the ordinary differential d is now a "total" differential D which uses total instead of partial derivatives. Although this is the easier of the two halves to define, the proof of exactness is by far the more complicated and requires the machinery of "higher Euler operators" developed at the end of this section. The result on null divergences appears at the next-to-last stage of this half of the complex. In second half of the variational complex, the role of functions is taken by the functionals of the variational calculus, with "functional forms" being defined analogously. The differential now is similar to the variational derivative of a functional, and is hence called the variational differential. Although the objects in this half are less familiar, the proof of exactness relies on a relatively simple extension of the de Rham homotopy operator. Included here is the solution to Helmholtz' inverse problem of the calculus of variations. The Euler operator itself provides the link between the two halves, the characterization of null Lagrangians providing the remaining step in the full exactness of the variational complex.

The D-Complex

The first half of the variational complex is obtained by adapting the de Rham complex to the space of differential functions defined over $M \subset X \times U$. A *total differential r-form* will take the form

$$\omega = \sum_J P_J[u]\, dx^J$$

in which the coefficients $P_J \in \mathscr{A}$ are now differential functions, and $dx^J = dx^{j_1} \wedge \cdots \wedge dx^{j_r}$, $1 \leqslant j_1 < \cdots < j_r \leqslant p$ form the standard basis of $\bigwedge_r T^*X$. If we replace u by some function $u = f(x)$, then we recover an ordinary differential r-form on the space X. We differentiate ω treating the u's as functions of the x's, leading to the *total differential*

$$\mathrm{D}\omega = \sum_{i=1}^{p} \sum_J D_i P_J\, dx^i \wedge dx^J. \tag{5.107}$$

For example, if $p = 2$, then

$$\omega = yu_x\, dx + uu_{xy}\, dy$$

is a total one-form, with total differential

$$\mathrm{D}\omega = [D_x(uu_{xy}) - D_y(yu_x)]\,dx \wedge dy = [uu_{xxy} + u_x u_{xy} - u_x - yu_{xy}]\,dx \wedge dy.$$

Since when we specialize $u = f(x)$, the total differential agrees with the exterior derivative, it is easy to see that D defines a complex (called the "big D-complex") on the spaces of total differential forms, meaning that $\mathrm{D}(\mathrm{D}\omega) = 0$ for any form ω. Over suitable subdomains $M \subset X \times U$ this complex is exact. The precise requirement on M is that it be *totally star-shaped*, meaning that it be (a) *vertically star-shaped*, so each vertical slice $M_x = \{u: (x, u) \in M\}$ is a star-shaped subdomain of U, and (b) the base horizontal slice $\Omega = \{x: (x, 0) \in M\}$ is a star-shaped subdomain of X.

Theorem 5.80. *Let M be totally star-shaped. Then the* D*-complex*

$$0 \to \mathbb{R} \to \textstyle\bigwedge_0 \xrightarrow{\mathrm{D}} \bigwedge_1 \xrightarrow{\mathrm{D}} \cdots \xrightarrow{\mathrm{D}} \bigwedge_{p-1} \xrightarrow{\mathrm{D}} \bigwedge_p$$

is exact, where $\bigwedge_r$ denotes the space of total r-forms. In other words, if $\omega \in \bigwedge_r$ for $0 < r < p$, then ω is D*-closed: $\mathrm{D}\omega = 0$, if and only if ω is* D*-exact: $\omega = \mathrm{D}\eta$ for some total $(r-1)$-form η, while if $\omega \in \bigwedge_0$, so ω is just a differential function, then $\mathrm{D}\omega = 0$ if and only if ω is constant.*

Example 5.81. Exactness of the D-complex at the $\bigwedge_{p-1}$-stage is easily seen to be equivalent to the characterization of null divergences given in Theorem 4.24. Indeed, using the notation of Example 1.62, any $(p-1)$-form $\omega = \sum (-1)^{j-1} P_j\, dx^{\hat{j}}$ can be identified with its coefficients $P = (P_1, \dots, P_p) \in \mathscr{A}^p$. We have $\mathrm{D}\omega = (\mathrm{Div}\, P)\, dx^1 \wedge \cdots \wedge dx^p$, so ω is D-closed if and only if P is a null divergence. On the other hand, a $(p-2)$-form takes the form $\eta = \sum (-1)^{j+k-1} Q_{jk}\, dx^{\widehat{jk}}$ where $Q_{jk} = -Q_{kj}$, and $\mathrm{D}\eta = \omega$ if and only if $P_j = \sum D_k Q_{jk}$. (Explicit formulae for the Q's in terms of the P's will be found in the course of the proof of Theorem 5.80.)

If we specialize $u = f(x)$ everywhere, then the D-complex reduces to the ordinary de Rham complex, which by the Poincaré lemma (Theorem 1.61) is exact. However, this does *not* prove the exactness of the D-complex! To see why not, let $\omega[u]$ be a total r-form depending on u and its derivatives and $\tilde{\omega}_f(x) = \omega[f(x)]$ the corresponding r-form on $\Omega \subset X$ once we substitute $f(x)$ for u everywhere. We have $\mathrm{D}\omega = 0$ if and only if $d\tilde{\omega}_f = 0$ for each f, and hence $\tilde{\omega}_f = d\tilde{\eta}_f$ for some $(r-1)$-form $\tilde{\eta}_f(x)$. What is *not* clear from the Poincaré homotopy formula (1.69) for $\tilde{\eta}_f$ is that there is a total $(r-1)$-form $\eta[u]$, depending just on u and its derivatives, which specializes to the given $\tilde{\eta}_f$ in every case: $\eta[f(x)] = \tilde{\eta}_f(x)$ for all f, the reason being that (1.69) is not a local map.

Indeed, the de Rham complex is exact at the $\bigwedge_{p-1} T^*\Omega \xrightarrow{d} \bigwedge_p T^*\Omega \to 0$ stage, but this is most definitely not true for the D-complex. Every total p-form $\omega = L[u]\, dx^1 \wedge \cdots \wedge dx^p$ is trivially D-closed, but it is D-exact, $\omega = \mathrm{D}\eta$, if and only if L is a total divergence, $L = \mathrm{Div}\, P$, and, as we know, not every differential function is a total divergence. The proof of Theorem 5.80 will therefore require new methods, in particular a new "total homotopy operator". The proof will be deferred until the end of this section.

The next step in the variational complex is to continue the D-complex beyond the $\bigwedge_p$-stage. This is something we essentially already know how to do, since by Theorem 4.7, $\omega = L\, dx^1 \wedge \cdots \wedge dx^p$ is D-exact, meaning $L = \operatorname{Div} P$ for some $P \in \mathscr{A}^p$, if and only if $\mathsf{E}(L) = 0$, where E is the Euler operator. Thus $\mathsf{D}\colon \bigwedge_{p-1} \to \bigwedge_p$ should be followed by the Euler operator or variational derivative expressed, perhaps, in a more intrinsic way. This will be implemented, and the variational complex continued even further, through the introduction of "functional forms" and the "variational differential", which in a sense accomplish for the dependent variables what the D-complex did for the independent variables.

Vertical Forms

The total r-forms concentrated on the "horizontal" variables x in $M \subset X \times U$ in that only the differentials dx^i appeared. Vertical forms are constructed by similarly concentrating on the "vertical" variables, which consist of the u's and all their derivatives.† Specially, a *vertical k-form* is a *finite* sum

$$\hat{\omega} = \sum P_{\mathbf{J}}^{\boldsymbol{\alpha}}[u]\, du_{J_1}^{\alpha_1} \wedge \cdots \wedge du_{J_k}^{\alpha_k}, \tag{5.108}$$

in which the coefficients $P_{\mathbf{J}}^{\boldsymbol{\alpha}} \in \mathscr{A}$ are differential functions. Since only the differentials du_J^α appear in these forms, the analogue of the differential of the ordinary de Rham complex is the *vertical differential*

$$\hat{d}\hat{\omega} = \sum \frac{\partial P_{\mathbf{J}}^{\boldsymbol{\alpha}}}{\partial u_K^\beta}\, du_K^\beta \wedge du_{J_1}^{\alpha_1} \wedge \cdots \wedge du_{J_k}^{\alpha_k}. \tag{5.109}$$

For example, if $p = q = 1$, a typical vertical two-form might be $\hat{\omega} = x u_{xx}\, du \wedge du_x$. Its vertical differential is then $\hat{d}\hat{\omega} = x\, du \wedge du_x \wedge du_{xx}$, the independent variable x only appearing parametrically.

Since any given vertical form $\hat{\omega}$ can depend on only finitely many of the variables u_J^α, and hence lives on a finite jet space $M^{(n)}$, the vertical differential $\hat{d}\hat{\omega}$ is in reality the same as the de Rham differential in these variables, the remaining independent variables playing the role of parameters. Thus the vertical differential is readily seen to have the usual bilinearity, anti-derivation and closure properties of the ordinary differential:

$$\hat{d}(c\hat{\omega} + c'\hat{\omega}') = c\,\hat{d}\hat{\omega} + c'\,\hat{d}\hat{\omega}',$$
$$\hat{d}(\hat{\omega} \wedge \hat{\eta}) = (\hat{d}\hat{\omega}) \wedge \hat{\eta} + (-1)^k \hat{\omega} \wedge \hat{d}\hat{\eta},$$
$$\hat{d}(\hat{d}\hat{\omega}) = 0,$$

† One can, of course, construct "hybrid" forms in both sets of variables, leading to the important "variational bicomplex". However, this would take us too far afield.

for $\hat{\omega}, \hat{\omega}' \in \hat{\bigwedge}^k$, the space of vertical k-forms over M, $\hat{\eta} \in \hat{\bigwedge}^l$ and c, c' constants. Moreover, the proof of the Poincaré lemma immediately extends to this situation to prove exactness of the "vertical complex" over suitable subdomains $M \subset X \times U$.

Theorem 5.82. *Let $M \subset X \times U$ be vertically star-shaped. Then the vertical complex*

$$\hat{\bigwedge}^0 \xrightarrow{\hat{d}} \hat{\bigwedge}^1 \xrightarrow{\hat{d}} \hat{\bigwedge}^2 \xrightarrow{\hat{d}} \cdots$$

is exact. In other words, for $k > 0$ a vertical k-form $\hat{\omega}$ is closed: $\hat{d}\hat{\omega} = 0$, if and only if it is exact: $\hat{\omega} = \hat{d}\hat{\eta}$ for some $(k-1)$-form $\hat{\eta}$. For $k = 0$, a 0-form or differential function is $\hat{d}$-closed if and only if it is a function of x only.

Note that although any given vertical form depends on only finitely many variables, the entire vertical complex never terminates since we can keep bringing in higher and higher order derivatives of u to construct nonzero vertical k-forms for any $k \geqslant 0$.

The proof of Theorem 5.82 uses the same homotopy operator as was used in the ordinary Poincaré lemma, but adapted to the infinity of variables u_J^α. The basic scaling vector field is $\text{pr}\, \mathbf{v}_u = \sum u_J^\alpha \partial/\partial u_J^\alpha$, the infinite prolongation of the evolutionary vector field $\mathbf{v}_u = \sum u^\alpha \partial/\partial u^\alpha$. There is a well-defined interior product between such vector fields and vertical forms, with $\{\partial/\partial u_J^\alpha\}$ and $\{du_J^\alpha\}$ being the dual bases of the relevant tangent and cotangent spaces. (Note that since vertical forms are required to be finite sums (5.108), we can allow infinite sums in our vector fields, since in computing $\text{pr}\, \mathbf{v}_u \,\lrcorner\, \hat{\omega}$, say, only finitely many terms in the full prolongation of $\mathbf{v}_u$ are needed.) The formula for the homotopy operator corresponding to (1.69) is then

$$\hat{h}(\hat{\omega}) = \int_0^1 \{\text{pr}\, \mathbf{v}_u \,\lrcorner\, \hat{\omega}[\lambda u]\} \frac{d\lambda}{\lambda}, \tag{5.110}$$

and we find for $\hat{\omega} \in \hat{\bigwedge}^k$, $k > 0$,

$$\hat{\omega} = \hat{d}\hat{h}(\hat{\omega}) + \hat{h}(\hat{d}\hat{\omega}).$$

In (5.110), $\hat{\omega}[u]$ indicates the dependence of $\hat{\omega}$ on u and all its derivatives, so to find $\hat{\omega}[\lambda u]$ we replace each u_J^α appearing in $\hat{\omega}$ (either explicitly or as a differential) by λu_J^α. Taking the interior product and then integrating out the λ's completes the determination of $\hat{h}(\hat{\omega})$. (In particular, there is no singularity in the integrand at $\lambda = 0$.)

Example 5.83. Let $p = q = 1$. If $\hat{\omega} = x u_x\, du \wedge du_x$, then $\hat{d}\hat{\omega} = x\, du_x \wedge du \wedge du_x = 0$, so $\hat{\omega}$ is closed. To find a one-form $\hat{\eta}$ such that $\hat{\omega} = \hat{d}\hat{\eta}$ we need only evaluate

$$\begin{aligned} \hat{\eta} = \hat{h}(\hat{\omega}) &= \int_0^1 \{\text{pr}\, \mathbf{v}_u \,\lrcorner\, [x(\lambda u_x)\, d(\lambda u) \wedge d(\lambda u_x)]\} \frac{d\lambda}{\lambda} \\ &= \int_0^1 \lambda^2 \{x u u_x\, du_x - x u_x^2\, du\}\, d\lambda \\ &= \tfrac{1}{3}(x u u_x\, du_x - x u_x^2\, du), \end{aligned}$$

which is correct.

Each vertical k-form determines an alternating k-linear map from the space of *vertical vector fields* $\mathbf{v}^* = \sum Q_\alpha^J \partial/\partial u_J^\alpha$ to the space $\mathscr{A}$ of differential functions; in particular, it determines an alternating multi-linear map on the space T_0 of evolutionary vector fields. The precise formula is written using determinants, as in (1.49), so if $\hat{\omega}$ is given by (5.108), then

$$\langle \hat{\omega}; \operatorname{pr} \mathbf{v}_1, \ldots, \operatorname{pr} \mathbf{v}_k \rangle = \sum_{\alpha, \mathbf{J}} P_{\mathbf{J}}^{\boldsymbol{\alpha}} \det(D_{J_i} Q_{\alpha_i}^j), \tag{5.111}$$

where $Q^j \in \mathscr{A}^q$ is the characteristic of $\mathbf{v}_j$ and the determinant is of the $k \times k$ matrix with the indicated (i, j) entry. For instance,

$$\begin{aligned} \langle x u_{xx}\, du \wedge du_x; \operatorname{pr} \mathbf{v}_Q, \operatorname{pr} \mathbf{v}_R \rangle &= x u_{xx} \det \begin{pmatrix} Q & R \\ D_x Q & D_x R \end{pmatrix} \\ &= x u_{xx}(Q D_x R - R D_x Q). \end{aligned}$$

Total Derivatives of Vertical Forms

For each $i = 1, \ldots, p$, the total derivative D_i can be thought of as a kind of vector field on the infinite jet space. As such, we can allow it to act on vertical forms as a "Lie derivative", which is determined by the following rules:

(a) *Linearity*:

$$D_i(c\hat{\omega} + c'\hat{\omega}') = c D_i \hat{\omega} + c' D_i \hat{\omega}', \qquad c, c' \in \mathbb{R}, \tag{5.112a}$$

(b) *Derivation*:

$$D_i(\hat{\omega} \wedge \hat{\eta}) = (D_i \hat{\omega}) \wedge \hat{\eta} + \hat{\omega} \wedge (D_i \hat{\eta}), \tag{5.112b}$$

(c) *Commutation with the Vertical Differential*:

$$D_i(\hat{d}\hat{\omega}) = \hat{d}(D_i \hat{\omega}), \tag{5.112c}$$

together with its well-established action on differential functions. (See (1.59), (1.60), (1.61).) In particular, D_i acts on the basic forms by $D_i\, du_J^\alpha = d(D_i u_J^\alpha) = du_{J,i}^\alpha$. The action is easy to reconstruct from these properties. For example,

$$\begin{aligned} D_x(x u_{xx}\, du \wedge du_x) &= D_x(x u_{xx})\, du \wedge du_x + x u_{xx} D_x(du) \wedge du_x \\ &\qquad + x u_{xx}\, du \wedge D_x(du_x) \\ &= (x u_{xxx} + u_{xx})\, du \wedge du_x + x u_{xx}\, du \wedge du_{xx}, \end{aligned} \tag{5.113}$$

the middle term vanishing since $D_x(du) = du_x$. The proof that (5.112) determine a well-defined action of D_i is not difficult; in essence, it is a direct consequence of the same uniqueness property of the ordinary Lie derivative. One key property is that the total derivative is compatible with the evaluation of vertical forms on *evolutionary* vector fields:

$$D_i \langle \hat{\omega}; \operatorname{pr} \mathbf{v}_1, \ldots, \operatorname{pr} \mathbf{v}_k \rangle = \langle D_i \hat{\omega}; \operatorname{pr} \mathbf{v}_1, \ldots, \operatorname{pr} \mathbf{v}_k \rangle, \tag{5.114}$$

whenever $1 \leqslant i \leqslant p$, $\hat{\omega} \in \hat{\bigwedge}^k$, $\mathbf{v}_i = \mathbf{v}_{Q^i}$, $Q^i \in \mathscr{A}^q$. Thus, for example, the D_x derivative of $xu_{xx}(QD_xR - RD_xQ)$ agrees with the evaluation of the two-form (5.113) on $\mathrm{pr}\,\mathbf{v}_Q$ and $\mathrm{pr}\,\mathbf{v}_R$. The proof of (5.114) rests on the Lie derivative formulae in Exercise 1.35 together with the fact (5.19) that total derivatives commute with evolutionary vector fields.

Functionals and Functional Forms

Actually, what we are really interested in are the "functional versions" of our vertical forms, which are related to them just as functionals are related to differential functions. Although the basic notion of a functional appeared in its traditional guise in Chapter 4, subsequent developments necessitate a more algebraic approach to these fundamental objects of the calculus of variations. Each differential function $L \in \mathscr{A}$ determines a functional $\mathscr{L}[u] = \int_\Omega L[u]\,dx$ defined over any region $\Omega \subset X$ in its domain of definition. Provided we ignore boundary contributions (say by considering only functions $u = f(x)$ vanishing sufficiently rapidly near the boundary) a second function $\tilde{L} \in \mathscr{A}$ will determine the same functional, i.e. $\int_\Omega L[u]\,dx = \int_\Omega \tilde{L}[u]\,dx$ for all such u, if and only if it differs from L by a total divergence:

$$\tilde{L} = L + \mathrm{Div}\,P, \quad \text{for some} \quad P \in \mathscr{A}^p. \tag{5.115}$$

This is the essential content of Theorem 4.7 in the case that $P[u] = 0$ on $\partial\Omega$. Condition (5.115) does not any longer depend on the domain Ω, and determines an equivalence relation on the space of differential functions. Specifically, L and $\tilde{L}$ are *equivalent*, and determine the same functional, provided (5.115) holds. Each functional is thereby uniquely determined by an equivalence class of differential functions and conversely. It is reasonable, therefore, to *define* the space of *functionals*, denoted $\mathscr{F}$, as the set of equivalence classes of the space $\mathscr{A}$ of differential functions under the equivalence relation (5.115). Put another way, $\mathscr{F} = \mathscr{A}/\mathrm{Div}(\mathscr{A}^p)$ is the quotient vector space of $\mathscr{A}$ under the subspace of total divergences, i.e. the "cokernel" of the total divergence map $\mathrm{Div}\colon \mathscr{A}^p \to \mathscr{A}$. The natural projection from $\mathscr{A}$ to $\mathscr{F}$ which associates to each differential function L its equivalence class or functional will be denoted suggestively by an integral sign, so $\int L\,dx \in \mathscr{F}$ is the functional, or equivalence class, corresponding to $L \in \mathscr{A}$. In particular, $\int L\,dx = 0$ if and only if $L = \mathrm{Div}\,P$ for some P. This allows us the freedom of "integrating functionals by parts":

$$\int (P \cdot D_i Q)\,dx = -\int (Q \cdot D_i P)\,dx, \qquad P, Q \in \mathscr{A}.$$

(From our earlier standpoint, the image of the total divergence can be identified with the image of $\mathrm{D}\colon \bigwedge_{p-1} \to \bigwedge_p$, where $L[u]$ corresponds to the p-form $L[u]\,dx = L[u]\,dx^1 \wedge \cdots \wedge dx^p$. We can identify $\mathscr{F}$, the space of functionals, with the cokernel $\mathscr{F} \simeq \bigwedge_p/\mathrm{D}\bigwedge_{p-1}$, the projection of $\hat{\omega} = L\,dx$ being the functional $\int \hat{\omega} = \int L\,dx$. Indeed, if we were pursuing a truly coordinate-free presentation, we should be working with $\bigwedge_p$, the space of total p-forms, rather than $\mathscr{A}$, the space of differential functions. Note also that we could thus complete the D-complex by appending the trivially exact piece $\bigwedge_{p-1} \xrightarrow{\mathrm{D}} \bigwedge_p \to \mathscr{F} \to 0$, but this is not as interesting as the full variational complex.)

One important point is that whereas the space $\mathscr{A}$ of differential functions is an algebra, the same is no longer true of the space $\mathscr{F}$ of functionals, since we cannot multiply functionals together in any natural way. For example, the differential functions u_x and u_{xxx} both determine trivial functionals: $\int u_x\,dx = 0 = \int u_{xxx}\,dx$, but their product $u_x u_{xxx}$ is *not* a divergence, and hence $\mathscr{L} = \int u_x u_{xxx}\,dx \neq 0$ is not a trivial functional. Indeed, $\delta\mathscr{L} = -2u_{xxxx} \neq 0$, hence by Theorem 4.7, $\mathscr{L} \neq 0$. Of course we can still take constant coefficient linear combinations of functionals, so $\mathscr{F}$ is a vector space.

Similarly, we define an equivalence relation on the space $\hat{\bigwedge}^k$ of vertical k-forms, with $\hat{\omega}$ equivalent to $\hat{\omega}'$ if they differ by a total divergence

$$\hat{\omega} = \hat{\omega}' + \mathrm{Div}\,\hat{\eta} = \hat{\omega}' + \sum_{i=1}^{p} D_i \hat{\eta}_i, \qquad \hat{\eta}_i \in \hat{\textstyle\bigwedge}^k,$$

where D_i acts on $\hat{\eta}_i$ according to (5.112). The space of equivalence classes is the space of *functional k-forms*, denoted

$$\textstyle\bigwedge_*^k = \hat{\bigwedge}^k / \mathrm{Div}(\hat{\bigwedge}^k)^p.$$

The natural projection from $\hat{\bigwedge}^k$ to $\bigwedge_*^k$ is again denoted by an integral sign, so $\int \hat{\omega}\,dx$ stands for the equivalence class containing $\hat{\omega} \in \hat{\bigwedge}^k$. In particular, $\int \mathrm{Div}\,\hat{\eta}\,dx = 0$ for any p-tuple of vertical k-forms $\hat{\eta}$. Coupled with the derivational property of the total derivative (5.112b), this gives the integration by parts formula

$$\int \hat{\omega} \wedge D_i \hat{\eta}\,dx = -\int (D_i \hat{\omega}) \wedge \hat{\eta}\,dx, \qquad \hat{\omega} \in \hat{\textstyle\bigwedge}^k, \quad \hat{\eta} \in \hat{\textstyle\bigwedge}^l. \tag{5.116}$$

Example 5.84. Let $p = q = 1$ and consider the functional two-form

$$\omega = \int \{u_x\,du \wedge du_{xx}\}\,dx.$$

We can integrate this by parts using the fact that $du_{xx} = D_x(du_x)$, so by (5.116)

$$\begin{aligned}\omega &= -\int \{D_x(u_x\,du) \wedge du_x\}\,dx = -\int \{(u_{xx}\,du + u_x\,du_x) \wedge du_x\}\,dx \\ &= -\int \{u_{xx}\,du \wedge du_x\}\,dx.\end{aligned}$$

It doesn't help, though, to try a second integration by parts, since we get

$$\omega = +\int \{D_x(u_{xx}\,du) \wedge du\}\,dx = +\int \{u_{xx}\,du_x \wedge du\}\,dx,$$

which is exactly the same form as before.

Just as we are not allowed to multiply functionals, there is *no* well-defined wedge product between functional forms, since if

$$\hat{\omega} = \tilde{\omega} + \text{Div}\,\eta \quad \text{and} \quad \hat{\theta} = \tilde{\theta} + \text{Div}\,\zeta,$$

are equivalent forms, there is no guarantee that $\hat{\omega} \wedge \hat{\theta}$ and $\tilde{\omega} \wedge \tilde{\theta}$ are equivalent. In the above example, $du_{xx} = D_x(du_x)$ is trivial, but the functional two-form ω is not trivial. (See Proposition 5.88.)

Each functional form is an alternating multi-linear map from the space of evolutionary vector fields to the space of functionals, defined so that

$$\langle \omega; \mathbf{v}_1, \ldots, \mathbf{v}_k \rangle = \int \langle \hat{\omega}; \text{pr}\,\mathbf{v}_1, \ldots, \text{pr}\,\mathbf{v}_k \rangle \, dx, \qquad \mathbf{v}_i \in T_0, \tag{5.117}$$

whenever $\omega = \int \hat{\omega}\, dx$, $\hat{\omega} \in \hat{\bigwedge}^k$. This is well defined by virtue of (5.114). For example, if $\omega = \int \{u_x \, du \wedge du_{xx}\}\, dx$ as above, then

$$\langle \omega; \mathbf{v}_Q, \mathbf{v}_R \rangle = \int u_x(QD_x^2 R - RD_x^2 Q)\, dx.$$

What is slightly less obvious is that this action uniquely determines ω:

Lemma 5.85. *If ω and ω' are functional k-forms, then $\omega = \omega'$ if and only if $\langle \omega; \mathbf{v}_1, \ldots, \mathbf{v}_k \rangle = \langle \omega'; \mathbf{v}_1, \ldots, \mathbf{v}_k \rangle$ for every set of evolutionary vector fields $\mathbf{v}_1, \ldots, \mathbf{v}_k$.*

The proof rests on a more basic result:

Lemma 5.86. *Suppose $u \in \mathbb{R}^q$ and $v \in \mathbb{R}^r$ are both dependent variables depending on $x \in \mathbb{R}^p$. Suppose $\mathscr{L}[u, v] = \int L(x, u^{(n)}, v^{(n)})\, dx$ is a functional with the property that $\mathscr{L}[u, Q[u]] = 0$ for all differential r-tuples $Q \in \mathscr{A}^r$ depending on x, u and derivatives of u. Then $\mathscr{L}[u, v] = 0$ as a functional in u and v.*

Proof. An equivalent way of stating this result is to say that if for every $Q \in \mathscr{A}^r$

$$L[u, Q[u]] = \text{Div}\, P_Q[u]$$

for some $P_Q \in \mathscr{A}^p$ depending on x, u and derivatives of u, then

$$L[u, v] = \text{Div}\, P^*[u, v]$$

for some p-tuple P^* depending on x, u, v and derivatives of u and v. In particular,

$$L[u, Q[u]] = \text{Div}\, P^*[u, Q[u]],$$

where P^* depends on Q and its total derivatives alone. (The same is not necessarily true of P_Q, especially if it was constructed using the method of proof of Theorem 4.7.)

To prove this result, let $Q, R \in \mathscr{A}^r$. Then by the methods used to determine the variational derivative

$$0 = \frac{d}{d\varepsilon}\bigg|_{\varepsilon=0} \mathscr{L}[u, Q + \varepsilon R] = \int \mathsf{E}_v(L)[u, Q] \cdot R \, dx,$$

where $\mathsf{E}_v(L)$ denotes the variational derivative of L with respect to v. By Corollary 5.68, $\mathsf{E}_v(L)[u, Q[u]] \equiv 0$ for all Q, hence $\mathsf{E}_v(L)[u, v] = 0$ for all u, v. Similarly, differentiating $\mathscr{L}[u + \varepsilon P[u], Q[u + \varepsilon P[u]]]$ with respect to ε at $\varepsilon = 0$ and using the vanishing of $\mathsf{E}_v(L)$, we find $\mathsf{E}_u(L) \equiv 0$. Theorem 4.7 immediately implies $L[u, v] = \operatorname{Div} P^*[u, v]$ for some P^*, proving the lemma. □

To prove Lemma 5.85, we need only show that

$$\mathscr{L}[u; Q^1, \ldots, Q^k] \equiv \langle \omega; \mathbf{v}_1, \ldots, \mathbf{v}_k \rangle = 0$$

for all $Q^\nu \in \mathscr{A}^q$, $\nu = 1, \ldots, k$, if and only if $\hat{\omega} = \operatorname{Div} \hat{\eta}$ for some p-tuple of vertical forms $\hat{\eta}$. Lemma 5.86 implies that

$$\langle \hat{\omega}; \operatorname{pr} \mathbf{v}_1, \ldots, \operatorname{pr} \mathbf{v}_k \rangle = \operatorname{Div} P^*[u; Q^1, \ldots, Q^k],$$

where P^* depends on $Q^1, \ldots, Q^k$ and their total derivatives only. As it stands, the components P_j^* of P^* can certainly be chosen to be linear in each Q^ν, but may not be alternating functions thereof. However, if we replace P^* by its "skew-symmetrization",

$$\hat{P}^*[u; Q^1, \ldots, Q^k] = \frac{1}{k!} \sum_\pi (-1)^\pi P^*[u; Q^{\pi 1}, \ldots, Q^{\pi k}],$$

the sum being over all permutations π of $\{1, \ldots, k\}$, we maintain the condition

$$\langle \hat{\omega}; \operatorname{pr} \mathbf{v}_1, \ldots, \operatorname{pr} \mathbf{v}_k \rangle = \operatorname{Div} \hat{P}^*[u; Q^1, \ldots, Q^k].$$

Moreover, each component of $\hat{P}^*$ is an alternating, multi-linear function of the Q^ν's and their total derivatives, and hence can be identified with a vertical k-form

$$\hat{P}_j^*[u; Q^1, \ldots, Q^k] = \langle \hat{\eta}_j; \operatorname{pr} \mathbf{v}_1, \ldots, \operatorname{pr} \mathbf{v}_k \rangle.$$

Since this holds for all such $Q^1, \ldots, Q^k$, we conclude that $\hat{\omega} = \operatorname{Div} \hat{\eta}$, and the lemma is proved. □

Let us look in more detail at the cases of functional one- and two-forms. Any one-form

$$\omega = \int \left\{ \sum_{\alpha, J} P_\alpha^J[u] \, du_J^\alpha \right\} dx$$

is determined by a finite collection of differential functions P_α^J, but the P_α^J are not uniquely determined by ω. Indeed, since $du_J^\alpha = D_J du^\alpha$, we can integrate each summand by parts, leading to the simpler expression

$$\omega = \int \left\{ \sum_{\alpha=1}^{q} P_\alpha[u]\, du^\alpha \right\} dx \equiv \int \{P \cdot du\}\, dx, \quad \text{where} \quad P_\alpha = \sum_J (-D)_J P_\alpha^J, \tag{5.118}$$

called the *canonical* form of ω. It is not hard to see that each functional one-form *does* have a uniquely determined canonical form.

Proposition 5.87. *Let $\omega = \int \{P \cdot du\}\, dx$ and $\tilde{\omega} = \int \{\tilde{P} \cdot du\}\, dx$ be functional one-forms in canonical form, so $P, \tilde{P} \in \mathscr{A}^q$. Then $\omega = \tilde{\omega}$ if and only if $P = \tilde{P}$.*

Proof. It suffices to show that a functional one-form $\omega = 0$ if and only if the p-tuple P appearing in its canonical form vanishes identically. Evaluating (5.118) on an arbitrary vector field, we have

$$\langle \omega; \mathbf{v}_Q \rangle = \int (P \cdot Q)\, dx.$$

According to Lemma 5.85, $\omega = 0$ if and only if this vanishes for all such $\mathbf{v}_Q$, but by Corollary 5.68 this occurs if and only if $P = 0$, proving the result. □

Next consider the case of functional two-forms, the most general one of which is

$$\omega = \int \left\{ \sum_{\substack{\alpha,\beta \\ J,K}} P_{\alpha\beta}^{JK}[u]\, du_J^\alpha \wedge du_K^\beta \right\} dx,$$

the sum as usual being finite. To simplify the vertical two-form in the integrand, we rewrite $du_J^\alpha = D_J\, du^\alpha$ and integrate by parts. This leads to an expression of the form

$$\omega = \int \left\{ \sum_{\alpha,\beta,I} P_{\alpha\beta}^I[u]\, du^\alpha \wedge du_I^\beta \right\} dx,$$

where the $P_{\alpha\beta}^I$ are determined from the $P_{\alpha\beta}^{JK}$ and their derivatives. Define the differential operators

$$\tilde{\mathscr{D}}_{\alpha\beta} = \sum_I P_{\alpha\beta}^I[u] D_I,$$

whereby the above expression can be written as

$$\omega = \int \left\{ \sum_{\alpha,\beta=1}^{q} du^\alpha \wedge \tilde{\mathscr{D}}_{\alpha\beta}\, du^\beta \right\} dx, \tag{5.119}$$

or, using a more compact matrix notation,

$$\omega = \int \{du \wedge \tilde{\mathscr{D}}\, du\}\, dx.$$

As it stands, though, the matrix differential operator $\tilde{\mathscr{D}} = (\tilde{\mathscr{D}}_{\alpha\beta})$ is *not* uniquely determined by ω. Indeed, (5.119) can be integrated by parts, leading to an equivalent expression

$$\omega = \int \left\{ \sum_{\alpha,\beta} \tilde{\mathscr{D}}^*_{\alpha\beta}(du^\alpha) \wedge du^\beta \right\} dx = -\int \left\{ \sum_{\alpha,\beta} du^\beta \wedge \tilde{\mathscr{D}}^*_{\alpha\beta}(du^\alpha)\, dx \right\}$$

involving the adjoint $\tilde{\mathscr{D}}^* = (\tilde{\mathscr{D}}^*_{\beta\alpha})$ of $\tilde{\mathscr{D}}$. If we set $\mathscr{D} = \tilde{\mathscr{D}} - \tilde{\mathscr{D}}^*$, so $\mathscr{D}: \mathscr{A}^q \to \mathscr{A}^q$ is a skew-adjoint differential operator: $\mathscr{D}^* = -\mathscr{D}$, then ω has the *canonical form*

$$\omega = \tfrac{1}{2} \int \{du \wedge \mathscr{D}\, du\}\, dx, \qquad \mathscr{D}^* = -\mathscr{D}. \tag{5.120}$$

Its value on a pair of evolutionary vector fields is then

$$\langle \omega; \mathbf{v}_Q, \mathbf{v}_R \rangle = \tfrac{1}{2} \int \{Q \cdot \mathscr{D}R - R \cdot \mathscr{D}Q\}\, dx = \int \{Q \cdot \mathscr{D}R\}\, dx$$

since $\mathscr{D}$ is skew-adjoint. *This* canonical form is uniquely determined by ω.

Proposition 5.88. *Let* $\omega = \frac{1}{2}\int \{du \wedge \mathscr{D}(du)\}\, dx$, $\tilde{\omega} = \frac{1}{2}\int \{du \wedge \tilde{\mathscr{D}}(du)\}\, dx$ *be functional two-forms in canonical form, so* $\mathscr{D}$ *and* $\tilde{\mathscr{D}}$ *are skew-adjoint* $q \times q$ *matrix differential operators. Then* $\omega = \tilde{\omega}$ *if and only if* $\mathscr{D} = \tilde{\mathscr{D}}$.

Proof. By Lemma 5.85, it suffices to prove that if $\mathscr{D}: \mathscr{A}^q \to \mathscr{A}^q$ is skew-adjoint, then $\int (Q \cdot \mathscr{D}R)\, dx = 0$ for all q-tuples $Q, R \in \mathscr{A}^q$ if and only if $\mathscr{D} = 0$. Corollary 5.68 implies that $\mathscr{D}R = 0$ for all R, which implies that $\mathscr{D} = 0$. (See Exercise 5.39.) □

The Variational Differential

Definition 5.89. Let $\omega = \int \hat{\omega}\, dx$ be a functional k-form corresponding to the vertical k-form $\hat{\omega}$. The *variational differential* of ω is the functional $(k+1)$-form corresponding to the vertical differential of ω:

$$\delta\omega = \int (\hat{d}\hat{\omega})\, dx. \tag{5.121}$$

The commutativity relation (5.112c) assures us that this operator is well defined on the spaces of functional forms. The basic properties follow at once from those of the vertical differential, so we immediately have an exact *variational complex.*

Theorem 5.90. *Let* $M \subset X \times U$ *be vertically star-shaped. The variational differential determines an exact complex*

$$0 \to \Lambda^0_* \xrightarrow{\delta} \Lambda^1_* \xrightarrow{\delta} \Lambda^2_* \xrightarrow{\delta} \Lambda^3_* \xrightarrow{\delta} \cdots$$

on the spaces of functional forms on M. In other words, a functional form is closed: $\delta\omega = 0$, if and only if it is exact: $\omega = \delta\eta$.

PROOF. The homotopy formula (5.110) immediately projects to a homotopy formula for the variational differential: if ω is any functional k-form, $k > 0$, then

$$\omega = \delta h(\omega) + h(\delta\omega),$$

where, for $\omega = \int \hat{\omega}\, dx$,

$$h(\omega) = \int \hat{h}(\hat{\omega})\, dx = \int \left\{ \int_0^1 (\mathrm{pr}\, \mathbf{v}_u \,\lrcorner\, \hat{\omega}[\lambda u]) \frac{d\lambda}{\lambda} \right\} dx. \tag{5.122}$$

This also extends to the case when $k = 0$, i.e. ω is a functional, since $\hat{\omega}$ only differs from $\hat{d}\hat{h}(\hat{\omega}) + \hat{h}(\hat{d}\hat{\omega})$ by a function of x alone, and any such function determines a trivial functional. This suffices to prove Theorem 5.90 in all cases. □

Example 5.91. Consider the functional two-form

$$\omega = \int \{u_{xxx}\, du \wedge du_x\}\, dx.$$

(Note that ω is not quite in canonical form, which would be

$$\tfrac{1}{2} \int \{du \wedge (2u_{xxx}\, du_x + u_{xxxx}\, du)\}\, dx$$

corresponding to the skew-adjoint operator $2u_{xxx}D_x + u_{xxxx}$.) The variational derivative is the functional three-form

$$\delta\omega = \int \{du_{xxx} \wedge du \wedge du_x\}\, dx.$$

This form is trivial: integrating by parts, we see

$$\begin{aligned} \delta\omega &= -\int \{du_{xx} \wedge D_x(du \wedge du_x)\}\, dx \\ &= -\int \{du_{xx} \wedge du_x \wedge du_x + du_{xx} \wedge du \wedge du_{xx}\}\, dx = 0. \end{aligned}$$

Equivalently, $du_{xxx} \wedge du \wedge du_x = D_x(du_{xx} \wedge du \wedge du_x)$ is a total x-derivative. (Another way to see this is to note that the evaluation of the corresponding vertical three-form on a triple of evolutionary vector fields is an x-derivative:

$$\begin{aligned} \langle du \wedge du_x \wedge du_{xxx}; \mathrm{pr}\, \mathbf{v}_P, \mathrm{pr}\, \mathbf{v}_Q, \mathrm{pr}\, \mathbf{v}_R \rangle &= \det \begin{bmatrix} P & Q & R \\ D_x P & D_x Q & D_x R \\ D_x^3 P & D_x^3 Q & D_x^3 R \end{bmatrix} \\ &= D_x \left\{ \det \begin{bmatrix} P & Q & R \\ D_x P & D_x Q & D_x R \\ D_x^2 P & D_x^2 Q & D_x^2 R \end{bmatrix} \right\}.) \end{aligned}$$

To compute a one-form η whose variational differential is ω, we use the homotopy formula (5.122),

$$\eta = h(\omega) = \int \left\{ \int_0^1 \lambda^2 (u u_{xxx}\, du_x - u_x u_{xxx}\, du)\, d\lambda \right\} dx$$

$$= \int \{\tfrac{1}{3} u u_{xxx}\, du_x - \tfrac{1}{3} u_x u_{xxx}\, du\}\, dx.$$

This has the canonical form

$$\eta = \int \{(-\tfrac{1}{3} u u_{xxxx} - \tfrac{2}{3} u_x u_{xxx})\, du\}\, dx,$$

and, indeed

$$\delta\eta = \int \{-\tfrac{1}{3} u\, du_{xxxx} \wedge du - \tfrac{2}{3} u_x\, du_{xxx} \wedge du - \tfrac{2}{3} u_{xxx}\, du_x \wedge du\}\, dx$$

can be shown to be equal to ω through a couple of integrations by parts.

The exactness of the variational complex at the $\bigwedge_*^1$-stage is of especial importance since it provides the afore-mentioned solution to the inverse problem of the calculus of variations. To see this, we need to first relate the variational differential to the variational derivative. If $\mathscr{L} = \int L\, dx$ is a functional, which we regard as an element of $\bigwedge_*^0$, then its variational differential is the functional one-form

$$\delta\mathscr{L} = \int \{\hat{d}L\}\, dx = \int \left\{ \sum_{\alpha=1}^{q} \sum_J \frac{\partial L}{\partial u_J^\alpha}\, du_J^\alpha \right\} dx.$$

As in (5.118), we can integrate this latter form by parts, leading to the canonical form

$$\delta\mathscr{L} = \int \left\{ \sum_{\alpha=1}^{q} \left(\sum_J (-D)_J \frac{\partial L}{\partial u_J^\alpha} \right) du^\alpha \right\} dx = \int \{\mathsf{E}(L) \cdot du\}\, dx,$$

cf. (4.3). Proposition 5.87 implies that $\delta\mathscr{L}$ can be uniquely identified with the Euler–Lagrange expression $\mathsf{E}(L)$, and this provides the connection between the variational differential and our previous notation for the variational derivative. (Indeed, if we interpret the differentials du^α as infinitesimal variations in the u^α, with corresponding variations $du_J^\alpha = D_J\, du^\alpha$ in the derivatives, then the above computation is the same as the traditional determination of the Euler–Lagrange equations from the definition of the variational derivative.) Exactness of the variational complex at the $\bigwedge_*^0$-stage, then, is equivalent to Theorem 4.7 that a functional is trivial if and only if its variational derivative vanishes identically.

We can thus "glue" the D-complex to the complex determined by the variational differential to obtain the full *variational complex*

$$0 \to \mathbb{R} \to \textstyle\bigwedge_0 \xrightarrow{\mathrm{D}} \bigwedge_1 \xrightarrow{\mathrm{D}} \cdots \xrightarrow{\mathrm{D}} \bigwedge_{p-1} \xrightarrow{\mathrm{D}} \bigwedge_p \xrightarrow{\mathsf{E}} \bigwedge_*^1 \xrightarrow{\delta} \bigwedge_*^2 \xrightarrow{\delta} \cdots,$$

which is exact over totally star-shaped domains $M \subset X \times U$.

Next consider the variational differential of a functional one-form, which we take in canonical form $\omega = \int \{P \cdot du\}\, dx$. We find

$$\delta\omega = \int \left\{ \sum_\alpha \sum_{\beta, J} \frac{\partial P_\alpha}{\partial u_J^\beta} du_J^\beta \wedge du^\alpha \right\} dx = \int \{\mathsf{D}_P(du) \wedge du\}\, dx,$$

where D_P is the Fréchet derivative of P, cf. (5.32). As in (5.120), we can integrate by parts a second time, leading to the canonical form

$$\delta\omega = \tfrac{1}{2} \int \{du \wedge (\mathsf{D}_P^* - \mathsf{D}_P)\, du\}\, dx.$$

In particular, ω is closed if and only if D_P is a self-adjoint differential operator. Exactness of the variational complex, coupled with the explicit form for the homotopy operator (5.122), thus gives the complete solution to the problem of characterizing the image of the Euler–Lagrange operator.

Theorem 5.92. *Let $P[u] \in \mathscr{A}^p$ be defined over a vertically star-shaped region $M \subset X \times U$. Then P is the Euler–Lagrange expression for some variational problem $\mathscr{L} = \int L\, dx$, i.e. $P = \mathsf{E}(L)$, if and only if the Fréchet derivative D_P is self-adjoint: $\mathsf{D}_P^* = \mathsf{D}_P$. In this case, a Lagrangian for P can be explicitly constructed using the homotopy formula*

$$L[u] = \int_0^1 u \cdot P[\lambda u]\, d\lambda. \tag{5.123}$$

Example 5.93. Let $p = q = 1$. The functional

$$\mathscr{L}[u] = \int (\tfrac{1}{2}u_{xx}^2 - uu_x^2)\, dx$$

has Euler–Lagrange expression

$$\mathsf{E}(L)[u] = P[u] = u_{xxxx} + 2uu_{xx} + u_x^2.$$

The Fréchet derivative of P is the ordinary differential operator

$$\mathsf{D}_P = D_x^4 + 2uD_x^2 + 2u_x D_x + 2u_{xx},$$

which is easily seen to be self-adjoint. If, on the other hand, we were just given P, we could reconstruct a variational problem using (5.123),

$$\mathscr{L}[u] = \int \left\{ \int_0^1 u(\lambda u_{xxxx} + 2\lambda^2 uu_{xx} + \lambda^2 u_x^2)\, d\lambda \right\} dx$$

$$= \int \{\tfrac{1}{2}uu_{xxxx} + \tfrac{2}{3}u^2 u_{xx} + \tfrac{1}{3}uu_x^2\}\, dx.$$

The Lagrangian, while not the same as the original one, is still equivalent, since

$$\tfrac{1}{2}uu_{xxxx} + \tfrac{2}{3}u^2u_{xx} + \tfrac{1}{3}uu_x^2 = \tfrac{1}{2}u_{xx}^2 - uu_x^2 + D_x(\tfrac{1}{2}uu_{xxx} - \tfrac{1}{2}u_xu_{xx} + \tfrac{2}{3}u^2u_x).$$

We have thus solved Helmholtz' version of the inverse problem of the calculus of variations: characterizing those q-tuples $P[u] \in \mathscr{A}^q$ which are Euler–Lagrange expressions. (The conditions requiring the self-adjointness of the operator D_P are often referred to as the *Helmholtz conditions.*) Although the solution is very neat, from the wider perspective of determining which systems of differential equations $\Delta = 0$ arise from variational principles, it is somewhat unsatisfactory. If one happens to write the equations in the "wrong" order, say $\Delta_1 = \mathsf{E}_2(L)$, $\Delta_2 = \mathsf{E}_1(L)$, etc., then the Helmholtz conditions for Δ will not hold, and the variational structure of the system will remain undiscovered. Even more difficult to detect will be when Δ is *equivalent* to a set of Euler–Lagrange equations, so $\Delta = A \cdot \mathsf{E}(L)$ for some invertible $q \times q$ matrix of differential functions A, or, even more generally, $\Delta = \mathscr{D}\mathsf{E}(L)$ for some differential operator $\mathscr{D}$. The solution to the general equivalence problem is unknown, even in the case when A is constant matrix! (Some special cases have been considered—see the notes at the end of the chapter.)

Higher Euler Operators

Although the D-complex was perhaps simpler to write down, the construction of a suitable homotopy operator is considerably more complicated. The usual de Rham formula no longer works, and we are forced to introduce the so-called "higher Euler operators". These arise most naturally through a detailed analysis of the fundamental integration by parts formula (4.39) used in the proof of Noether's theorem.

Definition 5.94. For each $1 \leqslant \alpha \leqslant q$ and each multi-index J, the *higher Euler operators* E_α^J are defined so that the formula

$$\operatorname{pr} \mathbf{v}_Q(P) = \sum_{\alpha=1}^{q} \sum_J D_J(Q_\alpha \cdot \mathsf{E}_\alpha^J(P)) \tag{5.124}$$

holds for every evolutionary vector field $\mathbf{v}_Q$ and every differential function $P \in \mathscr{A}$.

The fact that (5.124) serves to uniquely determine these operators can perhaps best be appreciated through an example.

Example 5.95. Let $p = q = 1$, so there are Euler operators $\mathsf{E}^{(0)}$, $\mathsf{E}^{(1)}$, $\mathsf{E}^{(2)}$, etc., satisfying

$$\operatorname{pr} \mathbf{v}_Q(P) = Q\,\mathsf{E}^{(0)}(P) + D_x(Q\,\mathsf{E}^{(1)}(P)) + D_x^2(Q\,\mathsf{E}^{(2)}(P)) + \cdots \tag{5.125}$$

for general $P = P(x, u, u_x, \ldots)$. Suppose $P = P(x, u, u_x, u_{xx})$ depends only on second order derivatives, so

$$\operatorname{pr} \mathbf{v}_Q(P) = Q\frac{\partial P}{\partial u} + D_x Q\frac{\partial P}{\partial u_x} + D_x^2 Q\frac{\partial P}{\partial u_{xx}}.$$

To rewrite this in the form (5.125), we must integrate the second and third terms by parts:

$$D_x Q \cdot \frac{\partial P}{\partial u_x} = -Q \cdot D_x \frac{\partial P}{\partial u_x} + D_x\left(Q \frac{\partial P}{\partial u_x}\right),$$

$$D_x^2 Q \cdot \frac{\partial P}{\partial u_{xx}} = Q \cdot D_x^2 \frac{\partial P}{\partial u_{xx}} - 2D_x\left(Q \cdot D_x \frac{\partial P}{\partial u_{xx}}\right) + D_x^2\left(Q \frac{\partial P}{\partial u_{xx}}\right).$$

Comparing with (5.125), we see that for such P,

$$\mathsf{E}^{(0)}(P) = \frac{\partial P}{\partial u} - D_x \frac{\partial P}{\partial u_x} + D_x^2 \frac{\partial P}{\partial u_{xx}},$$

$$\mathsf{E}^{(1)}(P) = \frac{\partial P}{\partial u_x} - 2D_x \frac{\partial P}{\partial u_{xx}}, \qquad \mathsf{E}^{(2)}(P) = \frac{\partial P}{\partial u_{xx}}.$$

If we carry through the same procedure for general P, we find that (5.125) holds provided we set

$$\mathsf{E}^{(k)}(P) = \sum_{l=k}^{\infty} \binom{l}{k} (-D_x)^{l-k} \frac{\partial P}{\partial u_l},$$

so that

$$\mathsf{E}^{(0)}(P) = \frac{\partial P}{\partial u} - D_x \frac{\partial P}{\partial u_x} + D_x^2 \frac{\partial P}{\partial u_{xx}} - D_x^3 \frac{\partial P}{\partial u_{xxx}} + \cdots$$

agrees with the usual Euler operator, while

$$\mathsf{E}^{(1)}(P) = \frac{\partial P}{\partial u_x} - 2D_x \frac{\partial P}{\partial u_{xx}} + 3D_x^2 \frac{\partial P}{\partial u_{xxx}} - 4D_x^3 \frac{\partial P}{\partial u_{xxxx}} + \cdots,$$

$$\mathsf{E}^{(2)}(P) = \frac{\partial P}{\partial u_{xx}} - 3D_x \frac{\partial P}{\partial u_{xxx}} + 6D_x^2 \frac{\partial P}{\partial u_{xxxx}} - 10D_x^3 \frac{\partial P}{\partial u_{xxxxx}} + \cdots,$$

and so on.

To state the general formula for the higher Euler operators, we need some further multi-index notation. Let I, J be unordered multi-indices of the type introduced in Chapter 2. We say $J \subset I$ if all the indices in J appear in I. We write $I \backslash J$ for the remaining indices in I. For example, if $p = 4$, $J = (1, 1, 2, 4)$ is contained in $I = (1, 1, 1, 2, 4, 4)$ and $J \backslash I = (1, 4)$. Given $I = (i_1, \ldots, i_n)$, let $\tilde{I} = (\tilde{\imath}_1, \ldots, \tilde{\imath}_p)$ be the "transposed" ordered multi-index, where $\tilde{\imath}_j$ equals the number of occurrences of the integer j in I; for the above example, $\tilde{I} = (3, 1, 0, 2)$ since there are three 1's, one 2, no 3's and two 4's in I. Set $I! = \tilde{I}! = \tilde{\imath}_1!\tilde{\imath}_2!\cdots\tilde{\imath}_p!$, and define the multinomial coefficient $\binom{I}{J} = I!/(J!(I \backslash J)!)$ when $J \subset I$; 0 otherwise. In the above example, $I! = 3! \cdot 1! \cdot 0! \cdot 2! = 12$, $\binom{I}{J} = 12/(2 \cdot 1) = 6$.

Proposition 5.96. *Let $1 \leqslant \alpha \leqslant q$, $\# J \geqslant 0$. Then*

$$\mathsf{E}_\alpha^J(P) = \sum_{I \supset J} \binom{I}{J} (-D)_{I \setminus J} \frac{\partial P}{\partial u_I^\alpha} \tag{5.126}$$

for all $P \in \mathscr{A}$.

Proof. First note that

$$R \cdot D_I Q = \sum_{J \subset I} \binom{I}{J} D_J(Q \cdot (-D)_{I \setminus J} R) \tag{5.127}$$

for any I, a formula which is easy to prove by induction starting with the Leibniz rule $RD_i Q = D_i(QR) - QD_i R$. Evaluating the left-hand side of (5.124), and using (5.127), we find

$$\operatorname{pr} \mathbf{v}_Q(P) = \sum_{\alpha, I} D_I Q_\alpha \cdot \frac{\partial P}{\partial u_I^\alpha} = \sum_{\alpha, I} \sum_{J \subset I} D_J \left(Q_\alpha \cdot \binom{I}{J} (-D)_{I \setminus J} \frac{\partial P}{\partial u_I^\alpha} \right).$$

Interchanging the order of summation proves (5.126) and hence the uniqueness of the Euler operators. In particular, for $J = 0$, $\mathsf{E}_\alpha^0 = \mathsf{E}_\alpha$ agrees with the usual Euler operator. □

Example 5.97. Let $p = 2$, $q = 1$ and let x, y denote the independent variables. Then, for instance,

$$\mathsf{E}^{(x)}(P) = \frac{\partial P}{\partial u_x} - 2D_x \frac{\partial P}{\partial u_{xx}} - D_y \frac{\partial P}{\partial u_{xy}} + 3D_x^2 \frac{\partial P}{\partial u_{xxx}} + 2D_x D_y \frac{\partial P}{\partial u_{xxy}} + D_y^2 \frac{\partial P}{\partial u_{xyy}} - \cdots,$$

$$\mathsf{E}^{(xy)}(P) = \frac{\partial P}{\partial u_{xy}} - 2D_x \frac{\partial P}{\partial u_{xxy}} - 2D_y \frac{\partial P}{\partial u_{xyy}} + 3D_x^2 \frac{\partial P}{\partial u_{xxxy}} + 4D_x D_y \frac{\partial P}{\partial u_{xxyy}} + 3D_y^2 \frac{\partial P}{\partial u_{xyyy}} - \cdots,$$

and so on.

Actually, for theoretical purposes, the precise formula for the E_α^J is not important; what is important is that they are uniquely determined by the integration by parts formula (5.124). As a first application, we find the explicit expression for the divergence in the key formula (4.39) used in Noether's theorem.

Proposition 5.98. *Let $Q \in \mathscr{A}^q$, $L \in \mathscr{A}$. Then*

$$\operatorname{pr} \mathbf{v}_Q(L) = Q \cdot \mathsf{E}(L) + \operatorname{Div} A, \tag{5.128}$$

where

$$A_k = \sum_{\alpha=1}^{q} \sum_{\#I \geq 0} \frac{\tilde{\imath}_k + 1}{\#I + 1} D_I[Q_\alpha \mathsf{E}_\alpha^{I,k}(L)], \qquad k = 1, \ldots, p. \tag{5.129}$$

PROOF. We compute

$$\operatorname{Div} A = \sum_{\alpha=1}^{q} \sum_{\#I \geq 0} \sum_{k=1}^{p} \frac{\tilde{\imath}_k + 1}{\#I + 1} D_{I,k}[Q_\alpha \mathsf{E}_\alpha^{I,k}(L)].$$

Now change the summation variable to be $J = (I, k)$, so $\tilde{\imath}_k + 1 = \tilde{\jmath}_k$ and $\#I + 1 = \#J = \sum \tilde{\jmath}_k$. Thus the coefficient of $D_J[Q_\alpha \mathsf{E}_\alpha^J(L)]$ is unity. Comparing this with (5.124), we see that only the terms $Q_\alpha \mathsf{E}_\alpha(L)$ corresponding to $\#J = 0$ are missing. Thus (5.128) follows immediately. □

The higher Euler operators are also intimately connected with the total derivatives.

Proposition 5.99. *Let* $1 \leq \alpha \leq q$, $1 \leq i \leq p$, $\#J \geq 0$. *Then*

$$\mathsf{E}_\alpha^J(D_i P) = \begin{cases} \mathsf{E}_\alpha^{J \setminus i}(P) & \text{if } i \subset J, \\ 0 & \text{if } i \not\subset J, \end{cases} \tag{5.130}$$

for any $P \in \mathscr{A}$.

PROOF. Although this can be proved directly from (5.126), it is simpler to use the uniqueness properties of (5.124). We have

$$\operatorname{pr} \mathbf{v}_Q(D_i P) = \sum_{\alpha, J} D_J[Q_\alpha \mathsf{E}_\alpha^J(D_i P)].$$

On the other hand, by (5.19), this equals

$$D_i \operatorname{pr} \mathbf{v}_Q(P) = \sum_{\alpha, K} D_i D_K[Q_\alpha \mathsf{E}_\alpha^K(P)].$$

Replacing K by $J = (K, i)$ and comparing the two expressions immediately gives (5.130) by uniqueness. □

Corollary 5.100. *A differential function* P *is an "n-th order divergence", i.e. there exist* $Q_I \in \mathscr{A}$, $\#I = n$, *such that* $P = \sum D_I Q_I$, *if and only if* $\mathsf{E}_\alpha^J(P) = 0$ *for all* $\alpha = 1, \ldots, q$, $0 \leq \#J \leq n - 1$.

The Total Homotopy Operator

As in our proof of the Poincaré lemma in Section 1.5, the construction of the homotopy operator for the D-complex rests on a formula for the Lie derivative of a total differential form with respect to an evolutionary vector field. To establish this result, we begin by noting that any operator, such as a total derivative, higher Euler operator or prolonged vector field, which acts on the space $\mathcal{A}$ of differential functions, can be made to act coefficient-wise on the total differential forms. For example, if $\omega = \sum P_I\, dx^I$, $P_I \in \mathcal{A}$, then

$$\operatorname{pr} \mathbf{v}_Q(\omega) = \sum \operatorname{pr} \mathbf{v}_Q(P_I)\, dx^I. \tag{5.131}$$

In particular, the total differential can be written as

$$\mathrm{D}\omega = \sum_{i=1}^{p} D_i(dx^i \wedge \omega) = \sum_{i=1}^{p} dx^i \wedge D_i\omega, \tag{5.132}$$

the D_i's acting only on the coefficients of ω.

The first goal in our construction is to establish a formula mimicking (1.65), but for the total differential. Thus we need to find "interior product" operators $\mathsf{I}_Q\colon \bigwedge_k \to \bigwedge_{k-1}$, $k = 1, \dots, p$, $Q \in \mathcal{A}^q$, such that

$$\operatorname{pr} \mathbf{v}_Q(\omega) = \mathrm{D}\mathsf{I}_Q(\omega) + \mathsf{I}_Q(\mathrm{D}\omega) \tag{5.133}$$

for any $\omega \in \bigwedge_r$, $0 < r < p$. It turns out that this *total interior product* can be written most succinctly in terms of the higher Euler operators:

$$\mathsf{I}_Q(\omega) = \sum_{\alpha=1}^{q} \sum_{\#I \geq 0} \sum_{k=1}^{p} \frac{\tilde{\imath}_k + 1}{p - r + \#I + 1} D_I \left\{ Q_\alpha \mathsf{E}_\alpha^{I,k} \left(\frac{\partial}{\partial x^k} \lrcorner\, \omega \right) \right\}, \qquad \omega \in \textstyle\bigwedge_r. \tag{5.134}$$

Before proving that this does satisfy (5.133), we look at a couple of special cases.

Example 5.101. If $\omega = L\, dx^1 \wedge \cdots \wedge dx^p$, then $\mathsf{I}_Q(\omega) \in \bigwedge_{p-1}$ and hence has the form

$$\mathsf{I}_Q(\omega) = \sum_{k=1}^{p} (-1)^{k-1} A_k\, dx^{\hat{k}}.$$

Since $(-1)^{k-1}\, dx^{\hat{k}} = \partial_{x^k} \lrcorner\, (dx^1 \wedge \cdots \wedge dx^p)$, (5.134) implies that

$$A_k = \sum_{\alpha, I} \frac{\tilde{\imath}_k + 1}{\#I + 1} D_I[Q_\alpha \mathsf{E}_\alpha^{I,k}(L)],$$

which recovers the divergence terms (5.129) in (5.128), which we can rewrite in "homotopy form"

$$\operatorname{pr} \mathbf{v}_Q(\omega) = \mathrm{D}(\mathsf{I}_Q(\omega)) + Q \cdot \mathsf{E}(\omega). \tag{5.135}$$

Example 5.102. Let $r = p - 1$, so ω is of the form

$$\omega = \sum_{k=1}^{p} (-1)^{k-1} P_k\, dx^{\hat{k}}.$$

The $(p-2)$-form $\mathsf{I}_Q(\omega)$ has the form

$$\mathsf{I}_Q(\omega) = \sum_{j<k} (-1)^{j+k-1} R_{jk}\, dx^{\widehat{jk}},$$

where

$$R_{jk} = \sum_{\alpha=1}^{q} \sum_{\#I \geq 0} D_I \left\{ Q_\alpha \left(\frac{\tilde{\imath}_j + 1}{\#I + 2} \mathsf{E}_\alpha^{I,j}(P_k) - \frac{\tilde{\imath}_k + 1}{\#I + 2} \mathsf{E}_\alpha^{I,k}(P_j) \right) \right\}.$$

The Lie derivative formula (5.133) takes the form

$$\mathrm{pr}\, \mathbf{v}_Q(P_k) = \sum_{j=1}^{p} D_j R_{jk} + A_k, \tag{5.136}$$

where A_k is given by (5.129) when $L = \mathrm{Div}\, P$, which, using (5.130), is

$$A_k = \sum_{\alpha, I} \sum_{l \subset I} \frac{\tilde{\imath}_k + 1}{\#I + 1} D_I [Q_\alpha \mathsf{E}_\alpha^{I,k \setminus l}(P_l)]. \tag{5.137}$$

(We leave to the reader the direct verification of (5.136).)

The proof of (5.133) is perhaps the most complex calculation of this book. (However, the present proof of the exactness of the D-complex is much easier than previous computational proofs!) We begin by analyzing the right-hand side using (5.132):

$$\begin{aligned} \mathsf{I}_Q(\mathrm{D}\omega) &= \sum_{l=1}^{p} \mathsf{I}_Q[D_l(dx^l \wedge \omega)] \\ &= \sum_{\alpha, I} \sum_{k,l=1}^{p} \frac{\tilde{\imath}_k + 1}{p - r + \#I} D_I \left\{ Q_\alpha \mathsf{E}_\alpha^{I,k} \left[\frac{\partial}{\partial x^k} \lrcorner D_l(dx^l \wedge \omega) \right] \right\}, \end{aligned} \tag{5.138}$$

since $\mathrm{D}\omega$ is an $(r+1)$-form. The principal constituent in (5.138) is the interior summation

$$\begin{aligned} &\sum_{k,l=1}^{p} (\tilde{\imath}_k + 1) \mathsf{E}_\alpha^{I,k} \left[\frac{\partial}{\partial x^k} \lrcorner D_l(dx^l \wedge \omega) \right] \\ &\quad = \sum_{k,l=1}^{p} (\tilde{\imath}_k + 1) \mathsf{E}_\alpha^{I,k} \left[D_l \left(\frac{\partial}{\partial x^k} \lrcorner (dx^l \wedge \omega) \right) \right], \end{aligned} \tag{5.139}$$

which we break into two pieces according to whether $k = l$ or $k \neq l$. If $k \neq l$, then by (5.130), $\mathsf{E}_\alpha^{I,k} \cdot D_l = \mathsf{E}_\alpha^{I \setminus l, k}$, where, by convention, this operator is 0 if l does not appear in I. Also, according to Exercise 1.37,

$$\frac{\partial}{\partial x^k} \lrcorner (dx^l \wedge \omega) = -dx^l \wedge \left(\frac{\partial}{\partial x^k} \lrcorner \omega \right), \qquad k \neq l.$$

We conclude that

$$\mathsf{E}_\alpha^{I,k} \left[\frac{\partial}{\partial x^k} \lrcorner D_l(dx^l \wedge \omega) \right] = -\mathsf{E}_\alpha^{I \setminus l, k} \left[dx^l \wedge \left(\frac{\partial}{\partial x^k} \lrcorner \omega \right) \right] \quad \text{whenever} \quad k \neq l. \tag{5.140}$$

The case $k = l$ is a bit more delicate. First note that $\mathsf{E}_\alpha^{I,k} \cdot D_k = \mathsf{E}_\alpha^I$, so the relevant sum is

$$\sum_{k=1}^{p} (\tilde{\imath}_k + 1)\mathsf{E}_\alpha^{I,k}\left[\frac{\partial}{\partial x^k} \lrcorner D_k(dx^k \wedge \omega)\right]$$
$$= \sum_{k=1}^{p} \mathsf{E}_\alpha^I\left[\frac{\partial}{\partial x^k} \lrcorner (dx^k \wedge \omega)\right] + \sum_{k=1}^{p} \tilde{\imath}_k \mathsf{E}_\alpha^I\left[\frac{\partial}{\partial x^k} \lrcorner (dx^k \wedge \omega)\right]. \quad (5.141)$$

On the right-hand side of (5.141), we use the two further identities of Exercise 1.37:

$$\sum_{k=1}^{p} \frac{\partial}{\partial x^k} \lrcorner (dx^k \wedge \omega) = (p - r)\omega$$

in the first summation, and

$$\frac{\partial}{\partial x^k} \lrcorner (dx^k \wedge \omega) = \omega - dx^k \wedge \left(\frac{\partial}{\partial x^k} \lrcorner \omega\right)$$

in the second. This yields

$$\sum_{k=1}^{p} (\tilde{\imath}_k + 1)\mathsf{E}_\alpha^{I,k}\left[\frac{\partial}{\partial x^k} \lrcorner D_k(dx^k \wedge \omega)\right]$$
$$= (p - r + \# I)\mathsf{E}_\alpha^I(\omega) - \sum_{k=1}^{p} \tilde{\imath}_k \mathsf{E}_\alpha^I\left[dx^k \wedge \left(\frac{\partial}{\partial x^k} \lrcorner \omega\right)\right] \quad (5.142)$$

since $\sum_{k=1}^{p} \tilde{\imath}_k = \# I$. Combining (5.139), (5.140) and (5.142), we conclude that

$$\sum_{k,l=1}^{p} (\tilde{\imath}_k + 1)\mathsf{E}_\alpha^{I,k}\left[\frac{\partial}{\partial x^k} \lrcorner D_l(dx^l \wedge \omega)\right]$$
$$= (p - r + \# I)\mathsf{E}_\alpha^I(\omega) - \sum_{k,l=1}^{p} (\tilde{\imath}_k + 1 - \delta_l^k)\mathsf{E}_\alpha^{I\setminus l,k}\left[dx^l \wedge \left(\frac{\partial}{\partial x^k} \lrcorner \omega\right)\right]. \quad (5.143)$$

This is our key identity in the proof of (5.133).

We now have

$$\mathsf{I}_Q(\mathsf{D}\omega) = \sum_{\alpha,I} D_I(Q_\alpha \mathsf{E}_\alpha^I(\omega))$$
$$- \sum_{\alpha,I}\sum_{k,l} \frac{\tilde{\imath}_k + 1 - \delta_l^k}{p - r + \# I} D_I\left\{Q_\alpha \mathsf{E}_\alpha^{I\setminus l,k}\left[dx^l \wedge \left(\frac{\partial}{\partial x^k} \lrcorner \omega\right)\right]\right\}.$$

By (5.124), the first summation on the right-hand side is just $\operatorname{pr} \mathbf{v}_Q(\omega)$, so to complete the proof of (5.133), we need only identify the second summation with

$$-\mathsf{D}\mathsf{I}_Q(\omega) = -\sum_{l=1}^{p} D_l\left\{dx^l \wedge \sum_{\alpha,J}\sum_{k} \frac{\tilde{\jmath}_k + 1}{p - r + \# J + 1} D_J\left[Q_\alpha \mathsf{E}_\alpha^{J,k}\left(\frac{\partial}{\partial x^k} \lrcorner \omega\right)\right]\right\}$$
$$= -\sum_{\alpha,J}\sum_{k,l} \frac{\tilde{\jmath}_k + 1}{p - r + \# J + 1} D_{J,l}\left[Q_\alpha \mathsf{E}_\alpha^{J,k}\left(dx^l \wedge \left(\frac{\partial}{\partial x^k} \lrcorner \omega\right)\right)\right].$$

But these two summations agree upon changing the multi-index summation variable from J to $I = (J, l)$, noting that $\tilde{\imath}_k = \tilde{\jmath}_k + \delta_l^k$, $\#I = \#J + 1$. This completes the proof of (5.133).

We now specialize (5.133) to the case of the scaling vector field $\mathrm{pr}\,\mathbf{v}_u$ introduced earlier in the proof of the exactness of the variational complex. Note that if $P[u] = P(x, u^{(n)})$ is any smooth differential function defined on a vertically star-shaped domain, then

$$\frac{d}{d\lambda} P[\lambda u] = \sum_{\alpha, J} u_J^\alpha \frac{\partial P}{\partial u_J^\alpha}[\lambda u] = \frac{1}{\lambda} \mathrm{pr}\,\mathbf{v}_u(P)[\lambda u],$$

where the notation means that we first apply $\mathrm{pr}\,\mathbf{v}_u$ to P and then evaluate at λu. Integrating, we find

$$P[u] - P[0] = \int_0^1 \mathrm{pr}\,\mathbf{v}_u(P)[\lambda u] \frac{d\lambda}{\lambda},$$

where $P[0] = P(x, 0)$ is a function of x alone. Since $\mathrm{pr}\,\mathbf{v}_u$ acts coefficient-wise on a total differential form $\omega(x, u^{(n)})$, we have the analogous formula

$$\omega[u] - \omega[0] = \int_0^1 \mathrm{pr}\,\mathbf{v}_u(\omega)[\lambda u] \frac{d\lambda}{\lambda}, \tag{5.144}$$

where $\omega[0] = \omega(x, 0)$ is an ordinary differential form on the base space X. If we now use (5.133) in the case $Q = u$, whereby

$$\mathsf{I}_u(\omega) = \sum_{\alpha=1}^{q} \sum_{I} \sum_{k=1}^{p} \frac{\tilde{\imath}_k + 1}{p - r + \#I + 1} D_I \left\{ u^\alpha \mathsf{E}_\alpha^{I,k} \left(\frac{\partial}{\partial x^k} \lrcorner\, \omega \right) \right\}, \tag{5.145}$$

we obtain the homotopy formula

$$\omega[u] - \omega[0] = \mathsf{DH}(\omega) + \mathsf{H}(\mathsf{D}\omega), \tag{5.146}$$

where the *total homotopy operator* is

$$\mathsf{H}(\omega) = \int_0^1 \mathsf{I}_u(\omega)[\lambda u] \frac{d\lambda}{\lambda}, \tag{5.147}$$

meaning that we first evaluate $\mathsf{I}_u(\omega)$ and then replace u by λu wherever it occurs. Except for the extra term $\omega[0]$ this would suffice to prove the exactness of the D-complex. However, $\omega[0]$ is an ordinary differential form on $\Omega = M \cap \{u = 0\}$, so provided Ω is also star-shaped we can use the ordinary Poincaré homotopy operator (1.69), with

$$\omega[0] - \omega_0 = dh(\omega[0]) + h(d\omega[0]), \tag{5.148}$$

where $\omega_0 = 0$ if $r > 0$, while $\omega_0 = f(0)$ if $\omega[0] = f(x)$ is a function, $r = 0$. For such forms, the total derivatives D_i and the partial derivatives $\partial/\partial x^i$ are the same, so we can replace the differential d by the total differential D. Combining (5.146) and (5.148), we obtain

$$\omega - \omega_0 = \mathsf{DH}^*(\omega) + \mathsf{H}^*(\mathsf{D}\omega), \tag{5.149}$$

for $\omega \in \bigwedge_r$, $0 \leqslant r < p$, where

$$\mathsf{H}^*(\omega) = \mathsf{H}(\omega) + h(\omega[0])$$

is the combined homotopy operator. Now that we have established (5.149), the proof of Theorem 5.80 is straightforward.

The homotopy formula (5.146) also extends to total p-forms $\omega = L[u]\, dx^1 \wedge \cdots \wedge dx^p$ if we utilize the modified formula (5.135) for the Lie derivative. Translating the differential form language, we see that if $L[u] = L(x, u^{(n)})$ is defined over a totally star-shaped domain $M \subset X \times U$, then

$$L[u] = \operatorname{Div} B^*[u] + \int_0^1 u \cdot \mathsf{E}(L)[\lambda u]\, d\lambda, \tag{5.150}$$

where B^* is the sum of the p-tuples $B[u] \in \mathscr{A}^p$ and $b(x)$ with entries

$$\begin{aligned} B_k(u) &= \int_0^1 \sum_{\alpha=1}^{q} \sum_I \frac{\tilde{\imath}_k + 1}{\#I + 1} D_I(u^\alpha \mathsf{E}_\alpha^{I,k}(L)[\lambda u])\, d\lambda, \\ & \qquad\qquad\qquad\qquad\qquad\qquad k = 1, \ldots, p. \\ b_k(x) &= \int_0^1 x^k L(\lambda x, 0)\, d\lambda, \end{aligned} \tag{5.151}$$

In particular, if $\mathsf{E}(L) = 0$, then $L = \operatorname{Div} B^*$ with B^* as above. Thus we have an explicit formula expressing any null Lagrangian as a divergence.

Example 5.103. Let $p = 2$, $q = 1$. Consider the null Lagrangian $L = u_x u_{yy}$. According to (5.150), $L = D_x A + D_y B$, where

$$A = \int_0^1 \{u \mathsf{E}^{(x)}(L) + D_x(u \mathsf{E}^{(xx)}(L)) + \tfrac{1}{2} D_y(u \mathsf{E}^{(xy)}(L)) + \cdots\}\, d\lambda,$$

$$B = \int_0^1 \{u \mathsf{E}^{(y)}(L) + \tfrac{1}{2} D_x(u \mathsf{E}^{(xy)}(L)) + D_y(u \mathsf{E}^{(yy)}(L)) + \cdots\}\, d\lambda,$$

where the differential functions $\mathsf{E}^{(x)}(L)$, $\mathsf{E}^{(y)}(L)$, etc. are to be evaluated at λu. In the case $L = u_x u_{yy}$, we have

$$\mathsf{E}^{(x)}(L) = u_{yy}, \qquad \mathsf{E}^{(y)}(L) = -2u_{xy}, \qquad \mathsf{E}^{(yy)}(L) = u_x,$$

and all the other terms in A and B vanish. (See Example 5.97.) Thus

$$A = \int_0^1 u(\lambda u_{yy})\, d\lambda = \tfrac{1}{2} u u_{yy},$$

$$B = \int_0^1 u(-2\lambda u_{xy}) + D_y[u(\lambda u_x)]\, d\lambda = -\tfrac{1}{2} u u_{xy} + \tfrac{1}{2} u_x u_y,$$

satisfy the above divergence identity. Even from this relatively simple example, it is easy to see how the homotopy formula (5.151) can rapidly become unmanageable. In practice, it is often easier to determine the divergence form directly by inspection, using (5.151) only as a last resort.

We conclude this section by specializing (5.149) to the case of total $(p-1)$-forms.

Theorem 5.104. *Let $P \in \mathscr{A}^q$ and let $L = \operatorname{Div} P$. Then*

$$P_k = \sum_{j=1}^{p} D_j Q_{jk}^* + B_k^*, \qquad k = 1, \ldots, p, \tag{5.152}$$

where $B^* = B + b$ is the p-tuple determined by (5.151), and $Q_{jk}^* = Q_{jk} + q_{jk}$, where

$$Q_{jk}[u] = \int_0^1 \sum_{\alpha=1}^{q} \sum_I D_I \left\{ u^\alpha \left[\frac{\tilde{\imath}_j + 1}{\#I + 2} \mathsf{E}_\alpha^{I,j}(P_k)[\lambda u] - \frac{\tilde{\imath}_k + 1}{\#I + 2} \mathsf{E}_\alpha^{I,k}(P_j)[\lambda u] \right] \right\} d\lambda, \tag{5.153}$$

$$q_{jk}(x) = \int_0^1 \{ x^j P_k(\lambda x, 0) - x^k P_j(\lambda x, 0) \} \, d\lambda.$$

In particular, if P is a null divergence, then (5.153) shows how to write P explicitly as a "total curl" $P_k = \sum D_j Q_{jk}^*$, where $Q_{jk}^* = -Q_{kj}^*$. For example, in the case $p = 2$, we have

$$D_x P + D_y \tilde{P} = 0$$

if and only if

$$P = D_y Q, \qquad \tilde{P} = -D_x Q,$$

where

$$Q = \int_0^1 \{ \tfrac{1}{2} u \mathsf{E}^{(y)}(P) + \tfrac{1}{3} D_x[u \mathsf{E}^{(xy)}(P)] + \tfrac{2}{3} D_y[u \mathsf{E}^{(yy)}(P)] + \cdots$$
$$- \tfrac{1}{2} u \mathsf{E}^{(x)}(\tilde{P}) - \tfrac{2}{3} D_x[u \mathsf{E}^{(xx)}(\tilde{P})] - \tfrac{1}{3} D_y[u \mathsf{E}^{(xy)}(\tilde{P})] - \cdots \} \, d\lambda, \tag{5.154}$$

where $\mathsf{E}^{(x)}(\tilde{P})$, $\mathsf{E}^{(y)}(P)$, etc. are all evaluated at λu. As a specific example, let $P = u_y u_{xy} + u_x u_{yy}$, $\tilde{P} = -u_y u_{xx} - u_x u_{xy}$, which do form a null divergence. The only nonzero Euler expressions appearing in (5.154) are

$$\mathsf{E}^{(y)}(P) = -2u_{xy}, \qquad \mathsf{E}^{(xy)}(P) = u_y, \qquad \mathsf{E}^{(yy)}(P) = u_x,$$
$$\mathsf{E}^{(x)}(\tilde{P}) = 2u_{xy}, \qquad \mathsf{E}^{(xx)}(\tilde{P}) = -u_y, \qquad \mathsf{E}^{(xy)}(\tilde{P}) = -u_x.$$

Thus

$$Q = \int_0^1 \{ u(-\lambda u_{xy}) + \tfrac{1}{3} D_x[u(\lambda u_y)] + \tfrac{2}{3} D_y[u(\lambda u_x)]$$
$$- u(\lambda u_{xy}) + \tfrac{2}{3} D_x[u(\lambda u_y)] + \tfrac{1}{3} D_y[u(\lambda u_x)] \} \, d\lambda = u_x u_y$$

satisfies $P = D_y Q$, $\tilde{P} = -D_x Q$.

Notes

Generalized symmetries first made their appearance in their present form in the fundamental paper of Noether, [1], in which their role in the construction of conservation laws was clearly enunciated. Anderson and Ibragimov, [1], and Ibragimov, [1], try to make the case that they date back to the investigations of Lie and Bäcklund, hence their choice of the term "Lie–Bäcklund transformation" for these objects. As far as I can determine, Lie only allows dependence of the group transformations on derivatives of the dependent variables in his theory of contact transformations, which are much more restrictive than generalized symmetries. He further, in Lie, [1; § 1.4], proposes the problem of looking at higher order generalizations of these contact transformations. Bäcklund, [1], in response, does write down transformations depending on derivatives of the dependent variables of arbitrary order, and so in a sense anticipates the theory of generalized symmetries, but he and Lie always require that the corresponding prolongations "close off" to define genuine geometrical transformations on some jet space. Bäcklund concludes that the only such transformations are the prolongations of ordinary point transformations or of Lie's contact transformations, hence fails to make the jump to true generalized symmetries. More telling is the fact that Bäcklund requires his transformations to depend on only finitely many derivatives of the dependent variables, while for true generalized symmetries, this is only the case for the infinitesimal generators (which Bäcklund never considers); the group transformations determined by the solutions of the associated evolution equation (5.14) are truly nonlocal, and are not determined solely by the values of finitely many derivatives of the dependent variables at a single point. In particular, in the case of one dependent variable, every contact transformation is equivalent to a first order generalized symmetry, whereas for more than one dependent variable, every contact transformation is the prolongation of a point transformation.

Since their introduction by Noether, generalized symmetries have been rediscovered many times, including the papers of Johnson, [1], [2], in differential geometry, Steudel, [1], in the calculus of variations, and Anderson, Kumei and Wulfman, [1], among others. Recent applications to differential equations can be found in Anderson and Ibragimov, [1], Kosmann-Schwarzbach, [1], [2], Fokas, [3], and Ibragimov, [1]; the last reference includes an extensive discussion of those second and third order evolution equations, as well as general second order equations in two independent variables, admitting generalized symmetries. Steudel, [1], was the first to note the importance of placing a generalized vector field in its evolutionary form. Recent investigations into the symmetry properties of systems of linear equations, including those of field theory (Fushchich and Nikitin, [2], and Kalnins, Miller and Williams, [1]) and elasticity (Olver, [9], [14]) have uncovered new generalized symmetries depending on first order derivatives of the dependent variables, whose significance is not yet fully understood, although they appear to play a role in the separation of variables for such systems.

An outstanding problem in the theory of generalized symmetries is whether a system of partial differential equations can admit only a finite-dimensonal space of generalized symmetries. Exercise 5.3 in the first edition, based on a paper of Ibragimov and Shabat, [1], was not correct; see Exercise 5.16 for a candidate system. Sokolov, [1; Proposition 11], proves that any evolution equation with an infinite-dimensional algebra of generalized symmetries always has an infinite-dimensional commutative subalgebra. The Theorem of Tu presented in the first edition is not correct; however see Exercise 5.4 for what can be proven in this direction, and Exercise 5.15 for a counterexample to the theorem as stated in Tu, [1], and the first edition. A more complete discussion of the properties of the heat equation mentioned in Example 5.11 can be found in Kovalevskaya, [1], Forsyth, [1; Vol. 5, §26] and Copson, [1; §12.4, 12.5].

Various further extensions of the concept of symmetry have appeared since the first edition came out. Nonlocal symmetries and conservation laws have received a fair amount of recent attention—see Krasilshchik and Vinogradov, [1], Vladimorov and Volovich, [1], [2], and Akhatov, Gazizov and Ibragimov, [1]. Also see Bluman and Kumei, [2], for a development of the concept of a potential symmetry. (However, there is not yet, as far as I know, an adequate computational calculus for determining symmetries of nonlocal (integro-differential) equations.) Anderson, Kamran and Olver, [1], discuss internal symmetries of differential equations, including a generalization of Bäcklund's theorem showing that for most systems every internal symmetry is equivalent to a first order generalized symmetry. Fushchich and Shtelen, [1], and Baikov, Gazizov and Ibragimov, [1], discuss approximate symmetries of perturbed differential equations.

The use of recursion operators for constructing infinite families of generalized symmetries is based on the recursive construction of the higher order Korteweg–de Vries equations due to Lenard (see Gardner, Greene, Kruskal and Miura, [1]) and was first presented in the general form in Olver, [1]. The extension of the theory of recursion operators to nonlinear partial differential equations in more than two independent variables was the source of considerable difficulty in the 1980's. Zakharov and Konopelchenko, [1], proved that recursion operators naturally generalizing those discussed here do not exist. However, Fokas and Santini, [1], and Santini and Fokas, [1], finally discovered the correct form of these operators for soliton equations; see also the review paper by Fokas, [2]. The concept of a hereditary operator was introduced in Fuchssteiner, [1]. See Li Yi-Shen, [1], for examples of recursion operators which do not satisfy the hereditary property. Master symmetries were introduced by Chen, Lee and Lin, [1], and Fuchssteiner, [2]; see also Oevel, [1].

For linear partial differential equations, higher order symmetries have been directly applied to the method of separation of variables by Miller, Kalnins, Boyer, Winternitz and others using the operator-theoretic approach mentioned in the text; see Miller, [3], and the references therein. The results concerning the number of independent symmetries of a given order of Laplace's equation and the wave equation appear in Delong's thesis, [1]. Later, Shapovalov and Shirokov, [1], proved that every generalized symmetry of these two equation is a linear symmetry, which is a polynomial in the first order symmetries. However, the latter statement is not true for more general linear equations; see Exercise 5.2.

The calculus of pseudo-differential operators discussed in the section on formal symmetries originates in studies of Lax representations of soliton equations due to Gel'fand and Dikii, [3], Adler, [1], who proved Proposition 5.73, and Wilson, [1]. In Adams, Ratiu and Schmidt, [1], [2], an attempt was made to make the formal calculus rigorous, utilizing the analytical theory of pseudo-differential operators, cf. M. Taylor, [1]. Formal symmetries were developed in the work of Shabat and collaborators; see Sokolov and Shabat, [1], Mikhailov, Shabat and Yamilov, [1], [2], Sokolov, [1], and, especially, the survey paper by Mikhailov, Shabat and Sokolov, [1].

If we omit the part concerning trivial symmetries and conservations laws, the version of Noether's Theorem 5.58 stated here dates back to Bessel-Hagen, [1]. (See Exercise 5.33 for Noether's original version, which does not use divergence symmetries.) The correspondence between nontrivial conservation laws and nontrivial variational symmetry groups proved here can be found in Alonso, [1], Olver, [11], and Vinogradov, [5]. Proposition 5.56 concerning the geometric interpreation of the group transformations of a variational symmetry can be found in Edelen, [1; p. 149]. The existence of infinite families of conservation laws for self-adjoint linear systems of differential equations was the cause of some astonishment in the mid-1960's with the discovery of the "zilch tensor" and related objects by Lipkin, [1], T. A. Morgan, [1] and Kibble, [1], in their work on field theories. An explanation using generalized symmetries and the full version of Noether's theorem similar to Proposition 5.62 was soon proposed by Steudel, [3]. Proposition 5.64 discussing the action of symmetries on conservation laws also appears in Khamitova, [1]. See Abellanas and Galindo, [1], for further results on conservation laws of linear systems.

The statement and proof of Noether's Second Theorem 5.66 on infinite-dimensional symmetry groups is from Noether's paper, [1]. The connections with the abnormality of the underlying system of Euler–Lagrange equations, however, is new; see Olver, [11]. One outstanding problem here is to complete the classification of symmetries and conservation laws for over-determined systems of Euler–Lagrange equations. In particular, does there exist an over-determined system for which a trivial variational symmetry group gives rise to a nontrivial conservation law? Such a system must be quite complicated; for instance, Exercise 5.51 shows that it cannot be homogeneous in u and its derivatives. (Fokas, [2], refers to an example of Ibragimov where this occurs, but the cited paper does not contain the purported example.)

The history of the variational complex and, in particular, the inverse problem in the calculus of variations is quite interesting. Helmholtz, [1], first proposed the problem of determining which systems of differential equations are the Euler–Lagrange equations for some variational problem and found necessary conditions in the case of a single second order ordinary differential equation. Mayer, [1], generalized Helmholtz' conditions to the case of first order Lagrangians involving one independent variable and several dependent variables, and also proved they sufficed to guarantee the existence of a suitable functional. In two incisive papers on this subject Hirsch, [1], [2], extended these results to the cases of higher order Lagrangians involving either one independent and several dependent variables, or two or three independent and one dependent variable. Hirsch's papers also include further results on what order derivatives can appear in the Lagrangian, as well as the "multiplier problem": when can one multiply a differential equation by a differential function so as to make it an Euler–Lagrange equation? However, the general self-adjointness condition and the homotopy formula (5.123) were independently discovered by Volterra, [1; pp. 43, 48]; see also Vainberg, [1], for a modern version. The next major work on the inverse problem was the profound paper of Douglas, [1], in which he states and solves the problem of determining when a system of two second order ordinary differential equations is *equivalent* to the Euler–Lagrange equations for some functional depending on at most first order derivatives of the dependent variables; the complexity of his solution no doubt hindered further research in this direction. Further recent work on this more difficult version of the inverse problem—when a system of differential equations is equivalent to a system of Euler–Lagrange equations—can be found in Anderson and Duchamp, [2], and Henneaux, [1]. The general case, however, remains unsolved to this day. See also Atherton and Homsy, [1], and Anderson and Thompson, [1], for further references on the inverse problem.

In the early 1970's, the inverse problem was seen to be part of a much larger machine—the variational complex and, more generally, the variational bicomplex—which has evolved into a pre-eminent position in the geometric theory of the calculus of variations. Intimations of this machinery can be found in Dedecker's work, [1], on the applications of algebraic topology to the calculus of variations. This complex first appears explicitly in the work of Vinogradov, [1], where deep methods from algebraic topology are used to prove exactness. A closely related complex appears in contemporaneous work of Tulczyjew, [1], [2]. Further developments, including applications to conservation laws, Cartan forms and characteristic classes, are to be found in Kupershmidt, [1], Takens, [1], Anderson and Duchamp, [1], Tsujishita, [1], [2], and Anderson, [1]. (A different complex including the solution to the inverse problem can be found in Olver and Shakiban, [1], and Shakiban, [1].) The formal variational calculus methods used in the development of this complex presented in Section 5.4, and in particular the abstract definition of a functional, owes much to the work of Gel'fand and Dikii, [1], [2], on the Korteweg–de Vries equation. Further developments of this complex can be found in the comprehensive papers of Vinogradov, [2], [3], [4], and in Olver, [4]. The new proof of exactness of the variational complex presented here was discovered by I. Anderson; the homotopy operators (5.147) serve to considerably simplify the earlier computational proofs of exactness of the D-complex of Takens, [1], and Anderson and Duchamp, [1]. The higher order Euler operators themselves first appeared in work of Kruskal, Miura, Gardner and Zabusky, [1], on the Korteweg–de Vries equation, and were subsequently developed by Aldersley, [1], Galindo and Martinez-Alonso, [1], and Olver, [3]. The present presentation is due to I. Anderson.

Exercises

5.1. Prove that the full symmetry group of the Kepler problem in $\mathbb{R}^3$, including those symmetries giving the Runge–Lenz vector, is locally isomorphic to the group SO(3, 1) of "rotations" in $\mathbb{R}^4$ preserving the Lorentz metric $(dx^1)^2 + (dx^2)^2 + (dx^3)^2 - (dx^4)^2$. (Goldstein, [1; p. 422].)

*5.2. The Schrödinger equation for a hydrogen atom is the quantized version of the Kepler problem and takes the form

$$\Delta u + |\mathbf{x}|^{-1} u = \lambda u,$$

where λ is a constant, $\mathbf{x} \in \mathbb{R}^3$ and $u \in \mathbb{R}$.

(a) Find the geometric symmetries of this equation for different values of λ.

(b) Prove that the "Runge–Lenz vector"

$$\mathbf{R}[u] = (\mathbf{x} \times \nabla) \times \nabla u - \nabla \times (\mathbf{x} \times \nabla u) + 2|\mathbf{x}|^{-1} \mathbf{x} u$$

provides three second order generalized symmetries, their characteristics being the three components of $\mathbf{R}$. Further show that these symmetries are *not* derivable from the first order symmetries of the equation coming from the evolutionary forms of the geometrical symmetries of part (a).

(Miller, [2; p. 376], Kalnins, Miller and Winternitz, [1].)

5.3. Let $u_t = K[u]$ be an n-th order evolution equation, and let $Q[u]$ be an m-th order symmetry. Prove that $(\partial K/\partial u_n)^m = c(\partial Q/\partial u_m)^n$ for some constant c.

5.4. Consider an evolution equation of the form

$$u_t = c u_n + \tilde{K}(u, u_x, \ldots, u_{n-1}),$$

where $c \neq 0$ is a constant and $u_n = D_x^n u$. (These include the Korteweg–de Vries equation and Burgers' equation.)

(a) Prove that every m-th order, x, t-independent, generalized symmetry has a characteristic of the same form: $Q[u] = \hat{c} u_m + \tilde{Q}(u, u_x, \ldots, u_{m-1})$, $\hat{c} \neq 0$. (Compare with Exercise 5.3.)

(b) Use this fact to prove that any two x, t-independent generalized symmetries $\mathbf{v}_Q$, $\mathbf{v}_R$ of such an evolution equation necessarily commute: $[\mathbf{v}_Q, \mathbf{v}_R] = 0$. (Tu, [1].)

5.5. Let Δ be a linear system of differential equations and $\mathscr{D}$ a linear recursion operator. Prove that whenever $u = f(x)$ is a solution to Δ, so is $\tilde{u} = \mathscr{D} f(x)$. How is $\exp(t\mathscr{D})u = u + t\mathscr{D}u + \cdots$ related to the flow generated by the symmetry with characteristic $\mathscr{D}u$? Does this result generalize to recursion operators for nonlinear systems?

5.6. Let $u_t = \mathscr{D}u$ be a linear evolution equation and $(\mathscr{E}u)\partial_u$ a linear symmetry. Prove that $(\mathscr{E}^*u)\partial_u$ is a linear symmetry of the adjoint equation $u_t = -\mathscr{D}^*u$.

5.7. Prove that, under appropriate existence and uniqueness assumptions, two evolutionary vector fields commute, $[\mathbf{v}_Q, \mathbf{v}_R] = 0$, if and only if their one-parameter groups $\exp(\varepsilon \mathbf{v}_Q)$, $\exp(\tilde{\varepsilon} \mathbf{v}_R)$ commute. Interpret.

5.8. An alternative approach to the definition of the flow associated with a generalized vector field $\mathbf{v} = \sum \xi^i \partial_{x^i} + \sum \phi_\alpha \partial_{u^\alpha}$ would be as follows. Consider the infinite prolongation (5.3), and write down the infinite system of ordinary differential equations

$$\frac{dx^i}{d\varepsilon} = \xi^i, \qquad \frac{du^\alpha}{d\varepsilon} = \phi_\alpha, \qquad \frac{du_J^\alpha}{d\varepsilon} = \phi_\alpha^J.$$

Define the flow of $\mathbf{v}$ on the infinite jet space to be the solution of this system with given initial values $(x, u^{(\infty)}) = (x^i, u^\alpha, u_J^\alpha)$:

$$\exp[\varepsilon \operatorname{pr} \mathbf{v}](x, u^{(\infty)}) = (x(\varepsilon), u^{(\infty)}(\varepsilon)).$$

For the "heat vector field" $\mathbf{v} = u_{xx}\partial_u$, compare this method with the evolutionary method (5.18) in the case of analytic initial data. Does this carry over to more general vector fields? (Anderson and Ibragimov, [1].)

5.9. (a) What happens if one applies the recursion operators for Burgers' equation to the symmetry with characteristic $\rho(x, t)e^{-u}$, ρ a solution to the heat equation? (See Example 5.30.)
(b) How are the recursion operators for Burgers' equation and the heat equation related through the Hopf–Cole transformation of Example 2.42?

5.10. (a) Prove that the nonlinear diffusion equation $u_t = D_x(u^{-2}u_x)$ admits the recursion operator $\mathscr{R} = D_x^2 \cdot u^{-1} D_x^{-1}$. How is this related to Exercise 2.22(d)?
(b) Prove that a general nonlinear diffusion equation $u_t = D_x(K(u)u_x)$ admits generalized symmetries if and only if K is either constant or a multiple of $(u + c)^{-2}$. (Bluman and Kumei, [1].)

5.11. The modified Korteweg–de Vries equation is $u_t = u_{xxx} + u^2 u_x$.
(a) Compute the geometrical symmetry group of this equation.
(b) Prove that the operator $\mathscr{R} = D_x^2 + \frac{2}{3}u^2 + \frac{2}{3}u_x D_x^{-1} \cdot u$ is a recursion operator. (The last summand is the operator that takes a differential function, multiplies it by u, then takes D_x^{-1} (if possible) and finally multiplies the result by $\frac{2}{3}u_x$.) What are the first few generalized symmetries of this equation?
(c) Let $v_x = u$, so v will be a solution of the "potential modified Korteweg–de Vries equation" $v_t = v_{xxx} + \frac{1}{3}v_x^3$. Find a recursion operator for this equation.
(d) Prove that if u is any solution to the modified Korteweg–de Vries equation, then its Miura transformation $w = u^2 + \sqrt{-6}u_x$ is a solution to the Korteweg–de Vries equation. How do the symmetries and recursion operators of these two equations match up? (Miura, [1], Olver, [1].)

*5.12. Prove that the operator $\mathscr{R} = D_x^2 + u_x^2 - u_x D_x^{-1} \cdot u_{xx}$ is a recursion operator for the sine–Gordon equation $u_{xt} = \sin u$. (*Hint*: In (5.42), $\tilde{\mathscr{R}} = \mathscr{R}^*$.) What are some generalized symmetries? Can you relate them to those of the potential modified Korteweg–de Vries equation in the previous example? (*Hint*: Try scaling v.) (Olver, [1].)

5.13. Discuss the conservation laws and linear recursion operators for the following linear equations:
(a) The telegraph equation $u_{tt} = u_{xx} + u$. (See Exercise 2.9.)
(b) The axially symmetric wave equation $u_{tt} - u_{xx} + (u/x) = 0$. (See Exercise 3.1.)

5.14. Find a recursion operator for the generalized Burgers' equation

$$u_t = u_{xx} + uu_x + h(x).$$

5.15. Prove that the operator $u^2 D_x^2 - uu_x D_x + uu_{xx} + u^3 u_{xxx} D_x^{-1} u^{-2}$ is a recursion operator for the alternative form $u_t = u^3 u_{xxx}$ of the Harry Dym equation. Is it hereditary? (Leo, Leo, Soliani, Solombrino, and Mancarella, [1].)

**5.16. (a) Prove that $\mathbf{v} = (u_{xxx} + 3vv_x)\partial_u + 4v_{xxx}\partial_v$ is a generalized symmetry for the evolutionary system

$$u_t = u_{xx} + \tfrac{1}{2}v^2, \qquad v_t = 2v_{xx}.$$

Find a higher order generalized symmetry and a recursion operator.
(b) Prove that the system

$$u_t = u_{xxxx} + v^2, \qquad v_t = \tfrac{1}{5}v_{xxxx}$$

has a sixth order generalized symmetry. (It seems likely that this is an example of an equation with only one generalized symmetry, but a rigorous proof of this fact seems to be quite difficult. Computer experiments of Bakirov, [1], have shown that the system has no other generalized symmetries of order $\leqslant 53$.)

5.17. Prove directly that the two types of Lie derivatives of differential operators (5.37), (5.40) are invariant under a change of variables $v = \varphi(x, u)$ (with appropriate action on the differential operator). Are changes of independent variables allowed?

5.18. Prove that the recursion operators for the physical form of Burgers' equation and the Korteweg–de Vries equation are hereditary operators. Discuss the same question for the recursion operators in Exercises 5.14–5.16. (Fuchssteiner, [1].)

5.19. Prove that a second order evolution equation of the form $u_t = \rho(x, u)u_{xx} + \sigma(x, u)u_x^2$ admits a formal symmetry of rank 3 if and only if either $\sigma = (\rho\rho_{uu}/\rho_u) - \tfrac{1}{2}\rho_u$ or $\rho_u = 0$. Find a recursion operator for the first type of equation. (This equation arises in the study of flow in porous media, Fokas and Yortsos, [1].)

5.20. Given a (formal) Laurent series $P[u, \xi] = \sum_{i=-\infty}^{n} P_i[u]\xi^i$ whose coefficients are differential functions, we can define an associated pseudo-differential operator $\mathscr{D} = P[u, D_x]$ by formally substituting the operator D_x for the variable ξ. Prove that the Leibniz rule for multiplying pseudo-differential operators can be restated in the following convenient form. If $P[u, \xi]$ and $Q[u, \xi]$ are Laurent series determining pseudo-differential operators $\mathscr{D} = P[u, D_x]$, $\mathscr{E} = Q[u, D_x]$, then the product operator $\mathscr{D} \cdot \mathscr{E} = R[u, D_x]$ is determined by the Laurent series

$$R[u, \xi] = \sum_{i=0}^{\infty} \frac{1}{i!} \frac{\partial^i P}{\partial \xi^i} D_x^i Q.$$

(Gel'fand and Dikii, [3].)

5.21. Prove that the coefficient of D_x^{-n} in $(D_x + u)^{-1}$ is given by $(-1)^{n-1}(D_x + u)^{n-2}u$ for $n \geqslant 2$.

5.22. Prove that any pseudo-differential operator of order $-m$, $m > 0$, has an m-th root.

*5.23. Suppose $L = D_x^2 + u$. Using the (positive) square root pseudo-differential operator $\sqrt{L} = D_x + \frac{1}{2}uD_x^{-1} + \cdots$, define the differential operators $B_m = (L^{m/2})_+$. Here the $+$ subscript means differential operator truncation, so that if $\mathscr{D} = \sum_{i=-\infty}^{n} P_i D_x^i$, then $\mathscr{D} = \mathscr{D}_+ + \mathscr{D}_-$, where $\mathscr{D}_+ = \sum_{i=0}^{n} P_i D_x^i$ and $\mathscr{D}_- = \sum_{i<0} P_i D_x^i$.
(a) Prove that the commutator $[B_m, L]$ is zero if m is even and is a nonzero multiplication operator if m is odd. (*Hint*: Use the fact that the power $L^{m/2} = (L^{m/2})_+ + (L^{m/2})_-$ commutes with L and that $(L^{m/2})_-$ has order strictly less than 0.)
(b) Prove that the *Lax representation* $L_t = [B_m, L]$ for m odd defines an evolution equation $u_t = K_m[u]$ of order m. Show that the case $m = 3$ produces the Korteweg–de Vries equation (up to are scaling).
(c) Prove that the higher order equations are the higher order Korteweg–de Vries equations.
(d) Perform an analogous construction for the third order differential operator $L = D_x^3 + uD_x + v$. The case $B_2 = (L^{2/3})_+$ yields the Boussinesq equation, which is, like the Korteweg–de Vries equation, a model for shallow water waves—see Example 7.28.
(Gel'fand and Dikii, [1], Wilson, [1], Dickey, [1].)

*5.24. Discuss the symmetries and conservation laws of the Helmholtz equation $\Delta u + \lambda u = 0$, $x \in \mathbb{R}^3$.

**5.25. Discuss the generalized symmetries of Maxwell's equations. (See Exercise 2.16.) What about conservation laws? (Pohjanpelto, [1].)

*5.26. (a) Derive the conservation laws for the two-dimensional wave equation listed in Example 5.63. Compare the direct method with the method from Example 5.65. Continue the list to find new second order conserved densities for the wave equation.
(b) Let $x \in \mathbb{R}^p$, $t \in \mathbb{R}$, $u \in \mathbb{R}^q$, and consider an autonomous system of partial differential equations $\Delta(x, u^{(n)}) = 0$ involving u and its x and t derivatives in which t does not explicitly appear. Prove that if $T(x, t, u^{(m)})$ is a conserved density, so are the partial derivatives $\partial T/\partial t$, $\partial^2 T/\partial t^2$, etc. Use this result to check your work in part (a).

5.27. (a) Let $u_t = \mathscr{D}u$ be a linear evolution equation. Prove that $\int q(x, t)u\, dx$ is conserved if and only if $q(x, t)$ is a solution to the adjoint equation $q_t = -\mathscr{D}^* q$.
(b) Prove that if $q(x, t) = (x - 2t\partial_x)^m(1)$, then $\int q(x, t)u\, dx$ is a conservation law for the heat equation $u_t = u_{xx}$. What does this imply about the time evolution of the moment $\int x^m u(x, t)\, dx$ when u is a solution to the heat equation?
(c) Do the same as part (b) for the Fokker–Planck equation of Exercise 2.8. (Steinberg and Wolf, [1].)

5.28. Try to generalize Example 5.50 by discussing the validity of the following statement: If $u_t = P(x, u, \ldots, u_{2m})$ is an evolution equation in only one spatial variable and $\partial P/\partial u_{2m} \neq 0$, then the only nontrivial conservation laws have characteristics independent of u and its derivatives.

5.29. Hamilton's principle in geometrical optics requires the minimization of the integral $\int_a^b N(\mathbf{x})|d\mathbf{x}/dt|\,dt$ in which $\mathbf{x}(t) \in \mathbb{R}^3$, $N(\mathbf{x})$ is the optical index of the material and $|\cdot|$ is the usual length on $\mathbb{R}^3$. What are the Euler–Lagrange equations? Prove that the variational symmetries of space translations and rotations lead, respectively, to Snell's law in the form $N\mathbf{n} =$ constant, where $\mathbf{n} = |d\mathbf{x}/dt|^{-1}\,d\mathbf{x}/dt$ is the unit velocity vector, and the "Optical sine theorem" $N\mathbf{n} \times \mathbf{x} =$ constant. Further, prove that the time translational symmetry leads to a trivial conservation law. What does this imply about the Euler–Lagrange equations? (Baker and Tavel, [1].) (Hint: see Exercise 5.34.)

5.30. Let $p = q = 3$. Prove that any functional $\mathscr{L}[u] = \int L(\operatorname{div}\mathbf{u})\,dx$ depending only on $\operatorname{div}\mathbf{u} = u_x + v_y + w_z$ admits an infinite-dimensional variational symmetry group. Discuss the consequences of Noether's second theorem in this case.

5.31. In Kumei, [1], the author finds "new" conservation laws of the sine–Gordon equation $u_{xt} = \sin u$ by starting with the elementary conservation law

$$D_t(\tfrac{1}{2}u_x^2) + D_x(\cos u) = 0$$

and applying generalized symmetries to it. For instance, the evolutionary symmetry with characteristic $Q = u_{ttt} + \frac{1}{2}u_t^3$ is shown to lead to the conservation law

$$D_t[u_x u_{xttt} + \tfrac{3}{2}u_x u_t^2 u_{xt}] - D_x[(u_{ttt} + \tfrac{1}{2}u_t^3)\sin u] = 0.$$

Prove that this conservation law is trivial! (What is its characteristic?) More generally, prove that any conservation law derived by this method from a symmetry whose characteristic does not explicitly depend on x is necessarily trivial.

5.32. Let $\mathscr{L}[u] = \int \frac{1}{2}u_x^2\,dx$. Show that the vector field $\tilde{\mathbf{v}} = u_x\partial_u$ is a variational symmetry, but the equivalent vector field (for the Euler–Lagrange equation $u_{xx} = 0$) $\tilde{\mathbf{v}} = (u_x + u_{xx})\partial_u$ is *not* a variational symmetry. Thus the equivalence relation on symmetries does *not* respect the variational property.

*5.33. *Noether's Version of Noether's Theorem.* A generalized vector field $\mathbf{v}$ will be called a *strict variational symmetry* of $\mathscr{L} = \int L\,dx$ if

$$\operatorname{pr}\mathbf{v}(L) + L\operatorname{Div}\xi = 0,$$

i.e. there is no divergence term in (5.84), as we had in our original discussion of variational symmetries in Chapter 4.

(a) Prove that for each conservation law of the Euler–Lagrange equation $\mathsf{E}(L) = 0$ there is a corresponding strict variational symmetry which gives rise to it via Noether's theorem.

(b) Prove that such a strict variational symmetry is unique up to addition of a null divergence in the sense that $\mathbf{v} = \sum \xi^i\partial_{x^i} + \sum \phi_\alpha\partial_{u^\alpha}$ and $\tilde{\mathbf{v}} = \sum \tilde{\xi}^i\partial_{x^i} + \sum \tilde{\phi}_\alpha\partial_{u^\alpha}$ are both strict variational symmetries giving rise to the same conservation law if and only if

$$\tilde{\xi}^i = \xi^i + \frac{1}{L}\sum_j D_j Q_{ij} \quad \text{where} \quad Q_{ij} = -Q_{ji}.$$

(c) What are the strict variational symmetries corresponding to the inversional symmetries of the wave equation?

(Noether, [1].)

*5.34. Prove that if the Euler–Lagrange equations $\mathrm{E}(L) = 0$ admit a trivial conservation law corresponding to a nontrivial variational symmetry, then they are necessarily under-determined and admit an infinite family of such laws. (Compare Exercise 5.29.)

5.35. Let $\mathscr{L}[u] = \int L[u]\,dx$ be a functional, and let $\mathbf{v}$ generate a one-parameter (generalized) group which does not leave $\mathscr{L}$ invariant but rather multiplies it by a scalar factor. (For example, the scaling symmetries for the wave equation are of this type.) Prove that there is a divergence identity of the form $\operatorname{Div} P[u] = L[u]$ which holds whenever u is a solution of the Euler–Lagrange equations $\mathrm{E}(L) = 0$. (Steudel, [2].)

*5.36. One trick used to construct variational principles for arbitrary systems of differential equations is the following. If $\Delta_\nu[u] = 0$, $\nu = 1, \ldots, l$ is the system, we let $\mathscr{L}[u] = \int \frac{1}{2}|\Delta[u]|^2\,dx = \int \frac{1}{2}\sum_\nu (\Delta_\nu[u])^2\,dx$ (called a "weak Lagrangian structure" by Anderson and Ibragimov, [1; §14.3]).

(a) Prove that the Euler–Lagrange equations for $\mathscr{L}$ take the form $\delta\mathscr{L} = \mathrm{D}_\Delta^*(\Delta) = 0$. Thus solutions of $\Delta = 0$ are solutions of the Euler–Lagrange equation for $\mathscr{L}$, but the converse is not true in general. What is $\mathscr{L}$ and $\delta\mathscr{L}$ in the case of the heat equation $u_t = u_{xx}$?

(b) Prove that if $\mathbf{v}_Q$ is any generalized symmetry of Δ, then one can construct a conservation law for Δ with characteristic $\mathrm{D}_\Delta(Q)$ using the techniques of Noether's theorem, but the conservation law is always trivial! Thus *this* method for finding variational principles in practice leads only to trivial conservation laws.

*5.37. A second method for finding variational principles for arbitrary systems of differential equations $\Delta = (\Delta_1, \ldots, \Delta_l) = 0$ is to introduce auxiliary variables $v = (v^1, \ldots, v^l)$ and consider the problem $\mathscr{L}[u, v] = \int v \cdot \Delta[u]\,dx$.

(a) Prove that the Euler–Lagrange equations for $\mathscr{L}$ are $\Delta = 0$ and $\mathrm{D}_\Delta^*(v) = 0$.

(b) Find variational symmetries and conservation laws for the heat equation $u_t = u_{xx}$ by this method. How does one intepret these results physically?

(Bateman, [2], Atherton and Homsy, [1].)

5.38. Define a pseudo-variational symmetry $\mathbf{v}$ to be a generalized vector field that satisfies (5.84) only on solutions u of the Euler–Lagrange equations. Prove that to every pseudo-variational symmetry of a normal variational problem there corresponds a conservation law, but that there is also always a true variational symmetry giving rise to the same law. How is the true variational symmetry related to the pseudo-variational symmetry?

5.39. Let $\mathscr{D}: \mathscr{A}^r \to \mathscr{A}^s$ be a matrix differential operator. Prove that $\mathscr{D}Q = 0$ for all $Q \in \mathscr{A}^r$ if and only if $\mathscr{D} = 0$ is the zero operator.

5.40. Let $\mathscr{D}$ be a scalar differential operator.

(a) Prove that $\mathscr{D}$ is self-adjoint if and only if $\mathscr{D}$ can be written in the form

$$\mathscr{D} = \sum_{\#J\ \text{even}} (D_J \cdot P_J + P_J \cdot D_J)$$

for certain differential functions P_J.

(b) Prove that $\mathscr{D}$ is skew-adjoint if and only if

$$\mathscr{D} = \sum_{\#J\ \text{odd}} (D_J \cdot P_J + P_J \cdot D_J)$$

for certain P_J.

(c) In the case $p = 1$, prove that $\mathscr{D}$ is self-adjoint if and only if

$$\mathscr{D} = \sum_j D_x^j Q_j D_x^j$$

for certain differential functions Q_j. Does this result generalize to the case $p \geqslant 2$?

5.41. Let $p = q = 1$, and let $P(x, u^{(2n+1)})$ be a differential function. Prove that the Fréchet derivative D_P is skew-adjoint, $\mathsf{D}_P^* = -\mathsf{D}_P$, if and only if P is linear in u, $u_x, \ldots, u_{2n+1}$. Is this true if $p > 1$?

*5.42. *The Substitution Principle.* For subsequent applications in Chapter 7, we will require a slight generalization of the technical vanishing results such as those in Corollary 5.68 and Lemma 5.86. The basic problem is that one has an expression depending on one or more q-tuples of differential functions $Q^1, \ldots, Q^k \in \mathscr{A}^q$ and, possibly, their *total* derivatives, with the property that it vanish whenever each $Q^i = \delta \mathscr{Q}_i$ is a variational derivative of some functionals $\mathscr{Q}_1, \ldots, \mathscr{Q}_k \in \mathscr{F}$. The conclusion to be drawn is that the same expression will vanish no matter what the Q^i are, variational derivatives or otherwise.

More specifically, the reader should prove the following.

(a) Let $P \in \mathscr{A}^q$ be a fixed q-tuple of differential functions. Then $\int P \cdot Q \, dx = 0$ whenever $Q = \delta \mathscr{Q} \in \mathscr{A}^q$ is a variational derivative if and only if $P = 0$, and hence $\int P \cdot Q \, dx = 0$ for all $Q \in \mathscr{A}^q$.

(b) Let $\mathscr{D}: \mathscr{A}^q \to \mathscr{A}^r$ be an $r \times q$ matrix of differential operators. Then $\mathscr{D}Q = 0$ whenever $Q = \delta \mathscr{Q} \in \mathscr{A}^q$ is a variational derivative if and only if $\mathscr{D} = 0$, and hence $\mathscr{D}Q = 0$ for all $Q \in \mathscr{A}^q$.

(c) Suppose $\mathscr{D}: \mathscr{A}^q \to \mathscr{A}^q$ is a differential operator. Then $\int Q \cdot \mathscr{D}R \, dx = 0$ for all variational derivatives $Q = \delta \mathscr{Q}$, $R = \delta \mathscr{R} \in \mathscr{A}^q$ if and only if $\mathscr{D} = 0$, and hence $\int Q \cdot \mathscr{D}R \, dx = 0$ for all $Q, R \in \mathscr{A}^q$.

*5.43. Let $p = q = 1$. Prove that if $L = L(u^{(n)})$ does not explicitly depend on x, then $\int u_x \mathsf{E}(L) \, dx = 0$. This shows that one must be careful with the above "substitution principle", since the following "slight" generalization of part (a) is *not* true: If $P \in \mathscr{A}^q$, and $\int P \cdot Q \, dx = 0$ for all $Q(u^{(n)}) = \delta \mathscr{Q} \in \mathscr{A}^q$ which do not depend explicitly on x, then $P = 0$. The above ($P = u_x$) is a definite counterexample. Is this the only such counterexample?

5.44. Let $P(u^{(m)})$ be a q-tuple of homogeneous differential functions of degree $\alpha \neq -1$: $P(\lambda u^{(m)}) = \lambda^\alpha P(u^{(m)})$. Prove that $P = \mathsf{E}(L)$ for some Lagrangian L if and only if $\mathsf{E}(u \cdot P) = (\alpha + 1)P$. Is this true if $\alpha = -1$? (Olver and Shakiban, [1], Shakiban, [1].)

*5.45. Prove the Helmholtz Theorem 5.45 directly, without using variational forms: If $P \in \mathscr{A}^q$ has self-adjoint Fréchet derivative, then $P = \mathsf{E}(L)$ where L is given by (5.123). Conversely, if $P = \mathsf{E}(L)$ for some L, then D_P is self-adjoint.

5.46. (a) Show that a single evolution equation $u_t = P[u]$, $u \in \mathbb{R}$, is never the Euler–Lagrange equation for a variational problem. Is the same true for systems of evolution equations?

(b) One common trick to put a single evolution equation into variational form is to replace u by a potential function v with $u = v_x$, yielding $v_{xt} = P[v_x]$. Show that the Korteweg–de Vries equation becomes the Euler–Lagrange equation of some functional in this way. Which of the geometrical and generalized symmetries of the Korteweg–de Vries equation yield conservation laws via Noether's theorem?

(c) Find necessary and sufficient conditions on P that the trick of part (b) yields an Euler–Lagrange equation.

(d) A second trick to convert such an evolution equation is to just differentiate with respect to x: $u_{xt} = D_x P[u]$. Prove that this yields an Euler–Lagrange equation if and only if $D_x P$ depends only on x, u_x, u_{xx}, ... (not u), and the equation is equivalent to an evolution equation $w_t = Q[w]$, $w = u_x$, for which the trick in part (b) is applicable.

*5.47. (a) Prove that if $Q(x, u^{(n)})$ is any system of differential functions which satisfies the Helmholtz conditions $\mathsf{D}_Q = \mathsf{D}_Q^*$, and $P(\lambda, x, u^{(m)})$ any one-parameter family of q-tuples of differential functions such that

$$P(0, x, u^{(m)}) = f(x), \qquad P(1, x, u^{(m)}) = u,$$

for some fixed $f(x)$, then

$$L = \int_0^1 \frac{\partial P}{\partial \lambda} \cdot Q(x, P^{(n)})\, d\lambda$$

is a Lagrangian for Q: $\mathsf{E}(L) = Q$.

(b) This method is useful for finding variational principles for systems of differential equations not defined on the entire jet space $M^{(n)}$. For example, let $p = q = 1$, $Q = u_x^{-2} u_{xx} + u_{yy}$, and use $P(\lambda) = (1 - \lambda)x + \lambda u$ to find a variational principle for Q. Why does the classical construction (5.123) break down in this case? (Horndeski, [1].)

*5.48. Given a differential equation $\Delta[u] = 0$, the ***multiplier problem*** of the calculus of variations is to determine a nonvanishing differential function $Q[u]$ such that $0 = Q \cdot \Delta = \mathsf{E}(L)$ is the Euler–Lagrange equation for some variational problem.

(a) Prove that if $\Delta[u] = u_{xx} - H(x, u, u_x)$ is a second order ordinary differential equation, then $Q(x, u, u_x)$ is such a multiplier if and only if Q satisfies the partial differential equation

$$\frac{\partial Q}{\partial x} + u_x \frac{\partial Q}{\partial u} + \frac{\partial}{\partial u_x}(QH) = 0.$$

Conclude that any second order ordinary differential equation is always locally equivalent to an Euler–Lagrange equation of some first order variational problem. (See Exercises 4.8 and 4.9.)

(b) Find all multipliers and corresponding variational problems for the equation $u_{xx} - u_x = 0$.

(c) Discuss the case of a higher order differential equation. (Hirsch, [1]; see Douglas, [1], and Anderson and Thompson, [1], for generalizations to systems of ordinary differential equations, and Anderson and Duchamp, [2], for second order partial differential equations.)

**5.49. Here we generalize the formulae in Theorem 4.8 for the action of the Euler operator under a change of variables, where now the new variables can depend on derivatives of the old variables. Let $x = (x^1, \ldots, x^p)$, $u = (u^1, \ldots, u^q)$ be the original variables and $\mathscr{L}[u] = \int L(x, u^{(n)})\, dx$ be a variational problem with Euler–Lagrange expression $\mathsf{E}(L)$. Suppose $y = (y^1, \ldots, y^p)$ and $w = (w^1, \ldots, w^q)$ are new variables, with $y = P(x, u^{(m)})$, $w = Q(x, u^{(m)})$, for certain differential functions $P \in \mathscr{A}^p$, $Q \in \mathscr{A}^q$. Let $\mathscr{L}[w] = \int \tilde{L}(y, w^{(l)})\, dy$ be the corresponding variational problem. Prove that

$$\mathsf{E}_{u^\alpha}(L) = \sum_{\beta=1}^{q} \frac{\mathsf{D}(P_1, \ldots, P_p, Q_\beta)}{\mathsf{D}(x^1, \ldots, x^p, u^\alpha)} \mathsf{E}_{w^\beta}(\tilde{L}).$$

Here the coefficient of E_{w^β} is a differential operator, given by a determinantal formula

$$\frac{\mathsf{D}(P_1, \ldots, P_p, Q_\beta)}{\mathsf{D}(x^1, \ldots, x^p, u^\alpha)} = \det \begin{bmatrix} D_1 P_1 & \cdots & D_p P_1 & \mathsf{D}^*_{P_1,\alpha} \\ \vdots & & \vdots & \vdots \\ D_1 P_p & \cdots & D_p P_p & \mathsf{D}^*_{P_p,\alpha} \\ D_1 Q_\beta & \cdots & D_p Q_\beta & \mathsf{D}^*_{Q_\beta,\alpha} \end{bmatrix},$$

in which

$$\mathsf{D}_{P,\alpha} = \sum_J \frac{\partial P}{\partial u^\alpha_J} D_J$$

is the Fréchet derivative of P with respect to u^α, $\mathsf{D}^*_{P,\alpha}$ its adjoint, and in the expansion of the determinant, the differential operators are written *first* in any product. For example,

$$\frac{\mathsf{D}(P, Q)}{\mathsf{D}(x, u)}(R) = \det \begin{pmatrix} D_x P & \mathsf{D}^*_P \\ D_x Q & \mathsf{D}^*_Q \end{pmatrix} R = \mathsf{D}^*_Q(D_x P \cdot R) - \mathsf{D}^*_P(D_x Q \cdot R).$$

Discuss how (4.7) is a special case. Try some specific examples, e.g. $y = x$, $w = u_x$, to see how this works in practice.

5.50. An n-th *order divergence* is a differential function $P[u]$ such that

$$P = \sum_{\#I=n} D_I Q_I$$

for certain differential functions Q_I. For example,

$$u_{xx}u_{yy} - u_{xy}^2 = D_x^2(-\tfrac{1}{2}u_y^2) + D_x D_y(u_x u_y) + D_y^2(-\tfrac{1}{2}u_x^2)$$

is a second order divergence.

(a) Prove that P is an n-th order divergence if and only if $\mathsf{E}^I_\alpha(P) = 0$ for all $\alpha = 1, \ldots, q$, $0 \leqslant \#I \leqslant n - 1$. (See Corollary 5.100.)

(b) Show that

$$u_{xx}v_{yy} - 2u_{xy}v_{xy} + u_{yy}v_{xx}$$

is a second order divergence, and

$$u_{xxx}v_{yyy} - 3u_{xxy}v_{xyy} + 3u_{xyy}v_{xxy} - u_{yyy}v_{xxx}$$

is a third order divergence. Can you generalize this result? (Olver, [6].)

*5.51. Suppose $\Delta[u] = 0$ is a homogeneous system of differential equations in the sense that $\Delta[\lambda u] = \lambda^\alpha \Delta[u]$ for all $\lambda \in \mathbb{R}$, where α is the degree of homogeneity. Prove that if Div $P = 0$ is a conservation law with *trivial* characteristic for such a system, then P itself is a trivial conservation law. (*Hint*: First reduce to the case of a homogeneous conservation law $P[\lambda u] = \lambda^\beta P[u]$. Use the homotopy formula (5.151) to reconstruct P from the characteristic form Div $P = Q \cdot \Delta$, where $Q = 0$ whenever $\Delta = 0$, Q homogeneous.) (Olver, [11].)

5.52. (a) If $\mathscr{L}[u] = \int L[u]\,dx$ is a functional, and $\mathbf{v}_Q$ an evolutionary vector field, prove that the prolonged action

$$\operatorname{pr} \mathbf{v}_Q(\mathscr{L}) \equiv \int \operatorname{pr} \mathbf{v}_Q(L)\,dx$$

gives a well-defined map on the space $\mathscr{F}$ of functionals.

(b) Prove that the action is effective, i.e., $\operatorname{pr} \mathbf{v}_Q(\mathscr{L}) = 0$ for all $\mathscr{L} \in \mathscr{F}$ if and only if $Q = 0$. Similarly prove that $\operatorname{pr} \mathbf{v}_Q(\mathscr{L}) = 0$ for all $Q \in \mathscr{A}^q$ if and only if $\mathscr{L} = 0$ in $\mathscr{F}$.

(c) Generalize this to define the Lie derivative of a functional form with respect to an evolutionary vector field. Prove a homotopy formula generalizing (1.67) or (5.122) to functional forms.

5.53. Let $\hat{\omega}$ be a vertical k-form and $\mathbf{v}_Q$ an evolutionary vector field with flow $\exp(\varepsilon \mathbf{v}_Q)$ determined by (5.14). Define a suitable action $\exp(\varepsilon \mathbf{v}_Q)_* \hat{\omega}$ of this flow on $\hat{\omega}$ and prove the Lie derivative formula

$$\operatorname{pr} \mathbf{v}_Q(\hat{\omega}) = \frac{d}{d\varepsilon}[\exp(\varepsilon \mathbf{v}_Q)_* \hat{\omega}].$$

(As always, assume existence and uniqueness for the relevant initial value problem.) Can the same be done if we use the definition of Exercise 5.8 for the flow generated by $\mathbf{v}_Q$?

CHAPTER 6

Finite-Dimensional Hamiltonian Systems

The guiding concept of a Hamiltonian system of differential equations forms the basis of much of the more advanced work in classical mechanics, including motion of rigid bodies, celestial mechanics, quantization theory and so on. More recently, Hamiltonian methods have become increasingly important in the study of the equations of continuum mechanics, including fluids, plasmas and elastic media. In this book, we are concerned with just one aspect of this vast subject, namely the interplay between symmetry groups, conservation laws and reduction in order for systems in Hamiltonian form. The Hamiltonian version of Noether's theorem has a particularly attractive geometrical flavour, which remains somewhat masked in our previous Lagrangian framework.

No previous knowledge of Hamiltonian mechanics will be assumed, so our first order of business will be to make precise the concept of a Hamiltonian system of differential equations. In this chapter, we concentrate on the more familiar, and conceptually easier case of systems of ordinary differential equations. Once we have mastered these, the generalizations to systems of evolution equations to be taken up in Chapter 7 will be quite natural. There are, at the outset, several different approaches to Hamiltonian mechanics, and that adopted here is slightly novel. It is important to realize the necessity of a coordinate-free treatment of "Hamiltonian structures" which does not assume the introduction of special canonical coordinates (the p's and q's of the elementary classical mechanics texts). Admittedly, one always has the temptation to simplify matters as much as possible, and, for finite-dimensional systems of constant rank, Darboux' theorem says that we could, if desired, always introduce such coordinates, with the attendant simplification in the formulae, but this may not always be the most natural or straightforward approach to the problem. Besides, in the infinite-dimensional version of this theory to be discussed later, such a result is no longer available; hence, if we are to receive a proper grounding in the finite-dimensional theory to make the ascension to infinite dimensions and evolution equations, we must cast aside the crutch of canonical coordinates and approach the Hamiltonian structure from a more intrinsic standpoint.

Even so, there are still several coordinate-free approaches to Hamiltonian mechanics. The one that requires the least preparatory work in differential geometric foundations concentrates on the Poisson bracket as the fundamental object of study. This has the advantage of avoiding differential forms almost entirely, and proceeding directly to the heart of the subject. In addition, the Poisson bracket approach admits Hamiltonian structures of varying rank (in a sense to be defined shortly), which have proved important in recent work on collective motion and stability. It includes as an important special case the Lie–Poisson bracket on the dual to a Lie algebra which plays a key role in representation theory and geometric quantization, as well as providing the theoretical basis for the general theory of reduction of Hamiltonian systems with symmetry.

6.1. Poisson Brackets

Given a smooth manifold M, a Poisson bracket on M assigns to each pair of smooth, real-valued functions $F, H: M \to \mathbb{R}$ another smooth, real-valued function, which we denote by $\{F, H\}$. There are certain basic properties that such a bracket operation must satisfy in order to qualify as a Poisson bracket. We state these properties initially in the simple, coordinate-free manner. Subsequently, local coordinate versions will be found, which, especially if M is an open subset of some Euclidean space, could equally well be taken as the defining properties for a Poisson bracket.

Definition 6.1. A *Poisson bracket* on a smooth manifold M is an operation that assigns a smooth real-valued function $\{F, H\}$ on M to each pair F, H of smooth, real-valued functions, with the basic properties:

(a) *Bilinearity*:

$$\{cF + c'P, H\} = c\{F, H\} + c'\{P, H\}, \quad \{F, cH + c'P\} = c\{F, H\} + c'\{F, P\},$$

for constants $c, c' \in \mathbb{R}$,

(b) *Skew-Symmetry*:

$$\{F, H\} = -\{H, F\},$$

(c) *Jacobi Identity*:

$$\{\{F, H\}, P\} + \{\{P, F\}, H\} + \{\{H, P\}, F\} = 0,$$

(d) *Leibniz' Rule*:

$$\{F, H\cdot P\} = \{F, H\}\cdot P + H\cdot\{F, P\}.$$

(Here $\cdot$ denotes the ordinary multiplication of real-valued functions.) In all these equations, F, H and P are arbitrary smooth real-valued functions on M.

A manifold M with a Poisson bracket is called a *Poisson manifold*, the bracket defining a *Poisson structure* on M. The notion of a Poisson manifold is slightly more general than that of a symplectic manifold, or manifold with Hamiltonian structure; in particular, the underlying manifold M need not be even-dimensional. This is borne out by the standard examples from classical mechanics.

Example 6.2. Let M be the even-dimensional Euclidean space $\mathbb{R}^{2n}$ with coordinates $(p, q) = (p^1, \ldots, p^n, q^1, \ldots, q^n)$. (In physical problems, the p's represent momenta and the q's positions of the mechanical objects.) If $F(p, q)$ and $H(p, q)$ are smooth functions, we define their Poisson bracket to be the function

$$\{F, H\} = \sum_{i=1}^{n} \left\{\frac{\partial F}{\partial q^i}\frac{\partial H}{\partial p^i} - \frac{\partial F}{\partial p^i}\frac{\partial H}{\partial q^i}\right\}. \tag{6.1}$$

This bracket is clearly bilinear and skew-symmetric; the verifications of the Jacobi identity and the Leibniz rule are straightforward exercises in vector calculus which we leave to the reader. We note the particular bracket identities

$$\{p^i, p^j\} = 0, \qquad \{q^i, q^j\} = 0, \qquad \{q^i, p^j\} = \delta^i_j, \tag{6.2}$$

in which i and j run from 1 to n, and δ^i_j is the Kronecker symbol, which is 1 if $i = j$ and 0 otherwise. (In (6.2) we are viewing the coordinates themselves as functions on M.)

More generally, we can determine a Poisson bracket on any Euclidean space $M = \mathbb{R}^m$. Just let $(p, q, z) = (p^1, \ldots, p^n, q^1, \ldots, q^n, z^1, \ldots, z^l)$ be the coordinates, so $2n + l = m$, and define the Poisson bracket between two functions $F(p, q, z)$, $H(p, q, z)$ by the same formula (6.1). In particular, if the function $F(z)$ depends on the z's only, then $\{F, H\} = 0$ for all functions H. Such functions, in particular the z^k's themselves, are known as *distinguished functions* or *Casimir functions* and are characterized by the property that their Poisson bracket with any other function is always zero. We must supplement the basic coordinate bracket relations (6.2) by the additional relations

$$\{p^i, z^k\} = \{q^i, z^k\} = \{z^j, z^k\} = 0 \tag{6.3}$$

for all $i = 1, \ldots, n$, and $j, k = 1, \ldots, l$. Although this example appears to be somewhat special, Darboux' Theorem 6.22 will show that locally, except at singular points, every Poisson bracket looks like this. We therefore call (6.1) the *canonical Poisson bracket*.

Definition 6.3. Let M be a Poisson manifold. A smooth, real-valued function $C: M \to \mathbb{R}$ is called a *distinguished function* if the Poisson bracket of C with any other real-valued function vanishes identically, i.e. $\{C, H\} = 0$ for all $H: M \to \mathbb{R}$.

In the case of the canonical Poisson bracket (6.1) on $\mathbb{R}^{2n}$, the only distinguished functions are the constants, which always satisfy the requirements of the definition. At the other extreme, if the Poisson bracket is completely trivial, i.e. $\{F, H\} = 0$ for every F, H, then every function is distinguished.

Hamiltonian Vector Fields

Let M be a Poisson manifold, so the Poisson bracket satisfies the basic requirements of Definition 6.1. Concentrating for the moment on just the bilinearity and Leibniz rule, note that given a smooth function H on M, the map $F \mapsto \{F, H\}$ defines a derivation on the space of smooth functions F on M, and hence by (1.20), (1.21) determines a vector field on M. This observation leads to a fundamental definition.

Definition 6.4. Let M be a Poisson manifold and $H: M \to \mathbb{R}$ a smooth function. The *Hamiltonian vector field* associated with H is the unique smooth vector field $\hat{v}_H$ on M satisfying

$$\hat{v}_H(F) = \{F, H\} = -\{H, F\} \tag{6.4}$$

for every smooth function $F: M \to \mathbb{R}$. The equations governing the flow of $\hat{v}_H$ are referred to as *Hamilton's equations* for the "Hamiltonian" function H.

Example 6.5. In the case of the canonical Poisson bracket (6.1) on $\mathbb{R}^m$, $m = 2n + l$, the Hamiltonian vector field corresponding to $H(p, q, z)$ is clearly

$$\hat{v}_H = \sum_{i=1}^{n} \left(\frac{\partial H}{\partial p^i} \frac{\partial}{\partial q^i} - \frac{\partial H}{\partial q^i} \frac{\partial}{\partial p^i} \right). \tag{6.5}$$

The corresponding flow is obtained by integrating the system of ordinary differential equations

$$\frac{dq^i}{dt} = \frac{\partial H}{\partial p^i}, \qquad \frac{dp^i}{dt} = -\frac{\partial H}{\partial q^i}, \qquad i = 1, \ldots, n, \tag{6.6}$$

$$\frac{dz^j}{dt} = 0, \qquad j = 1, \ldots, l, \tag{6.7}$$

which are Hamilton's equations in this case. In the nondegenerate case $m = 2n$, we have just (6.6), which is the canonical form of Hamilton's equations in classical mechanics. More generally, (6.7) just adds in the constancy of the distinguished coordinates z^j under the flow. In particular, if H depends only on the distinguished coordinates z, its Hamiltonian flow is completely trivial. This remark holds in general: A function C on a Poisson manifold is distinguished if and only if its Hamiltonian vector field $\hat{v}_C = 0$ vanishes everywhere.

There is a fundamental connection between the Poisson bracket of two functions and the Lie bracket of their associated Hamiltonian vector fields, which forms the basis of much of the theory of Hamiltonian systems.

Proposition 6.6. *Let M be a Poisson manifold. Let $F, H: M \to \mathbb{R}$ be smooth functions with corresponding Hamiltonian vector fields $\hat{v}_F, \hat{v}_H$. The Hamiltonian vector field associated with the Poisson bracket of F and H is, up to sign, the Lie bracket of the two Hamiltonian vector fields:*

$$\hat{v}_{\{F,H\}} = -[\hat{v}_F, \hat{v}_H] = [\hat{v}_H, \hat{v}_F]. \tag{6.8}$$

PROOF. Let $P: M \to \mathbb{R}$ be any other smooth function. Using the commutator definition of the Lie bracket, we find

$$\begin{aligned}
[\hat{v}_H, \hat{v}_F]P &= \hat{v}_H \cdot \hat{v}_F(P) - \hat{v}_F \cdot \hat{v}_H(P) \\
&= \hat{v}_H\{P, F\} - \hat{v}_F\{P, H\} \\
&= \{\{P, F\}, H\} - \{\{P, H\}, F\} \\
&= \{P, \{F, H\}\} \\
&= \hat{v}_{\{F,H\}}(P),
\end{aligned}$$

where we have made use of the Jacobi identity, the skew-symmetry of the Poisson bracket, and the definition (6.4) of a Hamiltonian vector field. Since P is arbitrary, this suffices to prove (6.8). □

Example 6.7. Let $M = \mathbb{R}^2$ with coordinates (p, q) and canonical Poisson bracket $\{F, H\} = F_q H_p - F_p H_q$. For a function $H(p, q)$, the corresponding Hamiltonian vector field is $\hat{v}_H = H_p \partial_q - H_q \partial_p$. Thus for $H = \frac{1}{2}(p^2 + q^2)$ we have $\hat{v}_H = p\partial_q - q\partial_p$, whereas for $F = pq$, $\hat{v}_F = q\partial_q - p\partial_p$. The Poisson bracket of F and H is $\{F, H\} = p^2 - q^2$, which has Hamiltonian vector field $\hat{v}_{\{F,H\}} = 2p\partial_q + 2q\partial_p$. This agrees with the commutator $[\hat{v}_H, \hat{v}_F]$, as the reader can verify.

The Structure Functions

To determine the general local coordinate picture for a Poisson manifold, we first look at the Hamiltonian vector fields. Let $x = (x^1, \ldots, x^m)$ be local coordinates on M and $H(x)$ a real-valued function. The associated Hamiltonian vector field will be of the general form $\hat{v}_H = \sum_{i=1}^m \xi^i(x)\partial/\partial x^i$, where the coefficient functions $\xi^i(x)$, which depend on H, are to be determined. Let $F(x)$ be a second smooth function. Using (6.4), we find

$$\{F, H\} = \hat{\mathbf{v}}_H(F) = \sum_{i=1}^{m} \xi^i(x) \frac{\partial F}{\partial x^i}.$$

But, again by (6.4),

$$\xi^i(x) = \hat{\mathbf{v}}_H(x^i) = \{x^i, H\},$$

so this formula becomes

$$\{F, H\} = \sum_{i=1}^{m} \{x^i, H\} \frac{\partial F}{\partial x^i}. \tag{6.9}$$

On the other hand, using the skew-symmetry of the Poisson bracket, we can turn this whole procedure around and compute the latter set of Poisson brackets in terms of the particular Hamiltonian vector fields $\hat{\mathbf{v}}_i = \hat{\mathbf{v}}_{x^i}$ associated with the local coordinate functions x^i; namely

$$\{x^i, H\} = -\{H, x^i\} = -\hat{\mathbf{v}}_i(H) = -\sum_{j=1}^{m} \{x^j, x^i\} \frac{\partial H}{\partial x^j},$$

the last equality following from a second application of (6.9), with H replacing F and x^i replacing H. Thus we obtain the basic formula

$$\{F, H\} = \sum_{i=1}^{m} \sum_{j=1}^{m} \{x^i, x^j\} \frac{\partial F}{\partial x^i} \frac{\partial H}{\partial x^j} \tag{6.10}$$

for the Poisson bracket. In other words, to compute the Poisson bracket of any pair of functions in some given set of local coordinates, it suffices to know the Poisson brackets between the coordinate functions themselves. These basic brackets,

$$J^{ij}(x) = \{x^i, x^j\}, \qquad i, j = 1, \ldots, m, \tag{6.11}$$

are called the *structure functions* of the Poisson manifold M relative to the given local coordinates, and serve to uniquely determine the Poisson structure itself. For convenience, we assemble the structure functions into a skew-symmetric $m \times m$ matrix $J(x)$, called the *structure matrix* of M. Using ∇H to denote the (column) gradient vector for H, the local coordinate form (6.10) for the Poisson bracket can be written as

$$\{F, H\} = \nabla F \cdot J \nabla H. \tag{6.12}$$

For example, in the case of the canonical bracket (6.1) on $\mathbb{R}^m$, $m = 2n + l$, the structure matrix has the simple form

$$J = \begin{pmatrix} 0 & -\mathrm{I} & 0 \\ \mathrm{I} & 0 & 0 \\ 0 & 0 & 0 \end{pmatrix}$$

relative to the (p, q, z)-coordinates, where I is the $n \times n$ identity matrix.

The Hamiltonian vector field associated with $H(x)$ has the form

$$\hat{\mathbf{v}}_H = \sum_{i=1}^{m} \left(\sum_{j=1}^{m} J^{ij}(x) \frac{\partial H}{\partial x^j} \frac{\partial}{\partial x^i} \right), \tag{6.13}$$

or, in matrix notation, $\hat{\mathbf{v}}_H = (J\nabla H) \cdot \partial_x$, ∂_x being the "vector" with entries $\partial/\partial x^i$. Therefore, in the given coordinate chart, Hamilton's equations take the form†

$$\frac{dx}{dt} = J(x) \nabla H(x). \tag{6.14}$$

Alternatively, using (6.9), we could write this in the "bracket form"

$$\frac{dx}{dt} = \{x, H\},$$

the i-th component of the right-hand side being $\{x^i, H\}$. A system of first order ordinary differential equations is said to be a *Hamiltonian system* if there is a Hamiltonian function $H(x)$ and a matrix of functions $J(x)$ determining a Poisson bracket (6.12) whereby the system takes the form (6.14). Of course, we need to know which matrices $J(x)$ are the structure matrices for Poisson brackets.

Proposition 6.8. *Let $J(x) = (J^{ij}(x))$ be an $m \times m$ matrix of functions of $x = (x^1, \ldots, x^m)$ defined over an open subset $M \subset \mathbb{R}^m$. Then $J(x)$ is the structure matrix for a Poisson bracket $\{F, H\} = \nabla F \cdot J \nabla H$ over M if and only if it has the properties of*:

(a) Skew-Symmetry:

$$J^{ij}(x) = -J^{ji}(x), \qquad i, j = 1, \ldots, m,$$

(b) Jacobi Identity:

$$\sum_{l=1}^{m} \{J^{il} \partial_l J^{jk} + J^{kl} \partial_l J^{ij} + J^{jl} \partial_l J^{ki}\} = 0, \qquad i, j, k = 1, \ldots, m, \tag{6.15}$$

for all $x \in M$. (Here, as usual, $\partial_l = \partial/\partial x^l$.)

Proof. In its basic form (6.12) the Poisson bracket is automatically bilinear and satisfies Leibniz' rule. The skew-symmetry of the structure matrix is clearly equivalent to the skew-symmetry of the bracket. Thus we need only verify the equivalence of (6.15) with the Jacobi identity. Note that by (6.10), (6.11)

$$\{\{x^i, x^j\}, x^k\} = \sum_{l=1}^{m} J^{lk}(x) \partial_l J^{ij}(x),$$

† More generally, we can allow $H(x, t)$ to depend on t as well, which leads to a time-dependent Hamiltonian vector field; see Section 6.3.

so (6.15) is equivalent to the Jacobi identity for the coordinate functions x^i, x^j and x^k. More generally, for $F, H, P: M \to \mathbb{R}$,

$$\{\{F, H\}, P\} = \sum_{k,l=1}^{m} J^{lk} \frac{\partial}{\partial x^l}\left\{\sum_{i,j=1}^{m} J^{ij} \frac{\partial F}{\partial x^i}\frac{\partial H}{\partial x^j}\right\}\frac{\partial P}{\partial x^k}$$

$$= \sum_{i,j,k,l}\left\{J^{lk}\frac{\partial J^{ij}}{\partial x^l}\frac{\partial F}{\partial x^i}\frac{\partial H}{\partial x^j}\frac{\partial P}{\partial x^k}\right.$$

$$\left. + J^{lk}J^{ij}\left(\frac{\partial^2 F}{\partial x^l \partial x^i}\frac{\partial H}{\partial x^j}\frac{\partial P}{\partial x^k} + \frac{\partial F}{\partial x^i}\frac{\partial^2 H}{\partial x^l \partial x^j}\frac{\partial P}{\partial x^k}\right)\right\}.$$

Summing cyclically on F, H, P, we find that the first set of terms vanishes by virtue of (6.15), while the remaining terms cancel due to the skew-symmetry of the structure matrix. □

Note that we could just as well take the requirements of Proposition 6.8 on the structure matrix as the definition of a Poisson bracket (6.12) in a local coordinate chart. The conditions (6.15) guaranteeing the Jacobi identity form a large system of *nonlinear* partial differential equations which the structure functions must satisfy. In particular, any constant skew-symmetric matrix J trivially satisfies (6.15) and thus determines a Poisson bracket.

The Lie–Poisson Structure

One of the most important examples of a Poisson structure is that associated with an r-dimensional Lie algebra $\mathfrak{g}$. Let c_{ij}^k, $i, j, k = 1, \dots, r$, be the structure constants of $\mathfrak{g}$ relative to a basis $\{\mathbf{v}_1, \dots, \mathbf{v}_r\}$. Let V be another r-dimensional vector space, with coordinates $x = (x^1, \dots, x^r)$ determined by a basis $\{\omega_1, \dots, \omega_r\}$. Define the *Lie-Poisson bracket* between two smooth functions $F, H: V \to \mathbb{R}$ to be

$$\{F, H\} = \sum_{i,j,k=1}^{r} c_{ij}^k x^k \frac{\partial F}{\partial x^i}\frac{\partial H}{\partial x^j}. \tag{6.16}$$

This clearly takes the form (6.10) with linear structure functions $J^{ij}(x) = \sum_{k=1}^r c_{ij}^k x^k$. The verification of the properties of Proposition 6.8 for the structure matrix follows easily from the basic properties (1.43), (1.44) of the structure constants; in particular, (6.15) reduces to the Jacobi identity (1.44), as the reader can easily verify.

There is a more intrinsic characterization of the Lie–Poisson bracket. First, recall that if V is any vector space and $F: V \to \mathbb{R}$ a smooth, real-valued function, then the gradient $\nabla F(x)$ at any point $x \in V$ is naturally an element of the dual vector space V^* consisting of all (continuous) linear functions on V. Indeed, by definition,

$$\langle \nabla F(x); y \rangle = \lim_{\varepsilon \to 0} \frac{F(x + \varepsilon y) - F(x)}{\varepsilon}$$

for any $y \in V$, where $\langle\ ;\ \rangle$ is the natural pairing between V and its dual V^*. Keeping this in mind, we identify the vector space V used in our initial construction of the Lie–Poisson bracket with the dual space $\mathfrak{g}^*$ to the Lie algebra $\mathfrak{g}$, $\{\omega_1, \ldots, \omega_r\}$ being the dual basis to $\{\mathbf{v}_1, \ldots, \mathbf{v}_r\}$. If $F: \mathfrak{g}^* \to \mathbb{R}$ is any smooth function, then its gradient $\nabla F(x)$ is an element of $(\mathfrak{g}^*)^* \simeq \mathfrak{g}$ (since $\mathfrak{g}$ is finite-dimensional). Then the Lie–Poisson bracket has the coordinate-free form

$$\{F, H\}(x) = \langle x; [\nabla F(x), \nabla H(x)]\rangle, \qquad x \in \mathfrak{g}^*, \tag{6.17}$$

where $[\ ,\ ]$ is the ordinary Lie bracket on the Lie algebra $\mathfrak{g}$ itself; the proof is left to the reader. If $H: \mathfrak{g} \to \mathbb{R}$ is any function, the associated system of Hamilton's equations takes the form

$$\frac{dx^i}{dt} = \sum_{j,k=1}^{r} c_{ij}^k x^k \frac{\partial H}{\partial x^j}, \qquad i = 1, \ldots, r,$$

in which the coordinates x^k themselves appear explicitly.

Example 6.9. Consider the three-dimensional Lie algebra $\mathfrak{so}(3)$ of the rotation group $SO(3)$. Using the basis $\mathbf{v}_1 = y\partial_z - z\partial_y$, $\mathbf{v}_2 = z\partial_x - x\partial_z$, $\mathbf{v}_3 = x\partial_y - y\partial_x$ of infinitesimal rotations around the x-, y- and z-axes of $\mathbb{R}^3$ (or their matrix counterparts), we have the commutation relations $[\mathbf{v}_1, \mathbf{v}_2] = -\mathbf{v}_3$, $[\mathbf{v}_3, \mathbf{v}_1] = -\mathbf{v}_2$, $[\mathbf{v}_2, \mathbf{v}_3] = -\mathbf{v}_1$. Let $\omega_1, \omega_2, \omega_3$ be a dual basis for $\mathfrak{so}(3)^* \simeq \mathbb{R}^3$ and $u = u^1\omega_1 + u^2\omega_2 + u^3\omega_3$ a typical point therein. If $F: \mathfrak{so}(3)^* \to \mathbb{R}$, then its gradient is the vector

$$\nabla F = \frac{\partial F}{\partial u^1}\mathbf{v}_1 + \frac{\partial F}{\partial u^2}\mathbf{v}_2 + \frac{\partial F}{\partial u^3}\mathbf{v}_3 \in \mathfrak{so}(3).$$

Thus from (6.17) we find the Lie–Poisson bracket on $\mathfrak{so}(3)^*$ to be

$$\begin{aligned}\{F, H\} &= u^1\left(\frac{\partial F}{\partial u^3}\frac{\partial H}{\partial u^2} - \frac{\partial F}{\partial u^2}\frac{\partial H}{\partial u^3}\right) + u^2\left(\frac{\partial F}{\partial u^1}\frac{\partial H}{\partial u^3} - \frac{\partial F}{\partial u^3}\frac{\partial H}{\partial u^1}\right)\\ &\quad + u^3\left(\frac{\partial F}{\partial u^2}\frac{\partial H}{\partial u^1} - \frac{\partial F}{\partial u^1}\frac{\partial H}{\partial u^2}\right)\\ &= -u\cdot\nabla F \times \nabla H,\end{aligned}$$

using the standard cross product on $\mathbb{R}^3$. Thus the structure matrix is

$$J(u) = \begin{bmatrix} 0 & -u^3 & u^2 \\ u^3 & 0 & -u^1 \\ -u^2 & u^1 & 0 \end{bmatrix}, \qquad u \in \mathfrak{so}(3)^*.$$

Hamilton's equations corresponding to the Hamiltonian function $H(u)$ are therefore

$$\frac{du}{dt} = u \times \nabla H(u).$$

For example, if

$$H(u) = \frac{(u^1)^2}{2I_1} + \frac{(u^2)^2}{2I_2} + \frac{(u^3)^2}{2I_3},$$

where I_1, I_2, I_3 are certain constants, then Hamilton's equations become the *Euler equations* for the motion of a rigid body

$$\frac{du^1}{dt} = \frac{I_2 - I_3}{I_2 I_3} u^2 u^3, \qquad \frac{du^2}{dt} = \frac{I_3 - I_1}{I_3 I_1} u^3 u^1, \qquad \frac{du^3}{dt} = \frac{I_1 - I_2}{I_1 I_2} u^1 u^2, \tag{6.18}$$

in which (I_1, I_2, I_3) are the moments of inertia about the coordinate axes and u^1, u^2, u^3 the corresponding body angular momenta. (The angular velocities are $\omega^i = u^i/I_i$.) The Hamiltonian function is the kinetic energy of the body.

6.2. Symplectic Structures and Foliations

In order to gain a more complete understanding of the geometry underlying a general Poisson structure on a smooth manifold, we need to look more closely at the structure matrix $J(x)$ which determines the local coordinate form of the Poisson bracket. The most important invariant of this matrix is its rank. If the rank is maximal everywhere, then, as we will see, we are in the more standard situation of a "symplectic structure" on a smooth manifold, treated in most books on Hamiltonian mechanics. In the more general case of nonmaximal rank, the Poisson manifold M will be seen to be naturally foliated into symplectic submanifolds in such a way that any Hamiltonian system on M naturally restricts to any one of the symplectic submanifolds and hence, by restriction, returns us to the more classical case of Hamiltonian mechanics. However, for many problems it is more natural to remain on the larger Poisson manifold itself, especially when one is interested in the collective behaviour of systems depending on parameters, with the underlying symplectic structure varying with the parameters themselves.

The Correspondence Between One-Forms and Vector Fields

As we saw in the previous section, a Poisson structure on a manifold M sets up a correspondence between smooth functions $H: M \to \mathbb{R}$ and their associated Hamiltonian vector field $\hat{v}_H$ on M. In local coordinates, this correspondence is determined by multiplication of the gradient ∇H by the structure matrix $J(x)$ determined by the Poisson bracket. This can be given a more intrinsic formulation if we recall that the coordinate-free version of the gradient of a real-valued function H is its differential dH. Thus the Poisson structure determines a correspondence between differential one-forms dH on M and their associated Hamiltonian vector fields $\hat{v}_H$, which in fact extends to general one-forms:

Proposition 6.10. *Let M be a Poisson manifold and $x \in M$. Then there exists a unique linear map*

$$\mathsf{B} = \mathsf{B}|_x : T^*M|_x \to TM|_x,$$

from the cotangent space to M at x to the corresponding tangent space, such that for any smooth real-valued function $H: M \to \mathbb{R}$,

$$\mathsf{B}(dH(x)) = \hat{\mathbf{v}}_H|_x. \tag{6.19}$$

PROOF. At any point $x \in M$, the cotangent space $T^*M|_x$ is spanned by the differentials $\{dx^1, \ldots, dx^m\}$ corresponding to the local coordinate functions near x. From (6.13), we see that at $x \in M$

$$\mathsf{B}(dx^j) = \sum_{i=1}^m J^{ij}(x) \frac{\partial}{\partial x^i}\bigg|_x, \qquad j = 1, \ldots, m.$$

By linearity, for any $\omega = \sum a_j \, dx^j \in T^*M|_x$,

$$\mathsf{B}(\omega) = \sum_{i,j=1}^m J^{ij}(x) a_j \frac{\partial}{\partial x^i}\bigg|_x$$

is essentially matrix multiplication by the structure matrix $J(x)$, proving the proposition. □

Example 6.11. In the case of $\mathbb{R}^m$ with canonical coordinates (p, q, z), as in Example 6.2, if

$$\omega = \sum_{i=1}^n [a_i dp^i + b_i dq^i] + \sum_{j=1}^l c_j \, dz^j$$

is any one-form, then

$$\mathsf{B}(\omega) = \sum_{i=1}^n \left\{ a_i \frac{\partial}{\partial q^i} - b_i \frac{\partial}{\partial p^i} \right\}.$$

In this particular case, the form of B does not vary from point to point. In particular, the kernel of B has the same dimension as the number of distinguished coordinates $z^1, \ldots, z^l$.

Rank of a Poisson Structure

Definition 6.12. Let M be a Poisson manifold and $x \in M$. The *rank* of M at x is the rank of the linear map $\mathsf{B}|_x : T^*M|_x \to TM|_x$.

In local coordinates, $\mathsf{B}|_x$ is the same as multiplication by the structure matrix $J(x)$, so the rank of M at x equals the rank of $J(x)$, independent of the choice of coordinates. Skew-symmetry of J immediately implies:

Proposition 6.13. *The rank of a Poisson manifold at any point is always an even integer.*

For example, the canonical Poisson structure (6.1) on $\mathbb{R}^m$, $m = 2n + l$, is of constant rank $2n$ everywhere. Later we will see that every Poisson structure of constant rank $2n$ looks locally like the canonical structure of this rank. In the case of the Lie–Poisson structure on $\mathfrak{so}(3)^*$, the rank is 2 everywhere except at the origin $u = 0$, where the rank is 0.

Since the rank of a linear mapping is determined by the dimension of its kernel, or of its range, we can compute the rank $2n$ of a Poisson manifold at a point either by looking at $\mathscr{K}|_x = \{\omega \in T^*M|_x \colon \mathsf{B}(\omega) = 0\}$, which has dimension $m - 2n$, or the image space $\mathscr{H}|_x = \{\mathbf{v} = \mathsf{B}(\omega) \in TM|_x \colon \omega \in T^*M|_x\}$, which has dimension $2n$. For instance, in the case of the canonical Poisson bracket (6.1), $\mathscr{K}|_x$ is spanned by the "distinguished differentials" $dz^1, \ldots, dz^l$, while $\mathscr{H}|_x$ is spanned by the elementary Hamiltonian vector fields $\partial/\partial q^i$, $\partial/\partial p^i$ corresponding to the coordinate functions p^i, $-q^i$ respectively. The image space $\mathscr{H}|_x$ is of particular significance; it can be characterized as the span of all the Hamiltonian vector fields on M at x:

$$\mathscr{H}|_x = \{\hat{\mathbf{v}}_H|_x \colon \quad H \colon M \to \mathbb{R} \text{ is smooth}\}.$$

Symplectic Manifolds

In classical mechanics, one usually imposes an additional nondegeneracy requirement on the Poisson bracket, which leads to the more restrictive notion of a symplectic structure on a manifold.

Definition 6.14. A Poisson manifold M of dimension m is *symplectic* if its Poisson structure has maximal rank m everywhere.

In particular, according to Proposition 6.13, a symplectic manifold is necessarily even-dimensional. The canonical example is the Poisson bracket (6.1) on $\mathbb{R}^m$ in the case $m = 2n$, so there are no extra distinguished coordinates. In terms of local coordinates, a structure matrix $J(x)$ determines a symplectic structure provided it satisfies the additional nondegeneracy condition $\det J(x) \neq 0$ everywhere. In this case, the complicated nonlinear equations (6.15) describing the Jacobi identity simplify to a *linear* system of differential equations involving the entries of the inverse matrix $K(x) = [J(x)]^{-1}$.

Proposition 6.15. *A matrix $J(x)$ determines a symplectic structure on $M \subset \mathbb{R}^m$ if and only if its inverse $K(x) = [J(x)]^{-1}$ satisfies the conditions*:

(a) Skew-Symmetry:

$$K_{ij}(x) = -K_{ji}(x), \qquad i, j = 1, \ldots, m,$$

(b) Closure (Jacobi Identity):

$$\partial_k K_{ij} + \partial_j K_{ki} + \partial_i K_{jk} = 0, \qquad i, j, k = 1, \ldots, m, \tag{6.20}$$

everywhere.

PROOF. The equivalence of the skew-symmetry of J to that of K is elementary. To prove the equivalence of (6.20) and (6.15), we use the formula for the derivative of a matrix inverse $\partial_k K = -K \cdot \partial_k J \cdot K$, where $K = J^{-1}$. Substituting into (6.20), we find

$$\sum_{l,n=1}^{m} \{K_{il}K_{jn}\partial_k J^{ln} + K_{kl}K_{in}\partial_j J^{ln} + K_{jl}K_{kn}\partial_i J^{ln}\} = 0.$$

Multiplying by $J^{\tilde{\imath}i} J^{\tilde{\jmath}j} J^{\tilde{k}k}$, and summing over i, j, k from 1 to m, leads to (6.15) with a slightly different labelling of indices. □

Maps Between Poisson Manifolds

If M and N are Poisson manifolds, a *Poisson map* is a smooth map $\phi: M \to N$ preserving the Poisson brackets:

$$\{F \circ \phi, H \circ \phi\}_M = \{F, H\}_N \circ \phi \qquad \text{for all } F, H: N \to \mathbb{R}.$$

In the case of symplectic manifolds these are the *canonical maps* of classical mechanics. A good example is provided by the flow generated by a Hamiltonian vector field.

Proposition 6.16. *Let M be a Poisson manifold and $\hat{v}_H$ a Hamiltonian vector field. For each t, the flow $\exp(t\hat{v}_H): M \to M$ determines a (local) Poisson map from M to itself.*

PROOF. Let F and P be real-valued functions, and let $\phi_t = \exp(t\hat{v}_H)$. If we differentiate the Poisson condition $\{F \circ \phi_t, P \circ \phi_t\} = \{F, P\} \circ \phi_t$ with respect to t and use (1.17), we find the infinitesimal version

$$\{\hat{v}_H(F), P\} + \{F, \hat{v}_H(P)\} = \hat{v}_H(\{F, P\})$$

at the point $\phi_t(x)$. By (6.4) this is the same as the Jacobi identity. At $t = 0$, ϕ_0 is the identity, and trivially Poisson, so a simple integration proves the Poisson condition for general t. □

For example, if $M = \mathbb{R}^2$ with canonical coordinates (p, q), then the function $H = \frac{1}{2}(p^2 + q^2)$ generates the group of rotations in the plane, determined by $\hat{v}_H = p\partial_q - q\partial_p$. Thus each rotation in $\mathbb{R}^2$ is a canonical map.

Since any Hamiltonian flow preserves the Poisson bracket on M, in particular it preserves its rank.

Corollary 6.17. *If $\hat{v}_H$ is a Hamiltonian vector field on a Poisson manifold M, then the rank of M at $\exp(t\hat{v}_H)x$ is the same as the rank of M at x for any $t \in \mathbb{R}$.*

For instance, the origin in $\mathfrak{so}(3)^*$, being the only point of rank 0, is a fixed point of any Hamiltonian system with the given Lie–Poisson structure. In fact, it is easy to see that any point of rank 0 on a Poisson manifold is a fixed point for any Hamiltonian system there.

Poisson Submanifolds

Definition 6.18. A submanifold $N \subset M$ is a *Poisson submanifold* if its defining immersion $\phi: \tilde{N} \to M$ is a Poisson map.

An equivalent way of stating this definition is that for any pair of functions $F, H: M \to \mathbb{R}$ which restrict to functions $\tilde{F}, \tilde{H}: N \to \mathbb{R}$ on N, their Poisson bracket $\{F, H\}_M$ naturally restricts to a Poisson bracket $\{\tilde{F}, \tilde{H}\}_N$. For example, the submanifolds $\{z = c\}$ of $\mathbb{R}^m$, $m = 2n + l$ corresponding to constant values of the distinguished coordinates are easily seen to be Poisson submanifolds, with the natural reduced Poisson bracket with respect to the remaining coordinates (p, q).

If $N \subset M$ is an arbitrary submanifold then there is a simple test that will tell whether or not it can be made into a Poisson submanifold, the reduced Poisson structure, if it exists, being uniquely determined by the above remark.

Proposition 6.19. *A submanifold N of a Poisson manifold M is a Poisson submanifold if and only if $TN|_y \supset \mathscr{H}|_y$ for all $y \in N$, meaning every Hamiltonian vector field on M is everywhere tangent to N. In particular, if $TN|_y = \mathscr{H}|_y$ for all $y \in N$, N is a symplectic submanifold of M.*

PROOF. Since a Poisson bracket is determined by its local character, we can without loss of generality assume that N is a regular submanifold of M and use flat local coordinates $(y, w) = (y^1, \ldots, y^n, w^1, \ldots, w^{m-n})$ with $N = \{(y, w): w = 0\}$. First suppose that N is a Poisson submanifold, and let $\tilde{H}: N \to \mathbb{R}$ be any smooth function. Then we can extend $\tilde{H}$ to a smooth function $H: M \to \mathbb{R}$ defined in a neighbourhood of N, with $\tilde{H} = H|N$. In our local coordinates, $\tilde{H} = \tilde{H}(y)$ and $H(y, w)$ is any function so that $H(y, 0) = \tilde{H}(y)$. If $\tilde{F}: N \to \mathbb{R}$ has a similar extension F, then by definition the Poisson bracket between $\tilde{F}$ and $\tilde{H}$ on N is obtained by restricting that of F and H to N:

$$\{\tilde{F}, \tilde{H}\}_N = \{F, H\}|N.$$

In particular, for any choice of $\tilde{F}, \tilde{H}$, the bracket $\{F, H\}|N$ cannot depend on the particular extensions F and H which are selected. Clearly, this is possible if and only if $\{F, H\}|N$ contains no partial derivatives of either F or H with respect to the normal coordinates w^i, so

$$\{F, H\}|N = \sum_{i,j} J^{ij}(y, 0)\frac{\partial F}{\partial y^i}\frac{\partial H}{\partial y^j} = \sum_{i,j} \tilde{J}^{ij}(y)\frac{\partial \tilde{F}}{\partial y^i}\frac{\partial \tilde{H}}{\partial y^j}. \tag{6.21}$$

But then the Hamiltonian vector field $\hat{v}_H$, when restricted to N, takes the form

$$\hat{v}_H|N = \sum_{i,j} \tilde{J}^{ij}(y)\frac{\partial H}{\partial y^j}\frac{\partial}{\partial y^i}, \tag{6.22}$$

and is thus tangent to N everywhere.

Conversely, if the tangency condition $\mathscr{H}|_y \subset TN|_y$ holds for all $y \in N$, any Hamiltonian vector field, when restricted to N, must be a combination of the tangential basis vectors $\partial/\partial y^i$ only, and hence of the form (6.22). If $F(w)$ depends on w alone, then $\{F, H\} = \hat{v}_H(F)$ must therefore vanish when restricted to N. In particular,

$$\{y^i, w^j\} = \{w^k, w^j\} = 0 \qquad \text{on } N \qquad \text{for all } i, j, k,$$

and hence the Poisson bracket on N takes the form (6.21) in which $\tilde{J}^{ij}(y) = J^{ij}(y, 0) = \{y^i, y^j\}|N$. The fact that the structure functions $\tilde{J}^{ij}(y)$ of the induced Poisson bracket on N satisfy the Jacobi identity easily follows from (6.15) since on restriction to N all the w-terms vanish. Thus N is a Poisson submanifold and the proposition is proved. Note that the rank of the Poisson structure on N at $y \in N$ equals the rank of the Poisson structure on M at the same point. □

Example 6.20. For the Lie–Poisson structure on $\mathfrak{so}(3)^*$, the subspace $\mathscr{H}|_u$ at $u \in \mathfrak{so}(3)^*$ is spanned by the elementary Hamiltonian vectors $\hat{v}_1 = u^3\partial_2 - u^2\partial_3$, $\hat{v}_2 = u^1\partial_3 - u^3\partial_1$, $\hat{v}_3 = u^2\partial_1 - u^1\partial_2$, $(\partial_i = \partial/\partial u^i)$, corresponding to the coordinate functions u^1, u^2, u^3 respectively. If $u \neq 0$, these vectors span a two-dimensional subspace of $T\mathfrak{so}(3)^*|_u$, which coincides with the tangent space to the sphere $S_\rho^2 = \{u \colon |u| = \rho\}$ passing through u: $\mathscr{H}|_u = TS_\rho^2|_u$, $|u| = \rho$. Proposition 6.19 therefore implies that each such sphere is a symplectic submanifold of $\mathfrak{so}(3)^*$. In terms of spherical coordinates $u^1 = \rho\cos\theta\sin\phi$, $u^2 = \rho\sin\theta\sin\phi$, $u^3 = \rho\cos\phi$ on S_ρ^2, the Poisson bracket between $\tilde{F}(\theta, \phi)$ and $\tilde{H}(\theta, \phi)$ is computed by extending them to a neighbourhood of S_ρ^2, e.g. set $F(\rho, \theta, \phi) = \tilde{F}(\theta, \phi)$, $H(\rho, \theta, \phi) = \tilde{H}(\theta, \phi)$, computing the Lie–Poisson bracket $\{F, H\}$, and then restricting to S_ρ^2. However, according to (6.10), $\{\tilde{F}, \tilde{H}\} = \{\theta, \phi\}(\tilde{F}_\theta\tilde{H}_\phi - \tilde{F}_\phi\tilde{H}_\theta)$, so we only really need compute the Lie–Poisson bracket between the spherical angles θ, ϕ:

$$\{\theta, \phi\} = -u\cdot(\nabla_u\theta \times \nabla_u\phi) = -1/(\rho\sin\phi).$$

Thus

$$\{\tilde{F}, \tilde{H}\} = \frac{-1}{\rho\sin\phi}\left(\frac{\partial\tilde{F}}{\partial\theta}\frac{\partial\tilde{H}}{\partial\phi} - \frac{\partial\tilde{F}}{\partial\phi}\frac{\partial\tilde{H}}{\partial\theta}\right)$$

is the induced Poisson bracket on $S_\rho^2 \subset \mathfrak{so}(3)^*$.

Thus, if $N \subset M$ is a Poisson submanifold, any Hamiltonian vector field $\hat{\mathbf{v}}_H$ on M is everywhere tangent to N and thereby naturally restricts to a Hamiltonian vector field $\hat{\mathbf{v}}_{\tilde{H}}$ on N, where $\tilde{H} = H|N$ is the restriction of H to N and we are using the induced Poisson structure on N to compute $\hat{\mathbf{v}}_{\tilde{H}}$. If we are only interested in solutions to the Hamiltonian system corresponding to H on M with initial conditions x_0 in N, we can restrict to the Hamiltonian system corresponding to $\tilde{H}$ on N without loss of information, thereby reducing the order of the system. In particular, as far as finding particular solutions of the Hamiltonian system goes, we may as well restrict to the *minimal* Poisson submanifolds of M in which the initial data lies. According to the next theorem, these are always symplectic submanifolds, so every Hamiltonian system can be reduced to one in which the Poisson bracket is symplectic.

Theorem 6.21. *Let M be a Poisson manifold. The system of Hamiltonian vector fields $\mathscr{H}$ on M is integrable, so through each point $x \in M$ there passes an integral submanifold N of $\mathscr{H}$ satisfying $TN|_y = \mathscr{H}|_y$ for each $y \in N$. Each integral submanifold is a symplectic submanifold of M, and, collectively, these submanifolds determine the* symplectic foliation *of the Poisson manifold M. Moreover, if $H: M \to \mathbb{R}$ is any Hamiltonian function, and $x(t) = \exp(t\hat{\mathbf{v}}_H)x_0$ any solution to the corresponding Hamiltonian system, with initial data $x_0 \in N$, then $x(t) \in N$ remains in a single integral submanifold N for all t.*

Proof. This is a direct consequence of the variable rank Frobenius' Theorem 1.41. The involutiveness of $\mathscr{H}$ follows from the fact that the Lie bracket of two Hamiltonian vector fields is again a Hamiltonian vector field, (6.8). The rank-invariance of $\mathscr{H}$ is given in Corollary 6.17. □

Thus each Poisson manifold naturally splits into a collection of even-dimensional symplectic submanifolds—the *leaves* of the symplectic foliation. The dimension of any such leaf N equals the rank of the Poisson structure at any point $y \in N$, so if M has nonconstant rank, the symplectic leaves will be of varying dimensions. For example, in the case of $\mathfrak{so}(3)^*$ the leaves are just the spheres S^2_ρ centred at the origin, together with the singular point $u = 0$. Any Hamiltonian system on M naturally restricts to any symplectic leaf. If we are only interested in the dynamics of particular solutions, then, we could effectively restrict our attention to the single symplectic submanifold in which our solution lies. For instance, the solutions of the equations (6.18) of rigid body motion naturally live on the spheres $|u| = \rho$.

Darboux' Theorem

If we restrict attention to the places where the Poisson structure is of constant rank (in particular, on the open submanifold where its rank achieves its maximum) the geometric picture underlying the symplectic foliation simplifies considerably. In fact, as with the constant rank version of Frobenius' Theorem 1.43, we can introduce flat local coordinates which make the foliation of a particularly simple, canonical form. This is the content of Darboux' theorem.

Theorem 6.22. *Let M be an m-dimensional Poisson manifold of constant rank $2n \leqslant m$ everywhere. At each $x_0 \in M$ there exist canonical local coordinates $(p, q, z) = (p^1, \ldots, p^n, q^1, \ldots, q^n, z^1, \ldots, z^l)$, $2n + l = m$, in terms of which the Poisson bracket takes the form*

$$\{F, H\} = \sum_{i=1}^{n} \left(\frac{\partial F}{\partial q^i} \frac{\partial H}{\partial p^i} - \frac{\partial F}{\partial p^i} \frac{\partial H}{\partial q^i} \right).$$

The leaves of the symplectic foliation intersect the coordinate chart in the slices $\{z^1 = c_1, \ldots, z^l = c_l\}$ determined by the distinguished coordinates z.

PROOF. If the rank of the Poisson structure is 0 everywhere, there is nothing to prove. Indeed, the Poisson bracket is trivial: $\{F, H\} \equiv 0$ for all F, H, and any set of local coordinates $z = (z^1, \ldots, z^l)$, $l = m$, satisfies the conditions of the theorem. Otherwise, we proceed by induction on the "half-rank" n.

Since the rank at x_0 is nonzero, we can find real-valued functions F and P on M whose Poisson bracket does not vanish at x_0:

$$\{F, P\}(x_0) = \hat{\mathbf{v}}_P(F)(x_0) \neq 0.$$

In particular, $\hat{\mathbf{v}}_P|_{x_0} \neq 0$, so we can use Proposition 1.29 to straighten out $\hat{\mathbf{v}}_P$ in a neighbourhood U of x_0 and thereby find a function $Q(x)$ satisfying

$$\hat{\mathbf{v}}_P(Q) = \{Q, P\} = 1$$

for all $x \in U$. (In the notation of Proposition 1.29, Q would be the coordinate y^1.) Since $\{Q, P\}$ is constant, (6.8) and (6.13) imply that

$$[\hat{\mathbf{v}}_P, \hat{\mathbf{v}}_Q] = \hat{\mathbf{v}}_{\{Q, P\}} = 0$$

for all $x \in U$. On the other hand, $\hat{\mathbf{v}}_Q(Q) = \{Q, Q\} = 0$, so $\hat{\mathbf{v}}_P$ and $\hat{\mathbf{v}}_Q$ form a commuting, linearly independent pair of vector fields defined on U. If we set $p = P(x)$, $q = Q(x)$, then Frobenius' Theorem 1.43 allows us to complete p, q to form a system of local coordinates $(p, q, y^3, \ldots, y^m)$ on a possibly smaller neighbourhood $\tilde{U} \subset U$ of x_0 with $\hat{\mathbf{v}}_p = \partial_q$, $\hat{\mathbf{v}}_q = -\partial_p$ there. The bracket relations $\{p, q\} = 1$, $\{p, y^i\} = 0 = \{q, y^i\}$, $i = 3, \ldots, m$, imply that the structure matrix takes the form

$$J(p, q, y) = \begin{bmatrix} 0 & 1 & 0 \\ -1 & 0 & 0 \\ 0 & 0 & \tilde{J}(p, q, y) \end{bmatrix},$$

where $\tilde{J}$ has entries $\tilde{J}^{ij} = \{y^i, y^j\}$, $i, j = 3, \ldots, m$. Finally, we prove that $\tilde{J}$ is actually independent of p and q, and hence forms the structure matrix of a

Poisson bracket in the y variables of rank two less than that of J, from which the induction step is clear. To prove the claim, we just use the Jacobi identity and the above brackets relations; for instance

$$\frac{\partial \tilde{J}^{ij}}{\partial q} = \{\tilde{J}^{ij}, p\} = \{\{y^i, y^j\}, p\} = 0,$$

and similarly for the p derivative. □

Example 6.23. Let us compute the canonical coordinates for the Lie–Poisson bracket on $\mathfrak{so}(3)^*$. According to the proof of Darboux' theorem, we need only find functions $P(u)$, $Q(u)$ whose Poisson bracket is identically 1. Here the function $z = u^3$ generates the rotational vector field $\hat{\mathbf{v}}_3 = u^2\partial_1 - u^1\partial_2$, which can be straightened out using the polar angle $\theta = \arctan(u^2/u^1)$ provided $(u^1, u^2) \neq (0, 0)$. We find $\{\theta, z\} = \hat{\mathbf{v}}_3(\theta) = -1$, hence θ and z will provide canonical coordinates on the symplectic spheres $S^2_\rho = \{|u| = \rho\}$. Indeed, an easy calculation shows that if we re-express $F(u)$ and $H(u)$ in terms of θ, z and ρ, then the Lie–Poisson bracket is simply $\{F, H\} = F_z H_\theta - F_\theta H_z$. In other words, while the symplectic leaves in $\mathfrak{so}(3)^*$ are spheres, canonical coordinates are provided by cylindrical coordinates z, θ!

The Co-adjoint Representation

In the case of a Lie–Poisson bracket on the dual to a Lie algebra $\mathfrak{g}^*$, the induced symplectic foliation has a particularly nice interpretation in terms of the dual to the adjoint representation of the underlying Lie group G on the Lie algebra $\mathfrak{g}$. (See Section 3.3.)

Definition 6.24. Let G be a Lie group with Lie algebra $\mathfrak{g}$. The *co-adjoint action* of a group element $g \in G$ is the linear map $\mathrm{Ad}^* g\colon \mathfrak{g}^* \to \mathfrak{g}^*$ on the dual space satisfying

$$\langle \mathrm{Ad}^* g(\omega); \mathbf{w} \rangle = \langle \omega; \mathrm{Ad}\, g^{-1}(\mathbf{w}) \rangle \tag{6.23}$$

for all $\omega \in \mathfrak{g}^*$, $\mathbf{w} \in \mathfrak{g}$. Here $\langle\ ;\ \rangle$ is the natural pairing between $\mathfrak{g}$ and $\mathfrak{g}^*$, and $\mathrm{Ad}\, g$ the adjoint action of g on $\mathfrak{g}$.

If we identify the tangent space $T\mathfrak{g}^*|_\omega$, $\omega \in \mathfrak{g}^*$, with $\mathfrak{g}^*$ itself, and similarly for $\mathfrak{g}$, then the infinitesimal generators of the co-adjoint action are determined by differentiating (6.23):

$$\langle \mathrm{ad}^*\, \mathbf{v}|_\omega; \mathbf{w} \rangle = -\langle \omega; \mathrm{ad}\, \mathbf{v}|_\mathbf{w} \rangle = \langle \omega; [\mathbf{v}, \mathbf{w}] \rangle, \tag{6.24}$$

for $\mathbf{v}, \mathbf{w} \in \mathfrak{g}$, $\omega \in \mathfrak{g}^*$ (cf. (3.21)).

The fundamental result that connects the co-adjoint action with the Lie–Poisson bracket is the following:

Theorem 6.25. *Let G be a connected Lie group with co-adjoint representation* Ad^*G *on* $\mathfrak{g}^*$. *Then the orbits of* Ad^*G *are precisely the leaves of the symplectic foliation induced by the Lie–Poisson bracket on* $\mathfrak{g}^*$. *Moreover, for each* $g \in G$, *the co-adjoint map* Ad^*g *is a Poisson mapping on* $\mathfrak{g}^*$ *preserving the leaves of the foliation.*

PROOF. Let $\mathbf{v} \in \mathfrak{g}$ and consider the linear function $H(\omega) = H_{\mathbf{v}}(\omega) = \langle \omega; \mathbf{v} \rangle$ on $\mathfrak{g}^*$. Note that for $\omega \in \mathfrak{g}^*$, the gradient $\nabla H(\omega)$, considered as an element of $T^*\mathfrak{g}^*|_\omega \simeq \mathfrak{g}$, is just $\mathbf{v}$ itself. Using the intrinsic definition (6.17) of the Lie–Poisson bracket, we find

$$\begin{aligned}\hat{\mathbf{v}}_H(F)(\omega) &= \{F, H\}(\omega) = \langle \omega; [\nabla F(\omega), \nabla H(\omega)] \rangle \\ &= \langle \omega; [\nabla F(\omega), \mathbf{v}] \rangle = \langle \omega; \mathrm{ad}\, \mathbf{v}(\nabla F(\omega)) \rangle \\ &= -\langle \mathrm{ad}^*\, \mathbf{v}(\omega); \nabla F(\omega) \rangle\end{aligned}$$

for any $F: \mathfrak{g}^* \to \mathbb{R}$. On the other hand,

$$\hat{\mathbf{v}}_H(F)(\omega) = \langle \hat{\mathbf{v}}_H|_\omega; \nabla F(\omega) \rangle$$

is uniquely determined by its action on all such functions. We conclude that the Hamiltonian vector field determined by the linear function $H = H_{\mathbf{v}}$ coincides, up to sign, with the infinitesimal generator of the co-adjoint action determined by $\mathbf{v} \in \mathfrak{g}$: $\hat{\mathbf{v}}_H = -\mathrm{ad}^*\, \mathbf{v}$. Thus the corresponding one-parameter groups satisfy

$$\exp(t\hat{\mathbf{v}}_H) = \mathrm{Ad}^*[\exp(-t\mathbf{v})].$$

Proposition 6.16 and the usual connectivity arguments show that $\mathrm{Ad}^*\, g$ is a Poisson mapping for each $g \in G$.

Moreover, the subspace $\mathscr{H}|_\omega$, $\omega \in \mathfrak{g}^*$, is spanned by the Hamiltonian vector fields $\hat{\mathbf{v}}_H$ corresponding to all such linear functions $H = H_{\mathbf{v}}$, $\mathbf{v} \in \mathfrak{g}$, hence $\mathscr{H}|_\omega = \mathrm{ad}^*\, \mathfrak{g}|_\omega$ coincides with the space spanned by the corresponding infinitesimal generators $\mathrm{ad}^*\, \mathbf{v}|_\omega$. Since $\mathrm{ad}^*\, \mathfrak{g}|_\omega$ is precisely the tangent space to the co-adjoint orbit of G through ω, which is connected, we immediately conclude that this co-adjoint orbit is the corresponding integral submanifold of $\mathscr{H}$. □

Corollary 6.26. *The orbits of the co-adjoint representation of G are even-dimensional submanifolds of* $\mathfrak{g}^*$.

Example 6.27. In the case of the rotation group SO(3), the co-adjoint orbits are the spheres $S^2_\rho \subset \mathfrak{so}(3)^*$ determined in Example 6.20. Indeed, according to Example 3.9, the adjoint representation of a rotation matrix $R \in \mathrm{SO}(3)$ on the Lie algebra $\mathfrak{so}(3) \simeq \mathbb{R}^3$ coincides with the rotation R itself relative to the standard basis: $\mathrm{Ad}\, R(\mathbf{v}) = R\mathbf{v}$, $\mathbf{v} \in \mathfrak{so}(3)$. Thus the co-adjoint action Ad^*R of R on $\mathfrak{so}(3)^*$ has matrix representation $\mathrm{Ad}^*R = (R^{-1})^T = R$ relative to the corresponding dual basis on $\mathfrak{so}(3)^* \simeq \mathbb{R}^3$, and the co-adjoint representation of SO(3) coincides with its usual action on $\mathbb{R}^3$ under the above identifications. In particular, the co-adjoint orbits are precisely the spheres S^2_ρ, $\rho \geqslant 0$.

6.3. Symmetries, First Integrals and Reduction of Order

For a system of ordinary differential equations in Lagrangian form, i.e. the Euler–Lagrange equations associated to some variational problem, Noether's theorem provides a connection between one-parameter variational symmetry groups of the system and conservation laws or first integrals. Moreover, the knowledge of such a first integral allows us to reduce the order of the system by two in the case of a one-parameter symmetry group, indicating that we only need find half as many symmetries as the order of the system in order to integrate it entirely by quadrature. All of these statements carry over to the Hamiltonian framework, and, in fact, arise in a far more natural geometric setting than our original Lagrangian results. In this section we discuss the general theory of symmetry and reduction for finite-dimensional Hamiltonian systems.

First Integrals

Consider a system of ordinary differential equations in Hamiltonian form

$$\frac{dx}{dt} = J(x)\nabla H(x, t), \tag{6.25}$$

where $H(x, t)$ is the Hamiltonian function and $J(x)$ the structure matrix determining the Poisson bracket. In this case, first integrals are readily characterized using the Poisson bracket.

Proposition 6.28. *A function $P(x, t)$ is a first integral for the Hamiltonian system (6.25) if and only if*

$$\frac{\partial P}{\partial t} + \{P, H\} = 0 \tag{6.26}$$

for all x, t. In particular, a time-independent function $P(x)$ is a first integral if and only if $\{P, H\} = 0$ everywhere.

PROOF. Let $\hat{\mathbf{v}}_H$ be the Hamiltonian vector field determining (6.25). Then, by (1.17), if $x(t)$ is any solution to Hamilton's equations,

$$\frac{d}{dt}P(x(t), t) = \frac{\partial P}{\partial t}(x(t), t) + \hat{\mathbf{v}}_H(P)(x(t), t).$$

Thus $dP/dt = 0$ along solutions if and only if (6.26) holds everywhere. □

Some first integrals are immediately apparent from the form of (6.26).

Corollary 6.29. *If $x_t = J\nabla H$ is any Hamiltonian system with time-independent Hamiltonian function $H(x)$, then $H(x)$ itself is automatically a first integral.*

Corollary 6.30. *If $x_t = J\nabla H$ is a Hamiltonian system, then any distinguished function $C(x)$ for the Poisson bracket determined by J is automatically a first integral.*

The first integrals supplied by the distinguished functions arise from degeneracies in the Poisson bracket itself and are *not* governed by any intrinsic symmetry properties of the particular Hamiltonian system under investigation. If the Poisson bracket is symplectic, only the constants are distinguished functions and Corollary 6.30 provides no new information. For a Poisson structure of constant rank, the common level sets of the distinguished functions are the leaves of the symplectic foliation, so Corollary 6.30 is just a restatement of Theorem 6.21 that any solution is contained in a single symplectic leaf.

Hamiltonian Symmetry Groups

For systems of Euler–Lagrange equations, first integrals arise from variational symmetry groups; for Hamiltonian systems this role is played by the one-parameter *Hamiltonian symmetry groups* whose infinitesimal generators (in evolutionary form) are Hamiltonian vector fields. First, it is easy to show that any first integral leads to such a symmetry group.

Proposition 6.31. *Let $P(x, t)$ be a first integral of a Hamiltonian system. Then the Hamiltonian vector field $\hat{\mathbf{v}}_P$ determined by P generates a one-parameter symmetry group of the system.*

PROOF. Note first that since the structure matrix $J(x)$ does not depend on t, the Hamiltonian vector field associated with $\partial P/\partial t$ is just the t-derivative $\partial \hat{\mathbf{v}}_P/\partial t$ of that associated with P. Thus the Hamiltonian vector field associated with the combination $\partial P/\partial t + \{P, H\}$ occurring in (6.26) is, using (6.8),

$$\partial \hat{\mathbf{v}}_P/\partial t + [\hat{\mathbf{v}}_H, \hat{\mathbf{v}}_P].$$

If P is a first integral, this last vector field vanishes, which is just the condition (5.26) that $\hat{\mathbf{v}}_P$ generate a symmetry group. □

In particular, if $H(x)$ is time-independent, the associated symmetry group is generated by $\hat{\mathbf{v}}_H$, which is equivalent to the generator ∂_t of the symmetry group of time translations reflecting the autonomy of the Hamiltonian system. For a distinguished function $C(x)$, the corresponding symmetry is trivial: $\hat{\mathbf{v}}_C \equiv 0$.

Example 6.32. Consider the equations of a harmonic oscillator $p_t = -q$, $q_t = p$, which form a Hamiltonian system on $M = \mathbb{R}^2$ relative to the canonical Poisson bracket. The Hamiltonian function $H(q, p) = \frac{1}{2}(p^2 + q^2)$ is thus a first integral, reflecting the fact that the solutions move on the circles $p^2 + q^2 = \text{constant}$.

Not every Hamiltonian symmetry group corresponds directly to a first integral. For example, on $\tilde{M} = M \backslash \{(p, 0): p \leqslant 0\}$, the vector field $\mathbf{w} = -(p^2 + q^2)^{-1}(p\partial_p + q\partial_q)$ generates a symmetry group. Moreover, $\mathbf{w} = \hat{\mathbf{v}}_{\tilde{P}}$ is Hamiltonian for $\tilde{P}(p, q) = \arctan(q/p)$. But $\tilde{P}$ is *not* a first integral; in fact $\tilde{P}(p(t), q(t)) = t + \theta_0$, a linear function of t, whenever $(p(t), q(t))$ solves the system.

The problem here, and more generally, is that there is *not* a one-to-one correspondence between Hamiltonian vector fields and their corresponding Hamiltonian functions. For example, the function $P(p, q, t) = \arctan(q/p) - t$, which *is* a first integral for the oscillator, has the same Hamiltonian vector field $\hat{\mathbf{v}}_P = \mathbf{w} = \hat{\mathbf{v}}_{\tilde{P}}$ as $\tilde{P}$. More generally, we can add any *time-dependent distinguished function* $C(x, t)$ (meaning that for each fixed t, C is a distinguished function) to a given function P without changing the form of its Hamiltonian vector field. Once we recognize the possibility of modifying the function determining a Hamiltonian symmetry group, we can readily prove a converse to the preceding proposition. This forms the Hamiltonian version of Noether's theorem.

Theorem 6.33. *A vector $\mathbf{w}$ generates a Hamiltonian symmetry group of a Hamiltonian system of ordinary differential equations if and only if there exists a first integral $P(x, t)$ so that $\mathbf{w} = \hat{\mathbf{v}}_P$ is the corresponding Hamiltonian vector field. A second function $\tilde{P}(x, t)$ determines the same Hamiltonian symmetry if and only if $\tilde{P} = P + C$ for some time-dependent distinguished function $C(x, t)$.*

PROOF. The second statement follows immediately from Definition 6.3 of a distinguished function applied to the difference $P - \tilde{P}$. To prove the first part, let $\mathbf{w} = \hat{\mathbf{v}}_{\tilde{P}}$ for some function $\tilde{P}(x, t)$. The symmetry condition (5.26) implies that the Hamiltonian vector field associated with the function $\partial\tilde{P}/\partial t + \{\tilde{P}, H\}$ vanishes everywhere, and hence this combination must be a time-dependent distinguished function $\tilde{C}(x, t)$:

$$\frac{d\tilde{P}}{dt} = \frac{\partial\tilde{P}}{\partial t} + \{\tilde{P}, H\} = \tilde{C}.$$

Set $C(x, t) = \int_0^1 \tilde{C}(x, \tau)\, d\tau$, so that C is also distinguished. Moreover, for solutions $x(t)$ of the Hamiltonian system,

$$\frac{dC}{dt} = \frac{\partial C}{\partial t} + \{C, H\} = \tilde{C}.$$

It is now easy to see that the modified function $P = \tilde{P} - C$ has the same Hamiltonian vector field, $\hat{\mathbf{v}}_P = \mathbf{w}$, and provides a first integral: $dP/dt = 0$ on solutions. □

In particular, if the Poisson bracket is symplectic, the only time-dependent distinguished functions are functions $C(t)$ which depend only on t. In this case the theorem states that the Hamiltonian vector field $\hat{\mathbf{v}}_{\tilde{P}}$ generates a symmetry group if and only if there is a function $C(t)$ such that $P(x, t) = \tilde{P}(x, t) - C(t)$ is a first integral. Note that even though both $H(x)$ and $\tilde{P}(x)$ might be t-independent, the first integral $P(x, t) = \tilde{P}(x) - C(t)$ may be required to depend on t! (Indeed, this was precisely the case in Example 6.32.) See Exercise 6.2 for further information on this case.

Example 6.34. The equations of motion of n point masses subject to pairwise potential interactions discussed in Example 4.31 can be put into a canonical Hamiltonian form. We use the positions $\mathbf{q}_i = (x^i, y^i, z^i)$ and momenta $\mathbf{p}_i = (\xi^i, \eta^i, \zeta^i) = m_i\dot{\mathbf{q}}_i$, $i = 1, \ldots, n$, as canonical coordinates. The Hamiltonian function is the total energy

$$H(p, q) = K(p) + U(q) = \sum_{i=1}^{n} \frac{|\mathbf{p}_i|^2}{2m_i} + \sum_{i<j} m_i m_j V(|\mathbf{q}_i - \mathbf{q}_j|),$$

where the potential $V(r)$ depends only on the distance between the two masses. Thc equations of motion are thus

$$\frac{d\mathbf{p}_i}{dt} = -\frac{\partial U}{\partial \mathbf{q}_i}, \quad \frac{d\mathbf{q}_i}{dt} = \frac{\mathbf{p}_i}{m_i}, \qquad i = 1, \ldots, n.$$

Several geometrical symmetry groups are immediately apparent. Simultaneous translation of all the masses in a given direction $\mathbf{a} = (a^1, a^2, a^3)$ is generated by the Hamiltonian vector field

$$\hat{\mathbf{v}}_P = \sum_{i=1}^{n} \left(a^1 \frac{\partial}{\partial x^i} + a^2 \frac{\partial}{\partial y^i} + a^3 \frac{\partial}{\partial z^i} \right),$$

and corresponds to the first integral $P = \sum_i \mathbf{a} \cdot \mathbf{p}_i$ representing the linear momentum in the given direction. Similarly, the group $SO(3)$ of simultaneous rotations of the masses about the origin leads to the integrals of angular momentum. For example, $Q = \sum_i (x^i\eta^i - y^i\xi^i)$, the angular momentum about the z-axis, generates the symmetry group

$$\hat{\mathbf{v}}_Q = \sum_{i=1}^{n} \left(x^i \frac{\partial}{\partial y^i} - y^i \frac{\partial}{\partial x^i} + \xi^i \frac{\partial}{\partial \eta^i} - \eta^i \frac{\partial}{\partial \xi^i} \right)$$

of simultaneous rotations about the z-axis. Besides the six momentum integrals, the constancy of the Hamiltonian function itself implies conservation of energy. Three further first integrals are provided by the uniform motion of the centre of mass, and these lead to three further Hamiltonian symmetry groups. For example, in the x-direction we have

$$R = \sum_{i=1}^{n} m_i x^i - t \sum_{i=1}^{n} \xi^i = \text{constant},$$

and hence

$$\hat{\mathbf{v}}_R = -\sum_{i=1}^{n} \left(t \frac{\partial}{\partial x^i} + m_i \frac{\partial}{\partial \xi^i} \right)$$

generates a one-parameter symmetry group of Galilean boosts.

Reduction of Order in Hamiltonian Systems

The use of symmetry groups to effect a reduction in order of a Hamiltonian system of ordinary differential equations parallels the methods for Euler–Lagrange equations of Section 4.3, but with the added advantage of an immediate geometrical interpretation. We first remark that if the underlying Poisson bracket is degenerate, we can always restrict attention to a single symplectic leaf. Thus each nonconstant distinguished function will reduce the order of the system by one. Other kinds of first integrals, which generate nontrivial Hamiltonian symmetry groups, can then be used to reduce the order by *two*. For simplicity, we restrict our attention to time-independent first integrals.

Theorem 6.35. *Suppose $\hat{\mathbf{v}}_P \neq 0$ generates a Hamiltonian symmetry group of the Hamiltonian system $\dot{x} = J\nabla H$ corresponding to the time-independent first integral $P(x)$. Then there is a reduced Hamiltonian system involving two fewer variables with the property that every solution of the original system can be determined using a single quadrature from those of the reduced system.*

PROOF. The construction is the same as the initial step in the proof of Darboux' Theorem 6.22. We introduce new variables $p = P(x)$, $q = Q(x)$, $y = (y^1, \ldots, y^{m-2}) = Y(x)$ which straighten out the symmetry, so $\hat{\mathbf{v}}_P = \partial_q$ in the (p, q, y)-coordinates. In terms of these coordinates, the structure matrix has the form

$$J(p, q, y) = \begin{bmatrix} 0 & 1 & 0 \\ -1 & 0 & a \\ 0 & -a^T & \tilde{J} \end{bmatrix},$$

where $a(p, q, y)$ is a row vector of length $m - 2$ and $\tilde{J}(p, y)$ is an $(m - 2) \times (m - 2)$ skew-symmetric matrix, which is independent of q, and for each fixed value of p is the structure matrix for a Poisson bracket in the y variables. (If $y = (y^1, \ldots, y^{m-2})$ are chosen as flat coordinates as in the proof of Darboux' theorem, then $a = 0$ and $\tilde{J}(y)$ is independent of p also, as we saw earlier. However, to effect the reduction procedure this is not necessary, and, indeed, may be impractical to achieve.) The proofs of the above statements on the form of the structure matrix follow as in the "flat" case.

The reduced system will be Hamiltonian with respect to the reduced structure matrix $\tilde{J}(p, y)$ for any fixed value of the first integral $p = P(x)$. Note that in terms of the (p, q, y) coordinates

$$0 = \{p, H\} = -\hat{\mathbf{v}}_P(H) = -\partial H/\partial q,$$

hence $H = H(p, y)$ also only depends on p and y. Therefore Hamilton's equations take the form

$$\frac{dp}{dt} = 0, \tag{6.27a}$$

$$\frac{dq}{dt} = -\frac{\partial H}{\partial p} + \sum_{j=1}^{m-2} a^j(p, y)\frac{\partial H}{\partial y^j}, \tag{6.27b}$$

$$\frac{dy^i}{dt} = \sum_{j=1}^{m-2} \tilde{J}^{ij}(p, y)\frac{\partial H}{\partial y^j}, \qquad i = 1, \ldots, m-2. \tag{6.27c}$$

The first equation says that p is a constant (as it should be). Fixing a value of p, we see that the $(m-2)$ equations (6.27c) form a Hamiltonian system relative to the reduced structure matrix $\tilde{J}(p, y)$ and the Hamiltonian function $H(p, y)$; this is the reduced system referred to in the statement of the theorem. Finally, (6.27b), which governs the time evolution of the remaining coordinate q, can be integrated by a single quadrature once we know the solution to the reduced system (6.27c), since the right-hand side does not depend on q. □

Example 6.36. Let $M = \mathbb{R}^4$ with canonical Poisson bracket and consider a Hamiltonian function of the form

$$H(p_1, p_2, q_1, q_2) = \tfrac{1}{2}(p_1^2 + p_2^2) + V(q_1 - q_2).$$

The corresponding Hamiltonian system

$$\frac{dq_1}{dt} = p_1, \quad \frac{dq_2}{dt} = p_2, \quad \frac{dp_1}{dt} = -V'(q_1 - q_2), \quad \frac{dp_2}{dt} = V'(q_1 - q_2), \tag{6.28}$$

determines the motion of two particles of unit mass on a line whose interaction comes from a potential $V(r)$ depending on their relative displacements. This system admits an obvious translational invariance $\mathbf{v} = \partial_{q_1} + \partial_{q_2}$; the corresponding first integral is the linear momentum $p_1 + p_2$. According to the theorem, we can reduce the order of the system by two if we introduce new coordinates

$$p = p_1 + p_2, \qquad q = q_1, \qquad y = p_1, \qquad r = q_1 - q_2,$$

which straighten out $\mathbf{v} = \partial_q$. In these variables, the Hamiltonian function is

$$H(p, y, r) = y^2 - py + \tfrac{1}{2}p^2 + V(r), \tag{6.29}$$

and the Poisson bracket is

$$\{F, H\} = \frac{\partial F}{\partial q}\frac{\partial H}{\partial y} + \frac{\partial F}{\partial r}\frac{\partial H}{\partial y} + \frac{\partial F}{\partial q}\frac{\partial H}{\partial p} - \frac{\partial F}{\partial y}\frac{\partial H}{\partial q} - \frac{\partial F}{\partial y}\frac{\partial H}{\partial r} - \frac{\partial F}{\partial p}\frac{\partial H}{\partial q}.$$

Further, the Hamiltonian system splits into

$$\frac{dp}{dt} = -\frac{\partial H}{\partial q} = 0, \qquad \frac{dq}{dt} = \frac{\partial H}{\partial p} + \frac{\partial H}{\partial y} = y,$$

and

$$\frac{dy}{dt} = -\frac{\partial H}{\partial q} - \frac{\partial H}{\partial r} = -V'(r), \qquad \frac{dr}{dt} = \frac{\partial H}{\partial y} = 2y - p. \tag{6.30}$$

The solution to the first pair,

$$p = a, \qquad q = \int y(t)\, dt + b,$$

(a, b constant) can be determined from the solutions to the second pair (6.30). These form a reduced Hamiltonian system relative to the reduced Poisson bracket $\{\tilde{F}, \tilde{H}\} = \tilde{F}_r\tilde{H}_y - \tilde{F}_y\tilde{H}_r$ for functions of y and r, with the energy (6.29) obtained by fixing $p = a$. Presently, we will see how the two-dimensional system (6.30) can be explicitly integrated, thereby solving the original two-particle system (6.28).

As with the general reduction method for ordinary differential equations, if the vector field $\hat{\mathbf{v}}_P$ associated with the first integral P is too complicated, it may not be possible to explicitly find the change of variables that straightens it out, and so the reduction method cannot be completed. (Of course, the fact that P is a first integral certainly allows a reduction in order by one in all cases.) For example, if the Hamiltonian $H(x)$ is time-independent, it provides a first integral, but straightening out its corresponding vector field $\hat{\mathbf{v}}_H$ is the same problem as solving the Hamiltonian system itself! In this special case, however, the fact that $\hat{\mathbf{v}}_H$ is equivalent to the time translational symmetry generator ∂_t allows us to reduce the order by two provided we are willing to go to a time-dependent Hamiltonian framework.

Proposition 6.37. *Let $\dot{x} = J\nabla H$ be a Hamiltonian system in which $H(x)$ does not depend on t. Then there is a reduced, time-dependent Hamiltonian system in two fewer variables, from whose solutions those of the original system can be found by quadrature.*

PROOF. The reduction in order by two *per se* is easy. First, since H is constant, we can restrict to a level set $H(x) = c$, reducing the order by one. Furthermore, the resulting system remains autonomous and so can be reduced in order once more using the method in Example 2.67. The problem is that unless we choose our coordinates more astutely, the system resulting from this reduction will not be of Hamiltonian form in any obvious way.

The easiest way to proceed is to first introduce the coordinates (p, q, y) used in the proof of Darboux' Theorem 6.22, relative to which the original system takes the form

$$\frac{dp}{dt} = -\frac{\partial H}{\partial q}, \qquad \frac{dq}{dt} = \frac{\partial H}{\partial p}, \qquad \frac{dy^i}{dt} = \sum_{j=1}^{m-2} \tilde{J}^{ij}(y)\frac{\partial H}{\partial y^j}, \qquad i = 1, \ldots, m-2.$$

Assume that $\partial H/\partial p \neq 0$, so that we can solve the equation $w = H(p, q, y)$ locally for $p = K(w, q, y)$. (If $\partial H/\partial p = 0$ everywhere, q is a first integral and we can use the previous reduction procedure!) We take t, w and y to be the new dependent variables and q the new independent variable, in terms of which the system takes the form

$$\frac{dt}{dq} = \frac{1}{\partial H/\partial p} = \frac{\partial K}{\partial w}, \qquad \frac{dw}{dq} = 0, \tag{6.31a}$$

$$\frac{dy^i}{dq} = \sum_{j=1}^{m-2} \tilde{J}^{ij}(y)\frac{\partial H/\partial y^j}{\partial H/\partial p} = \sum_{j=1}^{m-2} \tilde{J}^{ij}(y)\frac{\partial K}{\partial y^j}. \tag{6.31b}$$

The system (6.31b) is Hamiltonian using the reduced Poisson bracket corresponding to the structure matrix $\tilde{J}(y)$ and Hamiltonian function $K(w, q, y)$. For each fixed value of w, once we have solved (6.31b) we can determine the remaining variable $t(q)$ from (6.31a) by a single quadrature. This completes the procedure. □

Example 6.38. In the case of an autonomous Hamiltonian system

$$q_t = \partial H/\partial p, \qquad p_t = -\partial H/\partial q,$$

in the plane, we can use this method to explicitly integrate it. We first solve $w = H(p, q)$ for one of the coordinates, say p, in terms of q and w, which is constant. The first equation, then, leaves an autonomous equation for q, which we can solve by quadrature. For example, in the case of a simple pendulum $H(p, q) = \frac{1}{2}p^2 + (1 - \cos q)$, so on the level curve $H = \omega + 1$, $p = \sqrt{2(\omega + \cos q)}$. The remaining equation

$$dq/dt = p = \sqrt{2(\omega + \cos q)}$$

can be solved in terms of Jacobi elliptic functions

$$q(t) = 2 \sin^{-1}\{\mathrm{sn}(k^{-1}(t + \delta), k)\},$$

where sn has modulus $k = \sqrt{2/(\omega + 1)}$.

Similarly, in the case of the two-particle system on the line from Example 6.36, setting $H(y, r) = \omega + \frac{1}{4}p^2$, we find

$$y = \tfrac{1}{2}p \pm \sqrt{\omega - V(r)}.$$

Thus we recover the solution just by integrating

$$\frac{dr}{dt} = 2y - p = \pm 2\sqrt{\omega - V(r)}.$$

Example 6.39. Consider the equations of rigid body motion (6.18), which were realized as a Hamiltonian system on $\mathfrak{so}(3)^*$. The distingiushed function $C(u) = |u|^2$ naturally reduces the order by one by restriction to a level set or co-adjoint orbit. Provided the moments of inertia I_1, I_2, I_3 are not all equal, the Hamiltonian itself provides a second independent first integral. We conclude that the integral curves of this Hamiltonian vector field are determined by the intersection of a sphere $\{C(u) = |u|^2 = c\}$ and an ellipsoid $\{H(u) = \omega\}$ forming the common level set of these two first integrals. The explicit solutions can be determined by eliminating two of the variables, say u^2 and u^3, from the pair of equations $C(u) = c$, $H(u) = \omega$. Proposition 6.37 then guarantees that the one remaining equation for $u^1 = y$ is autonomous, and hence can be integrated. It turns out to be of the form

$$\frac{dy}{dt} = \sqrt{\alpha(\beta^2 - y^2)(\gamma^2 - y^2)},$$

and hence the solutions can be written in terms of elliptic functions. See Whittaker, [1; Chap. 6] for the explicit formula and a geometric interpretation.

Reduction Using Multi-parameter Groups

As we already saw in the case of Euler–Lagrange equations (cf. Exercise 4.11), it is not true that a Hamiltonian system which admits an r-parameter symmetry group can be fully reduced in order by $2r$, even if the group is solvable. In the Hamiltonian case, however, we can actually determine the degree of reduction which can be effected. Interestingly, this question is closely tied with the structure of the co-adjoint action of the symmetry group on its Lie algebra. We begin by considering an example.

Example 6.40. Let $M = \mathbb{R}^4$ with canonical coordinates $(p, \tilde{p}, q, \tilde{q})$ and consider a Hamiltonian function of the form $H(\tilde{p}, pe^{\tilde{q}}, t)$. Hamilton's equations are

$$\frac{dp}{dt} = 0, \qquad \frac{dq}{dt} = e^{\tilde{q}} H_r, \qquad \frac{d\tilde{p}}{dt} = -pe^{\tilde{q}} H_r, \qquad \frac{d\tilde{q}}{dt} = H_{\tilde{p}}, \tag{6.32}$$

where $r = pe^{\tilde{q}}$ is the second argument of H. They admit a *two*-parameter solvable symmetry group, generated by

$$\mathbf{v} = \partial_q, \qquad \mathbf{w} = -p\partial_p + q\partial_q + \partial_{\tilde{q}},$$

which correspond to the two first integrals $P = p$, $Q = pq + \tilde{p}$. Nevertheless, we can in general only reduce the order of (6.32) by two! There are, in fact, four distinct ways in which this reduction can be effected, and we examine them in turn.

(1) The simplest approach is to use the two first integrals directly and restrict to a common level set. Let $s = pq + \tilde{p}$, so s and p are constant. If we treat $p, \tilde{p}, r, s$ as the new variables (which is valid provided $p \neq 0$) then the reduced system is

$$\frac{d\tilde{p}}{dt} = -rH_r, \qquad \frac{dr}{dt} = rH_{\tilde{p}}, \tag{6.33}$$

which is Hamiltonian using the reduced Poisson bracket $\{F, H\} = r(F_r H_{\tilde{p}} - F_{\tilde{p}} H_r)$. However, there is no residual symmetry property of (6.33) reflecting the invariance of (6.32) under $\mathbf{v}$ and $\mathbf{w}$, so barring any special structure of the Hamiltonian function $H(\tilde{p}, r, t)$ (e.g. time-independence) we cannot reduce any further.

(2) Alternatively, we can employ the reduction procedure of Theorem 6.35 using the Hamiltonian symmetry $\mathbf{v}$. Here the coordinates are already in the appropriate form, with $y = (\tilde{p}, \tilde{q})$. Fixing p, we see that once we have solved the third and fourth equation in (6.32) for $\tilde{p}$ and $\tilde{q}$, we can determine q by quadrature. The reduced system for $\tilde{p}$ and $\tilde{q}$ is canonically Hamiltonian, but there is no symmetry or first integral of it which comes from the second Hamiltonian symmetry group of the full system. Again the order can only be reduced by two.

(3) Reduction using the symmetry group generated by $\mathbf{w}$ leads to a similar conclusion. The relevant flat coordinates are $s = pq + \tilde{p}$, $\tilde{q}$, $r = pe^{\tilde{q}}$ and $z = qe^{-\tilde{q}}$ in terms of which $\mathbf{w} = \partial_{\tilde{q}}$, $H = H(s - rz, r, t) = \tilde{H}(r, s, z, t)$. The system is now

$$\frac{ds}{dt} = 0, \qquad \frac{d\tilde{q}}{dt} = \tilde{H}_s, \qquad \frac{dr}{dt} = -\tilde{H}_z, \qquad \frac{dz}{dt} = \tilde{H}_r. \tag{6.34}$$

Fixing s, the third and fourth equation form a Hamiltonian system, the solutions of which determine $\tilde{q}$ by quadrature. Again, no symmetry or integral reflecting the original invariance under $\mathbf{v}$ remains.

(4) The final possibility is to ignore the Hamiltonian structure of (6.32) entirely and reduce using the symmetry procedure of Section 2.5. Noting that $[\mathbf{v}, \mathbf{w}] = \mathbf{v}$, we first reduce using $\mathbf{v}$, which is trivial. Namely the first, third and fourth equations of (6.32), once solved, will determine $q(t)$ by quadrature. This third order system remains invariant under the reduced vector field $\tilde{\mathbf{w}} = -p\partial_p + \partial_{\tilde{q}}$. We set $r = pe^{\tilde{q}}$ and use $r, \tilde{p}, \tilde{q}$ as variables. The result is identical with (6.33), using which we can determine $\tilde{q}$ (and hence q) by quadrature. As in part (2), no further reduction is possible in general!

Finally, note that for certain special initial conditions, e.g. $p = 0$, we can actually compute the solution by quadrature alone. Thus the degree of reduction possible would appear to depend both on the structure of the symmetry group and the precise initial conditions desired for the solution.

Hamiltonian Transformation Groups

Throughout the following discussion, the underlying symmetry group will be assumed to be Hamiltonian in the following strict sense.

Definition 6.41. Let M be a Poisson manifold. Let G be a Lie group with structure constants c_{ij}^k, $i, j, k = 1, \ldots, r$, relative to some basis of its Lie algebra $\mathfrak{g}$. The functions $P_1, \ldots, P_r: M \to \mathbb{R}$ generate a *Hamiltonian action* of G on M provided their Poisson brackets satisfy the relations

$$\{P_i, P_j\} = -\sum_{k=1}^{r} c_{ij}^k P_k, \qquad i, j = 1, \ldots, r.$$

Note that by (6.8), the corresponding Hamiltonian vector fields $\hat{\mathbf{v}}_i = \hat{\mathbf{v}}_{P_i}$ satisfy the same commutation relations (up to sign)

$$[\hat{\mathbf{v}}_i, \hat{\mathbf{v}}_j] = \sum_{k=1}^{r} c_{ij}^k \mathbf{v}_k,$$

and therefore generate a local action of G on M by Theorem 1.57. Given a Hamiltonian system on M, we will say that G is a *Hamiltonian symmetry group* if each of its generating functions P_i is a first integral, $\{P_i, H\} = 0$, $i = 1, \ldots, r$, which implies that each $\hat{\mathbf{v}}_i$ generates a one-parameter symmetry group.

As we saw in Section 2.5 and Exercise 3.12, any first order system of differential equations on a manifold M which admits a regular symmetry group G reduces to a first order system on the quotient manifold M/G. (Of course, if G is not solvable, we will not be able to reconstruct the solutions to the original system from those of the reduced system by quadrature, but we ignore this point at the moment.) In the case M is a Poisson manifold, and G a Hamiltonian group of transformations, the quotient manifold naturally inherits a Poisson structure, relative to which the reduced system is Hamiltonian. Moreover, the degree of degeneracy of the Poisson bracket on M/G will determine how much further we can reduce the system using any distinguished functions on the quotient space.

Theorem 6.42. *Let G be a Hamiltonian group of transformations acting regularly on the Poisson manifold M. Then the quotient manifold M/G inherits a Poisson structure so that whenever $\tilde{F}, \tilde{H}: M/G \to \mathbb{R}$ correspond to the G-invariant functions $F, H: M \to \mathbb{R}$, their Poisson bracket $\{\tilde{F}, \tilde{H}\}_{M/G}$ corresponds to the G-invariant function $\{F, H\}_M$. Moreover, if G is a Hamiltonian symmetry group for a Hamiltonian system on M, then there is a reduced Hamiltonian system on M/G whose solutions are just the projections of the solutions of the system on M.*

PROOF. First note that the fact that the Poisson bracket $\{F, H\}$ of two G-invariant functions remains G-invariant is a simple consequence of the

Jacobi identity and the connectivity of G; we find, for $i = 1, \ldots, r$,

$$\hat{\mathbf{v}}_i(\{F, H\}) = \{\{F, H\}, P_i\} = \{\{F, P_i\}, H\} + \{F, \{H, P_i\}\} = 0$$

since F and H are invariant, verifying the infinitesimal invariance condition (2.1). Thus the Poisson bracket is well defined on M/G; the verification that it satisfy the properties of Definition 6.1 is trivial.

Now if $H: M \to \mathbb{R}$ has G as a Hamiltonian symmetry group, then H is automatically a G-invariant function: $\hat{\mathbf{v}}_i(H) = \{H, P_i\} = 0$ since each P_i is, by assumption, a first integral. Let $\tilde{H}: M/G \to \mathbb{R}$ be the corresponding function on the quotient manifold. To prove that the corresponding Hamiltonian vector fields are related, $d\pi(\hat{\mathbf{v}}_H) = \hat{\mathbf{v}}_{\tilde{H}}$, where $\pi: M \to M/G$ is the natural projection, it suffices to note that by (1.24)

$$d\pi(\hat{\mathbf{v}}_H)(\tilde{F}) \circ \pi = \hat{\mathbf{v}}_H[\tilde{F} \circ \pi] = \{\tilde{F} \circ \pi, H\}_M$$

for any $\tilde{F}: M/G \to \mathbb{R}$. But this equals

$$\{\tilde{F}, \tilde{H}\}_{M/G} \circ \pi = \hat{\mathbf{v}}_{\tilde{H}}(\tilde{F}) \circ \pi$$

by the definition of the Poisson bracket on M/G, and hence proves the correspondence. □

Example 6.43. Consider the Euclidean space $\mathbb{R}^6$ with canonical coordinates $(p, q) = (p^1, p^2, p^3, q^1, q^2, q^3)$. The functions

$$P_1 = q^2 p^3 - q^3 p^2, \qquad P_2 = q^3 p^1 - q^1 p^3, \qquad P_3 = q^1 p^2 - q^2 p^1,$$

satisfy the bracket relations

$$\{P_1, P_2\} = P_3, \qquad \{P_2, P_3\} = P_1, \qquad \{P_3, P_1\} = P_2,$$

and hence generate a Hamiltonian action of the rotation group SO(3) on $\mathbb{R}^6$, which is, in fact, given by $(p, q) \mapsto (Rp, Rq)$, $R \in \mathrm{SO}(3)$. This action is regular on the open subset $M = \{(p, q): p, q \text{ are linearly independent}\}$, with three-dimensional orbits and global invariants

$$\xi(p, q) = \tfrac{1}{2}|p|^2, \qquad \eta(p, q) = p \cdot q, \qquad \zeta(p, q) = \tfrac{1}{2}|q|^2.$$

We can thus identify the quotient manifold with the subset $M/G \simeq \{(x, y, z): x > 0, z > 0, y^2 < 4xz\}$ of $\mathbb{R}^3$, where $x = \xi$, $y = \eta$, $z = \zeta$ are the new coordinates.

How do we compute the reduced Poisson bracket on M/G? According to (6.10), we need only compute the basic Poisson brackets between the corresponding invariants ξ, η, ζ using the Poisson bracket on M, and re-expressing them in terms of the invariant themselves. For instance, since

$$\{\xi, \eta\} = \sum_{i=1}^{3} \left(\frac{\partial \xi}{\partial q^i} \frac{\partial \eta}{\partial p^i} - \frac{\partial \xi}{\partial p^i} \frac{\partial \eta}{\partial q^i} \right) = -\sum_{i=1}^{3} (p^i)^2 = -2\xi,$$

we have $\{x, y\}_{M/G} = -2x$. Similarly the bracket relations $\{\xi, \zeta\} = -\eta$, $\{\eta, \zeta\} = -2\zeta$ on M lead to the structure functions $\{x, z\}_{M/G} = -y$,

$\{y, z\}_{M/G} = -2z$ on M/G. The structure matrix on M/G is thus

$$J/G = \begin{bmatrix} 0 & -2x & -y \\ 2x & 0 & -2z \\ y & 2z & 0 \end{bmatrix},$$

with Poisson bracket

$$\{\tilde{F}, \tilde{H}\} = -2x(\tilde{F}_x\tilde{H}_y - \tilde{F}_y\tilde{H}_x) - y(\tilde{F}_x\tilde{H}_z - \tilde{F}_z\tilde{H}_x) - 2z(\tilde{F}_y\tilde{H}_z - \tilde{F}_z\tilde{H}_y).$$

Any Hamiltonian system on M admitting the angular momenta P_i as first integrals will reduce to a Hamiltonian system on M/G. For example, the general Kepler problem of a mass moving in a central force field with potential $V(r)$ is such a candidate. Here the Hamiltonian function is the energy $H(p, q) = \frac{1}{2}|p|^2 + V(|q|)$. The reduced system on M/G is obtained by rewriting H in terms of the invariants and then using the given Poisson bracket to reconstruct the Hamiltonian vector field. We find the reduced Hamiltonian $\tilde{H}(x, y, z) = x + \tilde{V}(z)$, where $\tilde{V}(z) = V(\sqrt{2z})$, and reduced system

$$x_t = -y\tilde{V}'(z), \qquad y_t = 2x - 2z\tilde{V}'(z), \qquad z_t = y. \tag{6.35}$$

(The reader may enjoy deriving this directly from Hamilton's equations on M.)

Now M/G is three-dimensional, so there is at least one distinguished function. This is easily seen to be $C(x, y, z) = 4xz - y^2$, which is an invariant of any Hamiltonian system on M/G. (In the original variables, $C = |p \times q|^2$.) The hyperboloids $4xz - y^2 = k^2$, being the level sets of C, are the leaves of the symplectic foliation, and hence we can restrict (6.35) to any such leaf. Using (x, z) as coordinates, we find the fully reduced system

$$x_t = -\sqrt{4xz - k^2}\,\tilde{V}'(z), \qquad z_t = \sqrt{4xz - k^2}, \tag{6.36}$$

which is Hamiltonian relative to the induced Poisson bracket $\{\tilde{F}, \tilde{H}\} = -\sqrt{4xz - k^2}(\tilde{F}_x\tilde{H}_z - \tilde{F}_z\tilde{H}_x)$ on the hyperboloid. This final two-dimensional system can be solved by method of Proposition 6.37, so we can solve the reduced system (6.35) by quadrature. However, at this stage we cannot use this solution to integrate the original central force problem because SO(3) is not a solvable group. But, as we will soon see, this difficulty can be circumvented by an alternative approach to the reduction rocedure.

The Momentum Map

The above approach to the reduction problem, while geometrically appealing, leaves something to be desired from a computational standpoint. The problem is that we are concentrating initially on the more complicated aspect of a Hamiltonian symmetry group, namely the group transformations, and ignoring the first integrals, which are also present, until after the symmetry reduction has been effected, at which point they manifest their presence as distinguished functions. A more logical approach would be to use the first integrals at the outset, restricting the system to a common level set thereof, and then completing the reduction by using any residual symmetry properties of the resulting system. This turns out to be equivalent to the above procedure, but now we stand a better chance of being able to reconstruct the solution to the original system by quadratures alone.

The first step here is to organize the first integrals furnished by a Hamiltonian group of symmetries in a more natural framework. It is here that the dual to the Lie algebra of the symmetry group and, subsequently, the co-adjoint action makes its appearance.

Definition 6.44. Let G be a Hamiltonian group of transformations acting on the Poisson manifold M, generated by the real-valued functions $P_1, \ldots, P_r$. The *momentum map* for G is the smooth map $P\colon M \to \mathfrak{g}^*$ given by

$$P(x) = \sum_{i=1}^{r} P_i(x)\omega_i,$$

in which $\{\omega_1, \ldots, \omega_r\}$ are the dual basis to $\mathfrak{g}^*$ for the basis $\{\mathbf{v}_1, \ldots, \mathbf{v}_r\}$ of $\mathfrak{g}$ relative to which the structure constants c_{ij}^k were computed.

The key property of the momentum map, which explains why we allowed it to take values in $\mathfrak{g}^*$, is its invariance (or, more correctly, "equivariance") with respect to the co-adjoint representation of G on $\mathfrak{g}^*$.

Proposition 6.45. *Let $P\colon M \to \mathfrak{g}^*$ be the momentum map determined by a Hamiltonian group action of G on the Poisson manifold M. Then*

$$P(g \cdot x) = \mathrm{Ad}^* g(P(x)) \tag{6.37}$$

for all $x \in M$, $g \in G$.

PROOF. As usual, it suffices to prove the infinitesimal form of this identity, which is

$$dP(\mathbf{v}_j|_x) = \mathrm{ad}^* \, \mathbf{v}_j|_{P(x)}, \qquad x \in M, \tag{6.38}$$

for any generator $\mathbf{v}_j \in \mathfrak{g}$, $j = 1, \ldots, r$, of G. If we identify $T\mathfrak{g}^*|_{P(x)}$ with $\mathfrak{g}^*$ itself, then

$$dP(\mathbf{v}_j|_x) = \sum_{i=1}^{r} \mathbf{v}_j(P_i)\omega_i = \sum_{i=1}^{r} \{P_i, P_j\}(x)\omega_i = -\sum_{i,k=1}^{r} c_{ij}^k P_k(x)\omega_i,$$

cf. (1.24), (6.4) and Definition 6.41. By (6.24) this expression is the same as the right-hand side of (6.38).

To prove (6.37), we note that if $g = \exp(\varepsilon\mathbf{v}_j)$ and we differentiate with respect to ε, then we recover (6.38) at $\tilde{x} = \exp(\varepsilon\mathbf{v}_j)x$. Since this holds at all $\tilde{x}$, the usual connectivity arguments prove that (6.37) holds in general. □

Example 6.46. Consider the Hamiltonian action of SO(3) on $\mathbb{R}^6$ presented in Example 6.43. The momentum map is

$$P(p, q) = (q^2 p^3 - q^3 p^2)\omega_1 + (q^3 p^1 - q^1 p^3)\omega_2 + (q^1 p^2 - q^2 p^1)\omega_3,$$

where $\{\omega_1, \omega_2, \omega_3\}$ are the basis of $\mathfrak{so}(3)^*$ of Example 6.9. Note that if we identify $\mathfrak{so}(3)^*$ with $\mathbb{R}^3$, $P(p, q) = q \times p$ is the same as the cross product of vectors in $\mathbb{R}^3$. In this case, SO(3) acts on $\mathfrak{so}(3)^*$ by rotations, and the equivariance of the momentum map is just a restatement of the rotational invariance of the cross product: $R(q \times p) = (Rq) \times (Rp)$ for $R \in \mathrm{SO}(3)$.

Now, as remarked earlier, any Hamiltonian system with G as a Hamiltonian symmetry group naturally restricts to a system of ordinary differential equations on the common level set $\{P_i(x) = c_i\}$ of the given first integrals. Note that these common level sets are just the level sets of the momentum map, denoted $\mathcal{S}_\alpha = \{x\colon P(x) = \alpha\}$ where $\alpha = \sum c_i\omega_i \in \mathfrak{g}^*$. Moreover, the reduced system will automatically remain invariant under the *residual symmetry group*

$$G_\alpha \equiv \{g \in G\colon g \cdot \mathcal{S}_\alpha \subset \mathcal{S}_\alpha\}$$

of group elements leaving the chosen level set invariant. There is an easy characterization of this residual group.

Proposition 6.47. *Let $P\colon M \to \mathfrak{g}^*$ be the momentum map associated with a Hamiltonian group action. Then the residual symmetry group of a level set $\mathcal{S}_\alpha = \{P(x) = \alpha\}$ is the isotropy subgroup of the element $\alpha \in \mathfrak{g}^*$:*

$$G_\alpha = \{g \in G\colon \mathrm{Ad}^* g(\alpha) = \alpha\}.$$

Moreover, if $g \in G$ has the property that it takes one point $x \in \mathcal{S}_\alpha$ to a point $g \cdot x \in \mathcal{S}_\alpha$, then $g \in G_\alpha$, and has this property for all $x \in \mathcal{S}_\alpha$.

PROOF. By definition, $g \in G_\alpha$ if and only if $P(g \cdot x) = \alpha$ whenever $P(x) = \alpha$. But, by the equivariance of P,

$$\alpha = P(g \cdot x) = \mathrm{Ad}^* g(P(x)) = \mathrm{Ad}^* g(\alpha),$$

so g is in the isotropy subgroup of α. The second statement easily follows from this identity. □

Note that the residual Lie algebra corresponding to G_α is the *isotropy subalgebra* $\mathfrak{g}_\alpha \equiv \{\mathbf{v} \in \mathfrak{g}\colon \mathrm{ad}^*\, \mathbf{v}|_\alpha = 0\}$, which is readily computable. In particular, the dimension of G_α can be computed as the dimension of its Lie algebra $\mathfrak{g}_\alpha$. For instance, if G is an abelian Lie group, its co-adjoint representation is trivial, $\mathrm{Ad}^* g(\alpha) = \alpha$ for all $g \in G$, $\alpha \in \mathfrak{g}^*$, hence $G_\alpha = G$ for every α. Therefore any Hamiltonian system admitting an abelian Hamiltonian symmetry group remains invariant under the full group, even on restriction to a common level set $\mathcal{S}_\alpha$. This will imply that we can always reduce such a system in order by $2r$, twice the dimension of the group. As a second example, consider the two-parameter solvable group of Example 6.40. Here the momentum map is

$$P(p, q, \tilde{p}, \tilde{q}) = p\omega_1 + (pq + \tilde{p})\omega_2,$$

where $\{\omega_1, \omega_2\}$ are a basis of $\mathfrak{g}^*$ dual to the basis $\{\mathbf{v}, \mathbf{w}\}$ of $\mathfrak{g}$. The co-adjoint representation of $g = \exp(\varepsilon_1 \mathbf{v} + \varepsilon_2 \mathbf{w})$ is found to be

$$\mathrm{Ad}^* g(c_1\omega_1 + c_2\omega_2) = e^{-\varepsilon_2} c_1 \omega_1 + (\varepsilon_1 \varepsilon_2^{-1}(e^{-\varepsilon_2} - 1)c_1 + c_2)\omega_2$$

(with appropriate limiting values if $\varepsilon_2 = 0$). Thus the isotropy subgroup of $\alpha = c_1\omega_1 + c_2\omega_2$ is just $\{e\}$ unless $c_1 = 0$, in which case it is all of G. Thus we expect that the restriction of a Hamiltonian system with symmetry group G to a level set $\mathcal{S}_\alpha = \{p = c_1, pq + \tilde{p} = c_2\}$ will retain no residual symmetry group unless $c_1 = 0$, in which case the entire group G will remain. This is precisely what we observed in Example 6.40.

Once we have restricted the Hamiltonian system to the level set $\mathcal{S}_\alpha$, the idea is then to utilize the methods of Section 2.5 to reduce further using the residual symmetry group G_α. Under certain regularity assumptions on the group action, the quotient manifold $\mathcal{S}_\alpha/G_\alpha$, on which the fully reduced system will live, has a natural identification as a Poisson submanifold of M/G. Thus the fully reduced system inherits a Hamiltonian structure itself. In particular, if the residual group G_α is solvable (rather than G itself being solvable) we can reconstruct the solutions to the original system on $\mathcal{S}_\alpha$ by quadrature from those of the fully reduced system on $\mathcal{S}_\alpha/G_\alpha$. The general result follows:

Theorem 6.48. *Let M be a Poisson manifold and G a regular Hamiltonian group of transformations. Let $\alpha \in \mathfrak{g}^*$. Assume that the momentum map $P: M \to \mathfrak{g}^*$ is of maximal rank everywhere on the level set $\mathcal{S}_\alpha = P^{-1}\{\alpha\}$, and that the residual symmetry group G_α acts regularly on the submanifold $\mathcal{S}_\alpha$. Then there is a natural immersion ϕ making $\mathcal{S}_\alpha/G_\alpha$ into a Poisson submanifold of M/G in such a way that the diagram*

$$\begin{array}{ccccc} & & M & & \\ & \overset{i}{\nearrow} & & \overset{\pi}{\searrow} & \\ \mathcal{S}_\alpha & & & & M/G \\ & \underset{\pi_\alpha}{\searrow} & & \underset{\phi}{\nearrow} & \\ & & \mathcal{S}_\alpha/G_\alpha & & \end{array} \tag{6.39}$$

commutes. (Here π and π_α are the natural projections and i the immersion realizing $\mathcal{S}_\alpha$ as a submanifold of M.) Moreover, any Hamiltonian system on M which admits G as a Hamiltonian symmetry group naturally restricts to systems on the other spaces in (6.39), which are Hamiltonian on M/G and $\mathcal{S}_\alpha/G_\alpha$, and which are related by the appropriate maps. In particular, we obtain a Hamiltonian system on $\mathcal{S}_\alpha/G_\alpha$ by first restricting to $\mathcal{S}_\alpha$ and then projecting using π_α.

PROOF. Assume G is a global group of transformation, although the proof is easily modified to incorporate the local case. According to the diagram, if $z = \pi_\alpha(x) \in \mathcal{S}_\alpha/G_\alpha$, then we should define $\phi(z) = \pi(x) \in M/G$. Note that

$\pi_\alpha(x) = \pi_\alpha(\hat{x})$ if and only if $x = g \cdot \hat{x}$ for some $g \in G_\alpha$, but this means $\pi(x) = \pi(\hat{x})$ and hence ϕ is well defined. Similarly, ϕ is one-to-one since if $x, \hat{x} \in \mathscr{S}_\alpha$ and $\pi(x) = \pi(\hat{x})$, then $x = g \cdot \hat{x}$ for some $g \in G$; according to Proposition 6.47, $g \in G_\alpha$, and hence $\pi_\alpha(x) = \pi_\alpha(\hat{x})$. Finally, ϕ is an immersion, meaning $d\phi$ has maximal rank everywhere, since $d\phi \circ d\pi_\alpha = d\pi \circ di$, and by Proposition 6.47,

$$\ker d\pi_\alpha = \mathfrak{g}_\alpha = \mathfrak{g} \cap T\mathscr{S}_\alpha = \ker(d\pi \circ di).$$

Let $\tilde{H}: M/G \to \mathbb{R}$ correspond to the G-invariant function $H: M \to \mathbb{R}$, so by Theorem 6.42 the corresponding Hamiltonian systems are related: $\hat{\mathbf{v}}_{\tilde{H}} = d\pi(\hat{\mathbf{v}}_H)$. We also know that $\hat{\mathbf{v}}_H$ is everywhere tangent to the level set $\mathscr{S}_\alpha$, and hence there is a reduced vector field $\tilde{\mathbf{v}}$ on $\mathscr{S}_\alpha$ with $\hat{\mathbf{v}}_H = di(\tilde{\mathbf{v}})$ there. Moreover, as $\hat{\mathbf{v}}_H$ has G as a symmetry group, $\tilde{\mathbf{v}}$ retains G_α as a residual symmetry group and there is thus a well-defined vector field $\mathbf{v}^* = d\pi_\alpha(\tilde{\mathbf{v}})$ on the quotient manifold $\mathscr{S}_\alpha/G_\alpha$. Furthermore, this vector field agrees with the restriction of $\hat{\mathbf{v}}_{\tilde{H}}$ to the submanifold $\phi(\mathscr{S}_\alpha/G_\alpha)$ since

$$d\phi(\mathbf{v}^*) = d\phi \circ d\pi_\alpha(\tilde{\mathbf{v}}) = d\pi \circ di(\tilde{\mathbf{v}}) = d\pi(\hat{\mathbf{v}}_H) = \hat{\mathbf{v}}_{\tilde{H}}$$

there.

This last argument proves that *every* Hamiltonian vector field on M/G is everywhere tangent to $\phi(\mathscr{S}_\alpha/G_\alpha)$. Proposition 6.19 then implies that ϕ makes $\mathscr{S}_\alpha/G_\alpha$ into a Poisson submanifold of M/G and, moreover, the restriction of a Hamiltonian vector field $\hat{\mathbf{v}}_{\tilde{H}}$ on M/G to $\mathscr{S}_\alpha/G_\alpha$ (i.e. $\mathbf{v}^*$) is Hamiltonian with respect to the induced Poisson structure. This completes the proof of the theorem and hence the reduction procedure. □

If M is symplectic, then it is not true that M/G is necessarily symplectic. However, it is possible to show that the submanifolds $\mathscr{S}_\alpha/G_\alpha$ form the leaves of the symplectic foliation of M/G! (See Exercise 6.14.)

Example 6.49. Consider the abelian Hamiltonian symmetry group G acting on $\mathbb{R}^6$, with canonical coordinates $(p, q) = (p^1, p^2, p^3, q^1, q^2, q^3)$, generated by the functions $P = p^3$, $Q = q^1p^2 - q^2p^1$. The corresponding Hamiltonian vector fields

$$\mathbf{v}_1 = \frac{\partial}{\partial q^3} \quad \text{and} \quad \mathbf{v}_2 = p^1\frac{\partial}{\partial p^2} - p^2\frac{\partial}{\partial p^1} + q^1\frac{\partial}{\partial q^2} - q^2\frac{\partial}{\partial q^1}$$

generate a two-parameter abelian group of transformations. Any Hamiltonian function of the form $H(\rho, \sigma, \gamma, \zeta, t)$, where $\rho = \sqrt{(q^1)^2 + (q^2)^2}$, $\sigma = \sqrt{(p^1)^2 + (p^2)^2}$, $\gamma = q^1p^2 - q^2p^1$, $\zeta = p^3$, has G as a symmetry group; in particular, $H = \frac{1}{2}|p|^2 + V(\rho)$, a cylindrically symmetrical energy potential, is such a function.

The method of Proposition 6.48 will allow us to reduce the order of such a Hamiltonian system by four. (And, if H does not depend on t, we can integrate the entire system by quadratures.) First we restrict to the level set

$\mathscr{S} = \{P = \zeta, Q = \gamma\}$ for ζ, γ constant. If we use cylindrical coordinates

$$q = (\rho\cos\theta, \rho\sin\theta, z), \qquad p = (\sigma\cos\psi, \sigma\sin\psi, \zeta),$$

for q and p, then

$$\gamma = \rho\sigma\sin(\psi - \theta) = \rho\sigma\sin\phi,$$

where $\phi = \psi - \theta$. In terms of the variables ρ, θ, ϕ, z,† the Hamiltonian system, when restricted to $\mathscr{S}$, takes the form

$$\rho_t = \cos\phi\cdot H_\sigma, \qquad \phi_t = \sin\phi(\sigma^{-1}H_\rho - \rho^{-1}H_\sigma) \tag{6.40a}$$

$$\theta_t = \rho^{-1}\sin\phi H_\sigma + H_\gamma, \qquad z_t = H_\zeta, \tag{6.40b}$$

the subscripts on H denoting partial derivatives. These variables are also designed so that on $\mathscr{S}$, $\mathbf{v}_1 = \partial_z$, $\mathbf{v}_2 = \partial_\theta$. Theorem 6.48 guarantees that (6.40) is invariant under the reduced symmetry group of $\mathscr{S}$, which, owing to the abelian character of G, is all of G itself. This is reflected in the fact that neither z nor θ appears explicitly on the right-hand sides of (6.40). Thus, once we have determined $\rho(t)$ and $\phi(t)$ to solve the first two equations, $\theta(t)$ and $z(t)$ are determined by quadrature.

Moreover, Theorem 6.48 says that (6.40a) forms a Hamiltonian system in its own right. Fixing γ and ζ, let

$$\hat{H}(\rho, \phi, t) = H(\rho, \gamma/(\rho\sin\phi), \gamma, \zeta, t)$$

be the reduced Hamiltonian. Note that

$$\{\rho, \phi\} = -\gamma\rho^{-1}\sigma^{-2} = -\gamma^{-1}\rho\sin^2\phi.$$

An easy computation using the chain rule shows that (6.40a) is the same as

$$\rho_t = -\gamma^{-1}\rho\sin^2\phi\cdot\hat{H}_\phi, \qquad \phi_t = \gamma^{-1}\rho\sin^2\phi\cdot\hat{H}_\rho, \tag{6.41}$$

which is indeed Hamiltonian. In particular, if H (and hence $\hat{H}$) is independent of t we can, in principle, integrate (6.41) by quadrature and hence solve the original system. (In practice, however, even for simple functions H, the intervening algebraic manipulations may prove to be overly complex.)

In general, if a Hamiltonian system is invariant under an r-parameter *abelian* Hamiltonian symmetry group, one can reduce the order by $2r$. This is because the residual symmetry group is always the entire abelian group itself owing to the triviality of the co-adjoint action. A $2n$-th order Hamiltonian system with an n-parameter abelian Hamiltonian symmetry group, or, equivalently possessing n first integrals $P_1(x), \ldots, P_n(x)$ which are *in involution*:

$$\{P_i, P_j\} = 0 \quad \text{for all } i, j,$$

† These, of course, are not universally valid local coordinates; if $\rho = 0$ we must use slightly different variables.

is called a *completely integrable* Hamiltonian system since, in principle, its solutions can be determined by quadrature alone. Actually, much more can be said about such completely integrable systems and the topic forms a significant chapter in the classical theory of Hamiltonian mechanics.

Example 6.50. Consider the group of simultaneous rotations $(p, q) \mapsto (Rp, Rq)$, $R \in SO(3)$ acting on $\mathbb{R}^6$. In Example 6.43 this was shown to be a Hamiltonian group action generated by the components of the angular momentum vector $\boldsymbol{\omega} = q \times p$. Any Hamiltonian function of the form $H(|p|, |q|, p \cdot q)$ will be rotationally-invariant and thereby generate a Hamiltonian system with SO(3) as a Hamiltonian symmetry group. On the subset $M = \{(p, q): q \times p \neq 0\}$, SO(3) acts regularly with three-dimensional orbits. According to Theorem 6.48, we will be able to reduce any such Hamiltonian system in degree by a total of four; three from the reduction to a common level set $\mathscr{S}_{\boldsymbol{\omega}} = \{q \times p = \boldsymbol{\omega}\}$ and one further degree from the residual symmetry group $G_{\boldsymbol{\omega}} \simeq SO(2)$ of rotations around the $\boldsymbol{\omega}$-axis.

Before charging ahead with the reduction, it will help to make a small observation. By the equivariance of the momentum map $P: \mathbb{R}^6 \to \mathfrak{so}(3)^* \simeq \mathbb{R}^3$, $P(p, q) = q \times p = \boldsymbol{\omega}$, we see that $R \in SO(3)$ maps the level set $\mathscr{S}_{\boldsymbol{\omega}}$ to the level set $R \cdot \mathscr{S}_{\boldsymbol{\omega}} = \mathscr{S}_{R\boldsymbol{\omega}}$. Thus we can choose R so as to make $\boldsymbol{\omega} = (0, 0, \omega)$, $\omega > 0$, point in the direction of the positive z-axis. All other solutions, except those of zero angular momentum, which must be treated separately, can be found by suitably rotating these solutions. If $\boldsymbol{\omega}$ is of this form, both p and q must lie in the xy-plane. We use polar coordinates (ρ, θ) for q and (σ, ψ) for p (as in the previous example). Choosing three of these as local coordinates on $\mathscr{S}_{\boldsymbol{\omega}}$ (and ignoring singular points) we obtain the reduced system

$$\rho_t = \cos\phi \cdot H_\sigma + \rho H_\tau, \quad \sigma_t = -\cos\phi \cdot H_\rho - \sigma H_\tau, \quad \theta_t = \rho^{-1} \sin\phi \cdot H_\sigma. \tag{6.42}$$

Here ϕ denotes the angle from q to p, so $\omega = \rho\sigma \sin\phi$, and $\tau = p \cdot q = \rho\sigma\cos\phi$; subscripts on H denote partial derivatives. The residual symmetry group of rotations around the z-axis is generated by

$$\mathbf{v} = -q^2 \frac{\partial}{\partial q^1} + q^1 \frac{\partial}{\partial q^2} - p^2 \frac{\partial}{\partial p^1} + p^1 \frac{\partial}{\partial p^2} = \frac{\partial}{\partial \theta},$$

and is reflected in the fact that θ does not appear on the right-hand side of the equations in (6.42). We can thus determine $\theta(t)$ by a single quadrature from the solutions to the fully reduced system

$$\rho_t = \cos\phi \cdot \hat{H}_\sigma, \qquad \sigma_t = -\cos\phi \cdot \hat{H}_\rho, \tag{6.43}$$

which are Hamilton's equations for

$$\hat{H}(\rho, \sigma) = H(\rho, \sigma, \rho\sigma\cos\phi), \quad \text{where} \quad \omega^2 = \rho\sigma\sin\phi.$$

We leave it to the reader to check that the appropriate Poisson bracket is

$$\{\hat{F}, \hat{H}\} = \cos\phi \left(\frac{\partial \hat{F}}{\partial \rho} \frac{\partial \hat{H}}{\partial \sigma} - \frac{\partial \hat{F}}{\partial \sigma} \frac{\partial \hat{H}}{\partial \rho} \right).$$

In particular, if $\hat{H}$ is independent of t, we can integrate (6.43) by quadrature, leading to a full solution to the original system. The reader might check that the present procedure is, more or less, equivalent to our integration of the Kepler problem in Example 4.19.

NOTES

Hamiltonian mechanics and the closely allied concept of a Poisson bracket have their origins in the original investigations of Poisson, Hamilton, Ostrogradskii and Liouville in the nineteenth century; see Whittaker, [1; p. 264] for details on the historical development of the classical theory, which relied exclusively on the canonical coordinates (p, q) of a symplectic structure on $\mathbb{R}^{2n}$. Besides the classical work of Whittaker, [1], good general references for the theory of Hamiltonian mechanics in the symplectic framework include the books of Abraham and Marsden, [1], Arnol'd, [3], Goldstein, [1], and Arnol'd and Novikov [1].

The more general notion of a Poisson structure first appears in Lie's theory of "function groups" (which predates his theory of Lie groups!) and the integration of systems of first order linear partial differential equations; see Lie, [4; Vol. 2, Chap. 8], Forsyth, [1; Vol. 5, § 137], and Carathéodory, [1; Chap. 9] for this theory. Lie already proved the general Darboux Theorem 6.22 for a Poisson structure of constant rank, and called the distinguished functions "ausgezeichnete functionen", which Forsyth translates as "indicial functions". In this book, I have chosen to use Carathéodory's translation of Lie's term. Recently, Weinstein, [3], proposed the less historically motivated term "Casimir function" for these objects, which has become the more popular terminology of late. Lie's theory was, by and large, forgotten by both the mathematics and physics communities. Poisson structures were re-introduced, more or less independently, by Dirac, [1], Jost, [1], Sudarshan and Mukunda, [1], and, in its present form, Lichnerowicz, [1], [2], Marsden and Weinstein, [2], and Weinstein, [3]. They are of considerable importance in both mathematical physics and differential geometry.

Lie was also well aware of the Poisson bracket associated with the dual of a Lie algebra and its connections with the co-adjoint representation. The explicit formula for this Lie–Poisson bracket can be found in Lie, [4; Vol. 2, p. 294]. This bracket too was forgotten until the 1960's, when it was rediscovered by Berezin, [1], and used by Kirillov, [1], Kostant, [1], and Souriau, [1], in connection with representation theory and geometric quantization. This bracket then bore the name of one or more of the above authors until Weinstein, [2], pointed out its much earlier appearance in Lie's work; the name "Lie–Poisson bracket" first appears in Marsden and Weinstein, [2]. The connection between rigid body motion and the Lie–Poisson bracket on SO(3) is due to Arnol'd, [2]. There is also, of course, a Lie–Poisson bracket corresponding to the left-invariant vector fields on a Lie group; perhaps surprisingly, it differs from the right-invariant version merely by a sign. In fact, to be geometrically accurate, the rigid body bracket of

Example 6.9 is really the left-invariant Lie–Poisson bracket—see Marsden, [2]. See Whittaker, [1; §69], Goldstein, [1; Chap. 4], for classical developments of these equations and Holmes and Marsden, [1], for a more detailed exposition along the lines of this chapter. See Weinstein, [4], for applications to stability theory. See Marsden, [2], for a survey of recent developments.

The reduction of Hamiltonian systems with symmetry has a long history, and most of the techniques, including Jacobi's "elimination of the node" appear in their classical form in Whittaker, [1]. The modern approach to this theory has its origins in the paper of Smale, [1], where the present version of the momentum map was introduced. Further developments due to Souriau, [1], and Meyer, [1], led to the fully developed Marsden and Weinstein, [1], approach to the reduction procedure. The treatment in this chapter is a slightly simplified and slightly less general version of the Marsden–Weinstein theory. Completely integrable Hamiltonian systems, which we've only touched on, have been a subject of immense importance throughout the history of classical (and quantum) mechanics. Most of these examples, such as rigid body motion in $\mathbb{R}^3$ and the Kepler problem, have been known for a long time, but the Toda lattice of Exercise 6.11 is of more recent origin. Manakov, [1], has shown the complete integrability of rigid body motion in $\mathbb{R}^n$. Generalizing the notion of complete integrability to include systems whose integrals are not in involution, as in Exercise 6.12, has been popularized in recent years by Mishchenko and Fomenko, [1], and Kozlov, [1].

Exercises

6.1. Suppose $P(x, t)$ is a first integral of a time-independent Hamiltonian system. Prove that the derivatives $\partial P/\partial t$, $\partial^2 P/\partial t^2$, etc. are also all first integrals. (Whittaker, [1; p. 336].)

6.2. Suppose $\dot{x} = J\nabla H(x)$ is a time-independent Hamiltonian system. Suppose $\hat{\mathbf{v}}_P$ is a Hamiltonian symmetry of the system corresponding to a time-independent function $P(x)$. Prove that for any solution $x(t)$ of the system, $P(x(t)) = at + b$ is a linear function of t. How does this compare with Theorem 6.33? Prove that if the Hamiltonian system has a fixed point x_0, then $a = 0$ and P is actually a first integral as it stands.

6.3. Suppose $\hat{\mathbf{v}}_P$ is a Hamiltonian symmetry of the Hamiltonian system $\dot{x} = J\nabla H$. Let $f(s)$ be any real-valued function of the real variable s. Prove that $f(P(x))\hat{\mathbf{v}}_P$ is again a Hamiltonian symmetry and find the corresponding first integral.

6.4. Let M be a Poisson manifold of constant rank. Prove that a function $C: M \to \mathbb{R}$ is a distinguished function if and only if C is constant on the leaves of the symplectic foliation of M. Does this generalize to the case of nonconstant rank? (Weinstein, [3].)

6.5. Discuss the Lie–Poisson bracket and co-adjoint orbits for the Lie algebra $\mathfrak{sl}(2)$.

6.6. Determine the Lie–Poisson bracket for the Euclidean groups E(2) and E(3). What does it look like when restricted to a co-adjoint orbit?

6.7. Suppose $\{F, H\}$ is a Poisson bracket on $\mathbb{R}^m$ whose structure functions $J^{ij}(x)$ depend linearly on $x \in \mathbb{R}^m$. Prove that this bracket determines a Lie–Poisson structure on $\mathbb{R}^m$.

6.8. Solve the Hamiltonian system corresponding to the Hamiltonian function $H(p, \tilde{p}, q, \tilde{q}) = \frac{1}{2}\tilde{p}^2 + \frac{1}{2}\tilde{q}^{-2}(p^2 - 1)$ on $\mathbb{R}^4$ with canonical Poisson bracket. (Whittaker, [1; p. 314].)

6.9. For a conservative mechanical system, discuss the process of choosing coordinates that fix the centre of mass and the angular momentum of the system in light of our general group-reduction procedure.

*6.10. The motion of n identical point vortices in the plane is governed by the canonical Hamiltonian system in $M = \mathbb{R}^{2n}$ corresponding to the Hamiltonian function

$$H(p, q) = \sum_{i \neq j} \gamma_i \gamma_j \log[(p^i - p^j)^2 + (q^i - q^j)^2]$$

in which (p^i, q^i) is the planar coordinates of the i-th vortex and γ_i its strength. Prove that the Euclidean group E(2) of simultaneous translations and rotations of the vortices forms a symmetry group of this system. Show that each infinitesimal generator of this group is a Hamiltonian vector field, and determine the corresponding conserved quantity. Show that, however, the entire group E(2) is *not* a Hamiltonian symmetry group in the strict sense of Definition 6.41. For what values of n is the vortex problem completely integrable? (Kozlov, [1; p. 15].)

6.11. The three-particle periodic *Toda lattice* is governed by the Hamiltonian system with Hamiltonian function

$$H(p, q) = \tfrac{1}{2}|p|^2 + y^1 + y^2 + y^3, \qquad p, q \in \mathbb{R}^3,$$

where

$$y^1 = e^{q^1 - q^2}, \qquad y^2 = e^{q^2 - q^3}, \qquad y^3 = e^{q^3 - q^1},$$

and we use the canonical Poisson structure on $\mathbb{R}^6$. Prove that the functions

$$P(p, q) = p^1 + p^2 + p^3, \qquad Q(p, q) = p^1 p^2 p^3 - p^1 y^2 - p^2 y^3 - p^3 y^1,$$

are first integrals, and hence the Toda lattice is a completely integrable Hamiltonian system. Is it possible to explicitly integrate it? (Toda, [1; §2.10]).

6.12. Suppose a Hamiltonian system on a $2n$-dimensional symplectic manifold is invariant under an n-parameter solvable Hamiltonian transformation group. Prove that the solutions whose initial conditions cause the n integrals to all *vanish* can (in principle) be found by quadrature, generalizing Example 6.40. (Mishchenko and Fomenko, [1], Kozlov, [1].)

6.13. Let $M = \mathbb{R}^{2n}$ with the canonical Poisson structure. Discuss the reduction Theorem 6.48 for the symmetry group $\mathfrak{so}(2)$ whose action is generated by the energy of a harmonic oscillator, $H(p, q) = \frac{1}{2}(|p|^2 + |q|^2)$. (Arnol'd, [3; p. 377].)

*6.14. Let G and M satisfy the hypotheses of Theorem 6.48. Prove that if M is symplectic, the quotient manifold $\mathscr{S}_\alpha/G_\alpha$ is also symplectic. (Marsden and Weinstein, [1].)

6.15. Corresponding to a Hamiltonian system

$$\frac{dq^i}{dt} = \frac{\partial H}{\partial p^i}, \qquad \frac{dp^i}{dt} = -\frac{\partial H}{\partial q^i}, \qquad H = H(p, q, t),$$

in canonical form is the Hamilton–Jacobi partial differential equation

$$\frac{\partial u}{\partial t} + H\left(\frac{\partial u}{\partial x}, x, t\right) = 0.$$

Prove that the vector field $\mathbf{v} = A(t, x, \partial u/\partial x)\partial_u$ is a generalized symmetry of the Hamilton–Jacobi equation if and only if $A(t, q, p)$ is a first integral of Hamilton's equations. (Fokas, [1].)

6.16. Suppose $\mathscr{L}[u] = \int L(t, u^{(n)})\, dt$ is a functional involving $t \in \mathbb{R}$, $u \in \mathbb{R}$. Define a change of variables

$$q^1 = u, \qquad q^2 = u_t, \ldots, \qquad q^n = u_{n-1} = d^{n-1}u/dt^{n-1},$$

$$p^1 = \frac{\partial L}{\partial u_t} - D_t \frac{\partial L}{\partial u_{tt}} + \cdots + (-D_t)^{n-1} \frac{\partial L}{\partial u_n},$$

$$p^2 = \frac{\partial L}{\partial u_{tt}} - D_t \frac{\partial L}{\partial u_{ttt}} + \cdots + (-D_t)^{n-2} \frac{\partial L}{\partial u_n},$$

$$\vdots$$

$$p^n = \frac{\partial L}{\partial u_n}.$$

Finally, let

$$H(p, q) = -L + p^1 q^2 + p^2 q^3 + \cdots + p^{n-1} q^n + p^n u_n,$$

where $u_n = d^n u/dt^n$ is determined implicitly from the equation for p^n. Prove that $u(t)$ satisfies the Euler–Lagrange equation for $\mathscr{L}$ if and only if $(p(t), q(t))$ satisfy Hamilton's equations for H relative to the canonical Poisson bracket. (Whittaker, [1; p. 266].)

*6.17. *Integral Invariants.* Let M be a Poisson manifold and $\hat{\mathbf{v}}_H$ a Hamiltonian vector field on M. If $S \subset M$ is any subset, let $S(t) = \{\exp(t\hat{\mathbf{v}}_H)x : x \in S\}$, assuming the Hamiltonian flow $\exp(t\hat{\mathbf{v}}_H)$ at time t is defined over all of S. A differential k-form ω on M is called an (absolute) *integral invariant* of the Hamiltonian system determined by $\hat{\mathbf{v}}_H$ if $\int_{S(t)} \omega = \int_S \omega$ for all k-dimensional compact submanifolds $S \subset M$ (with boundary).

(a) Prove that ω is an integral invariant if and only if $\hat{\mathbf{v}}_H(\omega) = 0$ on all of M.

(b) Prove that if $F(x)$ is any first integral of the Hamiltonian system, the one-form dF is an integral invariant. Does the converse hold?

(c) Suppose M is symplectic and $K(x) = J(x)^{-1}$ is as in Proposition 6.15 in terms of local coordinates x. Prove that the differential two-form

$$\Omega = \tfrac{1}{2} \sum_{i,j=1}^{m} K_{ij}(x)\, dx^i \wedge dx^j$$

is independent of the choice of local coordinates, and is an integral invariant of *any* Hamiltonian system with the given Poisson structure.

(d) Conversely, prove that a two-form Ω on M determines a symplectic Poisson structure if and only if it is closed: $d\Omega = 0$, and of maximal rank.
(e) Prove that if ω and ζ are absolute integral invariants for $\hat{\mathbf{v}}_H$, so is $\omega \wedge \zeta$.
(f) Prove *Liouville's theorem* that any Hamiltonian system in $\mathbb{R}^{2n}$ with canonical Poisson bracket (6.1) is volume preserving:

$$\mathrm{Vol}(S(t)) = \mathrm{Vol}(S), \qquad S \subset \mathbb{R}^{2n}.$$

(See Exercise 1.36.) (Cartan, [1], Arnol'd, [3].)

*6.18. Let $\hat{\mathbf{v}}_H$ be a Hamiltonian vector field on the Poisson manifold M.
(a) Prove that if the k-form ω is an integral invariant and $\mathbf{w}$ generates a symmetry group, then the $(k-1)$-form $\mathbf{w} \lrcorner \omega$ is an integral invariant. Also show that the Lie derivative $\mathbf{w}(\omega)$ is an integral invariant.
(b) If M is symplectic, prove that any Hamiltonian system with two *non-*Hamiltonian symmetries has a first integral. What about Hamiltonian symmetries?
(c) Discuss the Hamiltonian vector field corresponding to this first integral. (Rosencrans, [1].)

*6.19. Let N be a smooth manifold, and $M = T^*N$ its cotangent bundle. Then there is a natural symplectic structure on $M = T^*N$ which can be described in any of the following equivalent ways.
(a) Let $q = (q^1, \ldots, q^n)$ be local coordinates on N. Then $T^*N|_q$ is spanned by the basic one-forms $dq^1, \ldots, dq^n$, so that $\omega \in T^*N|_q$ can be written as $\omega = \sum p^i\, dq^i$; hence (q, p) determine the local coordinates on T^*N. Define

$$\{F, H\} = \sum_{i=1}^{n} \frac{\partial F}{\partial q^i}\frac{\partial H}{\partial p^i} - \frac{\partial F}{\partial p^i}\frac{\partial H}{\partial q^i}.$$

Prove that $\{\ ,\ \}$ is a Poisson bracket, which is well-defined on all of T^*N.
(b) Let $\pi\colon T^*N \to N$ be the projection. The canonical one-form θ on $M = T^*N$ is defined so that for any tangent vector $\mathbf{v} \in TM|_\omega$ at $\omega \in M = T^*N$,

$$\langle \theta; \mathbf{v} \rangle = \langle \omega; d\pi(\mathbf{v}) \rangle.$$

Prove that in the local coordinates of part (a), $\theta = \sum p^i\, dq^i$. Therefore, its exterior derivative $\Omega = d\theta$, as in Exercise 6.17(c), defines the Poisson bracket on T^*M.
(c) Let $\mathbf{v}$ be a vector field on N with flow $\exp(\varepsilon\mathbf{v})\colon N \to N$. Prove that the pull-back $\exp(\varepsilon\mathbf{v})^*$ defines a flow on $M = T^*N$. What is its infinitesimal generator? If $H\colon T^*N \to \mathbb{R}$ is any function, prove that there is a unique vector field $\hat{\mathbf{v}}_H$ on $M = T^*N$ such that

$$\hat{\mathbf{v}}_H(\langle \omega; \mathbf{v} \rangle|_x) = \left.\frac{d}{d\varepsilon}\right|_{\varepsilon=0} H[\exp(\varepsilon\mathbf{v})^*(x, \omega)]$$

for all $(x, \omega) \in T^*N$, $\mathbf{v}$ a vector field on N, and that this vector field is the Hamiltonian vector field associated with H relative to the above symplectic structure.

*6.20. *Multi-vectors.* The dual objects to differential forms on a manifold are called "multi-vectors" and are defined as alternating, k-linear, real-valued maps on the cotangent space $T^*M|_x$, varying smoothly from point to point.

(a) Prove that a "uni-vector" (i.e. $k = 1$) is the same as a vector field.
(b) Prove that in a local coordinate chart, any k-vector is of the form

$$\theta = \sum_I h_I(x)\partial_{i_1} \wedge \cdots \wedge \partial_{i_k},$$

where the sum is over all strictly increasing multi-indices I, $\partial_i = \partial/\partial x^i$ form a basis for the tangent space $TM|_x$, and h_I depends smoothly on x.
(c) Let $J(x)$ be the structure matrix for a Poisson bracket on M. Prove that

$$\Theta = \tfrac{1}{2} \sum_{i,j=1}^{m} J^{ij}(x)\partial_i \wedge \partial_j$$

determines a bi-vector, defined so that

$$\langle \Theta; dH, dF \rangle = \{H, F\} \qquad (*)$$

for any pair of real-valued functions H, F.
(d) Prove that if θ is a k-vector and ζ an l-vector, then there is a uniquely defined $(k + l - 1)$-vector $[\theta, \zeta]$, called the *Schouten bracket* of θ and ζ, determined by the properties that $[\cdot,\cdot]$ is bilinear, skew-symmetric: $[\theta, \zeta] = (-1)^{kl}[\zeta, \theta]$, satisfies the Leibniz rule

$$[\theta, \zeta \wedge \eta] = [\theta, \zeta] \wedge \eta + (-1)^{kl+l}\zeta \wedge [\theta, \eta],$$

and agrees with the ordinary Lie bracket in the case θ and ζ are vector fields (uni-vectors). What is the analogue of the Jacobi identity for the Schouten bracket?
(e) Prove that if Θ is a bi-vector then the bracket between functions H and F defined by $(*)$ is necessarily bilinear and alternating; it satisfies the Jacobi identity if and only if the tri-vector $[\Theta, \Theta]$ obtained by bracketing Θ with itself vanishes identically:

$$[\Theta, \Theta] \equiv 0. \qquad (**)$$

Thus the definition of a Poisson structure on a manifold M is equivalent to choosing a bi-vector satisfying $(**)$. (Lichnerowicz, [1], [2].)

6.21. Are the orbits of the adjoint representation of a Lie group on its Lie algebra always even dimensional? Discuss.

*6.22. Suppose M is a Poisson manifold and G a local group of transformations each of which is a Poisson map.
(a) Are the infinitesimal generators of G necessarily Hamiltonian vector fields?
(b) Is G a Hamiltonian transformation group?
(c) Discuss the reduction procedure of Theorem 6.48 for such symmetry groups.

CHAPTER 7

Hamiltonian Methods for Evolution Equations

The equilibrium solutions of the equations of nondissipative continuum mechanics are usually found by minimizing an appropriate variational integral. Consequently, smooth solutions will satisfy the Euler–Lagrange equations for the relevant functional and one can employ the group-theoretic methods in the Lagrangian framework discussed in Chapters 4 and 5. However, when presented with the full dynamical problem, one encounters systems of evolution equations for which the Lagrangian viewpoint, even if applicable, no longer is appropriate or natural to the problem. In this case, the Hamiltonian formulation of systems of evolution equations assumes the natural variational role for the system.

Historically, though, the recognition of the correct general form for an infinite-dimensional generalization to evolution equations of the classical concept of a finite-dimensional Hamiltonian system has only recently been acknowledged. Part of the problem was the excessive reliance on canonical coordinates, guaranteed by Darboux' theorem in finite dimensions, but no longer valid for evolutionary systems. The Poisson bracket approach adopted here, however, does readily generalize in this context. The principal innovations needed to convert a Hamiltonian system of ordinary differential equations (6.14) to a Hamiltonian system of evolution equations are:

(i) replacing the Hamiltonian function $H(x)$ by a Hamiltonian functional $\mathscr{H}[u]$,

(ii) replacing the vector gradient operation ∇H by the variational derivative $\delta\mathscr{H}$ of the Hamiltonian functional, and

(iii) replacing the skew-symmetric matrix $J(x)$ by a skew-adjoint differential operator $\mathscr{D}$, which may depend on u.

The resulting Hamiltonian system will take the form

$$\frac{\partial u}{\partial t} = \mathscr{D} \cdot \delta \mathscr{H}[u].$$

There are certain restrictions, based on the Jacobi identity for the associated Poisson bracket, that a differential operator $\mathscr{D}$ must satisfy in order to qualify as a true Hamiltonian operator. These are described in Section 7.1; in their original form they are hopelessly complicated to work with but, using the theory of "functional multi-vectors", an efficient, simple computational algorithm for determining when an operator $\mathscr{D}$ is Hamiltonian is devised. The second section explores the standard applications of symmetry groups and conservation laws to Hamiltonian systems of evolution equations. The main tool is the Hamiltonian form of Noether's theorem. Applications to the Korteweg–de Vries equation and the Euler equations of ideal fluid flow are presented.

The final section deals with the recent theory of bi-Hamiltonian systems. Occasionally, as in the case of the Korteweg–de Vries equation, one runs across a system of evolution equations which can be written in Hamiltonian form in two distinct ways. In this case, subject to a mild compatibility condition, the system will necessarily have an infinite hierarchy of mutually commuting conservation laws and consequent Hamiltonian flows, generated by a recursion operator based on the two Poisson brackets, and hence can be viewed as a "completely integrable" Hamiltonian system. Such systems have many other remarkable properties, including soliton solutions, linearization by inverse scattering and so on. A new proof of the basic theorem on bi-Hamiltonian systems is given here, along with some applications.

7.1. Poisson Brackets

Recall first the basic set-up of the formal variational calculus presented in Section 5.4. Let $M \subset X \times U$ be an open subset of the space of independent and dependent variables $x = (x^1, \ldots, x^p)$ and $u = (u^1, \ldots, u^q)$. The algebra of differential functions $P(x, u^{(n)}) = P[u]$ over M is denoted by $\mathscr{A}$, and its quotient space under the image of the total divergence is the space $\mathscr{F}$ of functionals $\mathscr{P} = \int P\, dx$.

The main goal of this section is to make precise what we mean by a system of evolution equations

$$u_t = K[u] = K(x, u^{(n)}), \qquad K \in \mathscr{A}^q,$$

being a Hamiltonian system. Here K depends only on spatial variables x and spatial derivatives of u; t is singled out to play a special role. To do this, we need to pursue analogies to the various components of (6.14) in the present context. Firstly, the role of the Hamiltonian function in (6.14) should be played by a Hamiltonian *functional* $\mathscr{H} = \int H\, dx \in \mathscr{F}$. Therefore, we must replace the gradient operation by the "functional gradient" or variational derivative $\delta\mathscr{H} \in \mathscr{A}^q$ of $\mathscr{H}$. The remaining ingredient is the analogue of the skew-symmetric matrix $J(x)$ which serves to define the Poisson bracket. Here we need a linear operator $\mathscr{D}: \mathscr{A}^q \to \mathscr{A}^q$ on the space of q-tuples of differential functions, which, in most instances, will be a linear $q \times q$ matrix differential operator, and may depend on x, u and derivatives of u. To qualify as a Hamiltonian operator, $\mathscr{D}$ must enjoy further properties, which are found by looking at the corresponding Poisson bracket.

In finite dimensions, the Poisson brackets of two functions is a function which depends bilinearly on the respective gradients, the coefficients being determined by the Hamiltonian matrix $J(x)$, cf. (6.12). Thus, for evolution equations, the Poisson bracket of two functionals must be a *functional* depending bilinearly on the respective variational derivatives. Clearly, for a candidate Hamiltonian operator $\mathscr{D}$, the correct expression for the corresponding Poison bracket has the form

$$\{\mathscr{P}, \mathscr{Q}\} = \int \delta\mathscr{P} \cdot \mathscr{D}\delta\mathscr{Q}\, dx, \tag{7.1}$$

whenever $\mathscr{P}, \mathscr{Q} \in \mathscr{F}$ are functionals. Of course, the Hamiltonian operator $\mathscr{D}$ must satisfy certain further restrictions in order that (7.1) be a true Poisson bracket.

Definition 7.1. A linear operator $\mathscr{D}: \mathscr{A}^q \to \mathscr{A}^q$ is called *Hamiltonian* if its Poisson bracket (7.1) satisfies the conditions of *skew-symmetry*

$$\{\mathscr{P}, \mathscr{Q}\} = -\{\mathscr{Q}, \mathscr{P}\}, \tag{7.2}$$

and the *Jacobi identity*

$$\{\{\mathscr{P}, \mathscr{Q}\}, \mathscr{R}\} + \{\{\mathscr{R}, \mathscr{P}\}, \mathscr{Q}\} + \{\{\mathscr{Q}, \mathscr{R}\}, \mathscr{P}\} = 0, \tag{7.3}$$

for all functionals $\mathscr{P}, \mathscr{Q}, \mathscr{R} \in \mathscr{F}$.

If we compare Definition 7.1 with the finite-dimensional version of Definition 6.1, we see that two of the earlier conditions have been dropped. Of these, bilinearity is apparent from the form (7.1) of the bracket. The Leibniz rule has no counterpart in this situation, since, as we saw in Section 5.4, there is no well-defined multiplication between functionals. However, the principal use of Leibniz' rule was to deduce the existence of a Hamiltonian vector field from a real-valued function H, satisfying (6.4). This *does* carry over to the functional case:

Proposition 7.2. *Let $\mathscr{D}$ be a Hamiltonian operator with Poisson bracket* (7.1). *To each functional $\mathscr{H} = \int H\, dx \in \mathscr{F}$, there is an evolutionary vector field $\hat{\mathbf{v}}_{\mathscr{H}}$, called the* Hamiltonian vector field *associated with $\mathscr{H}$, which satisfies*

$$\operatorname{pr} \hat{\mathbf{v}}_{\mathscr{H}}(\mathscr{P}) = \{\mathscr{P}, \mathscr{H}\} \tag{7.4}$$

for all functionals $\mathscr{P} \in \mathscr{F}$. Indeed, $\hat{\mathbf{v}}_{\mathscr{H}}$ has characteristic $\mathscr{D}\delta\mathscr{H} = \mathscr{D}\mathrm{E}(H)$.

PROOF. Let $\mathscr{P} = \int P\,dx$, $P \in \mathscr{A}$. Then, using the integration by parts formula (5.128), we find

$$\{\mathscr{P}, \mathscr{H}\} = \int \mathsf{E}(P) \cdot \mathscr{D}\mathsf{E}(H)\,dx = \int \operatorname{pr} \mathbf{v}_{\mathscr{D}\mathsf{E}(H)}(P)\,dx = \operatorname{pr} \mathbf{v}_{\mathscr{D}\mathsf{E}(H)}\left(\int P\,dx\right).$$

(See Exercise 5.52). Thus (7.4) holds provided $\hat{\mathbf{v}}_{\mathscr{H}} = \mathbf{v}_{\mathscr{D}\mathsf{E}(H)}$. □

The Hamiltonian flow corresponding to a functional $\mathscr{H}[u]$ is obtained by exponentiating the corresponding Hamiltonian vector field $\hat{\mathbf{v}}_{\mathscr{H}}$. According to (5.14), then, a Hamiltonian system of evolution equations takes the form

$$\frac{\partial u}{\partial t} = \mathscr{D} \cdot \delta\mathscr{H}, \tag{7.5}$$

where δ is the variational derivative, and $\mathscr{D}$ the Hamiltonian operator. Note the complete analogy with the finite-dimensional Hamiltonian system (6.14). Before proceeding to examples of Hamiltonian systems, we need to have some reasonably straightforward means of determining when a given operator $\mathscr{D}$ is Hamiltonian. To begin with, the requirement imposed by the skew-symmetry of the Poisson bracket is immediate, being the infinite-dimensional verison of the skew-symmetry of the matrix J in (6.14).

Proposition 7.3. *Let $\mathscr{D}$ be a $q \times q$ matrix differential operator with bracket (7.1) on the space of functionals. Then the bracket is skew-symmetric, i.e.* (7.2) *holds, if and only if $\mathscr{D}$ is skew-adjoint: $\mathscr{D}^* = -\mathscr{D}$.*

PROOF. If $\mathscr{P} = \int P\,dx$, $\mathscr{Q} = \int Q\,dx$, then (7.2) can be written as

$$\int \mathsf{E}(P) \cdot \mathscr{D}\mathsf{E}(Q)\,dx = -\int \mathsf{E}(Q) \cdot \mathscr{D}\mathsf{E}(P)\,dx. \tag{7.6}$$

Using the definition (5.76) of the adjoint $\mathscr{D}^*$, we find (7.6) is equivalent to

$$\int \mathsf{E}(P) \cdot (\mathscr{D} + \mathscr{D}^*)\mathsf{E}(Q)\,dx = 0. \tag{7.7}$$

If (7.7) holds for all $P, Q \in \mathscr{A}$, then, as in the proof of Proposition 5.88, using the "substitution principle" which was enunciated in Exercise 5.42, we must have $\mathscr{D} + \mathscr{D}^* = 0$. □

The Jacobi Identity

At first sight, the direct verification of the Jacobi identity (7.3), even for the simplest skew-adjoint operators, appears a hopelessly complicated computational task. However, a considerable simplification is effected by utilizing some of our basic results from the formal variational calculus, bringing this problem within the realm of feasibility. An even further simplification can be made by introducing a version of the functional forms of Section 5.4 (although here they are, in a sense, "dual" objects), after which the verification of the Jacobi identity becomes a more or less routine computation.

Let $\mathscr{P}$, $\mathscr{Q}$, $\mathscr{R}$ be functionals, with variational derivatives $\delta\mathscr{P} = P$, $\delta\mathscr{Q} = Q$, $\delta\mathscr{R} = R$ in $\mathscr{A}^q$. (Note the change of notation: P is no longer the integrand for $\mathscr{P}$!) With this notation, the first term in the Jacobi identity (7.3) becomes

$$\{\{\mathscr{P}, \mathscr{Q}\}, \mathscr{R}\} = \operatorname{pr} \hat{\mathbf{v}}_{\mathscr{R}}\left(\int P \cdot \mathscr{D}Q \, dx\right) = \int \operatorname{pr} \mathbf{v}_{\mathscr{D}R}(P \cdot \mathscr{D}Q) \, dx.$$

Using Leibniz' rule, and the Lie derivative formula (5.38), this is

$$\{\{\mathscr{P}, \mathscr{Q}\}, \mathscr{R}\} = \int \{\operatorname{pr} \mathbf{v}_{\mathscr{D}R}(P) \cdot \mathscr{D}Q + P \cdot \operatorname{pr} \mathbf{v}_{\mathscr{D}R}(\mathscr{D})Q + P \cdot \mathscr{D}[\operatorname{pr} \mathbf{v}_{\mathscr{D}R}(Q)]\} \, dx. \tag{7.8}$$

From the formula (5.33) connecting the Lie derivative and Fréchet derivative, the first term in this expression is

$$\int \operatorname{pr} \mathbf{v}_{\mathscr{D}R}(P) \cdot \mathscr{D}Q \, dx = \int \mathsf{D}_P(\mathscr{D}R) \cdot \mathscr{D}Q \, dx.$$

Similarly, if we use the fact that $\mathscr{D}$ is skew-adjoint, the third term has an analogous form

$$\int P \cdot \mathscr{D}[\operatorname{pr} \mathbf{v}_{\mathscr{D}R}(Q)] \, dx = -\int \mathscr{D}P \cdot \mathsf{D}_Q(\mathscr{D}R) \, dx. \tag{7.9}$$

The second and third components in the Jacobi identity contribute similar expressions; for example, $\{\{\mathscr{Q}, \mathscr{R}\}, \mathscr{P}\}$ contains the terms

$$\int \mathsf{D}_Q(\mathscr{D}P) \cdot \mathscr{D}R \, dx \quad \text{and} \quad -\int \mathscr{D}Q \cdot \mathsf{D}_R(\mathscr{D}P) \, dx. \tag{7.10}$$

But according to Theorem 5.92, the Fréchet derivative of $Q = \delta\mathscr{Q}$ is a self-adjoint differential operator, so the first integral in (7.10) equals

$$\int \mathscr{D}P \cdot \mathsf{D}_Q(\mathscr{D}R) \, dx,$$

and cancels the integral (7.9) when substituted into the Jacobi identity. Thus, once we have expanded the Jacobi identity in this way, six of the terms cancel and we are left with the equivalent form

$$\int [P \cdot \operatorname{pr} \mathbf{v}_{\mathscr{D}R}(\mathscr{D})Q + R \cdot \operatorname{pr} \mathbf{v}_{\mathscr{D}Q}(\mathscr{D})P + Q \cdot \operatorname{pr} \mathbf{v}_{\mathscr{D}P}(\mathscr{D})R] \, dx = 0, \tag{7.11}$$

which must vanish for all P, Q, R which are variational derivatives of functionals.

At this stage, a further simplification arises. Note that the integrand in (7.11) depends only on P, Q and R and their *total* derivatives. According to our general "substitution principle" announced in Exercise 5.42, this expression vanishes for all variational derivatives $P = \delta\mathscr{P}$, $Q = \delta\mathscr{Q}$, $R = \delta\mathscr{R}$ if and only if it vanishes for *arbitrary* q-tuples $P, Q, R \in \mathscr{A}^q$. We have thus proved

Proposition 7.4. *Let $\mathscr{D}$ be a skew-adjoint $q \times q$ matrix differential operator. Then the bracket* (7.1) *satisfies the Jacobi identity if and only if* (7.11) *vanishes for all q-tuples $P, Q, R \in \mathscr{A}^q$.*

Corollary 7.5. *If $\mathscr{D}$ is a skew-adjoint $q \times q$ matrix differential operator whose coefficients do not depend on u or its derivatives, then $\mathscr{D}$ is automatically a Hamiltonian operator.*

In fact, in this case $\mathrm{pr}\, \mathbf{v}_Q(\mathscr{D}) = 0$ for any evolutionary vector field $\mathbf{v}_Q$, so (7.11) is trivially satisfied. □

Example 7.6. The Korteweg–de Vries equation

$$u_t = u_{xxx} + uu_x$$

can in fact be written in Hamiltonian form in two distinct ways. Firstly, we see

$$u_t = D_x(u_{xx} + \tfrac{1}{2}u^2) = \mathscr{D}\delta\mathscr{H}_1,$$

where $\mathscr{D} = D_x$ and

$$\mathscr{H}_1[u] = \int [-\tfrac{1}{2}u_x^2 + \tfrac{1}{6}u^3]\, dx$$

is one of the classical conservation laws. Note that $\mathscr{D}$ is certainly skew-adjoint, and hence by Corollary 7.5 is automatically Hamiltonian. The Poisson bracket is

$$\{\mathscr{P}, \mathscr{Q}\} = \int \delta\mathscr{P} \cdot D_x(\delta\mathscr{Q})\, dx. \tag{7.12}$$

(To gain a true appreciation of the efficacy of our formal variational methods, the reader might try verifying the Jacobi identity for (7.12) directly!).

The second Hamiltonian form is a bit less obvious. We find

$$u_t = (D_x^3 + \tfrac{2}{3}uD_x + \tfrac{1}{3}u_x)u = \mathscr{E}\delta\mathscr{H}_0,$$

where

$$\mathscr{H}_0[u] = \int \tfrac{1}{2}u^2\, dx$$

is another of the conserved quantities, and

$$\mathscr{E} = D_x^3 + \tfrac{2}{3}uD_x + \tfrac{1}{3}u_x.$$

It is easy to prove that $\mathscr{E}$ is skew-adjoint; to prove the Jacobi identity we look at (7.11). The first term there is

$$\int P \operatorname{pr} \mathbf{v}_{\mathscr{E}(R)}(\mathscr{E})Q\, dx = \int P[\tfrac{2}{3}(\mathscr{E}R)Q_x + \tfrac{1}{3}(\mathscr{E}R)_x Q]\, dx$$

$$= \int [\tfrac{2}{3}PR_{xxx}Q_x + \tfrac{1}{3}PR_{xxxx}Q + \tfrac{4}{9}uPR_xQ_x + \tfrac{2}{9}u_xPRQ_x$$

$$+ \tfrac{2}{9}uPR_{xx}Q + \tfrac{1}{3}u_xPR_xQ + \tfrac{1}{9}u_xPRQ]\, dx,$$

where we are using x-subscripts as abbreviations for *total* derivatives: $P_x = D_xP$, $P_{xx} = D_x^2P$, etc. We must add in the corresponding expressions stemming from the other two terms in (7.11) and then prove that the resulting integrand is a null Lagrangian, i.e. a total D_x-derivative, no matter what P, Q and R are. This is true, and we leave the proof to the reader, since later we will find a much simpler proof of this fact. We conclude that

$$\{\mathscr{P}, \mathscr{Q}\} = \int [\delta\mathscr{P}\cdot(D_x^3 + \tfrac{2}{3}uD_x + \tfrac{1}{3}u_x)\delta\mathscr{Q}]\, dx \tag{7.13}$$

does define a Poisson bracket on the space of functionals.

Although the above computation is quite a bit more tractable than the direct verification of the Jacobi identity, it still requires quite a lot of computational stamina even in such a relatively simple example. An even more radical simplification can be effected if we employ a theory of multilinear maps similar to that developed in Section 5.4. (An applications-oriented reader may wish to skip ahead to Section 7.2 at this stage.)

Functional Multi-vectors

In finite dimensions, multivectors are the dual objects to differential forms. In Exercise 6.20 it was shown how to develop the theory of finite-dimensional Poisson structures based on the theory of multivectors. Here we introduce the analogous objects for infinite-dimensional Hamiltonian systems of evolution equations. Since we are working with open subsets of Euclidean space $M \subset X \times U$, the theory of functional multi-vectors is identical with that of functional forms developed in Section 5.4. The only reason that we employ different terminology and notation is that, from a more global standpoint, the transformation rules for these objects under changes of variables are *not* the same; functional forms transform like Euler–Lagrange expressions whereas functional multi-vectors are more like evolutionary vector fields. Except for this distinction (which will not actually occur in this book) these objects are the same.

Recall that each functional k-form determines an alternating k-linear map from the space T_0 of evolutionary vector fields to the space $\mathscr{F}$ of functionals.

Similarly, a *functional k-vector* will be determined by an alternating k-linear map from the "dual" space $\bigwedge_*^1$ of *functional* one-forms to $\mathscr{F}$. Since each evolutionary vector field is uniquely determined by its characteristic, we can identify T_0 with $\mathscr{A}^q$, the space of q-tuples of differential functions on M. Similarly, according to Proposition 5.87, each functional one-form is uniquely determined by its canonical form, and hence we can also identify $\bigwedge_*^1$ with $\mathscr{A}^q$. Under these two identifications (which depend on the precise Euclidean coordinates on M), we obtain the identification of functional multi-vectors and forms.

Each functional form arises from a vertical form, so correspondingly each functional multi-vector arises from a vertical multi-vector. To preserve the notational distinction between the two, we use the notation θ_J^α for the "uni-vector" corresponding to the one-form du_J^α; thus

$$\langle \theta_J^\alpha; P \rangle = D_J P_\alpha \quad \text{whenever} \quad P = (P_1, \ldots, P_q) \in \mathscr{A}^q.$$

(From now on we replace $\bigwedge_*^1$ by $\mathscr{A}^q$.) Note that we could identify θ_J^α with the derivation $\partial/\partial u_J^\alpha$, which could be adopted as an alternative notation for multi-vectors, but one that is heavier and, later, slightly confusing. A general functional k-vector is thus a finite sum

$$\Theta = \int \left\{ \sum_{\alpha, \mathbf{J}} R_{\mathbf{J}}^{\boldsymbol{\alpha}}[u] \theta_{J_1}^{\alpha_1} \wedge \cdots \wedge \theta_{J_k}^{\alpha_k} \right\} dx,$$

with coefficients $R_{\mathbf{J}}^{\boldsymbol{\alpha}} \in \mathscr{A}$; it defines the k-linear map

$$\langle \Theta; P^1, \ldots, P^k \rangle = \int \left[\sum_{\alpha, \mathbf{J}} R_{\mathbf{J}}^{\boldsymbol{\alpha}} \det(D_{J_i} P_{\alpha_i}^j) \right] dx, \qquad P^j \in \mathscr{A}^q,$$

cf. (5.111). The total derivatives act as Lie derivatives on the vertical k-vectors, with $D_i(\theta_J^\alpha) = \theta_{J,i}^\alpha$, cf. (5.112). The space $\bigwedge_k^*$ of functional k-vectors is then the quotient space of the space of *vertical k-vectors* (i.e. finite sums of wedge products of the θ_J^α with coefficients in $\mathscr{A}$) under the image of total divergence. By Lemma 5.85, every functional k-vector is uniquely determined by its values on the space of q-tuples of differential functions. In this way, all of the theorems and examples of functional forms discussed in Section 5.4 carry over to functional multi-vectors once we replace du_J^α by its counterpart θ_J^α and the vector fields pr $\mathbf{v}_Q$ by their characteristics $Q \in \mathscr{A}^q$.†

For example, any functional uni-vector

$$\gamma = \int \left\{ \sum_{\alpha, J} R_J^\alpha \theta_J^\alpha \right\} dx$$

† One tricky point in this theory is that while the spaces $\bigwedge_*^k$ and $\bigwedge_k^*$ of functional forms and multi-vectors are defined in a dual manner, they are *not* naturally dual vector spaces for any $k > 1$! This is a reflection of our inability to define a multiplication on the space $\mathscr{F}$ of functionals.

can be put into *canonical form*

$$\gamma = \int \{R \cdot \theta\}\,dx = \int \left\{ \sum_{\alpha=1}^{q} R_\alpha \theta^\alpha \right\} dx, \qquad R_\alpha = \sum_J (-D)_J R_J^\alpha,$$

by an integration by parts. (Thus we can identify $\bigwedge_1^*$ with T_0, the space of evolutionary vector fields so γ corresponds to $\mathbf{v}_R$!) Similarly, any functional bi-vector has the *canonical form*

$$\Theta = \tfrac{1}{2} \int \{\theta \wedge \mathscr{D}\theta\}\,dx = \tfrac{1}{2} \int \left\{ \sum_{\alpha,\beta=1}^{q} \theta^\alpha \wedge \mathscr{D}_{\alpha\beta}\theta^\beta \right\} dx, \tag{7.14}$$

where $\mathscr{D} = (\mathscr{D}_{\alpha\beta})$ is a skew-adjoint $q \times q$ matrix differential operator; see Proposition 5.88. Such a bi-vector defines the bilinear map

$$\langle \Theta; P, Q \rangle = \tfrac{1}{2} \int (P \cdot \mathscr{D}Q - Q \cdot \mathscr{D}P)\,dx = \int (P \cdot \mathscr{D}Q)\,dx, \qquad P, Q \in \mathscr{A}^q,$$

where we used the skew-adjoint nature of $\mathscr{D}$. In particular, if P and Q are variational derivatives (or differentials if we revert back to $\bigwedge_*^1$),

$$\langle \Theta; \delta\mathscr{P}, \delta\mathscr{Q} \rangle = \int (\delta\mathscr{P} \cdot \mathscr{D}\delta\mathscr{Q})\,dx$$

reproduces the bracket $\{\mathscr{P}, \mathscr{Q}\}$ determined by the skew-adjoint operator $\mathscr{D}$. For example, the Poisson bracket for the second Hamiltonian operator $\mathscr{E}$ of the Korteweg–de Vries example is represented by the bi-vector

$$\Theta = \tfrac{1}{2} \int \{\theta \wedge \mathscr{E}(\theta)\}\,dx = \tfrac{1}{2} \int \{\theta \wedge \theta_{xxx} + \tfrac{2}{3}u\theta \wedge \theta_x\}\,dx, \tag{7.15}$$

the term involving $\theta \wedge \theta$ trivially vanishing.

The Jacobi identity provides a natural example of a functional tri-vector. Note that in its original form, the left-hand side of (7.3) is clearly an alternating, tri-linear function of the variational derivatives $\delta\mathscr{P}$, $\delta\mathscr{Q}$, $\delta\mathscr{R}$. Therefore (7.11), although it may not appear to be, is an alternating tri-linear function of the q-tuples P, Q, R and hence determines a functional tri-vector, which we denote by

$$\Psi = \tfrac{1}{2} \int \{\theta \wedge \operatorname{pr} \mathbf{v}_{\mathscr{D}\theta}(\mathscr{D}) \wedge \theta\}\,dx, \tag{7.16}$$

so that $\langle \Psi; P, Q, R \rangle$ is the left-hand side of (7.11). (See also Exercise 7.12.) It remains to explain the notation in (7.16).

The "vector field" $\mathbf{v}_{\mathscr{D}\theta}$ is a formal evolutionary vector field whose characteristic is the q-tuple

$$(\mathscr{D}\theta)_\alpha = \sum_{\beta=1}^{q} \mathscr{D}_{\alpha\beta}\theta^\beta$$

of vertical uni-vectors; thus formally, using (5.6) as a model,

$$\mathrm{pr}\,\mathbf{v}_{\mathcal{D}\theta} = \sum_{\alpha,J} D_J\left(\sum_\beta \mathcal{D}_{\alpha\beta}\theta^\beta\right)\frac{\partial}{\partial u_J^\alpha}.$$

In particular, if $R \in \mathcal{A}$ is any differential function,

$$\mathrm{pr}\,\mathbf{v}_{\mathcal{D}\theta}(R) = \sum_{\alpha,J}\frac{\partial R}{\partial u_J^\alpha} D_J\left(\sum_\beta \mathcal{D}_{\alpha\beta}\theta^\beta\right)$$

is a vertical uni-vector. For example, in the case of the second Hamiltonian operator for the Korteweg–de Vries equation, we have

$$\mathrm{pr}\,\mathbf{v}_{\mathcal{E}\theta}(u) = \mathcal{E}\theta = \theta_{xxx} + \tfrac{2}{3}u\theta_x + \tfrac{1}{3}u_x\theta,$$
$$\mathrm{pr}\,\mathbf{v}_{\mathcal{E}\theta}(u_x) = D_x(\mathcal{E}\theta) = \theta_{xxxx} + \tfrac{2}{3}u\theta_{xx} + u_x\theta_x + \tfrac{1}{3}u_{xx}\theta,$$

and so on.

Secondly, we can let $\mathrm{pr}\,\mathbf{v}_{\mathcal{D}\theta}$ act on a differential operator, say $\mathcal{D}$, as a Lie derivative; the resulting object will be a differential operator whose coefficients are functional uni-vectors in that they involve θ_J^α's. For example,

$$\begin{aligned}\mathrm{pr}\,\mathbf{v}_{\mathcal{E}\theta}(\mathcal{E}) &= \mathrm{pr}\,\mathbf{v}_{\mathcal{E}\theta}(D_x^3 + \tfrac{2}{3}uD_x + \tfrac{1}{3}u_x)\\ &= \tfrac{2}{3}\,\mathrm{pr}\,\mathbf{v}_{\mathcal{E}\theta}(u)D_x + \tfrac{1}{3}\,\mathrm{pr}\,\mathbf{v}_{\mathcal{E}\theta}(u_x)\\ &= \tfrac{2}{3}(\theta_{xxx} + \tfrac{2}{3}u\theta_x + \tfrac{1}{3}u_x\theta)D_x + \tfrac{1}{3}(\theta_{xxxx} + \tfrac{2}{3}u\theta_{xx} + u_x\theta_x + \tfrac{1}{3}u_{xx}\theta).\end{aligned}$$

Finally, we apply $\mathrm{pr}\,\mathbf{v}_{\mathcal{D}\theta}(\mathcal{D})$ to θ itself by a combination of differentiating and wedging in the obvious manner. For instance, the tri-vector for the Jacobi identity corresponding to $\mathcal{E}$ is

$$\begin{aligned}&\tfrac{1}{2}\int \{\theta \wedge \mathrm{pr}\,\mathbf{v}_{\mathcal{E}\theta}(\mathcal{E}) \wedge \theta\}\,dx\\ &= \int \{\tfrac{1}{3}\theta \wedge \theta_{xxx} \wedge \theta_x + \tfrac{2}{9}u\theta \wedge \theta_x \wedge \theta_x + \tfrac{1}{9}u_x\theta \wedge \theta \wedge \theta_x + \tfrac{1}{6}\theta \wedge \theta_{xxxx} \wedge \theta\\ &\qquad + \tfrac{1}{9}u\theta \wedge \theta_{xx} \wedge \theta + \tfrac{1}{6}u_x\theta \wedge \theta_x \wedge \theta + \tfrac{1}{18}u_{xx}\theta \wedge \theta \wedge \theta\}\,dx\\ &= -\tfrac{1}{3}\int \{\theta \wedge \theta_x \wedge \theta_{xxx}\}\,dx\end{aligned}$$

by the basic properties of the wedge product. This final tri-vector is also trivial, as can be seen by a simple integration by parts:

$$\begin{aligned}\int\{\theta \wedge \theta_x \wedge \theta_{xxx}\}\,dx &= -\int\{D_x(\theta \wedge \theta_x) \wedge \theta_{xx}\}\,dx\\ &= -\int\{\theta_x \wedge \theta_x \wedge \theta_{xx} + \theta \wedge \theta_{xx} \wedge \theta_{xx}\}\,dx = 0.\end{aligned}$$

Note that according to this notation, if we evaluate Ψ, as given in (7.16), on $P, Q, R \in \mathscr{A}^q$ we obtain six terms, the first two of which are

$$\tfrac{1}{2}\int [P \operatorname{pr} \mathbf{v}_{\mathscr{D}R}(\mathscr{D})Q - Q \operatorname{pr} \mathbf{v}_{\mathscr{D}R}(\mathscr{D})P]\, dx.$$

By Exercise 7.12, since $\mathscr{D}$ is a skew-adjoint differential operator, so is $\operatorname{pr} \mathbf{v}_Q(\mathscr{D})$ for any evolutionary vector field $\mathbf{v}_Q$. Thus these two particular terms are equal and combine to give the first term in the Jacobi identity (7.11). Thus, as claimed above, $\langle \Psi; P, Q, R\rangle$ is the same as the left-hand side of (7.11). Moreover, using Lemma 5.85 (or, rather, its counterpart for functional multi-vectors), we see that vanishing of (7.11) is equivalent to the triviality of the tri-vector Ψ.

Proposition 7.7. *Let $\mathscr{D}$ be a skew-adjoint $q \times q$ matrix differential operator. Then $\mathscr{D}$ is Hamiltonian if and only if the functional tri-vector* (7.16) *vanishes:* $\Psi = 0$.

There is one final simplification available. Let us extend the definition of the prolonged "vector field" $\operatorname{pr} \mathbf{v}_{\mathscr{D}\theta}$ to the space of vertical uni-vectors by setting

$$\operatorname{pr} \mathbf{v}_{\mathscr{D}\theta}(\theta_J^\alpha) = 0$$

for all α, J and extending it to act as a derivation on vertical multi-vectors. Thus we can write the integrand of (7.16) as

$$-\operatorname{pr} \mathbf{v}_{\mathscr{D}\theta}(\theta \wedge \mathscr{D}\theta) = \theta \wedge \operatorname{pr} \mathbf{v}_{\mathscr{D}\theta}(\mathscr{D}) \wedge \theta, \tag{7.17}$$

the minus sign coming from the fact that we have interchanged a wedge product of θ's. Moreover, $\operatorname{pr} \mathbf{v}_{\mathscr{D}\theta}$, being evolutionary, commutes with total differentiation:

$$\operatorname{pr} \mathbf{v}_{\mathscr{D}\theta} \cdot D_k = D_k \cdot \operatorname{pr} \mathbf{v}_{\mathscr{D}\theta}, \qquad k = 1, \ldots, p,$$

cf. (5.19), even when it acts on vertical multi-vectors. Therefore, if $\Phi = \int \tilde{\Phi}\, dx$ is any functional k-vector, we can unambiguously define the $(k+1)$-vector

$$\operatorname{pr} \mathbf{v}_{\mathscr{D}\theta}(\Phi) = \int \operatorname{pr} \mathbf{v}_{\mathscr{D}\theta}(\tilde{\Phi})\, dx.$$

(See Exercise 5.52.) In particular, the bi-vector Θ determining the Poisson bracket, (7.14), can be acted on this way, and by (7.17) we find

$$\operatorname{pr} \mathbf{v}_{\mathscr{D}\theta}(\Theta) = \tfrac{1}{2}\int \{\operatorname{pr} \mathbf{v}_{\mathscr{D}\theta}(\theta \wedge \mathscr{D}\theta)\}\, dx = -\Psi,$$

agrees, up to sign, with the tri-vector corresponding to the Jacobi identity. We have thus proved

Theorem 7.8. *Let $\mathcal{D}$ be a skew-adjoint $q \times q$ matrix differential operator and $\Theta = \frac{1}{2} \int \{\theta \wedge \mathcal{D}\theta\}\, dx$ the corresponding functional bi-vector. Then $\mathcal{D}$ is Hamiltonian if and only if*

$$\operatorname{pr} \mathbf{v}_{\mathcal{D}\theta}(\Theta) = 0. \tag{7.18}$$

Example 7.9. Let us return one final time to the Hamiltonian operator $\mathscr{E}$ associated with the Korteweg–de Vries equation. According to (7.15) and (7.18) we need only check the vanishing of

$$\begin{aligned}
\operatorname{pr} \mathbf{v}_{\mathscr{E}\theta} \int \{\tfrac{1}{2}\theta \wedge \theta_{xxx} + \tfrac{1}{3} u\theta \wedge \theta_x\}\, dx &= \tfrac{1}{3} \int \{\mathscr{E}(\theta) \wedge \theta \wedge \theta_x\}\, dx \\
&= \tfrac{1}{3} \int (\theta_{xxx} \wedge \theta \wedge \theta_x + \tfrac{2}{3} u\theta_x \wedge \theta \wedge \theta_x \\
&\qquad + \tfrac{1}{3} u_x \theta \wedge \theta \wedge \theta_x)\, dx \\
&= 0,
\end{aligned}$$

by our earlier computation. Thus we have proved, in a completely elementary fashion, the fact that $\mathscr{E}$ is Hamiltonian.

Example 7.10. Consider the Euler equatons of inviscid ideal fluid flow

$$\frac{\partial \mathbf{u}}{\partial t} + \mathbf{u} \cdot \nabla \mathbf{u} = -\nabla p, \qquad \nabla \cdot \mathbf{u} = 0.$$

(See Example 2.45 for the notation.) As these stand, they cannot take the form of a Hamiltonian system since, in particular, we have no equation governing the temporal evolution of the pressure p. The easiest way to circumvent this difficulty is to rewrite the equations in terms of the *vorticity* $\boldsymbol{\omega} = \nabla \times \mathbf{u}$. Taking the curl of the first set of equations, we find the vorticity equation

$$\frac{\partial \boldsymbol{\omega}}{\partial t} = \boldsymbol{\omega} \cdot \nabla \mathbf{u} - \mathbf{u} \cdot \nabla \boldsymbol{\omega}, \tag{7.19}$$

which we will put into Hamiltonian form

$$\frac{\partial \boldsymbol{\omega}}{\partial t} = \mathcal{D} \frac{\delta \mathscr{H}}{\delta \boldsymbol{\omega}} \tag{7.20}$$

for a suitable Hamiltonian operator $\mathcal{D}$. The Hamiltonian functional is the energy

$$\mathscr{H} = \int \tfrac{1}{2} |\mathbf{u}|^2\, dx,$$

but for (7.20) we need to compute its variational derivative with respect to $\boldsymbol{\omega}$, not $\mathbf{u}$! This is done (formally) by introducing the vector stream function

$\boldsymbol{\psi}$ satisfying $\nabla \times \boldsymbol{\psi} = \mathbf{u}$, $\nabla \cdot \boldsymbol{\psi} = 0$. Let $\boldsymbol{\eta}(\mathbf{x})$ have compact support, with $\nabla \times \boldsymbol{\eta} = \boldsymbol{\zeta}$. Then

$$\frac{d}{d\varepsilon}\bigg|_{\varepsilon=0} \mathscr{H}[\boldsymbol{\omega} + \varepsilon\boldsymbol{\zeta}] = \int \mathbf{u}\cdot\boldsymbol{\eta}\, d\mathbf{x} = \int (\nabla \times \boldsymbol{\psi})\cdot\boldsymbol{\eta}\, d\mathbf{x} = \int \boldsymbol{\psi}\cdot(\nabla \times \boldsymbol{\eta})\, d\mathbf{x}$$

$$= \int \boldsymbol{\psi}\cdot\boldsymbol{\zeta}\, d\mathbf{x},$$

hence $\delta\mathscr{H}/\delta\boldsymbol{\omega} = \boldsymbol{\psi}$.

In the two-dimensional case, $\mathbf{u} = (u, v)$ depends on (x, y, t) and there is a single vorticity $\omega = v_x - u_y$. The vorticity equation is then

$$\frac{\partial \omega}{\partial t} = -u\omega_x - v\omega_y = \omega_x\psi_y - \omega_y\psi_x, \tag{7.21}$$

where ψ is the stream function, with $\psi_x = v$, $\psi_y = -u$. If we set

$$\mathscr{D} = \omega_x D_y - \omega_y D_x,$$

then we see that (7.21) is of the form (7.20) using the energy as the Hamiltonian functional.

To prove that $\mathscr{D}$ is a Hamiltonian operator, note first that

$$\mathscr{D}^* = -D_y\cdot\omega_x + D_x\cdot\omega_y = -\mathscr{D},$$

so $\mathscr{D}$ is skew-adjoint. The Jacobi identity is proved by checking (7.18):

$$0 = \operatorname{pr} \mathbf{v}_{\mathscr{D}\theta} \int \{\omega_x\theta \wedge \theta_y - \omega_y\theta \wedge \theta_x\}\, dx\, dy$$

$$= \int \{D_x(\omega_x\theta_y - \omega_y\theta_x) \wedge \theta \wedge \theta_y - D_y(\omega_x\theta_y - \omega_y\theta_x) \wedge \theta \wedge \theta_x\}\, dx\, dy$$

$$= \int \{\omega_x(\theta_{xy} \wedge \theta \wedge \theta_y - \theta_{yy} \wedge \theta \wedge \theta_x)$$

$$+ \omega_y(\theta_{xy} \wedge \theta \wedge \theta_x - \theta_{xx} \wedge \theta \wedge \theta_y)\}\, dx\, dy.$$

Integrating the second and fourth terms by parts, we find this equals

$$\int \{\omega_x(\theta_{xy} \wedge \theta \wedge \theta_y + \theta_y \wedge \theta \wedge \theta_{xy}) + \omega_{xy}\theta_y \wedge \theta \wedge \theta_x$$

$$+ \omega_y(\theta_{xy} \wedge \theta \wedge \theta_x + \theta_x \wedge \theta \wedge \theta_{xy}) + \omega_{xy}\theta_x \wedge \theta \wedge \theta_y\}\, dx\, dy = 0,$$

since the wedge product is alternating. Thus $\mathscr{D}$ verifies the conditions of Theorem 7.8 and defines a true Poisson bracket, relative to which the two-dimensional Euler equations are Hamiltonian. The three-dimensional version is left to the reader; see Exercise 7.5.

7.2. Symmetries and Conservation Laws

In outline, the correspondence between Hamiltonian symmetry groups and conservation laws for systems of evolution equations in Hamiltonian form proceeds exactly as in the finite-dimensional case discussed in Section 6.3. First, we need to investigate the "distinguished functionals" arising from degeneracies of the Poisson bracket itself; these will provide conservation laws for *any* system having the given Hamiltonian structure. Further conservation laws, particular to the symmetry properties of the individual Hamiltonian functionals, can then be deduced from generalized symmetries which are themselves Hamiltonian.

Distinguished Functionals

Definition 7.11. Let $\mathscr{D}$ be a $q \times q$ Hamiltonian differential operator. A *distinguished functional* for $\mathscr{D}$ is a functional $\mathscr{C} \in \mathscr{F}$ satisfying $\mathscr{D}\delta\mathscr{C} = 0$ for all x, u.

In other words, the Hamiltonian system corresponding to a distinguished functional is completely trivial: $u_t = 0$. From (7.4) we conclude that a functional $\mathscr{C}$ is distinguished if and only if its Poisson bracket with every other functional is trivial:

$$\{\mathscr{C}, \mathscr{H}\} = 0 \quad \text{for all} \quad \mathscr{H} \in \mathscr{F}.$$

This immediately implies the conservative nature of such functionals.

Proposition 7.12. *Let $\mathscr{D}$ be a Hamiltonian operator. If $\mathscr{C}$ is a distinguished functional for $\mathscr{D}$, then $\mathscr{C}$ determines a conservation law for every Hamiltonian system $u_t = \mathscr{D}\delta\mathscr{H}$ relative to $\mathscr{D}$.*

Example 7.13. For the first Hamiltonian operator $\mathscr{D} = D_x$ of the Korteweg–de Vries equation, a distinguished functional must satisfy $D_x\delta\mathscr{C} = 0$, or, equivalently, $\delta\mathscr{C}$ is constant. Every such functional is a constant multiple of the mass $\mathscr{M}[u] = \int u\,dx$. Thus, according to Proposition 7.12, the L^1 solutions to any evolution equation of the form $u_t = D_x\delta\mathscr{H}$ automatically satisfy the constraint of mass conservation $\int u\,dx =$ constant. (Actually, this can be generalized, see Exercise 7.8.) The second Hamiltonian operator $\mathscr{E}$, on the other hand, has no nontrivial distinguished functionals, and thus might be regarded as "symplectic".

Lie Brackets

As in the finite-dimensional set-up, the main result required for establishing a Noether-type theorem relating symmetry groups and conservation laws is the correspondence between the Poisson bracket of functionals and the commutator of their corresponding Hamiltonian vector fields.

Proposition 7.14. *Let $\{\cdot, \cdot\}$ be a Poisson bracket determined by a differential operator $\mathscr{D}$. Let $\mathscr{P}, \mathscr{Q} \in \mathscr{F}$ be functionals, with corresponding Hamiltonian vector fields $\hat{\mathbf{v}}_{\mathscr{P}}, \hat{\mathbf{v}}_{\mathscr{Q}}$. Then the Hamiltonian vector field corresponding to the Poisson bracket $\{\mathscr{P}, \mathscr{Q}\}$ is the Lie bracket of the two vector fields;*

$$\hat{\mathbf{v}}_{\{\mathscr{P}, \mathscr{Q}\}} = -[\hat{\mathbf{v}}_{\mathscr{P}}, \hat{\mathbf{v}}_{\mathscr{Q}}] = [\hat{\mathbf{v}}_{\mathscr{Q}}, \hat{\mathbf{v}}_{\mathscr{P}}]. \tag{7.22}$$

(*The Lie bracket is that given by Definition* 5.14.)

PROOF. Let $\mathscr{R}$ be an arbitrary functional. Applying the prolongation of $\hat{\mathbf{v}}_{\{\mathscr{P}, \mathscr{Q}\}}$ to $\mathscr{R}$, and using (7.4) and the Jacobi identity, we find

$$\begin{aligned} \operatorname{pr} \hat{\mathbf{v}}_{\{\mathscr{P}, \mathscr{Q}\}}(\mathscr{R}) &= \{\mathscr{R}, \{\mathscr{P}, \mathscr{Q}\}\} \\ &= \{\{\mathscr{R}, \mathscr{P}\}, \mathscr{Q}\} - \{\{\mathscr{R}, \mathscr{Q}\}, \mathscr{P}\} \\ &= \operatorname{pr} \hat{\mathbf{v}}_{\mathscr{Q}}(\{\mathscr{R}, \mathscr{P}\}) - \operatorname{pr} \hat{\mathbf{v}}_{\mathscr{P}}(\{\mathscr{R}, \mathscr{Q}\}) \\ &= (\operatorname{pr} \hat{\mathbf{v}}_{\mathscr{Q}} \cdot \operatorname{pr} \hat{\mathbf{v}}_{\mathscr{P}} - \operatorname{pr} \hat{\mathbf{v}}_{\mathscr{P}} \cdot \operatorname{pr} \hat{\mathbf{v}}_{\mathscr{Q}})\mathscr{R}. \end{aligned}$$

By (5.21), this latter expression is the prolongation of the Lie bracket between $\hat{\mathbf{v}}_{\mathscr{P}}, \hat{\mathbf{v}}_{\mathscr{Q}}$, so

$$\operatorname{pr} \hat{\mathbf{v}}_{\{\mathscr{P}, \mathscr{Q}\}}(\mathscr{R}) = -\operatorname{pr}([\hat{\mathbf{v}}_{\mathscr{P}}, \hat{\mathbf{v}}_{\mathscr{Q}}])\mathscr{R}$$

for every $\mathscr{R} \in \mathscr{F}$. Exercise 5.52 says that this can happen only if the two generalized vector fields are equal, which proves (7.22). □

Conservation Laws

As we remarked in Chapter 4, any conservation law of a system of evolution equations takes the form

$$D_t T + \operatorname{Div} X = 0,$$

in which Div denotes spatial divergence and the conserved density $T(x, t, u^{(n)})$ can be assumed without loss of generality to depend only on x-derivatives of u. Equivalently, for $\Omega \subset X$, the functional

$$\mathscr{T}[t; u] = \int_\Omega T(x, t, u^{(n)})\, dx$$

is a constant, independent of t, for all solutions u such that $T(x, t, u^{(n)}) \to 0$ as $x \to \partial\Omega$.

Note that if $T(x, t, u^{(n)})$ is any such differential function, and u is a solution to the evolutionary system $u_t = P[u]$, then

$$D_t T = \partial_t T + \operatorname{pr} \mathbf{v}_P(T),$$

where $\partial_t = \partial/\partial t$ denotes the partial t-derivative. Thus T is the density for a conservation law of the system if and only if its associated functional $\mathscr{T}$ satisfies

$$\partial\mathscr{T}/\partial t + \mathrm{pr}\,\mathbf{v}_P(\mathscr{T}) = 0. \tag{7.23}$$

In the case our system is of Hamiltonian form, the bracket relation (7.4) immediately leads to the Noether relation between Hamiltonian symmetries and conservation laws.

Theorem 7.15. *Let $u_t = \mathscr{D}\delta\mathscr{H}$ be a Hamiltonian system of evolution equations. A Hamiltonian vector field $\hat{\mathbf{v}}_{\mathscr{P}}$ with characteristic $\mathscr{D}\delta\mathscr{P}$, $\mathscr{P} \in \mathscr{F}$, determines a generalized symmetry group of the system if and only if there is an equivalent functional $\tilde{\mathscr{P}} = \mathscr{P} - \mathscr{C}$, differing only from $\mathscr{P}$ by a time-dependent distinguished functional $\mathscr{C}[t; u]$, such that $\tilde{\mathscr{P}}$ determines a conservation law.*

Proof. By a time-dependent distinguished functional we mean, in analogy with Chapter 6, a functional $\mathscr{C}[t; u] = \int C(t, x, u^{(n)})\,dx$, with C depending on t, x, u and x-derivatives of u, and with the property that for each fixed t_0, $\mathscr{C}[t_0; u]$ is a distinguished functional: $\mathscr{D}\delta\mathscr{C} = 0$. Now, according to Proposition 5.19, $\hat{\mathbf{v}}_{\mathscr{P}}$ is a symmetry of the Hamiltonian system if and only if

$$\partial\hat{\mathbf{v}}_{\mathscr{P}}/\partial t + [\hat{\mathbf{v}}_{\mathscr{H}}, \hat{\mathbf{v}}_{\mathscr{P}}] = 0, \tag{7.24}$$

$\hat{\mathbf{v}}_{\mathscr{H}}$ being the associated Hamiltonian vector field. Since $\mathscr{D}$ does not explicitly depend on t, $\partial\hat{\mathbf{v}}_{\mathscr{P}}/\partial t$ is the Hamiltonian vector field corresponding to the functional $\partial\mathscr{P}/\partial t$, while by the previous proposition, $[\hat{\mathbf{v}}_{\mathscr{H}}, \hat{\mathbf{v}}_{\mathscr{P}}]$ is the Hamiltonian vector field for the Poisson bracket of $\mathscr{P}$ and $\mathscr{H}$. Thus (7.24) says that the Hamiltonian vector field for the combined functional $\partial_t\mathscr{P} + \{\mathscr{P}, \mathscr{H}\}$ is zero, and hence

$$\frac{\partial\mathscr{P}}{\partial t} + \{\mathscr{P}, \mathscr{H}\} = \tilde{\mathscr{C}}$$

for some time-dependent distinguished functional

$$\tilde{\mathscr{C}}[t; u] = \int \tilde{C}(t, x, u^{(n)})\,dx.$$

Now set

$$\mathscr{C}[t; u] = \int_{t_0}^{t} \tilde{\mathscr{C}}[s; u]\,ds \equiv \int\left(\int_{t_0}^{t} \tilde{C}(s, x, u^{(n)})\,ds\right)dx,$$

and let $\tilde{\mathscr{P}} = \mathscr{P} - \mathscr{C}$. Then

$$\frac{\partial\tilde{\mathscr{P}}}{\partial t} = \frac{\partial\mathscr{P}}{\partial t} - \tilde{\mathscr{C}},$$

while by the definition of distinguished functional,

$$\{\mathscr{P}, \mathscr{H}\} = \{\tilde{\mathscr{P}}, \mathscr{H}\}.$$

Thus $\tilde{\mathscr{P}}$ satisfies the condition (7.23) that it be conserved, and the theorem is proved. □

Example 7.16. Consider the Korteweg–de Vries equation

$$u_t = u_{xxx} + uu_x, \tag{7.25}$$

whose two Hamiltonian structures were discussed in Example 7.6. Let's investigate which of the classical symmetry groups of Example 2.44 are Hamiltonian and hence lead to conservation laws. The symmetries are generated by

$$\mathbf{v}_1 = \partial_x, \qquad \mathbf{v}_2 = \partial_t, \qquad \mathbf{v}_3 = t\partial_x - \partial_u, \qquad \mathbf{v}_4 = x\partial_x + 3t\partial_t - 2u\partial_u,$$

cf. (2.68) with x replaced by $-x$, with corresponding characteristics

$$Q_1 = u_x, \;\; Q_2 = u_{xxx} + uu_x, \;\; Q_3 = 1 + tu_x, \;\; Q_4 = 2u + xu_x + 3t(u_{xxx} + uu_x),$$

(up to sign).

For the first Hamiltonian operator $\mathscr{D} = D_x$, there is one independent nontrivial distinguished functional, the mass $\mathscr{P}_0 = \mathscr{M} = \int u \, dx$, which is therefore conserved. Of the above four characteristics, the first three are Hamiltonian

$$Q_i = D_x \delta \mathscr{P}_i, \qquad i = 1, 2, 3, \tag{7.26}$$

with conserved functionals

$$\mathscr{P}_1 = \int \tfrac{1}{2}u^2 \, dx, \qquad \mathscr{P}_2 = \int (\tfrac{1}{6}u^3 - \tfrac{1}{2}u_x^2) \, dx, \qquad \mathscr{P}_3 = \int (xu + \tfrac{1}{2}tu^2) \, dx.$$

Note that $\mathscr{P}_2$ is just the Hamiltonian functional for (7.25) with $\mathscr{D}$ as the Hamiltonian operator. Invariance of $\mathscr{P}_3$, when combined with that of $\mathscr{P}_1$, proves that the first moment of u is a linear function of t,

$$\int xu \, dx = at + b, \qquad a, b \text{ constant},$$

with $a = -\int \frac{1}{2}u^2 \, dx$, for any solution u decaying sufficiently rapidly as $|x| \to \infty$, or any periodic solution, where the integral is now over one period. The fourth characteristic Q_4 is not of the form (7.26), and hence does not lead to a conservation law.

What about the second Hamiltonian operator $\mathscr{E} = D_x^3 + \frac{2}{3}uD_x + \frac{1}{3}u_x$? Here there are no longer any distinguished functionals. In this case Q_1, Q_2 and Q_4 (but not Q_3) are Hamiltonian,

$$Q_i = \mathscr{E} \delta \tilde{\mathscr{P}}_i, \qquad i = 1, 2, 4, \tag{7.27}$$

where

$$\tilde{\mathscr{P}}_1 = 3\mathscr{P}_0 = 3 \int u \, dx, \qquad \tilde{\mathscr{P}}_2 = \mathscr{P}_1 = \int \tfrac{1}{2}u^2 \, dx,$$

$$\tilde{\mathscr{P}}_4 = 3\mathscr{P}_3 = 3 \int (xu + \tfrac{1}{2}tu^2) \, dx,$$

are the corresponding conservation laws. In this case, nothing new is obtained. Note that the other conservation law $\mathcal{P}_2$ did *not* arise from one of the geometrical symmetries. According to Theorem 7.15, however, there *is* a Hamiltonian symmetry which gives rise to it, namely $\hat{\mathbf{v}}_{\mathcal{P}_2}$. The characteristic of this generalized symmetry is

$$Q_5 = \mathcal{E}\delta\mathcal{P}_2 = (D_x^3 + \tfrac{2}{3}uD_x + \tfrac{1}{3}u_x)(u_{xx} + \tfrac{1}{2}u^2)$$
$$= u_{xxxxx} + \tfrac{5}{3}uu_{xxx} + \tfrac{10}{3}u_xu_{xx} + \tfrac{5}{6}u^2u_x.$$

We have thus rediscovered the fifth order generalized symmetry of Section 5.2! Pressing on, we note that Q_5 happens to satisfy the Hamiltonian condition (7.26) for $\mathcal{D}$ with the functional

$$\mathcal{P}_5 = \int (\tfrac{1}{2}u_{xx}^2 - \tfrac{5}{6}uu_x^2 + \tfrac{5}{72}u^4)\,dx$$

providing a further conservation law for the Korteweg–de Vries equation. By now, the signs of a recursive procedure of generating conservation laws and corresponding Hamiltonian symmetries of the Korteweg–de Vries equation are starting to appear. We take the new conservation law $\mathcal{P}_5$, determine its Hamiltonian vector field relative to the operator $\mathcal{E}$, which, by Theorem 7.15 is necessarily a symmetry, and then try to find a further functional $\mathcal{P}_6$ for which it is the Hamiltonian vector field relative to the other Hamiltonian operator $\mathcal{D}$, and so on. The rigorous implementation of this recursion scheme for general equations with two Hamiltonian structures will be the subject of Section 7.3.

Example 7.17. The two-dimensional Euler equations were cast into Hamiltonian form in Example 7.10. Let us investigate what type of conservation laws arise as a result. First we need to look at the distinguished functionals for the Hamiltonian operator $\mathcal{D} = \omega_x D_y - \omega_y D_x$. A straightforward computation proves that a differential function $P[\omega]$ lies in the kernel of $\mathcal{D}$, so $\mathcal{D}P = 0$, if and only if $P = P(\omega)$ is a function of ω (but not x or y, nor any derivatives of ω). We conclude that the functionals

$$\mathcal{C}[\omega] = \int C(\omega)\,dx\,dy,$$

where $C(\omega)$ is *any* smooth function of the vorticity ω, are all distinguished and hence are conserved for solutions of the Euler equations. These are the well-known "area integrals" and reflect the point-wise conservation of vorticity for two-dimensional incompressible fluid flow.

Conservation laws arising from the Euclidean symmetries of the Euler equations found in Example 2.45 are constructed next. Note first that we need to find the "ω-characteristic" of each of the symmetry generators, i.e. rewrite it as the prolongation of an evolutionary vector field in the form $\mathbf{v} = Q(x, y, t, u, v, p, \omega, \ldots)\partial_\omega$. If $\mathbf{v}$ is a Hamiltonian vector field, we may then deduce the existence of a conservation law $\mathscr{P}[\omega]$ with $\mathscr{D}\delta\mathscr{P} = Q$. For instance, the translational symmetry

$$\mathbf{v}_\alpha = \alpha\partial_x + \alpha_t\partial_u - \alpha_{tt}x\partial_p, \qquad \alpha = \alpha(t),$$

has evolutionary form

$$\tilde{\mathbf{v}}_\alpha = (\alpha_t - \alpha u_x)\partial_u - \alpha v_x\partial_v - (\alpha_{tt}x + \alpha p_x)\partial_p.$$

Prolonging $\tilde{\mathbf{v}}_\alpha$, we see that its ω-coefficient is

$$Q = -\alpha\omega_x = -\mathscr{D}(\alpha y) = -\mathscr{D}\delta\mathscr{P}_\alpha,$$

where

$$\mathscr{P}_\alpha = \int \alpha(t)y\omega\, dx\, dy = \int \alpha(t)u\, dx\, dy$$

(integrating by parts) is the associated conserved functional. Similarly, the translational symmetries $\mathbf{v}_\beta$ lead to conservation laws

$$\mathscr{P}_\beta = -\int \beta(t)x\omega\, dx\, dy = \int \beta(t)v\, dx\, dy,$$

where $\beta(t)$ is also an arbitrary function of t. The fact that $\mathscr{P}_\alpha$ and $\mathscr{P}_\beta$ are conservation laws for any functions $\alpha(t)$ and $\beta(t)$ appears to be paradoxical, but this is resolved by looking at the boundary contributions. In vector form, if $\boldsymbol{\alpha}(t) = (\alpha(t), \beta(t))$, then we have the divergence identity

$$D_t(\boldsymbol{\alpha}\cdot\mathbf{u}) + \operatorname{Div}[(\boldsymbol{\alpha}\cdot\mathbf{u} - \boldsymbol{\alpha}_t\cdot\mathbf{x})\mathbf{u} + p\boldsymbol{\alpha}] = 0$$

valid for all solutions $\mathbf{u} = (u, v)$ of the Euler equations. This integrates to the generalized momentum relations

$$\frac{d}{dt}\int_\Omega (\boldsymbol{\alpha}(t)\cdot\mathbf{u})\, d\mathbf{x} = -\int_{\partial\Omega} [(\boldsymbol{\alpha}\cdot\mathbf{u} - \boldsymbol{\alpha}_t\cdot\mathbf{x})\mathbf{u} + p\boldsymbol{\alpha}]\cdot dS$$

valid over an arbitrary subdomain Ω. It is in this sense that the above functionals $\mathscr{P}_\alpha$, $\mathscr{P}_\beta$ are "conserved".

The rotational symmetry

$$y\partial_x - x\partial_y + v\partial_u - u\partial_v$$

has ω-evolutionary form

$$(y\omega_x - x\omega_y)\partial_\omega,$$

which is Hamiltonian. We find

$$(y\omega_x - x\omega_y) = \mathscr{D}(\tfrac{1}{2}x^2 + \tfrac{1}{2}y^2) = \mathscr{D}\delta\mathscr{T},$$

where

$$\mathscr{T} = \int \tfrac{1}{2}(x^2 + y^2)\omega\, dx\, dy = \int (yu - xv)\, dx\, dy$$

is the conserved angular momentum of the fluid. For the two-dimensional Euler equations, then, there are three infinite families of conservation laws—two coming from the generalized translational symmetries and one, the area integrals, reflecting the degeneracy of the underlying Poisson bracket—together with the individual conservation laws of angular momentum and energy. The three-dimensional case is markedly different—see Exercise 7.5.

7.3. Bi-Hamiltonian Systems

In this final section, we discuss the remarkable properties of systems of evolution equations which, like the Korteweg–de Vries equation, can be written in Hamiltonian form in not just one but *two* different ways. We will thus be interested in systems of the form

$$\frac{\partial u}{\partial t} = K_1[u] = \mathscr{D}\delta\mathscr{H}_1 = \mathscr{E}\delta\mathscr{H}_0 \tag{7.28}$$

in which both $\mathscr{D}$ and $\mathscr{E}$ are Hamiltonian operators, and $\mathscr{H}_0[u]$ and $\mathscr{H}_1[u]$ appropriate Hamiltonian functionals. Subject to a compatibility condition between the two Poisson structures determined by $\mathscr{D}$ and $\mathscr{E}$, we will be able to recursively construct an infinite hierarchy of symmetries and conservation laws for the system in the following manner.

According to Theorem 7.15, if $\mathscr{P}[u]$ is any conserved functional for (7.28), then *both* of the Hamiltonian vector fields $\mathbf{v}_{\mathscr{D}\delta\mathscr{P}}$ and $\mathbf{v}_{\mathscr{E}\delta\mathscr{P}}$ are symmetries. In particular, since both $\mathscr{H}_0$ and $\mathscr{H}_1$ are conserved, not only is the original vector field $\mathbf{v}_{K_1} = \mathbf{v}_{\mathscr{D}\delta\mathscr{H}_1} = \mathbf{v}_{\mathscr{E}\delta\mathscr{H}_0}$ a symmetry of (7.28), but so are the two additional vector fields $\mathbf{v}_{\mathscr{D}\delta\mathscr{H}_0}$ and $\mathbf{v}_{\mathscr{E}\delta\mathscr{H}_1}$. The recursion algorithm proceeds on the assumption that one of these new symmetries, say $\mathbf{v}_{\mathscr{E}\delta\mathscr{H}_1}$, is a Hamiltonian vector field for the other Hamiltonian structure, so

$$\mathscr{E}\delta\mathscr{H}_1 = \mathscr{D}\delta\mathscr{H}_2$$

for some functional $\mathscr{H}_2$. Again, by Theorem 7.15, $\mathscr{H}_2$ (or some equivalent functional) is conserved, and so we obtain yet a further symmetry, this time with characteristic $\mathscr{E}\delta\mathscr{H}_2$. At this stage, the recursion pattern is clear. At the n-th stage we determine a new functional $\mathscr{H}_{n+1}$ satisfying the recursion relation

$$K_n \equiv \mathscr{D}\delta\mathscr{H}_n = \mathscr{E}\delta\mathscr{H}_{n-1}. \tag{7.29}$$

This will provide both a further conservation law for the original system (7.28) plus a further symmetry $\mathbf{v}_{n+1}$ with characteristic $K_{n+1} = \mathscr{E}\delta\mathscr{H}_n$. Note that if we define the operator $\mathscr{R} = \mathscr{E}\cdot\mathscr{D}^{-1}$ then, formally,

$$K_{n+1} = \mathscr{R}K_n,$$

and, as will be the case, we suspect that $\mathscr{R}$ will define a recursion operator for our system.

Example 7.18. Consider the Korteweg–de Vries equation, which was shown to have two Hamiltonian structures in Example 7.6, with

$$\mathscr{D} = D_x, \qquad \mathscr{E} = D_x^3 + \tfrac{2}{3}uD_x + \tfrac{1}{3}u_x.$$

The operator connecting our hierarchy of Hamiltonian symmetries is thus

$$\mathscr{R} = \mathscr{E}\cdot\mathscr{D}^{-1} = D_x^2 + \tfrac{2}{3}u + \tfrac{1}{3}u_x D_x^{-1},$$

which is nothing but the Lenard recursion operator of Section 5.2! Thus our results on bi-Hamiltonian systems will provide ready-made proofs of the existence of infinitely many conservation laws and symmetries for the Korteweg–de Vries equation.

To proceed rigorously, however, we need to ensure that the two Hamiltonian structures be "compatible" in the following precise sense:

Definition 7.19. A pair of skew-adjoint $q \times q$ matrix differential operators $\mathscr{D}$ and $\mathscr{E}$ is said to form a *Hamiltonian pair* if every linear combination $a\mathscr{D} + b\mathscr{E}$, $a, b \in \mathbb{R}$, is a Hamiltonian operator. A system of evolution equations is a *bi-Hamiltonian system* if it can be written in the form (7.28) where $\mathscr{D}$, $\mathscr{E}$ form a Hamiltonian pair.

Lemma 7.20. *If $\mathscr{D}$, $\mathscr{E}$ are skew-adjoint operators, then they form a Hamiltonian pair if and only if $\mathscr{D}$, $\mathscr{E}$ and $\mathscr{D} + \mathscr{E}$ are all Hamiltonian operators.*

PROOF. Note that the Jacobi identity, in any of its forms (7.3), (7.11) or (7.18) is a *quadratic* expression in $\mathscr{D}$. The lemma is then an easy consequence of the fact that any quadratic polynomial vanishing at three distinct points must vanish identically. More specifically, given $P, Q, R \in \mathscr{A}^q$, let $\mathscr{J}(\mathscr{D}, \mathscr{D}; P, Q, R)$ denote the left-hand side of (7.11). The corresponding symmetric bilinear form is

$$\begin{aligned}\mathscr{J}(\mathscr{D}, \mathscr{E}; P, Q, R) = {} & \\ & \tfrac{1}{2}\int [P\cdot \operatorname{pr} \mathbf{v}_{\mathscr{D}R}(\mathscr{E})Q + R\cdot \operatorname{pr} \mathbf{v}_{\mathscr{D}Q}(\mathscr{E})P + Q\cdot \operatorname{pr} \mathbf{v}_{\mathscr{D}P}(\mathscr{E})R \\ & + P\cdot \operatorname{pr} \mathbf{v}_{\mathscr{E}R}(\mathscr{D})Q + R\cdot \operatorname{pr} \mathbf{v}_{\mathscr{E}Q}(\mathscr{D})P + Q\cdot \operatorname{pr} \mathbf{v}_{\mathscr{E}P}(\mathscr{D})R]\, dx. \qquad (7.30)\end{aligned}$$

If $\mathscr{D}$, $\mathscr{E}$ and $\mathscr{D} + \mathscr{E}$ are Hamiltonian, then

$$\mathscr{J}(\mathscr{D}, \mathscr{D}; P, Q, R) = \mathscr{J}(\mathscr{E}, \mathscr{E}; P, Q, R) = 0$$

and

$$\begin{aligned}\mathscr{J}(\mathscr{D} + \mathscr{E}, \mathscr{D} + \mathscr{E}; P, Q, R) &= \mathscr{J}(\mathscr{D}, \mathscr{D}; P, Q, R) + 2\mathscr{J}(\mathscr{D}, \mathscr{E}; P, Q, R) \\ &\quad + \mathscr{J}(\mathscr{E}, \mathscr{E}; P, Q, R) \\ &= 0,\end{aligned}$$

hence

$$\mathscr{J}(\mathscr{D}, \mathscr{E}; P, Q, R) = 0. \tag{7.31}$$

From this it is easy to check that

$$\mathscr{J}(a\mathscr{D} + b\mathscr{E}, a\mathscr{D} + b\mathscr{E}; P, Q, R) = 0,$$

for any $a, b \in \mathbb{R}$. □

Corollary 7.21. *Let $\mathscr{D}$ and $\mathscr{E}$ be Hamiltonian differential operators. Then $\mathscr{D}, \mathscr{E}$ form a Hamiltonian pair if and only if*

$$\operatorname{pr} \mathbf{v}_{\mathscr{D}\theta}(\Theta_{\mathscr{E}}) + \operatorname{pr} \mathbf{v}_{\mathscr{E}\theta}(\Theta_{\mathscr{D}}) = 0, \tag{7.32}$$

where

$$\Theta_{\mathscr{D}} = \tfrac{1}{2} \int \{\theta \wedge \mathscr{D}\theta\}\, dx, \qquad \Theta_{\mathscr{E}} = \tfrac{1}{2} \int \{\theta \wedge \mathscr{E}\theta\}\, dx$$

are the functional bi-vectors representing the respective Poisson brackets.

Indeed, we have

$$\mathscr{J}(\mathscr{D}, \mathscr{D}; P, Q, R) = \langle \operatorname{pr} \mathbf{v}_{\mathscr{D}\theta}(\Theta_{\mathscr{D}}); P, Q, R \rangle,$$

so evaluating (7.32) at P, Q, R reproduces (7.31). □

Example 7.22. Consider the Hamiltonian operators $\mathscr{D}, \mathscr{E}$ connected with the Korteweg–de Vries equation. We have

$$\operatorname{pr} \mathbf{v}_{\mathscr{E}\theta}(\Theta_{\mathscr{D}}) = \operatorname{pr} \mathbf{v}_{\mathscr{E}\theta} \int \tfrac{1}{2}\{\theta \wedge \theta_x\}\, dx = 0$$

trivially, while,

$$\operatorname{pr} \mathbf{v}_{\mathscr{D}\theta}(\Theta_{\mathscr{E}}) = \operatorname{pr} \mathbf{v}_{\mathscr{D}\theta} \int \{\tfrac{1}{2}\theta \wedge \theta_{xxx} + \tfrac{1}{3}u\theta \wedge \theta_x\}\, dx = \int \{\tfrac{1}{3}\theta_x \wedge \theta \wedge \theta_x\}\, dx$$
$$= 0$$

by the properties of the wedge product. Thus $\mathscr{D}$ and $\mathscr{E}$ form a Hamiltonian pair.

Incidentally, when we discuss Hamiltonian pairs, we are always excluding the trivial case when one operator is a constant multiple of the other. In the case of systems (as opposed to scalar equations) we must impose an additional constraint on one of the operators, $\mathscr{D}$, owing to the appearance of its inverse in the form of the recursion operator.

Definition 7.23. A differential operator $\mathscr{D}: \mathscr{A}^r \to \mathscr{A}^s$ is *degenerate* if there is a nonzero differential operator $\tilde{\mathscr{D}}: \mathscr{A}^s \to \mathscr{A}$ such that $\tilde{\mathscr{D}} \cdot \mathscr{D} \equiv 0$.

For example, the matrix operator

$$\mathscr{D} = \begin{pmatrix} D_x^3 & -D_x^2 \\ D_x^2 & -D_x \end{pmatrix}$$

is degenerate (and Hamiltonian), since if $\tilde{\mathscr{D}} = (1, -D_x)$, then $\tilde{\mathscr{D}} \cdot \mathscr{D} \equiv 0$. It is not difficult to see that degeneracy is strictly a matrix phenomenon; any nonzero scalar operator $\mathscr{D}: \mathscr{A} \to \mathscr{A}$ is automatically nondegenerate. (A useful criterion for nondegeneracy is given in Exercise 7.14.)

We are now in a position to state the main theorem on bi-Hamiltonian systems.

Theorem 7.24. *Let*

$$u_t = K_1[u] = \mathscr{D}\delta\mathscr{H}_1 = \mathscr{E}\delta\mathscr{H}_0$$

be a bi-Hamiltonian system of evolution equations. Assume that the operator $\mathscr{D}$ of the Hamiltonian pair is nondegenerate. Let $\mathscr{R} = \mathscr{E} \cdot \mathscr{D}^{-1}$ be the corresponding recursion operator, and let $K_0 = \mathscr{D}\delta\mathscr{H}_0$. Assume that for each $n = 1, 2, \ldots$ we can recursively define

$$K_n = \mathscr{R}K_{n-1}, \qquad n \geq 1,$$

meaning that for each n, K_{n-1} lies in the image of $\mathscr{D}$. Then there exists a sequence of functionals $\mathscr{H}_0, \mathscr{H}_1, \mathscr{H}_2, \ldots$ such that

(i) *for each $n \geq 1$, the evolution equation*

$$u_t = K_n[u] = \mathscr{D}\delta\mathscr{H}_n = \mathscr{E}\delta\mathscr{H}_{n-1} \tag{7.33}$$

is a bi-Hamiltonian system;

(ii) *the corresponding evolutionary vector fields $\mathbf{v}_n = \mathbf{v}_{K_n}$ all mutually commute:*

$$[\mathbf{v}_n, \mathbf{v}_m] = 0, \qquad n, m \geq 0;$$

(iii) *the Hamiltonian functionals $\mathscr{H}_n$ are all in involution with respect to either Poisson bracket:*

$$\{\mathscr{H}_n, \mathscr{H}_m\}_{\mathscr{D}} = 0 = \{\mathscr{H}_n, \mathscr{H}_m\}_{\mathscr{E}}, \qquad n, m \geq 0, \tag{7.34}$$

and hence provide an infinite collection of conservation laws for each of the bi-Hamiltonian systems (7.33).

Before proving the theorem, some remarks are in order. Although as it stands, the result is quite powerful, there is one annoying defect; namely the fact that we must assume at each stage that we can apply the recursion operator to K_{n-1} to produce K_n, i.e. prove that K_{n-1} lies in the image of $\mathscr{D}$. In most examples known to date, this always seems to be the case, but it would be nice to have a general proof of this. (The argument given in Theorem 5.31 seems to be fairly specific to the Korteweg–de Vries equation.) At the moment, though, except in some special instances, this seems to be the best that we can do. A second problem is that of determining whether all the Hamiltonian functionals $\mathscr{H}_n$ are independent; in practice this is usually easy to see from the leading terms of the corresponding evolution equations.

The proof itself rests on the following technical lemma:

Lemma 7.25. *Suppose $\mathscr{D}$, $\mathscr{E}$ form a Hamiltonian pair with $\mathscr{D}$ nondegenerate. Let $P, Q, R, \in \mathscr{A}^q$ satisfy*

$$\mathscr{E}P = \mathscr{D}Q, \qquad \mathscr{E}Q = \mathscr{D}R. \tag{7.35}$$

If $P = \delta\mathscr{P}$, $Q = \delta\mathscr{Q}$ are variational derivatives of functionals $\mathscr{P}, \mathscr{Q} \in \mathscr{F}$, then so is $R = \delta\mathscr{R}$ for some $\mathscr{R} \in \mathscr{F}$.

Before proving the lemma, let us see how it is used to prove the theorem. For each $n \geq 0$, we let $K_n = \mathscr{D}Q_n$ where, by assumption, $Q_n \in \mathscr{A}^q$ is a well-defined q-tuple of differential functions. By the lemma, if $Q_{n-1} = \delta\mathscr{H}_{n-1}$, $Q_n = \delta\mathscr{H}_n$ are variational derivatives, so is $Q_{n+1} = \delta\mathscr{H}_{n+1}$ for some $\mathscr{H}_{n+1} \in \mathscr{F}$. Since we already know $Q_0 = \delta\mathscr{H}_0$, $Q_1 = \delta\mathscr{H}_1$ are of this form, the existence of the functionals $\mathscr{H}_n$, $n \geq 0$ follows by an easy induction. This proves part (i).

Part (ii) follows from part (iii) using (7.22), so we concentrate on (iii). According to (7.4)

$$\{\mathscr{H}_n, \mathscr{H}_m\}_{\mathscr{D}} = \operatorname{pr} \mathbf{v}_m(\mathscr{H}_n),$$

and

$$\{\mathscr{H}_n, \mathscr{H}_m\}_{\mathscr{E}} = \operatorname{pr} \mathbf{v}_{m+1}(\mathscr{H}_n),$$

so

$$\{\mathscr{H}_n, \mathscr{H}_m\}_{\mathscr{D}} = \{\mathscr{H}_n, \mathscr{H}_{m-1}\}_{\mathscr{E}}.$$

We now use the skew-symmetry of the Poisson brackets to work our way down. If $n < m$, then

$$\{\mathscr{H}_n, \mathscr{H}_m\}_{\mathscr{D}} = \{\mathscr{H}_n, \mathscr{H}_{m-1}\}_{\mathscr{E}} = \{\mathscr{H}_{n+1}, \mathscr{H}_{m-1}\}_{\mathscr{D}} = \cdots = \{\mathscr{H}_k, \mathscr{H}_k\} = 0,$$

where k is the integer part of $(m - n)/2$ and the final bracket is the $\mathscr{D}$-Poisson bracket if $m - n$ is even, or the $\mathscr{E}$-Poisson bracket if $m - n$ is odd. This proves (7.34) and completes the proof of the theorem. □

Proof of Lemma 7.25. Referring back to the derivation of (7.11) from the Jacobi identity (7.3) we recall that the large number of cancellations resulted from the fact that, at that stage, the q-tuples P, Q, R were assumed to be variational derivatives of functionals and, consequently, their Fréchet derivatives were all self-adjoint operators. If we were to drop this initial assumption and carry through the computation for arbitrary q-tuples $P, Q, R \in \mathscr{A}^q$, we would find an identity of the form

$$\mathscr{K}(\mathscr{D}, \mathscr{D}; P, Q, R) = \mathscr{L}(\mathscr{D}, \mathscr{D}; P, Q, R) + \mathscr{J}(\mathscr{D}, \mathscr{D}; P, Q, R),$$

where

$$\mathscr{K}(\mathscr{D}, \mathscr{E}; P, Q, R) =$$

$$\tfrac{1}{2}\left\{\operatorname{pr} \mathbf{v}_{\mathscr{D}R} \int P\cdot\mathscr{E}Q\, dx + \operatorname{pr} \mathbf{v}_{\mathscr{D}P} \int Q\cdot\mathscr{E}R\, dx + \operatorname{pr} \mathbf{v}_{\mathscr{D}Q} \int R\cdot\mathscr{E}P\, dx\right.$$

$$\left. + \operatorname{pr} \mathbf{v}_{\mathscr{E}R} \int P\cdot\mathscr{D}Q\, dx + \operatorname{pr} \mathbf{v}_{\mathscr{E}P} \int Q\cdot\mathscr{D}R\, dx + \operatorname{pr} \mathbf{v}_{\mathscr{E}Q} \int R\cdot\mathscr{D}P\, dx\right\}, \quad (7.36)$$

$\mathscr{J}(\mathscr{D}, \mathscr{D}; P, Q, R)$ is as in (7.30) and $\mathscr{L}$ is the quadratic version of the bilinear expression

$$\mathscr{L}(\mathscr{D}, \mathscr{E}; P, Q, R) =$$

$$-\tfrac{1}{2} \int \{\mathscr{D}P\cdot(\mathsf{D}_Q - \mathsf{D}_Q^*)\mathscr{E}R + \mathscr{D}Q\cdot(\mathsf{D}_R - \mathsf{D}_R^*)\mathscr{E}P + \mathscr{D}R\cdot(\mathsf{D}_P - \mathsf{D}_P^*)\mathscr{E}Q$$

$$+ \mathscr{E}P\cdot(\mathsf{D}_Q - \mathsf{D}_Q^*)\mathscr{D}R + \mathscr{E}Q\cdot(\mathsf{D}_R - \mathsf{D}_R^*)\mathscr{D}P + \mathscr{E}R\cdot(\mathsf{D}_P - \mathsf{D}_P^*)\mathscr{D}Q\}\, dx.$$

The above identity has a bilinear counterpart

$$\mathscr{K}(\mathscr{D}, \mathscr{E}; P, Q, R) = \mathscr{L}(\mathscr{D}, \mathscr{E}; P, Q, R) + \mathscr{J}(\mathscr{D}, \mathscr{E}; P, Q, R), \quad (7.37)$$

which holds for arbitrary $P, Q, R \in \mathscr{A}$ and arbitrary skew-adjoint operators $\mathscr{D}, \mathscr{E}$. In particular, if $\mathscr{D}, \mathscr{E}$ form a Hamiltonian pair, then the $\mathscr{J}$ term in (7.37) vanishes.

Now replace P by $S = \delta\mathscr{S}$, R by $T = \delta\mathscr{T}$ in (7.37) and assume that P, Q, R are related by (7.35). Since Q, S and T are all variational derivatives of functionals, Theorem 5.92, (7.37) and (7.31) imply that

$$0 = \mathscr{L}(\mathscr{D}, \mathscr{E}; Q, S, T) = \mathscr{K}(\mathscr{D}, \mathscr{E}; Q, S, T). \quad (7.38)$$

Moreover using (7.35), (7.36) and the skew-adjointness of $\mathscr{D}$ and $\mathscr{E}$ we find (upon rearranging terms)

$$\mathscr{K}(\mathscr{D}, \mathscr{E}; Q, S, T) = \tfrac{1}{2}\{\mathscr{K}(\mathscr{E}, \mathscr{E}; P, S, T) + \mathscr{K}(\mathscr{D}, \mathscr{D}; R, S, T)\}.$$

Now $\mathscr{E}$ is Hamiltonian and P, S, T are all variational derivatives of functionals, so $\mathscr{K}(\mathscr{E}, \mathscr{E}; P, S, T)$ is just the Jacobi identity for $\mathscr{E}$ and hence vanishes. We still don't know that R is the variational derivative of some functional, so we cannot make the same claim for $\mathscr{K}(\mathscr{D}, \mathscr{D}; R, S, T)$. However, by (7.37), (7.38) and Theorem 5.92 (for S and T) we find

$$0 = \mathscr{K}(\mathscr{D}, \mathscr{D}; R, S, T) = \mathscr{L}(\mathscr{D}, \mathscr{D}; R, S, T)$$

$$= \int \mathscr{D}T\cdot(\mathsf{D}_R - \mathsf{D}_R^*)\mathscr{D}S\, dx = -\int T\cdot\mathscr{D}(\mathsf{D}_R - \mathsf{D}_R^*)\mathscr{D}S\, dx.$$

Note that the operator $\mathscr{D}(\mathsf{D}_R - \mathsf{D}_R^*)\mathscr{D}$ is skew-adjoint. Since this identity holds for arbitrary variational derivatives $S = \delta\mathscr{S}$, $T = \delta\mathscr{T}$, Proposition 5.88

and the substitution principle of Exercise 5.42 imply that

$$\mathscr{D}\cdot(\mathsf{D}_R - \mathsf{D}_R^*)\cdot\mathscr{D} = 0.$$

Finally, the nondegeneracy hypothesis on $\mathscr{D}$ allows us to conclude that

$$\mathscr{D}\cdot(\mathsf{D}_R - \mathsf{D}_R^*) = 0.$$

Taking adjoints, we have

$$-(\mathsf{D}_R^* - \mathsf{D}_R)\cdot\mathscr{D} = 0.$$

One further application of the nondegeneracy of $\mathscr{D}$ allows us to conclude that R satisfies the Helmholtz conditions $\mathsf{D}_R^* = \mathsf{D}_R$, and hence, by Theorem 5.92, is the variational derivative of some functional. □

Recursion Operators

We have seen that given a bi-Hamiltonian system, the operator $\mathscr{R} = \mathscr{E}\cdot\mathscr{D}^{-1}$, when applied successively to the initial equation $K_0 = \mathscr{D}\delta\mathscr{H}_0$, produces an infinite sequence of generalized symmetries of the original system (subject to the technical assumptions contained in Theorem 7.24). It is still not clear that $\mathscr{R}$ is a true recursion operator for the system, in the sense that whenever $\mathbf{v}_Q$ is a generalized symmetry, so is $\mathbf{v}_{\mathscr{R}Q}$. So far we only know it for symmetries with $Q = K_n$ for some n. In order to establish this more general result, we need a formula for the infinitesimal change of the Hamiltonian operator itself under a Hamiltonian flow.

Lemma 7.26. *Let $u_t = K = \mathscr{D}\delta\mathscr{H}$ be a Hamiltonian system of evolution equations with corresponding vector field $\mathbf{v}_K = \hat{\mathbf{v}}_{\mathscr{H}}$. Then*

$$\operatorname{pr}\hat{\mathbf{v}}_{\mathscr{H}}(\mathscr{D}) = \mathsf{D}_K\cdot\mathscr{D} + \mathscr{D}\cdot\mathsf{D}_K^*,$$

PROOF. Let $L = \delta\mathscr{H}$, so $K = \mathscr{D}L$. Let $P = \delta\mathscr{P}$, $Q = \delta\mathscr{Q}$ be arbitrary variational derivatives. By the Jacobi identity for $\mathscr{D}$, in the form (7.11), and (7.7), (5.33),

$$\int [P\cdot\operatorname{pr}\hat{\mathbf{v}}_{\mathscr{H}}(\mathscr{D})Q]\,dx = \int [P\cdot\operatorname{pr}\hat{\mathbf{v}}_{\mathscr{Q}}(\mathscr{D})L - Q\cdot\operatorname{pr}\hat{\mathbf{v}}_{\mathscr{P}}(\mathscr{D})L]\,dx$$

$$= \int \{P\cdot[\operatorname{pr}\hat{\mathbf{v}}_{\mathscr{Q}}(K) - \mathscr{D}\operatorname{pr}\hat{\mathbf{v}}_{\mathscr{Q}}(L)]$$

$$- Q\cdot[\operatorname{pr}\hat{\mathbf{v}}_{\mathscr{P}}(K) - \mathscr{D}\operatorname{pr}\hat{\mathbf{v}}_{\mathscr{P}}(L)]\}\,dx$$

$$= \int \{P \cdot [\mathsf{D}_K(\mathscr{D}Q) - \mathscr{D}\mathsf{D}_L(\mathscr{D}Q)]$$

$$- Q \cdot [\mathsf{D}_K(\mathscr{D}P) - \mathscr{D}\mathsf{D}_L(\mathscr{D}P)]\}\, dx$$

$$= \int [P \cdot \mathsf{D}_K(\mathscr{D}Q) - Q \cdot \mathsf{D}_K(\mathscr{D}P)]\, dx$$

$$= \int [P \cdot (\mathsf{D}_K \mathscr{D} + \mathscr{D}\mathsf{D}_K^*)Q]\, dx.$$

The next to last equality follows from the fact that $\mathscr{D}\mathsf{D}_L\mathscr{D}$ is self-adjoint since $\mathscr{D}$ is skew-adjoint, while D_L is self-adjoint since $L = \delta\mathscr{H}$ is a variational derivative. The result now follows by the substitution principle. □

Theorem 7.27. *Let $u_t = K = \mathscr{D}\delta\mathscr{H}_1 = \mathscr{E}\delta\mathscr{H}_0$ be a bi-Hamiltonian system of evolution equations. Then the operator $\mathscr{R} = \mathscr{E} \cdot \mathscr{D}^{-1}$ is a recursion operator for the system.*

Proof. We must verify the infinitesimal criterion (5.43). Using the previous lemma, we have

$$\mathscr{R}_t = \mathrm{pr}\, \mathbf{v}_K(\mathscr{R}) = \mathrm{pr}\, \mathbf{v}_K(\mathscr{E}) \cdot \mathscr{D}^{-1} - \mathscr{E} \cdot \mathscr{D}^{-1} \cdot \mathrm{pr}\, \mathbf{v}_K(\mathscr{D}) \cdot \mathscr{D}^{-1}$$

$$= (\mathsf{D}_K\mathscr{E} + \mathscr{E}\mathsf{D}_K^*) \cdot \mathscr{D}^{-1} - \mathscr{E} \cdot \mathscr{D}^{-1}(\mathsf{D}_K\mathscr{D} + \mathscr{D}\mathsf{D}_K^*)\mathscr{D}^{-1}$$

$$= \mathsf{D}_K\mathscr{R} - \mathscr{R}\mathsf{D}_K.$$

This verifies (5.42). □

Notice that in this theorem, we did not require that $(\mathscr{D}, \mathscr{E})$ form a Hamiltonian pair—only that each individual operator be Hamiltonian. Thus the recursion operator remains valid in more general situations. However, without the assumption that $(\mathscr{D}, \mathscr{E})$ form a Hamiltonian pair, it is unclear whether the symmetries $\mathbf{v}_n$, $n = 0, 1, 2, \ldots$, determined by recursion, are Hamiltonian or not.

Example 7.28. The *Boussinesq equation*, which we take in the form

$$u_{tt} = \tfrac{1}{3}u_{xxxx} + \tfrac{4}{3}(u^2)_{xx},$$

arises in a model for uni-directional propagation of long waves in shallow water (despite the fact that it admits waves travelling in both directions!). It can be converted into an equivalent evolutionary system

$$u_t = v_x, \qquad v_t = \tfrac{1}{3}u_{xxx} + \tfrac{8}{3}uu_x, \tag{7.39}$$

which turns out to be bi-Hamiltonian. The first Hamiltonian formulation is easy to discern. We take

$$\mathscr{D} = \begin{pmatrix} 0 & D_x \\ D_x & 0 \end{pmatrix}$$

as the Hamiltonian operator (which trivially satisfies the Jacobi identity since it has constant coefficients) and Hamiltonian functional

$$\mathscr{H}_1[u, v] = \int (-\tfrac{1}{6}u_x^2 + \tfrac{4}{9}u^3 + \tfrac{1}{2}v^2)\, dx.$$

The second Hamiltonian structure is not so obvious. The Hamiltonian functional is

$$\mathscr{H}_0[u, v] = \int \tfrac{1}{2}v\, dx,$$

the Hamiltonian operator being

$$\mathscr{E} = \begin{pmatrix} D_x^3 + 2uD_x + u_x & 3vD_x + 2v_x \\ 3vD_x + v_x & \tfrac{1}{3}D_x^5 + \tfrac{5}{3}(uD_x^3 + D_x^3 \cdot u) - (u_{xx}D_x + D_x \cdot u_{xx}) + \tfrac{16}{3}uD_x \cdot u \end{pmatrix}.$$

Even the proof that $\mathscr{E}$ is Hamiltonian is a rather laborious computation. The associated bi-vector is

$$\Theta_{\mathscr{E}} = \tfrac{1}{2}\int \{\theta \wedge \theta_{xxx} + 2u\theta \wedge \theta_x + 2v\theta \wedge \zeta_x - 4v\theta_x \wedge \zeta$$
$$+ \tfrac{1}{3}\zeta \wedge \zeta_{xxxxx} + \tfrac{4}{3}u\zeta \wedge \zeta_{xxx} - 2u\zeta_x \wedge \zeta_{xx} + \tfrac{16}{3}u^2\zeta \wedge \zeta_x\}\, dx,$$

where $\boldsymbol{\theta} = (\theta, \zeta)$, and θ and ζ are the basic uni-vectors corresponding to u and v respectively. Evaluating (7.18) (for $\mathscr{E}$), we use the fact that

$$\operatorname{pr} \mathbf{v}_{\mathscr{E}\boldsymbol{\theta}}(u) = \theta_{xxx} + 2u\theta_x + u_x\theta + 3v\zeta_x + 2v_x\zeta,$$
$$\operatorname{pr} \mathbf{v}_{\mathscr{E}\boldsymbol{\theta}}(v) = 3v\theta_x + v_x\theta + \tfrac{1}{3}\zeta_{xxxxx} + \tfrac{10}{3}u\zeta_{xxx} + 5u_x\zeta_{xx} + (3u_{xx} + \tfrac{16}{3}u^2)\zeta_x$$
$$+ (\tfrac{2}{3}u_{xxx} + \tfrac{16}{3}uu_x)\zeta,$$

and a lot of integration by parts—the reader may wish to try his or her skill at this! The proof that $\mathscr{D}$, $\mathscr{E}$ form a Hamiltonian pair is easier; since $\mathscr{D}$ has constant coefficients, $\operatorname{pr} \mathbf{v}_{\mathscr{E}\boldsymbol{\theta}}(\Theta_{\mathscr{D}}) = 0$, so we only need to verify

$$\operatorname{pr} \mathbf{v}_{\mathscr{D}\boldsymbol{\theta}}(\Theta_{\mathscr{E}}) = 0, \quad \text{where} \quad \operatorname{pr} \mathbf{v}_{\mathscr{D}\boldsymbol{\theta}}(u) = \zeta_x, \qquad \operatorname{pr} \mathbf{v}_{\mathscr{D}\boldsymbol{\theta}}(v) = \theta_x.$$

There is thus a whole hierarchy of conservation laws and commuting flows for the Boussinesq equation. The recursion operator is

$$\mathscr{R} = \mathscr{E} \cdot \mathscr{D}^{-1} =$$
$$\begin{pmatrix} 3v + 2v_xD_x^{-1} & D_x^2 + 2u + u_xD_x^{-1} \\ \tfrac{1}{3}D_x^4 + \tfrac{10}{3}uD_x^2 + 5u_xD_x + 3u_{xx} + \tfrac{16}{3}u^2 + (\tfrac{2}{3}u_{xxx} + \tfrac{16}{3}uu_x)D_x^{-1} & 3v + v_xD_x^{-1} \end{pmatrix}$$

and we can apply $\mathscr{R}$ successively to the right-hand side of (7.39) to obtain the symmetries. The first stage in this recursion is the flow

$$\begin{pmatrix} u_t \\ v_t \end{pmatrix} = \mathscr{E}\delta\mathscr{H}_1 = \mathscr{D}\delta\mathscr{H}_2$$

$$= \begin{pmatrix} \frac{1}{3}u_{xxxxx} + \frac{10}{3}uu_{xxx} + \frac{25}{3}u_x u_{xx} + \frac{20}{3}u^2 u_x + 5vv_x \\ \frac{1}{3}v_{xxxxx} + \frac{10}{3}uv_{xxx} + 5u_x v_{xx} + \frac{10}{3}u_{xx}v_x + \frac{5}{3}u_{xxx}v + \frac{20}{3}u^2 v_x + \frac{40}{3}uu_x v \end{pmatrix},$$

with consequent conservation law

$$\mathscr{H}_2[u, v] = \int \left(\tfrac{1}{3}u_{xx}v_{xx} + \tfrac{10}{3}uu_{xx}v + \tfrac{5}{2}u_x^2 v + \tfrac{20}{9}u^3 v + \tfrac{5}{6}v^3\right) dx.$$

Alternatively, one can start out with the translational Hamiltonian symmetry

$$\begin{pmatrix} u_t \\ v_t \end{pmatrix} = \begin{pmatrix} u_x \\ v_x \end{pmatrix} = \mathscr{E}\delta\hat{\mathscr{H}}_0 = \mathscr{D}\delta\hat{\mathscr{H}}_1,$$

where

$$\hat{\mathscr{H}}_0[u, v] = \int u \, dx, \qquad \hat{\mathscr{H}}_1[u, v] = \int uv \, dx,$$

are both conserved. By Theorem 7.27, applying $\mathscr{R}$ to this symmetry leads to a *second* hierarchy of commuting flows and consequent conservation laws, the first of which is

$$\begin{pmatrix} u_t \\ v_t \end{pmatrix} = \mathscr{E}\delta\hat{\mathscr{H}}_1 = \mathscr{D}\delta\hat{\mathscr{H}}_2 = \begin{pmatrix} v_{xxx} + 4uv_x + 4u_x v \\ \frac{1}{3}u_{xxxxx} + 4uu_{xxx} + 8u_x u_{xx} + \frac{32}{3}u^2 u_x + 4vv_x \end{pmatrix},$$

where

$$\hat{\mathscr{H}}_2[u, v] = \int \left(\tfrac{1}{6}u_{xx}^2 - 2uu_x^2 + \tfrac{8}{9}u^4 + 2uv^2 - \tfrac{1}{2}v_x^2\right) dx$$

is yet another conservation law. (At each stage, one needs to know that the operator $\mathscr{D}$ can be inverted, but this is proved in a similar fashion as the Korteweg–de Vries case.)

Notes

Although Hamiltonian systems of ordinary differential equations have been of paramount importance in the theory of both classical and quantum mechanics, the extension of these ideas and techniques to infinite-dimensional systems governed by systems of evolution equations has been very slow in maturing. The main reason for this delay has been the insistence on using canonical coordinates for finite-dimensional systems, which always exist by virtue of Darboux' theorem; these coordinates, however, do not appear to exist for the evolutionary systems of interest, and hence familiarity with the more general concept of a Poisson structure is a prerequisite here. (The infinite-dimensional version of Darboux' theorem proved by Weinstein, [1], does not seem to apply in this context.)

The correct formulation of a Hamiltonian structure for evolution equations was based on two significant developments. Arnol'd, [1], [2], showed that the Euler equations for ideal fluid flow could be viewed as a Hamiltonian system on the infinite-dimensional group of volume-preserving diffeomorphisms of the underlying space using the Lie–Poisson bracket (as generalized to the corresponding infinite-dimensional Lie algebra). Arnol'd wrote his Hamiltonian structure in Lagrangian (moving) coordinates; the Eulerian version was first discovered by Kuznetsov and Mikhailov, [1]. The version presented here, including the derivation of the conservation laws, is based on Olver, [5] and Ibragimov, [1; §25.3]. (The Poisson bracket presented here, while formally correct, fails to incorporate boundary effects, and needs to be slightly modified when discussing solutions over bounded domains; see Lewis, Marsden, Montgometry and Ratiu, [1], for a discussion of this point.) Subsequently, Marsden and Weinstein, [2], showed that the Lagrangian and Eulerian Poisson brackets were indeed the same. This method of Arnol'd has been applied with great success to determine the Hamiltonian structures of many of the systems of differential equations arising in fluid mechanics, plasma physics, etc. These have been used for proving new nonlinear stability results for these complicated systems; see Holm, Marsden, Ratiu and Weinstein, [1], and the references therein.

The second important development in the general theory was the discovery by Gardner, [1], that the Korteweg–de Vries equation could be written as a completely integrable Hamiltonian system. This idea was further developed by Zakharov and Fadeev, [1], Gel'fand and Dikii, [1], [2], [3], and Lax, [3]. Adler, [1], showed that the (first) Hamiltonian structure of the Korteweg–de Vries equation could be viewed as a formal Lie–Poisson structure on the infinite-dimensional Lie algebra of pseudo-differential operators on the real line, and extended these results to more general soliton equations having Lax representations, including the Boussinesq equation of Example 7.28. See Lax, [1], Dickey, [1], and Newell, [1].

Early versions of the theory of Hamiltonian systems of evolution equations were restricted by their insistence on introducing canonical coordinates; see Broer, [1], for a representative of this approach. The general concept of a Hamiltonian system of evolution equations first surfaces in the work of Magri, [1], Vinogradov, [2], Kupershmidt, [1], and Manin, [1]. Further developments, including the simplified techniques for verifying the Jacobi identity, appear in Gel'fand and Dorfman, [1], Olver, [4] and Kosmann-Schwarzbach, [3]. The computational methods based on functional multi-vectors presented here are a slightly modified version of the methods introduced in the second of these papers. The operator $\text{pr}\,\mathbf{v}_{\mathscr{D}\theta}$ in (7.18) is the same as the Schouten bracket with the bi-vector Θ determining the Poisson bracket, an approach favoured by Magri, [1], and Gel'fand and Dorfman, [1]. (See Exercise 6.20 for the finite-dimensional version of this bracket and Olver, [10], for a general infinite-dimensional form.) See Dubrovin and Novikov, [1], [2], [3], for a remarkable connection between first order Hamiltonian operators and Riemannian geometry. See Astashov and Vinogradov, [1], Cooke, [1], Dorfman, [1], Doyle, [1], and Olver, [12], for classification results on low order Hamiltonian operators.

The basic theorem on bi-Hamiltonian systems is due to Magri, [1], [2], who was also the first to publish the second Hamiltonian structure for the Korteweg–de Vries and other equations. Magri's method was developed by Gel'fand and Dorfman, [1], [2], and Fuchssteiner and Fokas, [1]. The second Hamiltonian structures of other soliton equations were found by Adler, [1], and Gel'fand and Dikii, [4], with further developments by Kupershmidt and Wilson, [1]. The concept of a bi-Hamiltonian system has also recently surfaced in work on finite-dimensional Hamiltonian systems, in which families of conservation laws are constructed using a related recursive procedure; see Hojman and Harleston, [1], Crampin, [1], Olver, [15], and the references therein. See Olver and Nutku, [1], and Olver, [15], for examples of bi-Hamiltonian systems which do not satisfy the compatability conditions. Strangely, these examples are "even more integrable" than the compatible ones.

The proof of the basic Theorem 7.24 on bi-Hamiltonian systems is based on that of Gel'fand and Dorfman, [1]. The annoying technical hypothesis on the invertibility of the operator $\mathscr{D}$ at each stage is not particularly satisfying. However, it is possible to drop this hypothesis if $\mathscr{D}$ happens to be a constant-coefficient operator; the proof relies on the exactness of an infinite-dimensional generalization of a complex due to Lichnerowicz, [1], based on the Schouten bracket; see Olver, [13]. The theory of bi-Hamiltonian systems played a key role in motivating the important subjects of R-matrices and quantum groups; see Semenov-Tian-Shanski, [1], Drinfel'd, [1], and Kosmann-Schwarzbach, [4].

Exercises

*7.1. Let $p = q = 1$. Find all Hamiltonian operators of the form

$$D_x^3 + PD_x + Q,$$

where P and Q are differential functions. (Try P, Q just depending on u and u_x first.) (Gel'fand and Dorfman, [1].)

7.2. Let $p = 1$, $q = 3$, with dependent variables u, v, w. Let

$$\mathscr{D} = \begin{bmatrix} 0 & 0 & D_x \\ 0 & D_x & 0 \\ D_x & 0 & D_x^3 + 2uD_x + u_x \end{bmatrix}.$$

Prove that $\mathscr{D}$ is Hamiltonian. (Adler, [1].)

7.3. Prove that Maxwell's equations in the physical form of Exercise 2.16(a) form a Hamiltonian system with Poisson bracket

$$\{\mathscr{F}, \mathscr{H}\} = \int \left(\frac{\delta \mathscr{F}}{\delta E} \cdot \nabla \times \frac{\delta \mathscr{H}}{\delta B} - \frac{\delta \mathscr{H}}{\delta E} \cdot \nabla \times \frac{\delta \mathscr{F}}{\delta B} \right) dx.$$

Discuss symmetries and conservation laws. (See also Exercises 4.6 and 5.25.) (Born and Infeld, [1], Marsden, [1].)

7.4. Derive the conservation laws $\mathscr{P}_\alpha$, $\mathscr{P}_\beta$ for the two-dimensional Euler equations found in Example 7.17 directly from the conservation law of energy using Proposition 5.64. (Ibragimov, [1; p. 357].)

*7.5. Prove that the three-dimensional Euler equations for incompressible fluid flow, when replaced by the corresponding vorticity equations for $\boldsymbol{\omega} = \nabla \times \mathbf{u}$, form a Hamiltonian system relative to the operator $\mathscr{D}$, where

$$\mathscr{D}P = \boldsymbol{\omega} \cdot \nabla P - (\nabla \boldsymbol{\omega}) \nabla \times P$$

(∇ denoting total gradient, curl or divergence). Find the conservation laws corresponding to known symmetry groups. Prove that the only nontrivial distinguished functional is the "total helicity" $\mathscr{C} = \int (\mathbf{u} \cdot \boldsymbol{\omega})\, d\mathbf{x}$. (Olver, [5]; see Serre, [1], and Khesin and Chekanov, [1], for the n-dimensional case.)

7.6. Let $\mathscr{L}[u]$ be a variational problem with Euler–Lagrange equations $\delta \mathscr{L} = 0$. Suppose $\mathbf{v}_Q$ generates a variational symmetry group with conservation law $\operatorname{Div} P = 0$. Prove that the corresponding dynamical Hamiltonian equations $u_t = \mathscr{D} \cdot \delta \mathscr{L}$ have a corresponding conservation law if and only if $\mathbf{v}_Q = \hat{\mathbf{v}}_{\mathscr{P}}$ is Hamiltonian with respect to the given Poisson bracket.

7.7. The dynamical equations of elasticity take the form

$$\frac{\partial^2 u^\alpha}{\partial t^2} = \sum_{i=1}^{p} D_i \left(\frac{\partial W}{\partial u_i^\alpha} \right), \qquad \alpha = 1, \ldots, q,$$

where $W(x, \nabla u)$ is the stored energy function, cf. Example 4.32. Prove that these can be put into Hamiltonian form using the total energy

$$\mathscr{H} = \int \left[\tfrac{1}{2} |u_t|^2 + W(x, \nabla u) \right] dx$$

as the Hamiltonian and u, $v = u_t$ as canonical variables. Discuss the conservation laws of this system in light of Exercise 7.6 and Example 4.32. (D. C. Fletcher, [1], Marsden and Hughes, [1; § 5.5].)

7.8. (a) Let $\mathscr{D} = \mathscr{A}^q \to \mathscr{A}^q$ be a differential operator. Prove that if $\mathscr{C}[u]$ is any functional satisfying $\mathscr{D}^* \cdot \delta \mathscr{C} = 0$, then $\mathscr{C}$ is a conservation law for any evolutionary system of the form $u_t = \mathscr{D}Q$ for $Q \in \mathscr{A}^q$.

(b) Prove that any evolution equation of the form $u_t = D_x^m Q$, with $x, u \in \mathbb{R}$, always conserves the first $m + 1$ moments $\mathscr{M}_j = \int x^j u \, dx$, $j = 0, 1, \ldots, m$, of any solution.

7.9. Prove that the operators

$$\mathscr{D} = D_x, \qquad \mathscr{E} = D_x^3 + \tfrac{2}{3} D_x \cdot u D_x^{-1} \cdot u D_x,$$

form a Hamiltonian pair making the modified Korteweg–de Vries equation $u_t = u_{xxx} + u^2 u_x$ into a bi-Hamiltonian system. Find the recursion operator and the first few symmetries. How do these relate to the Korteweg–de Vries equation under the Miura transformation of Exercise 5.11? (Magri, [2])

7.10. The Harry Dym equation is $u_t = D_x^3(u^{-1/2})$. Prove that this is a bi-Hamiltonian system with $\mathscr{D} = 2uD_x + u_x$, $\mathscr{E} = D_x^3$. Discuss distinguished functionals, symmetries and conservation laws for this equation. The change of variables $v = u^{-1/2}$ changes this equation to $v_t = -\frac{1}{2}v^3 v_{xxx}$. Discuss its effects on the bi-Hamiltonian structure. (Magri, [1], Leo, Leo, Soliani, Solombrino and Mancarella, [1]; Ibragimov, [1; p. 300], shows how this equation can be transformed into the Korteweg–de Vries equation.)

**7.11. The system of equations

$$u_t = uu_x + v_x - \tfrac{1}{2}u_{xx}, \qquad v_t = (uv)_x + \tfrac{1}{2}v_{xx},$$

is equivalent, under a change of variables, to a system of Boussinesq equations modelling the bi-directional propagation of long waves in shallow water, first found by Whitham, [1] and Broer, [1]. Prove that this system is *tri-Hamiltonian*, meaning that it can be written as a Hamiltonian system using any one of the three Hamiltonian operators

$$\mathscr{D}_0 = \begin{pmatrix} 0 & D_x \\ D_x & 0 \end{pmatrix}, \qquad \mathscr{D}_1 = \begin{pmatrix} 2D_x & D_x \cdot u - D_x^2 \\ uD_x + D_x^2 & 2vD_x + v_x \end{pmatrix},$$

$$\mathscr{D}_2 = \begin{pmatrix} 4uD_x + 2u_x & 4vD_x + 2v_x + D_x(D_x - u)^2 \\ 4vD_x + 2v_x + (D_x + u)^2 D_x & (D_x + u)(2vD_x + v_x) - (2vD_x + v_x)(D_x - u) \end{pmatrix},$$

and any two of these operators form a Hamiltonian pair. Discuss symmetries and conservation laws of the system. (Kupershmidt, [2].)

7.12. (a) Prove that if $\mathscr{D}$ is a self-adjoint (respectively, skew-adjoint) matrix differential operator, and $\mathbf{v}_Q$ is any evolutionary vector field, then the Lie derivative $\operatorname{pr} \mathbf{v}_Q(\mathscr{D})$ is self-adjoint (skew-adjoint).
(b) Prove directly that (7.11) is an alternating, trilinear function of P, Q, R.

7.13. Prove that if $\mathscr{D}: \mathscr{A} \to \mathscr{A}$ and $\mathscr{E}: \mathscr{A} \to \mathscr{A}$ are nonzero scalar differential operators, then $\mathscr{E} \cdot \mathscr{D}: \mathscr{A} \to \mathscr{A}$ is a nonzero differential operator. Deduce that any scalar differential operator is nondegenerate in the sense of Definition 7.23.

*7.14. Let $\mathscr{D}: \mathscr{A}^r \to \mathscr{A}^s$ be a differential operator, and let $\mathscr{K}^* = \{Q \in \mathscr{A}^s: \mathscr{D}^* Q = 0\}$ be the kernel of its adjoint. Prove that if $\mathscr{K}^*$ is a finite-dimensional vector space over $\mathbb{R}$, then $\mathscr{D}$ is nondegenerate in the sense of Definition 7.23. How many distinguished functionals does a nondegenerate Hamiltonian operator have?

*7.15. The equations of polytropic gas dynamics have the form

$$u_t + uu_x + v^\sigma v_x = 0, \qquad v_t + (uv)_x = 0,$$

in which u represents the velocity, v the density and $\sigma = \gamma - 2$, where γ is the physical ratio of specific heats appearing in the pressure-density relation. Show that this system can be written in Hamiltonian form in three distinct ways, using the Hamiltonian operators

$$\mathscr{D}_1 = \begin{pmatrix} 0 & D_x \\ D_x & 0 \end{pmatrix},$$

$$\mathscr{D}_2 = \begin{pmatrix} 2v^\sigma D_x + (v^\sigma)_x & (\sigma+1)uD_x + u_x \\ (\sigma+1)uD_x + \sigma u_x & 2vD_x + v_x \end{pmatrix},$$

$$\mathscr{D}_3 = \begin{pmatrix} 2uv^\sigma D_x + (uv^\sigma)_x & \left[\tfrac{1}{2}(\sigma+1)u^2 + \dfrac{2}{\sigma+1}v^\sigma\right]D_x + uu_x + v^\sigma v_x \\ \left[\tfrac{1}{2}(\sigma+1)u^2 + \dfrac{2}{\sigma+1}v^\sigma\right]D_x + \sigma uu_x + v^\sigma v_x & 2uvD_x + (uv)_x \end{pmatrix}.$$

Prove that each operator is Hamiltonian. Which pairs are compatible? Discuss the consequential recursion operators, symmetries and conservation laws. (Whitham, [2], Nutku, [1], Olver and Nutku, [1].)

7.16. Prove that the Toda lattice equations of Exercise 6.11 are a bi-Hamiltonian system for the two Poisson brackets with structure matrices

$$\begin{pmatrix} 0 & 0 & 0 & 1 & 0 & -1 \\ 0 & 0 & 0 & -1 & 1 & 0 \\ 0 & 0 & 0 & 0 & -1 & 1 \\ -1 & 1 & 0 & 0 & 0 & 0 \\ 0 & -1 & 1 & 0 & 0 & 0 \\ 1 & 0 & -1 & 0 & 0 & 0 \end{pmatrix}, \quad \begin{pmatrix} 0 & -e^{q^1} & e^{q^3} & p^1 & 0 & -p^1 \\ e^{q^1} & 0 & -e^{q^2} & -p^2 & p^2 & 0 \\ -e^{q^3} & e^{q^2} & 0 & 0 & -p^3 & p^3 \\ -p^1 & p^2 & 0 & 0 & 1 & -1 \\ 0 & -p^2 & p^3 & -1 & 0 & 1 \\ p^1 & 0 & -p^3 & 1 & -1 & 0 \end{pmatrix},$$

relative to the coordinates $(p^1, p^2, p^3, q^1, q^2, q^3)$. Are these two Poisson brackets compatible? (Arnol'd and Novikov, [1; p. 58], Leo, Leo, Soliani, Solombrino and Mancarella, [2].)

*7.17. Show that the structure matrices

$$J_1 = \begin{pmatrix} 0 & 0 & 1 & 0 \\ 0 & 0 & 0 & 1 \\ -1 & 0 & 0 & 0 \\ 0 & -1 & 0 & 0 \end{pmatrix}, \quad J_2 = \begin{pmatrix} 0 & 0 & e^{p_1} & -p_2 e^{p_1} \\ 0 & 0 & 0 & -e^{p_1} \\ -e^{p_1} & 0 & 0 & 0 \\ p_2 e^{p_1} & e^{p_1} & 0 & 0 \end{pmatrix},$$

using coordinates (p_1, p_2, q_1, q_2), are Hamiltonian, but do not form a Hamiltonian pair. Discuss the integrability of any associated "bi-Hamiltonian systems." (Olver, [15].)

7.18. Let $u_t = \mathscr{D}\delta\mathscr{H}$ be a scalar Hamiltonian evolution equation. Prove that the pseudo-differential operator $\mathscr{D}^{-1}$ is a formal conservation law of rank ∞.

References

Abellanas, L. and Galindo, A.
1. Conserved densities for linear evolution systems, *Commun. Math. Phys.* **79** (1981), 341–351.

Ablowitz, M. J. and Kodama, Y.
1. Note on asymptotic solutions of the Korteweg–de Vries equation with solitons, *Stud. Appl. Math.* **66** (1982), 159–170.

Ablowitz, M. J., Ramani, A. and Segur, H.
1. Nonlinear evolution equations and ordinary differential equations of Painlevé type, *Lett. Nuovo Cim.* **23** (1978), 333–338.
2. A connection between nonlinear evolution equations and ordinary differential equations of P-type. I, *J. Math. Phys.* **21** (1980), 715–721.

Abraham, R. and Marsden, J. E.
1. *Foundations of Mechanics*, 2nd ed., Benjamin-Cummings, Reading, Mass., 1978.

Abramowitz, M., and Stegun, I.
1. *Handbook of Mathematical Functions*, National Bureau of Standards Appl. Math. Series, No. 55, U.S. Govt. Printing Office, Washington, D.C., 1970.

Adams, M., Ratiu, T. and Schmidt, R.
1. A Lie group structure for pseudodifferential operators, *Math. Ann.* **273** (1986), 529–551.
2. A Lie group structure for Fourier integral operators, *Math. Ann.* **276** (1986), 19–41.

Adler, M.
1. On a trace functional for formal pseudo-differential operators and the symplectic structure of the Korteweg–de Vries type equations, *Invent. Math.* **50** (1979), 219–248.

Ado, I. D.
1. The representation of Lie algebras by matrices, *Bull. Soc. Phys.-Math. Kazan* **7** (1935), 3–43; also *Usp. Mat. Nauk.* **2** (1947), 159–173, and Amer. Math. Soc. Transl., No. 2, 1949.

Akhatov, I. S., Gazizov, R. K. and Ibragimov, N. H.
1. Nonlocal symmetries. Heuristic approach, *J. Sov. Math.* **55** (1991), 1401–1450.

Aldersley, S. J.
1. Higher Euler operators and some of their applications, *J. Math. Phys.* **20** (1979), 522–531.

Alonso, L. M.
1. On the Noether map, *Lett. Math. Phys.* **3** (1979), 419–424.

Ames, W. F.
1. *Nonlinear Partial Differential Equations in Engineering*, Academic Press, New York, 1965, 1972.

Anderson, I. M.
1. Introduction to the variational bicomplex, *Comtemp. Math.* **132** (1992), 51–73.

Anderson, I. M. and Duchamp, T. E.
1. On the existence of global variational principles, *Amer. J. Math.* **102** (1980), 781–868.
2. Variational principles for second-order quasi-linear scalar equations, *J. Diff. Eq.* **51** (1984), 1–47.

Anderson, I. M., Kamran, N. and Olver, P. J.
1. Internal, external and generalized symmetries, *Adv. Math.* **100** (1993), 53–100.

Anderson, I. and Thompson, G.
1. The inverse problem of the calculus of variations for ordinary differential equations, *Mem. Amer. Math. Soc.* **473** (1992), 1–110.

Anderson, R. L. and Ibragimov, N. H.
1. *Lie-Bäcklund Transformations in Applications*, SIAM Studies in Appl. Math., No. 1, Philadelphia, 1979.

Anderson, R. L., Kumei, S. and Wulfman, C. E.
1. Generalizations of the concept of invariance of differential equations. Results of applications to some Schrödinger equations, *Phys. Rev. Lett.* **28** (1972), 988–991.

Appell, P.
1. Sur l'équation $\partial^2 z/\partial x^2 - \partial z/\partial y = 0$ et la théorie de la chaleur, *J. de Math. Pures et Appl.* **8**(4), (1892), 187–216.

Arnol'd, V. I.
1. Sur la géométrie différentielle des groupes de Lie de dimension infinite et ses applications à l'hydrodynamique des fluides parfaits, *Ann. Inst. Fourier Grenoble* **16** (1966), 319–361.
2. The Hamiltonian nature of the Euler equations in the dynamics of a rigid body and an ideal fluid, *Usp. Mat. Nauk*, **24** (1969), 225–226 (in Russian).
3. *Mathematical Methods of Classical Mechanics*, Springer-Verlag, New York, 1978.

Arnol'd, V. I. and Novikov, S. P. (eds.)
1. *Dynamical Systems IV*, Encyclopaedia of Mathematical Sciences, vol. 4, Springer-Verlag, New York, 1990.

Astashov, A. M. and Vinogradov, A. M.
1. On the structure of Hamiltonian operators in field theory, *J. Geom. Phys.* **3** (1986), 263–287.

Atherton, R. W. and Homsy, G. M.
1. On the existence and formulation of variational principles for nonlinear differential equations, *Stud. Appl. Math.* **54** (1975), 31–60.

Bäcklund, A. V.
1. Ueber Flächentransformationen, *Math. Ann.* **9** (1876), 297–320.

Baikov, V. A., Gazizov, R. K. and Ibragimov, N. H.
1. Approximate symmetries, *Math. USSR Sbornik* **64** (1989), 427–441.

Baker, J. W. and Tavel, M. A.
1. The applications of Noether's theorem to optical systems, *Amer. J. Physics* **42** (1974), 857–861.

Bakirov, I. M.
1. On the symmetries of some system of evolution equations, preprint, 1991.

Barenblatt, G. I.
1. *Similarity, Self-similarity, and Intermediate Asymptotics*, Consultants Bureau, New York, 1979.

Bargmann, V.
1. Irreducible unitary representations of the Lorentz group, *Ann. Math.* **48** (1947), 568–640.

Bateman, H.
1. The conformal transformations of a space of four dimensions and their applications to geometrical optics, *Proc. London Math. Soc.* **7** (1909), 70–89.
2. On dissipative systems and related variables, *Phys. Rev.* **38** (1931), 815–819.

Benjamin, T. B.
1. The stability of solitary waves, *Proc. Roy. Soc. London* **A328** (1972), 153–183.

Benjamin, T. B. and Olver, P. J.
1. Hamiltonian structure, symmetries and conservation laws for water waves, *J. Fluid Mech.* **125** (1982), 137–185.

Berezin, F. A.
1. Some remarks about the associated envelope of a Lie algebra, *Func. Anal. Appl.* **1** (1967), 91–102.

Berker, R.
1. Intégration des équations du mouvement d'un fluide visqueux incompressible, in *Handbuch der Physik*, VIII/2, Springer-Verlag, Berlin, 1963, pp. 1–384.

Bessel-Hagen, E.
1. Über die Erhaltungssätze der Elektrodynamik, *Math. Ann.* **84** (1921), 258–276.

Beyer, W. A.
1. Lie-group theory for symbolic integration of first order ordinary differential equations, in *Proceedings of the 1979 Macsyma Users Conference*, V. E. Lewis, ed., MIT Laboratory for Computer Science, Cambridge, Mass., 1979, pp. 362–384.

Bianchi, L.
1. *Lezioni sulla teoria dei Gruppi Continui Finiti di Transformazioni*, Enrico Spoerri, Pisa, 1918.

Bilby, B. A., Miller, K. J. and Willis, J. R.
1. *Fundamentals of Deformation and Fracture*, Cambridge University Press, Cambridge, 1985.

Birkhoff, G.
1. Lie groups isomorphic with no linear group, *Bull. Amer. Math. Soc.* **42** (1936), 883–888.
2. *Hydrodynamics—A Study in Logic, Fact and Similitude*, 1st ed., Princeton University Press, Princeton, 1950.

Bluman, G. W. and Cole, J. D.
1. The general similarity solution of the heat equation, *J. Math. Mech.* **18** (1969), 1025–1042.
2. *Similarity Methods for Differential Equations*, Appl. Math. Sci., No. 13, Springer-Verlag, New York, 1974.

Bluman, G. W. and Kumei, S.
1. On the remarkable nonlinear diffusion equation $(\partial/\partial x)[a(u+b)^{-2}(\partial u/\partial x)] - (\partial u/\partial t) = 0$, *J. Math. Phys.* **21** (1980), 1019–1023.
2. *Symmetries and Differential Equations*, Springer-Verlag, New York, 1989.

Boltzmann, L.
1. Zur integration der Diffusionsgleichung bei variabeln Diffusionscoefficienten, *Ann. der Physik und Chemie* **53** (1894), 959–964.

Boothby, W. M.
1. *An Introduction to Differentiable Manifolds and Riemannian Geometry*, Academic Press, New York, 1975.

Born, M. and Infeld, L.
1. On the quantization of the new field theory. II, *Proc. Roy. Soc. London* **150A** (1935), 141–166.

Bott, R. and Tu, L. W.
1. *Differential Forms in Algebraic Topology*, Springer-Verlag, New York, 1982.

Bourlet, M. C.
1. Sur les équations aux dérivées partielles simultanées, *Ann. Sci. École Norm. Sup.* **8**(3) (1891), Suppl. S.3–S.63.

Boyer, C. P., Kalnins, E. G. and Miller, W., Jr.
1. Symmetry and separation of variables for the Helmholtz and Laplace equations, *Nagoya Math. J.* **60** (1976), 35–80.

Broer, L. J. F.
1. Approximate equations for long water waves, *Appl. Sci. Res.* **31** (1975), 377–395.

Brown, A. B.
1. Functional dependence, *Trans. Amer. Math. Soc.* **38** (1935), 379–394.

Buchnev, A. A.
1. The Lie group admitted by the motion of an ideal incompressible fluid, *Dinamika Sploch. Sredi* **7** (1971), 212–214 (in Russian).

Byrd, P. F. and Friedman, M. D.
1. *Handbook of Elliptic Integrals for Engineers and Scientists*, Springer-Verlag, New York, 1971.

Carathéodory, C.
1. *Calculus of Variations and Partial Differential Equations of the First Order*, Vol. 1, Holden-Day, New York, 1965.

Carmichael, R. D.
1. Transformations leaving invariant certain partial differential equations of physics, *Amer. J. Math.* **49** (1927), 97–116.

Cartan, E.
1. *Leçons sur les Invariants Intégraux*, Hermann, Paris, 1922.
2. *La Théorie des Groupes Finis et Continus et l'Analysis Situs*, Mém. Sci. Math. No. 42, Gauthier-Villars, Paris, 1930; also, *Oeuvres Complètes*, vol. 1, Gauthier-Villars, Paris, 1952, pp. 1165–1225.
3. *La Topologie des Groupes de Lie*, Exp. de Géométrie, vol. 8, Hermann, Paris, 1936; also, *Oeuvres Complètes*, vol. 1, Gauthier-Villars, Paris, 1952, pp. 1307–1330.

Cartan, E. and Einstein, A.
1. *Letters on Absolute Parallelism 1929–1932*, Princeton Univ. Press, Princeton, N.J., 1979.

Champagne, B., Hereman, W. and Winternitz, P.
1. The computer calculation of Lie point symmetries of large systems of differential equations, *Comp. Phys. Commun.* **66** (1991), 319–340.

Chen, H. H., Lee, Y. C. and Lin, J.-E.
1. On a new hierarchy of symmetries for the integrable nonlinear evolution equations, in *Advances in Nonlinear Waves*, Vol. 2, L. Debnath, ed., Research Notes in Math., Vol. 111, Pitman Publ. Inc., Marshfield, Mass., 1985, pp. 233–239.

Chevalley, C. C.
1. *Theory of Lie Groups*, vol. 1, Princeton University Press, Princeton, N.J., 1946.

Clarkson, P. and Kruskal, M.
1. New similarity reductions of the Boussinesq equation, *J. Math. Phys.* **30** (1989), 2201–2213.

Cohen, A.
1. *An Introduction to the Lie Theory of One-Parameter Groups, with Applications to the Solution of Differential Equations*, D. C. Heath & Co., New York, 1911.

Conn, J. F.
1. Normal forms for analytic Poisson structures, *Ann. Math.* **119** (1984), 577–601.

Cooke, D. B.
1. Classification results and the Darboux theorem for low order Hamiltonian operators, *J. Math. Phys.* **32** (1991), 109–119.

Copson, E. T.
1. *Partial Differential Equations*, Cambridge University Press, Cambridge, 1975.

Courant, R. and Hilbert, D.
1. *Methods of Mathematical Physics*, Interscience, New York, 1953.

Crampin, M.
1. A note on non-Noether constants of motion, *Phys. Lett.* **95A** (1983), 209–212.

Cunningham, E.
1. The principle of relativity in electrodynamics and an extension thereof, *Proc. London Math. Soc.* **8** (1909), 77–98.

Dedecker, P.
1. Calcul des variations et topologie algébrique, *Mém. Soc. Roy. Sci. de Liège* **29** (1957), 7–216.

Delassus, E.
1. Extension du théorème de Cauchy aux systèmes les plus généraux d'équations aux dérivées partielles, *Ann. Sci. École Norm. Sup.* **13**(3) (1896), 421–467.

Delong, R. P., Jr.

1. Killing tensors and the Hamilton–Jacobi equation, Ph.D. thesis, Univ. of Minnesota, 1982.

Dickey, L.

1. *Soliton Equations and Hamiltonian Systems*, World Scientific, Singapore, 1991.

DiPerna, R. J.

1. Decay of solutions of hyperbolic systems of conservation laws with a convex extension, *Arch. Rat. Mech. Anal.* **64** (1977), 1–46.
2. Uniqueness of solutions to hyperbolic conservation laws, *Indiana Univ. Math. J.* **28** (1979), 137–188.

Dirac, P. A. M.

1. Generalized Hamiltonian dynamics, *Canad. J. Math.* **2** (1950), 129–148.

Dorfman, I. Y.

1. On differential operators that generate Hamiltonian structures, *Phys. Lett. A* **140** (1989), 378–382.

Dorodnitsyn, V. A.

1. Transformation groups in net spaces, *J. Sov. Math.* **55** (1991), 1490–1517.

Douglas, J.

1. Solution of the inverse problem of the calculus of variations, *Trans. Amer. Math. Soc.* **50** (1941), 71–128.

Doyle, P.

1. *Differential Geometric Poisson Bivectors and Quasilinear Systems in One Space Variable*, Ph.D. Thesis, University of Minnesota, 1992.

Dresner, L.

1. *Similarity Solutions of Nonlinear Partial Differential Equations*, Research Notes in Math., No. 88, Pitman, Boston, 1983.

Drinfel'd, V. G.

1. Quantum groups, in *Proc. Int. Cong. Math., Berkeley*, Vol. 1, 1986, pp. 798–820.

Dubrovin, B. A. and Novikov, S. A.

1. Hamiltonian formalism of one-dimensional systems of hydrodynamic type and the Bogolyubov–Whitham averaging method, *Sov. Math. Dokl.* **27** (1983), 665–669.
2. On Poisson brackets of hydrodynamic type, *Sov. Math. Dokl.* **30** (1984), 651–654.
3. Hydrodynamics of weakly deformed soliton lattices. Differential geometry and Hamiltonian theory, *Russian Math. Surveys* **44**: 6 (1989), 35–124.

Edelen, D. G. B.

1. *Isovector Methods for Equations of Balance*, Sijthoff and Noordhoff, Germantown, Md., 1980.

Ehresmann, C.

1. Les prolongements d'une variété différentiable, *C.R. Acad. Sci. Paris* **233** (1951), 598–600, 777–779, 1081–1083; **234** (1952), 1028–1030, 1424–1425.
2. Introduction à la théorie des structures infinitésimales et des pseudo-groupes de Lie, in *Géométrie Différentielle*, Colloq. Inter. du Centre Nat. de la Recherche Scientifique, Strasbourg, 1953, 97–110.

Eisenhart, L. P.

1. *Riemannian Geometry*, Princeton University Press, Princeton, 1926.
2. *Continuous Groups of Transformations*, Princeton University Press, Princeton, 1933.

Elkana, Y.
1. *The Discovery of the Conservation of Energy*, Hutchinson Educational Ltd., London, 1974.

Engel, F.
1. Über die zehn allgemeinen Integrale der klassischen Mechanik, *Nachr. König. Gesell. Wissen. Göttingen, Math.-Phys. Kl.* (1916), 270–275.

Ericksen, J. L.
1. On the formulation of St.-Venant's problem, in *Nonlinear Analysis and Mechanics: Heriot–Watt Symposium*, R. J. Knops, ed., Research Notes in Math., No. 17, Pitman, San Francisco, 1977, pp. 158–186.

Eshelby, J. D.
1. The continuum theory of lattice defects, in *Solid State Physics*, Vol. 3, F. Seitz and D. Turnbull, eds., Academic Press, New York, 1956, pp. 79–144.

Finzi, A.
1. Sur les systèmes d'équations aux dérivées partielles qui, comme les systèmes normaux, comportent autant d'équations que de fonctions inconnues, *Proc. Kon. Neder. Akad. v. Wetenschappen* **50** (1947), 136–142, 143–150, 288–297, 351–356.

Fletcher, D. C.
1. Conservation laws in linear elastodynamics, *Arch. Rat. Mech. Anal.* **60** (1976), 329–353.

Fletcher, J. G.
1. Local conservation laws in generally covariant theories, *Rev. Mod. Phys.* **32** (1960), 65–87.

Fokas, A. S.
1. Group theoretical aspects of constants of motion and separable solutions in classical mechanics, *J. Math. Anal. Appl.* **68** (1979), 347–370.
2. Generalized symmetries and constants of motion of evolution equations, *Lett. Math. Phys.* **3** (1979), 467–473.
3. A symmetry approach to exactly solvable evolution equations, *J. Math. Phys.* **21** (1980), 1318–1325.
4. Symmetries and integrability, *Stud. Appl. Math.* **77** (1987), 253–299.

Fokas, A. S. and Fuchssteiner, B.
1. The hierarchy of the Benjamin–Ono equation, *Phys. Lett.* **86A** (1981), 341–345.

Fokas, A. S. and Santini, P.
1. Recursion operators and bi-Hamiltonian structures in multi-dimensions. II, *Commun. Math. Phys.* **116** (1988), 449–474.

Fokas, A. S. and Yortsos, Y. C.
1. On the exactly solvable equation $S_t = [(\beta S + \gamma)^{-2} S_x]_x + \alpha(\beta S + \gamma)^{-2} S_x$ occurring in two-phase flow in porous media, *SIAM J. Appl. Math.* **42** (1982), 318–332.

Forsyth, A. R.
1. *The Theory of Differential Equations*, Cambridge University Press, Cambridge, 1890, 1900, 1902, 1906.

Frobenius, G.
1. Über das Pfaffsche Probleme, *J. Reine Angew. Math.* **82** (1877), 230–315.

Fuchssteiner, B.

1. Application of hereditary symmetries to nonlinear evolution equations, *Nonlinear Anal. Theory Meth. Appl.* **3** (1979), 849–862.
2. Mastersymmetries, high order time-dependent symmetries and conserved densities of nonlinear evolution equations, *Prog. Theor. Phys.* **70** (1983), 1508–1522.

Fuchssteiner, B. and Fokas, A. S.

1. Symplectic structures, their Bäcklund transformations and hereditary symmetries, *Physica* **4D** (1981), 47–66.

Fujimoto, A. and Watanabe, Y.

1. Polynomial evolution equations of not normal type admitting nontrivial symmetries, *Phys. Lett. A* **136** (1989), 294–299.

Fushchich, V. I. and Nikitin, A. G.

1. New and old symmetries of the Maxwell and Dirac equations, *Sov. J. Part. Nucl.* **14** (1983), 1–22.
2. *Symmetries of Maxwell's Equations*, D. Reidel, Dordrecht, 1987.

Fuschchich, W. I. and Shtelen, W. M.

1. On approximate symmetry and approximate solutions of the non-linear wave equation with a small parameter, *J. Phys. A* **22** (1989), L887–L890.

Galaktionov, V. A.

1. On new exact blow-up solutions for nonlinear heat conduction equations with source and applications, *Diff. Int. Eq.* **3** (1990), 863–874.

Galindo, A. and Martínez Alonso, L.

1. Kernels and ranges in the variational formalism, *Lett. Math. Phys.* **2** (1978), 385–390.

Gardner, C. S.

1. Korteweg–de Vries equation and generalizations. IV. The Korteweg–de Vries equation as a Hamiltonian system, *J. Math. Phys.* **12** (1971), 1548–1551.

Gardner, C. S., Greene, J. M., Kruskal, M. D. and Miura, R. M.

1. Korteweg–de Vries equation and generalizations. VI. Methods for exact solution, *Comm. Pure Appl. Math.* **27** (1974), 97–133.

Gel'fand, I. M. and Dikii, L. A.

1. Asymptotic behaviour of the resolvent of Sturm–Liouville equations and the algebra of the Korteweg–de Vries equations, *Russ. Math. Surveys* **30**:5 (1975), 77–113.
2. A Lie algebra structure in a formal variational calculation, *Func. Anal. Appl.* **10** (1976), 16–22.
3. Fractional powers of operators and Hamiltonian systems, *Func. Anal. Appl.* **10** (1976), 259–273.
4. A family of Hamiltonian structures connected with integrable nonlinear differential equations, Inst. Appl. Math. Acad. Sci. USSR, No. 136 (1978) preprint (in Russian).

Gel'fand, I. M. and Dorfman, I. Ya.

1. Hamiltonian operators and algebraic structures related to them, *Func. Anal. Appl.* **13** (1979), 248–262.
2. Hamiltonian operators and infinite-dimensional Lie algebras, *Func. Anal. Appl.* **15** (1981), 173–187.

Gel'fand, I. M. and Fomin, S. V.

1. *Calculus of Variations*, Prentice-Hall, Englewood Cliffs, N.J., 1963.

Goff, J. A.
1. Transformations leaving invariant the heat equation of physics, *Amer. J. Math.* **49** (1927), 117–122.

Goldberg, J. N.
1. Invariant transformations, conservation laws and energy-momentum, in *General Relativity and Gravitation*, vol. 1, A. Held, ed., Plenum Press, New York, 1980, pp. 469–489.

Goldstein, H.
1. *Classical Mechanics*, 2nd ed., Addison-Wesley, Reading, Mass., 1980.

Golubitsky, M. and Guillemin, V.
1. *Stable Mappings and Their Singularities*, Springer-Verlag, New York, 1973.

González-Gascon, F. and González-López, A.
1. Symmetries of differential equations. IV, *J. Math. Phys.* **24** (1983), 2006–2021.

Gorenstein, D.
1. *Finite Simple Groups*, Plenum Press, New York, 1982.

Gragert, P. H. K., Kersten, P. H. M. and Martini, R.
1. Symbolic computations in applied differential geometry, *Acta Appl. Math.* **1** (1983), 43–77.

Günther, W.
1. Über einige Randintegrale der Elastomechanik, *Abh. Braunschweiger Wissen. Gesell.* **14** (1962), 53–72.

Hadamard, J. S.
1. *La Théorie des Équations aux Dérivées Partielles*, Editions Scientifiques, Peking, 1964.

Harrison, B. K. and Estabrook, F. B.
1. Geometric approach to invariance groups and solution of partial differential equations, *J. Math. Phys.* **12** (1971), 653–666.

Hawkins, T.
1. The *Erlanger Programm* of Felix Klein: reflections on its place in the history of mathematics, *Hist. Math.* **11** (1984), 442–470.
2. Line geometry, differential equations and the birth of Lie's theory of groups, in *The History of Modern Mathematics*, Vol. 2, J. McCleary and D. Rowe, eds., Academic Press, New York, 1989, pp. 275–327.
3. Jacobi and the birth of Lie's theory of groups, *Arch. Hist. Exact Sci.* **42** (1991), 187–278.

Hejhal, D. A.
1. Monodromy groups and linearly polymorphic functions, *Acta Math.* **135** (1975), 1–55.

Helgason, S.
1. *Differential Geometry and Symmetric Spaces*, Academic Press, New York, 1962.
2. *Groups and Geometric Analysis: Integral Geometry, Invariant Differential Operators and Spherical Functions*, Academic Press, New York, 1983.

Helmholtz, H.
1. Über der physikalische Bedeutung des Princips der kleinsten Wirkung, *J. Reine Angew. Math.* **100** (1887), 137–166.

Henneaux, M.

1. Equations of motion, commutation relations and ambiguities in the Lagrangian formalism, *Ann. Physics* **140** (1982), 45–64.

Hermann, R.

1. The differential geometry of foliations, II, *J. Math. Mech.* **11** (1962), 303–315.
2. Cartan connections and the equivalence problem for geometric structures, *Contrib. Diff. Eq.* **3** (1964), 199–248.

Hill, E. L.

1. Hamilton's principle and the conservation theorems of mathematical physics, *Rev. Mod. Phys.* **23** (1951), 253–260.

Hirsch, A.

1. Über eine charakteristische Eigenschaft der Differentialgleichungen der Variationsrechnung, *Math. Ann.* **49** (1897), 49–72.
2. Die Existenzbedingungen des verallgemeinterten kinetischen Potentialen, *Math. Ann.* **50** (1898), 429–441.

Hojman, S. and Harleston, H.

1. Equivalent Lagrangians: multi-dimensional case, *J. Math. Phys.* **22** (1981), 1414–1419.

Holm, D. D.

1. *Symmetry Breaking in Fluid Dynamics: Lie Group Reducible Motions for Real Fluids*, Ph.D. Thesis, University of Michigan, 1976.

Holm, D. D., Marsden, J. E., Ratiu, T. and Weinstein, A.

1. Nonlinear stability of fluid and plasma equilibria, *Phys. Rep.* **123** (1985), 1–116.

Holmes, P. J. and Marsden, J. E.

1. Horseshoes and Arnol'd diffusion for Hamiltonian systems on Lie groups, *Indiana Univ. Math. J.* **32** (1983), 273–309.

Horndeski, G. W.

1. Differential operators associated with the Euler–Lagrange operator, *Tensor* **28** (1974), 303–318.

Ibragimov, N. H.

1. *Transformation Groups Applied to Mathematical Physics*, Reidel, Boston, 1985.
2. *Methods of Group Analysis for Ordinary Differential Equations*, Znanie Publ., Moscow, 1991. (In Russian.)

Ibragimov, N. H. and Shabat, A. B.

1. Evolutionary equations with nontrivial Lie–Bäcklund group, *Func. Anal. Appl.* **14** (1980), 19–28.

Ince, E. L.

1. *Ordinary Differential Equations*, Longmans, Green and Co., London, 1926.

Jacobi, C. G. J.

1. *Vorlesungen Über Dynamik*, G. Reimer, Berlin, 1866 (also in *Gesammelte Werke*, end of vol. 7, G. Reimer, Berlin, 1891).

Jacobson, N.

1. *Lie Algebras*, Interscience, New York, 1962.

Johnson, H. H.

1. Bracket and exponential for a new type of vector field, *Proc. Amer. Math. Soc.* **15** (1964), 432–437.
2. A new type of vector field and invariant differential systems, *Proc. Amer. Math. Soc.* **15** (1964), 675–678.

Jost, R.
1. Poisson brackets (an unpedagogical lecture), *Rev. Mod. Phys.* **36** (1964), 572–579.

Kahn, D. W.
1. *Introduction to Global Analysis*, Academic Press, New York, 1980.

Kalnins, E. G. and Miller, W., Jr.
1. Lie theory and separation of variables 5. The equations $iU_t + U_{xx} = 0$ and $iU_t - U_{xx} - c/x^2 U = 0$, *J. Math. Phys.* **15** (1974), 1728–1737.
2. Related evolution equations and Lie symmetries, *SIAM J. Math. Anal.* **16** (1985), 221–232.

Kalnins, E. G., Miller, W., Jr. and Williams, G. C.
1. Matrix operator symmetries of the Dirac equation and separation of variables, *J. Math. Phys.* **27** (1986), 1893–1900.

Kalnins, E. G., Miller, W., Jr. and Winternitz, P.
1. The group O(4), separation of variables and the hydrogen atom, *SIAM J. Appl. Math.* **30** (1976), 630–664.

Kamke, E.
1. *Differentialgleichungen Lösungsmethoden und Lösungen*, Chelsea, New York, 1971.

Kapitanskii, L. V.
1. Group analysis of the Euler and Navier–Stokes equations in the presence of rotational symmetry and new exact solutions of these equations, *Sov. Phys. Dokl.* **23** (1978), 896–898.
2. Group-theoretic analysis of the Navier–Stokes equations in the rotationally symmetric case and some new exact solutions, *Sov. J. Math.* **21** (1980), 314–327.

Khamitova, R. S.
1. Group structure and the basis of conservation laws, *Theor. Math. Phys.* **52** (1982), 777–781.

Khesin, B. A. and Chekanov, Y. V.
1. Invariants of the Euler equations for ideal or barotropic hydrodynamics and superconductivity in *D* dimensions, *Physica D* **40** (1989), 119–131.

Kibble, T. W. B.
1. Conservation laws for free fields, *J. Math. Phys.* **6** (1965), 1022–1026.

Kirillov, A. A.
1. *Elements of the Theory of Representations*, Springer-Verlag, New York, 1976.

Knops, R. J. and Stuart, C. A.
1. Quasiconvexity and uniqueness of equilibrium solutions in nonlinear elasticity, *Arch. Rat. Mech. Anal.* **86** (1984), 234–249.

Knowles, J. K. and Sternberg, E.
1. On a class of conservation laws in linearized and finite elastostatics, *Arch. Rat. Mech. Anal.* **44** (1972), 187–211.

Kosmann–Schwarzbach, Y.
1. Vector fields and generalized vector fields on fibered manifolds, in *Geometry and Differential Geometry*, R. Artzy and I. Vaisman, eds., Lecture Notes in Math., No. 792, Springer-Verlag, New York, 1980, pp. 307–355.

2. Generalized symmetries of nonlinear partial differential equations, *Lett. Math. Phys.* **3** (1979), 395–404.
3. Hamiltonian systems on fibered manifolds, *Lett. Math. Phys.* **5** (1981), 229–237.
4. Poisson–Drinfel'd groups, in *Topics in Soliton Theory and Exactly Solvable Nonlinear Equations*, M. Ablowitz, B. Fuchssteiner and M. Kruskal, eds., World Scientific, Singapore, 1987, pp. 191–215.

Kostant, B.
1. Quantization and unitary representations; part I: Prequantization, in *Lectures in Modern Analysis and Applications III*, C. Taam, ed., Lecture Notes in Math., No. 170, Springer-Verlag, New York, 1970, pp. 87–208.

Kostin, V. M.
1. Some invariant solutions of Korteweg–de Vries equations, *J. Appl. Mech. Tech. Phys.* **10** (1969), 583–586.

Kovalevskaya, S.
1. Zur Theorie der Partiellen Differentialgleichungen, *J. Reine Angew. Math.* **80** (1875), 1–32.

Kozlov, V. V.
1. Integrability and nonintegrability in Hamiltonian mechanics, *Russ. Math. Surveys* **38** (1983), 1–76.

Krasilshchik, I. S. and Vinogradov, A. M.
1. Nonlocal symmetries and the theory of coverings: an addendum to A. M. Vinogradov's 'Local symmetries and conservation laws', *Acta Appl. Math.* **2** (1984), 79–96.

Krause, J. and Michel, L.
1. Classification of the symmetries of ordinary differential equations, in *Group Theoretical Methods in Physics*, V. V. Dodonov and V. I. Man'ko, eds., Lecture Notes in Physics, Vol. 382, Springer-Verlag, New York, 1991, pp. 251–262.

Kruskal, M. D., Miura, R. M., Gardner, C. S. and Zabusky, N. J.
1. Korteweg–de Vries equation and generalizations. V. Uniqueness and nonexistence of polynomial conservation laws, *J. Math. Phys.* **11** (1970), 952–960.

Kumei, S.
1. Invariance transformations, invariance group transformations, and invariance groups of the sine–Gordon equations, *J. Math. Phys.* **16** (1975), 2461–2468.
2. Group theoretic aspects of conservation laws of nonlinear dispersive waves: KdV-type equations and nonlinear Schrödinger equations, *J. Math. Phys.* **18** (1977), 256–264.

Kumei, S. and Bluman, G. W.
1. When nonlinear differential equations are equivalent to linear differential equations, *SIAM J. Appl. Math.* **42** (1982), 1157–1173.

Kupershmidt, B. A.
1. Geometry of jet bundles and the structure of Lagrangian and Hamiltonian formalisms, in *Geometric Methods in Mathematical Physics*, G. Kaiser and J. E. Marsden, eds., Lecture Notes in Math., No. 775, Springer-Verlag, New York, 1980, pp. 162–218.
2. Mathematics of dispersive waves, *Comm. Math. Phys.* **99** (1985), 51–73.

Kupershmidt, B. A. and Wilson, G.
1. Modifying Lax equations and the second Hamiltonian structure, *Invent. Math.* **62** (1981), 403–436.

Kuznetsov, E. A. and Mikhailov, A. V.
1. On the topological meaning of canonical Clebsch variables, *Phys. Lett.* **77A** (1980), 37–38.

Lax, P. D.
1. Integrals of nonlinear equations of evolution and solitary waves, *Comm. Pure Appl. Math.* **21** (1968), 467–490.
2. Shock waves and entropy, in *Contributions to Nonlinear Functional Analysis*, E. H. Zarantonello, ed., Academic Press, New York, 1971, pp. 603–634.
3. A Hamiltonian approach to the KdV and other equations, in *Group-Theoretical Methods in Physics*, R. T. Sharp and B. Kolman, eds., Academic Press, New York, 1977, pp. 39–57.

Leo, M., Leo, R. A., Soliani, G., Solombrino, L. and Mancarella, G.
1. Lie–Bäcklund symmetries for the Harry Dym equation, *Phys. Rev. D* **27** (1983), 1406–1408.
2. Symmetry properties and bi-Hamiltonian structure of the Toda lattice, *Lett. Math. Phys.* **8** (1984), 267–272.

Levi, D. and Winternitz, P.
1. Nonclassical symmetry reduction: example of the Boussinesq equation, *J. Phys. A* **22** (1989), 2915–2924.
2. Continuous symmetries of discrete equations, *Phys. Lett. A* **152** (1991), 335–338.

Lewis, D., Marsden, J. E., Montgomery, R. and Ratiu, T.
1. The Hamiltonian structure for dynamic free boundary problems, *Physica D* **18** (1986), 391–404.

Lewy, H.
1. An example of a smooth linear partial differential equation without solution, *Ann. Math.* **66** (1957), 155–158.

Lichnerowicz, A.
1. Les variétés de Poisson et leurs algèbres de Lie associées, *J. Diff. Geom.* **12** (1977), 253–300.
2. Variétés de Poisson et feuilletages, *Ann. Fac. Sci. Toulouse* **4** (1982), 195–262.

Li Yi-Shen
1. The algebraic structure associated with systems possessing non-hereditary recursion operators, in *Nonlinear Evolution Equations and Dynamical Systems*, V. G. Makhankov and O. K. Pashaev, eds., Springer-Verlag, New York, 1991, pp. 107–109.

Lie, S.
1. Begründung einer Invariantentheorie der Berührungstransformationen, *Math. Ann.* **8** (1874), 215–288; also *Gesammelte Abhandlungen*, vol. 4, B. G. Teubner, Leipzig, 1929; pp. 1–96.
2. Über die Integration durch bestimmte Integrale von einer Klasse linear partieller Differentialgleichung, *Arch. for Math.* **6** (1881), pp. 328–368; also *Gesammelte Abhandlungen*, vol. 3, B. G. Teubner, Leipzig, 1922, pp. 492–523.
3. Klassifikation und Integration von gewöhnlichen Differentialgleichungen zwischen *x*, *y*, die eine Gruppe von Transformationen gestatten, *Math. Ann.* **32** (1888), 213–281; also *Gesammelte Abhundlungen*, vol. 5, B. G. Teubner, Leipzig, 1924, pp. 240–310.
4. *Theorie der Transformationsgruppen*, B. G. Teubner, Leipzig, 1888, 1890, 1893.
5. *Vorlesungen über Differentialgleichungen mit Bekannten Infinitesimalen Transformationen*, B. G. Teubner, Leipzig, 1891.

6. Zur allgemeinen Theorie der partiellen Differentialgleichungen beliebeger Ordnung, *Leipz. Berichte* **47** (1895), 53–128; also *Gesammelte Abhandlungen*, vol. 4, B. G. Teubner, Leipzig, 1929, pp. 320–384.
7. Die Theorie der Integralinvarianten ist ein Korollar der Theorie der Differentialinvarianten, *Leipz. Berichte* **49** (1897), 342–357; also *Gesammelte Abhandlungen*, vol. 6, B. G. Teubner, Leipzig, 1927, pp. 649–663.

Lipkin, D. M.
1. Existence of a new conservation law in electromagnetic theory, *J. Math. Phys.* **5** (1964), 696–700.

Lisle, I.
1. *Equivalence Transformations for Classes of Differential Equations*, Ph.D. Thesis, University of British Columbia, Vancouver, 1992.

Lloyd, S. P.
1. The infinitesimal group of the Navier–Stokes equations. *Acta Mech.* **38** (1981), 85–98.

Logan, J. D.
1. *Invariant Variational Principles*, Academic Press, New York, 1977.

Maeda, S.
1. The similarity method for difference equations, *IMA J. Appl. Math.* **38** (1987), 129–134.

Magri, F.
1. A simple model of the integrable Hamiltonian equation, *J. Math. Phys.* **19** (1978), 1156–1162.
2. A geometrical approach to the nonlinear solvable equations, in *Nonlinear Evolution Equations and Dynamical Systems*, M. Boiti, F. Pempinelli and G. Soliani, eds., Lecture Notes in Physics, No. 120, Springer-Verlag, New York, 1980.

Manakov, S. V.
1. Note on integration of Euler's equations of dynamics of an n-dimensional rigid body, *Func. Anal. Appl.* **10** (1976), 328–329.

Manin, Yu. I.
1. Algebraic aspects of nonlinear differential equations, *J. Soviet Math.* **11** (1979) 1–122.

Markus, L.
1. Group theory and differential equations, Lecture Notes, University of Minnesota, Minneapolis, 1960.

Marsden, J. E.
1. A group theoretic approach to the equations of plasma physics, *Canad. Math. Bull.* **25** (1982), 129–142.
2. *Lectures on Mechanics*, Cambridge Univ. Press, London, 1992.

Marsden, J. E. and Hughes, T. J. R.
1. *Mathematical Foundations of Elasticity*, Prentice-Hall, Englewood Cliffs, N.J., 1983.

Marsden, J. E., Ratiu, T. and Weinstein, A.
1. Semi direct products and reduction in mechanics, *Trans. Amer. Math. Soc.* **281** (1984), 147–177.

Marsden, J. E. and Weinstein, A.

1. Reduction of symplectic manifolds with symmetry, *Rep. Math. Phys.* **5** (1974), 121–130.
2. Coadjoint orbits, vortices, and Clebsch variables for incompressible fluids, *Physica D* **7** (1983), 305–332.

Mayer, A.

1. Die Existenzbedingungen eines kinetischen Potentiales, *Berich. Verh. König. Sach. Gesell. Wissen. Leipzig, Math.-Phys. Kl.* **84** (1896), 519–529.

McLeod, J. B. and Olver, P. J.

1. The connection between partial differential equations soluble by inverse scattering and ordinary differential equations of Painlevé type, *SIAM J. Math. Anal.* **14** (1983), 488–506.

Meyer, K. R.

1. Symmetries and integrals in mechanics, in *Dynamical Systems*, M. M. Peixoto, ed., Academic Press, New York, 1973, pp. 259–272.

Michal, A. D.

1. Differential invariants and invariant partial differential equations under continuous transformation groups in normed linear spaces, *Proc. Nat. Acad. Sci.* **37** (1951), 623–627.

Mikhailov, A. V., Shabat, A. B. and Sokolov, V. V.

1. The symmetry approach to classification of integrable equations, in *What is Integrability?*, V. E. Zakharov, ed., Springer Verlag, New York, 1990, pp. 115–184.

Mikhailov, A. V., Shabat, A. B. and Yamilov, R. I.

1. The symmetry approach to classification of nonlinear equations. Complete lists of integrable systems, *Russ. Math. Surveys* **42**:4 (1987), 1–63.
2. Extension of the module of invertible transformations. Classification of integrable systems, *Commun. Math. Phys.* **115** (1988), 1–19.

Miller, W., Jr.

1. *Lie Theory and Special Functions*, Academic Press, New York, 1968.
2. *Symmetry Groups and Their Applications*, Academic Press, New York, 1972.
3. *Symmetry and Separation of Variables*, Addison-Wesley, Reading, Mass., 1977.

Milnor, J. W.

1. *Topology from the Differentiable Viewpoint*, The University Press of Virginia, Charlottesville, Va., 1965.

Mishchenko, A. S. and Fomenko, A. T.

1. Generalized Liouville method of integration of Hamiltonian systems, *Func. Anal. Appl.* **12** (1978), 113–121.

Miura, R. M.

1. Korteweg–de Vries equation and generalizations. I. A remarkable explicit nonlinear transformation, *J. Math. Phys.* **9** (1968), 1202–1204.

Moon, P. and Spencer, D. E.

1. *Field Theory Handbook*, Springer-Verlag, New York, 1971.

Morawetz, C. S.

1. *Notes on Time Decay and Scattering for Some Hyperbolic Problems*, CBMS–NSF Regional Conf. Series in Appl. Math. No. 19, SIAM, Philadelphia, 1975.

Morgan, A. J. A.
1. The reduction by one of the number of independent variables in some systems of partial differential equations, *Quart. J. Math. Oxford* **3** (1952), 250–259.

Morgan, T. A.
1. Two classes of new conservation laws for the electromagnetic field and for other massless fields, *J. Math. Phys.* **5** (1964), 1659–1660.

Morosi, C.
1. The R-matrix theory and the reduction of Poisson manifolds, *J. Math. Phys.* **33** (1992), 941–952.

Nagano, T.
1. Linear differential systems with singularities and an application to transitive Lie algebras, *J. Math. Soc. Japan* **18** (1966), 398–404.

Narasimhan, R.
1. *Analysis on Real and Complex Manifolds*, North-Holland Publ. Co., Amsterdam, 1968.

Newell, A.
1. *Solitons in Mathematics and Physics*, CBMS-NSF Conference Series in Applied Math., Vol. 48, SIAM, Philadelphia, 1985.

Nirenberg, L.
1. *Lectures on Linear Partial Differential Equations*, CBMS Regional Conf. Series in Math., No. 17, Amer. Math. Soc., Providence, R.I., 1973.

Noether, E.
1. Invariante Variationsprobleme, *Nachr. König. Gesell. Wissen. Göttingen, Math.-phys. Kl.* (1918), 235–257 (see *Transport Theory and Stat. Phys.* **1** (1971), 186–207 for an English translation).

Nucci, M. C. and Clarkson, P.
1. The nonclassical method is more general than the direct method for symmetry reductions. An example of the Fitzhugh–Nagumo equation, *Phys. Lett. A* **164** (1992), 49–56.

Nutku, Y.
1. On a new class of completely integrable nonlinear wave equations. II. Multi-Hamiltonian structure, *J. Math. Phys.* **28** (1987), 2579–2585.

Oevel, W.
1. Master symmetries: weak action/angle structure for hamiltonian and non hamiltonian systems, preprint, 1986.

Olver, P. J.
1. Evolution equations possessing infinitely many symmetries, *J. Math. Phys.* **18** (1977), 1212–1215.
2. Symmetry groups and group invariant solutions of partial differential equations, *J. Diff. Geom.* **14** (1979), 497–542.
3. Euler operators and conservation laws of the BBM equation, *Math. Proc. Camb. Phil. Soc.* **85** (1979), 143–160.
4. On the Hamiltonian structure of evolution equations, *Math. Proc. Camb. Phil. Soc.* **88** (1980), 71–88.
5. A nonlinear Hamiltonian structure for the Euler equations, *J. Math. Anal. Appl.* **89** (1982), 233–250.
6. Hyperjacobians, determinantal ideals, and weak solutions to variational problems, *Proc. Roy. Soc. Edinburgh* **95A** (1983), 317–340.

7. Conservation laws and null divergences, *Math. Proc. Camb. Phil. Soc.* **94** (1983), 529–540.
8. Conservation laws in elasticity I. General results, *Arch. Rat. Mech. Anal.* **85** (1984), 111–129.
9. Conservation laws in elasticity II. Linear homogeneous isotropic elastostatics, *Arch. Rat. Mech. Anal.* **85** (1984), 131–160.
10. Hamiltonian perturbation theory and water waves, *Contemp. Math.* **28** (1984), 231–249.
11. Noether's theorems and systems of Cauchy–Kovalevskaya type, in *Nonlinear Systems of PDE in Applied Math.*, B. Nicolaenko, D. D. Holm and J. Hyman, eds., Lectures in Applied Math., Vol. 23, Part 2, Amer. Math. Soc., Providence, R.I., 1986, pp. 81–104.
12. Darboux' theorem for Hamiltonian differential operators, *J. Diff. Eq.* **71** (1988), 10–33.
13. BiHamiltonian systems, in *Ordinary and Partial Differential Equations*, B. D. Sleeman and R. J. Jarvis, eds., Pitman Research Notes in Mathematics Series, No. 157, Longman Scientific and Technical, New York, 1987, pp. 176–193.
14. Conservation laws in elasticity. III. Planar linear anisotropic elastostatics, *Arch. Rat. Mech. Anal.* **102** (1988), 167–181.
15. Canonical forms and integrability of biHamiltonian systems, *Phys. Lett. A* **148** (1990), 177–187.
16. Symmetry and explicit solutions of partial differential equations, *App. Num. Math.* **10** (1992), 307–324.

Olver, P. J. and Nutku, Y.
1. Hamiltonian structures for systems of hyperbolic conservation laws, *J. Math. Phys.* **29** (1988), 1610–1619.

Olver, P. J. and Rosenau, P.
1. The construction of special solutions to partial differential equations, *Phys. Lett. A* **114** (1986), 107–112.
2. Group-invariant solutions of differential equations, *SIAM J. Appl. Math.* **47** (1987), 263–278.

Olver, P. J. and Shakiban, C.
1. A resolution of the Euler operator. I. *Proc. Amer. Math. Soc.* **69** (1978), 223–229.

Ondich, J. R.
1. *Partially Invariant Solutions of Differential Equations*, Ph.D. Thesis, University of Minnesota, 1989.

Ovsiannikov, L. V.
1. Groups and group-invariant solutions of differential equations, *Dokl. Akad. Nauk USSR* **118** (1958), 439–442 (in Russian).
2. *Group Properties of Differential Equations*, Novosibirsk, Moscow, 1962 (in Russian; translated by G. W. Bluman, unpublished).
3. *Group Analysis of Differential Equations*, Academic Press, New York, 1982.

Palais, R. S.
1. *A Global Formulation of the Lie Theory of Transformation Groups*, Memoirs of the Amer. Math. Soc., No. 22, Providence, R.I., 1957.

Patera, J., Winternitz, P. and Zassenhaus, H.
1. Continuous subgroups of the fundamental groups of physics. I. General method and the Poincaré group, *J. Math. Phys.* **16** (1975), 1597–1614.

Petrovskii, I. G.
1. *Lectures on Partial Differential Equations*, Interscience, New York, 1954.

Pohjanpelto, P. J.
1. *Symmetries of Maxwell's Equations*, Ph.D. Thesis, University of Minnesota, 1989.
2. First order generalized symmetries of Maxwell's equations, *Phys. Lett. A* **129** (1988), 148–150.

Poincaré, H.
1. *Les Méthodes Nouvelles de la Mécanique Céleste*, vol. 3, Gauthiers-Villars, Paris, 1892.

Pommaret, J. F.
1. *Systems of Partial Differential Equations and Lie Pseudogroups*, Gordon & Breach, New York, 1978.
2. *Differential Galois Theory*, Gordon & Breach, New York, 1983.

Pontryagin, L. S.
1. *Toplogical Groups*, 2nd ed., Gordon & Breach, New York, 1966.

Posluszny, J. and Rubel, L. A.
1. The motions of an ordinary differential equation, *J. Diff. Eq.* **34** (1979), 291–302.

Pucci, P. and Serrin, J.
1. A general variational identity, *Indiana Univ. Math. J.* **35** (1986), 681–703.

Ramakrishnan, V. and Schaettler, H.
1. Controlled invariant distributions and group invariance, *J. Math. Syst. Est. Control* **1** (1991), 263–278.

Riquier, Ch.
1. Sur une question fondamentale du calcul integrale, *Acta Math.* **23** (1900), 203–332.

Ritt, J. F.
1. *Differential Algebra*, Colloq. Publ., vol. 33, Amer. Math. Soc., Providence, R.I., 1950.

Rosen, G.
1. Nonlinear heat conduction in solid H_2, *Phys. Rev. B* **19** (1979), 2398–2399.
2. Restricted invariance of the Navier–Stokes equation, *Phys. Rev. A* **22** (1980), 313–314.

Rosenau, P.
1. A note on the integration of the Emden–Fowler equation, *Int. J. Nonlinear Mech.* **19** (1984), 303–308.

Rosenau, P. and Schwarzmeier, J. L.
1. Similarity solutions of systems of partial differential equations using MACSYMA, Courant Inst. of Math. Sci., Magneto-Fluid Dynamics Division, Report No. COO-3077-160 MF-94, 1979.

Rosencrans, S. I.
1. Conservation laws generated by pairs of non-Hamiltonian symmetries, *J. Diff. Eq.* **43** (1982), 305–322.
2. Computation of higher order symmetries using MACSYMA, *Comp. Physics Commun.* **38** (1985), 347–356.

Rubel, L. A.
1. The motions of a partial differential equation, *J. Diff. Eq.* **48** (1983), 177–188.

Santini, P. and Fokas, A. S.
1. Recursion operators and biHamiltonian structures in multi-dimensions. I, *Commun. Math. Phys.* **115** (1988), 375–419.

Sattinger, D. H.
1. *Group-Theoretic Methods in Bifurcation Theory*, Lecture Notes in Math., No. 762, Springer-Verlag, New York, 1979.

Schouten, J. A. and Struik, D. J.
1. *Einführung in die Neueren Methoden der Differentialgeometrie*, vol. 1, P. Noordhoff N.V., Groningen-Batavia, 1935.

Schreier, O.
1. Abstrakte kontinuerliche Gruppen, *Abh. Math. Seminar Hamburg Univ.* **4** (1926), 15–32.

Schütz, J. R.
1. Prinzip der absoluten Erhaltung der Energie, *Nachr. König. Gesell. Wissen. Göttingen, Math.-Phys. Kl.* (1897), 110–123.

Schwarz, F.
1. Automatically determining symmetries of partial differential equations, *Computing* **34** (1985), 91–106.

Sedov, L. I.
1. *Similarity and Dimensional Methods in Mechanics*, Academic Press, New York, 1959.

Semenov-Tian-Shanskii, M. A.
1. What is a classical *R*-matrix?, *Func. Anal. Appl.* **17** (1983), 259–272.

Serre, D.
1. Invariants et dégénérescence symplectique de l'équation d'Euler des fluids parfaits incompressibles, *C. R. Acad. Sci. Paris* **298** (1984), 349–352.

Seshadri, R. and Na, T. Y.
1. *Group Invariance in Engineering Boundary Value Problems*, Springer-Verlag, New York, 1985.

Shakiban, C.
1. A resolution of the Euler operator II, *Math. Proc. Camb. Phil. Soc.* **89** (1981) 501–510.

Shapovalov, A. V. and Shirokov, I. V.
1. Symmetry algebras of linear differential equations, *Theor. Math. Phys.* **92** (1992), 697–705.

Shokin, Yu. I
1. *The Method of Differential Approximation*, Springer-Verlag, New York, 1983.

Smale, S.
1. Topology and mechanics. I, *Invent. Math.* **10** (1970), 305–331.

Sokolov, V. V.
1. On the symmetries of evolution equations, *Russ. Math. Surveys* **43**:5 (1988), 165–204.

Sokolov, V. V. and Shabat, A. B.
1. Classification of integrable evolution equations, *Math. Phys. Rev.* **4** (1984), 221–280.

Souriau, J.-M.
1. *Structure des Systèmes Dynamiques*, Dunod, Paris, 1970.

Steinberg, S.
1. Symmetry operators, in *Proceedings of the 1979 MACSYMA User's Conference*, V. E. Lewis, ed., Washington, 1979, pp. 408–444.
2. Symmetries of differential equations, Univ. of New Mexico preprint, 1983.

Steinberg, S. and Wolf, K. B.
1. Symmetry, conserved quantities and moments in diffusive equations, *J. Math. Anal. Appl.* **80** (1981), 36–45.

Stephani, H.
1. *Differential Equations: Their Solution Using Symmetries*, Cambridge Univ. Press, New York, 1989.

Steudel, H.
1. Uber die Zuordnung zwischen Invarianzeigenschaften und Erhaltungssätzen, *Zeit. Naturforsch.* **17A** (1962), 129–132.
2. Eine Erweiterung des ersten Noetherschen Satzes, *Zeit. Naturforsch.* **17A** (1962), 133–135.
3. Die Struktur der Invarianzgruppe fur lineare Feldtheorien, *Zeit. Naturforsch.* **21A** (1966), 1826–1828.
4. Noether's theorem and higher conservation laws in ultrashort pulse propagation, *Ann. Physik* **32** (1975), 205–216.
5. Noether's theorem and the conservation laws of the Korteweg–de Vries equation, *Ann. Physik* **32** (1975), 445–455.

Strauss, W. A.
1. Nonlinear invariant wave equations, in *Invariant Wave Equations*, G. Velo and A. S. Wightman, eds., Lecture Notes in Physics, No. 73, Springer-Verlag, New York, 1978, pp. 197–249.

Sudarshan, E. C. G. and Mukunda, N.
1. *Classical Dynamics: A Modern Perspective*, Wiley, New York, 1974.

Sussmann, H. J.
1. Orbits of families of vector fields and integrability of systems with singularities, *Bull. Amer. Math. Soc.* **79** (1973), 197–199.

Svinolupov, S. I. and Sokolov, V. V.
1. Factorization of evolution equations, *Russ. Math. Surveys* **47**:3 (1992), 127–162.

Takens, F.
1. A global version of the inverse problem of the calculus of variations, *J. Diff. Geom.* **14** (1979), 543–562.

Taylor, G. I.
1. The formation of a blast wave by a very intense explosion. I. Theoretical discussion, *Proc. Roy. Soc. London* **201A** (1950), 159–174.
2. The formation of a blast wave by a very intense explosion. II. The atomic explosion of 1945, *Proc. Roy. Soc. London* **201A** (1950), 175–186.

Taylor, M.
1. *Pseudodifferential Operators*, Princeton Univ. Press, Princeton, N.J., 1981.

Thirring, W. E.
1. *A Course in Mathematical Physics*, vol. 1, Springer-Verlag, New York, 1978.

Toda, M.

1. *Theory of Nonlinear Lattices*, Springer-Verlag, New York, 1981.

Tsujishita, T.

1. On variation bicomplexes associated to differential equations, *Osaka J. Math.* **19** (1982), 311–363.
2. Formal geometry of systems of differential equations, *Sugaku Exp.* **3** (1990), 25–73.

Tu, G.-Z.

1. A commutativity theorem of partial differential operators, *Commun. Math. Phys.* **77** (1980), 289–297.

Tulczyjew, W. M.

1. The Lagrange complex, *Bull. Soc. Math. France* **105** (1977), 419–431.
2. The Euler–Lagrange resolution, in *Differential Geometric Methods in Mathematical Physics*, P. L. Garcia, A. Pérez-Rendón, and J.-M. Souriau, eds., Lecture Notes in Math., No. 836, Springer-Verlag, New York, 1980, pp. 22–48.

Vainberg, M. M.

1. *Variational Methods for the Study of Nonlinear Operators*, Holden-Day, San Francisco, 1964.

van der Schaft, A. J.

1. Symmetries in optimal control, *SIAM J. Control Optimization* **25** (1987), 245–259.

van der Vorst, R.C.A.M.

1. Variational identities and applications to differential systems, *Arch. Rat. Mech. Anal.* **116** (1991), 375–398.

Verdier, J.-L.

1. Groupes quantiques, *Astérisque* **152–153** (1987), 305–319.

Vessiot, E.

1. Sur l'intégration des systèmes différentiels qui admettent des groupes continus de transformations, *Acta. Math.* **28** (1904), 307–349.

Vilenkin, N. J.

1. *Special Functions and the Theory of Group Representations*, Amer. Math. Soc., Providence, R.I., 1968.

Vinogradov, A. M.

1. On the algebra-geometric foundations of Lagrangian field theory, *Sov. Math. Dokl.* **18** (1977), 1200–1204.
2. Hamilton structures in field theory, *Sov. Math. Dokl.* **19** (1978), 790–794.
3. The $\mathscr{C}$-spectral sequence, Lagrangian formalism and conservation laws. I. The linear theory, *J. Math. Anal. Appl.* **100** (1984), 1–40.
4. The $\mathscr{C}$-spectral sequence, Lagrangian formalism and conservation laws. II. The nonlinear theory, *J. Math. Anal. Appl.* **100** (1984), 41–129.
5. Local symmetries and conservation laws, *Acta Appl. Math.* **2** (1984), 21–78.

Vladimorov, V. S. and Volovich, I.V.

1. Conservation laws for non-linear equations, *Usp. Mat. Nauk* **40**:4 (1985), 17–26.
2. Local and nonlocal currents for nonlinear equations, *Theor. Math. Phys.* **62** (1985), 1–20.

Volterra, V.
1. *Leçons sur les Fonctions de Lignes*, Gauthier–Villars, Paris, 1913.

Wahlquist, H. D. and Estabrook, F. B.
1. Prolongation structures of nonlinear evolution equations, *J. Math. Phys.* **16** (1975), 1–7.

Warner, F. W.
1. *Foundations of Differentiable Manifolds and Lie Groups*, Scott, Foresman, Glenview, Ill., 1971.

Weinstein, A.
1. Symplectic manifolds and their Lagrangian submanifolds, *Adv. Math.* **6** (1971), 329–346.
2. Sophus Lie and symplectic geometry, *Expo. Math.* **1** (1983), 95–96.
3. The local structure of Poisson manifolds, *J. Diff. Geom.* **18** (1983), 523–557.
4. Stability of Poisson–Hamilton equilibria, *Contemp. Math.* **28** (1984), 3–13.

Weisner, L.
1. Generating functions for Hermite functions, *Canad. J. Math.* **11** (1959), 141–147.

Weiss, J., Tabor, M. and Carnevale, G.
1. The Painlevé property for partial differential equations, *J. Math. Phys.* **24** (1983), 522–526.

Weyl, H.
1. *Die Idee der Riemannschen Fläche*, B. G. Teubner, Berlin, 1923.

Whitham, G. B.
1. Variational methods and applications to water waves, *Proc. Roy. Soc. London* **299A** (1967), 6–25.
2. *Linear and Nonlinear Waves*, Wiley, New York, 1974.

Whittaker, E. T.
1. *A Treatise on the Analytical Dynamics of Particles and Rigid Bodies*, Cambridge University Press, Cambridge, 1937.

Widder, D. V.
1. *The Heat Equation*, Academic Press, New York, 1975.

Wilczynski, E. J.
1. An application of group theory to hydrodynamics, *Trans. Amer. Math. Soc.* **1** (1900), 339–352.

Wilson, G.
1. Commuting flows and conservation laws for Lax equations, *Math. Proc. Camb. Phil. Soc.* **86** (1979), 131–143.

Wussing, H.
1. *The Genesis of the Abstract Group Concept*, MIT Press, Cambridge, Mass., 1984.

Zakharov, V. E. and Fadeev, L. D.
1. Korteweg–de Vries equation: a completely integrable Hamiltonian system, *Func. Anal. Appl.* **5** (1971), 280–287.

Zakharov, V. E. and Konopelchenko, B. G.
1. On the theory of recursion operator, *Commun. Math. Phys.* **94** (1984), 483–509.

Zharinov, V.
1. *Geometrical Aspects of Partial Differential Equations*, World Scientific, Singapore, 1992.

Symbol Index

Author Index

Subject Index

Graduate Texts in Mathematics

(continued from page ii)

120 ZIEMER. Weakly Differentiable Functions: Sobolev Spaces and Functions of Bounded Variation.
121 LANG. Cyclotomic Fields I and II. Combined 2nd ed.
122 REMMERT. Theory of Complex Functions. *Readings in Mathematics*
123 EBBINGHAUS/HERMES et al. Numbers. *Readings in Mathematics*
124 DUBROVIN/FOMENKO/NOVIKOV. Modern Geometry—Methods and Applications. Part III.
125 BERENSTEIN/GAY. Complex Variables: An Introduction.
126 BOREL. Linear Algebraic Groups. 2nd ed.
127 MASSEY. A Basic Course in Algebraic Topology.
128 RAUCH. Partial Differential Equations.
129 FULTON/HARRIS. Representation Theory: A First Course. *Readings in Mathematics*
130 DODSON/POSTON. Tensor Geometry.
131 LAM. A First Course in Noncommutative Rings.
132 BEARDON. Iteration of Rational Functions.
133 HARRIS. Algebraic Geometry: A First Course.
134 ROMAN. Coding and Information Theory.
135 ROMAN. Advanced Linear Algebra.
136 ADKINS/WEINTRAUB. Algebra: An Approach via Module Theory.
137 AXLER/BOURDON/RAMEY. Harmonic Function Theory.
138 COHEN. A Course in Computational Algebraic Number Theory.
139 BREDON. Topology and Geometry.
140 AUBIN. Optima and Equilibria. An Introduction to Nonlinear Analysis.
141 BECKER/WEISPFENNING/KREDEL. Gröbner Bases. A Computational Approach to Commutative Algebra.
142 LANG. Real and Functional Analysis. 3rd ed.
143 DOOB. Measure Theory.
144 DENNIS/FARB. Noncommutative Algebra.
145 VICK. Homology Theory. An Introduction to Algebraic Topology. 2nd ed.
146 BRIDGES. Computability: A Mathematical Sketchbook.
147 ROSENBERG. Algebraic *K*-Theory and Its Applications.
148 ROTMAN. An Introduction to the Theory of Groups. 4th ed.
149 RATCLIFFE. Foundations of Hyperbolic Manifolds.
150 EISENBUD. Commutative Algebra with a View Toward Algebraic Geometry.
151 SILVERMAN. Advanced Topics in the Arithmetic of Elliptic Curves.
152 ZIEGLER. Lectures on Polytopes.
153 FULTON. Algebraic Topology: A First Course.
154 BROWN/PEARCY. An Introduction to Analysis.
155 KASSEL. Quantum Groups.
156 KECHRIS. Classical Descriptive Set Theory.
157 MALLIAVIN. Integration and Probability.
158 ROMAN. Field Theory.
159 CONWAY. Functions of One Complex Variable II.
160 LANG. Differential and Riemannian Manifolds.
161 BORWEIN/ERDÉLYI. Polynomials and Polynomial Inequalities.
162 ALPERIN/BELL. Groups and Representations.
163 DIXON/MORTIMER. Permutation Groups.
164 NATHANSON. Additive Number Theory: The Classical Bases.
165 NATHANSON. Additive Number Theory: Inverse Problems and the Geometry of Sumsets.
166 SHARPE. Differential Geometry: Cartan's Generalization of Klein's Erlangen Program.
167 MORANDI. Field and Galois Theory.
168 EWALD. Combinatorial Convexity and Algebraic Geometry.
169 BHATIA. Matrix Analysis.
170 BREDON. Sheaf Theory. 2nd ed.
171 PETERSEN. Riemannian Geometry.
172 REMMERT. Classical Topics in Complex Function Theory.
173 DIESTEL. Graph Theory.
174 BRIDGES. Foundations of Real and Abstract Analysis.
175 LICKORISH. An Introduction to Knot Theory.
176 LEE. Riemannian Manifolds.
177 NEWMAN. Analytic Number Theory.
178 CLARKE/LEDYAEV/STERN/WOLENSKI. Nonsmooth Analysis and Control Theory.
179 DOUGLAS. Banach Algebra Techniques in Operator Theory. 2nd ed.

180 SRIVASTAVA. A Course on Borel Sets.
181 KRESS. Numerical Analysis.
182 WALTER. Ordinary Differential Equations.
183 MEGGINSON. An Introduction to Banach Space Theory.
184 BOLLOBAS. Modern Graph Theory.
185 COX/LITTLE/O'SHEA. Using Algebraic Geometry.
186 RAMAKRISHNAN/VALENZA. Fourier Analysis on Number Fields.
187 HARRIS/MORRISON. Moduli of Curves.
188 GOLDBLATT. Lectures on the Hyperreals: An Introduction to Nonstandard Analysis.
189 LAM. Lectures on Modules and Rings.
190 ESMONDE/MURTY. Problems in Algebraic Number Theory.
191 LANG. Fundamentals of Differential Geometry.
192 HIRSCH/LACOMBE. Elements of Functional Analysis.
193 COHEN. Advanced Topics in Computational Number Theory.
194 ENGEL/NAGEL. One-Parameter Semigroups for Linear Evolution Equations.
195 NATHANSON. Elementary Methods in Number Theory.
196 OSBORNE. Basic Homological Algebra.
197 EISENBUD/ HARRIS. The Geometry of Schemes.

9 780387 950006